A NATURAL APPROACH TO CHEMISTRY

HSU
CHANIOTAKIS
CARLISLE
DAMELIN

FIRST EDITION

LAB-AIDS®

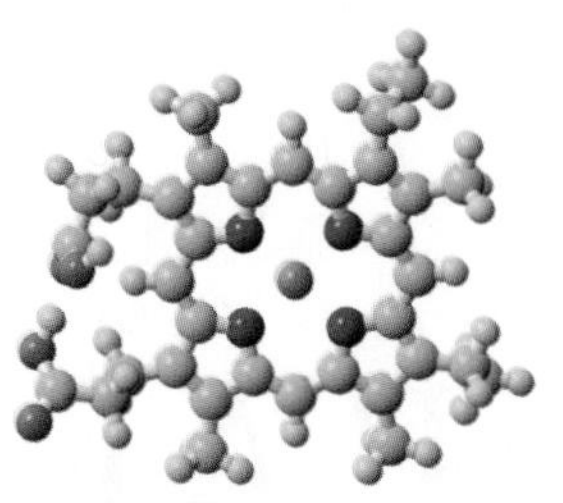

A Natural Approach to Chemistry

ISBN-13: 978-1-60301-313-0

2 3 4 5 6 7 8 9 - RRD - 15 14 13 12 11 10 09

Lab-Aids, Inc.
17 Colt Court
Ronkonkoma, NY 11779

This book was written, illustrated, published and printed in the United States of America

From the authors

We hope you find A Natural Approach to Chemistry to be successful and engaging for your students. As we developed the program, we strived for clarity and completeness within the context of a practical and useful chemistry course for all students. We drew inspiration and examples from the human body and from the environment to connect chemistry to students lives, today and in their future. We pruned and trimmed to make room to cover key ideas in greater depth and in repeated contexts. For example, you will find fundamental ideas like valence and molarity explained repeatedly over several chapters. In our experience, students need to see abstract but important ideas several times, in different contexts, before they truly learn them. After all, how many times did *you* parallel park before trying your driver's test?

As teachers ourselves, we place a lot of emphasis on effective, hands-on activities that bring students to the core of important concepts, such as the chemical formula. We also know that students have many strongly-held misconceptions, such as the difference between heat and temperature. We asked ourselves: what are the few most important concepts we want students to understand and use? We then took the most direct experimental path to get them there. You will see some novel, and maybe unfamiliar techniques in this program, such as using molecular models to learn the chemical formula, or using 0°C as a thermal energy reference. However, these techniques *really work*! Many traditional procedures, some of which date from Lavoisier, we found to be too abstract for learning at the level of an average high school student. We designed the unique lab equipment to ensure that it will do what the curriculum needs it to do, reliably, and in a way that maximizes student engagement and learning. In many ways the labs are the heart and soul of this course and we hope you make them a core part of your instruction as well.

Chemistry is a fascinating, beautiful, and extraordinarily important subject. Given the challenges of alternative energy, modified foods, new medicines, and environmental issues, we believe it is crucial that all thinking people have a basic understanding of chemistry. You will find a balance of deep conceptual treatment as well as rigorous but grade-appropriate quantitative problems and techniques.

We are writing this program for everyone who wants to know how this world of ours works.

Tom Hsu
Manos Chaniotakis
Debbie Carlisle
Dan Damelin

2009

A NATURAL APPROACH TO CHEMISTRY

Principal Authors

Dr. Tom Hsu is nationally known as an innovator in science equipment and curriculum. He is the author of six published middle and high school science programs in physics, physical science, and chemistry. Dr. Hsu did chemical engineering for Eastman Kodak, Xerox, and Dupont and also has taught grade 4 through graduate school. He holds a Ph.D. in applied plasma physics from MIT.

Dr. Manos Chaniotakis was a professor at MIT for 18 years where he is recognized for teaching innovative, hands-on courses. He is also the founder and president of MITOS Inc., which develops instruments for the analytical chemistry industry, including meters for pH, conductivity, and ion concentration. Dr. Chaniotakis holds a Ph.D. in Plasma Physics and Fusion Engineering from MIT.

Debbie Carlisle taught high school chemistry, biochemistry, and biology for 20 years, at Phillips Academy and other high schools. She has worked with students at all levels of ability from introductory to AP. She enhanced many important aspects of the laboratory program at Phillips that made it more exciting and accessible to students. Debbie has a Masters Degree in biological science, and technology.

Dan Damelin has taught high school chemistry since 1993. He is also the Principal Investigator on the NSF funded Molecular Workbench project, which has created extensive molecular modeling and computer simulations for teaching Physics, Chemistry, and Biology. Dan has a Master of Science degree in chemistry, environmental science, and computer science from Tufts University.

Consultants

Dr. Nick Chaniotakis
Professor of Chemistry
University of Crete, Greece

Technical

Michael Short
Electrical engineering and labs

Jared Sartee
Mechanical engineering

Adrian Culver
Laboratory experiments

Graphic Arts

James A. Travers
Lead graphic artist & illustrator

Linda L. Taylor
Taylor Graphics, Ashfield MA

Reviewers

Alana J. Nelson, MA
Chemistry Teacher
East Providence High School
Providence, RI

Amanda. Nycole Noble
Chemistry Teacher
Portsmouth High School
Portsmouth, RI

Amy L. Biagioni, M.Ed.
Chemistry Teacher
Cranston High School West
Cranston, RI

Brenda G. Weiser, Ed.D.
Clinical Assoc. Prof. Sci. Educ.
Univ. of Houston at Clear Lake
Clear Lake, TX

Brian Fortney, Ph.D.
Assistant Instructor
The University of Texas at Austin
Austin, TX

Britany W. Coleman, MA
Chemistry Teacher
Lincoln High School
Lincoln, RI

Christine M. Lawrence
Special Education Teacher
Portsmouth High School
Portsmouth, RI

David Bain
Science Department Chair
Lakes Community High School
Lake Villa, IL

David M Axelson
Chemistry Teacher
East Providence High School
Providence, RI

Reviewers

Deirdre D. London, M.Ed.
Chemistry Teacher
Cranston High School West
Cranston, RI

Dennis A. Nobrega, M.Ed. CAGS
Chemistry Teacher
East Providence High School
Providence, RI

Donald Lurgio
Chemistry Teacher
East Providence High School
Providence, RI

Donna M. Wise, M.Ed.
Science Specialist
Region 7 Education Service Center
Kilgore, Tx

Elaine Megyar, Ph.D.
Professor of Chemistry
Rhode Island College
Providence, RI

Ellen Will
Chemistry Teacher
East Providence High School
Providence, RI

Helaine Hager
Chemistry Teacher
E Cubed Academy
Lincoln, RI

James Megyar, Ph.D.
Professor of Chemistry
Rhode Island College
Providence, RI

Janet Dickinson
Chemistry Lead Teacher
Kingwood High School
Kingwood, TX

Janet M. Miele
Chemistry Teacher
Woonsocket High School
Woonsocket, RI

Janet R. Kasparian, MS
Biology and Chemistry Teacher
Portsmouth High School
Portsmouth, RI

Jeffrey Soares
Science Education Specialist
East Bay Educ.. Collaborative
Warren, RI

Jennifer P. Cameron
Chemistry Teacher
Lincoln High School
Lincoln, RI

Judy Grubbs
K-12 Science Specialist
Region 7 Educ. Service Center
Kilgore, Tx

Judy H. McGowan
Chemistry Teacher
Mount Pleasant High School
Providence, RI

Judy York, M.S.
Education Specialist
ESC Region 12
Waco, TX

Kathleen M. Beebe
Chemistry Teacher
Portsmouth High School
Portsmouth, RI

Kathleen Siok
Chemistry Specialist
East Bay Educ. Collaborative
Warren, RI

Keith D. Ward
Chemistry Teacher
Cranston High School West
Cranston, RI

Kimberly A. Laliberte, MS, PMP
Chemistry Teacher
East Providence High School
Providence, RI

Kristian Fischer Trampus M.Ed.
Dir. - East Texas STEM Center
Univ. of Texas at Tyler
Tyler, TX

Leanne M. Gordon Perry, MS
Biology and Chemistry Teacher
Portsmouth High School
Portsmouth, RI

Pamela R. Kahn
Chemistry Teacher
Cranston High School West
Cranston, RI

Peter Madeiros
Chemistry Teacher
Woonsocket High School
Woonsocket, RI

Ron DeFronzo
Science Education Specialist
East Bay Educ. Collaborative
Warren, RI

Ronald Kahn
Science Education Specialist
East Bay Educ. Collaborative
Warren, RI

Roxanne Minix-Wilkins, M.S.
Science Program Coordinator
Region 5 Educ. Serv. Ctr.
Silsbee, TX.

Suzanne Doucette
Chemistry Teacher
Woonsocket High School
Woonsocket, RI

Thomas J. Holstein, Jr, MS
Chemistry Teacher
Portsmouth High School
Portsmouth, RI

Thomas S. Morra
Chemistry Teacher
Mount Pleasant High School
Providence, RI

Timothy J. Brown
Chemistry Teacher
Woonsocket High School
Woonsocket, RI

A Natural Approach to Chemistry

Using the text

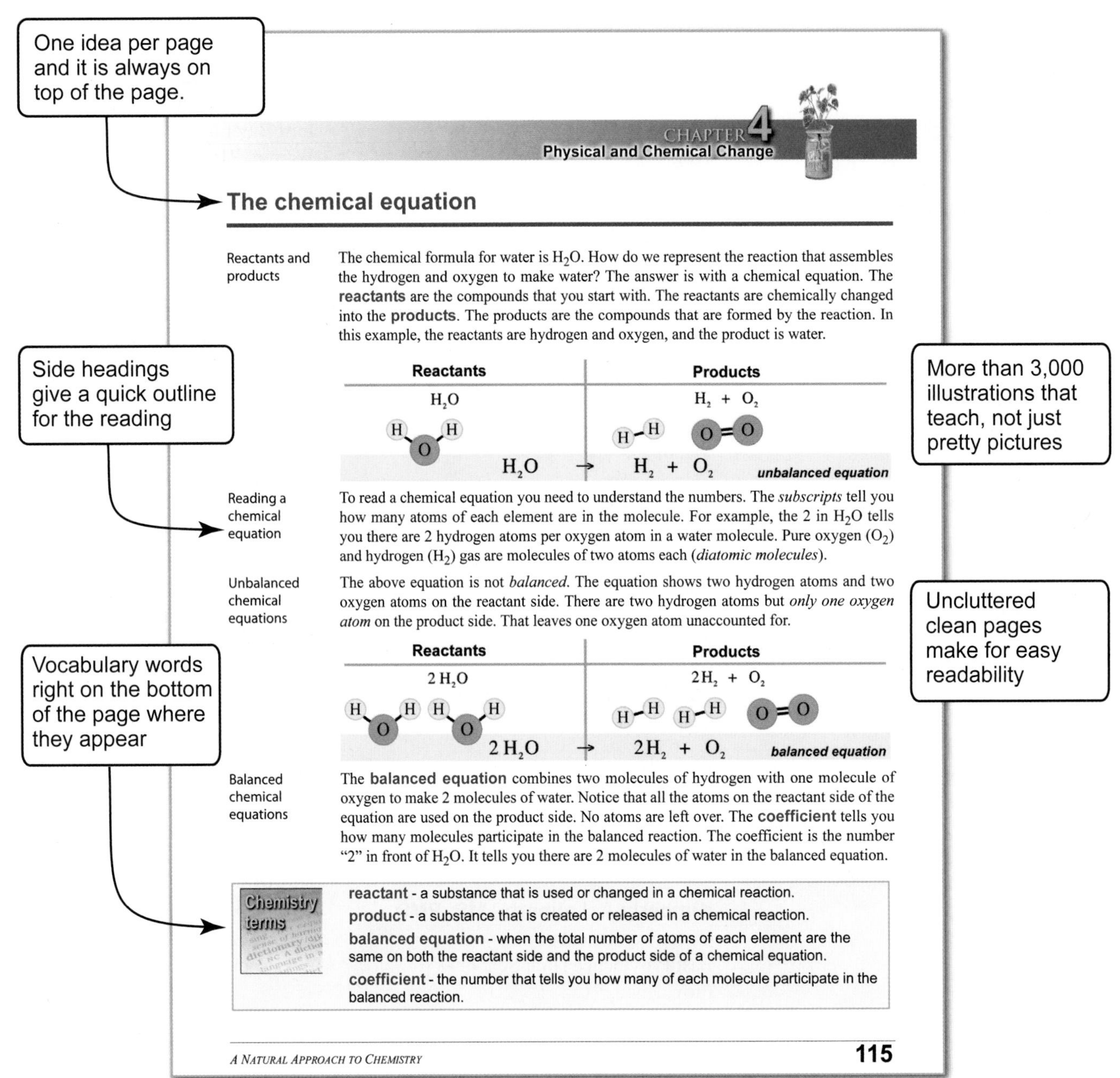

One idea per page and it is always on top of the page.
Side headings give a quick outline for the reading
Vocabulary words right on the bottom of the page where they appear
More than 3,000 illustrations that teach, not just pretty pictures
Uncluttered clean pages make for easy readability

The chemical equation

Reactants and products

The chemical formula for water is H_2O. How do we represent the reaction that assembles the hydrogen and oxygen to make water? The answer is with a chemical equation. The **reactants** are the compounds that you start with. The reactants are chemically changed into the **products**. The products are the compounds that are formed by the reaction. In this example, the reactants are hydrogen and oxygen, and the product is water.

Reactants	Products
H_2O	$H_2 + O_2$

$H_2O \rightarrow H_2 + O_2$ ***unbalanced equation***

Reading a chemical equation

To read a chemical equation you need to understand the numbers. The *subscripts* tell you how many atoms of each element are in the molecule. For example, the 2 in H_2O tells you there are 2 hydrogen atoms per oxygen atom in a water molecule. Pure oxygen (O_2) and hydrogen (H_2) gas are molecules of two atoms each (*diatomic molecules*).

Unbalanced chemical equations

The above equation is not *balanced*. The equation shows two hydrogen atoms and two oxygen atoms on the reactant side. There are two hydrogen atoms but *only one oxygen atom* on the product side. That leaves one oxygen atom unaccounted for.

Reactants	Products
$2\,H_2O$	$2H_2 + O_2$

$2\,H_2O \rightarrow 2H_2 + O_2$ ***balanced equation***

Balanced chemical equations

The **balanced equation** combines two molecules of hydrogen with one molecule of oxygen to make 2 molecules of water. Notice that all the atoms on the reactant side of the equation are used on the product side. No atoms are left over. The **coefficient** tells you how many molecules participate in the balanced reaction. The coefficient is the number "2" in front of H_2O. It tells you there are 2 molecules of water in the balanced equation.

Chemistry terms

reactant - a substance that is used or changed in a chemical reaction.

product - a substance that is created or released in a chemical reaction.

balanced equation - when the total number of atoms of each element are the same on both the reactant side and the product side of a chemical equation.

coefficient - the number that tells you how many of each molecule participate in the balanced reaction.

A NATURAL APPROACH TO CHEMISTRY

Problem solving

Concentration

Potassium concentration in your body

Potassium is critical to many of your body's functions, such as transmitting nerve signals. In a healthy person, potassium is dissolved in blood at a concentration of 140 to 200 milligrams (mg) per liter. A blood potassium concentration of less than 130 mg/L causes muscle weakness and heart rhythm instability (*hypokalemia*). A concentration of more than 215 mg/L may also result in heart instability (*hyperkalemia*). Since potassium is lost through excretion and sweat, you need to eat foods containing 4 - 5 grams of potassium (4,000 - 5,000 mg) each day. Fruits and vegetables are the best dietary sources.

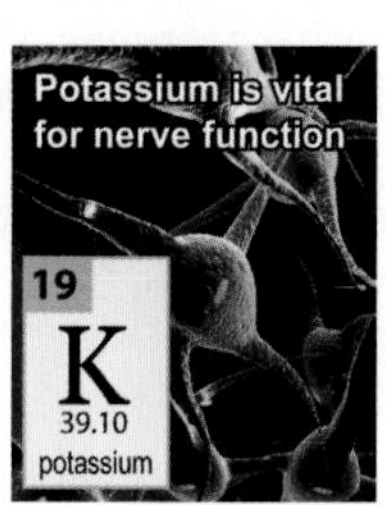

Quantitative relationships are clearly given with units

Different ways to measure concentration

Three common ways to express concentration are:

- As a percent, using mass of solute per liter of solution (g/L or mg/L)
- As a percent using mass of solute ÷ by total mass of solution
- In molarity, which is moles of solute per liter of solution (M)

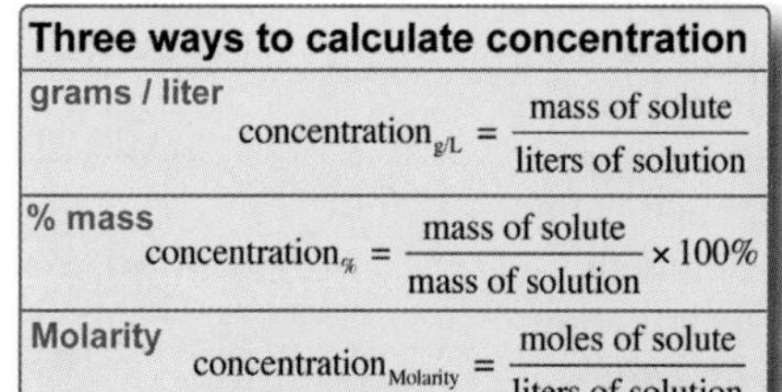

Three ways to calculate concentration

grams / liter

$$\text{concentration}_{g/L} = \frac{\text{mass of solute}}{\text{liters of solution}}$$

% mass

$$\text{concentration}_{\%} = \frac{\text{mass of solute}}{\text{mass of solution}} \times 100\%$$

Molarity

$$\text{concentration}_{\text{Molarity}} = \frac{\text{moles of solute}}{\text{liters of solution}}$$

Solved problems use consistent strategy

Very low concentrations

Environmental scientists often use a variation of percent concentration. Parts per million (ppm), parts per billion (ppb), and parts per trillion (ppt) are commonly used to describe very small concentrations of solutes in the environment. These terms are measures of the ratio (by mass) of one material in a much larger amount of another material. For example, a pinch (gram) of salt in 10 tons of corn chips is about 1 g salt per billion grams chips, or a concentration of 1 ppb.

Suppose you dissolve 10.0 grams of sugar in 90.0 grams of water. What is the mass percent concentration of sugar in the solution?

Asked: *What is the mass percent concentration?*

Given: *10 grams of solute (sugar) and 90 grams solvent (water)*

Relationships: $\text{concentration} = \frac{\text{mass of solute}}{\text{total mass of solution}} \times 100\%$

Solve:

$$\text{concentration} = \frac{\text{10 grams sugar}}{(10+90)\text{ grams of solution}} \times 100\% = 10\%\text{ sugar}$$

A NATURAL APPROACH TO CHEMISTRY

Organization of the book

Chemistry is important, relevant, and learnable by all students.

Themes

Energy is a unifying theme that explains why chemistry occurs
The atomic model of matter is consistently woven through every chapter
Understanding of "why" chemistry occurs is emphasized
Principles are illustrated with examples from the human body and the environment

Chapters 1 - 4: Fundamentals

Present a comprehensive overview of all the main ideas in chemistry such as the atomic nature of matter, systems, temperature and energy. Students don't get bogged down in the details, but see the conceptual "big picture."

Chapters 5 - 14: Core Concepts

These chapters present in-depth coverage of all major topic areas. They develop *usable* understanding of the big ideas laid out in the first four chapters. The treatment includes strong conceptual development as well as algebra-based quantitative problem solving. *All academic content and instruction standards for chemistry have been met by the end of Chapter 14.*

Each chapter begins with a real-lab example as a point of engagement for the content of the chapter.
3-5 additional labs per chapter develop more aspects of chemistry.

Chapters 15 - 21: Applications

The final seven chapters provide deeper exploration of significant areas of interest in chemistry. For example, Chapter 15 addresses the chemistry of rechargeable batteries. Chapter 16 addresses materials science, such as metal alloys. Chapter 21 looks at recent evidence for oceans of liquid water on Europa and Enceladus and the sulfuric atmosphere of Venus.

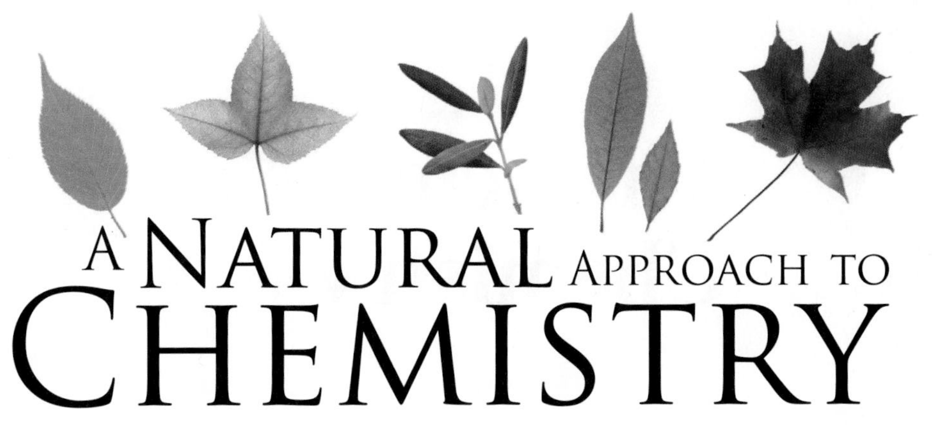
A NATURAL APPROACH TO
CHEMISTRY

CHAPTER 1

The Science of Chemistry

What is chemistry?

Why is chemistry important to know?

What does *chemistry* mean to you? What do you think of when you read the word *chemical*? Do you think of smelly concoctions in test tubes, mixed by a scientist in a white lab coat?

Would it surprise you to know that YOU are a mixture of chemicals? So is a tree, the ocean and even the cleanest, purest air. A glass of orange juice contains more than 100 chemicals. In fact, the word chemical describes any pure material around you, from the chlorophyll in a plant to the vitamins in your food. Water (H_2O) is a chemical. Sodium chloride, or table salt (NaCl), is another chemical. In being “alive”, your body is constantly changing chemicals into different chemicals, both using and releasing energy in the process.

Chemistry is the science of matter and its changes. The process that creates rain from evaporating water is part of chemistry. Digestion of food is part of chemistry. In short, chemistry describes the material world around you and the myriad of ways in which the material world changes. As you will see, not all chemicals are hazardous to your health! Our world is chemicals, as are we. Some chemicals are necessary for growth and survival, others are poisons. A wise person knows enough about chemistry to tell the difference!

Is this chemistry?
Is this chemistry?

The case of the disappearing sugar

Put 15 grams of warm water into a 20 mL graduated cylinder. The water level should come to 15 mL.

Measure out 5 grams of sugar on a weighing tin. Add this to a second small graduated cylinder.

- *What are the volumes of the water and sugar?*
- *What volume do you expect when you add them together?*

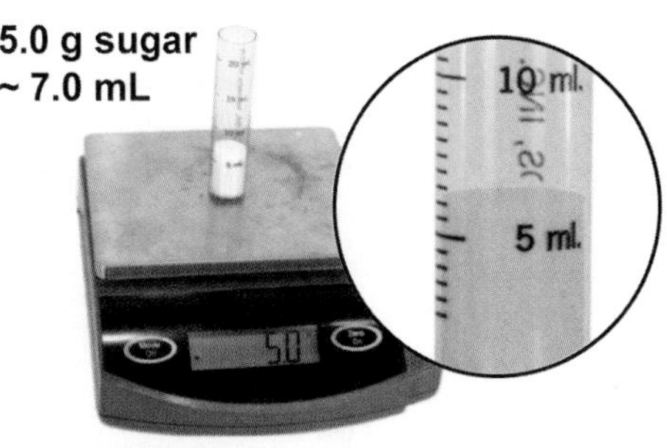

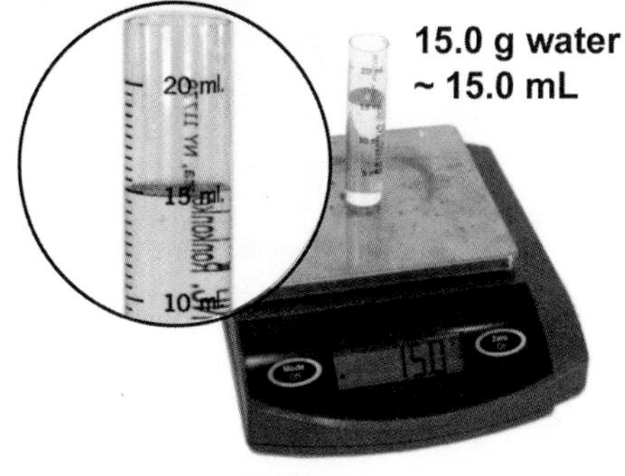

Add the sugar to the vial of warm water and stir until it is all dissolved.

A mixture of sugar dissolved in water is an example of a *solution*. Most of the sodas and juices you drink are solutions of sugar and other flavors in water.

When you do the experiment, here's what happens:

1) The sugar dissolves and appears to vanish

2) 5 g of sugar + 15 g of water = 20 g solution

3) 7 mL of sugar + 15 mL of water = *18 mL of solution!!!*

What happened to the sugar? You can't see it. The water looks just like it did before. Also, why didn't 7 mL + 15 mL add up to 22 mL? Whatever sugar is, some of it seems to have vanished!

Of course, the sugar didn't vanish. The graduated cylinder does contain 20.0 grams of matter. Why is the volume only 18 mL instead of 22 mL?

There are two parts to the explanation.

First, granulated sugar has some air trapped between the tiny grains of solid sugar. The air spaces get filled with water in the solution, but if this were the whole explanation, it would imply that 4 mL out of 7 mL, or 57% of powdered sugar is air. There is not *that* much air in powdered sugar.

Second, water really consists of tiny particles (molecules) whirling around with some space between them. Sugar also consists of tiny molecules. Under a magnifying glass you can see that the sugar is tiny solid crystals. However, individual sugar crystals are NOT the molecules I mean. While sugar crystals are small, the fundamental sugar molecules inside are a thousand, thousand times smaller still. When sugar dissolves, the sugar molecules partially fit in between the water molecules. That's the other half of why in this case, 7 + 15 = 18!

1.1 What Chemistry is About

Where does the mass of a tree come from?

When a tree grows, where does the matter come from? How does the tree get bigger? We know that matter cannot come from *nothing*. A big redwood tree might weigh 500,000 pounds. If all this weight came from the soil, we would see big holes in the ground around large trees. But, we do NOT see big holes around trees! Experiments reveal that only a tiny fraction of mass is lost from the soil around a tree. If the matter in a tree does not come from the soil, then where does all this matter come from?

Most of the solid matter in a tree comes from air

The answer is mostly *from the air*! A tree takes in carbon dioxide from the air and water from the ground. Carbon dioxide is a gas, but it is still matter. You don't "feel" the weight of a gas like air, because the matter is spread very thinly. However, there is a *lot* of air - more than enough to create majestic trees! An incredible chemical reaction called photosynthesis plucks carbon atoms from carbon dioxide in the air and compacts them into molecules of a sugar called glucose. The glucose is used to build the tree along with water and a tiny amount of minerals from the soil. In Chapter 10 we will learn a lot more about photosynthesis and other chemical reactions that are important to life on Earth.

Fire: a chemical reaction that changes matter from one form to another

What is fire? Is it a chemical? Is it matter? Is it energy? In early times it was believed that air, earth, water and fire were the major elements of the universe. You can feel fire, see it and smell it, so our senses tell us it is present. Water, air and earth are all made up of many atoms connected together, so they do represent types of matter! Fire is a form of heat and light energy, not matter. Chemical reactions transform one kind of matter, such as wood, into other kinds of matter, such as ashes, smoke and water. Fire is produced when matter changes its form. For example, when wood burns, the carbon reacts with oxygen producing energy and carbon dioxide. This energy is the heat and the light that we feel from a fire.

Chemistry in your body

Why should you eat foods high in antioxidants?

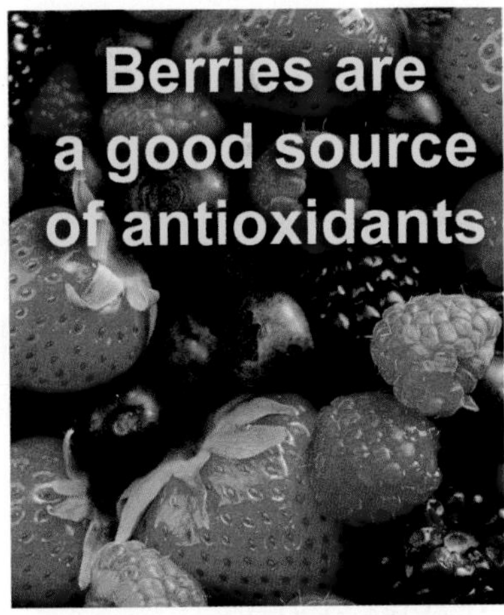
Berries are a good source of antioxidants

We are told to eat foods that are high in antioxidants, such as blueberries, cranberries, artichokes and beans. How does this help? The answer has to do with the chemistry of life. When energy in food is used by our cells, chemicals called *free radicals* are produced as a by-product. Free radicals are very *reactive*. Think of a pin-ball game where the ball is the "free radical". Anything the pinball hits may become damaged. The ball continues to bounce around and damage useful chemicals, until it is absorbed by something. In the pinball machine an "absorbed" ball falls to the bottom and is no longer "in-play" causing damage.

What do antioxidants do?

In your body, antioxidants absorb the free radicals before they can do much damage. Once neutralized by an antioxidant, free radicals are harmless. That is why it is important to have high levels of antioxidants in your body. Antioxidants are even more important with age because your metabolism becomes less efficient. This causes the number of free radicals to increase. Antioxidants have been shown to help prevent age-related diseases such as cancer.

What is the purpose of vitamins?

Vitamins are chemicals that the body needs but does not produce itself from the raw materials in food. For example, vitamin C is used by the body to make *collagen*. Collagen is in joints and tendons that connect muscles and bones. If you don't get enough vitamin C from fresh fruits and vegetables, you may develop *scurvy*. Scurvy is a disease that used to sicken or even kill sailors in Columbus' day.

Too much of certain vitamins can be as harmful as too little. All vitamin supplements list the Recommended Daily Allowance (RDA) for each vitamin

Different kinds of vitamins

Some vitamins dissolve in fat (vitamins A, D, E, K) and others dissolve in water (vitamin C and B-complex). The water soluble vitamins need to be consumed each day since they are excreted in urine. The fat soluble vitamins are stored in your liver, so they do not need to be replaced in your body every day.

Humans require vitamins in very small amounts; you can get sick from either too much (overdose) or too little (deficiency). That's why scientists came up with nutritional guidelines that give approximate daily recommended amounts of each one.

Water, an important chemical

Most of your body is water

A physically fit human body is about 60% water by weight. Take away the water and the rest is mostly different proteins and fats, with about 6% other chemicals and minerals such as calcium, phosphorus and iron. During one hour of exercise, you may lose as much as a half-gallon of water by sweating and breathing! You also lose small amounts of dissolved salts. If the lost water and dissolved salts are not replaced, your body stops working.

Human body
(physically fit)

60% water
17% proteins
17% fats
6%
calcium
phosphorus
glycogen
sodium
potassion
iron
zinc
sulfur
misc

Water is critical to life

The process of living requires your body to change the chemicals in food into other chemicals needed by your blood, muscles, nerves and the rest of your body. Solids, like sugar, cannot easily move around in your body. This is why most of the chemistry of living happens only in water. The two main reasons are:

1. Many substances dissolve in water, including sugars, salts and proteins.
2. Once dissolved, chemicals can circulate and interact with other chemicals.

Sugar is a chemical that dissolves in water

A good example of a chemical that dissolves in water is sugar. Sugar is made of tiny particles (molecules) that are so small that more than a million would fit on the head of a pin. Water is also made of tiny particles! A tablespoon of sugar is solid. The sugar molecules are stuck to each other. When sugar is mixed with water, the tiny molecules separate from each other. Each sugar molecule disperses among many water molecules. In fact, the sugar seems to visibly "disappear" once it has been dissolved. One taste will tell you the sugar is still there, but it has been dissolved.

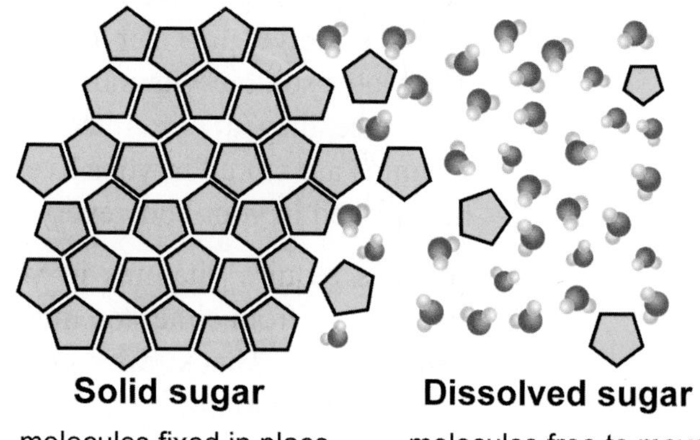

Measurements and units

Why units are necessary

In science, it is often not enough to say, "it's hot". Scientists want to communicate precisely how hot. To describe how hot it is you might say 82 degrees Fahrenheit (82°F). The value 82°F has two parts: a number (82) and a unit (°F). The number tells you how much and the unit tells you what the number means. One without the other can lead to big mistakes. If you say 82 degrees without specifying the unit of Celsius or Fahrenheit, a person from Canada might think you mean 82 Celsius (82°C) which is 180°F. 82°C would be fatal, while 82°F is comfortable for the beach!

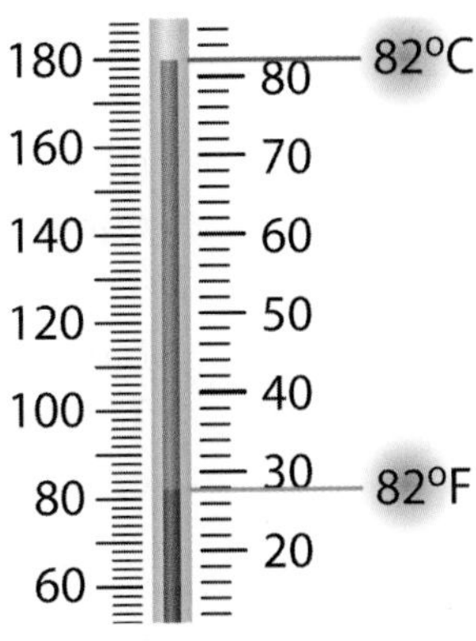

Kinds of units you use in chemistry

To understand chemistry, you need to speak the language of units. Just about everything that can be measured has its own unit. Below are some of the ones that are most useful.

TABLE 1.1. Some Units in Chemistry

Unit	Used for	Unit	Used for
Celsius degree (°C)	temperature	meter (m)	length
Fahrenheit degree (°F)	temperature	centimeter (cm)	length
kilogram (kg)	mass	mole (M)	counting atoms
gram (g)	mass	joule (J)	energy
liter (L)	volume	watt (W)	power
milliliter (mL)	volume	Pascal (P)	pressure
second (s)	time	atmosphere (atm)	pressure

Quantities have more than one unit

Notice that there is more than one unit for the same quantity. Your laboratory balance will measure mass in grams. A graduated cylinder measures volume in milliliters. A stopwatch measures time in seconds.

Definition of measurement

When you put a substance on a balance, you are making a **measurement** of its mass. A measurement is a specific kind of information that describes a physical quantity with both a number and a unit. The value 105.4 grams is a measurement because it has a number (104.5) and a unit (grams) that describe a real, physical quantity.

Chemistry terms

measurement - information that describes a physical quantity with both a number and a unit.

Mass and weight

Mass

Mass is a measure of how much matter there is. A grain of salt has very little mass and a planet has a great amount of mass. Note that size (or volume) does NOT tell you how much matter there is. A block of foam and a brick can be the same size, but one contains much more matter than the other. Don't be fooled by size. The definitive, reliable way to communicate a quantity of matter is by giving its mass.

Weight

A digital balance uses **weight** to measure mass. Weight is the *force* of gravity. The balance senses the weight of whatever you place on it. Then it calculates the mass by dividing by the number 9.8, which is the acceleration due to gravity. Pounds or ounces are units of weight, not mass, in the English system of units.

The kilogram

Mass is measured in **kilograms (kg)** and **grams (g)**. One kilogram is about the mass of a one liter bottle of water you buy in the grocery store. One gram is about the mass of a single peanut. There are 1,000 grams in 1 kilogram and 0.001 kilograms in 1 gram.

Equal masses means equal amounts of matter

When we say an object has a mass of one kilogram, we are saying that the object has the same amount of matter as one kilogram of water. Even if we do not know what the object is made of, we can measure how much matter it has. Air is very light, but a cubic meter of air at sea level has a mass of about 1 kg. If you put your hand out of the window in a moving car you can feel the mass of the air.

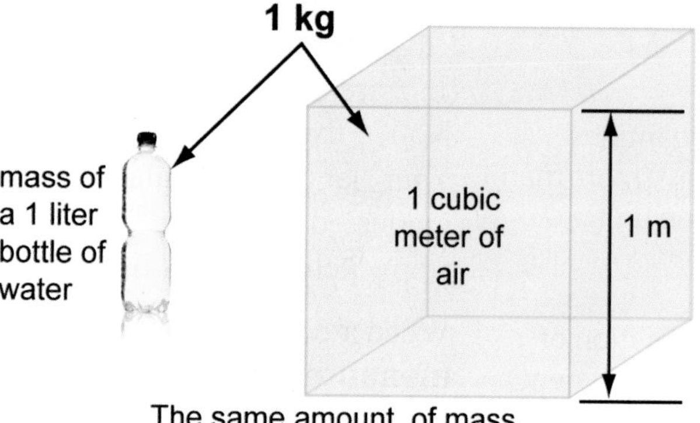

The same amount of mass

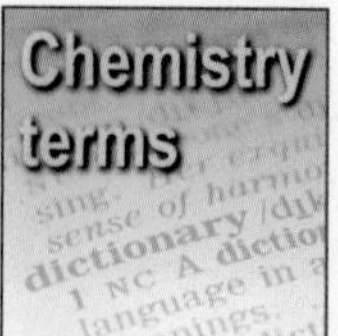

mass - measures how much matter there is, units of grams or kilograms (SI).

weight - a force (push or a pull) that results from gravity acting on mass, measured in newtons in the SI system of units and pounds or ounces in the English system of units.

kilogram, gram - the SI units of mass. There are 1,000 grams in a kilogram.

Volume

The definition of volume

Volume is an amount of space having length, width and height. A large object, like a box truck, takes up a lot of space and has a large volume. A tiny object, like a mouse, has a proportionally tiny volume.

The units of volume

In your chemistry lab, most of your measurements of volume will be in **liters** (L) or **milliliters** (mL). One liter is the volume of a cube ten centimeters on a side. One milliliter is the volume of a cube one centimeter on a side. A milliliter is a fairly small volume; there are five mL in a standard teaspoon. The chart below shows several volume units and their equivalent in liters and milliliters.

Volume units are derived from length units

Fundamentally, the unit for volume comes from the unit for length. A centimeter (cm) is a unit of length. The volume of a cubic centimeter is 1 cm × 1 cm × 1 cm = 1 cm^3. A milliliter is the same volume as a cubic centimeter. One liter, the volume of a soda bottle, is the same as 10 cm × 10 cm × 10 cm =1,000 cm^3. Empty a one liter bottle and it just fills a 10 centimeter cube. For larger volumes, cubic meters (m^3) are used. There are 1,000 liters in a cubic meter.

The graduated cylinder

The **graduated cylinder** is a tool for measuring the volume of liquids. Most graduated cylinders have markings that read in milliliters. The example shows a volume of 75 mL. Notice that you read the graduated cylinder at the flat (lowest) part of the surface. Water adheres slightly to glass and plastic, so the surface creeps up around the edges. This shape is called a *meniscus*. For a graduated cylinder you read the lowest point of the meniscus, NOT the highest point.

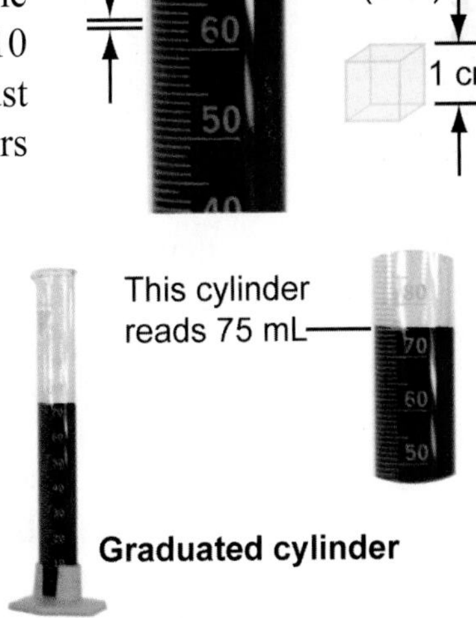

volume - an amount of space having length, width and height.
liter - an SI unit of volume equal to a cube 10 centimeters on a side, or 1,000 cm^3
milliliter - an SI unit of volume equal to a cube 1 cm on a side, or 1 cm^3
graduated cylinder - a measuring instrument used to measure volume.

Density

Equal size does not mean equal mass

You come into contact with many kinds of matter every day. Some matter is solid and hard, like iron and plastic. Some matter is liquid like water; some is a gas like air. Even within the category of solid matter, there are big differences. Think about the differences between polyethylene plastic, iron and glass. Imagine you had a solid cube of each substance, and all the cubes were the same size and painted black. Does each contain the same amount of matter? Could you tell which was plastic, iron or glass?

Equal size does not mean equal mass

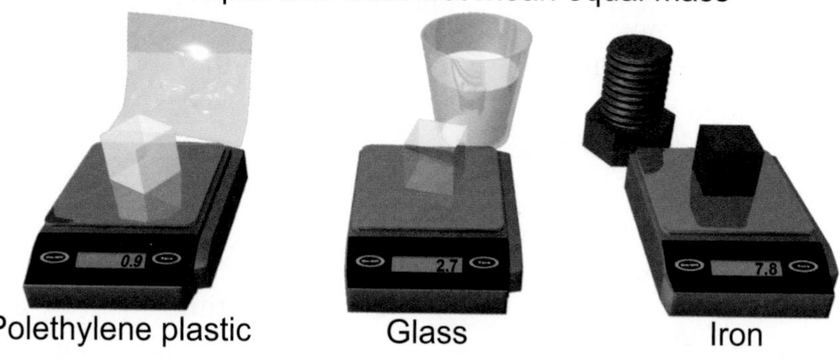

Density describes the mass per unit volume

A block of plastic and a block of steel may be the same size but one has a lot more *mass* than the other. Because of the difference, plastic floats in water and iron sinks. Whether an object floats or sinks in water is related to its *density*. **Density** describes how much mass is in a given volume of a material. The units of density are mass divided by volume, often grams per cubic centimeter (g/cm^3). Iron has a high density; it contains 7.8 grams of mass per cubic centimeter (7.8 g/cm^3). A one centimeter cube of polyethylene plastic contains only 0.94 grams of matter (0.94 g/cm^3).

Density

$$\underset{(g/cm^3)}{\text{density}} = \frac{\text{mass (g)}}{\text{volume }(cm^3)}$$

$$d = \frac{m}{V}$$

1. Density is a property of matter - independent of size or shape
2. Density is mass per unit volume.

The density of water and air

Solids range in density from cork (0.12 g/cm^3) to platinum, a precious metal with a density of 21.5 g/cm^3. The density of water is about one gram per cubic centimeter. The density of air is much lower, about 0.001 grams per cubic centimeter.

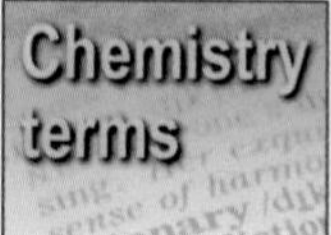

density - a property of a substance that describes how much matter the substance contains per unit volume - typical units are grams per cubic centimeter (g/cm^3)

Doing calculations with density

Measuring density

To find the density of a material, you need to know the mass and volume of a solid sample of the material. Mass is measured with a balance or a scale. For simple shapes you can calculate the volume. For irregular objects the *displacement method* is used to find the volume. Suppose you want to know the volume of five steel nuts. You record the volume of water in the graduated cylinder before inserting the nuts. Then you gently drop the five nuts in and record the volume again. The volume of the nuts is the change in volume.

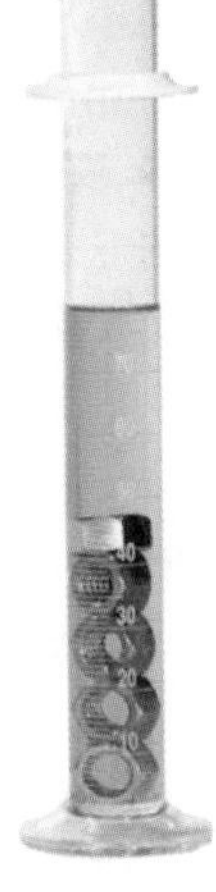

Densities of common substances

TABLE 1.2.
Densities of common substances

Material	Density (g/cm^3)	Material	Density (g/cm^3)
Platinum	21.5	Nylon Plastic	2.3
Lead	11.3	Rubber	1.2
Iron	7.8	Liquid water	1.0
Titanium	4.5	Polyethylene plastic	0.94
Aluminum	2.7	Ice	0.92
Glass	2.7	Oak (wood)	0.60
Granite	2.6	Pine (wood)	0.44
Concrete	2.3	Cork	0.12

45 grams of titanium are added to a graduated cylinder containing 50 mL of water. What will the cylinder read after the titanium has been added?

Asked: *Volume of graduated cylinder after adding 45 grams of titanium*

Given: *45 grams of titanium, density of titanium* $d=4.5\ g/cm^3$*, 50 mL of water*

Relationships: $d = m/V$

Solve: $d = m \div V$, *therefore* $V = m \div d$

$$V = 45\ g \div 4.5\ g/cm^3 = 10\ mL.$$

The titanium adds 10 mL to the cylinder which now reads 60 mL.

Answer: *60 mL*

Discussion: *This is an example of a displacement method measurement.*

Pressure

Force and fluids

Think about what happens when you pump up a bicycle tire. As you push down on the pump, you squeeze air in the tire. The tire expands and gets firm. The firmness of the inflated tire is an example of *pressure*. **Pressure** is a force per unit area exerted by matter, often when it is restrained from moving or expanding. For example, suppose the bicycle tire is inflated to a pressure of 60 pounds per square inch (60 psi). Each square inch of the inside of the tire feels a force of 60 pounds from the trapped air in the tire.

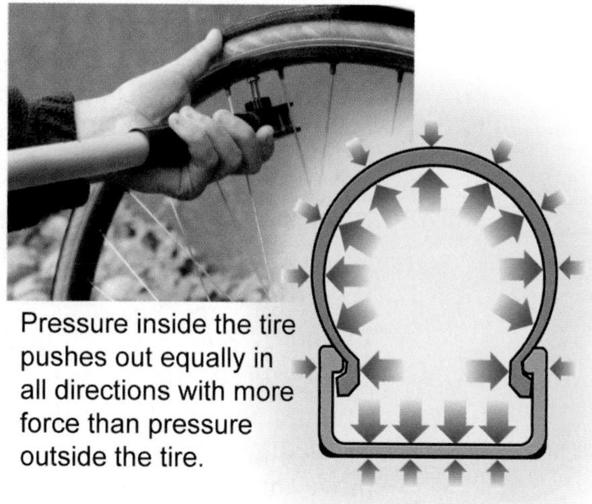

Pressure inside the tire pushes out equally in all directions with more force than pressure outside the tire.

Force

A **force** is an action, like a push or a pull, that has the ability to make things move (or stop them). In the SI system, force is measured in newtons. However, not many people use newtons in ordinary life. It takes about four and a half newtons to make a pound (4.448 N = 1 lb). In chemistry, we are not so much interested in forces, but in the ratio of force per unit area, which is pressure.

Pressure is most important in the understanding of gases (like air) or liquids (like water). For example, consider the water in a hose. If there is pressure in the water, the pressure exerts forces in all directions, on any surface touching the water. You can prove this is true by poking a hole in the hose. The water squirts out no matter whether you poke the hole in the top, bottom, or side of the hose.

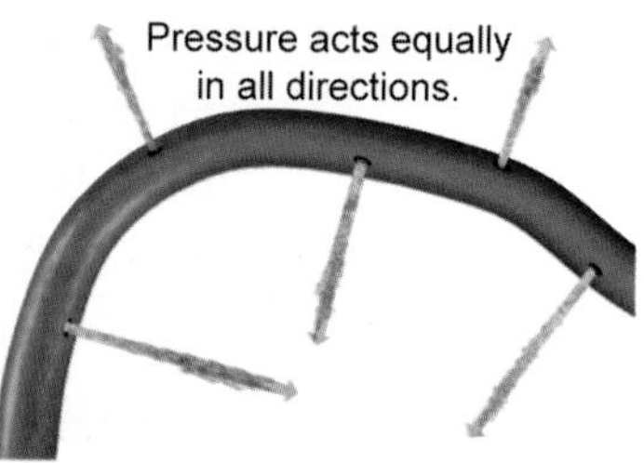
Pressure acts equally
in all directions.

Pressure in a liquid or gas pushes equally in all directions

You know that liquids flow to take the shape of a container. This happens because liquids have weight. Weight pulls down, but in a liquid or gas, the downward force of gravity force becomes pressure which pushes equally in all directions. Pressure pushes the boundary of a liquid outward until it fills its container in the precise shape that gets the most liquid closest to the ground.

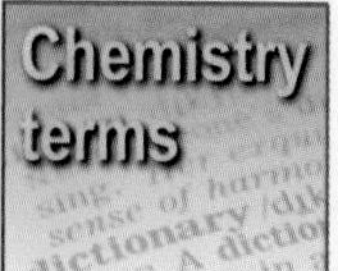
Chemistry terms

pressure - a expansive force per unit area that acts equally in all directions within a liquid or a gas.

force - an action such as a push or a pull that has the ability to change the motion of an object, such as to start it moving, stop it, or turn it.

Air pressure

Is an empty bottle really empty?

Hold up an empty plastic water bottle and think about what is inside. What comes to mind? Nothing? Is an empty bottle really full of nothing? To answer the question, consider what happens when you put the cap tightly on the empty bottle and try to squeeze it flat. You can't do it. Something in the "empty" bottle prevents you from squeezing it flat. If you open the cap, you can squeeze the bottle easily. Whatever is inside an empty bottle comes out if the cap is off, but is trapped inside if the cap is on.

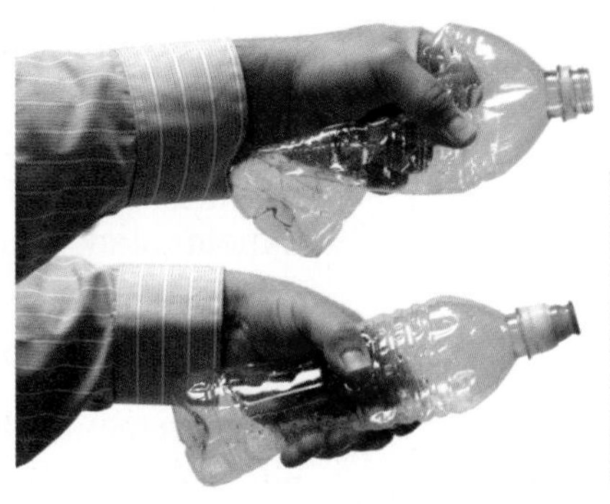

Air is matter

Air is matter! Air is transparent and its mass is spread so thin that you move through it easily. Put your hand out a moving car window and you can instantly feel that air has mass because it takes force to push it out of the way. In fact, a cubic meter of air has a mass of about 1 kilogram, about the same as a 1 liter bottle of water.

The weight of air creates air pressure

Air has weight. The weight of the atmosphere creates pressure that acts equally in all directions. A board with an area of one square meter feels a force from air pressure of 101,325 newtons on one side. That is an incredible 22,780 pounds! The board doesn't move because air pressure acts equally in all directions and the other side of the board feels an equal force in the opposite direction. The board doesn't move, but it's not because air is nothing! The board doesn't move because the air pressure is equal on all sides.

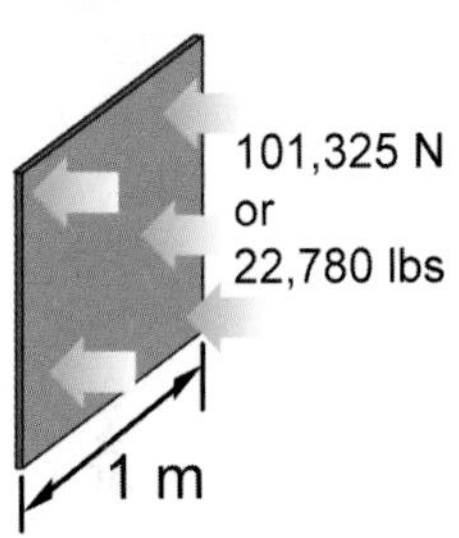

Demonstrating air pressure

To demonstrate air pressure, watch what happens when you suck the air out of a bottle. Why does the bottle collapse? You did not apply any force to it. The bottle collapses because removing some air from the inside makes the pressure inside lower than the pressure outside. The pressure of the *outside* air is what crushes the bottle.

Units of pressure

If you have a force of 1 newton distributed over an area of 1 square meter you have a pressure of 1 **Pascal** (Pa). The pascal is the SI unit of pressure, but it is very small. Many chemistry applications use *atmospheres* as a unit of pressure. One **atmosphere** (1 atm) is 101,235 Pa and is the average pressure of the air at sea level.

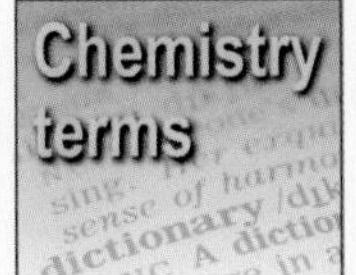

Pascal (Pa) - a very small unit of pressure equal to one newton of force per square meter of area.

Atmosphere (atm) - a large unit of pressure equal to 101,325 Pa, - the average pressure of air at sea level, also equal to 14.7 pounds per square inch (14.7 psi).

Accuracy and precision

Precision — When we take a number of measurements, **precision** tells us how close the measured values are to each other. A balance that can read to 0.1 grams has a precision of 0.1 g. Many of your lab experiments will be done with balances that have this precision.

Accuracy — The word **accuracy** tells us how close a measurement is to the true value. For example, a meter stick that has been stretched can make a measurement of length that is precise to one millimeter. However, measurement will not be accurate because the meter stick is no longer a meter long.

Measurements are never "perfect" — Measurements of real quantities in experiments are never exact. For example, you cannot determine that something has a mass of *exactly* 10 grams. Why not? Because measurements of mass are made with instruments, like balances. All real instruments have a limit to how small a quantity they can measure.

Is the mass exactly 10.0 g?

We don't know since any mass between 9.9 and 10.1 would round off to 10.0

Suppose you have an accurate balance that can measure mass to a precision of 0.1 grams. You claim the mass is exactly 10.0 grams because your balance shows 10.0. However, suppose the real mass was 10.02 grams. Your balance *rounded the measurement off to 10.0.* One way to show this is to write your measurement as 10.0 g +/- 0.1 g. This tells a reader that the actual mass could have been anything between 9.9 g and 10.1 g.

Why accuracy and precision are important — For many of your lab experiments, two masses of 10.01 g and 10.03 g can be considered the same because they differ by only 0.02 grams. This is smaller than the balance can measure. When analyzing the observations you make in the lab, you must consider both the accuracy and the precision of your measurements before making a conclusion. A measurement that is not *accurate* may give you the wrong conclusion. A measurement that is not *precise* may not be able to tell the difference between agreement or disagreement.

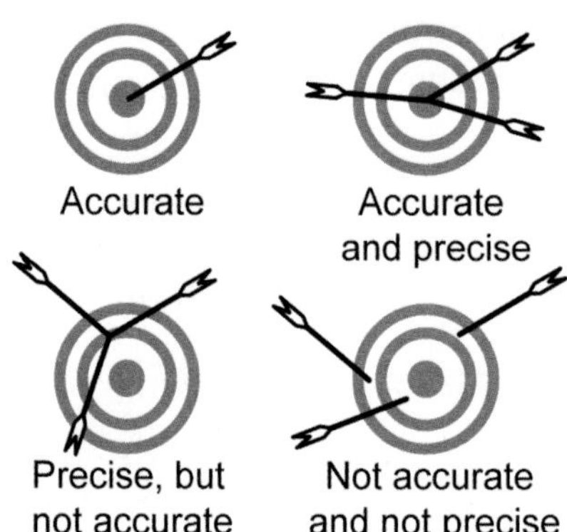

Chemistry terms

accuracy - describes how close a measurement is to the true value.

precision - describes how close measured values are to each other.

Significant figures

How many decimal places do you record?

Suppose you measure 10.0 grams of sugar on a balance. The balance has a precision of 0.1 grams. In your data table do you write down 10 grams or 10.0 grams? Does it make a difference?

Do you record 10 g or 10.0 g?

Significant figures tell the precision of a measurement

The answer is yes, it does make a difference. Scientists use the rule of **significant figures** as a code for telling the precision of a measurement without writing +/- after each value. The rule assumes *the last written digit is uncertain by plus or minus the value of the uncertainty.* For example, suppose you write the measurement as 10 grams. According to the rule of significant figures, a reader can "trust" the first digit completely (the 1) and assume the second digit (0) is plus or minus one, or +/- 1 grams. The actual mass of sugar could have been anything between 9 g and 11 g. This is an uncertainty of 1 out of 10, or 10% in the measurement!

10.0 g is a measurement with three significant figures

If you write the measurement as 10.0 grams, it now has three significant figures. A reader can assume they can "trust" the first two digits (10) and the uncertainty is only plus or minus one tenth, or +/- 0.1. The uncertainty is now 1 out of 100, or 1%. Significant figures are a way of recording data that tells a reader how precise a measurement was made. The word "significant" means you include the last digit that has any meaning when you record the result, even if that digit is a zero.

What value should be recorded for the volume measurement in the picture?

What is the volume?

20ml.

15 ml.

Asked: *for a value with the correct number of significant figures.*

Given: *you can estimate to a tenth of the graduation of a cylinder or ruler.*

Relationships: *the last digit on the right is assumed to be plus or minus one tenth.*

Solve: *The meniscus is right on 18, so estimate 18.0 mL.*

Answer: *18.0 mL*

Discussion: *The real value is confidently known to be between 17.9 and 18.1 mL.*

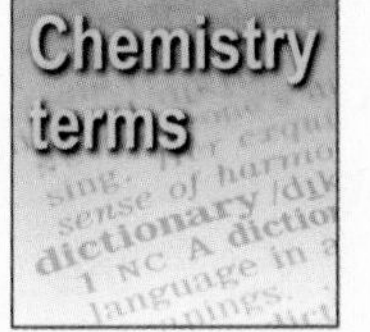

significant figures - a way of writing data that tells the reader how precise a measurement is. For example, 34.400 has 5 significant figures and we assume the uncertainty in the measurement is +/- 0.001. The number 34.4 has only three significant figures and the uncertainty is assumed to be +/- 0.1. Both numbers have the same value but have different precision.

Very large numbers

Describing large and small quantities

Atoms are so small that describing them requires extreme numbers. A single grain of sand contains 200,000 *million million* atoms. This number written out the usual way is

200,000 million million = 200,000,000,000,000,000

Don't worry if you cannot get your mind around this huge number—no one can without a reference. Fortunately, there is a shorthand method for writing and calculating with extremely large or small numbers. The method is called **scientific notation**.

Scientific notation for large numbers

Scientific notation works by expressing very small or very large numbers as the product of two smaller numbers. The first number is called the **mantissa**. The second number is a *power of ten*. Any number can be represented as a mantissa times a power of ten. As an example consider the number 1,500:

- 1,500 = 15 × 100. The number 15 is the mantissa. The number 100 is a power of ten.
- 100 = 10 squared, and is usually written 10^2.
- so 1,500 = 15 × 10^2 The small superscript number 2 in 10^2 is called the **exponent**.

Numbers larger than 1

10^1 = 10
10^2 = 100
10^3 = 1,000
10^4 = 10,000
10^5 = 100,000
10^6 = 1,000,000
10^7 = 10,000,000
10^8 = 100,000,000
10^9 = 1,000,000,000

Writing the base

The mantissa is usually written with only one digit in front of the decimal point. For example, 1,500 would be written 1.5 × 10^3 because 10^3 is 1,000. To make 1,500 into 1.5 you have to move the decimal three places so the correct power of ten is 10^3.

For the number 1,500, scientific notation does not seem to be very beneficial. However, 200,000 million million in scientific notation is 2 × 10^{17}. This number is much easier to write and calculate with when expressed in scientific notation. The number 6.02 × 10^{23} is very important in chemistry and it is very nicely presented in scientific notation.

Convert 34,500 to scientific notation.

Asked: *number in scientific notation*

Given: *34,500 as a decimal number*

Relationships: *10,000 =* 10^4

Answer: *3.45 ×* 10^4

scientific notation - a method of writing numbers as a base times a power of ten

mantissa - a decimal number that multiplies the power of ten in scientific notation.

exponent - the power of ten in scientific notation, e.g. in the number 1.5 × 10^2, the number 2 is the exponent.

Very small numbers

Chemistry may use very small numbers

Many of the important ideas in chemistry involve what happens on the size of individual atoms, the particles of matter. Atoms are extraordinarily small. A grain of sand is 0.01 centimeters across. A single atom is a million times smaller, or 0.00000001 centimeters across. Fortunately, scientific notation allows us to work with extremely small numbers as easily as with large numbers.

Scientific notation for small numbers

Powers of ten that are negative mean numbers smaller than one. Consider the number 0.0015. Let's write it in scientific notation:

- 0.0015 = 1.5 × 0.001.
- The number 0.001 is 1 ÷ 1000 = $1 \div 10^3 = 10^{-3}$.
- 0.0015 = 1.5×10^{-3} in scientific notation.

It is important to remember that a negative sign on the exponent of 10 *does not mean the whole number is negative*! Negative exponents mean a value that is less than one.

Numbers smaller than 1

10^{-9} = 0.000 000 001
10^{-8} = 0.000 000 01
10^{-7} = 0.000 000 1
10^{-6} = 0.000 001
10^{-5} = 0.000 01
10^{-4} = 0.000 1
10^{-3} = 0.001
10^{-2} = 0.01
10^{-1} = 0.1

Convert 0.00065 to scientific notation.

Asked: *number in scientific notation*

Given: *0.00065 as a decimal number*

Relationships: $0.0001 = 10^{-4}$

Answer: 6.5×10^{-4}

Using a calculator

Scientific calculators can work with numbers in scientific notation, IF you know how to enter them! On the specific calculator in the diagram, you use the following keystrokes to enter the number 1.5×10^{-3}. The keystrokes may be different on other calculators; however, the key labeled "EE" is usually the one that allows you to enter the exponent.

For this calculator (others may differ)

To enter the number 5.5×10^{-3} you press

5 . 5 2nd EE (-) 3

The calculator display reads

5.5 E -3

Converting between units

Two measurements can only be compared if both are in the same units

Suppose you drive to Mexico and gas at one filling station costs $1.25 per *liter*. Across the border in Texas, gas is $4.05 per *gallon*. Which is less expensive?

The answer is not obvious. Think of the different systems of units as different languages. To compare two measurements, both need to be in the same units. That means you must translate between language of dollars per liter and dollars per gallon. To translate Spanish to English you need a Spanish/English dictionary. To translate between units you need **conversion factors**. Conversion factors are ratios of two units. An example is that one gallon = 3.785 liters. To convert the gas prices, we use the conversion factor.

Convert $1.25/liter to dollars per gallon

Asked: *for a value in dollars per gallon.*

Given: *$1.25/L*

Relationships: *1 gallon = 3.785 liters*

Solve: $\left(\frac{\$1.25}{\cancel{\text{liter}}}\right)\left(\frac{3.785\ \cancel{\text{liter}}}{1\ \text{gallon}}\right) = \left(\frac{\$4.73}{\text{gallon}}\right)$

Answer: *$4.73/gallon*

Discussion: *$1.25/L is more expensive than $4.05/gallon*

Converting from one unit to another

To do a conversion, you arrange the conversion factors as ratios and multiply them so the units you don't want cancel out. You should be left with only the units you want. There are many conversion factors on the inside back cover of this book.

Another word for units is *dimensions*. The process of converting between units with conversion factors is also called **dimensional analysis**. Dimensional analysis is a very powerful tool and it is used extensively by scientists and engineers. It is also a very useful tool in our daily lives.

Some useful conversion factors

$\frac{1\ \text{meter}}{39.37\ \text{in}}$	$\frac{1\ \text{liter}}{1000\ \text{mL}}$	$\frac{1\text{kg}}{1000\text{g}}$
$\frac{1\ \text{inch}}{2.54\ \text{cm}}$	$\frac{1\ \text{gallon}}{3.785\ \text{liters}}$	$\frac{1\text{minute}}{60\text{sec}}$
$\frac{1\ \text{meter}}{100\ \text{cm}}$	$\frac{1\ \text{teaspoon}}{5\ \text{mL}}$	$\frac{1\text{hour}}{3600\text{sec}}$
$\frac{1\ \text{cm}}{10\ \text{mm}}$	$\frac{1\ \text{cm}^2}{1\ \text{mL}}$	$\frac{1\ ^{\text{g}}/_{\text{mL}}}{1000\ ^{\text{kg}}/_{\text{m}^3}}$

Chemistry terms

conversion factor - a ratio of two different units that has a value of 1. For example, 3.785 liters/1 gallon is a conversion factor. The numbers are different but the actual physical quantity is the same because 1 gallon is the same volume as 3.785 liters.

dimensional analysis - using conversion factors to convert between units.

1.2 Scientific Inquiry

The search for explanations

The ancient Greek scientist Aristotle thought that smoke rose because smoke was a form of fire and fire's natural place was in the sun. Rocks fell because rocks were made of earth and earth's natural place was on the ground. Aristotle was not stupid. His eyes told him the correct behavior of things. Smoke rises and rocks fall. However, today the explanation for *why* has changed. Today we explain the same observations using the ideas of gravity, density and buoyancy. Why do we think today's explanation is correct? If Aristotle was so smart (and he was) then why did he have the wrong explanation?

Why does smoke rise?
Why do rocks fall?

Natural laws

We believe that the universe obeys a set of rules that we call **natural laws**. We believe that everything that happens everywhere obeys the same natural laws. Unfortunately, the natural laws are not written down nor are we born knowing them. *The primary goal of science is to discover what the natural laws are*. Humans learned science by asking harder and harder questions. Aristotle's questions were among the earliest, and reflected a much smaller body of understanding than we have today.

Inquiry is learning by questioning

Learning through questions is called **inquiry**. An inquiry is like a crime investigation in that there is a mystery to solve. In a criminal inquiry, something illegal happened and the detective must figure out what it was. In a scientific inquiry, the mystery is how and why things happen the exact way they do.

Deduction and theories

Of course, the criminal detective did not see what happened. The detective must *deduce* what happened in the past from information collected in the present. In the process of inquiry, the detective asks lots of questions. The best questions lead to *evidence*. Eventually, the detective comes up with a **theory** about what happened.

When is a theory correct?

At first, the detective's theory is only one explanation among several of what *might have happened*. The detective must have proof that a theory describes what actually *did happen*. To be proven, a theory must pass three demanding tests:

1. It must be supported by significant evidence;
2. There can't be a *single piece* of evidence that disagrees with the theory;
3. The theory must be unique.

If two theories both fit the facts equally well, you cannot tell which is correct. Only when a theory passes all three tests does it become a potential solution to the mystery.

Chemistry terms

natural laws - the unwritten rules that govern everything in the universe.
inquiry - the process of learning through asking questions.
theory - an explanation that is supported by evidence.

Experiments and hypotheses

A hypothesis is a tentative explanation for something

Science is largely learned by trial and error over a long period of time and by many people. New scientific knowledge begins when you observe something interesting and wonder why or how it happens. For example, it takes sugar a long time to dissolve when you stir it into cold water. This leads to a question: does temperature affect how quickly sugar dissolves in water? You may make a guess, "I think sugar dissolves faster in hot water". This tentative answer, or educated guess, is called a **hypothesis**. A good hypothesis has outcomes you can test to see if they really happen the way you think they will.

I think sugar dissolves faster in hot water

Hypotheses are tested by experiments

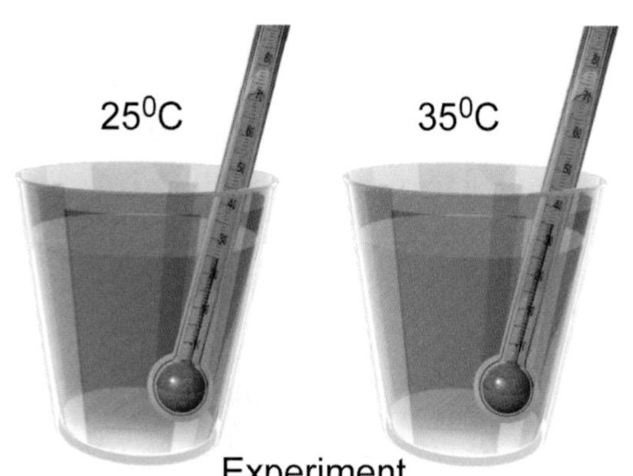
25^0C
35^0C
Experiment

An **experiment** is a situation you set up specifically to see what happens, often to test a hypothesis. You might try several different temperatures of water to see if the dissolving time is different. The results of your experiment help you change your hypothesis. You might even discard your first hypothesis altogether in favor of a new one that better fits the results of the experiment.

The truth is what actually happens, not the hypothesis!

The results of real experiments and observations are the test of what is correct and true in science. It does not matter what anyone thinks or says or even writes in a book. We believe the scientific ideas in this book only because they have been tested over and over with actual experiments, some for over a hundred years. You gain the benefit of that experience so you don't have to test every idea yourself. However, what is between these covers is just the beginning. Human knowledge of science is always growing. Someday we hope you add your own pages to a book of science! To unravel the mysteries of nature, your thinking must always be guided by what nature actually does, either through experiments or direct observations.

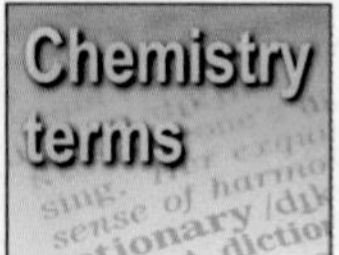
Chemistry terms

hypothesis - a tentative explanation for something, or a tentative answer to a question.

experiment - a situation specially set up to observe how something happens or to test a hypothesis.

Experimental and control variables

Designing good experiments

Suppose you do an experiment to see if sugar dissolves faster in hot water compared to cold water. You add different amounts of sugar to different cups of water at different temperatures. Cup B dissolves the fastest. What does the experiment tell you? Does adding more water makes sugar dissolve faster? Does temperature even matter?

Trial A	Trial B	Trial C
25^{0}C	45^{0} C	35^{0} C
5 g sugar	2 g sugar	8 g sugar
100 mL water	60 mL water	150 mL water
53 seconds to dissolve	35 seconds to dissolve	47 seconds to dissolve

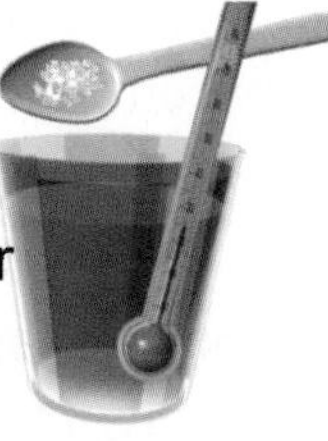

You cannot answer because you changed three things at once. A **variable** is a quantity, like temperature or volume, that you measure or change in an experiment. All you know is that the dissolving rate changed, but not *which variable* caused the change.

- Dissolving could be affected by the amount of water.
- Dissolving could be affected by temperature.
- Dissolving could be affected by the amount of sugar.

The experiment still *worked*. You just don't know how to understand the results.

The experimental variable

A better design for the experiment would be to add the same amount of sugar to the same amount of water in each of the three cups. Only the temperature is different. This way if you see a change in the time that it takes to dissolve then you know it came from changing the temperature. The single variable you change in an experiment is called the **experimental variable**. For this experiment, temperature is the experimental variable.

Trial A	Trial B	Trial C
25 ^{0}C	45 ^{0}C	35 ^{0}C
5 g sugar	5 g sugar	5 g sugar
100 mL water	100 mL water	100 mL water
53 seconds to dissolve	32 seconds to dissolve	44 seconds to dissolve

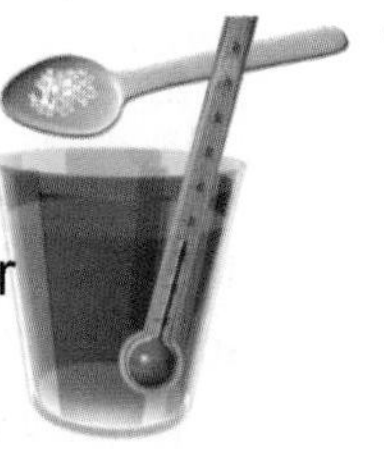

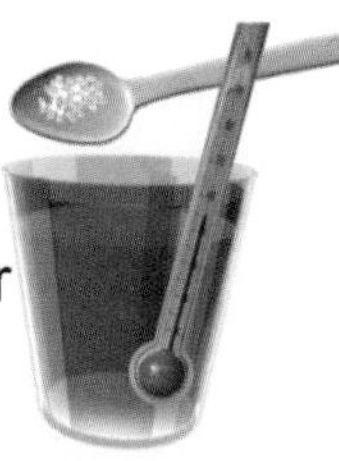

Control variables

If only one variable is allowed to change, then the others must be kept constant. Constant means NOT allowed to change. Variables that are kept constant are called **control variables**. The volume of water and the amount of sugar are control variables.

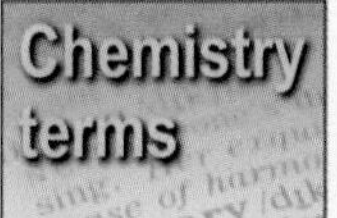

variable - a quantity that is measured or changed in an experiment or observation.
experimental variable - the single variable that is changed to test its effect.
control variables - variables that are kept constant.

Uncertainty and error

All measurements have error and ERROR IS NOT A MISTAKE!

Suppose you use a stopwatch to measure the dissolving time in the experiment with sugar and water. It is hard to tell the exact moment when all the sugar has been completely dissolved. If you repeat the experiment three times, each time you obtain a slightly different result *even if you do the exact same thing*. This is because any measurement always contains some uncertainty, or **error**. In the context of measurement, *error is not a mistake*. Error is the unavoidable difference between a measurement and the true value of what you are measuring. Errors can be small if the experiment is accurate and precise, but there is ALWAYS error in any measurement, no matter how careful you are.

Data from sugar dissolving experiment

Trial	Sugar	Water	Temp.	Time
1	5.0 g	100 mL	25.4 °C	53 sec
2	5.0 g	100 mL	25.1 °C	45 sec
3	5.0 g	100 mL	26.1 °C	48 sec
avg.	5.0 g	100 mL	25.5 °C	48.7 sec

The average

When you make many measurements of the same thing you should notice that they cluster around an **average** value. Some measurements are more than the average and some are less. To calculate the average, you add up all the measurements and divide by the number of measurements you have. For example, the average of the dissolving times is 48.7 seconds. Notice that the average has one more significant figure than any individual data point.

Taking the average

1 add up the values

$$\begin{array}{r} 53 \\ +\ 45 \\ +\ 48 \\ \hline 146 \end{array}$$

2 divide by the number of measurements

$$146 \div 3 = 48.7$$

Why taking the average is useful

There are two important reasons why you should always make several measurements and take the average:

1. The average of several measurements is usually more accurate than a single measurement. The average is more accurate because errors in the negative direction partially cancel errors in the positive direction. That's why there is one more significant figure.
2. The differences between the average and the actual measurements gives you an estimate of the possible error in your average result.

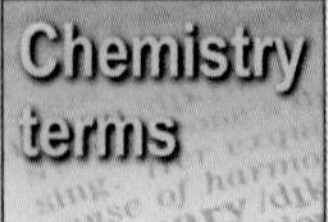

error - the unavoidable difference between a real measurement and the unknown true value of the quantity being measured.

average - you calculate the average of a set of measurements by adding up all their values and dividing by the total number of measurements.

Comparing data: when are two results the same?

What does it mean for results to be the same??

Earlier in this Chapter we said that one criteria for scientific evidence was that it be *reproducible*. Reproducibility means *two* things that we can now discuss more carefully:

1. Others who repeat the same experiment get the same result;
2. If you repeat the experiment the same way, you get the same result.

When different numbers are "the same"

According to rule #2, the experiment with dissolving fails the test for scientific evidence. Three repetitions gave three different measurements (53, 45, and 48 seconds). How can you tell if results are the *same* when all results have errors in them? The word "same" when applied to experimental measurements does not mean what it does in ordinary conversation. Two measurements are considered the same if their difference is less than or equal to the amount of error. *This is important to remember.*

Two measurements are considered the same if their difference is less than or equal to the amount of error

How to estimate the error

How can we know the error if we do not know the true value of a measurement? The way scientists estimate the error is to *assume the average is the true value.*

1. Find the difference between each measurement and the average
2. The error is roughly the largest difference between the average and a measured value.

Estimating the error

Time	Average	Difference
53 sec	48.7 sec	+4.3 sec
45 sec	48.7 sec	-3.7 sec
48 sec	48.7 sec	+0.7 sec

The largest difference between the data and the average is 4.3 seconds so the estimated error is about ±4 seconds

Later you may learn a more sophisticated way to estimate the error in a measurement, called the *standard deviation*. For now, the largest difference is 4.3 seconds so, by rounding, the error is +/- 4 seconds. If the largest value of the error was 4.6 then we would have rounded up to 5.

Significant figures in the result

The result of our experiment is the average value plus or minus the error. What is the correct number of significant figures? We know that we can't have more significant figures in our result than the number of significant figures that our measurements have. In other words the calculations performed on the data, like the average in this case, can not increase the number of significant figures. Our date have two significant figures and so the average should also have two significant figures. The calculated average of 48.7 is rounded up to 49. The error is +/- 4 seconds and the result is 49 +/- 4 seconds.

Drawing conclusions from data

We do experiments to reach conclusions

The point of experiments is to produce data that allows a scientist (like you) to come to a **conclusion**. A conclusion is a stated decision whether the results of experiments or observations confirm your idea, the hypothesis, is right or not. *You need to know about errors before you can make a conclusion.*

Are these results different?

Here is a good example. Suppose another group measures the dissolving time at 27°C and finds it *slower* than you did at 25°C. They use the same 5 grams of sugar and 100 mL of water, so the two experiments should be comparable. Do the experiments prove that sugar dissolves *slower* in warmer water? At first glance, that is what the results suggest. In fact, that conclusion is *not* supported once you consider the error.

Group A

25 °C
5 g sugar
100 mL water

Data from sugar dissolving experiment

Trial	Sugar	Water	Temp.	Time
1	5.0 g	100 mL	25.4 °C	53 sec
2	5.0 g	100 mL	25.1 °C	45 sec
3	5.0 g	100 mL	26.1 °C	48 sec
avg.	5.0 g	100 mL	25.5 °C	48.7 sec

Group B

27 °C
5 g sugar
100 mL water

Data from sugar dissolving experiment

Trial	Sugar	Water	Temp.	Time
1	5.0 g	100 mL	27.4 °C	52 sec
2	5.0 g	100 mL	28.1 °C	47 sec
3	5.0 g	100 mL	27.1 °C	49 sec
avg.	5.0 g	100 mL	27.5 °C	49.3 sec

When differences are significant

In both experiments the error is about plus or minus 4 seconds. The difference between the two groups is 0.6 seconds, which is *smaller than the estimated error in the results.* The conclusion is that there is no **significant** difference between the results. Both groups have the same time: 49 +/- 4 seconds. A difference between data is only *significant if* the difference is substantially greater than the error. This is an important consideration because experiments rarely produce exactly the same numbers twice in a row. Numbers that are different in a mathematical sense may not be *significantly* different in a scientific sense.

Chemistry terms

conclusion - a stated decision whether the results of experiments or observations confirm an idea or hypothesis, or not.

significant - a difference between two results is only significant when it is substantially greater than the error in either result.

Scientific evidence

Qualities of scientific evidence

Good experiments produce scientific evidence. Scientific evidence may be measurements, data tables, graphs, observations, or other information that describes what actually happens. Two important characteristics of scientific evidence are that it be **objective** and **repeatable**. "Objective" means the evidence should describe only what actually happened as exactly as possible. The personal opinions of the person doing the experiment do not count as scientific evidence. "Repeatable" means that others who repeat the same experiment observe the same results.

Evidence may come from observations

Scientific evidence may come from observing nature without doing an experiment. For example, Galileo used his telescope to observe the moon and recorded his observations by sketching. The sketches are *objective* because they represent what he actually saw. Others who looked through his telescope saw the same thing, therefore the sketches are also *repeatable*. The scientific evidence of Galileo's sketches convinced people that the moon was actually a world like the Earth with mountains and valleys. This was not what people believed prior to Galileo's time.

Galileo's sketches

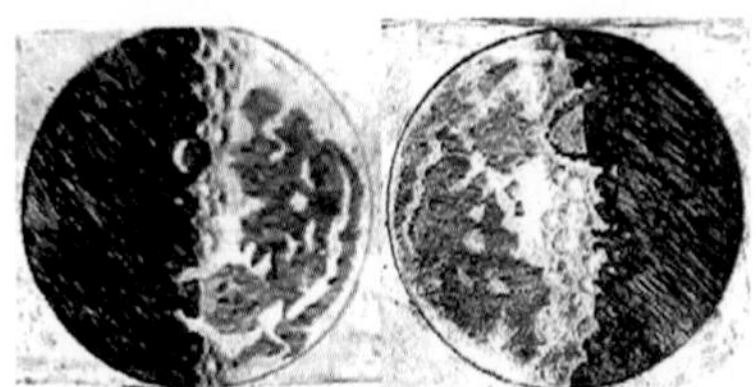

Scientific evidence must be **objective** and **repeatable**

The scientific meaning of words

It is important that scientific evidence be communicated clearly, with no room for misunderstanding. Scientists attach careful meanings to words like "weight" and "mass." Many "everyday" words are defined more precisely in science. For example, "heat" in chemistry means a certain type of *energy* that depends on temperature. This definition allows a quantity of heat to be defined with a precise measurement that tells someone else exactly how much energy is involved.

Procedures describe how scientific evidence is collected

The way scientific evidence is gathered is usually important to understanding what the evidence means. Any presentation of scientific evidence always includes a description of how the evidence was collected. For example, an experiment includes a **procedure**, which describes how the experiment was done. The procedure includes important details, such as what instruments were used to make any measurements. The procedure must include enough detail so that someone else familiar with chemistry can repeat the same experiment to verify the repeatability of the evidence.

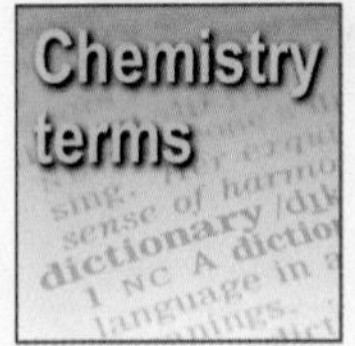

objective - describing only what actually occured without opinion or bias.

repeatable - others who do the experiment or make the observation the same way obtain the same results or observations.

procedure - detailed instructions on how to do an experiment or make an observation.

The Scientific Method

Learning by the scientific method

It takes a long time to learn by trying everything. Furthermore, you can never be sure you tried *everything*. The **scientific method** is a much more dependable way to learn.

The Scientific Method

1. Scientists observe nature, then invent or revise hypotheses about how things work;
2. The hypotheses are tested against evidence collected from observations and experiments;
3. Any hypothesis which correctly accounts for all of the evidence from the experiments is a potentially correct theory;
4. A theory is continually tested by collecting new and different evidence. Even a single piece of evidence that does not agree with a theory causes scientists to return to step one.

observe and think

make a hypothesis

compare to experiments & observations

revise and improve a theory

Why the scientific method works

The scientific method is the underlying logic of science. It is basically a careful and cautious way to build a supportable, evidence-based understanding of our natural world. Each theory is continually tested against the results of observations and experiments. Such testing leads to continued development and refinement of theories to explain more and more different things. The way people learned about many things great and small, to the solar system and beyond, can be traced through many hypotheses.

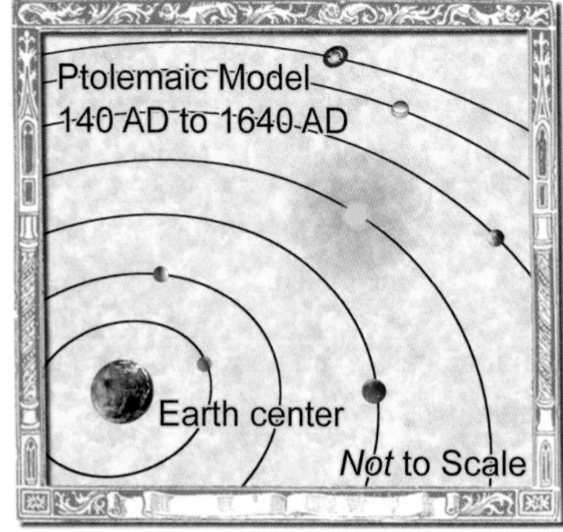

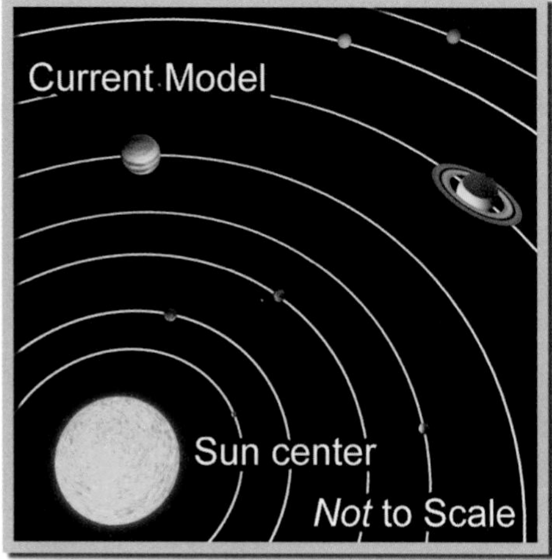

1.3 Matter and Energy

Chemistry describes the different forms of matter

Look at a bicycle and describe what the bicycle is made of. To a scientist, a bicycle is made of **matter**. Matter is everything that has mass and takes up space. This book is matter. You are matter. Even the air is matter. At the next level of detail, a bicycle contains many different kinds of matter, such as steel, aluminum, plastic, and rubber. Chemistry is about the different kinds of matter and how they are created. Chemistry also tells us how matter can be changed from one kind to another.

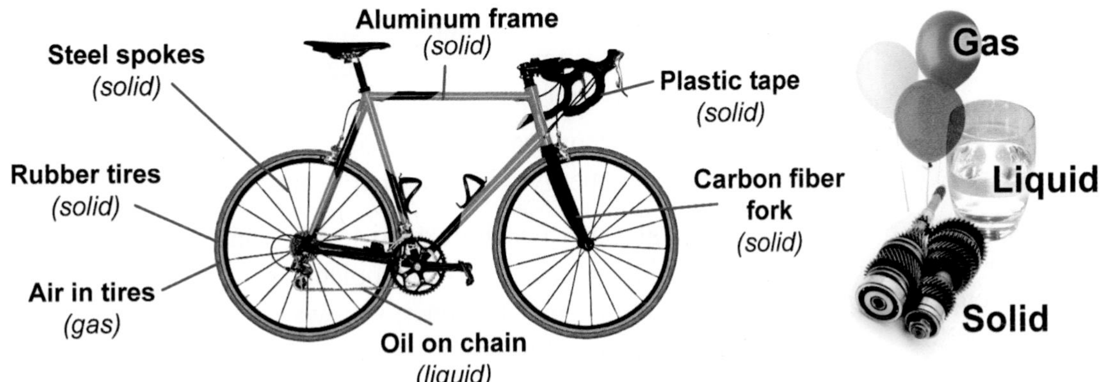

The phases of matter: solid, liquid and gas

The matter in a bicycle comes in three **phases** we call *solid*, *liquid* and *gas*. Solid matter, like the steel frame or rubber tires, is stiff, holds its shape, and may be strong. Liquid matter, like the oil in the wheel bearings, flows and does not hold its shape. The air in the tires is a gas. Gaseous matter flows like liquid but also expands to fill any container. You may not believe air has mass and is matter, but consider letting the air out of the bicycle tires. What happens? The air in the tires can easily support both bicycle and rider because air *does* have mass and under sufficient pressure can exert large forces.

The thermal energy in matter determines its phase

Whether matter is solid, liquid or gas depends on how much internal thermal energy is associated with its constituent molecules and atoms. The lowest thermal energy form of matter is solid. Matter that contains a moderate amount of thermal energy is liquid. Matter that contains a relatively high internal energy can be a gas. If you heat steel hot enough it can become a gas. There is a fourth state of matter, plasma, that appears at even higher energies. Lightning is a plasma.

Chemistry terms

matter - material, has mass and takes up space.

phases - the physical forms of matter: solid, liquid, and gas (and plasma).

Energy

Energy

Energy is an important concept, fundamental to all science, but one that is difficult to define. Energy is a measure of a system's ability to change or create change in other systems. A hot cup of coffee has more energy than a cold cup. As the coffee cools down, its energy warms the air around it.

$CH_4 + 2O_2 \rightarrow 2H_2O + CO_2$ + energy

Chemical energy

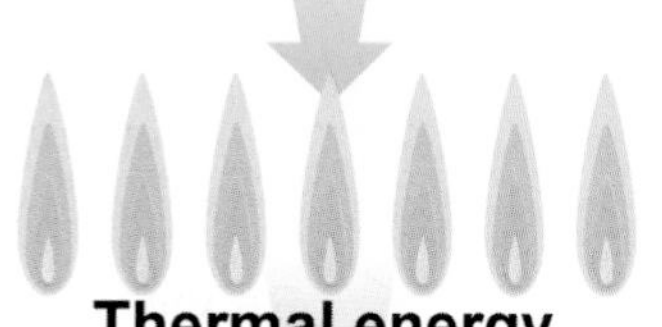

Thermal energy

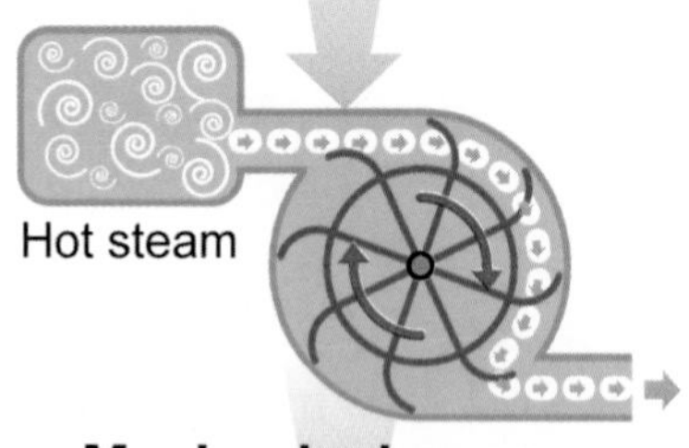

Mechanical energy

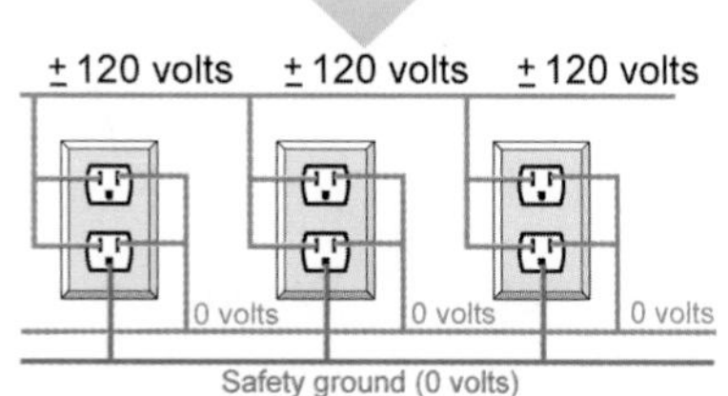

Electrical energy

Forms of energy

Energy appears in many forms, such as heat, motion, height, pressure, electricity and chemical bonds between atoms. Energy transforms during changes. For example, the chemical energy in fuel is transformed into heat energy when the fuel is burned.

Things tend to flow from higher energy to lower energy

When left alone, systems in nature tend to go from higher energy to lower energy. Warm objects have more energy than cold ones. The energy flows from warm to cold until there is no temperature difference any more. The same is true of chemical energy. We will see that many chemical reactions occur because the energy of the matter involved is lower after the reaction than it was before.

Conservation of energy

What happens to the energy in hot coffee as it cools? The energy lost by the coffee is *exactly* balanced by the energy gained by the room. The idea that energy converts from one form into another without a change in the total amount is called the **law of conservation of energy**. This very important natural law says that energy can never be created or destroyed, just converted from one form into another. The law of conservation of energy is one of the most important laws in all of science. It applies to all forms of energy, not just heat.

Energy can never be created or destroyed, just converted from one form into another.

Chemistry terms

energy - energy is a measure of a system's ability to change or create change in other systems, measured in joules

law of conservation of energy - energy can never be created or destroyed, just converted from one form into another

"Using" and "conserving" energy

"Conserving" energy

Almost everyone has heard that it is good to "conserve energy" and not waste it. This is good advice because energy from gasoline or electricity costs money and uses resources. What does it mean to "use energy" in the everyday sense? If energy can never be created or destroyed, how can it be "used up"? Why do people worry about "running out" of energy?

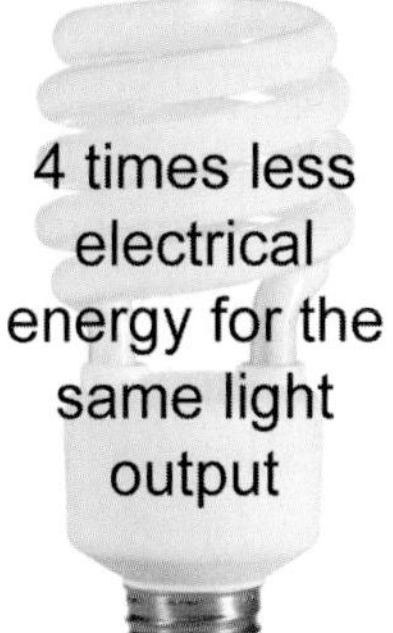

"Using" energy

When you "use" energy by turning on a light, you are really converting energy from one form (electricity) to other forms (light and heat). What gets "used up" is the amount of energy *in the form of electricity*. Electricity is a valuable form of energy because it is easy to move over long distances (through wires). In the "chemical" sense, the energy is not "used up", but converted into other forms. The total amount of energy stays constant.

Power plants

Electric power plants don't *make* electrical energy. Energy cannot be created. What power plants do is convert other forms of energy (chemical, solar, nuclear) into electrical energy. When someone advises you to turn out the lights to conserve energy, they are asking you to use less electrical energy. If people used less electrical energy, power plants would burn less oil, gas, or other fuels in "producing" the electrical energy they sell.

"Running out" of energy

Many people are concerned about "running out" of energy. What they worry about is running out of certain *forms* of energy that are easy to use, such as oil and gas. When you use gas in a car, the chemical energy in the gasoline mostly becomes heat energy. It is not practically possible to put the energy back into the form of gasoline, so we say the energy has been "used up" even though the energy itself is still there, only in a different form.

Gas pumps are inspected regularly to ensure that they accurately measure in full gallons

Learning how to measure and use units is an important part of chemistry, and also an important part of your life! Scientists use measurements to gather data and test theories. You use measurements for health and medicine, to purchase things you need, to build, and even to travel. For example, when you add gasoline to a car at the gas station, the pump measures the amount so that you know how much has been transferred into your car. Next time you are near a gas pump, look for the inspection sticker. Every state checks gas pumps to ensure they are measuring accurately. This is important since when you pay for a gallon of gas you should be getting a gallon in your car, not less.

Health is a field that requires numerous measurements. Besides height and weight, almost everything doctors use to monitor your health is based on measurements, referred to as "lab" work, or radiology (x-rays, ultrasound). When young children go to the doctor for a check-up, a lead test is often done. A blood sample is taken and the amount of lead in their blood is measured. Lead levels are measured in micrograms per deciliter (µg/dL). A microgram is one millionth of a gram. A deciliter is 0.1 liters or 100 mL. One µg in 100 mL is a *very* small amount yet it can be measured reliably.

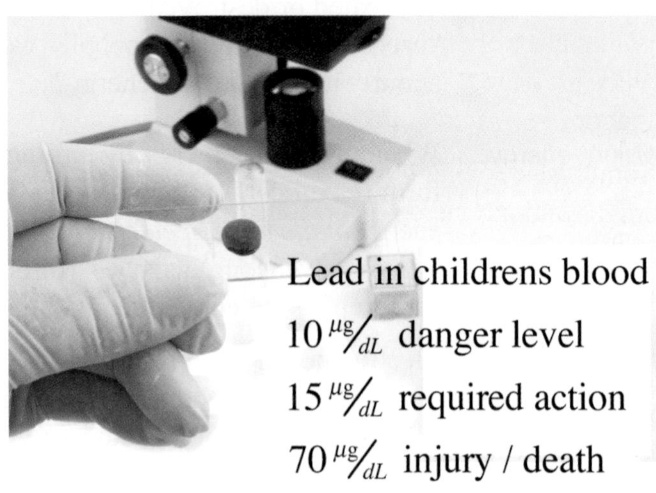

Gamma ray scanners measure airport cargo for excess nitrogen

Recently, there have been many advances in measuring devices. Scientists have had to come up with new ways to measure. Airport security has become very important and a unique measuring device was designed to help with the screening process.

Major airports now use an instrument that bombards cargo with high energy particles (gamma rays). These particles are able to detect high levels of nitrogen. Nitrogen is the element most common to explosives. This method of screening is effective and saves a good deal of time searching through large amounts of cargo and luggage.

Air quality is another important area requiring measurement. Environmental scientists use units of parts per million (ppm) to describe the concentration of compounds such as carbon monoxide (CO). A CO level of 1 ppm means there is 1 gram of carbon monoxide per million grams of air. Units of ppm are used to monitor harmful chemicals like nitrous oxides, and also harmless substances such as fragrances from perfumes! New sensor technology is capable of detecting very low levels of gases and pollutants, even less than one part per million. These new sensors allow us to warn people if unhealthy pollutants at high levels are in the air. People with breathing problems may choose to stay indoors.

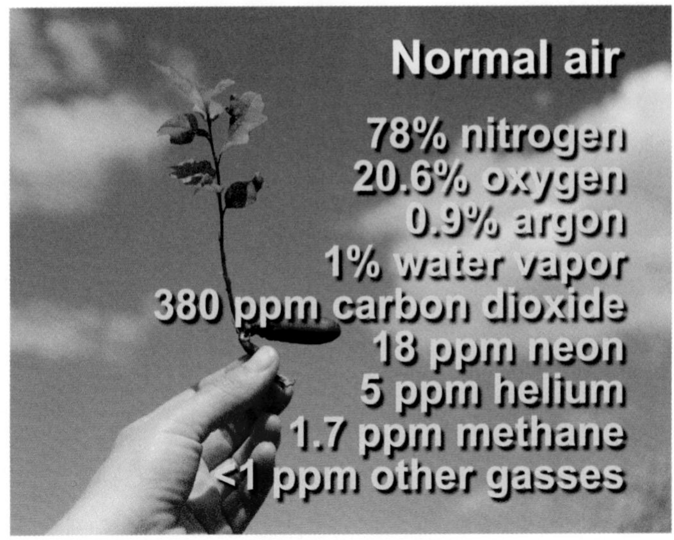

Normal air
78% nitrogen
20.6% oxygen
0.9% argon
1% water vapor
380 ppm carbon dioxide
18 ppm neon
5 ppm helium
1.7 ppm methane
<1 ppm other gasses

Sometimes new knowledge can be scary without knowing how to interpret it. Now that we can measure small amounts of chemicals more accurately, we are more aware of what is in our air, food and water. Sometimes we should worry, such when lead or arsenic is found in drinking water; other times, there is no problem at all, such as finding a few ppm of dissolved iron in drinking water. Iron is a necessary mineral nutrient. Knowing what is present allows us to better understand our environment. Just keep in mind that many chemicals, such ascorbic acid (vitamin C) and potassium, are necessary for health!

Learning to think like a scientist means that we learn to ask questions, perform experiments and gather information. When we perform experiments, our observations help us to collect information and make sense of it. Sometimes our "interpretation" of the information leads us to a new idea. This is the exciting part of science. Forensic scientists work together to try to solve mysteries and to provide useful information that may help people.

A forensic chemist has several ways to identify what type of molecules are present in materials. For example, during a criminal investigation samples of clothing fibers, medications, and perhaps blood stains are collected and sent to the forensic laboratory for analysis. The investigating detective knows that one suspect takes a certain medication. The forensic chemist uses a sophisticated measuring instrument called a *mass spectrometer* to identify the chemicals in the lab samples. If there is a match with the suspect's medication, that is a useful piece of evidence. As you might imagine, experience in the laboratory helps a chemist to decide what tests to run, and how to interpret what the results mean.

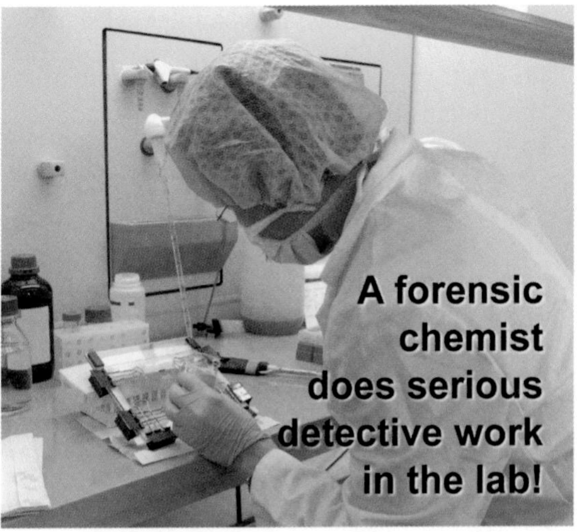

A forensic chemist does serious detective work in the lab!

Chapter 1 Review.

Vocabulary

Match each word to the sentence where it best fits.

Section 1.1

pressure	measurement
force	mass
Pascal	weight
milliliter	gram
graduated cylinder	volume
accurate	air
precise	density
significant figures	atmosphere
conversion factor	mantissa
scientific notation	exponent
kilogram	elements
	atom

1. As a tree grows its mass primarily comes from _____.
2. A bicycle tire expands because the _________ inside is greater than it is outside.
3. The _______ are pure forms of matter that cannot be broken down into simpler substances.
4. An ______ is the smallest particle of matter that retains the identity of an element.
5. The amount of matter in an objects is most accurately given by the object's ________.
6. A _______ could be a push or a pull.
7. Which of the above words describes the largest SI unit of mass?________
8. _______ is a measure of the amount of space something occupies.
9. The __________ is a small unit of pressure equal to one newton per square meter.
10. Chemists use a ________ to measure the volume of a liquid accurately.
11. How close a measured value is to the actual known value, tells you how______ your measurement is.
12. A volume measured to be 12 mL is less_______, than the volume measured to be 12.0 mL.
13. When changing the units of a measured value a student must use a _________, so that the units cancel properly.
14. The _______ tells you how much mass of a substance there is per unit volume.
15. One _________ is the average pressure at sea level on Earth.
16. The ______ is number that corresponds to N when a value is written $N \times 10^6$.
17. Scientists use _____________ to express very large or very small numbers.

Section 1.2

hypothesis	experiments
objective	inquiry
average	repeated
theory	procedure
natural law	error
significant	uncertainty
variable	control variable
conclusion	scientific method

18. An ________ is a sutuation specifically set up to observe what happens under certain conditions.
19. A tentative explanation for an observation is called a __________.
20. Learning through questioning is called ________, and it is the basis for all scientific discovery.
21. A good experiment must be able to be ________ by other scientists, who will obtain similar results.
22. When making observations in the laboratory a scientist must be _______ about what they record as data.

23. A ____________ is a quantity that is kept the same during an experiment where other things are allowed (or observed) to change.

24. A __________ is a statement of something learned or demonstrated by an experiment.

25. A _________ is a series of steps that outlines how to perform a particular experiment.

26. measurement ___________ can be small if the experiment is accurate and precise.

27. An explanation advances to become a _________ when it is supported by a substantial body of evidence.

28. All measurements contain some _________, and this is unavoidable.

29. A difference is __________ onlywhen it is larger than the estimated error in the measurements.

30. A _________ is a quantity that is observed and may change in an experiment.

Section 1.3

energy	phases
matter	conservation

31. ________ is the quantity that determines how much change is possible in a physical system.

32. The law of __________ of energy states that energy can never be created or destroyed.

33. Anything that takes up space and has mass is referred to as _______________.

34. Solid is one of the three ordinary _________of matter.

Conceptual Questions:

Section 1.1

35. Is fire a form of matter? Briefly explain your thinking.

36. What is the function of an antioxidant? Explain.

37. Describe how vitamins are useful to our bodies.

38. Write two sentences about one difference between solid sugar and sugar dissolved in water.

39. which is a better value; one gallon of milk for \$3.00 or one liter of milk for \$3.00.

40. The process of dissolving allows chemical reactions to take place more effectively in the body. Explain why this is so.

41. Which unit represents a larger amount of matter, a kilogram or a gram? Explain your reasoning.

42. If you have a solid chocolate Easter rabbit and a hollow chocolate Easter rabbit of the same size.:
    a. Which one contains more matter?
    b. Which one has a larger volume?

43. From the figure below, what is the best measurement?

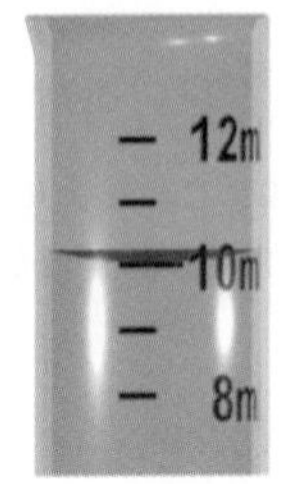

a) 10 mL
b) 10.0 mL
c) 10.2 mL
d) 10.5 mL
e) 10.00 mL

44. Your measurements are limited by the precision of the equipment you use. Which of the two choices below gives the more precise measurement? Explain.
    a. a) A top-loading balance with two pennies that reads 6 grams.
    b. b) A top-loading balance with the same two pennies that reads 6.30 grams.

45. Explain in words how you would determine how many dozen eggs are represented by 42 individual eggs. (12 eggs = 1 dozen)

Chapter 1 Review.

46. Explain the difference between accuracy and precision using the dart board pictures below.

47. Why should a measurement always be recorded to the correct number of significant figures?
48. When you squeeze an inflated balloon in one place, it bulges out in another place. Explain why this happens.
49. Why is an empty bottle sitting on your table not truly empty?
50. Give two observations that demonstrate that air has mass.

Section 1.2

51. What steps are involved in the scientific method?
52. Give an example of how an auto mechanic or a baker might use the scientific method in their daily lives.
53. When you are conducting an experiment, it is important to have only one experimental variable when possible. Explain why.
54. When performing experiments it is important to have several trials and therefore several measurements of the same process. This repetition makes it possible to say that the results are more "accurate" than another group that performed the experiment only once. Explain why this is true using the words "error" and "average" in your answer.
55. Chemists were asked to analyze an unknown sample. They were able to determine three physical properties about the sample. When run through a series of tests, a yellow solid was isolated that had a melting point of 402°C and a boiling point of 954°C. Use the Handbook of Chemistry and Physics to determine whether the substance is cadmium iodide, lead iodide or sodium iodide.

Section 1.3

56. Which of the following are examples of matter?
    a. the air
    b. a crayon
    c. a cat
    d. you
    e. a pencil
57. Design a simple experiment to test the temperature at which water boils. Define your system under study (use a simple sketch), the variable that you are testing, and the variables you will hold constant.
58. When we talk about "conserving" energy, what are we referring to?

Quantitative Problems

Section 1.1

59. Which number is larger: 0.001 or 1×10^{-4}?
60. What is the volume of a paperback book 20cm tall, 11cm wide, and 3.5cm thick?
61. Count the significant figures in each of the following:
    a. 10 swimmers
    b. 11,300 grams
    c. 0.0234 grams
    d. 6.00 liters
62. Calculate the numeric answer to the correct number of significant figures.
    a. 62.4 liters - 8.21 liters + 4.3 liters
    b. 6.43 grams - 2.10 grams
    c. 1.54 meters + 10.44 meters
63. Soda is often sold in 12 ounce cans. How many milliliters are in a 12 oz. can?

64. How many significant figures are in picture #2's measured length?
    a. Which measured value is the most precise?
    b. Given the calibrations on the rulers above are each of the measurements in relative agreement? Explain.
65. Perform the following calculations and report your answer to the correct number of significant figures.
    a. 6.25 meters x 0.51 meters
    b. 2.46 meters / 7.8 meters
    c. 3.20 meters x 2.5 meters
66. If a person weighs 82.5 kg, what is the persons weight in pounds?
67. Perform the following conversions:
    a. 350 mL to quarts (qt)
    b. 3.0 ft to meters (m)
    c. 80 km to miles (mi)
    d. 25 mL to cm^3
68. Write out the following numbers as decimal numbers.
    a. 4.5×10^{-2}
    b. 3.33×10^{-9}
    c. 2.5×10^{4}
69. Write each of the following numbers in scientific notation.
    a. 450,000
    b. 45
    c. 0.45
    d. 0.0045
70. A bar of aluminum has a volume of 1.45 mL, and a mass of 3.92 g. Calculate its density.
71. 54 grams of aluminum are added to a graduated cylinder containing 50 mL of water. What will be the level of water in the cylinder after the aluminum has been added?
72. Into a test tube which has 75 mL of water we add 30 grams of titanium (Ti) and 30 grams of zinc (Zn). What will be the level of water after the Ti and Zn have been added?
73. The density of silver is 10.5 g/mL.
    a. What volume would 2.86 g of silver occupy?
    b. What is the mass of a 16.3 cm^3 piece of silver?
74. A penny is measured to have a mass of 2.52 g and a volume of 0.35 cm^3. Is this penny made of pure copper?
75. Modern day pennies are made of zinc with a small covering of copper. The weight of a penny is 2.52 g and the volume it occupies is 0.35 cm^3. Calculate the amount of copper used in the penny.
76. An iceberg has a volume of 650 m^3. What is the mass of the iceberg in kg? (density of ice = 0.917 g/cm^3.)
77. An iceberg is floating in the ocean. A scientist using a specialized instrument measures the volume of the iceberg that is above the water surface and finds it to be 75 m^3. What is the volume of the iceberg below the water surface? (density of ice = 0.917 g/cm^3, density of water = 1.000 g/cm^3).
78. You are given three rings to test to determine if they are made of gold or if they are fake. You decide to measure the mass of each ring by weighing them on a balance and their volume by measuring the water they displace when immersed in water. The measurements for the three rings are:
    Ring A: Mass = 3.5 g. Volume = 0.18 cm^3.
    Ring B: Mass = 2.8 g. Volume = 0.18 cm^3.
    Ring C: Mass = 3.9 g. Volume = 0.20 cm^3.
    Using this data perform the appropriate calculations to determine which rings are gold.

CHAPTER 2
Matter and Atoms

What is matter made of?
What is the explanation for the diversity of matter?
How do we explain properties like solid or liquid?

You, the plant, and the coal
are mostly carbon

What are you made of? I mean, what are you *physically* made from? What sort of "stuff" makes up your skin, bones and muscles? Is it the same basic kind of "stuff" that makes up rocks, trees, air, plastic and the rest of the material world? The technical term for "stuff" is *matter* and matter is what makes up the physical universe.

Normal matter, including you, is made from tiny particles called *atoms*. The atoms of matter in your body are the same as the ones that make up rocks, water, plastic and even metals. Of course, the *details* are different! The blend and arrangement of different atoms in human muscle tissue is different from a log or a sheet of plastic wrap. The atoms themselves are mostly identical. Your body, wood and a sheet of plastic are mostly *carbon* atoms. In fact, it is possible to take *cellulose*, the main ingredient of wood, and turn it into a clear plastic sheet called *cellophane*. Many foods are wrapped with cellophane.

Chemistry is the science of matter and its changes. Chemistry tells us how matter is formed and how one kind of matter (wood) can be turned into a totally different kind of matter (plastic sheet).

A slimy chemical reaction

Take a small plastic cup and add 30 grams of white glue. Add 30 mL of water to the glue and stir it until well mixed. In a second plastic cup, combine 100 mL of warm water with 4g of borax powder. Stir this until the powder is dissolved.

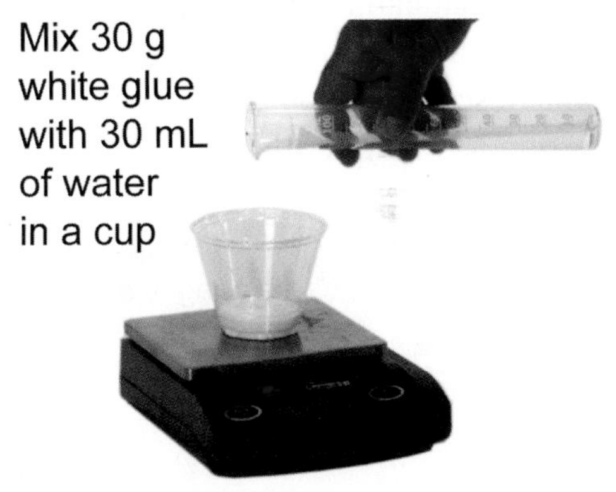

Mix 30 g white glue with 30 mL of water in a cup

Measure 4 g borax and mix with 100 mL of water in a second cup

Both cups are examples of *mixtures*. A mixture contains more than one kind of matter, which is the same as saying more than one chemical. In one cup the mixture is water and glue. In the other cup the mixture is water and borax.

- How are the mixtures alike?
- How are they different?
- Would you classify them as solid, liquid or gas?

Pour the colored glue mixture into the borax and water mixture while slowly stirring

To get some color, add a drop of food color to the glue/water and mix it in. Combine the two mixtures by slowly pouring the glue mixture into the cup with the water and borax while you gently keep stirring. What happens to the mixture now? Pick it up and work it between your fingers. Write down observations on what this new mixture is like.

- Is the new mixture a solid, liquid or gas?

The white glue is a mixture of water and polyvinyl acetate (PVA). PVA is a long molecule made of repeating units. Glue is so viscous, or thick, because the long strands of PVA stick to themselves like wet spaghetti. Borax is a compound of sodium, boron and oxygen that is "hydrated" (meaning it includes water molecules). Putting the two together creates a chemical reaction than combines them, linking the polyvinyl acetate molecules together and making the "slime". The more you work it, the more links you get and the stiffer the slime becomes.

What is it?

Why did it change from two liquids into this "slime"

2.1 Matter and the Elements

Explaining the diversity of matter

There are probably a million different kinds of matter in and around your classroom. Think about every different ingredient in the air, the soil, the walls and floor, the books and desks, and everything else. How do we make sense of such variety? Are the millions of different kinds of matter actually mixtures of a few simpler things?

Substances contain only a single kind of matter

The first step toward finding an answer is to search for pure *substances*. A **substance** cannot be separated into different kinds of matter by physical means such as sorting, filtering, drying, dissolving, heating or cooling. Corn oil is a pure substance. Salad dressing is not, because it contains oil, water, spices and other substances. In oterh words, something is a pure substance if it is a single chemical throughout.

Corn oil is a pure ***substance***

Oil and vinegar dressing is a ***mixture*** of substances

Mixtures contain more than one kind of matter

"Pure" is a much looser word in every day language. A container might say "pure orange juice". However, to a chemist, orange juice is a **mixture**. Orange juice can be separated into water, different flavoring chemicals, citric acid, sugars and fruit pulp. A mixture is matter that contains more than one substance. Wood is a mixture because there are many substances in wood including water, cellulose, tannic acid, lignin and other chemicals.

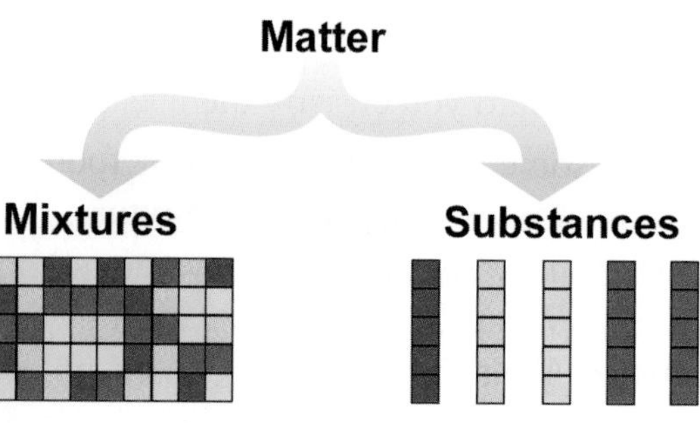

Most ordinary matter is mixtures

Mixtures can be separated by physical means

Mixtures are made of substances. They can be separated into their component substances by physical means such as sorting, filtering, heating or cooling. Chicken noodle soup, for example, could be separated into its components by using strainers and filters of different sizes. The separation process does not change the characteristics of each substance. You still have water, salt, fats, noodles and chicken.

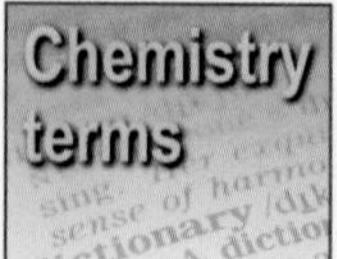

substance - a kind of matter that can't be separated into other substances by physical means such as heating, cooling, filtering, drying, sorting or dissolving.

mixture - matter that contains more than one substance.

Physical properties and physical change

Physical properties

Properties that you can measure or see through direct observation are called **physical properties**. For example, water is a colorless liquid at room temperature. The quality of "color" is a physical property. Temperature is another physical property.

Color and Temperature are physical properties

Brittleness and malleability are physical properties

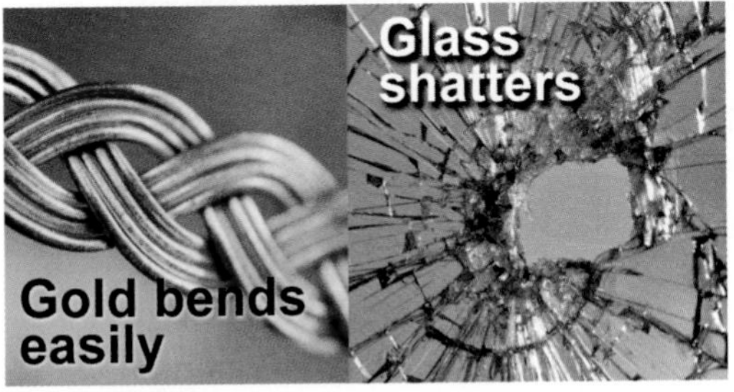

Gold is a shiny, *malleable* solid at room temperature. Malleable means gold can be hammered into thin sheets without cracking. The opposite is *brittle*. Glass is brittle. Brittle materials break if you hammer or bend them. Malleability and brittleness are examples of physical properties of solid materials.

Density and phase are physical properties

The density of a material is a physical property. Wood is less dense than water, which is less dense than rock or steel. Wood floats on water because its density is lower than the density of water. The phase (solid, liquid, gas) of a material is also a physical property.

Wood floats because its density is less than the density of water

Physical changes are reversible

Physical changes include
heating
cooling
melting
freezing
boiling
dissolving
bending
evaporating

Physical changes are changes in the physical properties of matter. Physical changes can be reversed without changing one substance into another. Heating, cooling, dissolving, melting, freezing, boiling or bending are all physical changes. When water freezes, it undergoes a physical change from a liquid to a solid. This does not change the water into a new substance. The change can easily be reversed by melting the water. Bending a steel bar is another physical change. Bending changes the shape of the bar, but it is still steel.

Chemistry terms

physical property - property such as mass, density or color that you can measure or see through direct observation.

physical change - a change in physical properties, such as shape, phase or temperature; for example, grinding, melting, boiling, dissolving, heating or cooling.

Chemical properties and chemical change

Chemical properties

Properties that can only be observed when one substance changes into a different substance are called **chemical properties**. For example, if you leave an iron nail outside, it will eventually rust. A chemical property of iron is that it reacts with oxygen in the air to form iron oxide (rust).

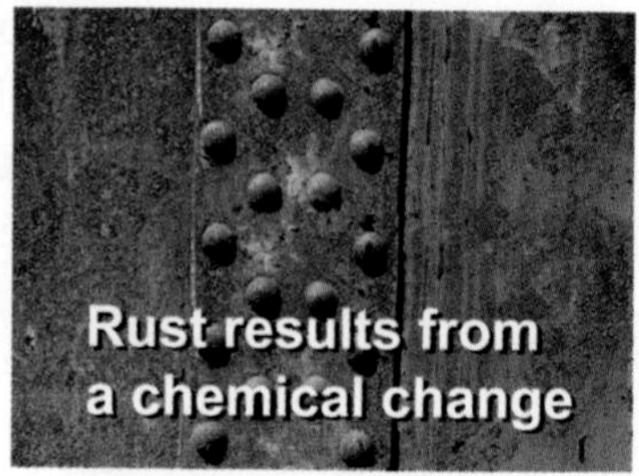

Chemical changes are hard to reverse

Any change that transforms one substance into a different substance is called a **chemical change.** The transformation of iron into rust is a chemical change. Chemical changes are not easily reversible. For example, rusted iron will not turn shiny again even if you take oxygen away.

Using chemical changes

We use chemical changes to create useful materials. The slime you made in an experiment is an example of a chemical change. The polyvinyl acetate (PVA) in the glue was a viscous liquid. Adding the borax links adjacent molecules together like rungs on a ladder. That made the liquid into a semi-solid mass, a polymer, that is more able to hold its shape.

The glue and borax together created a chemical change

Recognizing chemical change

Chemical changes are created by **chemical reactions**. A chemical reaction is any process in which one substance changes into a different substance.

A chemical reaction occurs when you mix baking soda with vinegar. The mixture bubbles violently as carbon dioxide gas, a new substance, is formed. The temperature of the mixture also gets noticeably colder. Bubbling, new substances, and temperature change can all be evidence of a chemical change.

A new gas is forming

Turns cloudy

A new solid is forming

Temperature change

Chemical bonds are changing

Color change

A new substance is forming

chemical property - property that can only be observed when one substance changes into a different substance - such as iron's tendency to rust.

chemical change - transforms one substance into another substance.

chemical reaction - the process that creates chemical changes.

The macroscopic and microscopic scales

An example of different scales

Is sand a substance or a mixture? Look at a pile of sand from across the room. It appears smooth, and shows no detail. Look close and you see that sand is many small grains. Look closer and you see that each grain has edges, and may even have a different color than the grain beside it. Sand is a *mixture* of different minerals, air and water.

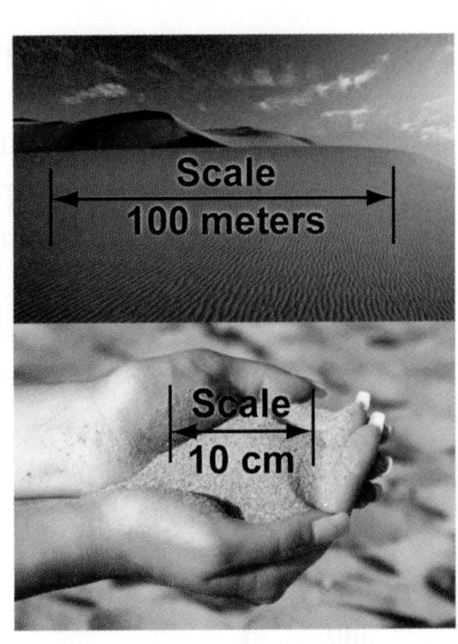

Like sand, all matter shows different and important structure at different *scales*. A **scale** is a typical length, or size, that shows a particular level of detail you wish to examine. For example, at a centimeter scale you can see the shape of the sand pile. At a scale of 0.01 millimeters you can see the details of each sand grain, but the shape of the pile is completely lost.

The macroscopic scale

The universe can be understood differently at different scales. It depends on what you are trying to understand. To calculate how much sand fills a bucket, you only need to know about sand's properties on the largest scale. We call this the **macroscopic** scale. Observations are macroscopic when they are large enough for us to see or directly measure in experiments, like mass in grams or the temperature of a pot of water.

The macroscopic scale can be observed and measured directly

The microscopic scale: The scale of atoms

Temperature is related to energy but you can't explain temperature on the macroscopic scale. To understand temperature, and many other aspects of matter, we must look on a scale a million times smaller than a grain of sand. This is the scale of atoms. Macroscopic properties like temperature ultimately depend on the behavior of atoms. We will use the term **microscopic** to mean "*on the scale of atoms.*" This is really a bit of a misnomer because normal microscopes do not have enough magnification to see the microscopic scale.

The scale of atoms is the microscopic scale

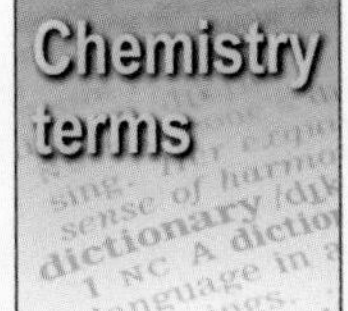

scale - a typical size that shows a certain level of detail.

macroscopic - on the scale that can be directly seen and measured, from a bacteria up to the size of a planet.

microscopic - to a chemist, on the scale of atoms, 10^{-9} meters and smaller.

Atoms and elements

Atoms

All the matter around you is made of atoms. Atoms make up everything that we see, hear, feel, smell and touch. We don't experience atoms directly because they are so incredibly small. A single grain of sand contains 200 million *million* atoms! A single atom is about 10^{-10} meters in diameter. That means you can lay ten billion, or 10,000,000,000 (10^{10}) atoms side by side in a one meter length.

Elements are the building blocks of matter

A mineral called *feldspar* is common in sand. Feldspar is a *substance*, because it is the same throughout. However, on the microscopic scale we see that feldspar contains *three kinds* of atoms. Feldspar contains oxygen atoms, silicon atoms and potassium atoms. Oxygen, silicon and potassium are **elements**. Elements are the fundamental "pure substances" from which all other matter is made.

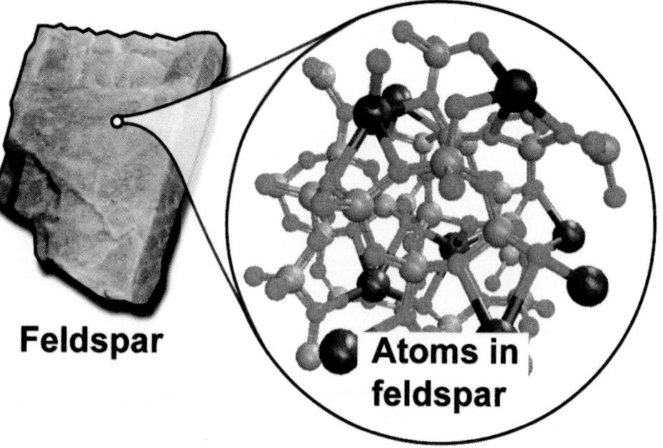

Each element is a unique type of atom

Think of an element as a unique type of atom. Oxygen atoms are different from silicon atoms, which are different from potassium atoms. All atoms of a given element are similar to each other and different from atoms of any other element.

Each element is a unique type of atom

Atoms of the same element are all similar to each other and different from atoms of any other element

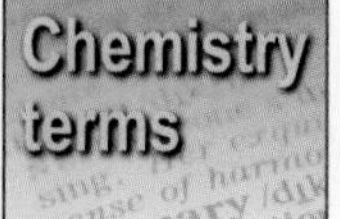

element - a unique type of atom. All atoms of the same element are similar to each other and different from atoms of other elements.

The periodic table

The modern periodic table

As early chemists worked on determining which substances were elements, they noticed that some elements acted like other elements. The soft metals lithium, sodium and potassium always combine with oxygen in a ratio of two atoms of metal per atom of oxygen. From these kinds of observations, scientists developed the *periodic table of the elements*. The **periodic table** organizes the elements according to how they combine with other elements (chemical properties). When this book was written, we knew of 118 elements, of which 88 occur naturally.

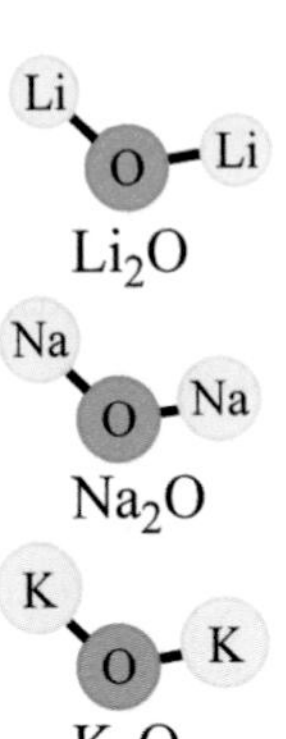

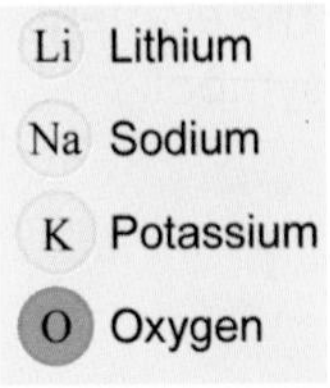

Li Lithium
Na Sodium
K Potassium
O Oxygen

The elements Li, Na, and K all form compounds in a 2:1 ratio with oxygen

Metals and non-metals

One big distinction is between metals and non-metals. Metals tend to be solids at room temperature and are shiny and dense. Metals are good conductors of electricity, like copper or aluminum. At room temperature, non-metals may be solid like carbon, liquid like bromine or gas like oxygen. Non-metals are generally poor conductors of electricity.

Periodic Table of the Elements

Non-metals

Metals

1	2	3	4	5	6	7	8	9	10	11	12	13	14	15	16	17	18
1 H 1.0079 hydrogen																	2 He 4.0028 helium
3 Li 6.941 lithium	4 Be 9.0122 beryllium											5 B 10.811 boron	6 C 12.011 carbon	7 N 14.007 nitrogen	8 O 15.999 oxygen	9 F 18.998 fluorine	10 Ne 20.180 neon
11 Na 22.990 sodium	12 Mg 24.305 magnesium											13 Al 26.982 aluminum	14 Si 28.086 silicon	15 P 30.974 phosphorous	16 S 32.065 sulfur	17 Cl 35.453 chlorine	18 Ar 39.948 argon
19 K 39.098 potassium	20 Ca 40.078 calcium	21 Sc 44.956 scandium	22 Ti 47.867 titanium	23 V 50.942 vanadium	24 Cr 51.996 chromium	25 Mn 54.938 manganese	26 Fe 55.845 iron	27 Co 58.933 cobalt	28 Ni 58.693 nickel	29 Cu 63.546 copper	30 Zn 65.38 zinc	31 Ga 69.723 galium	32 Ge 72.61 germanium	33 As 79.922 arsenic	34 Se 78.96 selenium	35 Br 79.904 bromine	36 Kr 83.80 krypton
37 Rb 85.468 rubidium	38 Sr 87.62 strontium	39 Y 88.906 yttrium	40 Zr 91.224 zirconium	41 Nb 92.906 niobium	42 Mo 95.96 molybdenum	43 Tc (98) technetium	44 Ru 101.07 ruthenium	45 Rh 102.91 rhodium	46 Pd 106.42 palladium	47 Ag 107.87 silver	48 Cd 112.41 cadmium	49 In 114.82 indium	50 Sn 118.71 tin	51 Sb 121.76 antimony	52 Te 127.60 tellurium	53 I 126.90 iodine	54 Xe 131.29 xenon
55 Cs 132.91 cesium	56 Ba 137.33 barium	71 Lu 174.97 lutetium	72 Hf 178.49 hafnium	73 Ta 180.95 tantalum	74 W 183.84 tungsten	75 Re 186.21 rhenium	76 Os 190.23 osmium	77 Ir 192.22 iridium	78 Pt 195.08 platinum	79 Au 196.97 gold	80 Hg 200.559 mercury	81 Tl 204.38 thallium	82 Pb 207.2 lead	83 Bi 208.98 bismuth	84 Po (209) polonium	85 At (210) astatine	86 Rn (222) radon
87 Fr (223) francium	88 Ra (226) radium	103 Lr (262) lawrencium	104 Rf (267) rutherfordium	105 Db (268) dubnium	106 Sg (271) seaborgium	107 Bh (272) bohrium	108 Hs (270) hassium	109 Mt (276) meitnerium	110 Ds (281) darmstadtium	111 Rg (280) roentgenium	112 Uub (285) ununbium	113 Uut (284) ununtrium	114 Uuq (289) ununquadium	115 Uup (288) ununpentium	116 Uuh (293) ununhexium		118 Uuo (294) ununoctium

57 La 138.91 lanthanum	58 Ce 140.12 cerium	59 Pr 140.91 praseodymium	60 Nd 144.24 neodymium	61 Pm (145) promethium	62 Sm 150.36 samarium	63 Eu 151.96 eruopium	64 Gd 157.25 gadolinium	65 Tb 158.93 terbium	66 Dy 162.50 dysprosium	67 Ho 164.93 holmium	68 Er 167.26 erbium	69 Tm 168.93 thulium	70 Yb 173.06 ytterbium
89 Ac (227) actinium	90 Th 232.04 thorium	91 Pa 231.04 protactinium	92 U 238.03 uranium	93 Np (237) neptunium	94 Pu (244) plutonium	95 Am (243) americium	96 Cm (247) curium	97 Bk (247) berkelium	98 Cf (251) californium	99 Es (252) einsteinium	100 Fm (257) fermium	101 Md (258) mendelevium	102 No (259) nobelium

Reading the periodic table

The element symbol

Each element in the periodic table has its own box. The largest item in the box is usually the **element symbol**. This is a one or two letter abbreviation of the element. For example, C in a chemical formula or diagram will always stand for carbon because C is the element symbol for carbon, Li for lithium and so on.

Element symbol

3 Li 6.941 lithium — 6 C 12.011 carbon

Li stands for *lithium*

C stands for *carbon*

The atomic number is unique for each element

Notice that the periodic table arranges the elements in increasing **atomic number** from left to right then top to bottom. The lightest element is hydrogen, atomic number 1. The heaviest naturally occurring element is uranium, atomic number 92. Elements heavier than uranium are not found naturally on earth but have been created in laboratories.

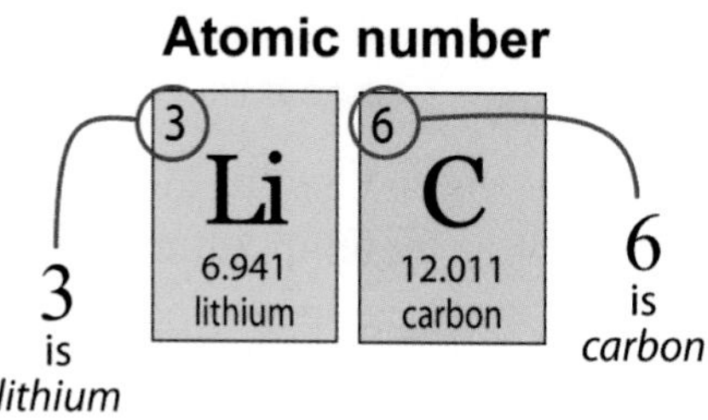

The rows and columns of the periodic table

Each row on the periodic table is also called a **period**. Each column is called a **group**.

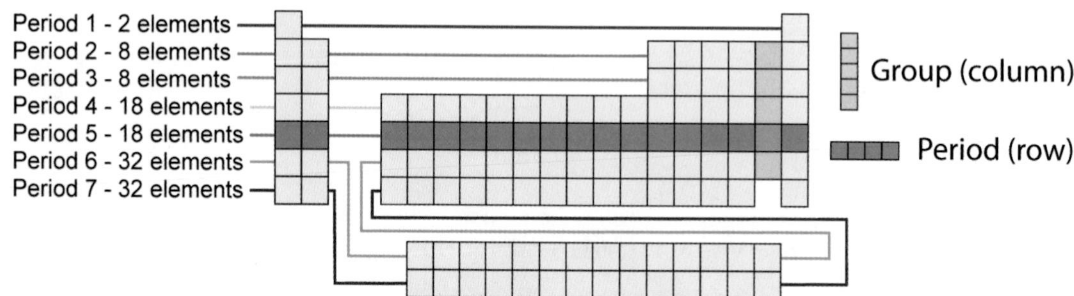

- The first row (period) has only two elements: hydrogen (1) and helium (2).
- The 2nd and 3rd rows have 8 elements each.
- The 4th and 5th rows have 18 elements.
- The 6th and 7th rows have 32 elements. To fit the table to a page, part of the 6th and 7th rows (elements 57-70 and 89-102) is usually dropped below the main table.

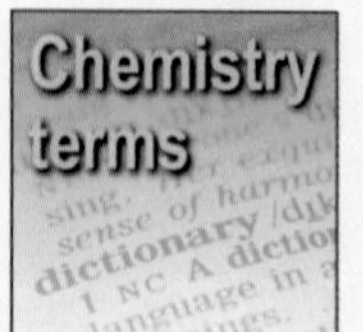

element symbol - a one or two letter abbreviation for each element.

atomic number - each element has a unique atomic number, the number of protons in the nucleus.

period - a row of the periodic table.

group - a column of the periodic table - all the elements in a group have similar chemical properties.

The mole

The atomic mass unit (amu)

A single atom is so small that you need an equally small unit of mass to describe it conveniently. One **atomic mass unit (amu)** is 1.66×10^{-24} grams. A single hydrogen atom has a mass of 1.01 amu. A carbon atom has a mass of 12.0 amu.

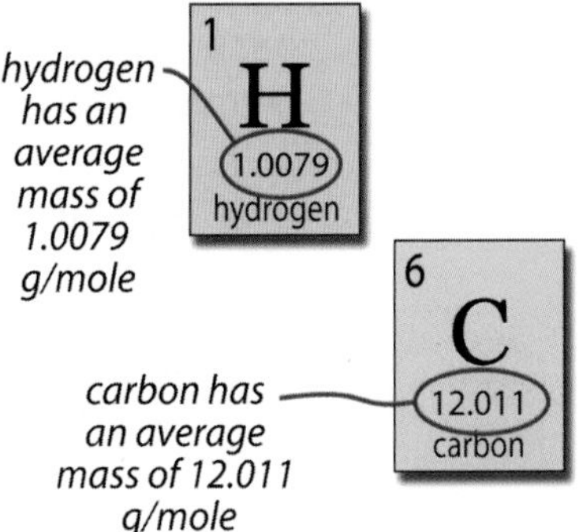

Written right below the element symbol is the average atomic mass in amu. You can see that the mass increases with atomic number (with a few exceptions). The atomic mass has two important interpretations.

1. It is the mass of a single atom in *amu.*
2. It is the mass of a *mole* of atoms in grams

Avogadro's number: 6.02×10^{23}

We can't measure masses as small as amu easily. Our average laboratory balances measure *grams*. How many atoms are in 1.01 grams of hydrogen? The answer is 6.02×10^{23} atoms. How about 12.0 grams of carbon? The answer is the same, 6.02×10^{23} atoms. The number 6.02×10^{23} is called **Avogadro's number** and it is very important in chemistry. The atomic mass of any element contains 6.02×10^{23} atoms. Scientists have developed the unit of **mole** that expresses the mass of 6.02×10^{23} atoms of any element. One mole of N weighs 14.0 grams and contains 6.02×10^{23} atoms of N.

mole

Hydrogen atoms
1.01 amu / atom

$$1.01\text{g}\left(\frac{1\text{ atom}}{1.01\text{ amu}}\right)\left(\frac{1\text{ amu}}{1.66 \times 10^{-24}\text{g}}\right) = 6.02 \times 10^{23}\text{ atoms}$$

1 mole of atoms

Carbon atoms
12.0 amu / atom

$$12.0\text{ g}\left(\frac{1\text{ atom}}{12.0\text{ amu}}\right)\left(\frac{1\text{ amu}}{1.66 \times 10^{-24}\text{g}}\right) = 6.02 \times 10^{23}\text{ atoms}$$

One mole is the number of atoms in the average atomic mass if the value is assumed to be in grams. The average atomic mass of magnesium (Mg) is 24.31 g. That means 6.02×10^{23} atoms of magnesium have a combined, total mass of 24.31 grams of Mg.

One mole contains 6.02×10^{23} atoms

The average atomic mass in grams is the mass of one mole of atoms

atomic mass unit (amu) - a mass unit equal to 1.66×10^{-24} grams.

mole - the number 6.02×10^{23}, the average atomic mass in grams is the mass of one mole of atoms.

Avogardo's number - the number of atoms contained in one mole. It is 6.02×10^{23}

Calculating moles

Moles have a measurable mass

Atoms are too small to weigh individually, so chemists often work in moles of atoms. A mole of atoms has a mass that is convenient to measure on a balance. For example, a mole of iron has a mass of 55.85 grams. This is about the mass of a few nails.

1 mole of iron

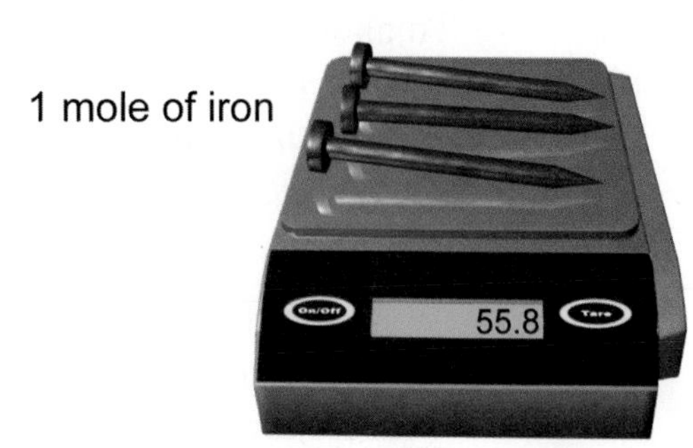

Mole problems tend to be of two kinds

1. How many moles are in a given mass of a substance?
2. What is the mass in grams of a specific number of moles of a substance?

In both types of problems you can use dimensional analysis to get the answer. Use the ratio of 1 mole to the atomic weight from the periodic table as a conversion factor. For example, one mole of calcium weighs 40.078 grams; one mole of sulfur weighs 32.065 grams; one mole of oxygen weighs 15.999 grams.

1 mole = the atomic weight from the periodic table

How many moles are in 100 grams of sulfur (S)?

Asked:	*number of moles*
Given:	*element is sulfur and there are 100 grams*
Relationships:	*One mole of sulfur has a mass of 32.065 grams.*
Solve:	$100g\ S \times \left(\frac{1\ mole}{32.065g\ S}\right) = 3.12\ moles$
Answer:	*100 grams of sulfur contains 3.12 moles of sulfur atoms.*

16
S
32.065
sulfur

How many grams of calcium (Ca) do you need to have 2.50 moles of calcium?

Asked:	*number of grams*
Given:	*element is calcium and there are 2.50 moles*
Relationships:	*One mole of calcium has a mass of 40.078 grams.*
Solve:	$2.50\ moles\ Ca \times \frac{40.078\ g}{1\ mole} = 100.20\ g$
Answer:	*2.50 moles of calcium has a mass of 100.20 grams.*

20
Ca
40.078
calcium

2.2 Molecules and Compounds

The smallest piece of matter

Think about a wax candle. Imagine carving off a tiny sliver of wax with your fingernail. Is the small piece still wax? Of course it is. Now think about cutting your sliver of wax in half, and in half again. Keep cutting and eventually you reach a particle of wax so small that if you cut it again, it is no longer wax but something else. That single smallest particle of wax that *retains the identity of wax* is called a **molecule**. Molecules are made of atoms. A single molecule is the smallest particle of a substance that can be.

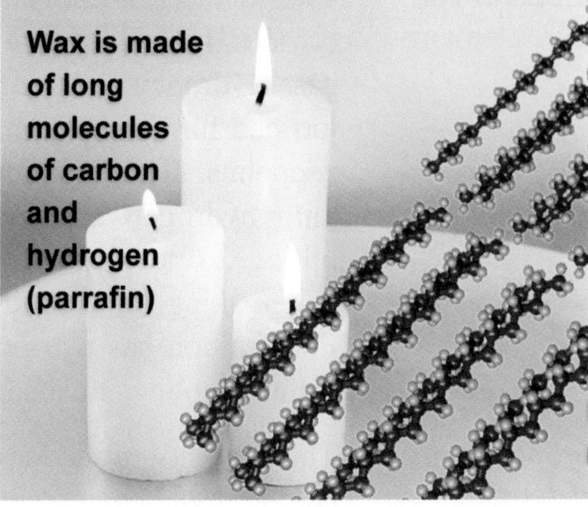

Wax is made of long molecules of carbon and hydrogen (parrafin)

Most matter is in compounds

Wax is a **compound** of the elements carbon and hydrogen. Pure elements are actually quite rare in nature. Most matter exists as compounds. With a few notable exceptions (like gold) pure elements quickly combine with other elements to make compounds. For example, water (H_2O) is a compound of hydrogen and oxygen. Virtually everything you eat and everything in your kitchen is a compound (or mixture of compounds). Salt is a compound of sodium and chlorine. Sugar is a compound of carbon, hydrogen and oxygen. "Pure sugar" is not a pure *element* in the chemical sense.

Most matter is in the form of compounds

Compounds versus mixtures

Compounds and mixtures are different. A mixture of hydrogen and oxygen is a gas at room temperature. The compound H_2O is a liquid. A compound contains two or more different elements that are *chemically bonded* together. In water, the oxygen atoms and hydrogen atoms are bonded together. A mixture contains two or more elements and/or compounds that are *not* chemically bonded together. In fact, most mixtures are mixtures of compounds! Sugar is a compound and soda is mostly a mixture of three compounds, water, sugar and carbon dioxide gas.

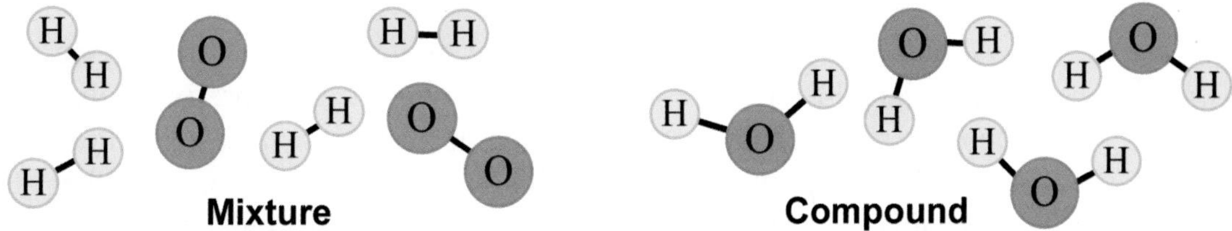

Chemistry terms

molecule - a group of atoms chemically bonded together.

compound - a substance containing more than one element in which atoms of different elements are chemically bonded together.

Millions of substances from 92 elements

Properties of pure elements

Consider three elements: hydrogen, oxygen and carbon. By itself, pure hydrogen is a light, colorless gas. Hydrogen is extremely flammable, which was too bad for the early makers of the airships called zeppelins. The infamous Hindenburg was filled with hydrogen gas and exploded May 6, 1937, killing 36 people. Pure oxygen is also a colorless gas, making up about 21% of Earth's atmosphere. Pure carbon has two forms: a black slippery solid called graphite and a hard crystal called diamond.

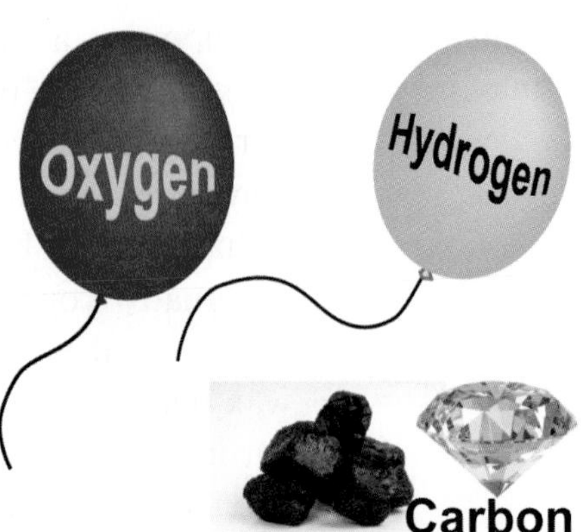

Properties are determined by the compound not the constituent elements

The diversity of matter comes from compounds. That's because *the properties of matter belong to the compound, not the constituent elements*. Consider just a few compounds of these three elements. Combine one carbon atom, one oxygen atom and four hydrogen atoms and you have methanol, commonly called wood alcohol. Nine carbon, four oxygen and eight hydrogen atoms make aspirin. Twenty one carbon atoms and 44 hydrogen atoms in a long chain make paraffin, or candle wax. Eight carbon and 18 hydrogen atoms make octane, a component of gasoline.

These very different substances contain just carbon, hydrogen and oxygen atoms

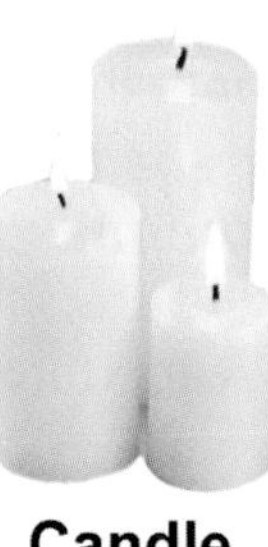

Candle wax

Aspirin

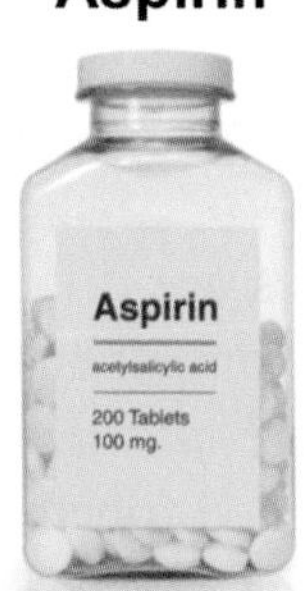

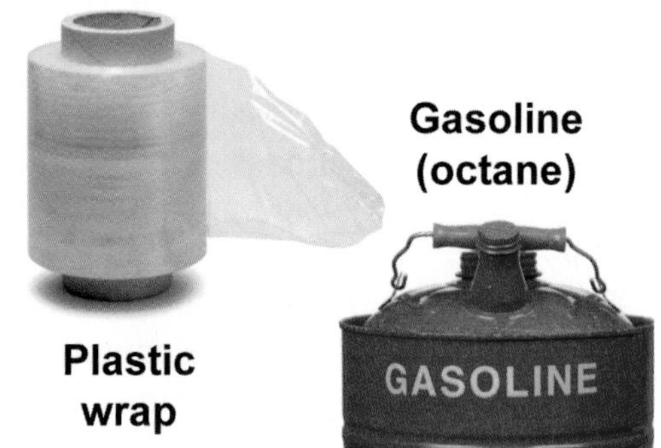

There are trillions of different combinations of 92 elements

This is the explanation for the diversity of matter

What are plastics?

Almost everything we own and use contains some form of the substance known as "plastic". Plastics are man-made polymers. The prefix "poly" means "many" and polymers are chain molecules made of many repeating units. Polyethylene is a polymer used for making bags, children's toys, bottles and many other items. Polyethylene contains only carbon and hydrogen atoms! Interestingly, the idea for polymers was found when early scientists were looking for a way to make silk. Years ago, silk was a very valuable fabric, used by royalty and the very wealthy. Scientists were hoping to create artificial silk and make lots of money! Rayon was the first artificial silk produced and it was made from cellulose polymers.

The chemical formula

The chemical formula

Compounds are represented by a **chemical formula.** The chemical formula tells you how many of each kind of atom are in the compound. For example, the chemical formula for water is H_2O. The subscript 2 indicates there are two hydrogen atoms for every oxygen atom in water. This is important to know if you wish to make water from the elements oxygen and hydrogen.

Reading a chemical formula

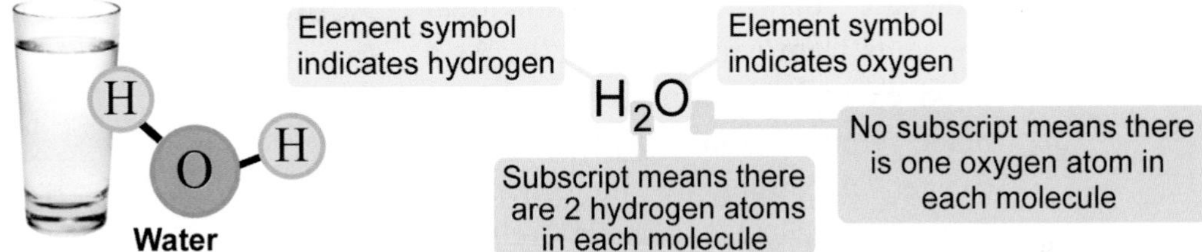

Examples of chemical formulas

The diagram below shows some other chemical formulas. Sodium bicarbonate ($NaHCO_3$) is commonly known as baking soda. This compound contains oxygen, carbon, hydrogen and sodium in a ratio of three oxygen atoms to one atom each of carbon, hydrogen and sodium. CH_3OH is the chemical formula for methanol, or wood alcohol. This formula is written to show that the carbon and three hydrogen atoms are together. The compound *paraffin* in candle wax is $C_{22}H_{44}$.

Sodium bicarbonate (baking soda)

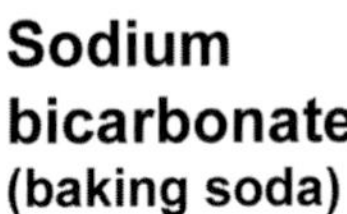

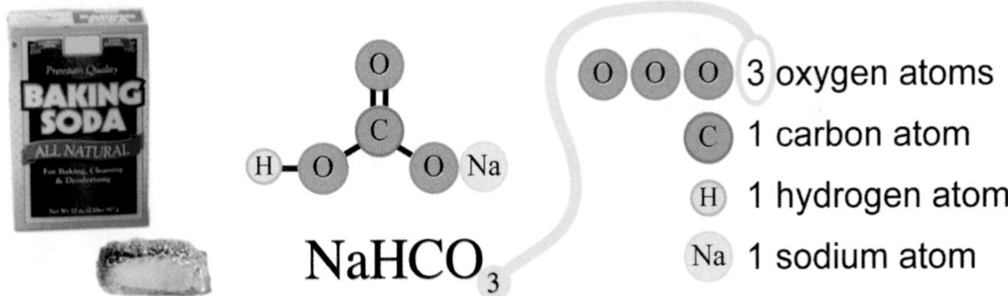

Write a chemical formula for a compound that has three hydrogen (H) atoms for each atom of nitrogen (N).

Asked: *Chemical formula*

Given: *3 hydrogen (H) and 1 nitrogen (N)*

Relationships: *The subscript tells the number of each element in the compound.*

Answer: NH_3

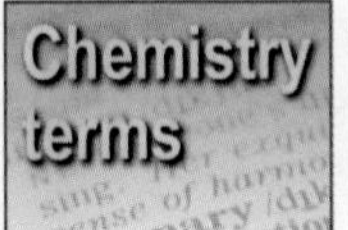

chemical formula - a combination of element symbols and subscripts that tells you the ratio of elements in a compound. For example, H_2O is the chemical formula for water and tells you there are 2 hydrogen (H) atoms for every one oxygen (O) atom.

Molecules

Molecular compounds

In a molecular compound, the atoms are bonded together in molecules. A pure molecular compound contains only identical molecules, each with the very same shape and chemical formula. Water is a molecular compound, as is methane (CH_4), ammonia (NH_3) and glucose ($C_6H_{12}O_6$).

Properties come from the molecule

The properties of a compound depend *much* more on the exact structure of its molecule than on the individual elements from which it is made. As a good example, aspirin is a molecule made from carbon, hydrogen and oxygen according to the chemical formula $C_9H_8O_4$. Aspirin relieves swelling and reduces pain in humans.

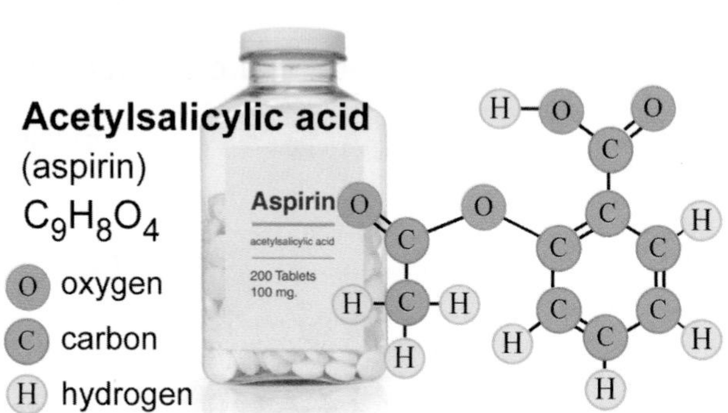

Properties depend on the exact chemical formula

By themselves, the elements (C, H, O) do not have the property of reducing pain. Other molecules made of the same elements have very different properties. For example, formaldehyde (a toxic preservative) is also made from just carbon, oxygen, and hydrogen. The beneficial properties of aspirin come from the specific combination of exactly eight hydrogen, nine carbon, and four oxygen atoms in the exact structure of the aspirin molecule. Remove even one hydrogen atom, and the resulting molecule would not have the properties of aspirin.

Three different molecules!
same chemical formula!

Acetylsalicylic acid (aspirin) $C_9H_8O_4$

Benzodioxole-5 carboxylic acid, methyl ester $C_9H_8O_4$

Acetyl benzoyl peroxide $C_9H_8O_4$

Properties also depend on molecular structure

The structure of a molecule is very important to the properties of a compound. The same 21 atoms in aspirin can be combined in other structures with the same chemical formula (above)! The resulting molecules are completely different chemicals and they do *not* have the beneficial properties of aspirin. *Both chemical formula and structure determine the properties of a compound.* The shape of a molecule is critical to its chemical and physical properties.

Describing molecules

What is a complete description of a molecule?

To completely describe a real molecule, you need to know both its composition and its structure. Consider acetic acid which has a chemical formula CH_3COOH. This is what gives the sour taste in vinegar and the fizzy reaction when mixed with baking soda. The chemical formula tells us the component atoms, but not much about the structure, or the exact *shape* in which the atoms combine. However, we need to know more to uniquely describe the chemical: *acetic acid*. The diagram below shows five different ways to represent the acetic acid molecule. Each kind of representation is best for certain applications.

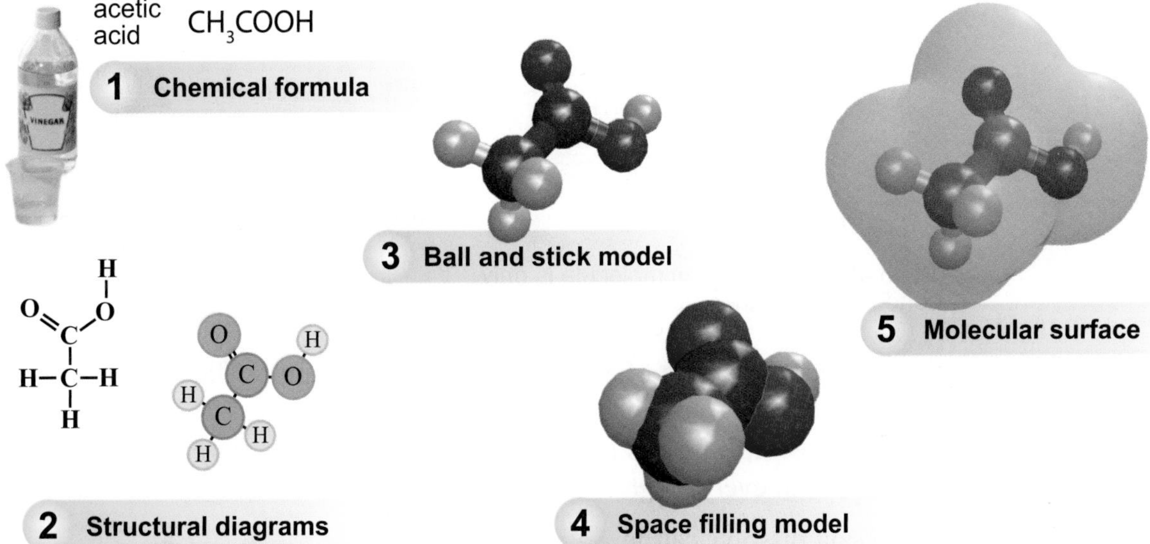

The structural diagram

The structural diagram shows the bonds, or connections, between atoms in the molecule. You can see that the two carbon atoms are bonded to each other. Structural diagrams are the easiest way to show how the atoms in a molecule are connected to each other. Notice that some of the lines are double. A double line means a *double bond* between the atoms.

The ball and stick model

The ball and stick model gives you the best sense of the three-dimensionality of the molecule. Real molecules are rarely flat like a structural diagram. The balls show the atoms and the sticks are the bonds between atoms.

The space-filling model

In reality, atoms are so close that they overlap when they bond. This is not shown by the ball and stick model. The space filling model more accurately shows how the atoms overlap each other in a molecule.

The molecular surface model

The molecular surface model combines the ball and stick view with a transparent shape that represents the "surface" of the molecule. Don't be fooled that this is "photorealistic" though. Molecules and atoms do not have hard, shiny surfaces as drawn in diagrams. The true boundary of an atom or molecule is very fuzzy and diffuse.

Ionic compounds

The compound sodium chloride (NaCl) has no molecules

Salt, or sodium chloride (NaCl), is a compound of sodium (Na) and chlorine (Cl) in a ratio of one sodium atom per chlorine atom. Salt is fundamentally different from aspirin in the atomic level. In aspirin, the atoms in a single molecule are chemically bonded *only* to other atoms in the same molecule. In salt *each* sodium atom is bonded to *every neighboring chlorine atom*. Each chlorine atom is likewise bonded to every neighboring sodium atom. The sodium and chlorine atoms in NaCl are always in a ratio of one sodium per chlorine. However, there are no *molecules* where a single sodium atom is bonded uniquely to a single chlorine atom.

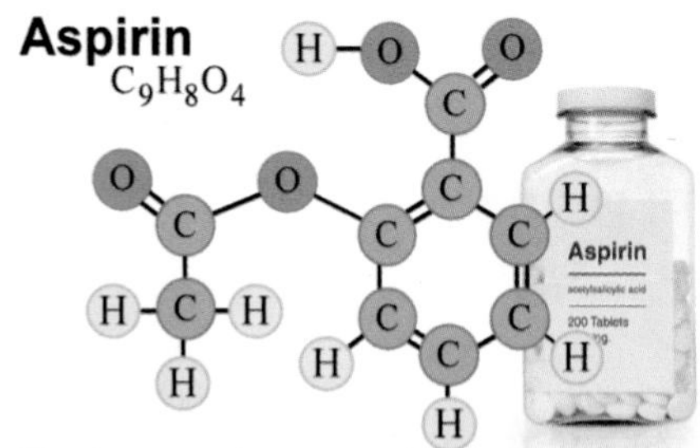

Each atom in an aspirin molecule is only bonded to one other atom in the molecule

In salt, every chlorine atom is attracted to all its neighboring sodium atoms and vice-versa

Ionic compounds

Salt is a good example of a whole class of compounds called **ionic compounds**. Salt still has a chemical formula, because there is always one sodium atom for every chlorine atom. However, salt does not have molecules because the bonds that hold salt together form between every neighboring atom, and not just the atoms in the molecule.

An ion is a charged atom

When you dissolve salt in water, the sodium and chlorine atoms come apart. They do not remain bonded together in a molecule. In fact, in salt each chlorine atom acquires a negative *electric charge* and becomes an *ion*. An **ion** is a charged atom. The sodium atoms also become ions, but with a positive charge.

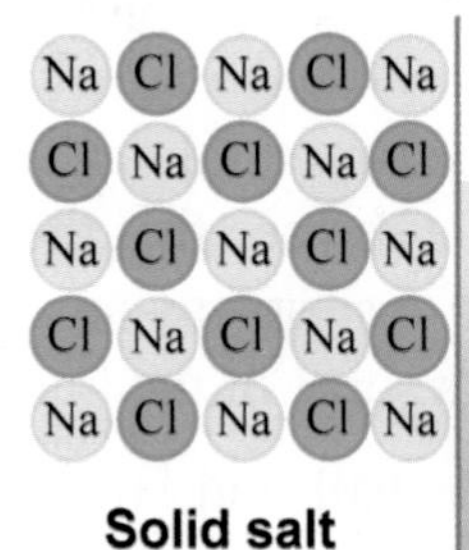

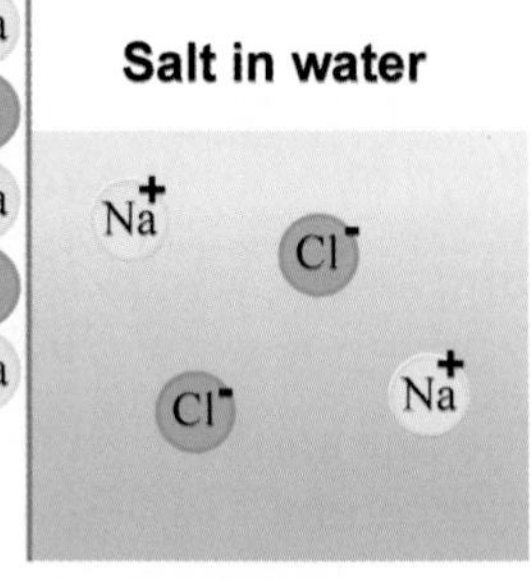

Positive and negative charges attract each other

Positive and negative charges attract each other. That is why every positive sodium ion is attracted to all the surrounding negative chlorine ions and vice versa. A compound becomes ionic when either atoms or groups of atoms become electrically charged.

ionic compound - a compound in which positive and negative ions attract each other to keep matter together. Salt (NaCl) is a good example.

ion - an “atom” that has acquired an electric charge, either positive or negative.

The formula mass

Moles of compounds

We know how to calculate a mole of any element from its average atomic mass. However, most matter exists in compounds. What about compounds? Most of chemistry deals with compounds. How do we calculate moles of a compound?

The formula mass

The answer is to use the chemical formula. The **formula mass** is the mass of one mole of a compound with a given chemical formula. As an example, start with water, H_2O. What is the mass (in grams) of one mole of water?

1. The chemical formula says one mole of water contains two moles of hydrogen and one mole of oxygen
2. Add up the mass of two moles of hydrogen (2 g) and one mole of oxygen (16 g)
3. The result (18 g) is the mass of 1 mole of water

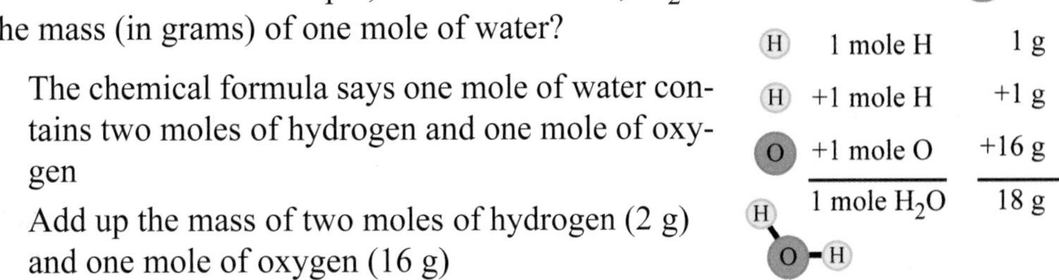

What is the mass of 1 mole of methane which has the chemical formula CH_4?

Asked: *mass of 1 mole*

Given: *Methane (CH_4) contains 1 carbon (C) and 4 hydrogen (H) atoms*

Relationships: *The formula mass is the sum of the atomic masses for each atom in the compound*

Solve: $4H + C = 4(1.0079) + 12.011 = 16.04$ *grams*

Answer: *One mole of methane (CH_4) has a mass of 16.04 grams*

How many moles are in 100 grams of water (H_2O)?

Asked: *moles in 100 grams*

Given: *Water H_2O contains 2 hydrogen (H) and 1 oxygen (O) atom.*

Relationships: *the formula mass is the sum of the atomic masses for each atom in the compound.*

Solve: $2H + O = 2(1.0079) + 15.999 = 18.015$ *grams per mole of* H_2O

$$100\text{ g} \times \left(\frac{1\text{ mole}}{18.015\text{ g}}\right) = 5.55\text{ moles}$$

Answer: *100 grams of water (H_2O) contains 5.55 moles.*

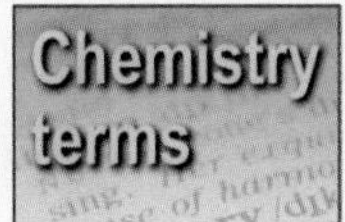

formula mass - the mass of one mole of a compound with a given chemical formula.

Solving problems with Moles

Converting back and forth between grams and moles is a very useful skill in chemistry. This is because we weigh substances in grams, but we need to know how many moles there are in order to determine the number of atoms or molecules in that mass. The periodic table allows us to obtain the formula mass for any compound that we might be experimenting with in the lab. You will find unit cancellation to be very important when you are doing this type of calculation. Keeping track of your units helps you to eliminate simple mistakes. To convert from grams to moles we divide by the molar mass.

How many moles are in 25.0 g of copper oxide (CuO)?

Asked: *How many moles of CuO?*

Given: *25.0 g of CuO*

Relationships: *Formula mass for CuO = 63.55 + 15.999 = 79.55* $\frac{\text{g}}{\text{mole}}$

Solve: $25.0\ \text{g CuO} \times \frac{1\text{mole CuO}}{79.55\ \text{g CuO}} = 0.314\ \text{moles of CuO}$

Answer: *25.0 g of CuO contains 0.314 moles of copper oxide (CuO)*

How many grams are in 2.30 moles of butane (C_4H_{10})? Butane is used as a lighter fluid in disposable lighters.

Given: *2.30 moles of* C_4H_{10}

Relationships: *Formula mass for* C_4H_{10} *= 4(C) + 10(H) = (4 × 12.011) + (10 × 1.0079) = 58.12 g/mole*

Solve: $2.30\ \text{moles}\ C_4H_{10} \times \frac{58.12\ \text{g}\ C_4H_{10}}{1\ \text{mole}\ C_4H_{10}} = 133.68\ \text{g}\ C_4H_{10}$

Answer: *133.68 g are in 2.30 moles of butane* (C_4H_{10}).

How many moles are in 30.0 g of baking soda ($NaHCO_3$)?

Given: *30.0 g of* $NaHCO_3$

Relationships: *Formula mass for* $NaHCO_3$ *= 22.99 + 1.0079 + 12.011 + 3×15.999 = 84.01 g/mole*

Solve: $30.0\ \text{g}\ NaHCO_3 \times \frac{1\ \text{mole}}{84.01\text{g}\ NaHCO_3} = 0.357\ \text{moles}$

Answer: *30.0 g are in 0.357 moles of baking soda* ($NaHCO_3$)

Using Avogadro's Number

We can use Avogadro's number to tell us how many atoms are in a given mass of a metal such as lead (Pb).

How many atoms of lead (Pb) are in a lead pipe that weighs 105 grams?

Given: *105 grams of Pb*

Relationships: *The atomic mass of lead is 207.2 g/mole.*

$$105\ \cancel{\text{g Pb}} \times \frac{1\ \cancel{\text{mole Pb}}}{207.2\ \cancel{\text{g Pb}}} \times \frac{6.022 \times 10^{23}\ \text{atoms Pb}}{1\ \cancel{\text{mole Pb}}} = 3.05 \times 10^{23}\ \text{atoms Pb}$$

Answer: *There are 3.05 x 10^{23} atoms of Pb in a lead pipe weighing 105 grams.*

Sometimes we want to know how many atoms of an element are in a given mass of a compound. For example:

How many oxygen atoms are in 200.0 g of glucose ($C_6H_{12}O_6$)?

Given: *200.0 grams of* $C_6H_{12}O_6$

Relationships: *The molar mass of glucose is = 6(C) + 12(H) + 6(O) = 6(12.011) + 12(1.0079) + 6(15.999) = 180.15 g/mole*

Solve: *First we find how many moles are in 200 grams*

$$200.0\ \cancel{\text{g}\ C_6H_{12}O_6} \times \frac{1\ \text{mole}\ C_6H_{12}O_6}{180.15\ \cancel{\text{g}\ C_6H_{12}O_6}} = 1.11\ \text{moles}\ C_6H_{12}O_6$$

Next we find how many molecules are contained in 1.11 moles

$$1.11\ \cancel{\text{mole}\ C_6H_{12}O_6} \times \frac{6.022 \times 10^{23}\ \text{molecules}\ C_6H_{12}O_6}{1\ \cancel{\text{mole}\ C_6H_{12}O_6}} = 6.68 \times 10^{23}\ \text{molecules}\ C_6H_{12}O_6$$

Then we find how many oxygen atoms are contained in 1.11 moles of glucose

$$6.68 \times 10^{23}\ \cancel{\text{molecules}\ C_6H_{12}O_6} \times \frac{6\ \text{O atoms}}{1\ \cancel{\text{molecule}\ C_6H_{12}O_6}} = 4.01 \times 10^{24}\ \text{atoms of O}$$

Answer: *There are 4.01 x 10^{24} atoms of O in 200.0 g of glucose ($C_6H_{12}O_6$).*

2.3 Mixtures and Solutions

Homogeneous mixture is the same throughout

Most matter is mixtures of many different compounds. A homogeneous mixture is the same throughout. All samples of a homogeneous mixture are the same. For example, stirring food color in water can produce a homogeneous mixture. The water near the top of the test tube has the same concentration of color as the water near the bottom. Brass is another example of a homogeneous mixture. Brass is made of 70 percent copper and 30 percent zinc. If you cut a brass candlestick into ten pieces, each piece would contain the same 70% copper and 30% zinc.

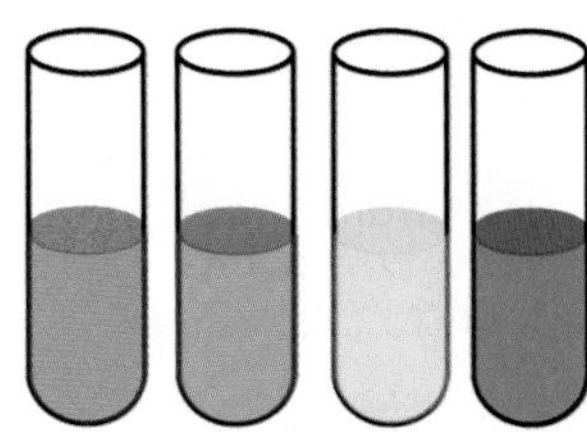

Well mixed food color in water is a ***homogeneous*** mixture

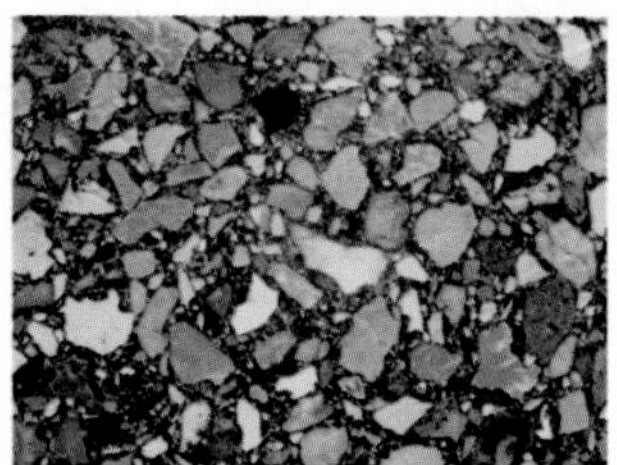

Concrete is a ***heterogeneous*** mixture

Heterogeneous mixtures

A heterogeneous mixture is one in which different samples are not necessarily made up of exactly the same proportions of matter. One common heterogeneous mixture is concrete. Concrete contains cement and rocks of many different types and sizes.

TABLE 2.1. Summary of the types of matter

Type of matter	Definition	Examples
Homogeneous mixture	A mixture that contains more than one type of matter and is the same throughout.	soda pop, air,
Heterogeneous mixture	A mixture that contains more than one type of matter and is not the same throughout.	chicken soup, soil, fudge ripple ice cream
Element	A substance that contains only one type of atom.	copper metal, oxygen gas, liquid nitrogen
Compound	A substance that contains more than one type of atom.	table salt, rust (iron oxide), carbon dioxide gas

Chemistry terms

homogenous mixture - is a mixture that is uniform throughout, any sample has the same composition as any other sample.

heterogeneous mixture - is a mixture that is not uniform, different samples may have different compositions.

Solutions

Defining a solution

A **solution** is a mixture that is *homogenous on the molecular level*. That means there are no clumps bigger than a molecule. The orange juice you drink is partly a solution. The sweetness comes from dissolved sugars. The "orange" taste comes from other dissolved molecules that provide flavor, scent and color. Orange juice is only *partly* a solution because some substances are *not* dissolved, such as small bits of orange pulp.

Solvent and solute

A solution always has a *solvent* and at least one *solute*. The **solvent** is the substance that makes up the biggest percentage of the mixture and it is usually a liquid. Water is the most important solvent for living things. For example, the solvent in orange juice is water. The **solutes** are the other substances in the solution. Sugar, citric acid and other flavoring chemicals are solutes in orange juice.

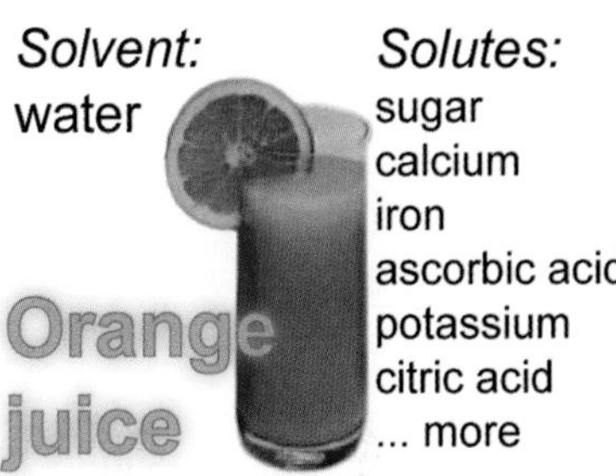

Dissolving

When the solute particles are evenly distributed throughout the solvent, we say that the solute has **dissolved**. The picture below shows a sugar and water solution being prepared. The solute (sugar) starts as a solid in the graduated cylinder on the left. Water is added and the mixture is carefully stirred until all the solid sugar has dissolved. Once the sugar has dissolved the solution is clear again.

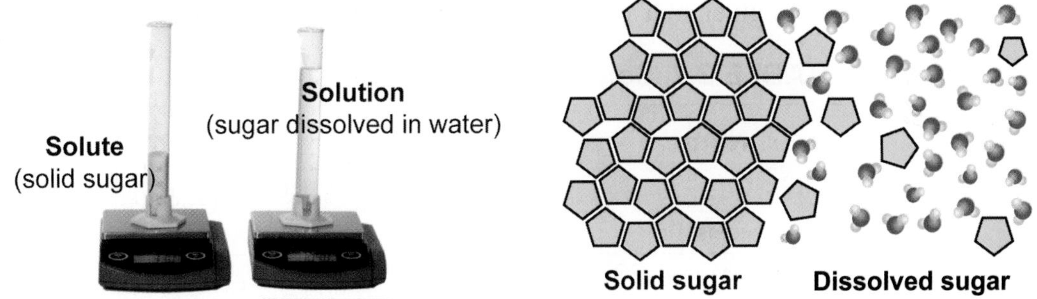

The molecular explanation for dissolving

On the molecular level, dissolving of a solid (like sugar) occurs when molecules of solvent collide with, and transfer energy to, molecules of solid solute. This explains why substances dissolve faster at higher temperatures. Hotter solvent molecules have more energy and are more effective at knocking molecules of solute into the solution. You may have noticed that sugar dissolves much faster in hot water than in cold water.

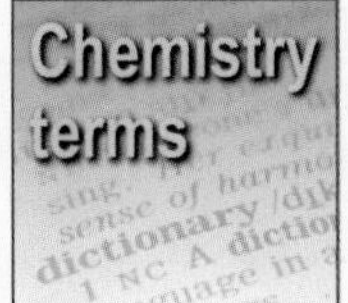

solution - a mixture that is homogeneous.

solvent - the substance that makes up the biggest percentage of the mixture and is usually a liquid.

solute - any substance in a solution other than the solvent.

dissolved - when molecules of solute are completely separated from each other and dispersed into a solution.

Concentration and solubility

Concentration and dilute solutions

The **concentration** of a solution describes how much of each solute there is compared to the total solution. A solution is **dilute** when there is very little solute. Mixing one gram of sugar with 99 mL of water makes 100 grams of a dilute sugar solution. A 10 gram sample of this solution contains only 0.1 grams of sugar compared to 9.9 grams of water.

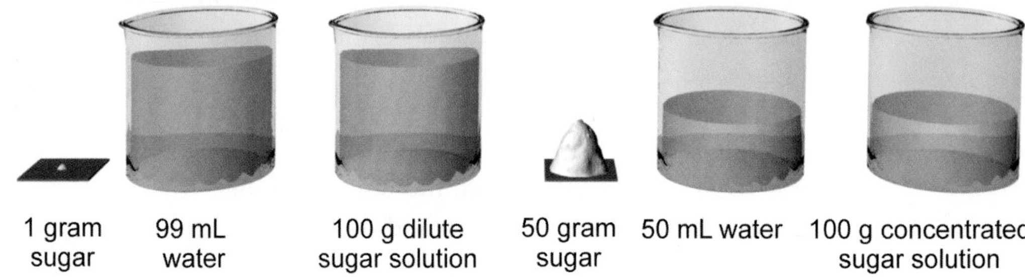

Concentrated solutions

A solution is **concentrated** when there is a lot of dissolved solute compared to solvent. Mixing 50 grams of sugar into 50 mL of water makes a concentrated sugar solution. A 10 gram sample of this solution contains 5 grams of sugar and 5 grams of water.

What is solubility?

Solubility means the amount of solute (if any) that can be dissolved in a quantity of solvent. Solubility is often listed in grams per 100 milliliters of solvent. Solubility is always given at a specific temperature since temperature strongly affects solubility.

TABLE 2.2. Solubility of common substances in water at 25°C

Name	Solubility (g/L)	Name	Solubility (g/L)
Salt (NaCl)	36	Chalk	0
Sugar (sucrose)	2000	Vegetable oil	0
Baking soda ($NaHCO_3$)	about 10	Oxygen	8.3×10^{-3}

Insoluble substances do not dissolve

Notice that chalk and talc have *zero* solubility. These substances are **insoluble** in water because they do not dissolve. You can mix chalk dust and water and stir it up all you want. You will just have a mixture of chalk dust and water. The water will not separate the chalk dust into individual molecules because chalk is insoluble in water.

concentration - the amount of each solute compared to the total solution.

dilute - a solution containing relatively little solute compared to solvent.

concentrated - a solution containing a lot of solute compared to solvent.

solubility - the amount of a solute that will dissolve in a particular solvent at a particular temperature and pressure.

insoluble - not dissolvable in a particular solvent.

Calculating solute and solution quantities

Working with concentration in g/L

Many experiments (and recipes) call for precise amounts of dissolved substances. For example, suppose you have an experiment that requires 10 grams of sugar. How much solution do you add if the concentration of the sugar solution is 75 g/L? This is a typical chemistry question. To find the answer, we rearrange the concentration formula to solve for liters of solution.

Calculating solution volume from solute mass and concentration in g/L

$$\text{concentration}_{g/L} = \frac{\text{mass of solute}}{\text{liters of solution}}$$

$$\text{liters of solution} = \frac{\text{mass of solute}}{\text{concentration}_{g/L}}$$

If the concentration of a sugar solution is 75 g/L, how much solution do you need if you want 10 g of sugar?

Asked: *Volume of solution*

Given: *10 grams of solute and concentration of 75 g/L*

Relationships: *liters of solution = mass of solute ÷ concentration in g/L*

Solve: *10 g ÷ 75 g/L = 0.133 liters, or 133 mL.*

Working with percent (%) concentration

Each of the concentration formulas can be rearranged to give one of the quantities in terms of the other two. For example, a mouthwash contains 0.05% menthol. Menthol ($C_{10}H_{20}O$) is a chemical derived from the peppermint plant, and it gives mouthwash its minty flavor. How much menthol do you need to make 10 kg of mouthwash?

Calculating solute mass from solution mass and % concentration

$$\text{concentration}_{\%} = \frac{\text{mass of solute}}{\text{mass of solution}} \times 100\%$$

$$\text{mass of solute} = \text{mass of solution} \times \frac{\text{concentration}_{\%}}{100}$$

How much menthol do you need to make 10 kg of mouthwash if the concentration of menthol is 0.05%?

Asked: *Mass of solute*

Given: *10 kilograms of solution, solute concentration of 0.05%*

Relationships: *mass of solute = mass of solution × (concentration ÷ 100)*

Solve: *10 kg × (0.05 ÷ 100) = 0.005 kg = 5 grams*

Molarity

What is molarity?

The **molarity** of a solution is the number of moles of solute per liter of solution. Molarity is useful because chemists need a way to control the ratios of different molecules in reactions. It's impossible to count molecules one by one! Instead, *chemists make solutions with a known molarity.* Once you know the molarity you can figure out how many moles of each solute there are in a given volume of solution. If you know how many moles you need, you can calculate how many milliliters of solution you have to add. Milliliters are easy to measure and many chemical experiments use solutions of standard molarity for this reason.

Molarity: The number of moles of solute per liter of solution

How to calculate molarity

To find molarity, you need to know how many moles are dissolved per liter of solution. Citric acid is what gives the sour taste to lemons and limes. Suppose you add ten grams of citric acid to a half liter of water. What is the molarity of the solution? The problem is solved in three steps:

Citric acid
$C_6H_8O_7$

1. Calculate the formula mass so you know the mass of one mole;
2. Use the formula mass to figure out how many moles there are;
3. Calculate molarity by dividing the number of moles by the volume of solution.

10 g of citric acid ($C_6H_8O_7$) is added to 500 mL of water. What is the molarity of the resulting solution?

Asked: *Find the molarity of a solution*

Given: *Amount of solute, citric acid and the volume of solution.*

Relationships: *Molarity = moles solute ÷ volume of solution*

Solve: *Start by calculating the formula mass of $C_6H_8O_7$.*

$$6 \times 12.011 + 8 \times 1.0079 + 7 \times 15.999 = 192.12\frac{g}{mole}$$

Next we calculate the number of moles in 10 g of $C_6H_8O_7$.

$$\text{\# moles} = \frac{10g}{192.12g/mole} = 0.052 \text{ moles}$$

Now we calculate the molarity

$$\text{Molarity} = \frac{0.052 \text{ moles}}{0.5 \text{ liters}} = 0.104 \text{ M}$$

Chemistry terms

molarity - the number of moles of solute per liter of solution.

Calculating solution quantities with molarity

Why molarity is useful

Chemical reactions rearrange reactant molecules to form product molecules. For this reason, the *quantity* of molecules is often the most important factor in determining what happens in a reaction. *Molarity* tells you the number of molecules (in moles), so molarity is the best way to describe concentration in many situations.

Vitamins

Vitamin C is a compound called *ascorbic acid* ($C_6H_8O_6$). The ascorbic acid molecule is a sugar (5-sided ring) with short carbon chain. Besides being a vitamin, ascorbic acid is also a food preservative! It works by accepting electrons (and losing hydrogens) to bind up reactive oxygen that might otherwise cause spoiling reactions in food.

Ascorbic acid
$C_6H_8O_6$

Reaction

$$2C_6H_8O_6 + O_2 \rightarrow 2C_6H_6O_6 + 2H_2O$$

Using molarity in a reaction: An ordinary bread bag holds about 2 liters of air, which contains about 0.02 moles of oxygen (O_2). Suppose you have a 1 molar (1M) solution of ascorbic acid. How much do you need to completely react with 0.02 moles of oxygen?

1. Use the balanced reaction to calculate the moles of ascorbic acid you need;
2. Use the concentration formula to find the required volume of 1M solution.

How much (volume) of a 1M ascorbic acid solution will completely react with 0.02 moles of oxygen (O_2)?

Asked: *volume of solution*

Given: *concentration (1M) and balanced reaction*

Relationships: *Molarity = moles solute ÷ volume of solution*

Solve: *According to the balanced reaction we need 2 moles of ascorbic acid for every mole of O_2.*
That means we need 0.04 moles of ascorbic acid.
volume = moles of solute ÷ molarity
= (0.04) ÷ 1 = 0.04 liters, or 40 mL
40 mL of the solution contains 0.04 moles of ascorbic acid, which is enough to react with 0.02 moles of oxygen (O_2).

Mixtures of gasses

Air is a mixture of gasses

Many mixtures are gases, like air. Gases like air can expand (or contract) to fill their container. For example, 1 gram of air can fill a 1 liter bottle at 1 atmosphere of pressure. The same one gram of air can expand to fill 1 cubic meter at 0.0001 atmospheres or contract to fill 1 mL at 1,000 atmospheres.

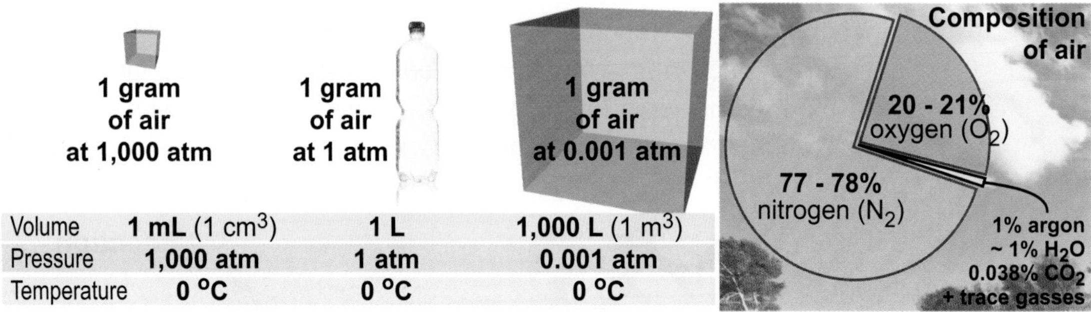

Volume	**1 mL** (1 cm^3)	**1 L**	**1,000 L** (1 m^3)
Pressure	**1,000 atm**	**1 atm**	**0.001 atm**
Temperature	**0 °C**	**0 °C**	**0 °C**

The molar volume and STP

Concentration in a mixture of gasses is usually given as a percent by volume. This is because gasses at the same temperature and pressure, have the same volume per mole *no matter what kind of gas it is*. For example, a mole of oxygen (O_2) gas at one atmosphere of pressure and 0°C has a volume of 22.4 liters. A mole of nitrogen gas at 1 atm. and 0°C has the same volume of 22.4 liters, and so does a mole of carbon dioxide gas (CO_2). The constant of 22.4 moles per liter is called the **molar volume**. The conditions of 0°C and 1 atmosphere of presssure are called **Standard Temperature and Pressure** (STP).

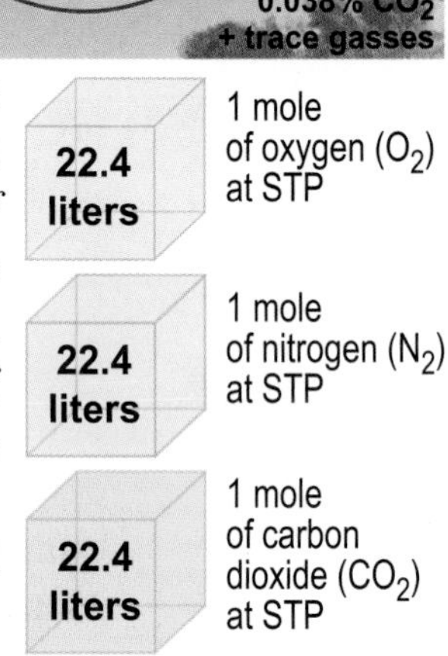

Using volume to determine quantity

Working with gases in terms of volume is necessary because the mass of a gas is hard to measure directly. When you put an "empty" bottle on a balance, you do not get the mass of the air inside the bottle. The most reliable way to determine or specify a quantity of gas is to use volume and pressure to calculate moles, then use moles to calculate mass.

How many moles of oxygen (O_2) are in a 1 liter bottle at STP (1 atm and 0°C)?

Asked: *moles of a gas (O_2)*

Given: *volume, and STP conditions*

Relationships: *The molar volume of a gas at STP is 22.4 liters / mole*

Solve: $1 \text{ liter} = \frac{1 \text{ L}}{} \times \frac{1 \text{ mol}}{22.4 \text{ L}} = 0.045 \text{ moles}$.

Answer: *There are 0.045 moles in 1 liter of O_2 at STP.*

Chemistry terms

molar volume - 22.7 liters per mole at 0°C and 1 atmosphere of pressure.

standard temperature and pressure - (STP) - conditions of one atmosphere of pressure and 0°C.

Partial pressures

Air is a mixture of gasses

Air is not an element, or even a compound but is a mixture of gasses. Dry air contains roughly (by volume) 78.08% nitrogen, 20.95% oxygen, 0.93% argon, 0.038% carbon dioxide, and trace amounts of other gases. Normal air you breathe is not completely dry and contains a some amount of water vapor, on average around 1%.

Looking at the components in air

The average air pressure, at sea level, is 101,325 Pa, which is defined to be one atmosphere (1 atm). Because air is a mixture, 101,325 Pa is the *total pressure* from all the gases in air. If you were to separate the different gases in 1 liter of air, there are two useful ways to think about the separation. Both are correct and useful.

Equal volumes - different pressures

Nitrogen (N_2) Oxygen (O_2) Argon (Ar)

1 liter 1 liter 1 liter

79,033 Pa (0.78 atm) 21,228 Pa (0.21 atm) 9,423 Pa (0.093 atm)

Equal pressures - different volumes

Nitrogen (N_2) Oxygen (O_2) Argon (Ar)

0.78 liters 0.21 liters 0.093 liters

101,325 Pa (1.00 atm) 101,325 Pa (1.00 atm) 101,325 Pa (1.00 atm)

Dalton's law of partial pressures

In a mixture of gasses, the total pressure is the sum of the partial pressures of each individual gas in the mixture. This rule is called **Dalton's law of partial pressures**. If you divide 1 liter of air into *three equal volumes*, the nitrogen gas has a partial pressure of 79,033 Pa. The oxygen gas has a partial pressure of 21,228 Pa. The argon gas has a partial pressure of 9,423 Pa.

How volume divides at equal pressure

If you divide 1 liter of air into three volumes of *equal pressure*, the nitrogen has a volume of 0.78 liters. The oxygen has a volume of 0.21 liters and the argon has a volume of 0.093 liters (93 mL).

1 liter of helium (75%) and neon (25%) is at STP. What is the partial pressure of helium?

Asked: *partial pressure of helium*

Given: *75% He and 25% Ne, and STP conditions*

Relationships: *The total pressure is the sum of the partial pressures of each gas.*

Solve: *0.75 × 101,325 Pa = 75,994 Pa or 0.75 atm.*

Answer: *The partial pressure of helium is 75,994 Pa or 0.75 atm.*

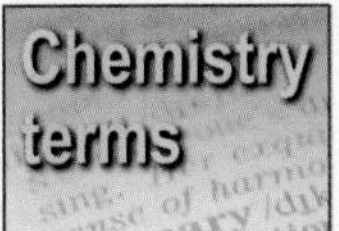

Dalton's law of partial pressures - the total pressure in a mixture of gasses is the sum of the partial pressures of each individual gas in the mixture

Perhaps the most magical thing about chemistry is the fact that when atoms of different elements and molecules combine a new compound is formed that has completely new properties! These new properties are nothing like the original properties of the pure elements.

One example of this involves sodium metal and chlorine gas. Did you know that sodium metal (Na), atomic number 11 on the periodic table, is so reactive that it catches on fire when placed in water? In the laboratory, sodium metal is stored in a glass jar and covered with oil, the oil coats the surface of the sodium metal so that it cannot react with any oxygen or moisture in the air.

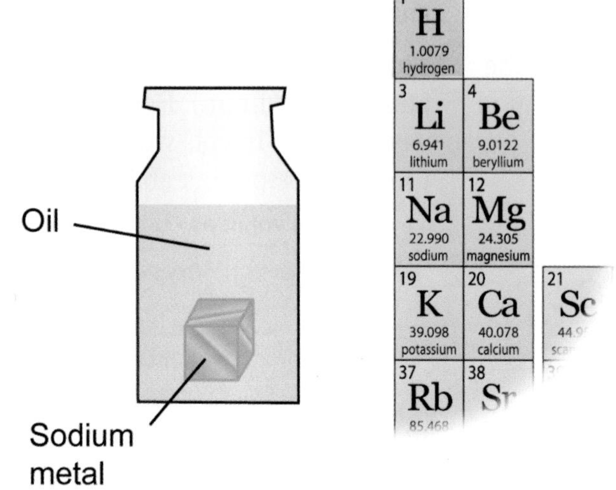

Sodium is a very reactive metal. It rapidly oxidizes, which means that it loses electrons in a process similar to rusting. Large amounts of sodium are explosive in water, because it rapidly generates hydrogen gas. This makes it extremely dangerous to handle. Even small amounts of hydrogen gas are highly flammable. The other product, sodium hydroxide, is caustic and can cause serious burns.

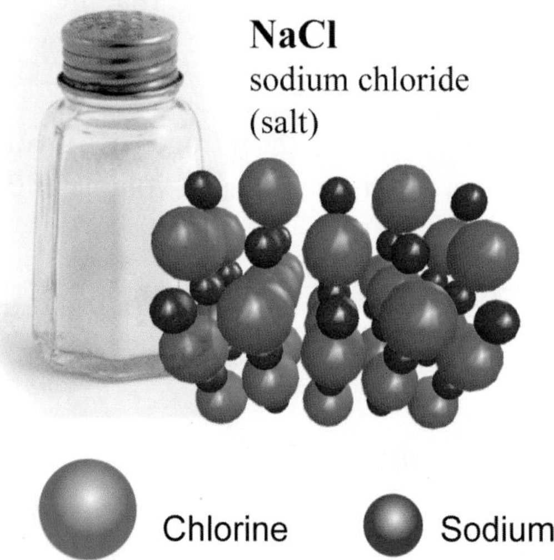

Sodium is the most abundant alkali metal making up about 3% of the earth's crust. In nature, sodium metal does not exist in its pure form, it needs to be extracted from compounds in the earth's crust, such as salt (NaCl).

.

The most common method used to obtain sodium metal is called electrolysis. Electrolysis uses electricity to separate substances. A voltage is applied to a molten liquid of NaCl, causing electrons to flow in such a manner that the sodium ion (Na^+) becomes Na(s), by gaining an electron. In this case, sodium chloride liquid, (NaCl(l)), is separated into sodium metal (Na(s)) and chlorine gas ($Cl_2(g)$).

$$2\ NaCl(l) \rightarrow 2\ Na(s) + Cl_2(g)$$

Chlorine gas can also be obtained from the electrolysis process. During electrolysis two chlorine ions $2Cl^-$ release electrons and becomes neutral Cl_2 gas.

Pure elemental sodium metal is toxic to ingest and hazardous to animals and humans. Elemental chlorine (Cl_2) is a light yellow-greenish gas at room temperature. It is also poisonous to animals and humans if inhaled. Chlorine is an example of a diatomic molecule, meaning its atoms occur in nature as a pair, or in units of two atoms joined together. Chlorine is not stable as an individual atom, much like oxygen is not stable alone. Chlorine is used to make many consumer products, such as: plastics, dry cleaning chemicals, textiles, insecticides and pharmaceuticals .

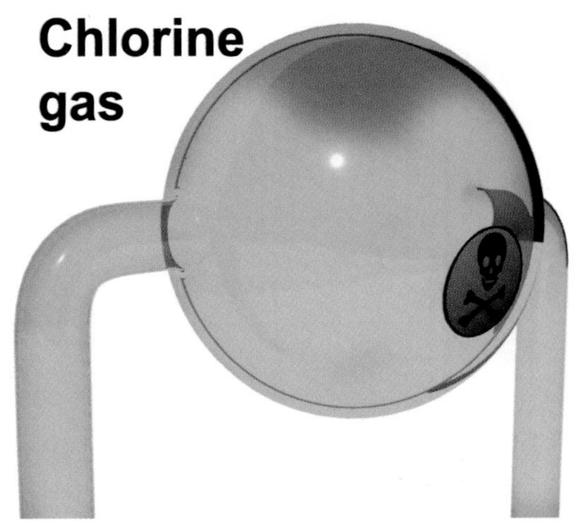

Both sodium metal and chlorine gas are toxic to us in their elemental form. When these elements react together they produce sodium chloride (NaCl) or table salt, which is a compound essential for life! How can this be? Two dangerous chemicals turn into something not only safe to consume but essential! This chemical reaction produces a compound, sodium chloride that is no longer dangerous!

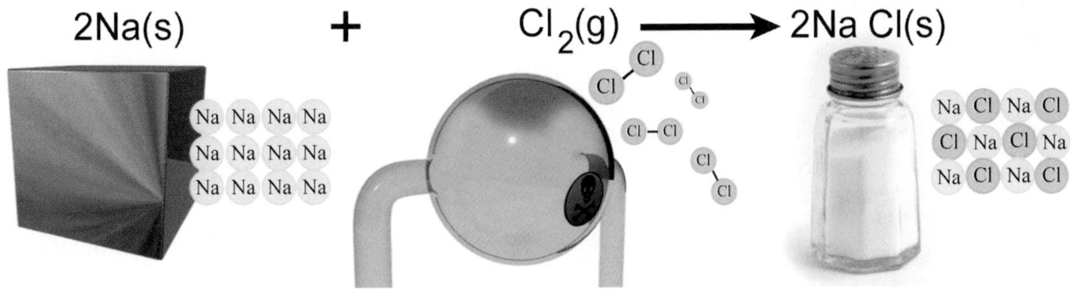

Once the atoms of sodium and chlorine recombine and form a compound, the result is a product that is common in each of our homes. Salt is required by the body for several important reasons: It helps to maintain the proper fluid concentrations in your tissues, transmits nerve signals, and it helps the heart muscle to contract.

Salt is also essential for animals. If you have visited a farm you may have seen salt licks, or salt blocks, that provide cattle and horses with the sodium chloride their bodies need. Some game preserves place them out for deer and other wild animals. Unlike humans who obtain salt in their daily diets, animals that eat mostly grass need to be supplemented with salt. Large quantities of sodium chloride (NaCl) are found dissolved in ocean water. Much of our salt is produced by evaporating seawater or by mining for "rock" salt in the earth.

Chapter 2 Review.

Vocabulary

Match each word to the sentence in which it best fits.

Section 2.1

scale	substance
mixture	chemical change
physical property	physical change
chemical property	macroscopic
element	microscopic
period	mole
group	atomic number

1. A typical size that shows a certain level of detail is called a ________.
2. A substance that cannot be broken down to a simpler substance, and one in which all the atoms have the same atomic number is called an _________.
3. A ________ is a kind of matter that can be separated by physical means.
4. When two or more substances are combined by physical means a __________ is formed.
5. The type of change that transforms one substance into another is _____________________.
6. Mass is an example of a __________ property.
7. The number which represents how many protons are in the nucleus is called the ___________.
8. The word ________ on the periodic table refers to the arrangement of the horizontal rows.
9. The _______ is a unit that chemists use to represent a large number of atoms or molecules.
10. A vertical row, such as 7A, on the periodic table is referred to as a ________.
11. The __________ is created or caused by the exact molecular structure of a compound.

Section 2.2

molecule	compound
chemical formula	structural diagram
ions	space filling model
ionic bond	formula mass

12. A _______ is a mixture of two or more different elements.
13. The _____________ is a way to represent the ratio of the different elements in a compound.
14. In a chemical substance, such as water, the single smallest unit, H_2O, is referred to as a _______________.
15. An ______ is an atom that has gained or lost an electron, and now carries a charge.
16. The _______________ is the weight of all of the atoms represented in the chemical formula added together.
17. A ___________is one way to show how all the atoms are connected to one another.
18. When Na^+ and Cl^- are attracted to each other by their opposing charges, they form an _____________.

Section 2.3

heterogeneous	solvent
homogeneous	solution
solute	concentrated
concentration	dilute
solubility	insoluble
dissolved	molarity
molar volume	partial pressue

19. The substance that is dissolved in a solution is called the _______________.
20. The substance which does the dissolution is called the ______________.

21. A mixture that does not "mix" in such a way that all substances are evenly distributed throughout is called _________________.

22. A ___________ solution is one in which there is a large quantity of solute particles present.

23. A mixture which has uniform (evenly distributed) particles throughout is called a _______________ mixture.

24. The value 22.7 liters per mole is known as the ___________ of a gas.

25. When a teaspoon of sugar is added to a cup of warm tea and stirred we can say that the sugar _________ in the tea.

26. A substances ability to dissolve is referred to as its _____________.

27. Chemists use the unit of ____________ to represent how many moles are in one liter of solution.

28. The ____________ of a gas in a mixture is the pressure exerted by that gas alone as if it were the only gas in the container.

29. When a substance does not dissolve in a solvent it is said to be _____________.

Conceptual Questions:

Section 2.1

30. List three examples of a physical property.

31. When sand is mixed with water it can be easily separated. What type of a mixture is this an example of?

32. Explain why breaking a piece of chalk is an example of a physical change and not a chemical change?

33. Compare and contrast the process of chemical versus physical change.

34. Give an example of a physical change that you observe in your everyday life.

35. Is the burning of magnesium ribbon an example of a chemical or physical change? How do you know?

36. Give an example of a chemical change that you observe in your everyday life.

37. For the following changes listed below label them as C (chemical) or P (physical).
    a. a match burning
    b. an ice cube melting
    c. metal rusting
    d. salt dissolving in water

38. Find the element Mg, magnesium on the periodic table.
    a. What period is it in?
    b. What group is it in?
    c. What is its atomic number?

39. An element has the atomic number of 18.
    a. What period and group is it in?
    b. List another element that will likely have similar chemical properties.

40. The mole is often referred to as the "chemists dozen", explain why the unit of a mole is useful to a chemist?

41. What is the mass of 1 mole of copper?

Section 2.2

42. Sketch an example of two different molecules that can be formed by carbon(C) and oxygen (O). Think of some you are familiar with in your everyday life (Hint: the gas you exhale.)

43. Using the chemical formula $C_6H_{12}O_6$, for the sugar glucose, what is the ratio of carbon atoms to oxygen atoms? Carbon atoms to hydrogen atoms?

Chapter 2 Review.

44. Write the chemical formula represented for the molecule in the picture.

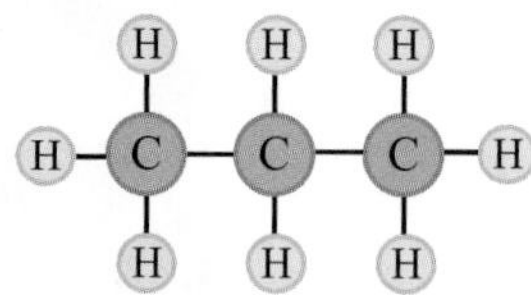

45. Explain why Na^+ is an ion of sodium and not an atom of sodium?

46. The formula mass of methane gas, CH_4, is 16 g/mole. Is this correct? Explain.

Section 2.3

47. Describe what happens when you mix 1 teaspoon of salt, NaCl, to 2 cups of water. Explain which substance is the solute and which is the solvent.

48. Compare and contrast a heterogeneous mixture with a homogeneous mixture. Use a table to organize the differences.

49. Given that you have 1 packet of gatorade mix and 1 liter of water describe how you would make a
    a. dilute solution of gatorade and
    b. concentrated solution of gatorade.

50. Describe how you would prepare 250 mL of a 0.850 M solution of NaCl.

51. Suppose a mixture of gasses contained equal masses of nitrogen and oxygen gas. Would the partial pressures be equal? Explain why or why not.

52. A laboratory experiment compares the volumes of one mole of hydrogen gas (H_2, 2 g/mole) with one mole of argon gas (Ar, 40 g/mole). The experiment should find which of the following is true?
    a. The hydrogen gas has a lesservolume because it has a lower molecular weight.
    b. The argon gas has a lesser volume because it has a higher molecular weight.
    c. The two gases have equal volumes.

Quantitative Problems

Section 2.1

53. How many oranges are in 1 mole of oranges?

54. If you have 9.1×10^{25} oranges how many moles of oranges are there?

55. What is the weight, in grams, of
    a. 1 mole of sulfur, S ?
    b. 1 mole of nitrogen, N ?

56. Calculate the mass in grams of
    a. 1 atom of gold Au
    b. 1 atom of copper, Cu

57. Calculate the number of atoms in
    a. 12.0 g of carbon, C
    b. 16.0 g of oxygen, O
    c. 7.0 g of lithium, Li
    d. 27.0 g of aluminum, Al

58. Calculate the number of moles in
    a. 40.5 grams of magnesium, Mg.
    b. 25.0 g of copper, Cu
    c. 10.0 g of aluminum
    d. 6.0 grams of helium, He

59. How many molecules of water, H_2O, are in 5.0 moles ?

60. Calculate the number of grams in 5.0 moles of argon, Ar.

61. How many gold, Au, atoms are in 45.0 grams of gold?

62. How may moles are in 1 gram of water?

63. How many hydrogen atoms and how many oxygen atoms are contained in 1 gram of water?

Section 2.2

64. Determine the formula mass of the following compounds
    a. sulfur dioxide, SO_2

b. methanol, CH_3OH,
c. ammonia, NH_3,
d. caffeine, $C_8H_{10}N_4O_2$.
e. aluminum oxide, Al_2O_3
f. lithium oxide, Li_2O

65. Determine the formula mass of the following compounds
    a. KNO_3
    b. Na_2CO_3
    c. Ca_3N_2
    d. H_2SO_4

66. Determine the number of moles in
    a. 20.0 g of ammonia, NH_3
    b. 30.0 grams of sodium chloride, NaCl
    c. 10.0 grams of carbon dioxide, CO_2
    d. 16.0 g of nitrogen dioxide, NO_2

67. Calculate the number of grams in
    e. 0.85 moles of CO_2
    f. 4.30 moles of NO_2
    g. 1.56 moles of NaCl

Section 2.3

68. If you have 25.0 g of sugar and you require a solution that has a concentration of 60 g/L of sugar, how much solution do you need to add to 25.0 g to achieve this concentration?

69. How many grams of alcohol do you need to make 400 g of an antiseptic solution, assuming the concentration of the alcohol is 0.10 % ?

70. How many moles of KOH are in 45.0 mL of a 6.50 M solution?

71. How many grams of KOH are in 45.0 mL of a 6.50 M solution?

72. How many moles of $CaCl_2$ are present in 60.0 mL of a 0.10 M $CaCl_2$ solution?

73. Calculate the mass of KBr in grams required to prepare 500 mL of a 2.50 M solution.

74. A recipe calls for 1 tablespoon of salt (5 grams) in 1 liter of water. What is the molarity of the solution?

75. Calculate the molarity of each of the following solutions:
    a. 25.0 g of ethanol (C_2H_5OH) in 455 mL of solution,
    b. 14.8 g of sucrose ($C_{12}H_{22}O_{11}$) in 78.0 mL of solution
    c. 9.00 g of sodium chloride (NaCl) in 95.0 mL of solution.

76. Calculate the volume in mL of a solution required to provide the following:
    a. 3.50 g of NaCl from a 0.45 M solution
    b. 5.0 g of C_2H_5OH from a 1.8 M solution
    c. 0.70 g of acetic acid(CH_3COOH) from a 0.50 M solution

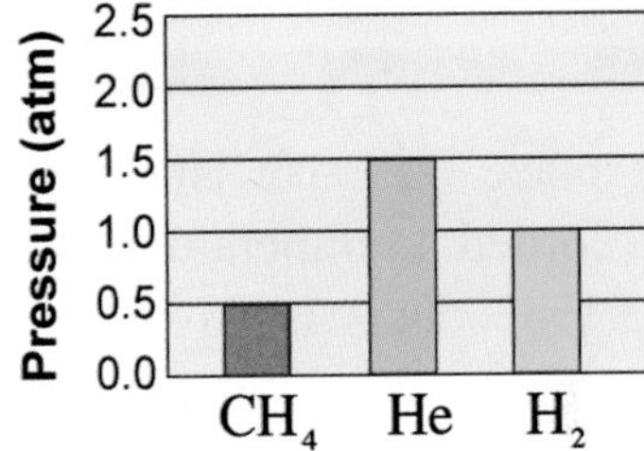

77. The diagram above shows the partial pressures for a mixture of 3 gasses. What is the total pressure in the container?

78. 10 grams of helium gas (4 g/mol) are mixed with 4 grams of neon gas (20.1 g/mol) at STP. What is the partial pressure of neon gas in the mixture?

Challenge Problems

79. A 50.0 mL of 0.560 M glucose ($C_6H_{12}O_6$) solution is mixed with 130.0 mL of 2.80 M glucose solution. What is the final concentration of the solution? Assume the volumes are additive.

80. Calculate the number of C, H, and O atoms in 2.60 grams of glucose, $C_6H_{12}O_6$.

CHAPTER 3 Temperature, Energy and Heat

What is temperature?

Why is temperature important in chemistry?

How is energy related to temperature?

Is there a difference between heat and temperature?

The coldest place on earth is in Antarctica where geologists recorded a temperature of -89°C (-129°F) on July 21, 1983. This is so bitterly cold that an unprotected human would perish in minutes. The Antarctic penguins and sea birds spend most of their lives on the water. The largest purely land-dwelling creature is an insect no bigger than your fingernail. This insect produces a chemical called *glycerol* in its body, which is a natural antifreeze!

The hottest temperature recorded on earth was 58°C (136°F) in Libya on September 13, 1922. This narrowly beat the previous record of 57°C set in Death Valley, California, on July 10, 1913. Fortunately the average temperature on the surface of our planet is 15°C, or 59°F. This is perfect for living things as it is comfortably within the range for which water is a liquid. Above 100°C, water boils and important chemicals break down, or react quickly, as they do in cooking. Below 0°C, pure water is a solid and the chemistry necessary for life cannot take place in a solid.

When an object is *hot*, what is different about its matter compared to the same object when it is *cold*? This chapter is about explaining the concept of temperature and what it really means.

In the lab

Dry ice is an amazing substance. Use some heavy gloves or tongs to hold a small piece of it. NEVER touch dry ice with bare skin. ALWAYS wear proper protection when working with dry ice.

- *How would you describe this material? Is it solid, liquid, or gas?*
- *What is the "smoke" you see? Is it hot or cold?*
- *Does the "smoke" go up or down?*

Place the chunk of dry ice on a balance. Record the mass every 30 seconds for 2 minutes.

- *Is dry ice matter? How do you know?*
- *Is the "smoke" matter? How do you know?*
- *What happens to the "lost" mass as the dry ice chunk gets smaller? Where did the mass go?*

Why does the mass decrease?
Where does the matter go?

Place a small piece of dry ice in a beaker of water. Observe what happens. You can touch the vapor with your finger however keep your skin away from any solid dry ice which may be floating on the water.

- *What is the vapor that you see?*
- *Is the vapor hot or cold?*
- *Does the vapor flow up or down? Why?*

Take about 100 mL of warm water and add a couple squirts of dish detergent. Stir the mixture up to thoroughly mix the soap into the warm water. Pour the mixture into a 250 mL graduated cylinder or other tall, thin container.

Add a small piece of dry ice to the mixture and stand back!

- *Explain what caused the explosion of soap suds.*
- *Do the bubbles that form float up or down? Why?*

What causes the rapid foaming action?

At room temperature of 22°C (72°F) air is a gas, and water is a liquid. Heat water up to 100°C and water becomes a gas, called steam. The water molecules are still together as H_2O. But the molecules have so much energy that they fly apart from each other to make steam. Carbon dioxide (CO_2) is a gas at room temperature, but becomes a solid at -79°C. Dry ice is frozen CO_2 gas. It does not melt into a liquid but *sublimates* directly back into gas as the temperature increases.

3.1 Temperature

Particles of matter are in constant motion

Milk looks like a uniform liquid, but it really isn't. Under the microscope, you see tiny globules of fat suspended in water. What is more, the fat particles are in constant motion! The particles jitter around in a very agitated way and *never slow down or stop*. This is strange when you think about it. Motion requires energy. What possible source of energy keeps the fat particles dancing around?

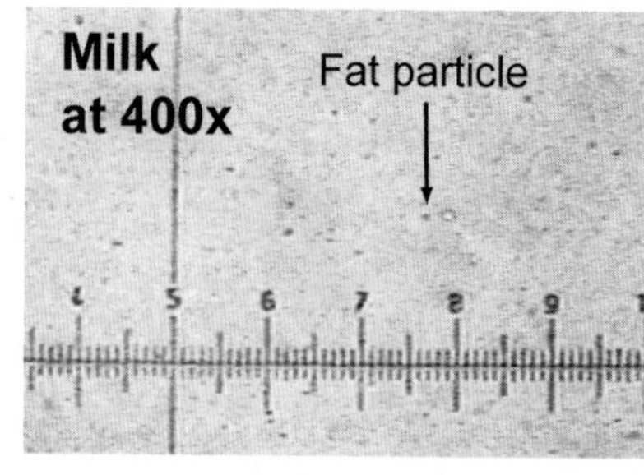

Brownian motion

If you look more carefully at the very smallest particles, you see they don't move smoothly, as they would if they were floating. Instead they move in a jerky, irregular way. The jerky movement of a very small particle in water is called **Brownian motion** and is a direct consequence of atoms and temperature. In 1905, Albert Einstein proved that matter was made of atoms by explaining Brownian motion.

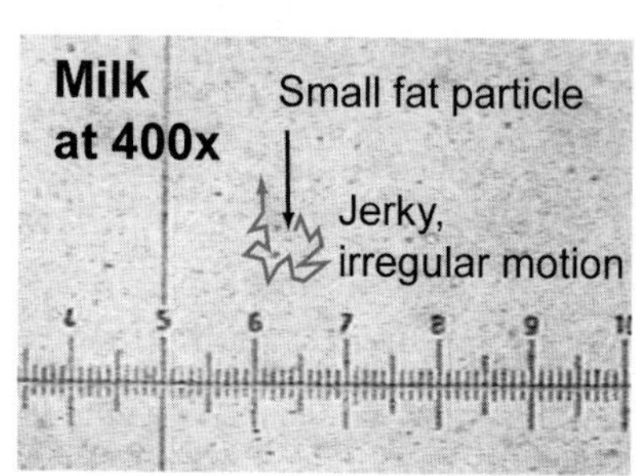

Why Brownian motion occurs

Brownian motion occurs for two important reasons:

1. Matter (including water) is made of atoms;
2. Atoms and molecules are in constant, agitated motion.

If the fat particle is *very* small, collisions with single molecules of water are visible because the mass of a water molecule is not that much smaller. The constant motion of individual water molecules causes Brownian motion.

A human-sized example

Imagine throwing marbles at a tire tube and a foam cup floating in a pool. The motion of the tube is smooth because each marble has a lot less mass than the tube. The foam cup jerks under the impact of each marble, like the fat particle in Brownian motion. This is because the mass of the cup is not much greater than the mass of a single marble. Brownian motion proves that matter exists in discrete atoms and molecules. It also proves that *at room temperature*, atoms and molecules are in constant, agitated motion.

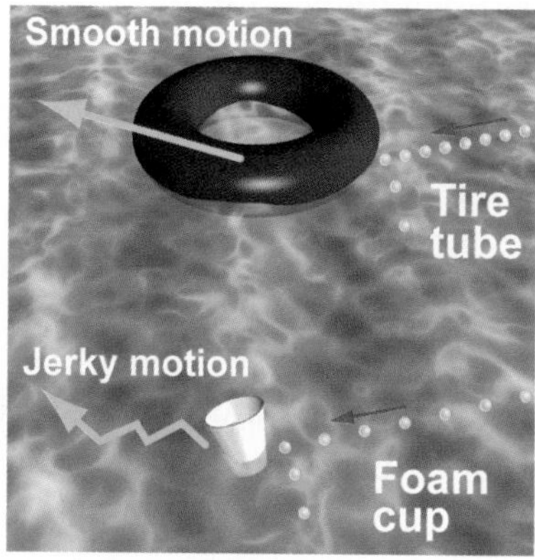

Chemistry terms

Brownian motion - the erratic, jerky movement of tiny particles suspended in water, due to the random impacts of individual molecules in thermal motion.

The explanation of temperature

Temperature and energy

Brownian motion demonstrates that atoms are constantly jiggling around. "Jiggling" implies motion, and motion always involves energy. Ordinary matter, even sitting still, contains energy in the microscopic motion of its constituent atoms. This "embedded" energy is what **temperature** measures. Temperature is a measure of the average **kinetic energy** of individual atoms (or molecules). When the temperature goes up, the energy of motion increases and the atoms jostle more vigorously.

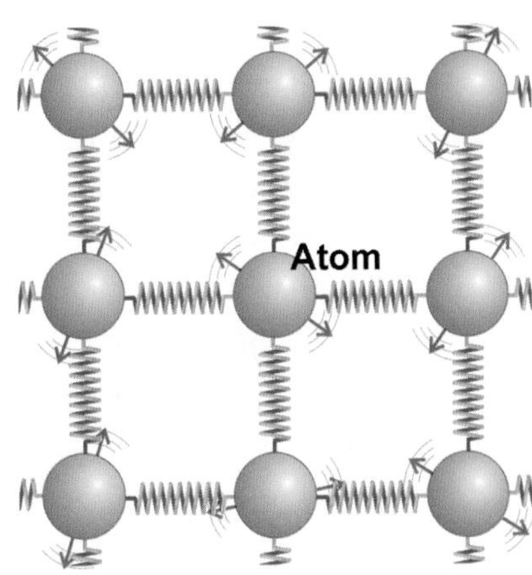

Average motion

If atoms are always moving, how can a grain of sand just stand still? The answer to this question has three parts:

1. Atoms are tiny and there are trillions of them even in a sand grain;
2. The motion due to temperature is fast, but very short, typically the width of a single atom in liquids and solids (more in gases);
3. Constant collisions cause the motion of individual atoms to be *random*.

Atoms are typically so close together that they constantly bang into each other and change direction. At any given moment there are as many atoms bouncing one way as there are the other way. The *average* speed of the whole group is zero. However, no individual atom is standing still.

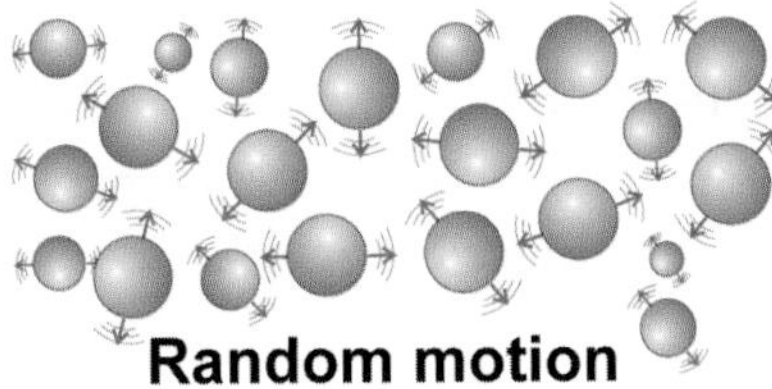

Random motion

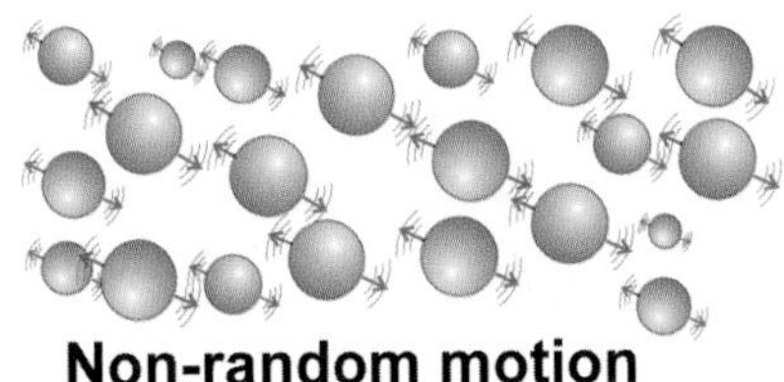

Non-random motion

Temperature and random motion

Random motion is motion that is scattered equally in all directions. Temperature only measures the energy in the *random* motion of atoms and molecules. Temperature is not affected by energy due to motion of the whole group. That is why throwing a rock does not make it hotter. When you throw a rock you give each atom in the rock the same average motion because all the atoms move together.

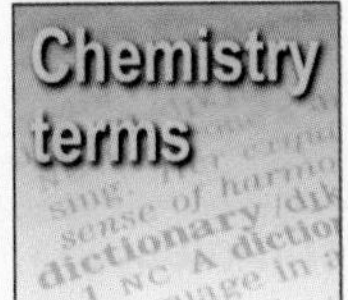

kinetic energy - energy of motion.

temperature - a measure of the average kinetic energy of atoms or molecules.

random - scattered equally among all possible choices with no organized pattern.

Temperature is an average

Molecules at a given temperature have a range of energies

In a given quantity of matter there are trillions of atoms. Some of them have more energy than the average and some have less energy than the average. This is easiest to see in a gas where molecules are not bound together. The graph below has thermal energy per molecule on the horizontal axis and the number of molecules on the vertical axis. The graph shows that molecules have a wide range of energy. A few have a lot more energy than average and a few have a lot less.

Distribution of molecular energies in a gas at different temperatures

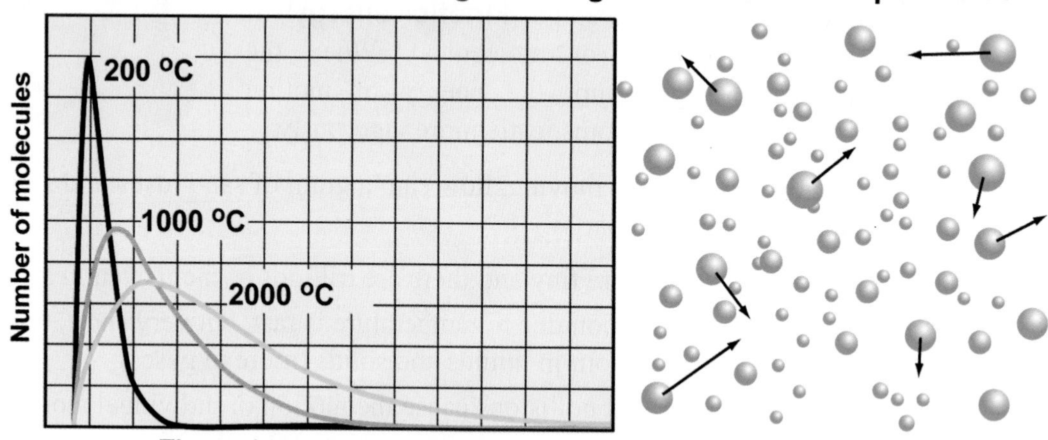

The energy changes with temperature

As the temperature increases, the graph changes in two ways:

1. The average energy of the molecules increases, so the peak of the graph shifts to the right:
2. The graph gets wider, so the range of energy of the molecules increases.

Molecular energy in liquids and solids

In a gas, the average distance between molecules is thousands of times the size of a single molecule. Molecules in a gas interact when they happen to hit each other. The weak interaction is why the graph of molecular energy of a gas spreads out so much.

In a liquid or solid, the molecules are very close together, separated by less than their own size. Molecules are in continuous motion making them bump together constantly. The constant bumping creates a strong interaction between molecules. In turn, the strong interaction means the energy curve is much narrower. Any molecule with significantly *more* energy quickly bounces into its neighbors and loses some. Any molecule with significantly *less* energy is bumped and jostled by its neighbors and sped up a little so it gains energy.

Distribution of molecular energy in liquids and solids

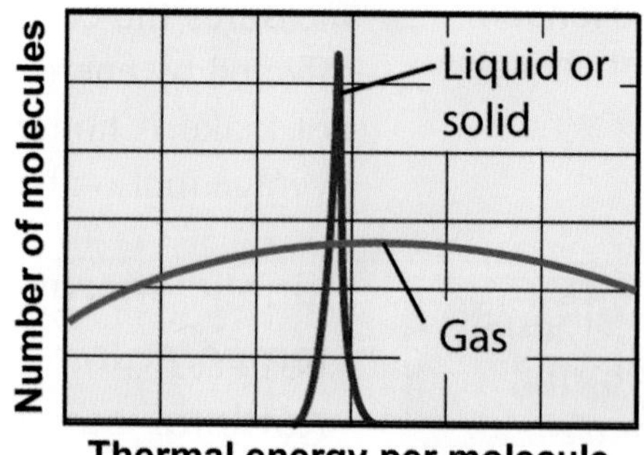

Temperature scales

Why is temperature important?

Humans can sense relative temperature through nerves in the skin. Ice feels cold because its temperature is *less* than the temperature of your skin. A cup of coffee feels hot because its temperature is *greater* than the temperature of your skin. Temperature is important to life because most of the internal processes in living organisms only work in a narrow range of temperatures.

The Fahrenheit scale

There are two commonly used temperature scales. The Fahrenheit scale and the Celsius scale. In the **Fahrenheit scale**, water freezes at 32 degrees and boils at 212 degrees. There are 180 Fahrenheit degrees between the freezing point and the boiling point of water. Temperature in the United States is commonly measured in Fahrenheit. For example, 81°F is the temperature of a warm summer day.

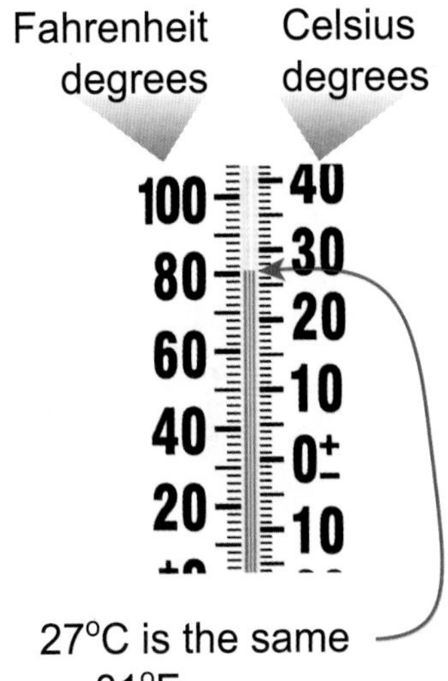

The Celsius scale

The **Celsius scale** divides the difference between the freezing and boiling points of water into 100 degrees (instead of 180). Water freezes at 0°C and boils at 100°C. Most science and engineering temperature measurement is in Celsius. Most other countries use the Celsius scale for all descriptions of temperature, including daily weather reports. The same 81°F warm summer day is 27°C.

Other countries use Celsius degrees for everyday temperature

If you travel to other countries, you will want to know the difference between the two temperature scales. A temperature of 21°C in Paris, France is a pleasant spring day, suitable for shorts and a T-shirt. A temperature of 21°F would be like a cold January day in Minneapolis, Minnesota! You would need a heavy winter coat, gloves and a hat to be comfortable outdoors in 21°F. The United States is one of the few countries still using the Fahrenheit scale.

Chemistry terms

Fahrenheit scale - a temperature scale with 180 degrees between the freezing point and the boiling point of water; water freezes at 32 °F and boils at 212 °F.

Celsius scale - a temperature scale with 100 degrees between the freezing point and the boiling point of water; water freezes at 0°C and boils at 100°C.

Converting Fahrenheit to Celsius

Differences between the two temperature scales

Fahrenheit degrees are smaller than Celsius degrees. For example a temperature change of 9°F is the same as a temperature difference of 5°C. Besides the different size of a degree between the two scales, there is also a 32°F *offset* (0°C is 32°F) that must be considered when converting from one scale to the other. Therefore, when we convert from °C to °F we must first scale the °C by 9/5 and then add 32 degrees. The formulas for the conversions are:

Converting between Celsius and Fahrenheit

$$T_{Fahrenheit} = \frac{9}{5}T_{Celsius} + 32 \qquad T_{Celsius} = \frac{5}{9}\left(T_{Fahrenheit} - 32\right)$$

Fahrenheit to Celsius

To convert from Fahrenheit to Celsius, subtract 32 then multiply by five-ninths. Subtracting 32 is necessary because water freezes at 32°F and 0°C. The factor of 5/9 is applied because the Celsius degree is larger than the Fahrenheit degree.

What temperature in Celsius is the same as 100°F?

Asked: *temperature in °C*

Given: *100°F*

Relationships: $T_C = \frac{5}{9}(T_F - 32)$

Solve: $T_C = \frac{5}{9}(100 - 32) = \frac{5}{9}(68) = 37.8$

Answer: *100°F is the same temperature as 37.8°C*

Celsius to Fahrenheit

To convert from Celsius to Fahrenheit, multiply by 9/5 then add 32. The factor of 9/5 accounts for the different size degrees and the 32 corrects for the zero offset.

What is the Fahrenheit equivalent of 15°C?

Asked: *temperature in °F*

Given: *15°C*

Relationships: $T_F = \frac{9}{5}T_C + 32$

Solve: $T_F = \frac{9}{5}(15) + 32 = 27 + (32) = 59^oF$

Answer: *15°C is the same temperature as 59°F*

Measuring temperature

Human temperature sense

Scientists use instruments to measure temperature because the human sense of temperature is not very accurate. You can feel when something is warm or cold, but not its exact temperature. For example, if you walk into a 65°F room from being outside on a winter day, the room feels warm. The same room will feel cool if you come in from outside on a hot summer day.

Thermometers

Our sense of temperature is thus very relative. We can easily compare the temperature of things that we touch, but we can't easily say what the actual temperature is. In science, we try to perform absolute measurements. This means that we need to know the actual value of the temperature. A **thermometer** is an instrument that measures temperature. The common alcohol thermometer uses the expansion of liquid alcohol. As the temperature increases, the alcohol expands and rises up a long, thin tube. The temperature is measured by the height the alcohol rises. The thermometer can read small changes in temperature because the bulb at the bottom has a much larger volume than the tube.

Molecules move around more as the temperature increases. So, the same number molecules takes up more space at higher temperature

alcohol molecules at 0°C

alcohol molecules at 100°C

How thermometers work

All thermometers are based on a physical property that changes with temperature. A **thermistor** is a temperature sensor that changes its electrical properties as the temperature changes. A **thermocouple** is another electrical sensor that measures temperature. The temperature probe you use in your laboratory experiments uses one of these two sensor types.

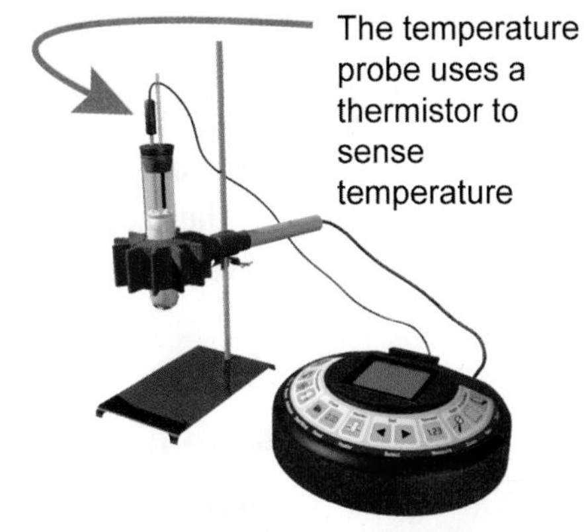

The temperature probe uses a thermistor to sense temperature

Some chemicals change color at different temperatures. One type of aquarium thermometer has a stripe that changes color when the water is too hot or cold.

Some aquarium thermometers use a color changing chemical to sense temperature

Chemistry terms

thermometer - an instrument that measures temperature.

thermistor, thermocouple - two types of electronic sensors for measuring temperature.

Absolute zero

Absolute zero

As the temperature gets lower, atoms have less and less thermal energy. This implies that a temperature can be reached at which the thermal energy is *zero*. This lowest possible temperature is called **absolute zero** where atoms have essentially zero thermal energy. Absolute zero is minus 273°C or -459°F. Think of absolute zero as the temperature where even atoms are completely frozen, like ice, with no motion. *It is not possible to have a temperature lower than absolute zero, or -273.15°C.*

No temperatures below absolute zero

Technically, we believe atoms can never stop moving *completely*. Even at absolute zero some tiny amount of thermal energy is left. For our purposes, however, this "zero point" energy might as well be truly zero because the rules of quantum physics prevent the energy from ever going any lower. Figuring out what happens when atoms are cooled to absolute zero is an area of active research.

The Kelvin scale

The **Kelvin** temperature scale is useful for many scientific calculations because it starts at absolute zero. For example, the pressure in a gas depends on how fast the atoms are moving. The Kelvin scale is used because it measures the actual thermal energy of atoms. A temperature in Celsius measures only the *relative* energy, relative to zero Celsius.

Converting to Kelvin

The Kelvin (K) unit of temperature is the same size as the Celsius unit. Add 273 to the temperature in Celsius to get the temperature in Kelvins. For example, a temperature of 21°C is equal to 294 K (21 + 273). Note that the word "degree" is not used with the Kelvin scale. A temperature of 300K is "300 Kelvin', and is abbreviated without the degree symbol.

$$T_{Kelvin} = T_{Celsius} + 273$$

Convert 27°C into Kelvin

Asked: *Temperature in Kelvin (K, note there is no degree symbol)*

Given: *27°C*

Relationships: $T_K = T_C + 273$

Solve: $T_K = 27 + 273 = 300K$

Answer: *300K is the same temperature as 27°C*

absolute zero - the lowest possible temperature, at which the energy of molecular motion is essentially zero, or as close to zero as allowed by quantum theory.

Kelvin scale - a temperature scale that starts at absolute zero and has the same size degrees as Celsius degrees. $T_{Kelvin} = T_{Celsius} + 273$.

3.2 Heat and Thermal Energy

Thermal energy is also called heat

Heat is another word for thermal energy. On the molecular level, thermal energy is the random kinetic energy of a collection of atoms and/or molecules. On a macroscopic level, thermal energy is the energy stored in matter that is *proportional to temperature*. To change the temperature of matter, you need to add or subtract heat. You add heat to warm your house in the winter. If you want to cool your house in summer, you remove heat.

Temperature
the average energy per molecule

Heat
the total energy in a collection of molecules

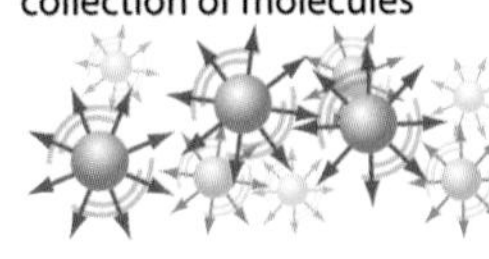

Joules

The **joule** (J) is the fundamental SI unit of energy and heat. A joule is a fairly small unit of energy. Heating 100 mL (100g) of water from room temperature to boiling requires about 33,000 joules of heat.

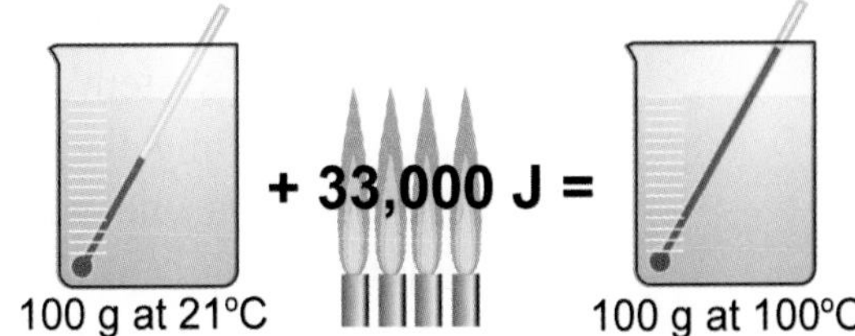

Calories

The **calorie** is an older unit of heat used in chemistry. One calorie is the amount of heat required to raise the temperature of one gram of water by one degree Celsius. There are 4.184 joules in one calorie so a calorie is more energy than a joule. To make things more confusing the Calories listed in foods (with capital "C") are really *kilocalories*. One food calorie = 1,000 calories = 4,184 joules.

British thermal units (Btu)

The air conditioner or furnace in your house is rated in **British thermal units (Btu)**. One Btu is the amount of heat required to raise the temperature of one pound of water by one degree Fahrenheit. A typical home-heating furnace can produce 10,000 to 100,000 Btu per hour. One Btu equals 1,055 joules.

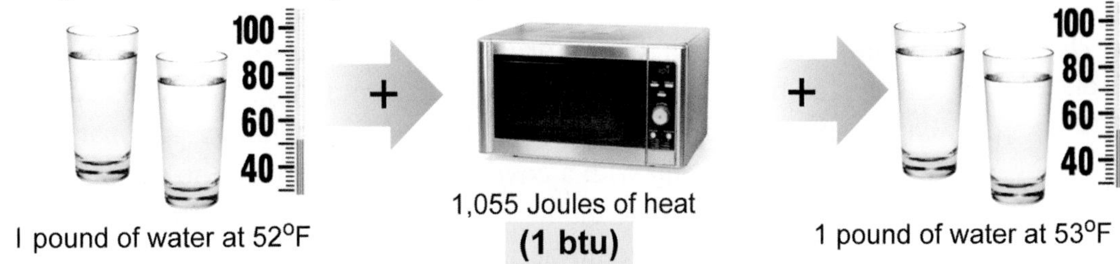

1 btu raises the temperature of 1 pound of water by 1°F

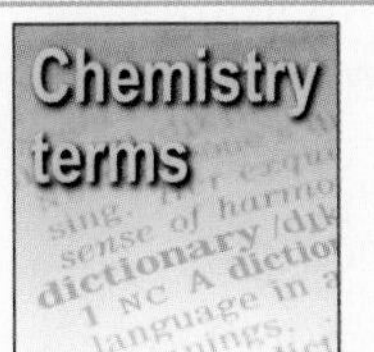

heat - thermal energy, energy due to temperature, the total energy in random molecular motion contained in matter.

joule - the fundamental SI unit of energy (and heat).

calorie - older unit of heat, 1 calorie = 4.184 joules.

British thermal unit (btu) - large unit of heat used in the U.S., 1 btu = 1055 joules.

Thermodynamics and systems

Heat flows from hot to cold (2nd law)

Imagine standing outside on a cold winter day without a coat. Your body feels cold because you are losing energy to your surroundings. That's because your body temperature is higher than the air temperature. Heat (energy) moves from high temperature to lower temperature. This common-sense rule is called the **second law of thermodynamics**.

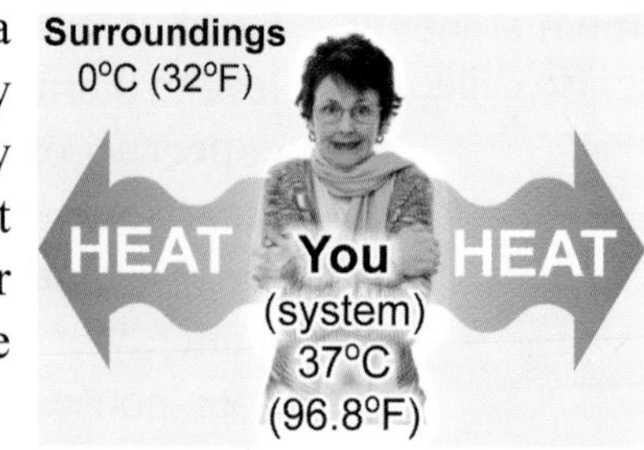

Choosing a system

To see how heat moves and what it does, we must look carefully at what we are studying. To a chemist the part of the "universe" under study is called the **system**. In the no-coat example, the system is your body; everything else is called the *surroundings*. By carefully defining the system and the surroundings, you can keep track of how much energy there is and where it goes. For example, if the system is defined as a beaker containing 200mL of water at 60°C, then the surroundings would be the lab bench that the beaker is resting on and the air surrounding the beaker.

Why heat flow is important

The flow of heat energy is important to virtually everything that occurs in both nature and technology. The weather on earth is a giant heat recycler between the oceans, the land and the atmosphere. A car engine converts heat from burning gasoline to energy of motion of the car.

Open, closed, and isolated systems

Systems can be open, closed or isolated. In an **open system**, matter and energy can be exchanged between the system and the surroundings. In a **closed system**, only energy can be exchanged between the system and surroundings. An example of a closed system would be a coffee cup with a cover on it. Heat can still be exchanged or lost between the air and the coffee, but no matter could be exchanged. In the third type of system which is **isolated**, the coffee would be inside a perfectly insulated thermos where no matter or heat could escape.

second law of thermodynamics - energy (heat) spontaneously flows from higher temperature to lower temperature (basic interpretation of 2nd law).

system -a group of interacting objects and effects that are selected for investigation.

open system - matter and energy can be exchanged with the surroundings.

closed system - only energy is allowed to be exchanged with the surroundings.

isolated system - neither matter nor energy can be exchanged with surroundings.

Thermal equilibrium

What stops heat from flowing?

Suppose we pour the hot and cold water into a divided cup so they can't mix, but heat can flow. Heat flows from the hot water to the cold water. The hot water gets cooler as it loses heat. The cold water gets warmer as it gains the same heat. *When does the heat flow stop?* Why doesn't the hot water keep losing heat until it gets *colder* than the cold water? Obviously, this does not happen; heat flow eventually stops.

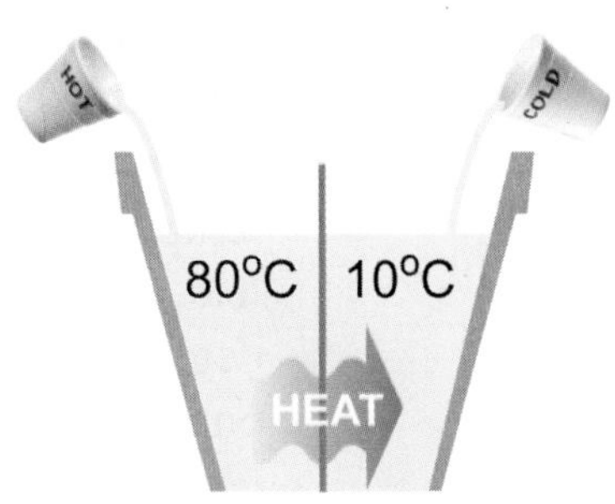

Thermal equilibrium

Two bodies are in **thermal equilibrium** when they have the same temperature. In thermal equilibrium, no heat flows because the temperatures are the same. In nature, heat always flows from hot to cold until thermal equilibrium is reached. Making the divider out of insulating foam only slows the process down. Heat still flows, only slower. The hot and cold water take *longer* to reach thermal equilibrium, but the end result is the same.

The rate of heat flow

The rate at which heat flows drops off gradually as the temperatures get closer together. A lot of heat flows quickly when the temperature difference is large. As the temperature difference gets smaller, the rate of heat flowing gets proportionally smaller, too. As two objects approach thermal equilibrium, the rate of heat flow between them becomes zero because both reach the same temperature.

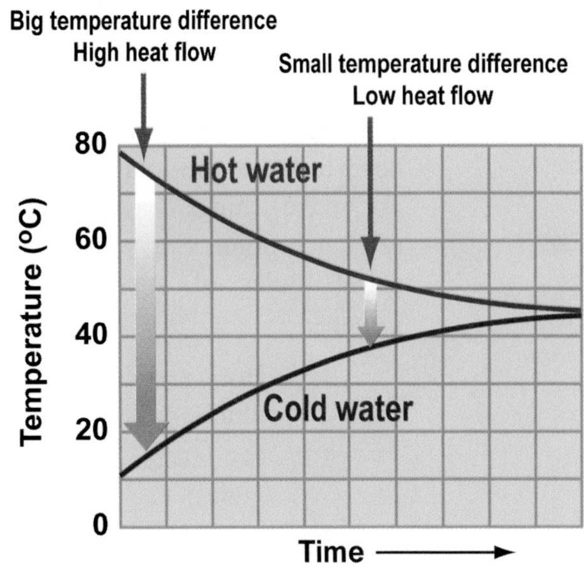

Heat transfer in living things

Heat flow is necessary for life because biological processes release energy. Your body regulates its temperature through the constant flow of heat. The inside of your body averages 37°C. Humans are most comfortable when the air is around 25°C because the rate of heat flow out of the body matches the rate at which the body generates heat internally. If the air is 10°C, you get cold because heat flows out of your body too rapidly. If the air is 40°C, you feel hot because your body cannot get rid of its internal heat.

Chemistry terms

thermal equilibrium - a condition where the temperatures are the same and heat no longer flows.

The first law: energy conservation

The first law

The **first law of thermodynamics** is the law of energy conservation. All the energy lost by one system must be gained by the surroundings (another system). Consider an experiment where you mix 100 grams each of hot and cold water. How can the temperature of the mixture be predicted?

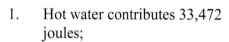

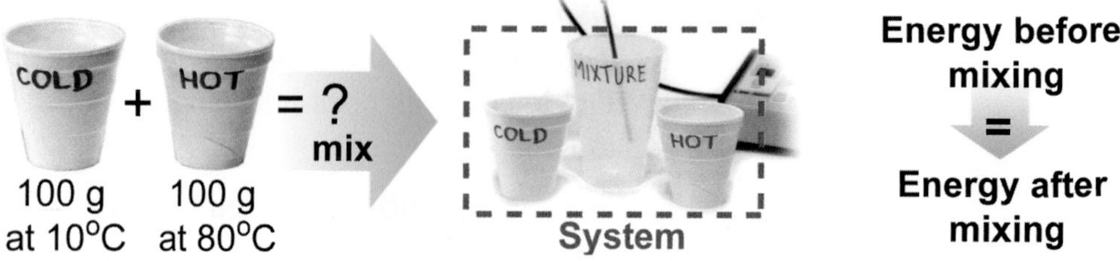

Assume the two cups are an isolated system; no energy or mass is allowed to cross the boundary. That means that the energy after mixing is the same as it was before mixing.

Energy depends on mass and temperature

The amount of heat energy in a quantity of matter, like a cup of water, is proportional to mass and temperature. More mass means more energy at any temperature and, the higher the temperature, the higher the energy.

Tracing the energy

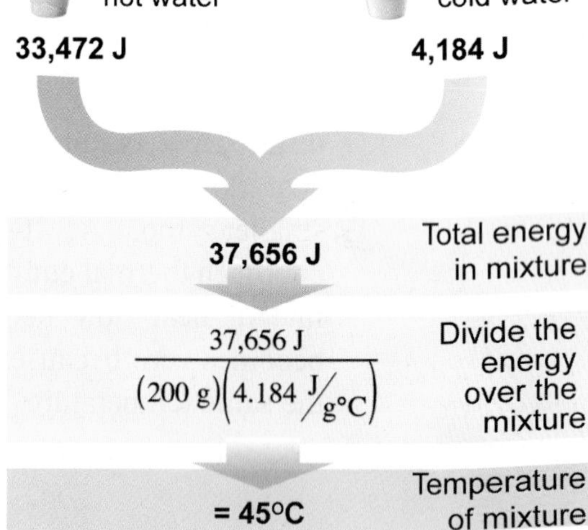

To find the final mixture temperature, you need to know the total energy of the system. To make things easy, assume 0°C as a reference point. For water, each gram stores 4.184 joules of energy for every degree Celsius.

1. Hot water contributes 33,472 joules;
2. Cold water contributes 4,184 joules;
3. The mixture has a total energy of 37,656 joules (hot + cold);
4. 200 g of water containing 37,656 joules of heat, is at 45°C.

How to solve the problem

How did we solve this problem? We added up all the energy in the system. We then analyzed where the energy went. In this case, all the energy went to changing the temperature of the mixture. The problem was solved by calculating what temperature the mixture *had to be* to contain all the energy in the system.

first law of thermodynamics - energy can neither be created nor destroyed. The total energy in an isolated system remains constant; all the energy lost by one system must be gained by the surroundings or another system.

Specific heat

Differences in materials

The same amount of heat causes a different change in temperature in different materials. For example, it takes 4.184 joules of heat energy to raise the temperature of one gram of water by one degree Celsius. If you add the same quantity of heat to one gram of gold, the temperature goes up by 32.4°C! The different temperature rise happens because different materials have different abilities to store thermal energy.

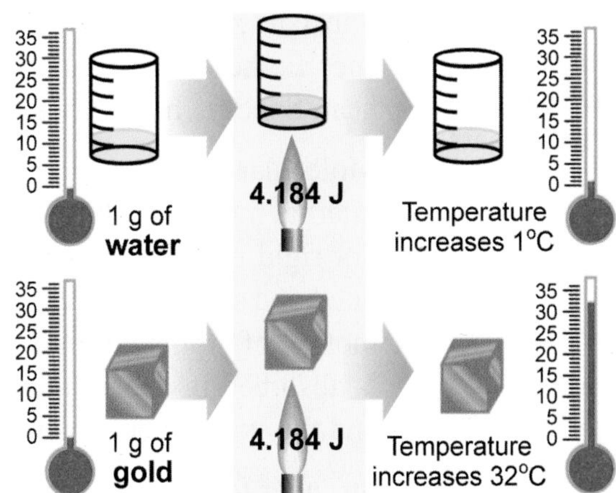

Specific heat

The **specific heat** in J/g°C is the quantity of energy it takes to raise the temperature of one gram of a material by one degree Celsius. Water is an important example; the specific heat of water is 4.184 J/g°C. It takes 4.184 joules to raise the temperature of one gram of water by one degree Celsius. The specific heat of gold is 0.129 J/g°C. It only takes 0.129 J to raise the temperature of 1 gram of gold by 1°C. The specific heat of water is 32 times higher than it is for gold.

TABLE 3.1. Specific heat of some common substances

Material	Specific heat (J/g°C)
Air at 1 atm	1.006
Water	4.184
Aluminum	0.900
Steel	0.470
Silver	0.235

Material	Specific heat (J/g°C)
Oil	1.900
Concrete	0.880
Glass	0.800
Gold	0.129
Wood	2.500

The heat equation

The heat equation is used to calculate how much energy (E) it takes to make a temperature change ($T_2 - T_1$) in a mass (m) of material with specific heat (c_p).

Heat equation

$$E = mc_p(T_2 - T_1)$$

Energy (J); Mass (g); Specific heat (J/g°C); Temperature change (°C)

Chemistry terms

specific heat - the quantity of energy, measured in J/g°C, it takes per gram to raise the temperature one degree Celsius.

Why specific heat varies

Different substances have different specific heats

Substances have a wide range of specific heats. Pure metals, like gold, tend to have a low specific heat. Molecular substances, like water and oil, tend to have a higher specific heat. Specific heat varies for many reasons.

Molecular substances can absorb energy in ways that don't increase temperature, such as internal motion of the atoms within a molecule. This is because bonds between atoms are not rigid rods, like the diagrams show. Rather, bonds are like flexible springs that can bend and stretch. Typically, only the motion of whole molecules affects temperature. The motion of atoms within a molecule, however, may not affect temperature. When energy is absorbed in ways other than motion of the whole molecule, temperature goes up less and the specific heat increases.

Stronger forces between molecules mean it takes more energy to cause a single molecule to move a given amount. This makes the specific heat higher. In general, strong bonds between molecules raise the specific heat because they limit thermal motion of individual molecules (or atoms).

Why specific heat varies

Materials with heavy atoms or molecules have low specific heat compared with materials with lighter atoms. This is because temperature measures the energy per atom. Heavy atoms mean fewer atoms per kilogram. Energy that is divided between fewer atoms means more energy per atom, and therefore more temperature change. Silver's specific heat is 235 J/kg°C and aluminum's specific heat is 900 J/kg°C. One gram of silver has fewer atoms than a gram of aluminum because silver atoms are heavier than aluminum atoms. When heat is added, each atom of silver gets more energy than each atom of aluminum because there are fewer silver atoms in a gram. Since the energy per atom is greater, the temperature increase in the silver is also greater.

Why is the specific heat of aluminum almost 4 times greater than the specific heat of silver?

1 gram

Silver
Specific heat: **235** J/kg°C
Heavier atoms, mean **fewer** atoms per gram.

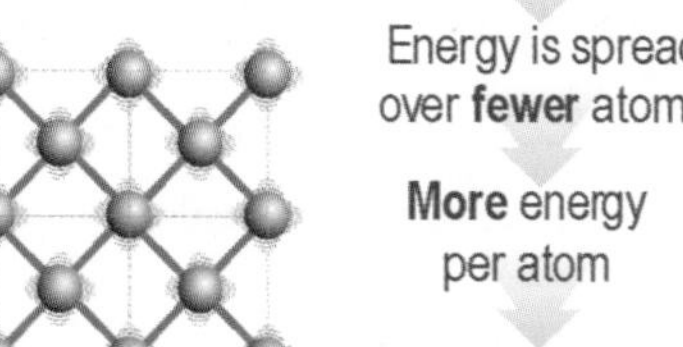
Energy is spread over **fewer** atoms
More energy per atom
Higher temperature gain per joule (lower specific heat)

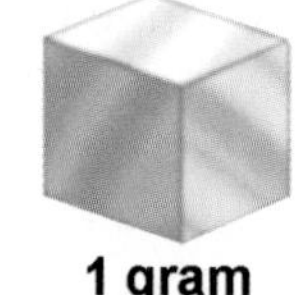
1 gram

Aluminum
Specific heat: **900** J/kg°C
Lighter atoms, mean **more** atoms per gram.

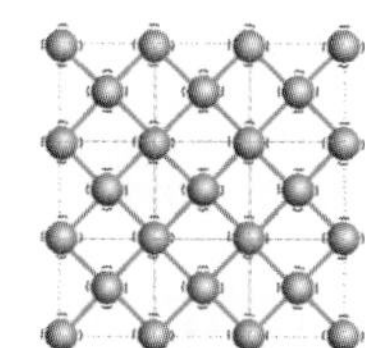
Energy is spread over **more** atoms
Less energy per atom
Lower temperature gain per joule (higher specific heat)

Calculating temperature and heat

Most specific heat problems include one or both of the following two calculations.

1. Calculate the temperature change from a given heat input.
2. Calculate how much heat is needed to reach a specified temperature.

On a sunny day, each square centimeter of the ocean absorbs 180 joules of energy from the sun each hour. Assume all the heat is absorbed in the first ten meters, which has a mass of 1,000 grams. How much does the water temperature increase?

180 J sunlight
1cm square
10m

Asked: *temperature increase*

Given: *energy input: 180 joules, mass of water: 1,000 grams*

Relationships: $E = m\, c_p\, (T_2 - T_1)$

Solve: $180\ J = (1000\ g) \times (4.184\ J/g°C) \times (T_2 - T_1)$

$T_2 - T_1 = 180 \div 4{,}184 = 0.04°C$

Answer: *The temperature rise in the top ten meters of the ocean is only 0.04°C per hour of full sunlight.*

A metal working process needs to heat steel from room temperature (20°C) up to 2000°C. If there are 100 grams of steel, how much heat is required?

Asked: *quantity of heat*

Given: *100 g of steel,* $c_p = 0.47\ J/g°C$ *temperature difference = 20°C to 2000°C*

Relationships: $E = m\, c_p\, (T_2 - T_1)$

Solve: $E = (100\ g) \times (0.47\ J/g°C) \times (2000 - 20)$

$= 93{,}060$ *joules*

Answer: *It takes 93,060 joules to raise the temperature of 100 grams of steel up to 2000°C, assuming no heat gets lost during the process (not a very good assumption!)*

Applying the first law

Heat flow out of a system equals heat absorbed by surroundings

When you leave a cup of hot coffee on the table, it cools down. Heat flows from the hot coffee to the cooler air in the room. The thermal energy of the hot coffee is decreased and the thermal energy of the air is increased by the same amount. Everything balances; the increase in thermal energy of the air is exactly the same as the decrease in thermal energy of the coffee.

Solving problems with the first law

The first law is useful because it allows you to solve problems quickly and with minimal work. In real life, a hot coffee cup heats up the nearby air, which rises and makes currents that change how fast the heat flows. It would be difficult to calculate precisely how much heat flows every second. Luckily, it does not matter. By looking at the energy at the start, and after the coffee has completely cooled down, we can get a pretty good answer without knowing the exact details of exactly how the heat went from cup to air.

TABLE 3.2. Applying the first law

Step1	Step2	Step3
• Identify the system and all the sources and uses of energy in the system before the change	• Set the total energy of the system after the change equal to the total energy before the change	• Account for all the uses of energy after the change and solve the problem

300 grams of water at 80°C cools down to 20°C. Assume all the heat from the water is absorbed by 100 cubic meters of air (a small room), with a mass of 100,000 grams. What is the temperature change in the air?

Asked: *temperature change in °C*

Given: *300 g of water (c_p = 4.18 J/g°C), change of 60°C (80°C - 20°C) and 100,000 grams of air (c_p = 1.006 J/g°C)*

Relationships: $E = mc_p(T_2-T_1)$

Solve: *Energy lost by the water = (300g)(4.184 J/g°C)(60°C) = 75,312 J*

Energy gained by the air = energy lost by water = 75,312 J

$$\text{Temperature change of air} = \frac{75{,}312\ J}{100{,}000\ g \times 1.006\ J/g\,°C} = 0.75\ °C$$

Answer: *The air in the room gets warmer by about 0.75°C, less than 1 degree*

Heat transfer

Conduction

Conduction is the transfer of heat *through* materials by the direct contact of matter. Imagine bringing in contact two pieces of material - one hot at 100°C and the other cold at 0°C. When the two pieces come in contact with each other, heat begins to flow. The heat flows from the hot piece to the cold piece. The cold piece warms up and the hot piece cools down. The heat will continue to flow until the two pieces are at the same temperature.

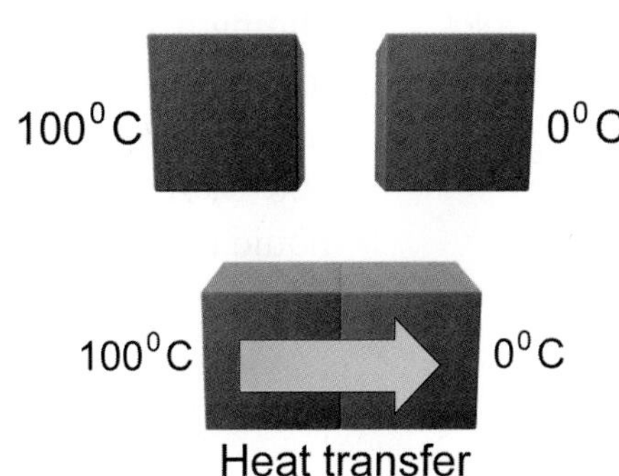

Thermal conductors

Glass and metal are **thermal conductors.** Think about holding a test tube of hot water. Thin glass conducts heat relatively well so heat flows rapidly through the glass from the hot water to your skin. You cannot hold on for long because the heat raises your skin temperature quickly.

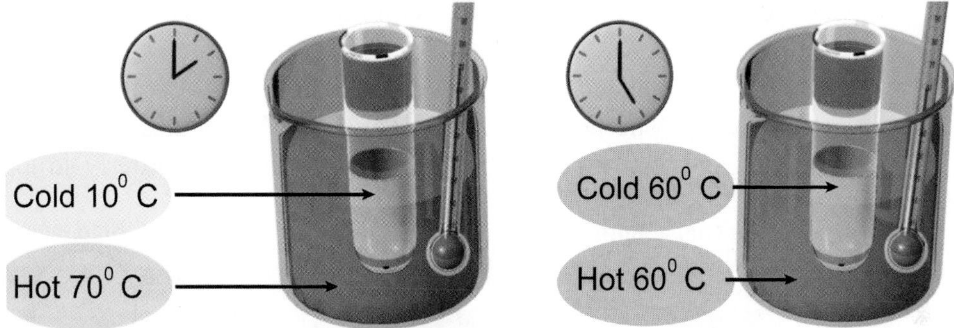

Thermal insulators

A **thermal insulator** is a material that conducts heat poorly. Styrofoam is a good example. You can comfortably hold a hot test pipe surrounded by a centimeter of foam insulation. Heat flows very slowly through the foam so that the temperature of your hand does not rise very much. Plastic foam gets its insulating ability by trapping spaces of air in bubbles. We use thermal insulators to maintain temperature differences without allowing much heat to flow.

The ability to conduct heat depends on many factors

All materials conduct heat at some rate. Solids usually are better heat conductors than liquids, and liquids are better conductors than gases. The ability to conduct heat often depends more on the structure of a material than on the material itself. For example, solid glass is a thermal conductor when it is made into windows. When glass is spun into fine fibers and made into insulation (fiberglass), the combination of glass fibers and trapped air makes a thermal insulator.

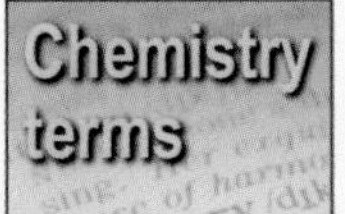

conduction - the flow of heat energy through the direct contact of matter.
thermal conductor - a material that conducts heat easily.
thermal insulator - a material that resists the flow of heat.

3.3 Phase Changes

Solid, liquid and gas

During a **phase change**, a substance rearranges the order of its particles (atoms or molecules). Examples of phase change include melting (solid to liquid) and boiling (liquid to gas). The most familiar example is water which melts at 0°C and boils at 100°C. All substances have phase changes, even metals such as iron. Iron melts into liquid at 1,535°C and boils into a gas at 2,750°C.

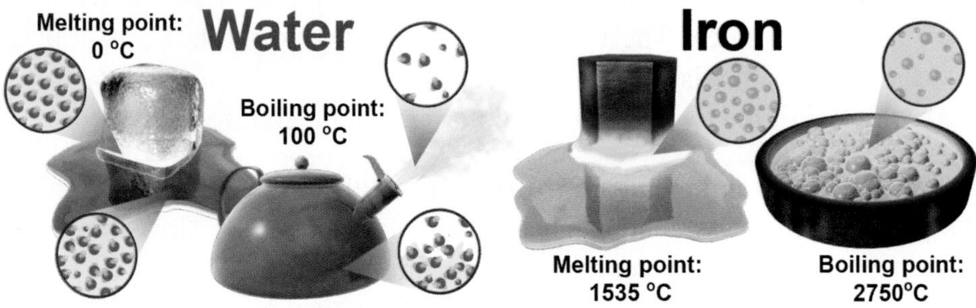

Why phase changes occur

Phase changes come from the competition between temperature and attractive intermolecular forces. On one side of this competition are intermolecular forces which tend to attract molecules together into rigid structures. On the opposite side is the action of temperature. Thermal energy is disruptive. Molecules with lots of thermal energy shake back and forth so much they cannot stay in a nice orderly structure, like a solid.

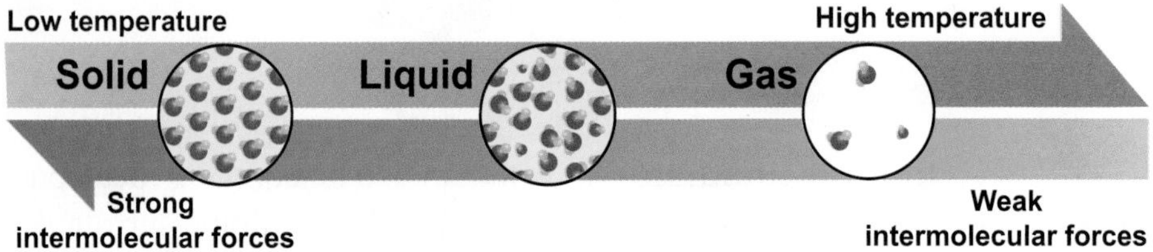

Melting point

The **melting point** is the temperature at which a substance changes from solid to liquid. Melting occurs when the thermal energy of individual atoms becomes comparable to the attractive force between atoms.

Different materials have different melting points because their intermolecular forces have different strengths. Water melts at 0°C (32°F). Iron melts at a much higher temperature, about 1,535°C (2,795°F). The difference in melting points tells us the attractive force between iron atoms is much greater than the attractive force between water molecules.

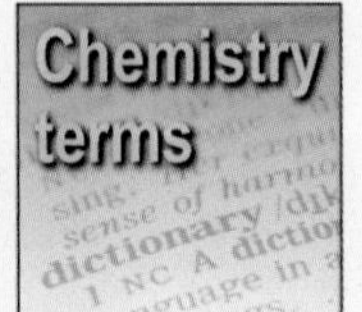

phase change - occurs when a substance changes how its molecules are organized without changing the individual molecules themselves. Examples are changing from solid to liquid or liquid to gas.

melting point - the temperature at which a substance changes phase from solid to liquid. For example, the melting point of water is 0°C.

Boiling and the heat of vaporization

Boiling point

The **boiling point** is the temperature at which the phase changes from liquid to gas. In a gas, all the bonds between one atom and its neighbors are completely broken. Water boils at 100°C (212°F) at a pressure of one atmosphere. The steam above the teapot is water molecules in the gas phase. Iron boils at 2,750°C (4,982°F).

Heat of vaporization

It takes energy for a molecule to completely break its bonds with its neighbors and go from liquid to gas. The **heat of vaporization** is the amount of energy it takes to convert one gram of liquid to one gram of gas at the boiling point. Some representative values are given below. A quick calculation shows it takes 2,256,000 joules to turn a kilogram of boiling water into steam! This explains why stoves require so much electric power.

TABLE 3.3. Heat of vaporization for some common substances

Substance	Heat of fusion. ΔH_f (J/g)	Heat of vaporization. ΔH_v (j/g)
Water	335	2,256
Alcohol	104	854
Liquid nitrogen	25.5	201
Iron	267	6,265
Silver	88	2,336

Comparing heats of vaporization and fusion

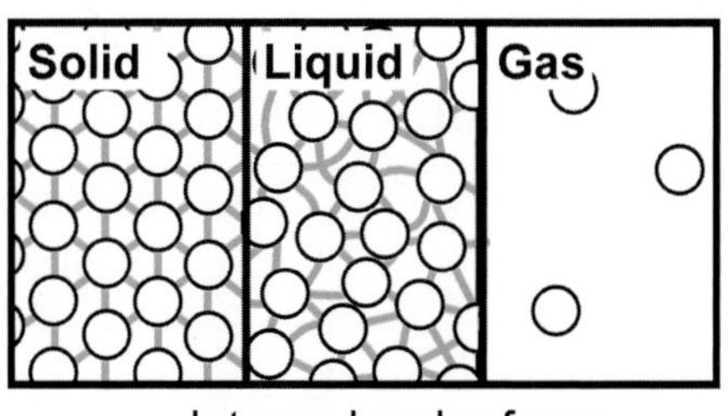

The heat of vaporization is much greater than the heat of fusion because breaking bonds between atoms or molecules takes much more energy than exchanging bonds. In a liquid, molecules move around by exchanging bonds with neighboring molecules. The energy needed to break one bond is recovered when the molecule forms a new bond with its neighbor.

Chemistry terms

heat of vaporization - the energy required to change the phase of one gram of a material from liquid to gas or gas to liquid at constant temperature, constant pressure, and at the boiling point.

boiling point - the temperature at which a substance changes phase from liquid to gas. For example, the boiling point of water is 100°C.

The heat of fusion

Temperature does not always rise when heat is added!

Think about heating a block of ice with an initial temperature of -20°C. As heat energy is added, the ice warms. However, once the ice reaches 0°C, *the temperature stops increasing!* This is because ice is melting to form liquid water. As heat is added, more ice becomes water but the temperature stays the same. This can easily be observed with an ordinary thermometer in a *well-stirred* experiment.

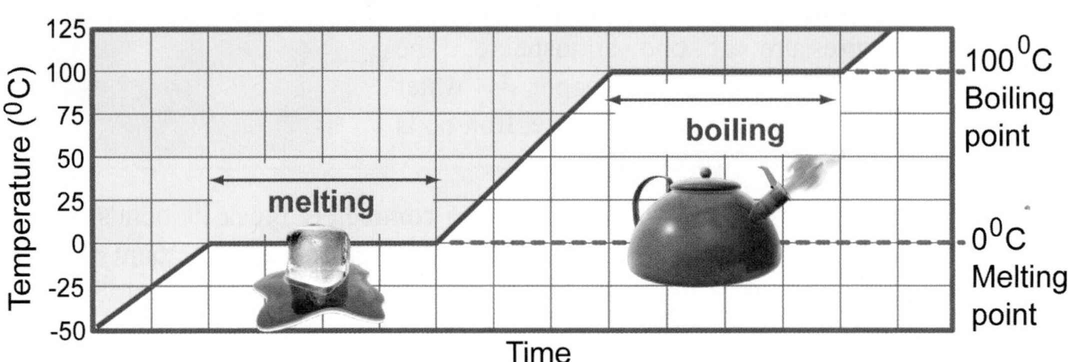

It takes energy to change phase

It takes energy to separate molecules in a solid and make them liquid, *even if the temperature stays the same.* The temperature does not change while the ice is melting because the added energy goes into loosening the bonds between neighboring water molecules. When thermal energy is added or subtracted from a material, either the temperature changes, or the phase changes, but usually not both at the same time. Once all the ice has become liquid, the temperature starts to rise again as more heat is added.

The heat of fusion

The **heat of fusion** is the amount of energy it takes to change one gram of material from solid to liquid or vice versa. The table below gives some values for the heat of fusion (Δh_f) for common materials. Note how large the values are. It takes 335,000 joules of energy to turn one kilogram of ice into liquid water.

TABLE 3.4. The heat of fusion for some common substances

Substance	Heat of fusion. ΔH_f (j/g)
Water	335
Aluminum	321
Iron	267
Paraffin (wax)	200-220

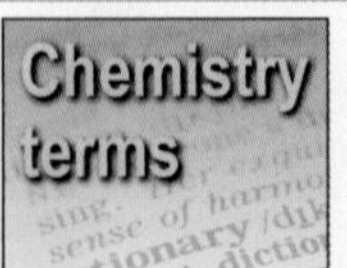

heat of fusion - the energy required to change the phase of one gram of a material from liquid to solid or solid to liquid at constant temperature, constant pressure, and at the melting point.

Solving phase change problems

Condensation releases the heat of fusion

Energy can be released in phase changes, as well as be absorbed. For example, condensation occurs when a gas turns back into a liquid again. When water vapor condenses, it gives up the heat of vaporization! Hot steam causes severe burns because each gram of water vapor delivers 2,256 J of energy to your skin from the heat of vaporization. By comparison, a gram of water at 100°C only releases 314 J as it cools down to skin temperature (25°C).

Solving phase change problems

Many problems involving phase changes can be solved by thinking about where the energy is absorbed or released.

1. Energy can be absorbed or released during phase changes;
2. Energy can be absorbed or released by changes in temperature.

The important idea is to realize that energy is conserved. Any energy used to change phase is no longer available to change temperature, and vice versa. For example, if 100 joules of energy is applied to a system and 40 joules are used to raise the temperature, then only 60 joules are available for changing phase.

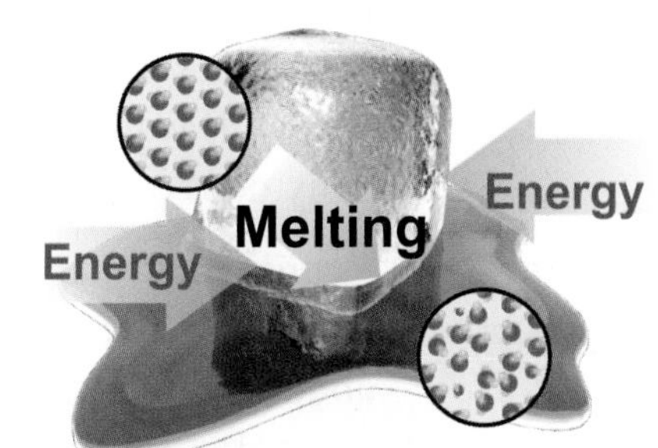

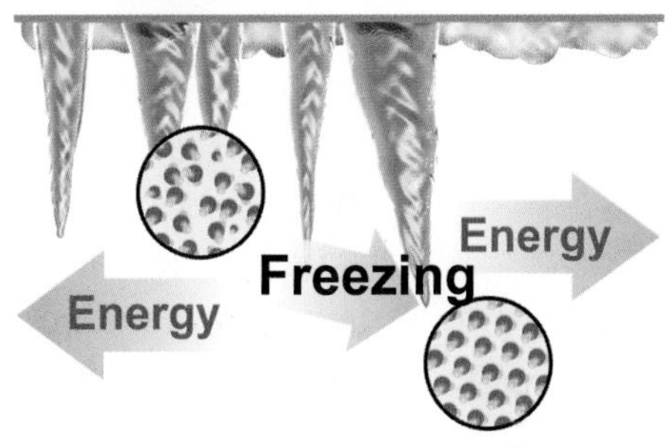

Ice cubes with a temperature of -25°C are used to cool off a glass of punch. Which absorbs more heat: warming up the ice or melting the ice into water? The specific heat of ice is 2.0 J/g°C.

Asked: *Which absorbs more heat, warming ice by 25°C or melting ice?*

Given: *The ice starts at -25°C. The specific heat of ice is 2.0 J/g°C.* $\Delta H_f(water) = 335\ J$

Relationships: $E = mc_p(T_2 - T_1)$ *and* $E = m\,\Delta h_f$

Solve: *First, let's calculate the energy that it takes to warm up a gram of ice from -25°C to 0°C.*
One gram of ice takes $mc_p(T_2 - T_1) = (1\ g)(2.0\ j/g°C)(25°C) = 50\ J$ *to warm up from -25°C to 0°C.*
The same gram of ice takes 335 J to melt into liquid water

Answer: *Changing phase (melting) absorbs 335 J per gram of ice while warming the ice only absorbs 50 J/g. The phase change is responsible for most of ice's cooling effect on drinks!*

Evaporation

Why does cold water evaporate?

Think about wiping a surface with a wet cloth? Over time the water *evaporates*. In **evaporation**, liquid molecules change phase into gas at a temperature far below the boiling point! Virtually all liquids evaporate, some much faster than others. Dampen a surface with alcohol and it evaporates almost immediately. A surface wet with water stays wet for much longer. Why does evaporation occur?

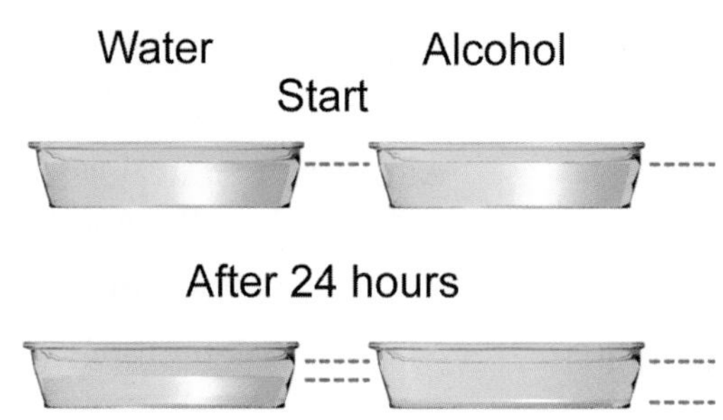

The average energy of molecules

Evaporation happens because there is a surface between liquid and gas. Think about water molecules at the surface compared to molecules deep below the surface. All molecules are in constant jostling motion. The deeper molecules are hit upward by their neighbors as often as they are hit downward. They stay in the water. The molecules on the surface however, have no molecules on top to hit them back down. When a surface molecule is jostled hard enough by a molecule below it, the surface molecule is ejected from the water and becomes a gas. This is the explanation for evaporation.

Explaining evaporation

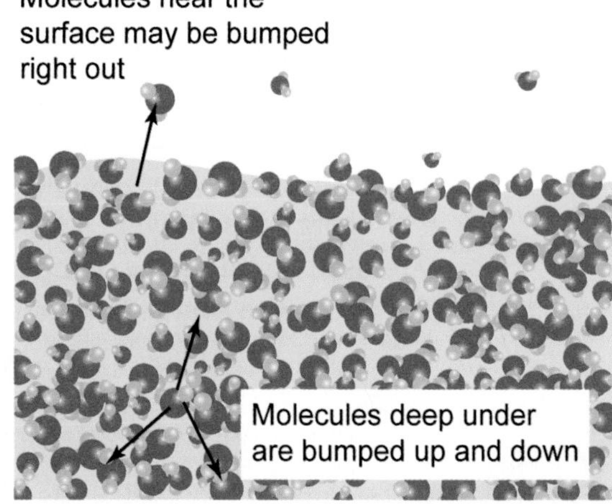

Evaporation cools liquids

How does an evaporating molecule get enough energy to break its connections with the other molecules and escape? The energy is taken away from the molecules just below that did the actual bumping! It actually takes the same amount of energy to evaporate a gram of water at room temperature as it does to boil a gram of water at 100°C! Evaporation transfers thermal energy from a liquid to a gas, and therefore *cools* the remaining liquid. That is why your wet skin feels cold when you get out of the shower. The evaporating water from your skin takes heat away from your body. This cooling effect is also why you sweat while exercising. The evaporation of sweat from your skin removes heat generated by your muscles.

Chemistry terms

evaporation - a phase change from liquid to gas at a temperature below the boiling point.

Condensation

Condensation

Have you ever noticed that the outside of a glass of cold water becomes wet on a humid summer day? Where did the water come from? Did it come through the glass?

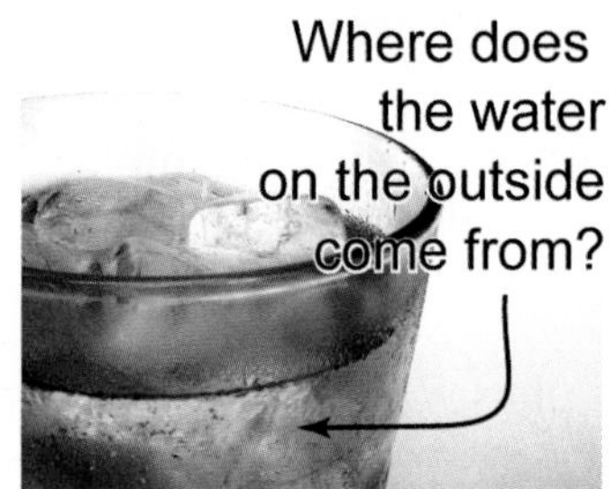

Where does the water on the outside come from?

The water on the outside of the glass comes out of the air in a process called *condensation*. **Condensation** is the changing of phase from gas to liquid. Water vapor in the air condenses on a cold glass to become liquid. Condensation occurs below a substance's boiling point.

Water cycle

While all liquids and gases can condense, the process is particularly important with water because evaporation and condensation create the water cycle of earth. Condensation may occur when a substance in its gas phase is cooled to below its boiling point. Clouds are an excellent example. Water vapor evaporates from the ocean and rises with warm air. The upper atmosphere is colder and the water vapor condenses into tiny droplets of liquid to make clouds. Clouds are not water vapor, a gas. Clouds are condensed liquid water. Evaporation and condensation create the water cycle in the atmosphere.

Condensation releases heat

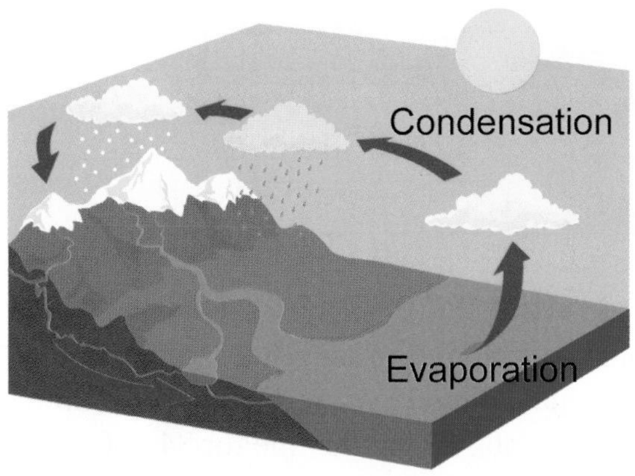

Condensation is the exact opposite process of evaporation. The energy transferred during condensation is exactly the same, but in the opposite direction, as in evaporation. From Table 3.3 we know that the heat of vaporization of water is 2,256 J/g. When one gram of water evaporates, it carries away 2,256 joules of energy, *cooling* any matter left behind. When a gram of water vapor condenses, it gives off 2,256 joules of energy, *warming* any matter it condenses on.

Latent heat

The heat energy given off during condensation is often called the *latent heat*. **Latent heat** is thermal energy associated with a phase change. For evaporation and condensation, the latent heat is the heat of vaporization. For melting and freezing, the latent heat is the heat of fusion. The latent heat released during cloud formation transfers heat from the oceans to the upper atmosphere, warming the atmosphere significantly. This is a very significant component of the earth's energy balance.

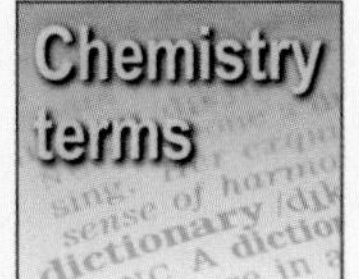

condensation - a phase change from gas to liquid, a substance in its gas phase may condense at a temperature below its boiling point.

latent heat - thermal energy that is absorbed or released by a phase change.

Pressure affects phase changes

Water boils at a lower temperature at high altitude

Cookbooks often have special instructions to cook food for longer at high altitudes. This is because water boils at different temperatures depending on the air pressure. For example, in Denver Colorado water boils at 95°C (203°F) instead of the usual 100°C (212°F). The boiling point is lower because Denver is at an altitide of 1,610 meters and the air pressure is lower that it is at sea level.

The phase diagram

The temperature at which a phase change occurs typically depends on pressure. The relationship between the phases of matter, temperature and pressure is shown on a *phase equilibrium diagram*. Point A is the normal melting/freezing point which is 0°C at 101,325 Pa (1 atm). Point B is the normal boiling point, which is 100°C at 1 atmosphere.

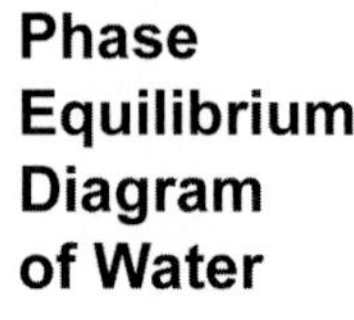

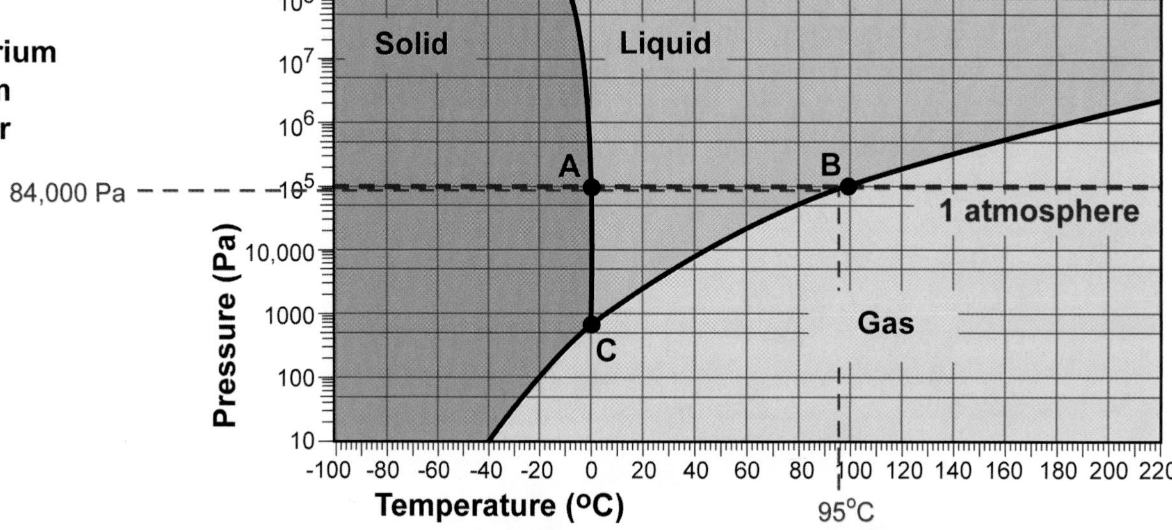

What the curves mean

Each curve represents a phase change. Crossing the liquid/gas curve means changing from liquid to gas or gas to liquid. This curve represents the boiling point. For Denver CO, the pressure is 84,000 Pa and notice that the line for 84,000 Pa crosses the liquid/gas curve at 95°C. Crossing the solid/liquid curve means changing phase from solid to liquid or vice versa. Water freezes or melts at the same temperature from 800 Pa to 200,000 Pa. At higher pressures water freezes at a temperature below 0°C.

The triple point

Point C on the phase equilibrium diagram is called the *triple point*. The **triple point** of water is at exactly 273.16 K (0.01 °C) and a pressure of 611.73 pascals (0.006037 atm). At the triple point, liquid, solid, and vapor can be present at the same time. A sample of water at its triple point can become solid, liquid, or gas with very small changes in pressure or temperature.

Chemistry terms

triple point - the temperature and pressure at which the solid, liquid, and gas phases of a substance can all exist in equilibrium together.

Relative humidity (Rh)

The meaning of humidity

A 90°F day in Houston, Texas feels much more uncomfortable than a 90°F day in Phoenix, Arizona. The reason is the *relative humidity*. **Relative humidity** describes how much water vapor is in the air compared to how much water vapor the air can hold when it is completely saturated.

Reading the boiling point

The diagram shows water's phase equilibrium diagram between 0°C (32°F) and 40°C (104°F). The curve represents the boiling point. The curve tells you that water boils at 20°C at a pressure of 2,340 Pa.

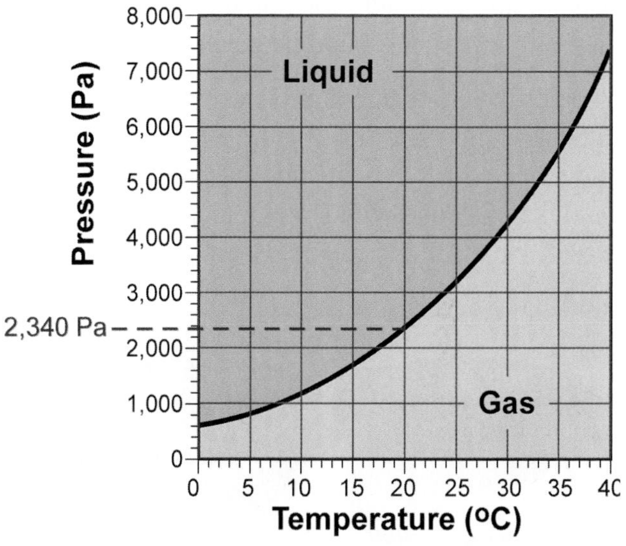

Reading the equilibrium vapor pressure

The curve also represents the equilibrium partial pressure of water vapor in air. For example, at 20°C water vapor has a partial pressure of 2,340 Pa. This is called the *saturation* vapor pressure. Like any saturated solution, if more water vapor is added to air at 20°C, it condenses into liquid. This interpretation is correct because a liquid boils when its vapor pressure is equal or greater than the ambient atmospheric pressure.

Calculating the relative humidity

Suppose 20°C air contains less than 2,340 Pa of water vapor. This is an *unsaturated* solution. It means the air could absorb more water vapor and not have it condense out. The relative humidity (Rh) is defined as the actual partial pressure of water vapor in air divided by the saturation vapor pressure at the same temperature. For example, the air in Phoenix, Arizona is very dry, and even at 20°C might contain only 1,000 Pa partial pressure of water vapor. The relative humidity is 1,000 ÷ 2,340 = 0.42 or 42%.

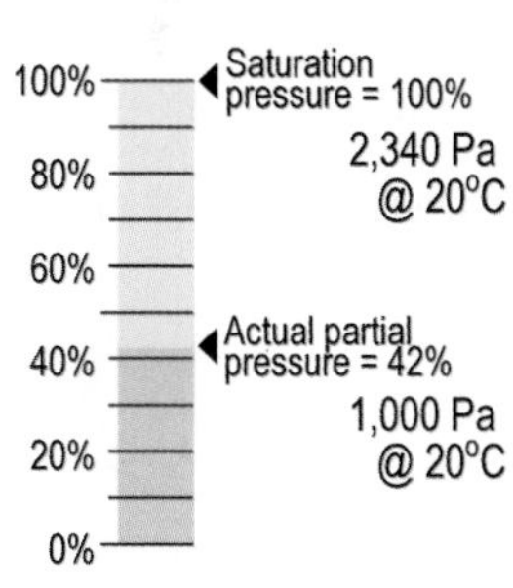

The dew point

Another important concept is the *dew point*. The **dew point** is the temperature at which air is saturated. For example, suppose air has a partial pressure of 1,000 Pa of water vapor. On the diagram, this corresponds to a dew point of 8°C. At 8°C the relative humidity is 100%. If the air gets cooler than 8°C the mixture becomes supersaturated and liquid water will condense. Greater than 8°C the relative humidity is less than 100%.

Chemistry terms

relative humidity - the actual partial pressure of water vapor in air divided by the saturation vapor pressure at the same temperature.

dew point - the temperature at which air is saturated with H_2O vapor (Rh = 100%)

Refrigeration by Evaporation

CHEMISTRY
CONNECTIONS

Mohammed Bah Abba is a smart, young African teacher who designed a clever refrigerator that needs no electricity! His invention uses earthenware clay pots and the principles of evaporation. As a young man Abba had a fascination with clay pots and their ability to absorb water. He wondered how the clay could absorb so much water, yet still hold together. He had observed that most things which absorb water become soft and lose their shape. He was also curious as to why the clay pots kept things cool. In college, Mohammed took some classes in biology, chemistry and geology and with this background he developed the "pot-in-pot refrigerator."

The pot-in-pot refrigerator is designed using two clay pots of different sizes, one large and the other small. The small pot is placed inside the larger pot and sand is added to fill the space between the walls of the two pots. The sand is moistened with water and a damp cloth is used to cover the top.

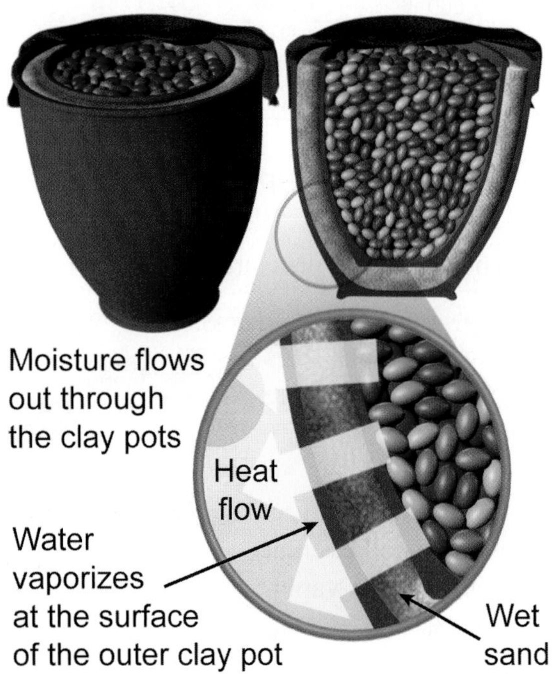
Moisture flows out through the clay pots
Heat flow
Water vaporizes at the surface of the outer clay pot
Wet sand

How does the pot-in-pot refrigerator keep things cool? To explain this, we will use the concepts you have learned about the phase changes of water and why they occur. When water molecules evaporate, they absorb energy, because they are going from the liquid phase to the gas phase. Do you recall why this requires energy? It is because the hydrogen bonds holding the water molecules together in the liquid phase must be *broken* to allow the water molecules to separate and become a vapor. H_2O(liquid) + Energy $\rightarrow$ H_2O(gas).

The hot climate in Africa provides warm, dry air outside the pot and this causes evaporation to occur toward the walls of the large outer pot. The moisture from the damp sand is pulled toward the outer pot as evaporation takes place. The porous clay allows the water molecules to travel from the moist interior to the outer pot where they are warmed and evaporate. The warmth outside the pots provides the liquid water molecules with the energy they need to evaporate and change into water vapor. The heat of vaporization, ΔH_V for water is 2256 j/g or 40.6 kJ/mole. This large amount of energy causes significant cooling that results in a temperature drop of several degrees! The more water molecules evaporate the cooler the temperature becomes. This causes the inner pot to become cool. The cooler temperature of the inner pot preserves foods that would otherwise spoil quickly in the heat.

Throughout human history, single clay pots have been used to keep things cool. Usually the clay pots are soaked in water for 20 minutes or so and then removed. Throughout the next few hours the water absorbed by the clay slowly evaporates. This process is used by street vendors to keep their produce and drinks cool for their customers. The single pot design works but it is less efficient and does not last as long as Mohammed's pot-in-pot design. His invention shows how important it is to use the ideas behind simple technology in an applied way! Mohammed wanted to help the poor people in the rural regions of Northern Nigeria and his inspiration gave way to this elegant, efficient pot-in-pot refrigerator.

To fully appreciate and understand the significance of Mohammed's invention it helps to understand the people of Northern Nigeria. The geography of this area is semi-desert scrubland and the rural people live in the poorest of conditions. There is no electricity and therefore no refrigeration. Even in the towns, the power supply is not dependable. Most people cannot afford to buy electricity even if it is available. The villagers' way of life is based on agriculture and young girls have the responsibility of rapidly selling the food that is grown each day before it spoils. Mohammed's invention allows these young girls to attend school more often instead of having to sell food.

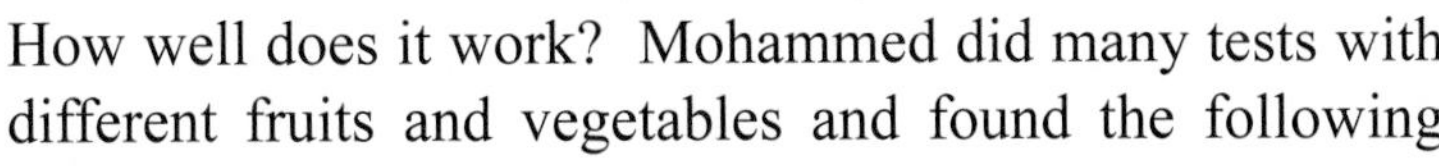

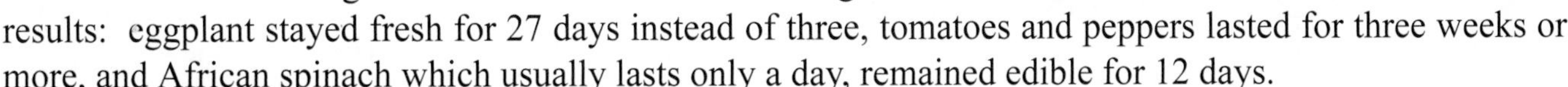

How well does it work? Mohammed did many tests with different fruits and vegetables and found the following results: eggplant stayed fresh for 27 days instead of three, tomatoes and peppers lasted for three weeks or more, and African spinach which usually lasts only a day, remained edible for 12 days.

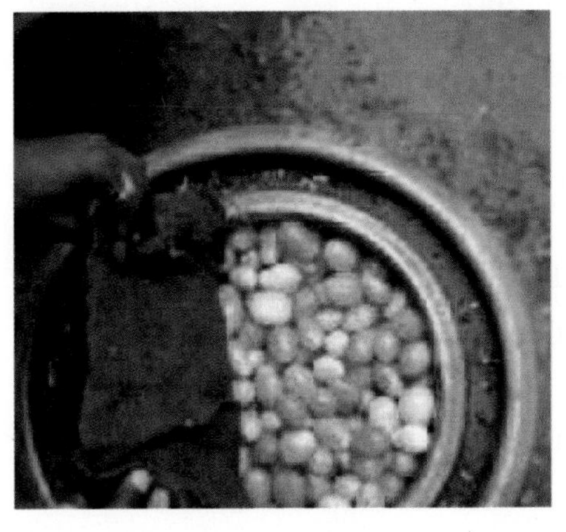

Mohammed worked to refine his invention for two years before beginning to distribute the pot-in-pot. He hired local pot makers to produce the first 5,000 pot-in-pots, which he distributed for free to five local villages in Jigawa. Later, with some limited financial backing, he distributed 7,000 more to another dozen local villages. Educating the illiterate local villagers on how to use the pot-in-pot was a challenge that Mohammed overcame by using a video recorded play of local actors who acted out how to use the pot-in-pot, and demonstrated how food stayed fresh, for a much longer time. As a result of the pot-in-pot, the farmers income levels have risen, because they can now sell their produce to consumers based on demand, where before they had to sell quickly, because the food would spoil. The number of girls attending local village schools has risen, because they no longer need to sell produce each day in order to survive. The pot-in-pot has also given rise to healthier food for all people to consume, therefore decreasing disease and illness in these areas.

Mohammed Bah Abba was only in his twenties when his invention won him the Rolex Award for enterprise in 2000. He won the prestigious award not only for his design of the pot-in-pot refrigerator, but for his efforts to make it affordable and available to the people who really needed it. These pots sell for thirty to fourty cents a set!

Chapter 3 Review.

Vocabulary

Match each word to the sentence where it best fits.

Section 3.1

kinetic energy	thermometer
Brownian motion	thermistor
Fahrenheit	absolute zero
random	Kelvin
temperature	Celsius

1. The erratic movement of a small particle in water due to constant collisions of molecules is called ________________.
2. ___________ is a measure of the average thermal energy of atoms or molecules.
3. The _________temperature scale is used in the United States where the boiling temperature of water is 212.
4. The energy of motion is referred to as __________.
5. A device used to measure temperature is called a ______________.
6. The theoretical temperature where all motion stops is called ______________ .
7. The SI unit of temperature is called the _________ and when using this scale water freezes at zero degrees.
8. Motion that is scattered equally in all directions is called _________ motion.
9. A ___________ is a temperature sensor that detects changes in electrical properties as the temperature changes.
10. The temperature scale that starts at absolute zero is called the _________ temperature scale.

Section 3.2

specific heat	heat
joule	hydrogen bonding
calorie	calorimetry
thermal equilibrium	first law
second law	conductor
system	insulator

11. ______ is a measure of the thermal energy of a substance
12. A group of interacting objects and effects that are selected for investigation is called a __________.
13. The ______________ of thermodynamics states that thermal energy moves from higher temperature to lower temperature.
14. The law of conservation of energy is also called the ____________ of thermodynamics.
15. The units of ________ and _______ are used to measure heat energy. The unit of the ________ is larger than the unit of the _______.
16. _______ is an attractive force between water molecules, that is responsible for the unusually high specific heat of water.
17. When the temperature of two systems is the same and heat no longer flows between them is called ____________________.
18. A material that easily permits the flow of heat across it is called a _______________.
19. The amount of heat needed to raise 1 gram of a substance by one degree is called _________.
20. This is a useful method for measuring the heat of a solution _________________.
21. A material that does not permits the flow of heat across it is called an ______________.

Section 3.3

gas	heat of fusion
solid	melting point
liquid	boiling point
triple point	ideal gas law
vaporization	phase change
evaporation	dew point

22. In the __________ phase of matter the atoms or molecules are able to move and change positions, but are still touching each other.
23. The process of _________ requires a great deal of energy, because it is breaking the bonds that hold molecules together in their liquid phase and separating them.
24. In the __________ phase of matter the atoms or molecules are not able to move and change positions.
25. The amount of heat required to melt one mole of ice is called the ________________.
26. Liquids kept in an uncovered container slowly lose volume over time due to the process of _____________.
27. A substance undergoes a ____________ when its molecules are rearranged without changing the individual molecules.
28. The temperature at which a substance changes from a solid to a liquid is called the ____________.
29. The ______________ allows us to calculate any variable that effects a gas under normal conditions.
30. The temperature at which a substance changes from a liquid to a gas is called the ____________.
31. The phase of matter that has no definite volume is called a ___________.
32. The ___________ is the temperature at which the partial pressure of water vapor in a sample of air equals its saturation vapor pressure.
33. The ____________ is the unique value of pressure and temperature at which solid, liquid and gas phases co-exist in equilibrium.

Conceptual Questions

Section 3.1

34. Explain why Brownian motion provides evidence that molecules are in constant random motion.
35. Describe the differences between the Fahrenheit and the Celsius temperature scales.
36. What units do we use to measure temperature?
37. How is the kinetic energy of a system related to its temperature?
38. Describe how random motion of atoms and molecules is related to temperature.

Section 3.2

39. To calculate the heat absorbed by an object, what information is needed, besides its specific heat?
40. Objects tend to exchange heat until "thermal equilibrium" is reached. Does this make sense? Explain your reasoning.
41. Does the specific heat change when you have a greater mass of a substance, say 10.0 g or 100.0 g? Explain.
42. If copper has a specific heat of 0.34 J/ g°C. What does this tell you about the structure of copper? (explain what the unit of specific heat means)
43. Adding heat to a substance can cause its temperature to increase. Explain why.
44. The specific heat of water is much higher than other substances. What does this mean?
45. You have a 20.0 g sample of two metals, one with a high specific heat and one with a low specific heat. Upon heating them with the same amount of heat which one will have the greatest change in temperature? Explain.
46. Suppose you have a flat hot bar of iron that weighs 25.0g at 90°C and you place a 10.0 g bar of copper

at 25°C on top of it. What will happen? Describe at the molecular level.

47. Tropical Islands often have what is referred to as temperate climates, which means the daily temperature is generally the same. It does not get really hot or really cold. Since islands are surrounded by water explain why the temperature remains relatively constant.

Section 3.3

48. Make a sketch of the "heating curve" for water. Label your axis appropriately, and label the solid, liquid and gas phases.

49. Why are there "flat" places on your sketch for the heating curve above?

50. Explain why the heat of fusion is always less than the heat of vaporization for any substance?

51. Explain why your skin feels "cool" when you sweat.

52. Some substances are liquids at room temperature. What does this tell you about the substances melting point?

53. Two identical glasses full of water are left in two different rooms. One room is maintained at a temperature of 10°C and the other room at 50°C. What would you find when you come back a week later?

54. What phase transition does the red arrow line represent in the phase equilibrium diagram shown here?

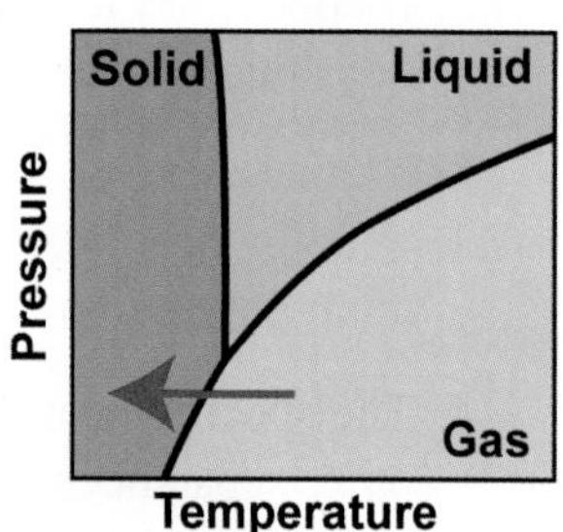

55. Describe with simple words the process of evaporation.

56. Water is the element with the highest value of heat of fusion and heat of vaporization (335 J/g). Describe the possible role that this large value has for life on earth.

57. Suppose the air is at a relative humidity of 100% then the temperature goes down.
    a. Does the partial pressure of water vapor change, why or why not?
    b. If the partial pressure changes, what happens to the "lost" water vapor?
    c. What is the relative humidity at the new temperature?

Quantitative Problems

Section 3.1

58. Normal human body temperature is about 98.7°C. Calculate this temperature in Celsius.

59. Ethylene glycol is a liquid organic compound, known as antifreeze, it freezes at -12°C. Calculate the freezing temperature on the Fahrenheit scale.

60. Paper catches on fire at about 451°F, what is this temperature on the Celsius scale.

61. Liquid nitrogen is used as a commercial refrigerant to flash freeze foods. It boils at -196°C. What is this temperature in Fahrenheit

62. Convert -20°C to Kelvin.

63. Convert 273K to Fahrenheit

Section 3.2

64. How many joules (J) are needed to increase the temperature of 15.0 g of lead from 20°C to 40°C? (Pb = 0.128 J/g°C)

65. An unknown metal substance absorbs 387 J of energy. If that substance weighs 50.0 g and changes temperature from 30°C to50°C, identify the metal? (refer to your table of specific heats)

66. A 5.0 g piece of gold, Au, changes temperature from 22°C to 80°C. How much heat does the gold absorb? (Au = 0.129J/g°C)

67. A 6.22 kg piece of copper metal is heated from 21.5°C to 310°C. Calculate the heat absorbed in kJ by the metal. (Cu = 0.34 J/g°C)

68. If 856 J of heat was absorbed by a block of iron that weighs 55.0 g, calculate the change in temperature of the block of iron.
(Fe = 0.44 J/g°C)

69. A 80.0 g mass of a metal was heated to 100°C and then plunged into 100g of water at 24°C. The temperature of the resulting mixture was 30°C. (use specific heat of water 4.18 J/g°C)
   a. How many joules of heat did the water absorb?
   b. How many joules of heat did the metal lose?
   c. What is the specific heat of the metal?

70. One beaker contains 145 g of water at 22°C and a second beaker contains 76.5 g of water at 96°C. What is the final temperature of the water after the beakers are mixed?

71. A 12.6 g piece of zinc was heated to 99.5°C in boiling water. It was then added to a beaker containing 50.0 g of water at 23°C. When the water and the metal come to thermal equilibrium, the temperature is 25.8°C. What is the specific heat of zinc?

72. A student mixed 100.0 mL of 0.500 M HCl with 100.0 mL of 0.500 M NaOH in a coffee cup calorimeter. Both solutions were initially at 23°C. After mixing the final temperature of the solution was 25.76°C. Calculate the heat change for the neutralization reaction in kJ per mole of NaOH.

   $HCl(aq) + NaOH(aq) \rightarrow NaCl(aq) + H_2O(l)$

   Assume that the densities and the specific heats of the solutions are the same as for water.
   (Density = 1.0 g/mL and Cp = 4.18 J/g°C)

73. Adding 6.43 g of $NH_4NO_3(s)$ to 150.0 g of water in a coffee-cup calorimeter, resulted in a decrease in temperature from 22.0 °C to 19.8°C. Calculate the heat change for dissolving $NH_4NO_3(s)$ in water, in kJ/mole. Assume the solution has the same specific heat as water.

Section 3.3

74. What quantity of heat is released (evolved) when 1.5 L of water at 0 °C solidifies to ice at 0°C?

75. The heat energy required to melt 1.0 g of ice at 0°C is 333 J. If one ice cube has a mass of 58.0 g, and you have 12 ice cubes, what quantity of energy is required to melt all of the ice cubes to form liquid water at 0°C?

76. What quantity of heat is required to vaporize 110 g of benzene, C_6H_6 at it's boiling point? Benzene's boiling point is 80.1°C and it's heat of vaporization is 30.8 kJ/mol.

77. What quantity of heat energy, in joules, is required to raise the temperature of 1.50 kg of ethanol from 22.0°C to it's boiling point, and then to change the liquid to vapor at that temperature? The boiling point of ethanol is 78.3°C. (The specific heat capacity of liquid ethanol is 2.44J/g°C and the heat of vaporization for ethanol is 855 J/g).

78. Nitrogen gas in an automobile air bag, with a volume of 65 L, exerts a pressure of 820 mm Hg at 24°C. How many moles of N_2 gas are in the air bag? R = 0.0821L·atm/ mol·K

79. A balloon holds 35.0 g of helium. What is the volume of the balloon if the pressure is 1.20 atm and the temperature is 25°C?

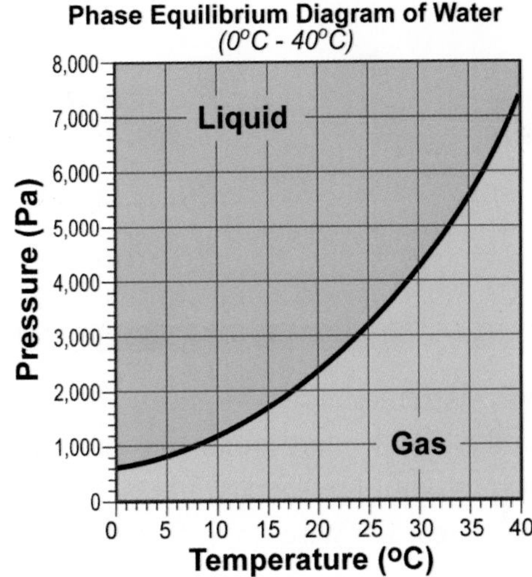

Use the phase equilibrium diagram above to solve problems 80 - 83.

80. What is the partial pressure of water vapor in saturated air at 15°C?

81. The water vapor in a sample of air has a partial pressure of 4,400 Pa. At what temperature is this air saturated, or at its dew point?

82. Some warm air has a temperature of 30°C and a partial pressure of water vapor of 2,000 Pa. This air is cooled to 10°C as night falls. Will some liquid condense or not? Explain how you get your answer.

83. Desert air at 35°C has a partial pressure of water vapor of 1,200 Pa. What is the relative humidity.

84. Which has more water vapor
   a. saturated air (100% Rh) at 15°C,
   b. air at 28°C with a relative humidity of 50%

CHAPTER 4
Physical and Chemical Change

What is the difference between physical change and chemical change?

What kinds of chemical changes are there?

How do we recognize chemical changes?

Memory wire is remarkable. Memory wire is made of an alloy, of nickel and titanium that looks like ordinary metal wire. You can bend it into complex shapes, like springs. What is special is that when you heat the wire up, it forcefully returns to the shape it originally had! Something about the metal "remembers" its original physical shape and tries to return once there is a heat source to provide energy. Bending is an example of physical change. Physical changes can be reversed by using physical means, such as changing temperature, filtering or drying.

Bending is a physical change

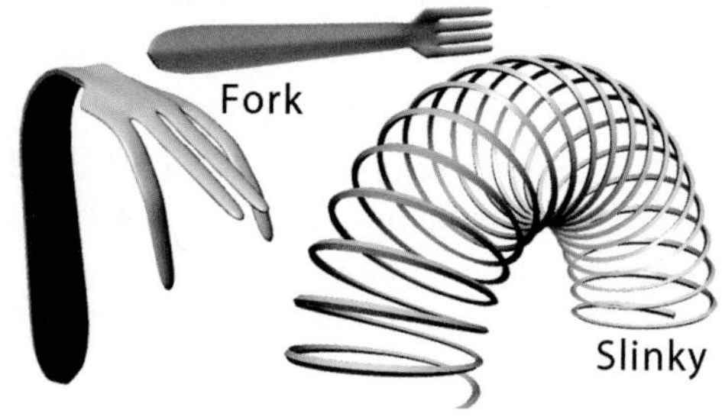

Consider burning a log in a fireplace. The matter in the log changes from wood into ashes, smoke and water vapor. When the fire cools down, the ashes do not spontaneously turn back into wood again! In fact, there is no technical way known to humans to turn the ashes, water and smoke back into a log again. Given enough time, nature can do it, but only by growing a new tree! Burning is an example of a chemical change. Chemical changes turn one substance into a completely different substance, or substances, which may have very different properties from the starting materials. Chemical changes are NOT reversible by physical means. Some chemical changes can be reversed, but only by inducing more chemical changes!

Observing the effects of chemical reactions.

Lets observe some chemical reactions. This will allow us to better appreciate chemical change. Some chemical changes are accompanied by brilliant visual differences when the chemicals are mixed. Here we will make some observations of the reactants we start with, followed by observations of the products formed after the chemical reaction takes place.

Measure about 20 mL of 0.50 M silver nitrate, $AgNO_3$ solution and pour this into a test tube. Silver nitrate looks clear like tap water, but it contains little silver ions, Ag^+ floating around that our eyes cannot see. Now take a 10.0 cm piece of copper wire and wrap it around a glass stirring rod (or a pencil). Place the copper wire in the test tube. Make initial observations and continue to make observations every 5 minutes. The pictures below show the system. (* t = 0, t = 2 min, t = 5 min, t = 10 min)

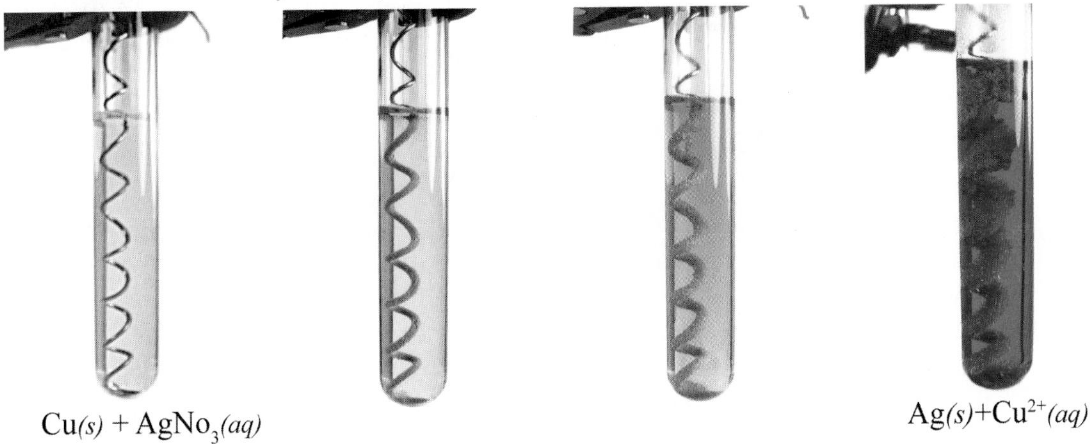

$Cu(s) + AgNo_3(aq)$ $Ag(s)+Cu^{2+}(aq)$

As you can see, the silver (Ag) forms on the copper wire and the solution begins to turn a greenish-blue color after 5+ minutes.

Why does this occur? Our observations allow us to see the grayish, silvery formation on the copper wire and the blue-green color of the solution. During this chemical change the silver ions (Ag^+) dissolved in the solution *gained electrons* and formed solid silver (Ag). You see the solid silver forming on the copper wire. Some of the solid copper (Cu(s)) *loses electrons* and dissolves in the solution forming Cu^{2+} ions, giving the solution the greenish-blue color. This type of reaction, where chemicals transfer electrons, is an important area of chemistry.

Another type of common chemical reaction is a precipitate reaction. To try this, you will need a dropper bottle of 1.0 M silver nitrate ($AgNO_3$) and a dropper bottle of 1.0 M sodium chloride (NaCl). Set up a test tube with 2.0 mL (or 20 drops) of NaCl, place this in a test tube holder. Slowly add $AgNO_3$. Make your observations!

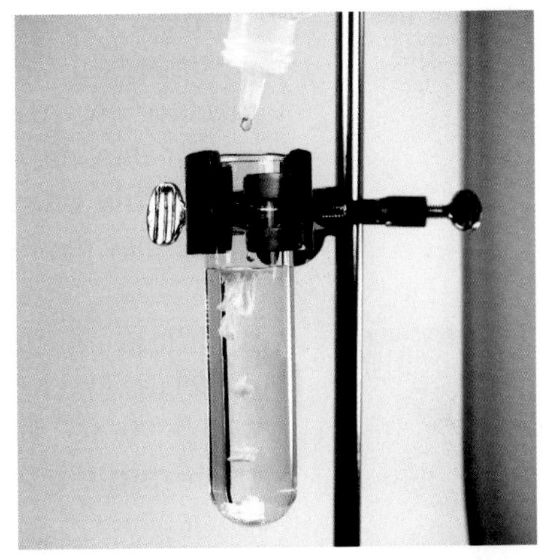

The white cloudiness formed is solid silver chloride (AgCl). If we filtered and dried the solid that settles to the bottom of the test tube, you would have a powdery white solid.

Isn't it amazing that such a brilliant yellow forms from two relatively clear solutions! Lets learn more about why chemicals do this.

4.1 Understanding Chemical Changes

Reversible changes

Physical and chemical change are fundamentally different. For example, consider a burning candle. Some of the wax in a candle is solid and some is melted. Melting (solid to liquid) and freezing (liquid to solid) are examples of *reversible* changes. If you cool the melted liquid wax down, it becomes solid again. Reversible changes are physical changes.

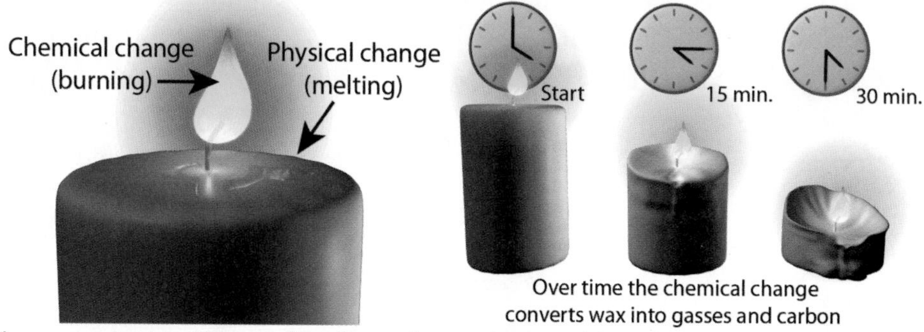

Irreversible changes

What happens inside the candle *flame* is not just melting, however. Over time, the wax vanishes! Burning wax is a very different, *irreversible* kind of change. In the flame, a chemical reaction changes wax into carbon, water vapor and carbon dioxide, three very different substances. If you cooled down the smoke from a burning candle, it would not become wax again. Wax disappears as a candle burns because water vapor and carbon dioxide are gases that float away into the air.

Physical changes are reversible

$C_{20}H_{42}$, a type of *paraffin* is a major component of candle wax. Each molecule of paraffin contains 20 carbon atoms and 42 hydrogen atoms. If you pour the wax into different shapes, or melt it, or cut it up in tiny pieces, each bit will still be made from molecules with 20 carbon and 42 hydrogen atoms. That's because a physical change leaves the molecules of a substance the same. *Physical changes are reversible*: melting, shaping, cutting, bending and freezing are all physical changes.

Chemical changes are irreversible

Wax burning in a flame is a chemical change. A **chemical change** is a change in the molecules themselves. In the candle flame, the atoms in paraffin molecules are rearranged into molecules of water and carbon dioxide. In burning, wax undergoes an **irreversible change**. Chemical changes are irreversible because they rearrange atoms into different substances.

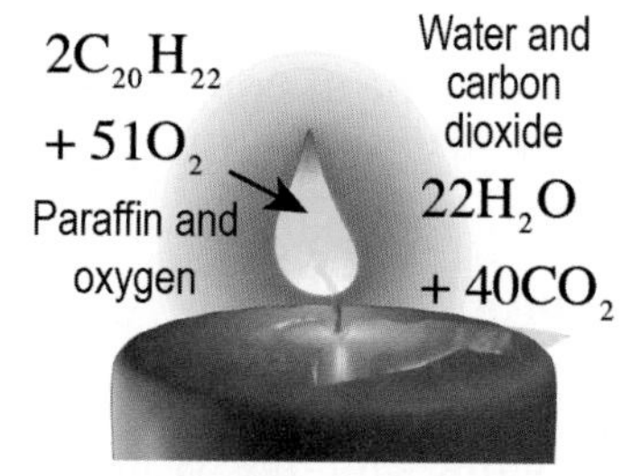

Chemistry terms

chemical change - a change that affects the structure or composition of the molecules that make up a substance, typically turning one substance into another substance with different physical properties.

irreversible change - is a chemical change that rearranges atoms into different substances

Energy and change

Energy determines change in matter

Changes in matter involve an exchange of energy. Whether a change is physical or chemical fundamentally depends on the amount of energy. If the energy is great enough to break a molecule apart, then chemical change is possible. If the energy is lower, then only physical change is possible. To understand why, we have to think about the forces between and within atoms and molecules. Strong forces take a lot of energy to change; weaker forces take less energy to change.

Interatomic forces act within a molecule

Consider water as a substance. **Interatomic forces** hold the two hydrogen atoms tightly to the one oxygen atom in a single water molecule. Interatomic forces are relatively strong: they take a lot of energy to break.

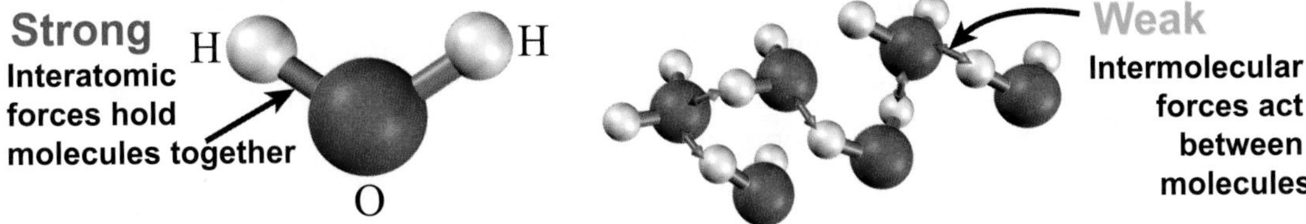

Intermolecular forces act between molecules

The fact that water molecules stick together to make ice or liquid means that there are other forces that act *between* molecules. These forces are called **intermolecular forces**. *Physical changes involve only intermolecular forces.*

Intermolecular forces are much weaker than interatomic forces

Chemical changes involve interatomic forces

The graph shows the relationship between the energy and the types of changes that occur in water. It takes 333 J of energy to melt 1 gram of solid ice into liquid water at 0°C. By comparison, it takes 51,000 J of energy to turn one gram of water into separated hydrogen and oxygen gas! To make a chemically change requires about 100 times more energy than a physical change.

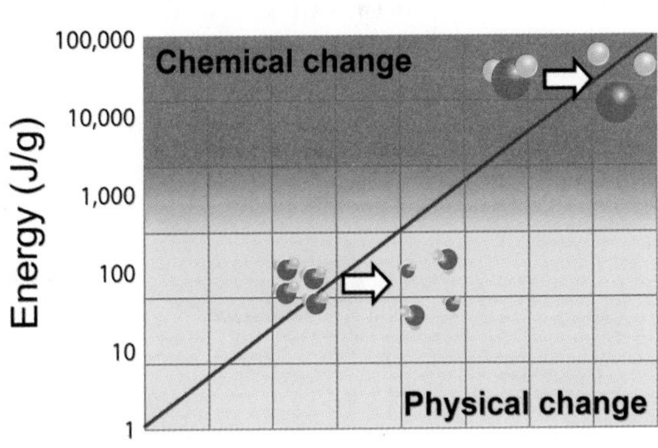

Chemical changes can release energy

The amount of energy needed to break water into hydrogen and oxygen is released if we let the hydrogen and oxygen come back together again and make water. In fact, that is the basic principle behind hydrogen fueled cars and trucks! A hydrogen car burns hydrogen gas along with oxygen from the air and creates harmless water.

interatomic forces - bond atoms together into molecules or ions.

intermolecular forces - act between molecules, typically much weaker than the forces acting within molecules (interatomic forces).

What is NOT a chemical change

Changes in size or shape are physical changes

Any process that changes the shape of a substance is usually a physical change. That means bending or deforming are physical changes. It also means grinding something up into powder is a physical change. Rock candy, granulated sugar and confectioners sugar are the same substance, but they have been ground into different textures.

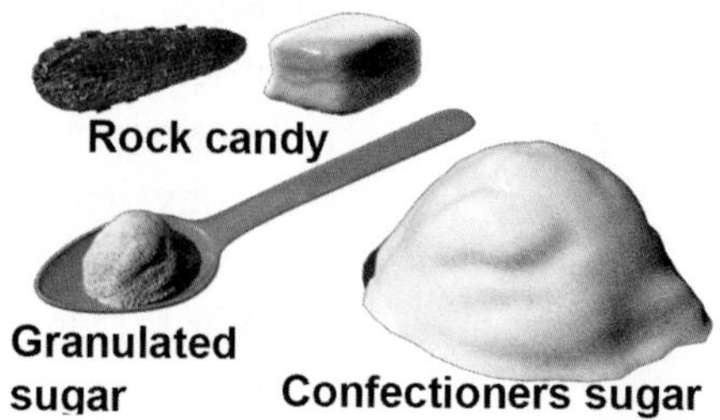

Mixing and dissolving are physical changes.

Adding food color to water spreads dissolved dye molecules evenly through the water but does not change the dye molecules into other molecules. This is evidence that dissolving is a physical change. Even vigorous mixing is still a physical change. For example, mix some corn oil in water and it floats. Whip it with a beater and the mixture turns cloudy and white, like milk. Under the microscope however, you still see a mixture of water and oil. The oil droplets have become very small, but each droplet still contains thousands or millions of atoms. Making bigger drops into smaller drops is definitely a physical change. Milk and mayonaisse are mixtures of oils or fats and water.

Mixing is a physical change

"Drying" may be a chemical or physical change

Drying is the opposite of dissolving. In drying, the water is removed from a mixture, leaving any solutes in their dry form. Like dissolving, drying is *usually* a physical change. Drying paint for example is not always just a physical change. Certain molecules in latex or acrylic paint react chemically with oxygen to link together. Dried latex or acrylic paint is a solid that does not become liquid again when you heat it, or add water back to it. In the sense of chemistry, "drying" means the purely physical process of removing liquid without chemical changes.

Atoms and chemical bonds

Chemistry is explained by the structure inside atoms

Before we can understand chemical change, we need to learn why nearly all ordinary matter exists as compounds and not single atoms. Why does one oxygen atom bond with two hydrogen atoms to make water? Why not three (H_3O) or even four (H_4O)? Why are pure hydrogen and oxygen in diatomic molecules (H_2, O_2) instead of single atoms? The answers to these questions can be found by looking *inside* the atom, at the structure within. The structure of atoms explains the reason for chemical bonds, and the chemical bonds are the source of chemical energy.

Atoms are not hard little balls

We draw atoms as hard, colored balls, but they are not like that at all! A better mental image of an atom is of an extremely tiny, hard core surrounded by a vast, thin cloud. Atoms really have no definite "edge" or "surface".

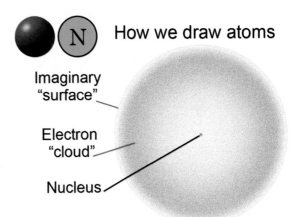

The nucleus

The core is called the **nucleus**, and it contains 99.8% of the mass of the atom. Compared to the size of a whole atom, the nucleus is extraordinarily tiny. If the atom were the size of your classroom, the nucleus would be the size of a single grain of sand in the center!

Electrons are in the space outside the nucleus

Around the nucleus are tiny particles called **electrons**. Hydrogen has one electron, helium has two electrons and lithium has three. The number of electrons corresponds to the atomic number of the element, as shown on the periodic table. Chapter 5 will explain more about the structure of atoms. For now, let's look at the big ideas to begin understanding chemical bonds.

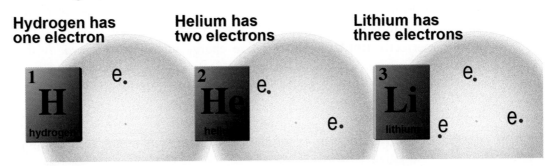

Chemistry terms

nucleus - the tiny, dense core of an atom which contains all the positive charge and 99.8% of the mass.

electron - a tiny particle that fills the outer volume of an atom. Electrons have negative charge and are responsible for chemical bonds.

Electric charge

Electric charge is a property of matter

Along with mass and volume, matter has a fundamental property called **electric charge**. Electric charge is important because it creates both the forces that hold the atoms together and the forces that cause atoms to combine into compounds and molecules.

There are only two kinds of electric charge: we call them **positive** charge, indicated by the + sign, and **negative** charge, indicated by the - sign. A positive and a negative charge attract each other. Two positive charges repel each other. Two negative charges also repel each other.

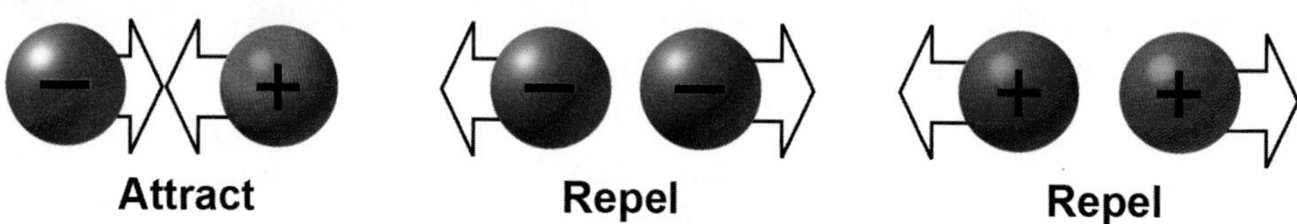

Protons have positive charge

Positive charge is a property of one of the particles in the nucleus called the **proton**. All of the protons, and therefore all the positive charge in an atom are in the nucleus. In fact, the atomic number is defined as the number of protons in the nucleus. Hydrogen has one proton in its nucleus, Helium has two protons, lithium has three protons and so on.

Electrons are attracted to the nucleus

Electrons are bound to the nucleus by the attractive force between electrons (-) and protons (+). The electrons don't fall into the nucleus because of their kinetic energy. Think of the earth orbiting the sun. Gravity creates a force that pulls the earth toward the sun. Earth's kinetic energy causes it to orbit the sun rather than fall straight in. While electrons don't really move in orbits, the energy analogy is approximately right.

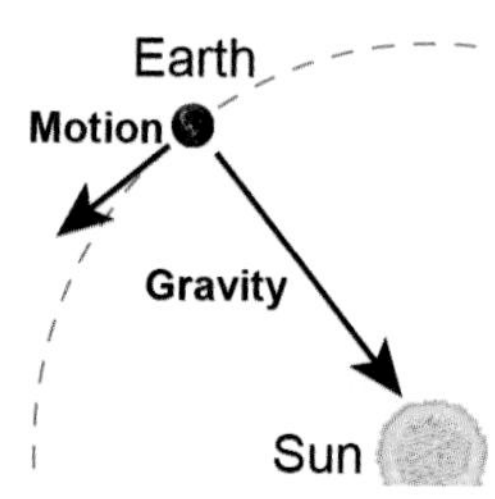

Neutral atoms have zero total charge

The force between electric charges is extremely strong. The electrical attraction between a proton and an electron is approximately 10^{40} times as strong as gravity. This is a ten with forty zeros after it! The reason you don't notice electric charge is that atoms are perfectly **neutral**. The positive charge of a proton is *exactly* the same amount as the negative charge on an electron. A carbon atom has six protons and six electrons. Its total electric charge is exactly zero because +6 from the protons and -6 from the electrons add up to zero.

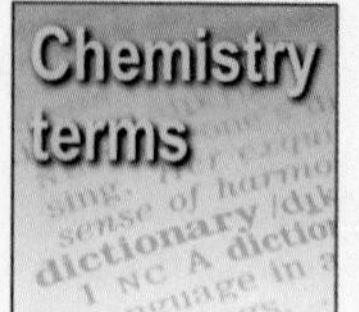

electric charge - a fundamental property of matter than comes in positive and negative.

positive, negative - the charge on a proton is defined to be positive and the charge on an electron is defined to be negative.

proton - a tiny particle in the nucleus that has a positive charge.

neutral - an atom or molecule is neutral when it has zero total electric charge.

Covalent bonds

Elements and electrons

Each element has a different number of electrons and a unique way that the electrons are arranged around the nucleus. This different number and arrangement of electrons is what creates each element's unique chemical properties. Hydrogen has a single electron, and that single electron is what makes hydrogen combine with oxygen in a two-to-one ratio to make H_2O. Nearly all the elements readily form chemical bonds. This is why most of the matter you experience is in the form of compounds.

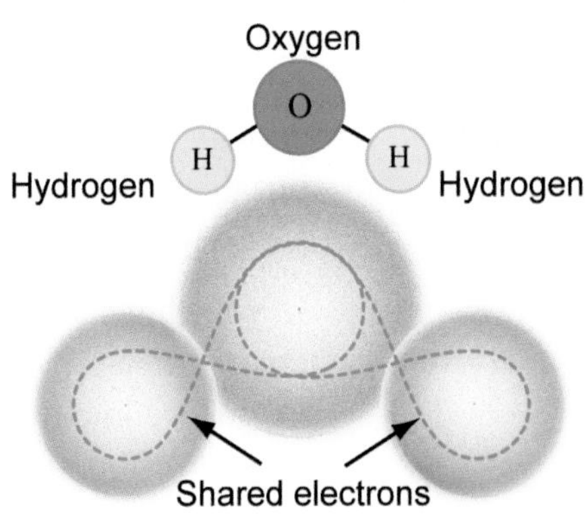

Electrons form chemical bonds

Chemical bonds are created by electrons. Two atoms that are sharing one or more electrons are chemically bonded and move together. In a water molecule, each hydrogen atom shares its single electron with the oxygen atom at the center.

A chemical bond is formed by sharing or transferring electrons

Covalent bonds

The strongest chemical bond is called a **covalent bond**. A covalent bond is formed when two atoms share a single electron. A water molecule contains two covalent bonds between oxygen and hydrogen. Each bond represents one electron. In a covalent bond, electrons are shared between atoms, not transferred.

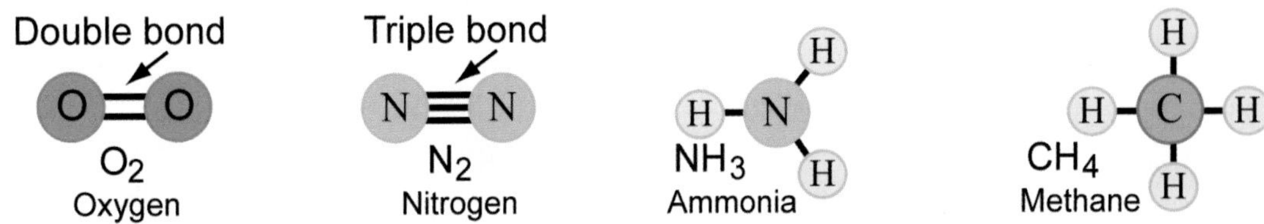

Molecules

An electrically neutral group of atoms held together by covalent bonds is called a **molecule**. Water is a molecule, and so is glucose. Other examples of molecules are methane (CH_4), ammonia (NH_3), and carbon dioxide (CO_2).

Some elements can share multiple electrons with the same atom. Good examples are oxygen (O_2) and nitrogen (N_2). An oxygen molecule contains a double covalent bond (2 shared electrons). A nitrogen molecule has a triple covalent bond (3 shared electrons). In diagrams, double and triple bonds are represented as double and triple lines connecting the atoms.

chemical bond - a relatively strong connection between two atoms.
covalent bond - a chemical bond that consists of one shared electron.
molecule - a neutral group of atoms that are covalently bonded together.

Why do chemical bonds form?

Bonds give an energy advantage to atoms

Chemical bonds form when attractive forces create an advantage in energy. For example, think about two strong magnets as they are brought near each other. When they get close enough the magnets snap quickly together. Clearly, a force pulls the magnets together, but there is another way to look at the situation. *The total energy of the two magnets is lower when they are together compared to when they are apart.*

Energy is released when bonds form

If you think about pulling the magnets apart, you need a force and it takes energy to create forces. If it takes energy to pull them apart, the same energy is *released* when the magnets come together. In fact, the force that pulls the magnets together is created by the energy difference between being apart and being together. The same is true of chemical bonds. *Energy is released when chemical bonds form.* Energy is released because chemically bonded atoms have less total energy than free atoms.

Atoms form bonds to reach a lower energy state

A general principle of chemistry is that atoms arrange themselves so they have the lowest possible energy. Like a ball rolling downhill, atoms form compounds because the atoms have lower energy when they are together in compounds compared to when they are separate. Consider water for example, one oxygen and two hydrogen atoms have more total energy apart than they do when combined in a water molecule.

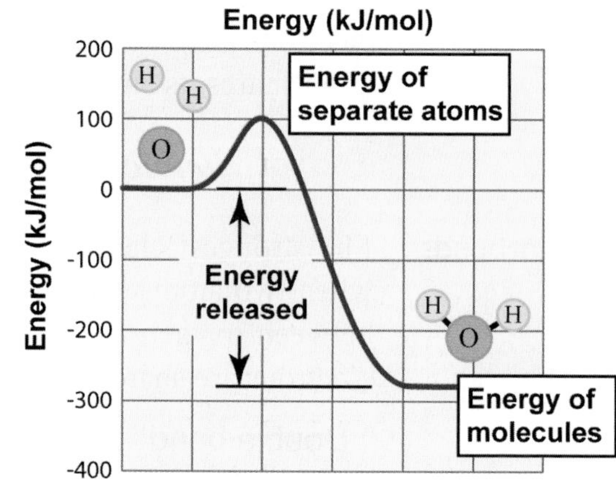

The enthalpy of formation

When hydrogen and oxygen combine to make water, 285,000 joules of energy is released for every mole of water molecules created. This energy is called the **enthalpy of formation** shown as ΔH_f. Table 4.1 lists the enthalpy of formation for water as negative because energy is given off instead of absorbed. By convention, pure elements are assigned an energy of zero.

TABLE 4.1. Enthalpy of formation (ΔH_f) for some common substances

Substance	ΔH_f (kJ/mol)	Substance	ΔH_f (kJ/mol)
Hydrogen (H_2)	0	H_2O	-285.5
Carbon (C)	0	CO_2	-393.5
Oxygen (O_2)	0	CH_4	-74.6
Nitrogen (N_2)	0	$C_6H_{12}O_6$	-1266.5

Chemistry terms

enthalpy of formation - the change in energy when one mole of a compound is assembled from pure elements.

Ionic bonds

An ion is a charged atom

Not all compounds are made of molecules. For example, sodium chloride (NaCl) is a compound of sodium (Na) and chlorine (Cl) in a ratio of one sodium atom per chlorine atom. The difference is that in sodium chloride, the electron is essentially *transferred* from the sodium atom to the chlorine atom. When atoms gain or lose an electron they become *ions*. An **ion** is an atom which either lost one or more electrons or gained one or more electrons. By losing an electron, the sodium atom becomes a sodium ion with a charge of +1. By gaining an electron, the chlorine atom becomes a chloride ion with a charge of -1 (when chlorine becomes an ion, the name changes to chloride).

An ionic bond forms when one or more electrons is transferred from one atom to another

Ionic bonds

Sodium and chlorine form an ionic bond because the positive sodium ion is attracted to the negative chloride ion. Ionic bonds are bonds in which electrons are transferred from one atom to another. In general, ionic bonds are slightly weaker than covalent bonds.

Ionic compounds do not form molecules

Ionic bonds are not limited to a single pair of atoms like covalent bonds. In sodium chloride, each positive sodium ion is attracted to all of the neighboring chloride ions. Likewise, each chloride ion is attracted to all the neighboring sodium atoms. Since the bonds are not just between pairs of atoms, ionic compounds do not form molecules! In an ionic compound, each atom bonds with all of its neighbors through attraction between positive and negative charge.

Sodium Chloride (NaCl) is an ionic compound

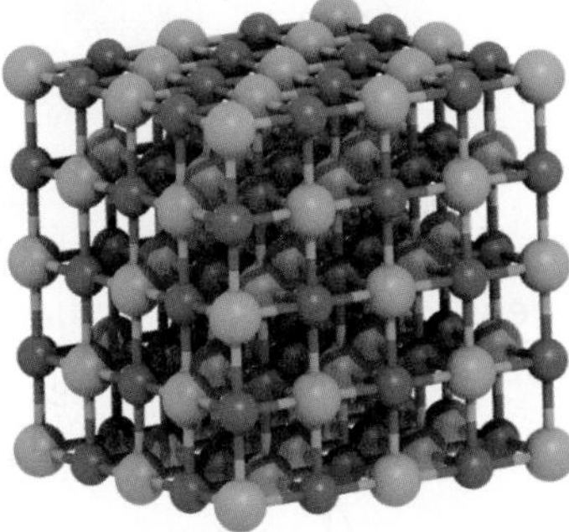

The chemical formula for ionic compounds

The chemical formula for an ionic compound like salt is used in the exact same way as the formula for a molecular compound like water. The chemical formula for salt (NaCl) means that there is one sodium atom per chlorine atom. You calculate the formula mass of salt (58.5 g/mole) by adding the atomic masses of sodium (23.0 g/mole) and chlorine (35.5 g/mole).

Ions may be multiply charged

Sodium chloride involves the transfer of one electron. However, ionic compounds may also be formed by the transfer of two or more electrons. A good example is magnesium chloride ($MgCl_2$). The magnesium atom gives up two electrons to become a magnesium ion with a charge of +2 (Mg^{2+}). Each chlorine atom gains one electron to become a chloride ion with a charge of -1 (Cl^-). The ion charge is written as a superscript after the element (Mg^{2+}, Fe^{3+},Cl^-, etc.).

Reactivity

Elements that react strongly are rarely found in pure form

Some elements react so strongly that they are never found in nature by themselves as pure elements. Good examples are lithium (Li), sodium (Na), and potassium (K). These are very common elements, but are always in compounds such as salt (NaCl), minerals in rocks such as petalite ($LiAlSi_4O_{10}$) and even chemicals in your body.

Some elements do not react

Other elements such as helium, neon and argon are always found as pure elements. These elements are not reactive and are called noble gases. The noble gases are not very reactive due to their electronic structure. Noble gases do not form chemical bonds with other elements.

Reactivity and the periodic table

What does it mean for an element to be "reactive?" In a literal sense it means that it forms bonds with anything it touches! If you look at the periodic table, you see that the three most reactive metals (Li, Na, K) are in the same group (column). The unreactive elements (He, Ne, Ar, Kr, Xe) are also in a group, but on the far right of the table.

← Electrons away from noble gas →

Not reactive
Moderately reactive
Very reactive

1	2	3	4										4	3	2	1	
H 1																	He 2
Li 3	Be 4											B 5	C 6	N 7	O 8	F 9	Ne 10
Na 11	Mg 12											Al 13	Si 14	P 15	S 16	Cl 17	Ar 18
K 19	Ca 20	Sc 21	Ti 22	V 23	Cr 24	Mn 25	Fe 26	Co 27	Ni 28	Cu 29	Zn 30	Ga 31	Ge 32	As 33	Se 34	Br 35	Kr 36
Rb 37	Sr 38	Y 39	Zr 40	Nb 41	Mo 42	Tc 43	Ru 44	Rh 45	Pd 46	Ag 47	Cd 48	In 49	Sn 50	Sb 51	Te 52	I 53	Xe 54

The halogens are reactive elements

To the immediate left of the noble gases is another group of very reactive elements - the halogens which include fluorine (F), chlorine (Cl) and bromine (Br).

If we look at the two example compounds of salt (NaCl) and the mineral petalite ($LiAlSi_4O_{10}$) we see an important pattern. Sodium and lithium from the left of the periodic table are bonded with elements from the right half of the periodic table: chlorine, aluminum, oxygen and silicon.

Elements on the far left and far right of the periodic table are more likely to form chemical bonds

Reactivity and the periodic table

The more reactive elements in the periodic table tend to form chemical bonds more easily. For example, sodium readily bonds with chlorine and the reaction releases a lot of energy. Sodium and chlorine are on opposite sides of the periodic table.

Chemistry terms

reactivity - the tendency of elements to form chemical bonds. A reactive element forms bonds easily therefore tends to have many reactions.

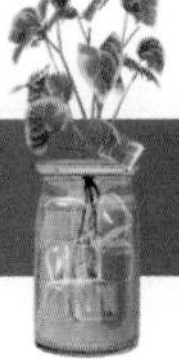

Reactivity and the periodic table

The two main classes of chemicals that we work with in general chemistry are ionic and molecular compounds. The periodic table helps us tell the difference between these two types of compounds.

Ionic compounds contain a metal and a nonmetal

Ionic compounds, which are often referred to as salts, form between metals and nonmetals. You have learned that ionic compounds transfer electrons. In the case of an ionic compound the metal loses the electron(s) and the nonmetal(s) gains the electrons. Lets look at the periodic table below to see how this pattern works.

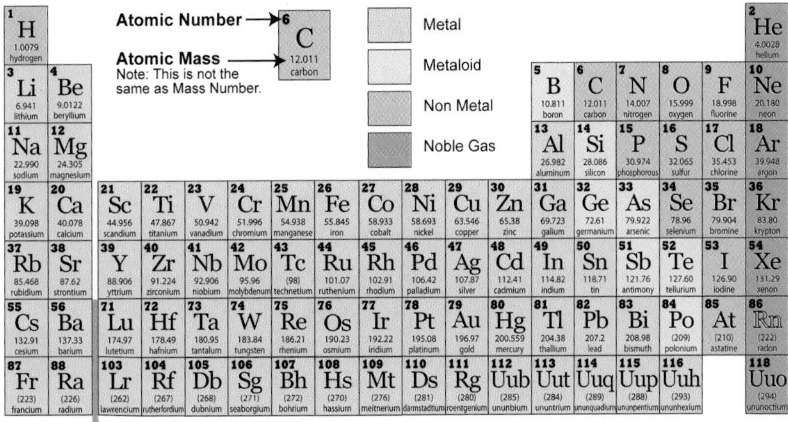

When dealing with a chemical compound, you can look at the periodic table and see if there is a metal bonded to a nonmetal; if so, you can classify the compound as ionic.

Molecular compounds contain two or more nonmetals

Molecular compounds are made up of two or more nonmetals bonded together. You can see from the periodic table that these elements appear on the far right upper corner. Molecular compounds share electrons. For example carbon dioxide, CO_2, is made of carbon and oxygen and both are nonmetals.

Use the periodic table to help determine whether a compound is ionic or molecular

Hydrogen (H) is ambidextrous and has metallic and nonmetallic behavior. The fact that it only has one electron and one proton makes hydrogen small in size. This is why it is a unique element on the periodic table. In most compounds, hydrogen shares it's one electron and forms a molecular compound. This is because hydrogen holds it's one electron very tightly to the nucleus.

Is the compound CF_4 ionic or molecular? We start by locating carbon (C) and fluorine (F) on the periodic table, and find that they are both nonmetals. This indicates that CF_4 is a molecule! We know it shares its electrons in the bonds it makes.

4.2 Chemical Reactions

Examples of chemical change

Chemical changes turn one substance into a different substance. We already mentioned that the digestion of food is a process of chemical changes. The substances that leave your body are definitely not the same substances that went in! Here is another (cleaner) example. The device in the diagram performs *electrolysis*. In electrolysis an electric current transfers a relatively large amount of energy to water molecules. The water molecules change slowly into gas.

Electrolysis is not the same as boiling

Over time the mass of water goes down and the volume of gas goes up. But, this is not boiling because *the gas is not water vapor.*

1. When you cool the gas, it does not condense back into droplets of water.
2. One of the gases burns! Water vapor (steam) does not burn.
3. The other gas causes flames to get much brighter! This is also not a characteristic of water vapor.

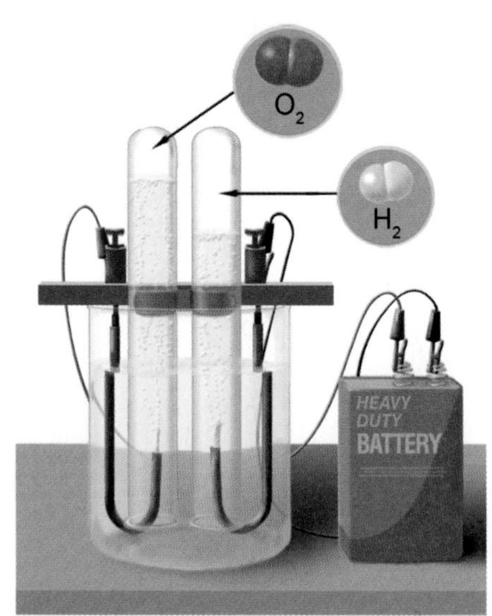

Water is different from hydrogen and oxygen

On the molecular level, each particle (or molecule) of water contains one oxygen atom bonded to two hydrogen atoms. The chemical reaction that occurs in electrolysis splits up the water molecule into separate atoms of hydrogen and oxygen. Hydrogen and oxygen are different substances from water. For example, both are gasses at room temperature while water is a liquid.

Chemical reactions

A chemical change is caused by a **chemical reaction**. A chemical reaction is usually written as an equation that has the molecules you start with on the left and the molecules you finish with on the right. The reaction for electrolysis starts with two water molecules and produces two hydrogen (H_2) molecules and one oxygen (O_2) molecule.

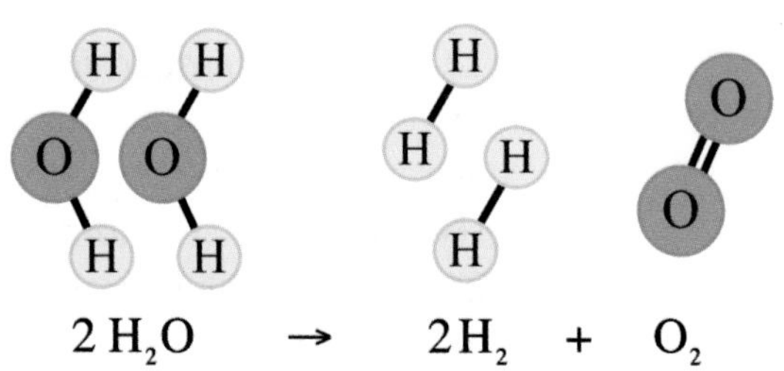

chemical reaction - a process that rearranges the atoms in any substance(s) to produce one or more different substances. Chemical reactions are described by chemical equations such as $2H_2O \rightarrow 2H_2 + O_2$.

chemical change - a result of chemical reaction.

The chemical equation

Reactants and products

The chemical formula for water is H_2O. How do we represent the reaction that assembles the hydrogen and oxygen to make water? The answer is with a chemical equation. The **reactants** are the compounds that you start with. The reactants are chemically changed into the **products**. The products are the compounds that are formed by the reaction. In this example, the reactants are hydrogen and oxygen, and the product is water.

Reactants	Products
H_2O	$H_2 + O_2$

$$H_2O \rightarrow H_2 + O_2$$

unbalanced equation

Reading a chemical equation

To read a chemical equation you need to understand the numbers. The *subscripts* tell you how many atoms of each element are in the molecule. For example, the 2 in H_2O tells you there are 2 hydrogen atoms per oxygen atom in a water molecule. Pure oxygen (O_2) and hydrogen (H_2) gas are molecules of two atoms each (*diatomic molecules*).

Unbalanced chemical equations

The above equation is not *balanced*. The equation shows two hydrogen atoms and two oxygen atoms on the reactant side. There are two hydrogen atoms but *only one oxygen atom* on the product side. That leaves one oxygen atom unaccounted for.

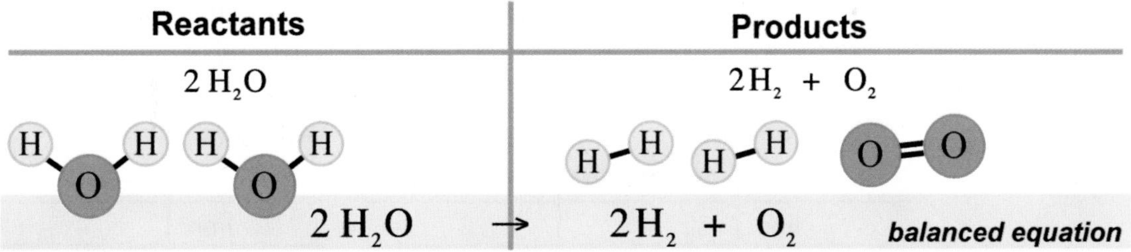

Balanced chemical equations

The **balanced equation** combines two molecules of hydrogen with one molecule of oxygen to make 2 molecules of water. Notice that all the atoms on the reactant side of the equation are used on the product side. No atoms are left over. The **coefficient** tells you how many molecules participate in the balanced reaction. The coefficient is the number "2" in front of H_2O. It tells you there are 2 molecules of water in the balanced equation.

reactant - a substance that is used or changed in a chemical reaction.

product - a substance that is created or released in a chemical reaction.

balanced equation - when the total number of atoms of each element are the same on both the reactant side and the product side of a chemical equation.

coefficient - the number that tells you how many of each molecule participate in the balanced reaction.

The balanced chemical equation

Rules for balancing equations

When balancing a chemical equation, you want to get the same number of each type of atom on both the reactant and product sides. Here are the rules for doing it.

1. *DO NOT* change the subscripts because this would change the identity of the substances in the reaction. For example H_2O is water but H_2O_2 is hydrogen peroxide.
2. *DO* change the coefficients to adjust how many molecules of each substance appear as either products or reactants.

Balancing the reaction for rust

The process is often trial and error. You start with an unbalanced reaction that shows the correct chemical formula for each substance in the reactants and products. You then try different coefficients to see what works. For example: Iron (Fe) reacts with oxygen (O_2) in the air to form rust (Fe_2O_3). The balanced and unbalanced reaction are shown below. It is useful to make a table showing atoms of each element on each side. Try different coefficients until the table shows the same numbers on both sides.

unbalanced reaction

$Fe + O_2 \rightarrow Fe_2O_3$

Reactants	Products
$Fe + O_2$	Fe_2O_3

Atoms (or moles)

	Reactants	Products
Iron	1	2
Oxygen	2	3

balanced reaction

$4Fe + 3O_2 \rightarrow 2Fe_2O_3$

Reactants	Products
$4Fe + 3O_2$	$2Fe_2O_3$

Atoms (or moles)

	Reactants	Products
Iron	4	4
Oxygen	6	6

When magnesium metal (Mg) reacts with oxygen (O_2) it forms magnesium oxide (MgO). Write the balanced equation for this reaction.

Given: *The unbalanced reaction is* $Mg + O_2 \rightarrow MgO$.

Relationships: *The same number of each type of atom must appear on each side*

Solve: *There are two oxygen atoms on the left side but only one on the right side. Oxygen is unbalanced.*
Adding a 2 in front of water (2 H_2O*) balances the oxygen atoms.*
Adding a 2 in front of magnesium (2 Mg) balances the Mg atoms.

Answer: *The balanced chemical equation is* $2\ Mg + O_2 \rightarrow 2\ MgO$

The meaning of a balanced chemical equation

The photosynthesis reaction

The basis of most life on earth is the *photosynthesis* reaction. This reaction takes place inside plants, and combines water from the soil with carbon dioxide from the air to make glucose (sugar) and oxygen. Plants use the sugar for energy and to build starch and cellulose, the larger molecules plants are made of. Animals breathe the oxygen and eat the plants to get the sugar and other plant molecules.

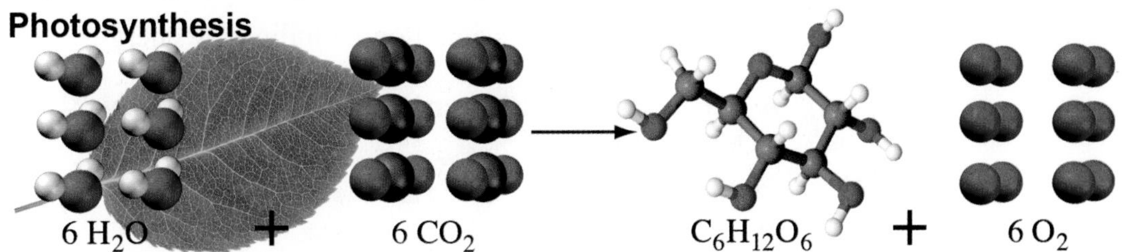

Reading the reaction

The equation tells us 6 molecules of water and six molecules of carbon dioxide react to produce one molecule of glucose and 6 molecules of oxygen. Molecules are so tiny though; how does the equation relate to the macroscopic (real) world?

The coefficients represent moles

The answer is that the coefficients also represent *moles* of each substance. Remember, a mole always has the same number of atoms. We can interpret the coefficients to mean *6 moles* of water combine with *6 moles* of carbon dioxide to produce *1 mole* of glucose and *6 moles* of oxygen. This is the real power of the balanced equation. It tells us precisely how many moles of each substance are used (reactants) or produced (products). A mole is a measurable quantity, often amounting to grams or even kilograms.

Converting from moles to grams

We use the formula mass to convert from moles to grams. For example, the formula mass for water is approximately 18 g/mole. That means 6 moles of water have a mass of $6 \times 18.16 = 108$ grams. The balanced reaction says 108 grams of water combine with 264 grams of carbon dioxide to produce 180 grams of glucose and 192 grams of water.

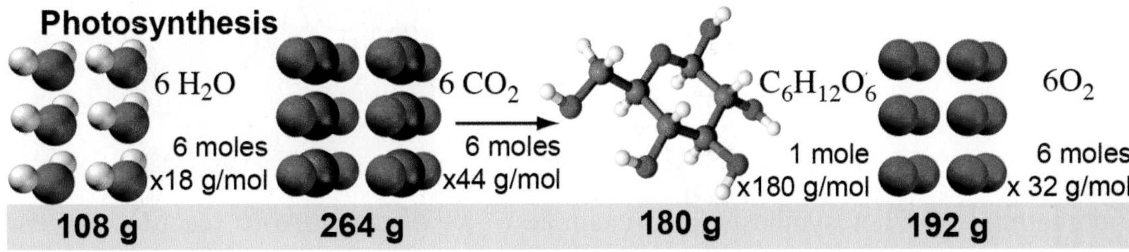

The conservation of mass

Note that both products and reactants add up to the same total mass. Because the same atoms appear on both the reactant and product side of the balanced equation, the total mass on each side must be the same. This is called the law of **conservation of mass**.

Chemistry terms

conservation of mass - in any chemical reaction, the total mass remains the same: the total mass of reactants equals the total mass of products.

Endothermic reactions

Energy is also part of reactions

If you mix water and carbon dioxide together, you do not produce sugar. There is carbon dioxide in the atmosphere, and there is plenty of water, yet, we do not see puddles spontaneously turn into sugar! That tells us there is something missing from the chemical equation. You may have guessed that the "something" is *energy*. Energy is both the cause of chemical change and also what limits change from happening.

Including energy in photosynthesis

Let's consider the photosynthesis equation including the energy.

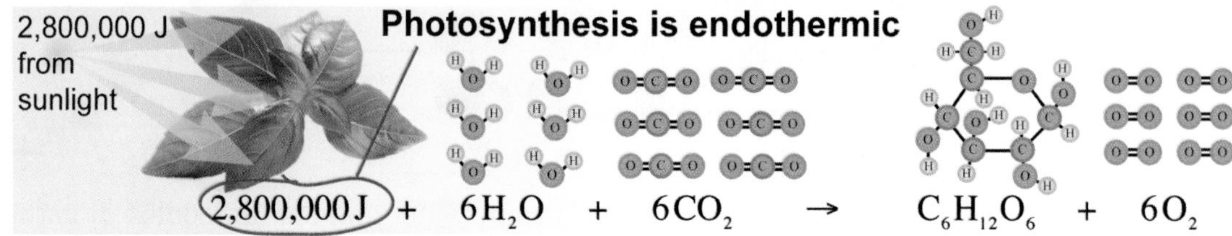

The photosynthesis reaction takes a lot of energy

This equation tells us that energy is one of the reactants. Energy needs to be supplied in order for this reaction to proceed. In fact, a lot of energy is required. One mole of glucose has a mass of 180 grams. It takes an *input* of 2,800,000 joules to make 180 grams of glucose from water and carbon dioxide. This is enough energy to lift a small car 300 meters in the air, almost 1,000 feet up. All that energy just to make a small cupful of sugar from water and carbon dioxide. Puddles of water do not become sugar because there is normally not enough energy to do it.

Energy for photosynthesis comes from the sun

Glucose is made in plants, so where does the energy come from? You already know the answer: the sun. Plants use the energy in sunlight to assemble glucose molecules from carbon dioxide and water in the environment. At night, plants do not make glucose because there is no adequate source of energy.

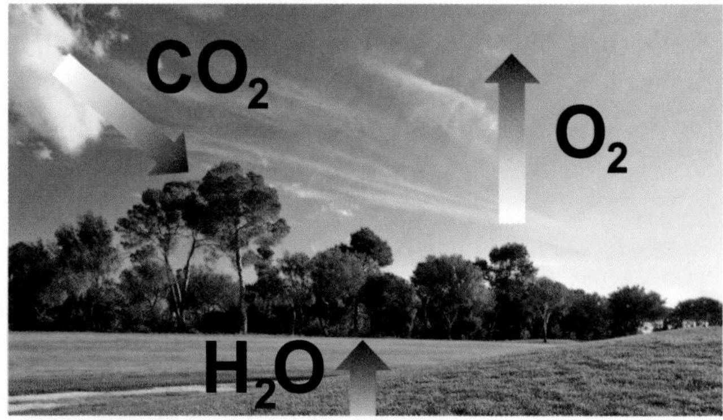

Endothermic reactions

Photosynthesis is an example of an **endothermic** reaction. An endothermic reaction absorbs energy in going from reactants to products. Energy needs to be supplied or the reaction does not work. Many of the reactions in your body are also endothermic reactions, as are many of the reactions that occur when food is cooked.

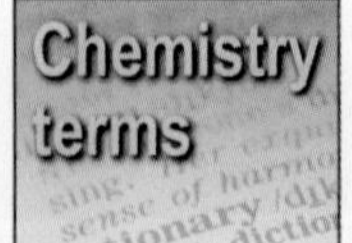

endothermic reaction - a chemical reaction that requires an input of energy to go from reactants to products. Endothermic reactions absorb energy.

Exothermic reactions

The cellular respiration reaction

The opposite reaction of photosynthesis is *cellular respiration*. In respiration, glucose is combined with oxygen to yield water and carbon dioxide. The chemical reaction is written exactly backwards, *including the energy.*

Respiration is exothermic

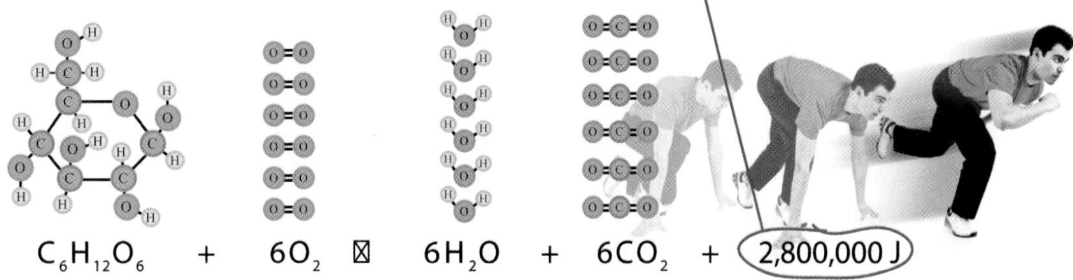

Exothermic reactions give off energy

In cellular respiration, oxygen is used and energy is released. A reaction that releases energy is called **exothermic**. The exothermic reaction that combines glucose with oxygen is fundamentally how animals, including you, derive energy from food. In every cell in your body, glucose is being broken down into carbon dioxide and water. The exact reactions are more complex, but the result is exactly what is written above. Every 1 mole (180 g) of glucose your body burns releases 2,800,000 joules of energy. The reaction requires 6 moles of oxygen to proceed, which must be obtained from the air you breathe.

Glucose stores chemical energy like a molecular "battery"

In many ways, the glucose molecule is a natural *battery*. Plants "charge" the battery with energy from sunlight by combining carbon dioxide and water. Animals (and plants) "use" the energy by breaking the molecule back down into carbon dioxide and water again. The energy is stored in the chemical bonds between atoms.

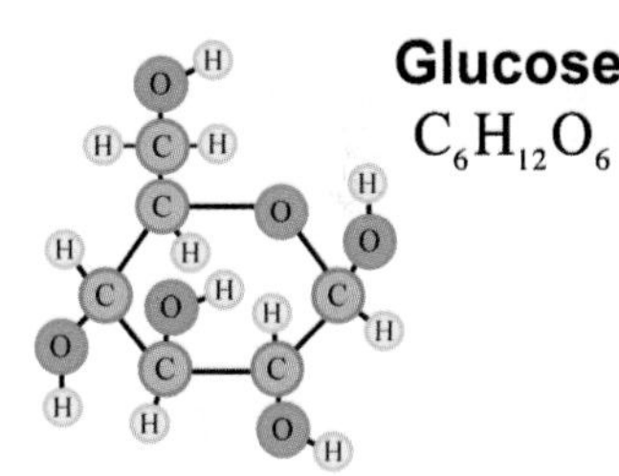

Enthalpy

The energy potential of a chemical reaction has a special name: **enthalpy**. Enthalpy is measured in joules per mole (J/mol) or kilojoules per mole (kJ/mol). If the enthalpy is positive, the reaction increases the total energy of the constituent atoms. Reactions with positive enthalpy are always endothermic, because they need to absorb energy to proceed. Photosynthesis has an enthalpy of +2,800 kJ/mol. Reactions with negative enthalpy release energy. Exothermic reactions always have negative enthalpy because they release energy. Respiration has an enthalpy of -2,800 kJ/mol.

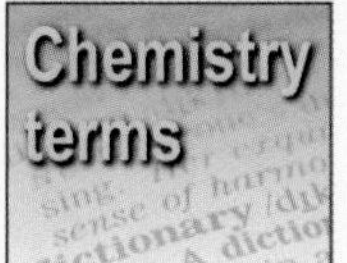

exothermic reaction - a chemical reaction that releases energy in going from reactants to products. Endothermic reactions have negative enthalpy.

enthalpy - the energy potential of a chemical reaction at standard temperature and pressure - measured in joules per mole (J/mol) or kilojoules per mole (kJ/mol).

System and surroundings for reactions

The big picture of energy flow

Consider the two reactions of photosynthesis and respiration together. The energy it takes to make the glucose molecule is exactly the same as the energy released when the molecule is broken down again. As mentioned earlier the First Law of Thermodynamics says that energy cannot be created or destroyed, it is always conserved. Another way to to say this is that the total energy of the universe is a constant. Energy that is stored in one place has to be taken away from some other place.

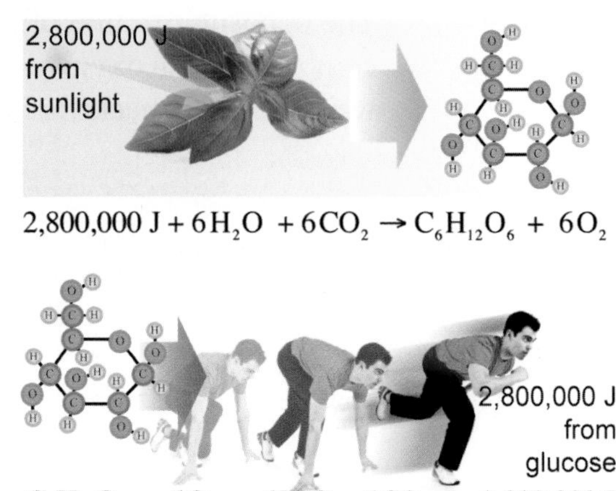

$$2{,}800{,}000\ \text{J} + 6H_2O + 6CO_2 \rightarrow C_6H_{12}O_6 + 6O_2$$

$$C_6H_{12}O_6 + 6O_2 \rightarrow 6H_2O + 6CO_2 + 2{,}800{,}000\ \text{J}$$

The system includes what we are interested in

It's impossible to do experiments using the whole universe, so we choose a small portion to be our **system**. We choose the system to include just the specific matter and energy we are interested in investigating. Very often the system will be the products and reactants of the reaction.

The surroundings are everything else

The rest of the universe all around the system is called the **surroundings**.

1. The total mass and energy of the system + surroundings is constant;
2. Any matter or energy *gained* by the system must be *lost* by the surroundings;
3. Any matter or energy *lost* by the system must be *gained* by the surroundings.

The "system" is often the reactants and products of a reaction

The surroundings are everything else, except the products and reactants

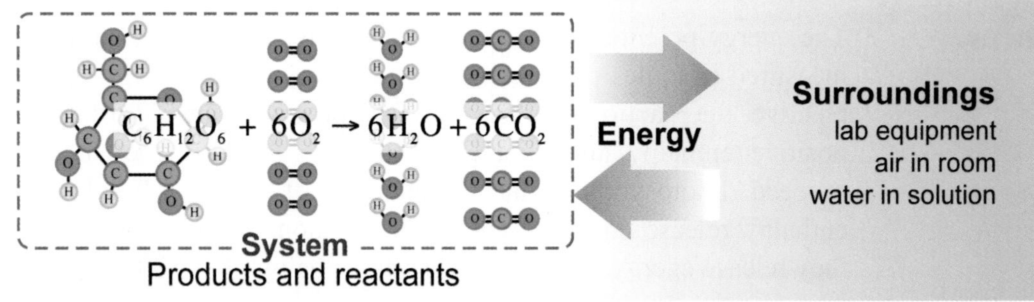

Chemistry terms

system - an interrelated group of matter and energy that we choose to investigate.
surroundings - everything outside the "system".

Activation energy

Why don't exothermic reactions occur spontaneously?

The respiration reaction releases energy by breaking glucose down into carbon dioxide and water. The reaction is not endothermic so why doesn't a tablespoon of glucose turn into CO_2 and H_2O by itself? The answer has to do with the microscopic details of how chemical reactions occur. The hydrogen, carbon and oxygen atoms in $C_6H_{12}O_6$ must first be separated before they can regroup to form the new molecules of CO_2 and H_2O. Separating the atoms means breaking the *chemical bonds* that hold the glucose molecule together.

It takes energy to break chemical bonds

Think of two chemically bonded atoms like two magnets that are stuck together. It takes energy to pull the magnets apart. Similarly, it takes energy to break the chemical bonds and separate the atoms in glucose.

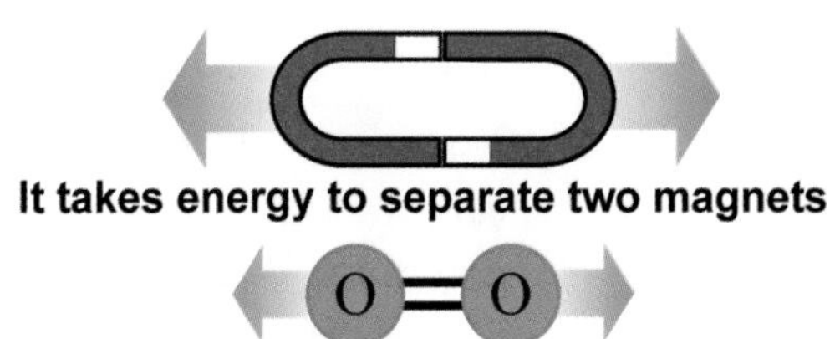

Activation energy

In an exothermic reaction, the atoms arranged as compounds of the products have lower energy than they had when arranged as compounds of the reactants. However, to get to the lower energy, the molecule needs enough input energy to get "over the hump". This energy is called **activation energy**. Activation energy is the energy needed to start a reaction and break chemical bonds in the reactants. Without enough activation energy, a reaction will not happen, even if it releases energy when it does happen. That is also why a flammable material, like gasoline, does not burn without a spark or flame. The spark supplies the activation energy to start the reaction.

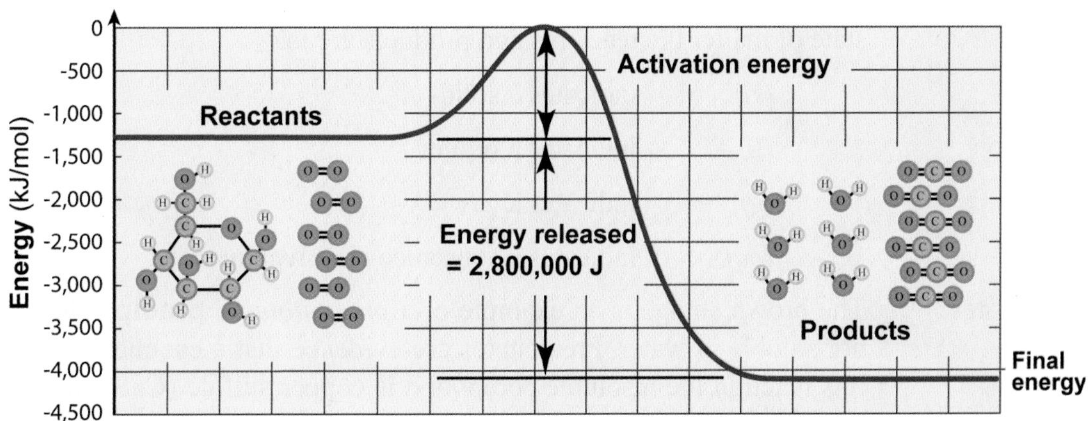

Chemistry terms	**activation energy** - the energy necessary to break bonds in the reactants so a chemical reaction can occur. Units are joules per mole (J/mol) or kilojoules per mole (kJ/mol).

4.3 Chemical Reactions in the Lab

Many reactions occur in solution

Many reactions occur between chemicals dissolved in water. This includes all of the reactions that sustain life in your body. Why are solutions so important in reactions?

The answer is not complicated. Reactions can only occur if *molecules can move around and touch each other.* For example, sodium sulfide (NaS) is a white powdery mineral salt. Copper sulfate ($CuSO_4$) is a bright blue powder. If you mix the two chemicals as powder, nothing really happens.

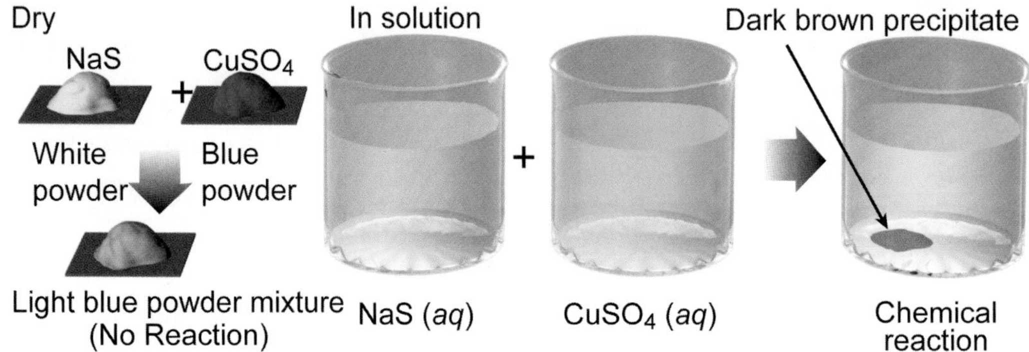

The result is quite different if you dissolve the powdered chemicals in water and then mix the solutions. The copper sulfate is a bright blue solution. The sodium sulfide is a colorless solution. When you mix the two, a deep brown, almost black sludge forms and settles at the bottom. If the quantities are right, the blue color may almost disappear.

Aqueous solutions

A solution with water as the solvent is called an **aqueous** solution. Aqueous solutions are so important that chemists consider being dissolved in water to be almost a fourth state of matter! In writing reactions we use the symbols *(s)*, *(l)*, *(g)* and *(aq)* to show what state of matter the reactants and products are in.

(s)	indicates a solid
(l)	indicates a liquid
(g)	indicates a gas
(aq)	indicates a substance dissolved in water

Precipitates

The brown sludge is an example of a *precipitate.* A **precipitate** is a compound that is not soluble in water. Precipitates are evidence that a chemical reaction has occurred. In this reaction the insoluble compound is copper sulfide (CuS).Aqueous means dissolved in water.

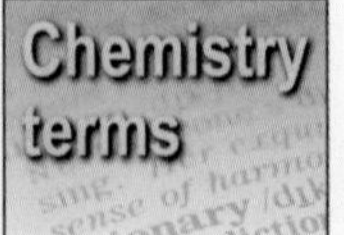

aqueous - dissolved in water- indicated by *(aq)*.

precipitate - an insoluble compound that forms in a chemical reaction in aqueous solution.

Oxidation and reduction

Salts are ionic compounds that dissolve to form ions

In the next few chapters, you will see how electrons in atoms are directly responsible for chemical properties. For example, when sugar is dissolved in water, the molecules become separated, but remain whole. This is not true for *salts* such as copper sulfate ($CuSO_4$). **Salts** are ionic compounds which form *ions* when dissolved in water. If you could look at the molecular level, you would find that in aqueous solution, copper sulfate dissociates into copper ions (Cu^{2+}) and sulfate ions (SO_4^{2-}). This happens because each single copper atom transfers two electrons to the sulfate ion.

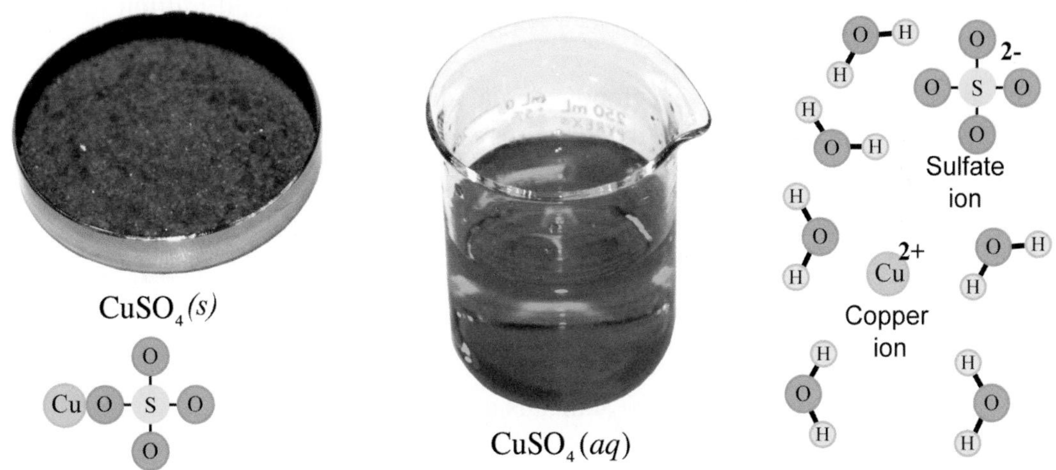

Oxidation is losing electron(s)

Pure zinc is a relatively soft, grey metal. Copper is also a soft metal but is reddish-gold in color. When zinc metal is placed in a solution of copper sulfate, the zinc slowly disappears! The zinc disappears because it is being oxidized. In chemistry, **oxidation** means losing an electron and becoming more positive. The zinc metal atoms become Zn^{2+} ions and go into the solution.

Reduction is gaining electron(s)

Zinc has to give up two electrons to become Zn^{2+} so where do these electrons go? They go to the copper ions. Cu^{2+} ions accept the two electrons to become copper (Cu) atoms. Copper atoms are no longer soluble in water, so they drop out of the solution as copper metal. The reddish-brown sludge that appears in the test tube is pure copper metal. In chemistry, **reduction** is the process of gaining electrons and becoming more negative. The copper ion (Cu^{2+}) accepts two electrons and loses its positive charge to become copper metal.

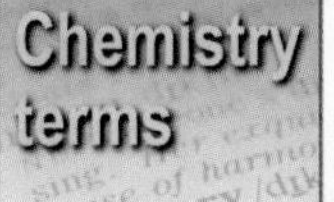

salt - an ionic compound that dissolves in water to produce ions.

oxidation - a chemical reaction that increases the charge of an atom or ion by giving up electrons.

reduction - a chemical reaction that decreases the charge of an atom or ion by accepting electrons.

Acids and bases

The dissociation of water

When pure water is carefully examined, we find that on average, one out of every 550 million water molecules is *dissociated,* or separated, into a hydrogen (H^+) ion and a hydroxide (OH^-) ion.

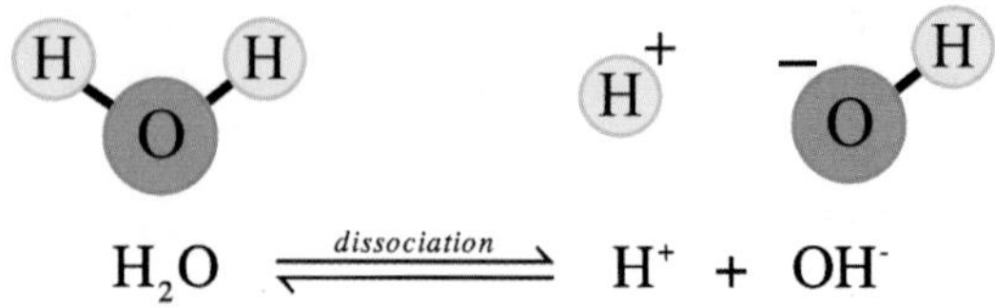

The dissociation reaction is written with the double arrow to show that the ions can re-combine to make a normal water molecule. In equilibrium, the constant dissociation and recombination leaves (on average) one out of 550 million molecules broken apart into into H^+ and OH^-.

Properties of acids

An **acid** is a compound that dissolves in water to make a solution that contains *more* H^+ ions than there are in pure water. Some properties of acids are listed below:

- Acids create the sour taste in food, like lemons;
- Acids react with metals to produce hydrogen gas (H_2);
- Acids can corrode metals and burn skin through chemical action.

Acids create the sour taste in food

When hydrochloric acid (HCl) dissolves in water it ionizes into hydrogen (H^+) and chlorine (Cl^-) ions. The fact that the H^+ ion is created is what makes HCl an acid.

Properties of bases

A **base** is a compound that dissolves in water to make a solution with more OH^- ions than there are in pure water. Some of the extra OH^- combines with H^+ to make water again, so another way to think about bases is that they reduce the concentration of H^+ ions. Some properties of bases are listed below:

- Bases create a bitter taste;
- Bases have a slippery feel, like soap;
- Bases can neutralize acids.

Common bases

Ammonia

Soap

Chemistry terms

acid - a chemical that dissolves in water to create more H^+ ions than there are in pure water.

base - a chemical that dissolves in water to create less H^+ ions than there are in pure water (or equivalently, more OH^- ions).

Acid / base reactions

Acids and bases react to form salts and water

When acids and bases are combined in aqueous solution they react to make water and salt. In fact, the chemical definition of *salt* is any compound formed from the positive ion from a base and the negative ion from an acid. For example, hydrochloric acid (HCl) is a common laboratory acid, also found in your stomach where it helps break down food. Sodium hydroxide (NaOH) is a common base. Sodium hydroxide is found in drain cleaners and industrial processes.

Acids and bases react to produce water and a salt

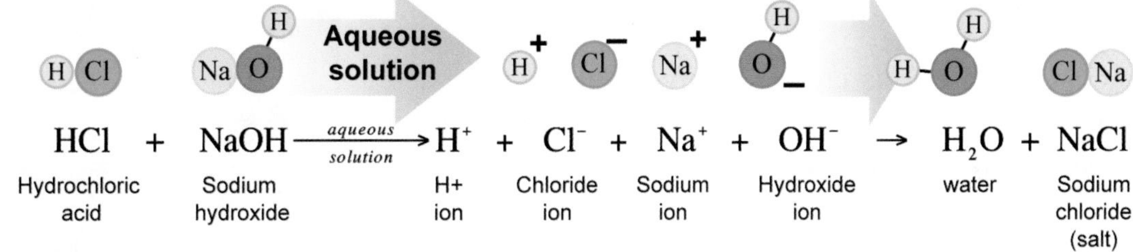

Table salt (NaCl) is formed by the positive sodium ion from sodium hydroxide (a base) and the negative chloride ion from hydrochloric acid.

Acids are important because of the H^+ ion

The H^+ ion is very important in chemistry. The main reason is that H^+ is a single bare nucleus, *with no electrons*. Earlier in the chapter we said that chemical bonds come from the interactions of electrons, in the outer part of the atom. This is true with one big exception - the H^+ ion. Since H^+ is chemically powerful, and it is *always* present in water, acids play a crucial role in chemistry. Many reactions that occur in aqueous solutions are sensitive to the concentration of H^+ ions, including nearly all the reactions in your body.

The pH scale

Because acids and bases are so important, they have a special measurement - the *pH scale*. The **pH scale** tells you whether a solution is acidic or basic. A pH less than 7 indicates an acid. An acid has higher concentration of H+ than pure water. A pH greater than 7 indicates a base. A base has H+ concentration less than pure water.

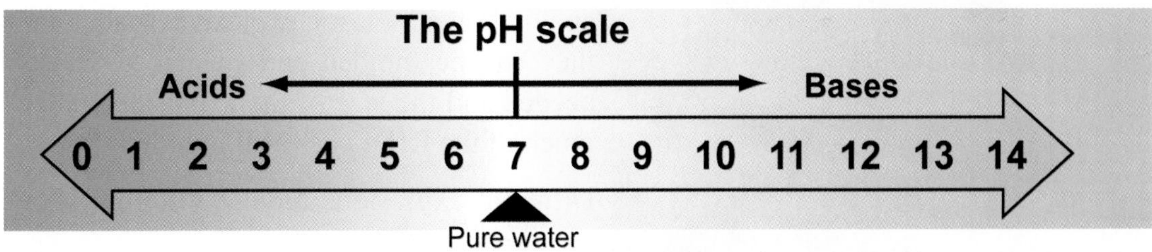

Chemistry terms

pH scale - a measurement of the H+ ion concentration that tells whether a solution is acid or base. Pure water has a pH of 7. Solutions with pH <7 are acidic. Solutions with pH >7 are basic.

Chemistry in the Movies

When you go to the movies, chances are there are chemical reactions involved in making the special effects you see. Computers are used to generate some effects, but chemicals are still used when a realistic or more dramatic effect is needed. Action movies, and movies that involve magic, seem to have the most chemical reactions.

Chemical reactions are often used to generate convincing explosions and fires. On a Hollywood movie set these chemical reactions need to be contained and easily extinguished. To accomplish this, just the right amount of chemical is added to achieve the desired effect. To carry out a chemical reaction with some precision requires skill and experience. For this reason, creating special effects is sometimes called an "art."

How are these effects created? Fires scenes are created using combustible gases such as propane. As you have learned fire is the result of a chemical reaction between two gases, the hydrocarbon made of C and H atoms and oxygen, O_2. When a spark or heat is added to this mixture it provides the energy necessary for some of the bonds to be broken between atoms. These atoms then recombine with oxygen atoms to form CO_2, H_2O and more heat, which then increases the combustion process. Sometimes other chemicals, such as lithium (Li) and strontium (Sr) are added to the flames to achieve a more reddish-colored flame. When these effects are required on a Hollywood film set, the amounts of fuel are small and contained. Fire extinguishers are handy just in case. Fire blazes can also be filmed separately and then "added in" to the film scene using computer technology.

Explosions seen in movies are created by *pyrotechnics*. The stunt crew in charge of pyrotechnics has special training and certification for using explosives. Small amounts of C-4 explosive are used and placed in areas that will provide the desired visual effect. C-4 explosives are made of approximately 90% RDX, which is a chemical composed of nitrogen, carbon and hydrogen. Plastic explosives made of C-4 are used because they can be molded and shaped to fit easily where they are needed and the size of the explosion can be controlled by how much of the RDX chemical is added.

Different explosions require different chemical mixtures, and this takes time to perfect!

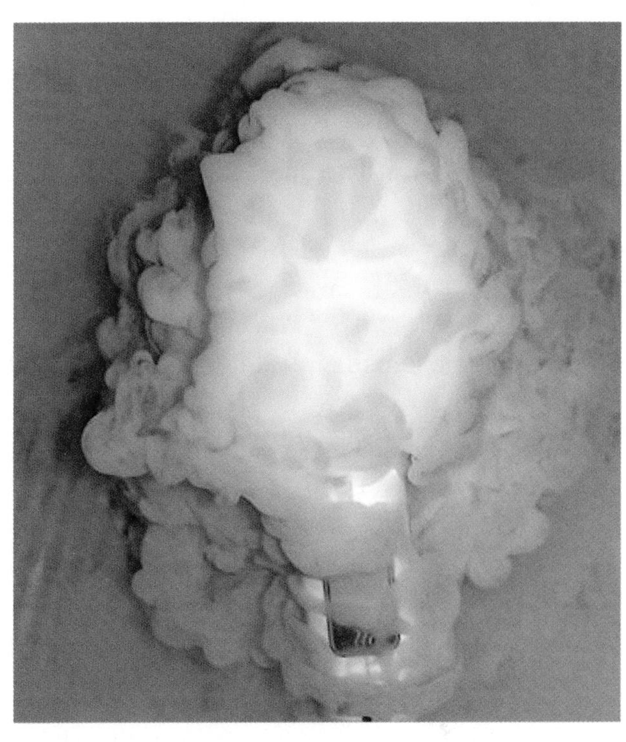

In a movie involving magic and sorcery, chemicals may be used to create an eerie smoke when entering a mysterious cave. The smoke can be white, grey or even colored. A variety of methods have been used in Hollywood to make smoke. The first method was to use dry ice or solid CO_2. Dry ice sublimes readily at room temperature giving a smoky effect. However, CO_2 gas from subliming dry ice displaces oxygen in the air. When too much dry ice is used, actors could experienced nausea and dizziness from lack of oxygen.

Liquid nitrogen (N_2) is also sometimes used for fog and smoke. Liquid nitrogen creates a "low lying" fog effect. Liquid nitrogen is pumped into a machine that contains hot water (near boiling) and this causes rapid condensation and a thick white fog. Today for the most part glycols are used. Glycols are also called "fog juice". Fog juice consists of glycol, glycerin, mineral oil and distilled water. The mixture is heated and forced into the air under pressure, creating a fog. Glycols are chemicals containing C, H and O; propylene glycol $C_3H_8O_2$ at 90% produces a thick dense fog. Glycerin has the chemical formula of $C_3H_5(OH)_3$ and, when mixed with distilled water, produces more of a haze. The mixture of these ingredients can produce a fog that is more or less dense.

Sparks flying and glowing potions are all part of the magic of chemistry. Potassium nitrate (KNO_3)and potassium perchlorate ($KClO_4$) along with metals such as steel and aluminum create a sparkling "flash" effect when burned or combusted with oxygen in the air. Glow-in-the-dark goop or slime contains hydrogen peroxide and phenyl oxalate ester (compound with C, H and O elements) and a fluorescent dye. When the hydrogen peroxide reacts with the oxalate ester causing oxidation, this releases energy that causes the electrons in the fluorescent dye to be excited. These excited fluorescent atoms release light upon returning to their ground state. The different colors are achieved by adding different dyes.

Next time you go to the movies and see a fire, explosion or even some fog, think about how chemistry has made these special effects real and believable. Special effect technicians have a very cool science job!

Chapter 4 Review.

Vocabulary:

Match each word to the sentence where it best fits.

Section 4.1

interatomic forces	intermolecular forces
irreversible	electric charge
physical change	electron
chemical change	neutral
covalent bond	proton
ionic bond	nucleus
molecule	enthalpy of formation
chemical bond	reactivity

1. ___________ comes from a competition between temperature and attractive intermolecular forces.
2. When _________________ happens the molecules of the substance do not change.
3. ____________________ exist between molecules.
4. When _________________ happens the molecules of the substance change.
5. _____________________ exist between atoms and are much stronger than intermolecular forces.
6. An _________________ change rearranges atoms to form different substances
7. The _________ of an atom contains 99.8% of its mass.
8. _____________ have negative charge and are responsible for chemical bonds
9. _____________ has a positive charge.
10. When a chemical bond between two atoms is formed by sharing a single electron it is called a _____________.
11. __________________ is a fundamental property of matter that comes in positive and negative.
12. An ____________ forms when one or more electrons is transferred from one atom to another.
13. A neutral group of atoms that are covalently bonded is called a ________________.
14. The connection between two atoms is called a __________________.
15. ______________________ is the change in energy when 1 mole of a compound is assembled from pure elements.

Section 4.2

chemical reaction	endothermic reaction
reactants	exothermic reaction
products	enthalpy
balanced equation	activation energy
coefficient	system
conservation of mass	surroundings

16. The process of chemical change is represented by a ________________.
17. The initial compounds involved in a chemical reaction are called the __________.
18. The compounds generated by a chemical reaction are called the ____________ of the reaction.
19. All chemical reactions must satisfy the property of ____________________.
20. The number that tells us how many of each molecule participate in a chemical reaction is called the __________________.
21. A chemical reaction that releases energy is called ______________________.
22. A chemical reaction that absorbs energy is called ______________________.

23. _____________ is the energy potential of a chemical reaction.

24. The energy required to start a reaction is called _______________________.

25. A chemical equaiton that consrves mass is called a ____________________.

26. A group of matter and energy that we choose to investigate is called a _________________.

27. Everything around a system is called the ____________________.

Section 4.3

aqueous solution	oxidation
precipitate	reduction
salt	acid
pH scale	base

28. A solution with water as the solvent is called an ______________________.

29. The insoluble compound that results from a chemical reaction that takes place in an aqueous solution is called a _______________.

30. A ___________ is an ionic compound that form ions when dissolved in water.

31. _____________happens when an atom loses electrons.

32. _____________ happens when an atom gains electrons.

33. When a compound dissolves in water forming an increased amount of H^+ ions it is called an __________.

34. A __________ is a compound that dissolves in water to make a solution with a large number of OH- ions.

35. The _____________ is a mesurement of the H+ ion concentration and it tells us whether a solution is an acid or a base.

Conceptual Questions

Section 4.1

36. Give examples of a reversible and an irreversible process.

37. Why are physical changes reversible?

38. What happens during a chemical change that makes them irreversible?

39. Describe the difference between interatomic and intermolecular forces.

40. Describe what happens when you take a glass of water and add one drop of blue food coloring at the top. Explain this at the molecular level.

41. An atom is electrically neutral. What does that mean?

42. Why do different elements have different chemical properties?

43. What particle plays the dominant role in chemical bonds?

44. Describe what is a covalent bond and how it is formed.

45. What is the primary reason why chemical bonds form?

46. What is the relation between energy and chemical bonds?

47. What is the difference between covalent and ionic bonds?

Section 4.2

48. Hydrogen and oxygen combine to form water. Describe the general process by which this happens.

Chapter 4 Review.

49. Describe the main components of a chemical equation.

50. Many chemical reactions release energy. Where does this energy come from?

51. Why do some reactions release more energy than others?

52. Describe the concept of mass conservation and use the chemical equation below to demonstrate how it is used.

$$4Fe + 3O_2 \rightarrow 2Fe_2O_3$$

53. An ice pack used by athletic trainers becomes cold when activated. Is this the result of an exothermic or an endothermic reaction? Explain.

54. Explain the photosynthesis reaction. Is it an exothermic or an endothermic reaction?

55. Give two simple examples of an exothermic and an endothermic chemical reaction.

56. Describe the concept of enthalpy. How is enthalpy realted to the energy of a chemical reaction?

57. Describe in your own words what is the activation energy of a chemical reaction.

58. Why do different chemical reactions have different activation energy?

59. Why is it important to specify what makes up a system when we study chemical reactions?

Section 4.3

60. What is an aqueous solution?

61. Describe which element groups in the periodic table have the highest reactivity?

62. What happens to an atom or ion when it is oxidized?

63. What happens to an atom or ion when it is reduced?

Quantitative Problems

Section 4.1

64. The melting of ice is a physical change. Separation of water into hydrogen and oxygen is a chemical change. How much energy is required to melt 10 g of ice and how much energy is required to break up 10 g of water into hydrogen and oxygen?

65. If you combine 10 g of oxygen gas and 10 g of hydrogen gas how many grams of water would you obtain?

Section 4.2

66. Acid rain contains sulfuric acid which is formed when sulful dioxide reacts with water. Write the balanced equation for this reaction.

67. Another polutant in acid rain is nitric acid which is formed when nitrogen dioxide reacts with oxygen and water. Write the balanced equation for this reaction.

68. When wood of gasoline is burned we may say that the following reaction occurs:
$C + O_2 \rightarrow CO_2$
If we burn 1 ton (1000 kg) of carbon how many kg of CO_2 is produced?

69. The burning of ethanol, the alcohol in alcoholic drinks, follows the reaction equation.
$C_2H_5OH + 3O_2 \rightarrow CO_2 + H_2O$
Is this reaction balanced?
If not balance it.

70. Balance the following chemical equations.
    a. $N_2O \rightarrow N_2 + O_2$
    b. $HBr \rightarrow H_2 + Br_2$
    c. $CH_4 + O_2 \rightarrow CO_2 + H_2O$

71. The following figures show graphical schematics of chemical equations. White the equation from the schematic.

    a. ______________________________

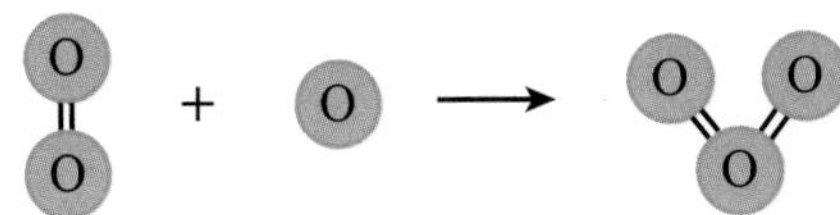

    b. ______________________________

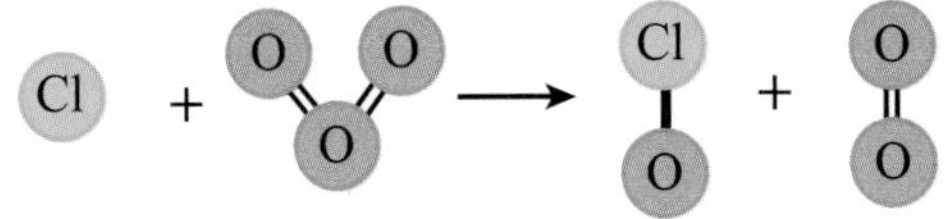

    c. ______________________________

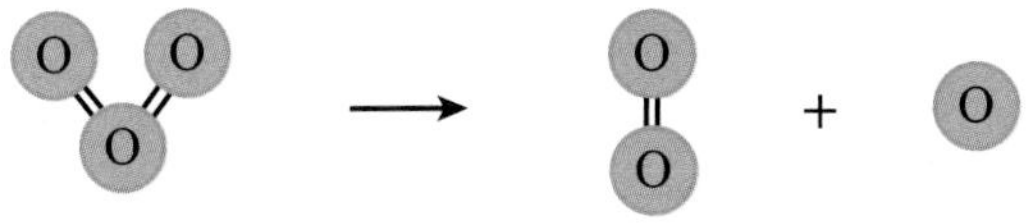

    d. ______________________________

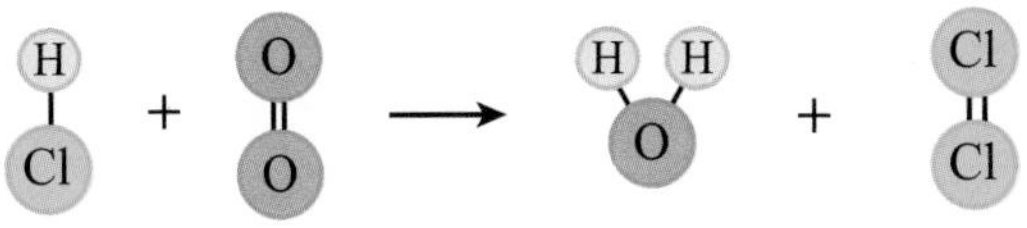

72. The following figures show graphical schematics of chemical equations that are not balanced. White the chemical equations from the schematic. and balance them.

    a. ______________________________

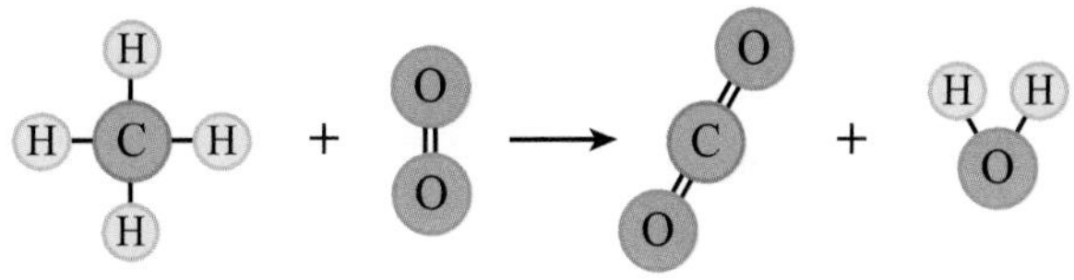

    b. ______________________________

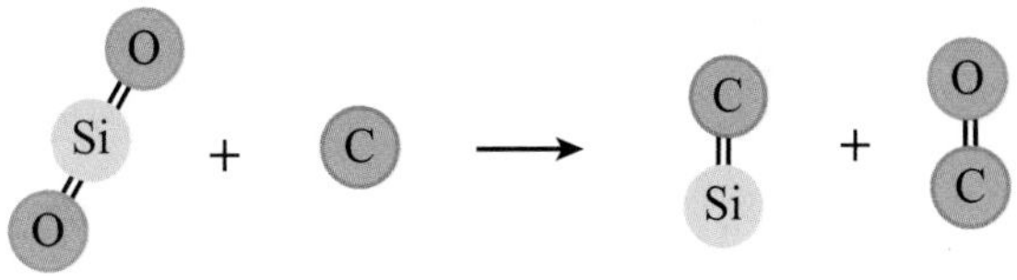

    c. ______________________________

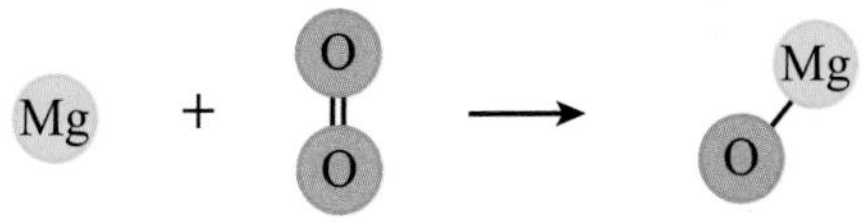

    d. ______________________________

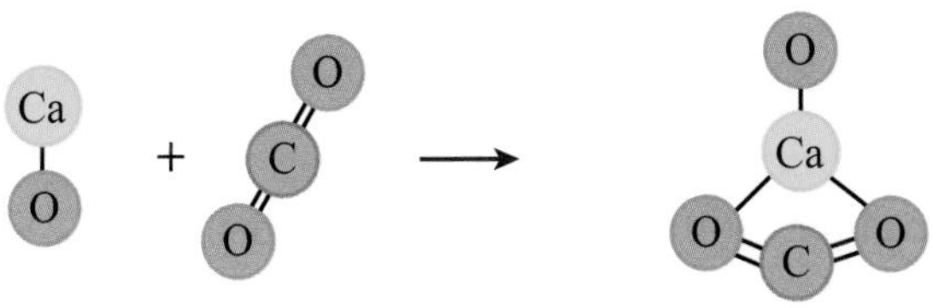

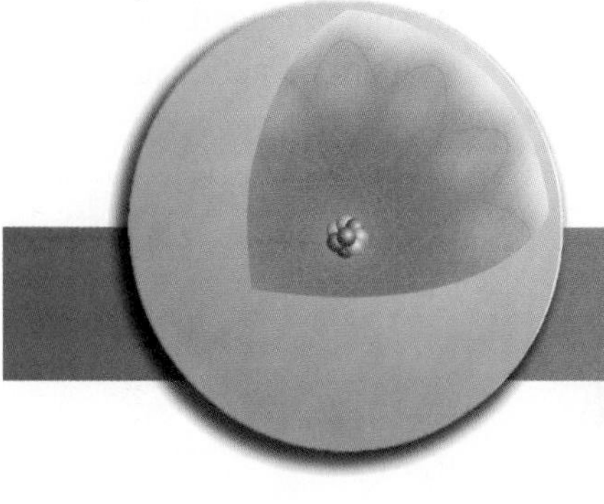

CHAPTER 5
The Structure of the Atom

What are atoms?

What are their properties?

Why are atoms important in chemistry?

What would happen if you took a piece of aluminum foil and tore it in half again and again? Then when that piece of aluminum became too small to hold, imagine that you could cut it in half, then in half again. Keep thinking of smaller and smaller pieces of aluminum. Could you keep cutting forever, or do you reach a limit, the smallest possible piece of aluminum?

The answer to this question has far reaching consequences. Either there is a smallest piece, which means the aluminum is made from tiny particles of aluminum (atoms of aluminum), or the aluminum is one smooth substance that can be cut in half forever.

Repeated Dividing of Aluminum Foil

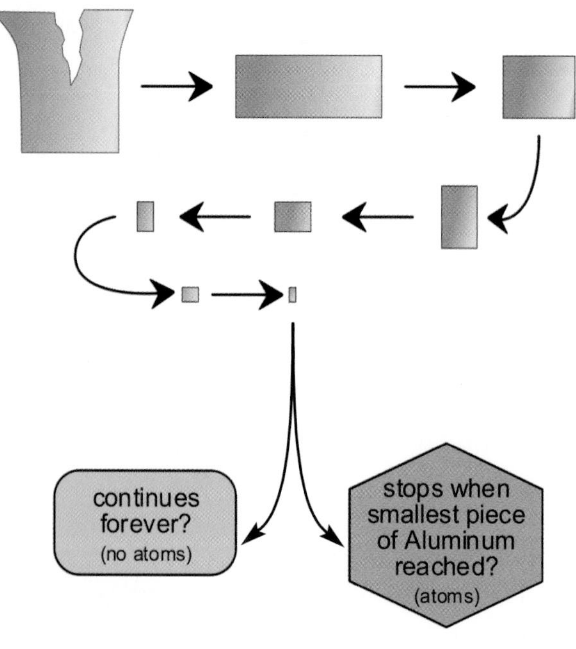

Over time, starting with the ancient Greeks, scientists struggled with this question, each building upon the work of others. Today we know that everything is made from atoms, that there **is** a smallest piece of aluminum. Our modern view of the atom is the culmination of 2,500 years of scientific thinking and experimentation. Our understanding of the properties of the atom has led to major chemical discoveries that have improved life from medicine to energy.

Light, color, atoms, electrons

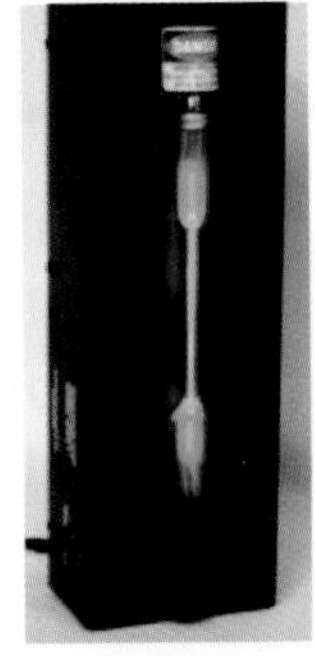
Gas discharge tube power supply

When you put high-voltage electricity through a gas, you can get light. This is how a neon sign works. In the lab we use glass tubes filled with gas in a high voltage power supply (picture at left).

Different elements have noticeably different colors. The most interesting thing happens when you observe the light given off through a special lens called *diffraction grating*. You may know that "pure white" light is really an equal mixture of all colors together. The light from a gas discharge tube is also a mixture of colors. A diffraction grating separates the light into its component colors.

To the eye, the light from nitrogen gas looks purple and helium looks pink. However, through the diffraction grating, the purple and pink are not so simple! Instead of a single color, the diffraction grating shows us that the light is really many colors. Each color shows up as a vertical line in what chemists call a "line spectra."

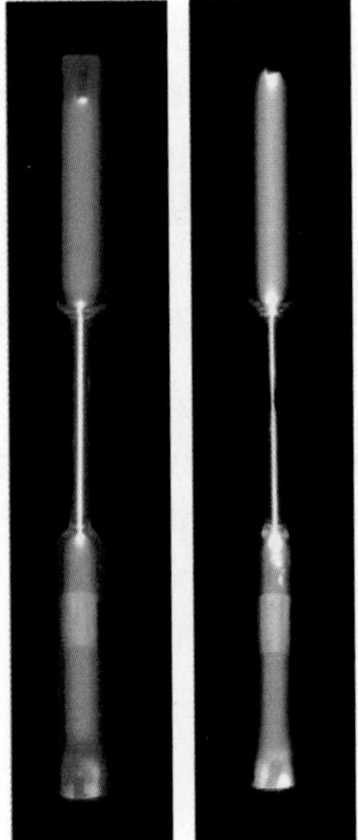
Nitrogen Helium

Notice how nitrogen and helium are different? Each element has its own unique line spectra. In fact, the unique patterns of spectral lines are often called the "finger prints" of the elements.

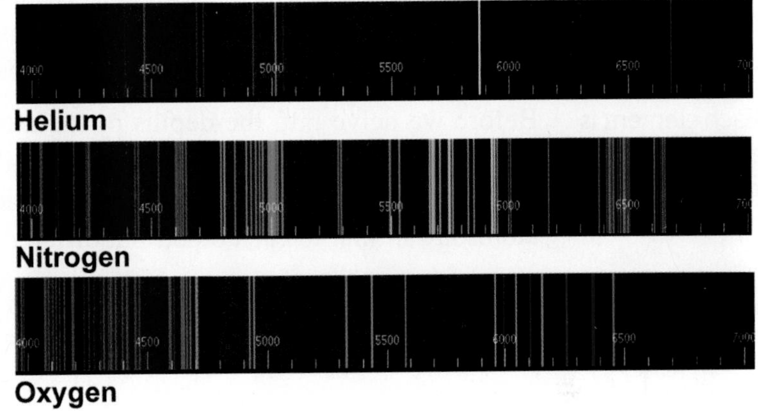

How are these line spectra different?

Do the line spectra remind you of a "bar code" used at the checkout counter of a store? It is a good analogy! Line spectra contain information about the structure of each element just as bar codes contain information about product numbers. Compare the line spectra of helium, nitrogen and oxygen. What differences do you see? What similarities?

- There are areas of the "rainbow" in which no color is emitted (black).
- Nitrogen and oxygen have more lines than helium.
- All three are different.
- Some of the lines are thicker than others.

The number of spectral lines is related to the number of electrons in each atom. Helium has 2 electrons and nitrogen has 7 electrons. Could it be that nitrogen has more spectral lines because it has more electrons? In this chapter we will learn how to categorize atoms according to their electron structure. We will organize the elements in a table called the periodic table, which is the major reference document of chemistry.

5.1 The Atom has a Structure

Elements and compounds

In the previous chapters, we learned that ordinary matter is made of atoms of the 92 naturally occurring elements. These atoms usually form compounds, such as salt (NaCl) or sodium bicarbonate ($NaHCO_3$). Compounds explain how we get millions of types of matter from 92 elements, but 92 is still a relatively large number. Are the 92 different types of atoms (i.e. the elements) made of even smaller things? The answer is: *yes*.

Atoms are made from smaller particles called protons, neutrons and electrons

How can three particles explain the universe?

This is an *extraordinary* fact! How can the incredible variety of matter in the universe come from only *three particles*? Think about making all the words in the dictionary with only 3 letters, or all the paintings in the world with only three colors. It may seem incredible, but it is true none theless. The atoms of all 92 elements (and more) are created from three basic particles: electrons, protons and neutrons. The beautiful variety of nature arises from how the three particles come together in rich and complex ways.

Each element is a unique type of atom

Before we delve into the depths of the atom, lets review what we know. There are 92 naturally occurring elements, and 20 to 30 other elements that have been created in a laboratory. Each element represents a unique type of atom. For example, all oxygen atoms are similar to each othe,r but different from carbon atoms or hydrogen atoms.

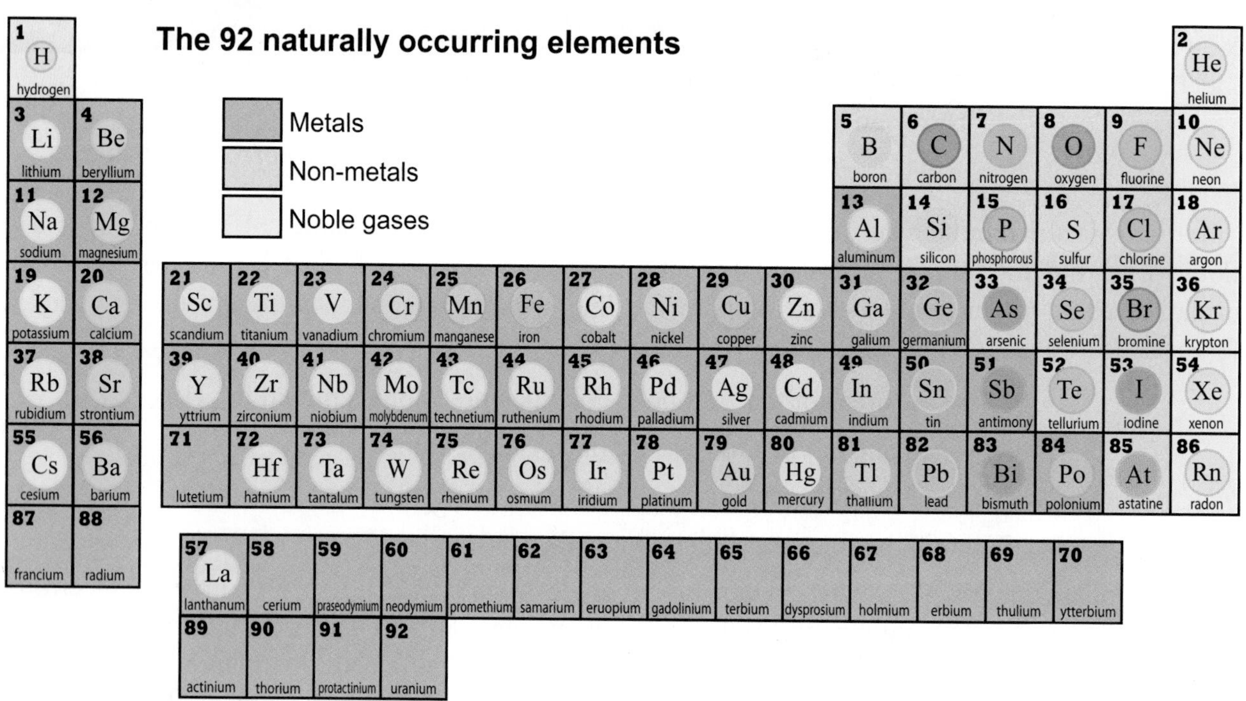

The beginning of atomic theory

Democritus (460 - 370 BC) - the beginning of atomic theory

Democritus, an ancient Greek philosopher who lived from 460BC - 370 BC, proposed the idea that you can't divide something in half forever. Eventually, he argued, you must reach a smallest indivisible part. He called this smallest piece of matter an *atom*. Democritus correctly deduced the existence of atoms, but he could go no further in discovering any of their properties. For the next 2000 years *atomism* was an interesting idea but there was no good scientific evidence to support its truth or falsehood.

John Dalton (1768 - 1828) - first "modern" atomic theory

In 1808, John Dalton, an English school teacher, put together many ideas in his four postulates of the atomic theory. Daltons four postulates were a brilliant synthesis based on what little evidence there was at that time. They remain true today.

1. All elements are made of tiny indivisible particles called atoms.
2. All atoms of the same element are alike but different from atoms of every other element.
3. Chemical reactions rearrange atoms but do not create, destroy or convert atoms from one element to another.
4. Compounds are made from combining atoms in simple whole number ratios

Cathode rays

In the mid 1800's it was discovered that high voltage made a "glow" in a sealed glass tube from which most of the air had been pumped out. In 1870 William Crookes invented a tube in which virtually ALL of the gas was removed. Now, the glow inside the tube disappeared, but *the glass at one end of the tube was glowing*. Some kind of invisible "ray" was being emitted from the cathode end of the tube and striking the glass at the other end. These rays were called *cathode rays*, and a great debate occurred over the nature of them. Were they another kind of light? Where they a stream of particles?.

high voltage electricity creating cathode rays inside a Crookes tube

J.J. Thomson (1856 - 1940) - the discovery of the electron

Thomson

In 1897, J.J. Thomson was able to definitively resolve the debate. His experiments showed that cathode rays were deflected toward a positively charged plate and away from a negatively charged plate. Thomson deduced that cathode rays must be negative. He found they could be deflected by magnetic fields. No ordinary ray of light would behave this way. He tried using different metals, or starting with the tube filled with different gasses. None of those factors mattered. He always got the same cathode rays and they always were deflected in the same way

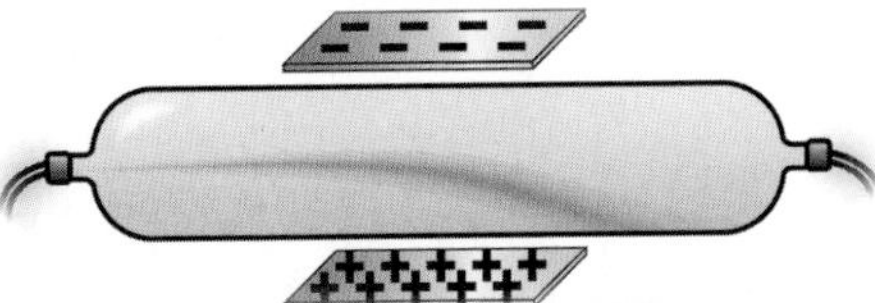
cathode rays deflected away from negative plate and toward positive plate

The discovery of the nucleus

Cathode rays are electrons

Thomson's discovery stunned the scientific world. Cathode rays were a stream of particles 2000 times lighter than the lightest known atom (H)! How there be a particle smaller than an atom? Because Thomson always got the same cathode rays regardless of whatever metals he used for the electrodes in his Crookes tube, he named the new particle an *electron* and proposed that electrons were inside ALL atoms.

The atom must have a structure inside

If electrons were inside atoms, then atoms could not be the most elementary particles of matter. Furthermore, electrons were *negative* and atoms were *neutral* so there had to also be something *positive* within atoms to cancel the charge of the electrons. The search was on to discover the *structure* inside the atom.

Ernest Rutherford - the gold foil experiment

In 1910, Ernest Rutherford designed and carried out the crucial experiments that provided the answer. Marie and Pierre Curie had discovered that uranium was radioactive and released energetic alpha particles at high velocity. Alpha particles were positively charged and had a mass about 8000 times that of an electron. Rutherford devised an experiment to shoot alpha particles through a thin gold foil and observe what happend as they collided with gold atoms. He expected most of the alpha particles to be deflected a little as they crashed through gold atoms.

Rutherford

The Gold Foil Experiment

α particle

very thin sheet of gold

Most α particles passed straight through, but a few were deflected to the side and once in a while one would bounce straight backwards!

Rutherford's discovery of the atomic nucleus

Rutherford's results were completely unexpected. Most of the alpha particles went straight through the gold foil with no deflection at all. A few were deflected slightly off their original path and about 1 of every 20,000 reversed direction, bouncing back from the foil! Rutherford determined that atoms have nearly all their mass concentrated in a very tiny, very dense, positively charged nucleus and this was his reasoning:

1. The deflected alpha particles were repelled by something with the same charge, so the nucleus must be positively charged.
2. Very few alpha particles were deflected so it must rare for one to come close to a nucleus. This meant the nucleus had to be tiny, about 1/10,000 the diameter of the atom.
3. The alpha particles were travelling at such high velocity that only something with significant mass could deflect them.

The interior of an atom

The big ideas

The three most important ideas in this chapter are:

1. Atoms are made of neutrons, protons and electrons, with the number of protons and electrons always being equal.
2. The number of protons determines the element; all atoms of hydrogen have one proton, all atoms of helium have 2, lithium has three and so on;
3. Most of the properties of atoms are determined by their electrons. Atoms interact with each other via their electrons.

Of course, there are many interesting details! Electrons are quirky particles and they behave in very strange ways, but that is what makes chemistry so interesting.

The nucleus

Neutrons and protons make up the *nucleus*. The **nucleus** is at the center of the atom. There are no electrons in the nucleus, only protons and neutrons. The nucleus is extremely small, even compared to an atom. If the atom were the size of your classroom, the nucleus would be the size of a single grain of sand in the center of the room!

	Mass (kg)	Charge (C)	*Relative mass*	*Relative charge*
Proton	1.673×10^{-27}	$+1.602 \times 10^{-19}$	1,835	+1
Neutron	1.675×10^{-27}	0	1,837	0
Electron	9.109×10^{-31}	-1.602×10^{-19}	1	-1

Mass of protons, neutrons and electrons

Look at the masses of the three particles (we will discuss charge on the next few pages). The masses are very small. The mass of a single electron in kilograms has 30 zeroes between the decimal point and the first non-zero digit. More important are the *relative* masses. The proton and neutron are much more massive than the electron. On the scale of atoms, the protons and neutrons have essentially all the mass.

Mass and the nucleus

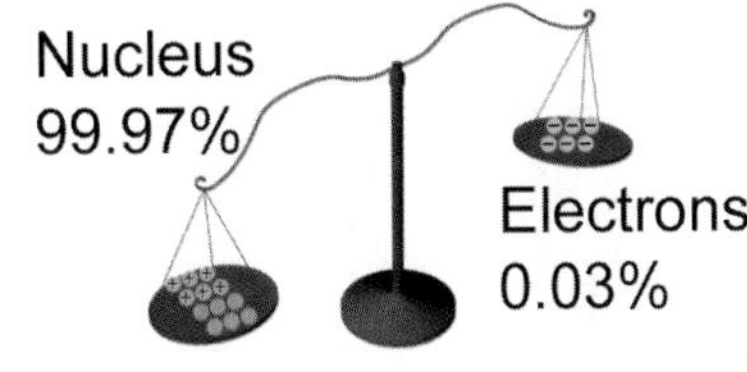

Most of an atom's mass is concentrated in the nucleus. The number of electrons and protons is the same, but electrons contribute very little mass. For example, a carbon atom has six protons, six electrons and six neutrons. 99.97 percent of the carbon atom's mass is in the nucleus and only 0.03 percent is electrons.

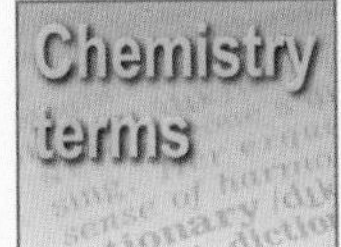

nucleus - the tiny core of an atom that contains all the protons and neutrons. The nucleus is extraordinarily small - about 1/10,000 the diameter of the atom.

Atomic number and atomic mass

The atomic number

The **atomic number** of each element is the number of protons in its nucleus. All atoms of the same element have the same number of protons in the nucleus. For example, every atom of helium has two protons in its nucleus. Every atom of carbon has six protons in its nucleus. The periodic table arranges the elements in increasing atomic number. Atomic number one is hydrogen with one proton. Atomic number 92 is uranium with 92 protons.

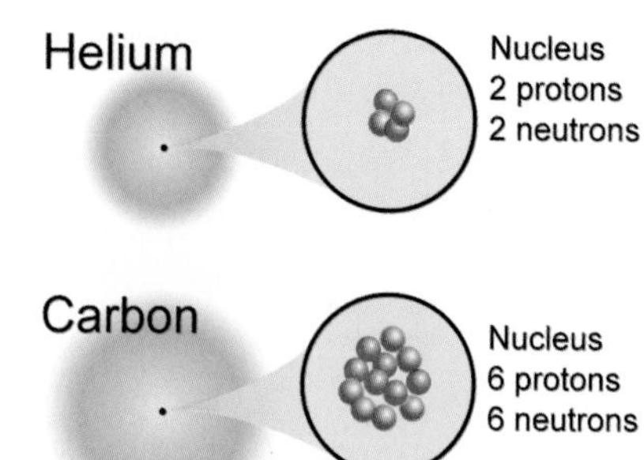

Neutrons act like "glue"

All protons have positive electric charge (more about that later). That means they repel each other. So how does the nucleus stay together? The answer is neutrons. Think about neutrons as "glue" particles that help the nucleus stay together. Every element heavier than helium has at least as many neutrons as protons in its nucleus.

Isotopes

All atoms of the same element have the same number of protons in the nucleus. However, they do not necessarily have the same number of neutrons. Three different *isotopes* of carbon are found on earth. **Isotopes** are atoms of the same element that have different numbers of neutrons in the nucleus.

Isotopes of Carbon

CARBON-12	CARBON-13	CARBON-14
$^{12}_{6}C$	$^{13}_{6}C$	$^{14}_{6}C$

The mass number

The most common isotope of carbon is carbon-12, written ^{12}C. A nucleus of carbon-12 contains 6 protons (making it carbon) and 6 neutrons. The superscript "12" before the symbol "C" tells you the *mass number* of the nucleus. The **mass number** is the total number of protons plus neutrons. The two other isotopes of carbon are carbon-13 (^{13}C) and carbon-14 (^{14}C). These isotopes are carbon because they have six protons in the nucleus but ^{13}C has seven neutrons and ^{14}C has eight neutrons.

Atomic mass unit (amu)

Because the mass of a proton is tiny by normal standards, scientists use **atomic mass units** (amu). One amu is 1.661×10^{-27} kg, slightly less than the mass of a proton.

atomic number - the number of protons in the nucleus - unique to each element.

isotopes - atoms or elements that have the same number of protons in the nucleus, but different number of neutrons.

mass number - the number of protons plus neutrons in the nucleus.

atomic mass unit (amu) - One amu is 1.661×10^{-27} kg.

Average atomic mass

The average atomic mass may not be a whole number

Consider the element lithium. Lithium has two isotopes that are found in nature. Lithium-6 has three protons and three neutrons. Lithium-7 has three protons and four neutrons. The periodic table lists the atomic mass of lithium as 6.94. How is that possible? Do lithium atoms have 0.94 neutrons?

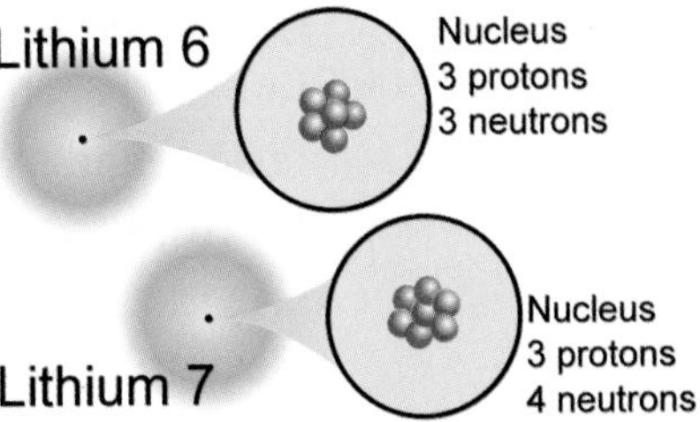

Elements in nature contain a mix of isotopes

It is not possible to split a proton or a neutron in ordinary matter. Lithium atoms have either 6 or 7 *whole* neutrons. The reason the atomic mass if 6.94 is that on average, 94 out of 100 atoms of lithium are ^{7}Li and 6 out of 100 atoms are ^{6}Li. The *average* atomic mass is 6.94 because of the mixture of isotopes. No lithium atom has a mass of 6.94 amu.

7 7 7 7 7 7 7 7 7 7 7 6 7 7 7 7 7 7 7 7 7 7 7 7 7
7 7 6 7 7 7 7 7 7 7 7 7 7 7 6 7 7 7 7 7 7 7 7 7 7
7 7 7 7 7 7 7 7 7 7 7 7 7 7 7 6 7 7 7 7 7 7 7 7 7
7 7 7 7 7 7 7 6 7 7 7 7 7 7 7 7 7 7 7 6 7 7 7 7 7

$$avg = \frac{(94 \times 7 + 6 \times 6)}{100}$$
$$= 6.94$$

Radioactivity

Not all isotopes exist in nature. For example, suppose scientists create a nucleus with 3 protons and 5 neutrons. This would have an atomic number of 3, making it lithium. The mass number would be 8. Lithium-8 is unstable and quickly decays into two atoms of helium instead! When an atomic nucleus decays or gives off energy, the process is called **radioactivity**. It means that the nucleus undergoes a spontaneous change, **decay**, often turning one element into a different element.

Lithium-8 is radioactive and decays into other elements

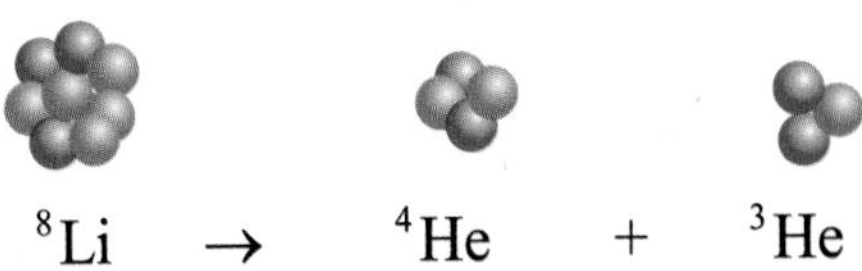

$^{8}\text{Li} \rightarrow {}^{4}\text{He} + {}^{3}\text{He}$

How many neutrons are in the nucleus of neon-21 (^{21}Ne)?

Asked: *number of neutrons*

Given: *mass number (21) and element (neon)*

Relationships: *mass number is protons plus neutrons, number of protons = atomic number*

Solve: *From the periodic table we find the atomic number of neon is 10. So, there are 10 protons in the nucleus: 21 - 10 = 11 neutrons*

Answer: *There are 11 neutrons in a nucleus of neon-21.*

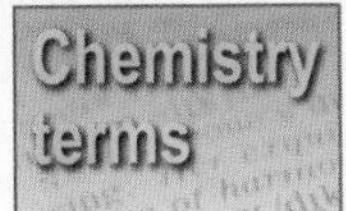

radioactivity - a process by which the nucleus of an atom spontaneously changes itself by emitting particles or energy.

decay - the process during which a nucleus undergoes spontaneous change.

The electron cloud

The electron cloud

Compared to protons and neutrons, an electron is much lighter. This has the effect of making the electron much faster and wider-ranging than either protons or neutrons. Because electrons are so fast and light, scientists call the region outside the nucleus the electron "cloud." Think about a swarm of bees buzzing in a "cloud" around a beehive. It is not easy to precisely locate any one bee, but you can easily see that, on average, the bees are confined to a cloud of a certain size around the hive. On average, electrons are confined to a similar cloud around the nucleus.

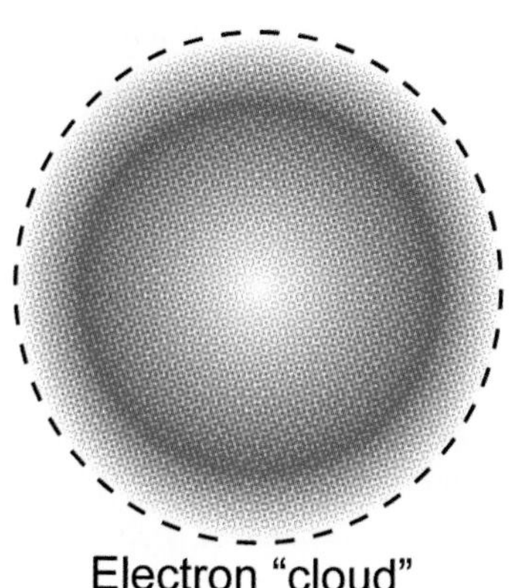

Electron "cloud"

Drawing electrons

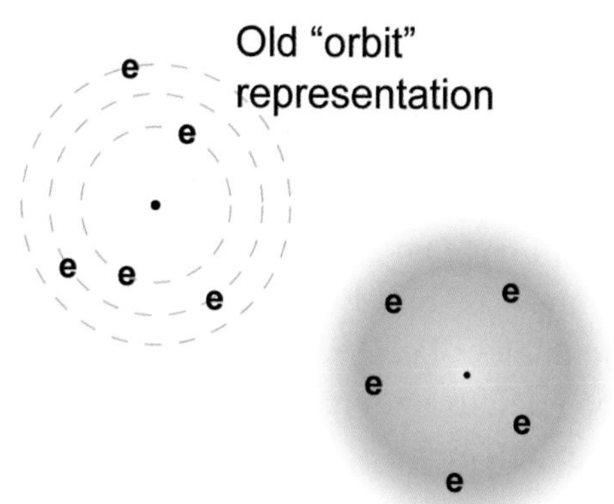

Electron cloud representation

Electrons are not really particles in the same way a dust particle is a particle. Although electrons have a definite mass, *they do not have a definite size.* The matter in an electron spreads out over a relatively large volume within an atom. In early drawings of atoms, scientists represent electrons like tiny planets orbiting the nucleus of an atom. Today we draw atoms with a tiny hard nucleus surrounded by the wispy electron cloud.

Electrons determine the size of atoms

The "size" of an atom is really the size of its electron cloud. When we talk about the size of atoms, what we really mean is how close atoms get to each other. Unless the atoms are chemically bonded together, the electron cloud of one atom does not normally overlap the electron cloud of another.

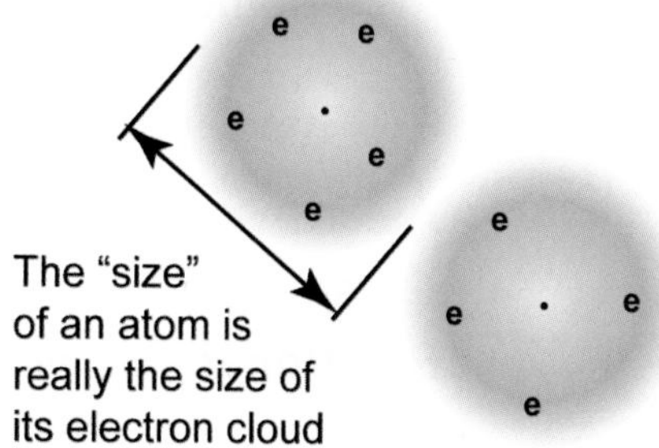

Except for mass, virtually every property of atoms is determined by electrons, including size and chemical bonding

The electrons determine virtually everything about how one atom interacts with another atom. For this reason, most of chemistry will be concerned with electrons and their unusual organization within atoms. The nucleus is buried deep inside, and contributes mass, but not much else.

Electric charge

Comparing charge and mass

Charge is a difficult idea to grasp so let's start by comparing mass and gravity with electric charge. Mass is a fundamental property of matter. Any two objects that have mass attract each other through a force called gravity. One way to think about it is that mass is the property that determines an object's response to the force of gravity. The more mass there is, the stronger the force of gravity.

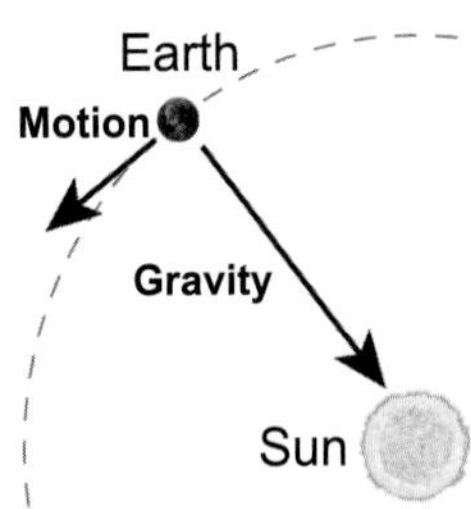

Positive and negative

Electric charge is another fundamental property of matter. However, unlike mass, there are two kinds of charge: positive and negative. Like gravity, there is a force that acts on electric charge called the electromagnetic force. Unlike gravity, however, the electromagnetic force can attract *or repel.*

The charge of the three particles

The electric charge on a proton is what scientists define to be *positive*. The charge on an electron is defined as *negative*. Neutrons are neutral and have zero charge. That means two protons repel each other, and so do two electrons. Protons and electrons attract each other. Neutrons feel no electromagnetic forces from either protons or electrons.

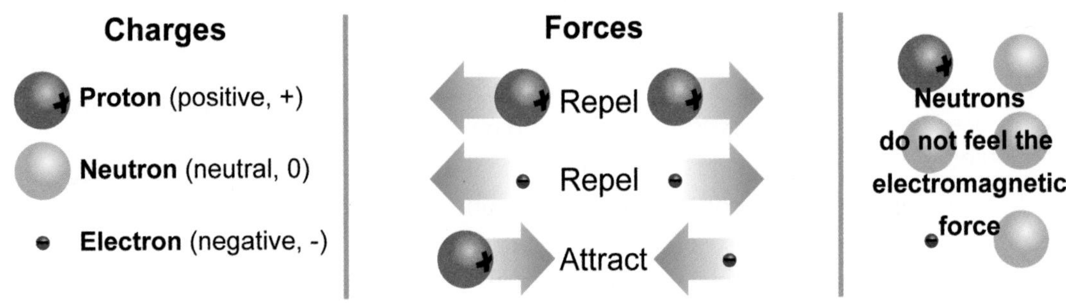

A complete atom has a charge of zero because the charge of the proton is exactly equal but opposite to the charge of the electron

Complete atoms have zero net charge

The charge on the electron and proton are *exactly* equal and opposite. If you put a proton and an electron together, the total effective charge is zero. For this reason, the charge on a complete atom is always zero because a complete atom contains the same number of protons as electrons. For example, the positive charge from six protons in a carbon nucleus is exactly cancelled by the six electrons in the electron cloud.

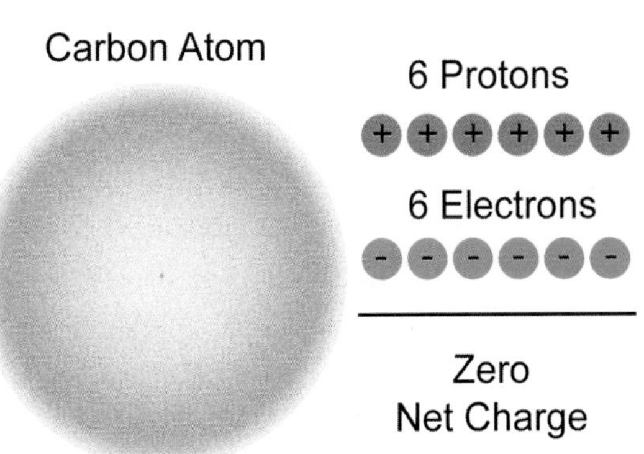

Forces in the atom

Why an atom stays together

The attractive electromagnetic force between protons in the nucleus and electrons is what holds an atom together. The electromagnetic force between electrons is also what creates chemical bonds between atoms, as we shall see. In fact, almost all of the chemistry we learn is driven by the electromagnetic force.

Forces in the nucleus

It does not affect chemistry directly, but there is a force in the nucleus which is even stronger than the electromagnetic force. It is called the *strong nuclear force* and it attracts protons to protons, neutrons to neutrons, and protons and neutrons to each other. If there are enough neutrons in the nucleus, the attractive strong nuclear force is able to overcome the repulsive electromagnetic force between protons. This is the reason most elements have equal numbers or slightly more neutrons than protons in the nucleus.

The **strong nuclear force** attracts protons to protons, neutrons to neutrons and protons to neutrons

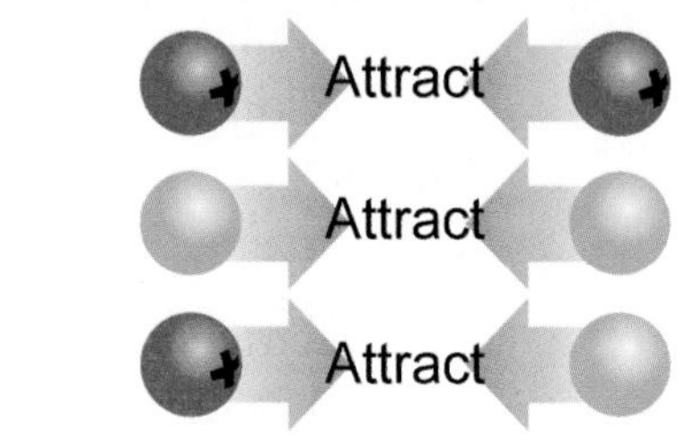

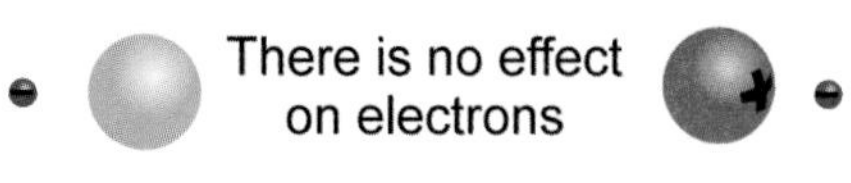

Electrons are both attracted and repelled

Let's think about the electron cloud in an atom like carbon. Electrons have energy, so they cannot just "fall in" to the nucleus, but must be constantly in motion. Each of carbon's six electrons is attracted to the nucleus, but repelled by all the other electrons! The combination of energy with attraction and repulsion is one reason behind the peculiar behavior of electrons in an atom.

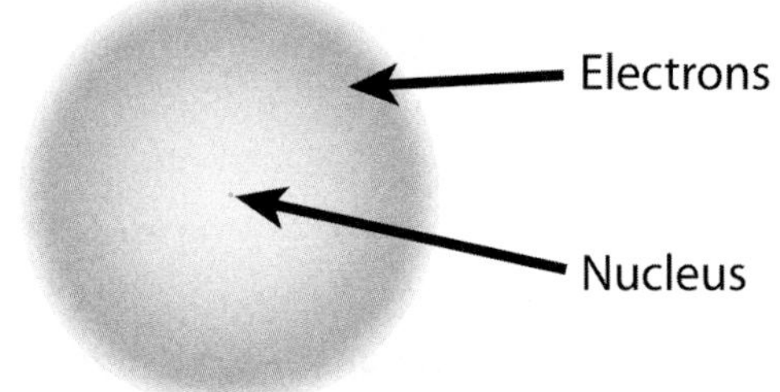

Electrons are responsible for the chemical bonds between atoms. When a water molecule forms, the oxygen atom shares electrons with two hydrogen atoms. Each shared electron is a chemical bond. The molecule has its "bent" shape because eight of the ten electrons in a water molecule repel each other in four pairs. Two pairs are shared with hydrogen atoms and two pairs are not shared.

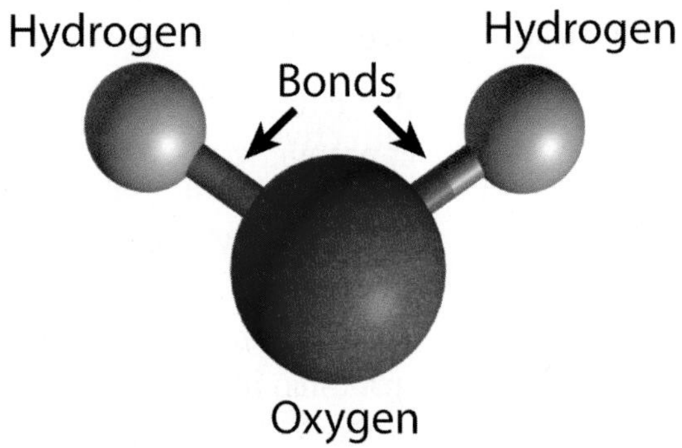

Ions

All matter contains charge

The fact that everything is made from atoms and atoms are made from electrically charged particles means that everything, even you, are a big bundle of electric charge. This fact becomes more obvious in the winter when you scuff your feet across a carpet and get shocked when touching a metal door knob. Static electricity and lightning are examples of how we are surrounded by, and made of, electrically charged particles.

Electric charges in the atoms of our bodies can cause sparks.

Neutral matter has equal positive and negative charge

If the number of electrons and protons in an atom are not equal, the atom will have an overall charge. Let's consider sodium as an example. Sodium has an atomic number of 11, so that means all sodium atoms have 11 protons. Each proton has a positive charge of +1. If these same atoms have 11 electrons, each with a -1 charge, then the -11 charge from the electrons exactly cancels out the +11 charge from the protons, and the atom is neutral.

A Neutral Sodium Atom

Na

The protons and electrons cancel each other out.

11 protons: ⊕⊕⊕⊕⊕⊕⊕⊕⊕⊕⊕
11 electrons: ⊖⊖⊖⊖⊖⊖⊖⊖⊖⊖⊖

Ions

However, if there were only 10 electrons, then the total charge from the electrons would be -10 while the charge from the protons would be +11. That means there is one proton which is not being cancelled out by an electron. This gives the atom an overall charge of +1. Charged atoms are called **ions**. Whenever the number of protons and electrons are not equal, an overall positive or negative charge will occur, and an ion will be formed. Ions can be single atoms or small molecules with an overall charge.

A Positive Sodium Ion

Na^{1+}

One proton is not neutralized by an electron, making this a +1 charged atom.

11 protons: ⊕⊕⊕⊕⊕⊕⊕⊕⊕⊕⊕
10 electrons: ⊖⊖⊖⊖⊖⊖⊖⊖⊖⊖

A Negative Oxygen Ion

O^{2-}

Two electrons are not neutralized by protons, making this a -2 charged atom.

8 protons: ⊕⊕⊕⊕⊕⊕⊕⊕
10 electrons: ⊖⊖⊖⊖⊖⊖⊖⊖⊖⊖

Ionic compounds

Positive and negative ions attract each other, just as protons and electrons do. The ionic compounds we introduced in Chapter 4 are examples. Sodium ions (Na^+) have one less electron and are attracted to chloride ions (Cl^-) which have one extra electron. The overall compound, sodium chloride (NaCl), is electrically neutral because there are equal numbers of sodium and chloride ions.

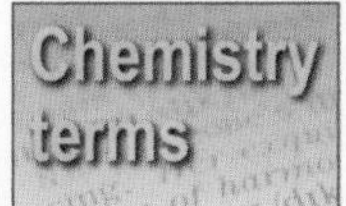

ion - an atom or small molecule with an overall positive or negative charge due to an imbalance of protons and electrons.

5.2 The Quantum Atom

Why is it that that the noble gases do not react?

How do the chemical properties of the elements arise from the structure of atoms? We have seen that electrons are the components in atoms that are near the "surface" and interact with other atoms. Why is it that the noble gases (helium, neon, argon, and krypton) do not react with other elements and among themselves? Note that these elements have 2, 10, 18 and 36 electrons respectively.

Form negative ions	Form no ions	Form positive ions
9 F 18.998 fluorine	2 He 4.0028 helium	3 Li 6.941 lithium
17 Cl 35.453 chlorine	10 Ne 20.180 neon	11 Na 22.990 sodium
35 Br 79.904 bromine	18 Ar 39.948 argon	19 K 39.098 potassium

Elements just before or just after the noble gasses are VERY reactive

The elements hydrogen, fluorine, chlorine and bromine have 1, 9, 17 and 35 electrons. So, they have one less electron than their respective noble gas group. These elements tend to form negative ions; they *accept* electrons from other atoms.

The elements with one more electron than the noble gas group are lithium, sodium, potassium and rubidium with 3, 11, 19 and 37 electrons respectively. These elements tend to form positive ions; they *donate* electrons to other atoms.

Some findings of quantum theory

The basic understanding of the electronic structure in the atom started around 1920 with a new theory called **quantum theory**. Championed by Neils Bohr, quantum theory provided a completely new way to look at things on the scale of atoms. The quantum theory finally gave us a way to understand why and how the elements make the bonds that they do. The quantum theory makes the following statements about matter and energy on the scale of atoms, and particularly about electrons in atoms. Don't worry if these statements seem quite strange. Quantum theory is accurate but not intuitive.

1. On the scale of atoms, a particle of matter such as an electron is not "solid" but is "smeared out" into a *wave* over a region of space;
2. When electrons are confined in an atom, their wave properties force them into a *pattern* that minimizes their energy;
3. Each unique "place" in the pattern is called a **quantum state** and can hold one single electron;
4. Electrons are always found to be in one quantum state or another, and are not found in-between states;
5. No two electrons can be in the same quantum state at the same time.

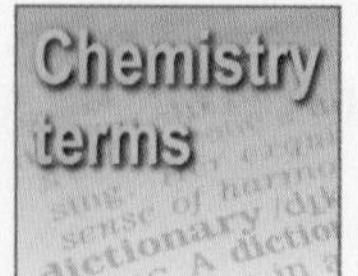

quantum theory - a theory of physics and chemistry that accurately describes the universe on very small scales, such as the inside of an atom.

quantum state - a specific combination of values of variables such as energy and position that is allowed by quantum theory.

Waves and particles

Our intuition is often wrong in the quantum world

The most basic idea of quantum theory is that our intuitive notion of a "particle" cannot be applied to the tiny world of the atom. To most of us, a particle is a tiny speck of matter that has a definite size, mass, and position, like a tiny ball. A "ball-like" particle can be either here or there but cannot be in two places at the same time. The quantum theory tells us *this intuition is wrong when things get as small as an atom*. In the quantum world, a particle is not like a tiny ball at all. Instead, the mass, size, and even the location is spread out into a wave.

Intuitive "particle"
mass and position are at a definite point

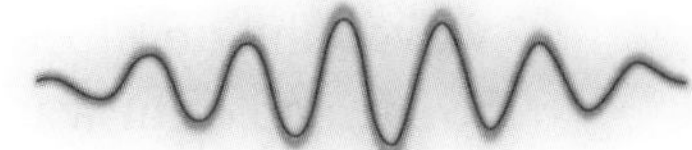

Wave "particle"
spreads the mass and position out over space

A wave is a travelling oscillation

To understand what this means we must first study the properties of waves. A wave is an organized flow of energy that *oscillates*, or goes up and down, rhythmically in time. It might be electricity that oscillates, or water, or air pressure. An oscillation of electricity might be a light wave. An oscillation of water is what you see at the beach. An oscillation of air pressure is a sound wave. The common property of all waves is a repeating oscillation of *something*.

Frequency

The oscillations of a wave have *frequency* and *wavelength*. The **frequency** is the number of times per second any point on the wave goes back and forth. A light wave has a very high frequency, 10^{12} times per second or more! A water wave might oscillate one or two times per second.

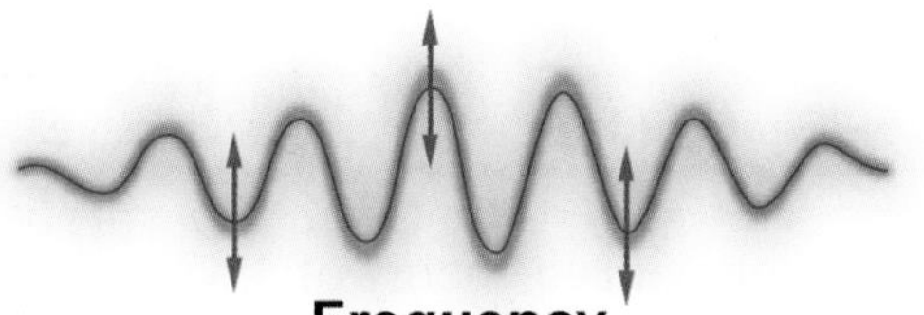

Frequency
The rate at which any point oscillates

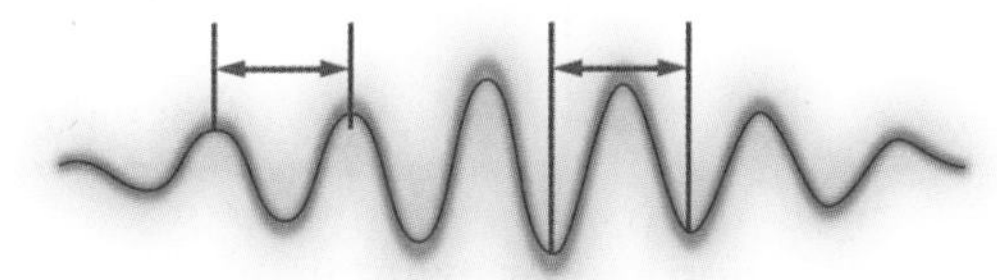

Wavelength
The separation of successive peaks (or valleys

Wavelength

When a wave moves through space, the successive peaks (or valleys) are separated by a distance called the **wavelength**. The wavelength is the same from one peak to the next and like the frequency, is a characteristic of a particular wave. The wavelength of light waves is very small, 10^{-8} meters or so.

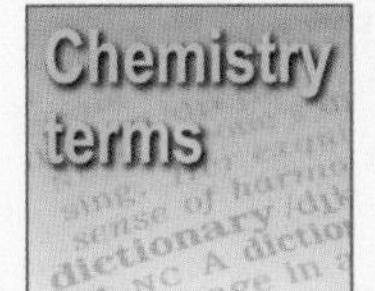

frequency - the rate at which an oscillation repeats - one hertz (Hz) is a frequency of 1 oscillation per second

wavelength - the distance (separation) between any two successive peaks (or valleys) of a wave.

Planck's constant

Wave energy increases with frequency

Waves in water carry energy, as anyone who has seen a storm crash on a beach can vividly remember. The same is true of light waves and particle-waves such as electrons. You might intuitively think the faster something oscillates, the more energy it has, and you would be right. The energy carried by a wave is proportional to how fast it oscillates, or its *frequency*. High frequency means faster oscillation, and more energy.

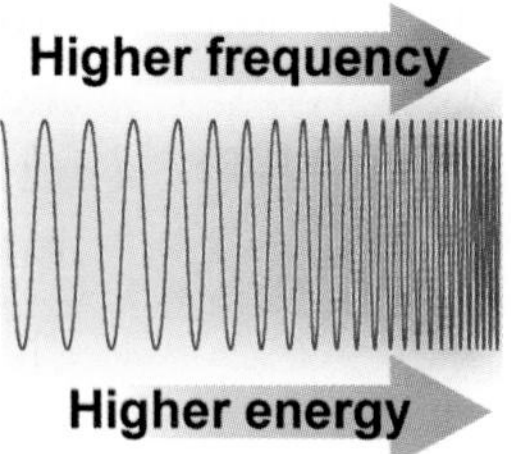

A photon is the smallest quantity of light energy

Before the quantum theory, the ideas of "particle" and "wave" were distinct. Light was a wave, and an electron was a particle. Today we know that an electron is also a *matter wave*, and light waves come in tiny bundles of energy called *photons*. A **photon** is like a particle in that it has a definite energy and moves with a certain speed and direction. However, a photon has no mass, just pure energy. You don't see light as a stream of photons for the same reason you don't see individual atoms. A small 3 watt flashlight beam emits 10^{19} photons per second!

Planck's constant

The scale at which the "granular" nature of matter and energy becomes evident is determined by *Planck's constant*. **Planck's constant** has the symbol, *h*, and a value of 6.626×10^{-34} joule-seconds (J s). The energy and wavelength of both electrons and photons are calculated from Planck's constant.

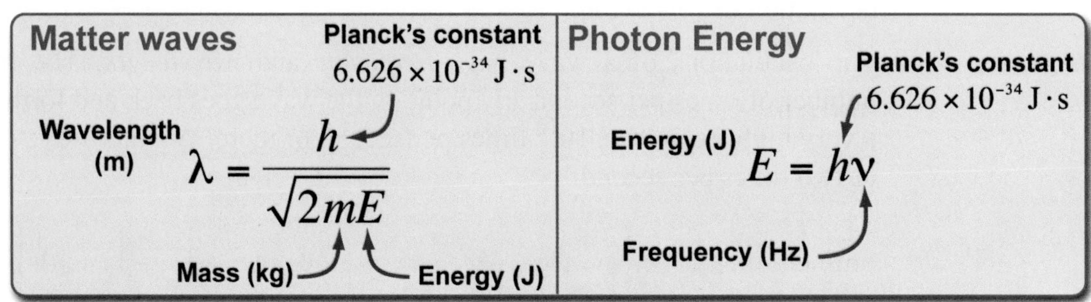

The connection between atoms, light, and electrons

For example, your eye "sees" a photon with an energy of 3.82×10^{-19} joules as green light. Suppose an electron has the same energy. The wavelength of the electron comes out to be 7.94×10^{-10} meters. *This is the same size as an atom!* There is an intimate relationship between atoms, electrons, and light that comes from the fact that visible light contains the same range of energy as electrons whose wavelengths are the size of atoms.

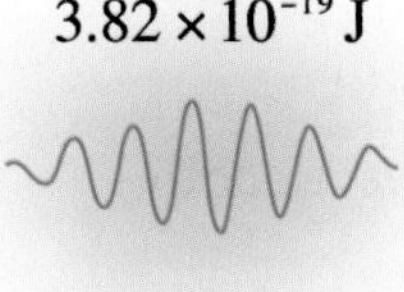

Energy of a green photon

Chemistry terms

Planck's constant - defines the scale of energy at which quantum effects must be considered. It is equal to h = 6.626×10^{-34} joule-seconds (J·s).

photon - the smallest possible quantity (or quanta) of light. Light exists in discrete bundles of energy called photons.

Electrons in the quantum atom

The wavelength of an electron in an atom

Most of the properties of the elements are caused by what happens to the wavelength of an electron when that electron is bound up inside an atom. To get a sense for how this might be, consider a ball in a box. The ball bounces off the walls of the box. At any instant there is one ball at one particular place in the box.

Waves versus balls

Like a ball, when a wave hits a wall it reflects. However, waves are very different from balls. Waves are extended in space, the wave and its reflection are in the box at the same time. One ball plus one ball is two balls, but one wave plus another wave could add up to *zero*. If one wave is "up" when the other is "down" then they cancel each other out! When averaged over time, the only waves that survive in a box are the ones whose wavelength fits the size of the box. All the others interfere with their own reflections and average out to zero.

The sum of two waves can be zero (no wave)

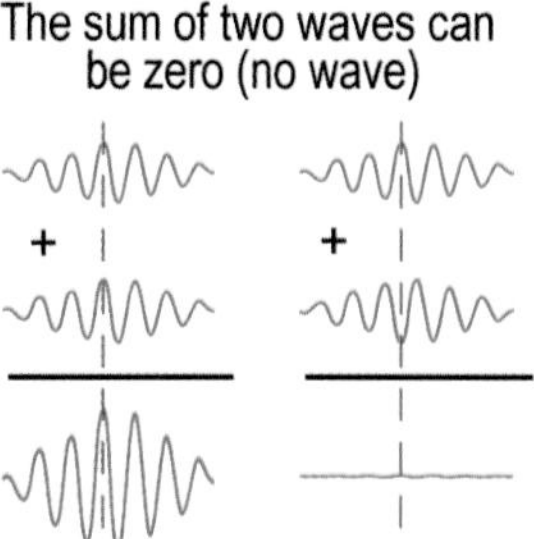

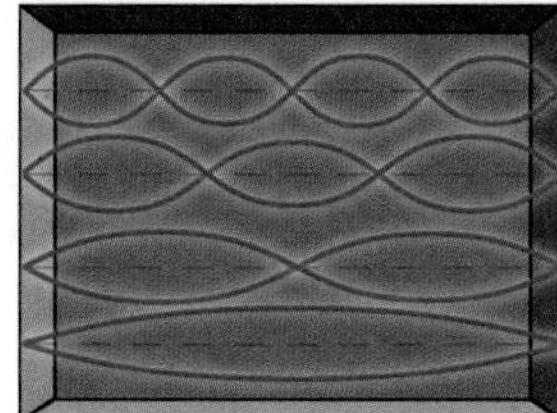

Waves that exactly "fit" in the box with their own reflections

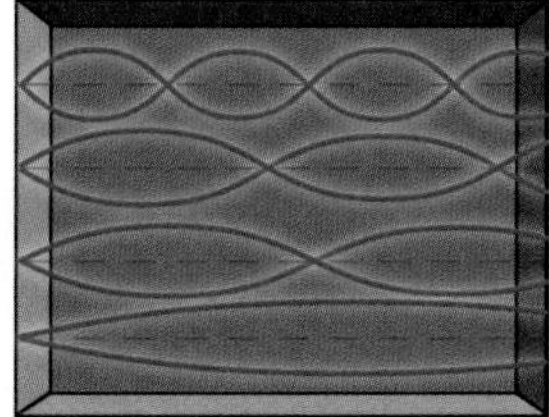

Waves that do NOT exactly "fit" in the box with their own reflections

The atom is a "box" for electrons

Now replace the "box" with an atom and let the "wave' be an electron. Electrons are "boxed" inside atoms by the attraction from the positive nucleus. The "walls" of the box are "soft" in that the electron never hits a hard surface, but needs more and more energy to get farther away from the nucleus. Given a limited amount of energy, an electron is confined to be within a certain distance from the nucleus and this confinement acts just like a "soft-walled" box.

The key idea of the quantum atom

Here is the key idea of the quantum atom. *The wavelength of the electron must be a multiple of the "size" of the atom.* If it is not, then the electron wave cancels with its own reflections over time. The diagram shows three "allowed" electron waves, and two that are "not allowed". To be stable inside an atom an electron must have one of the allowed wavelengths that exactly fits the size of the atom The elements differ because the "size" of the atom depends on the strength of attraction from the nucleus, which depends on the atomic number.

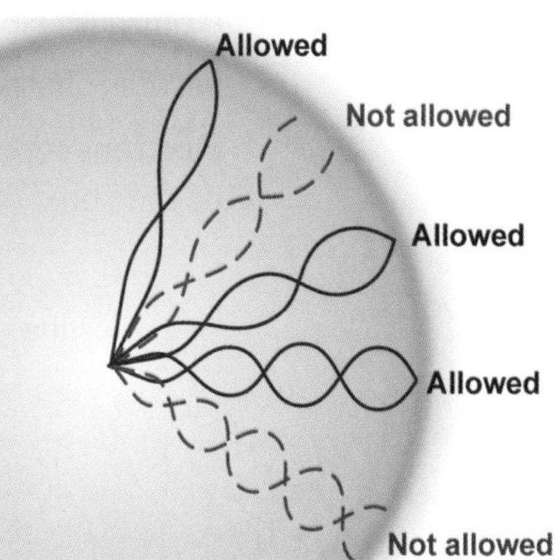

Quantum states

Consequence of restricting the electron wavelength

What does it mean that an electron can only have a wavelength that matches the "size" of an atom? The most important consequence comes from the relationship between wavelength and energy. If you know the energy of an electron, you also know its wavelength. Conversely, if the wavelength of the electron is fixed, you also fix its energy as well. *Electrons inside atoms can only have specific energies that match wavelengths they are allowed to have.*

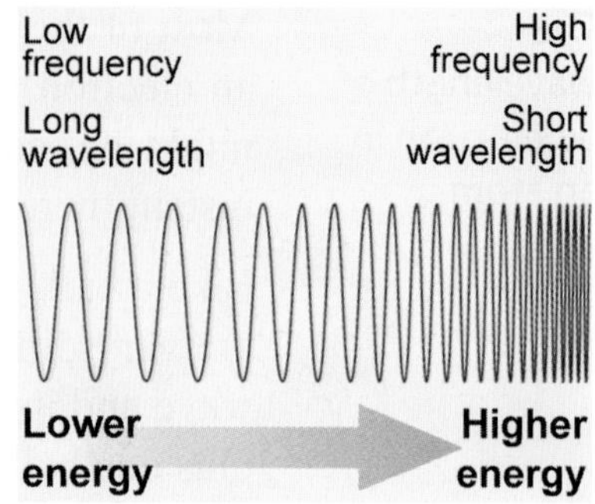

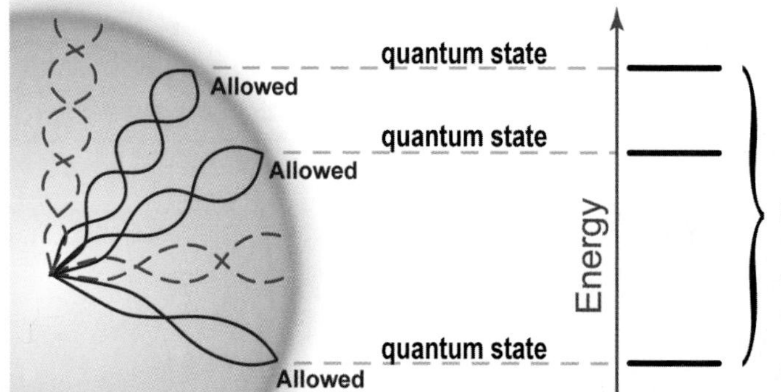

Quantization
Because only certain electron wavelengths are allowed, the energy of those electrons can only have the corresponding specific values.

Quantization

The restriction of energies to specific discrete values is called *quantization.* This is one of the most important consequences of quantum theory. An electron trapped inside an atom *cannot* have any value of energy. The electron can have only those specific values of energy that correspond to the allowed wavelengths.

We now have a good way to define a *quantum state*. A quantum state is one of the allowed wavelengths. Because of the connection between wavelength and energy, each quantum state has a specific energy. This is the "big idea" of the quantum atom.

Real quantum states

The details of the real quantum states inside an atom are complicated because,

- atoms are 3-dimensional and not simple boxes,
- besides charge and mass, electrons have a purely quantum property called *spin*,
- electrons repel each other, so in atoms with more than one electron the size and shape of the "box" depends on both the nucleus and on the other electrons in the atom.

Multiple states can have the same wavelength

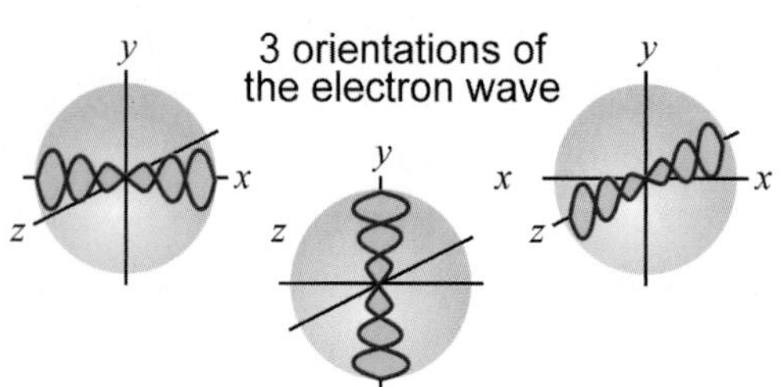

For example, because the electron wave can be aligned along any of the three coordinate axes (x, y, or z) there are 3 different quantum states that have the same wavelength, and therefore the same energy. This is a three dimensional effect.

Orbitals

The origin of s, p, d, and f

The quantum states are grouped in a peculiar, yet historically interesting way. When scientists first started using spectroscopy to explore the elements, they noticed that the spectra of the metals (Li, Na, K) had four characteristic groups of spectral lines. They named the groups *sharp*, *principal*, *diffuse*, and *fundamental* but had no real idea what caused the differences between the four groups. Today we know each group is associated with a particular "shape" of quantum states. In deference to history, the four types of shapes are called by the letters s, p, d, and f.

Orbitals

Orbitals are groups of quantum states that have similar shapes in space (spatial shapes). To understand what it means, consider the single electron in a hydrogen atom. With no other electrons to repel it, this electron has an equal chance to be at any angular position around the nucleus. Hydrogen's lone electron is in an "s" orbital. The "s" orbitals have spherical symmetry and can hold two electrons.

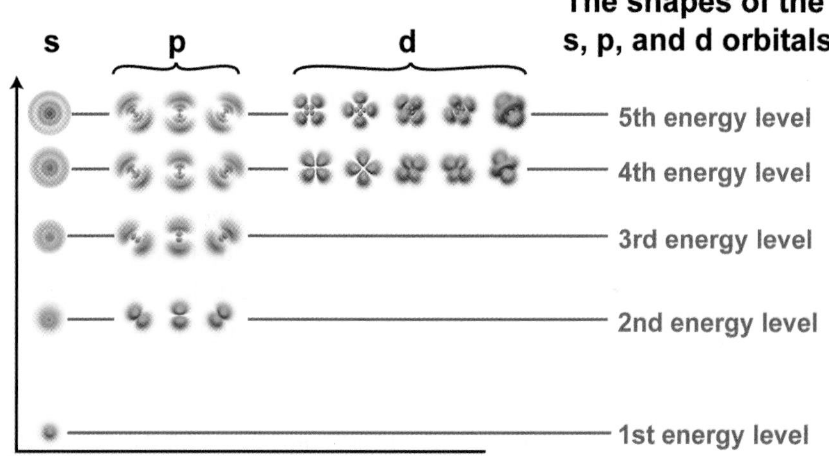

The shapes of the s, p, and d orbitals

Why some orbitals have strange shapes

Carbon has six electrons. With this many electrons, repulsion between electrons changes the shape of the electron cloud. The p orbitals are shaped like dumbbells along each of the three coordinate axes. The p orbitals can hold 6 electrons. The d orbitals are more elaborately shaped and can hold ten electrons.

Orbitals create 3-dimensional molecules

The three-dimensional shape of the orbitals represents the real three dimensional shape of the electron cloud. The orbital shapes are directly responsible for the three-dimensional shape of molecules. Methane (CH_4) is tetrahedral because of the way the s and p orbitals form the chemical bonds between carbon and hydrogen.

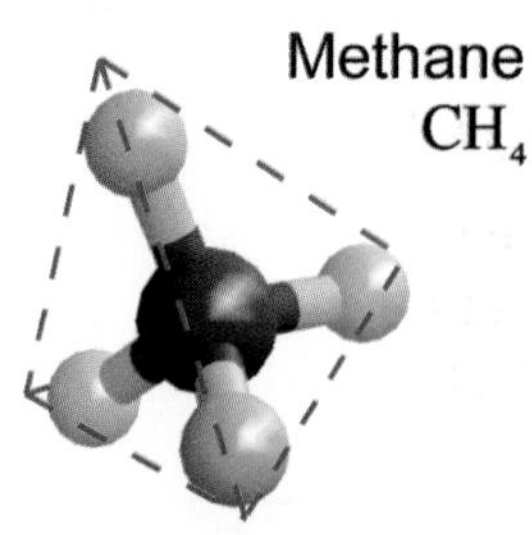

Chemistry terms

orbital - group of quantum states that have similar spatial shapes. The orbitals are labeled s, p, d and f.

Energy levels

The quantum states are grouped into energy levels

Since several quantum states have the same energy, the states are grouped into **energy levels**. The diagram below shows how the quantum states are arranged in the first five energy levels. The first energy level has two quantum states. The second and third energy levels have eight quantum states. The fourth and fifth levels have 18 states.

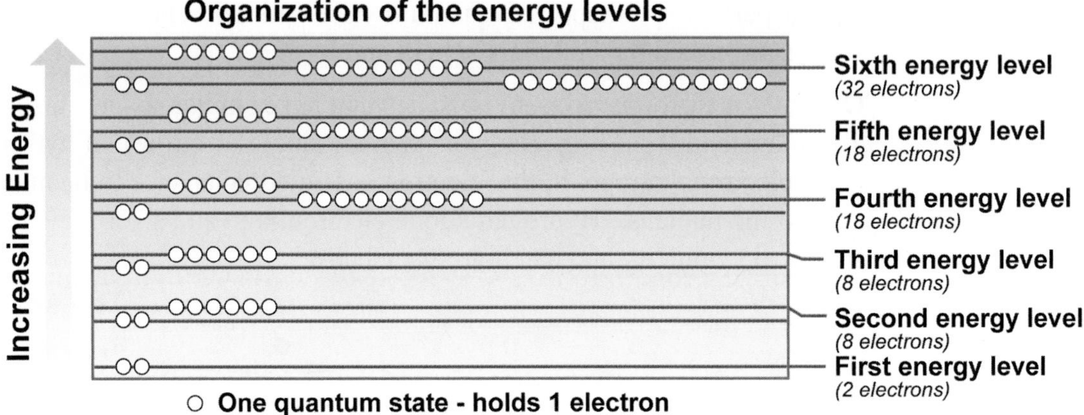

The Pauli exclusion principle

Electrons confined to the same atom obey a quantum rule called the *Pauli exclusion principle*, after Wolfgang Pauli, a famous physicist and chemist. The **Pauli exclusion principle** states that two electrons in the same atom may never be in the same quantum state. That means electrons fill the quantum states from lowest energy to highest. An atom with three electrons cannot put all three in the lowest energy level, which has only two quantum states. The third electron has to go up into the second energy level. This means that an atom with three electrons has intrinsically higher energy than an atom with two electrons.

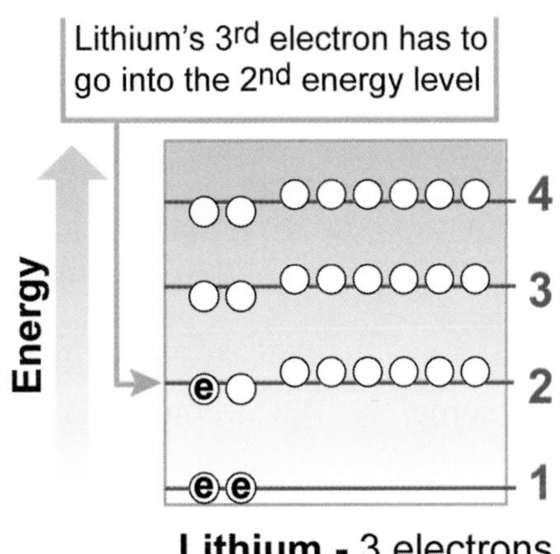

A good analogy for energy levels is a multilevel parking garage. Each floor of the garage has a limited number of parking spaces for cars. Each parking space can hold one car. Each energy level is like one floor of the garage. Each quantum state in an energy level is like a parking space for one electron.

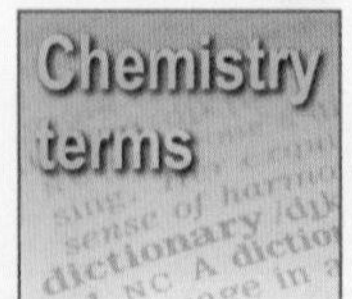

energy level - for an electron in an atom, an energy level is the set of quantum states that have approximately the same energy

Pauli exclusion principle - two electrons in the same atom may never be in the same quantum state

The periodic table

Rows correspond to energy levels

The rows of the periodic table correspond directly to the energy levels for electrons. The first energy level has two quantum states. Atomic hydrogen (H) has one electron and atomic helium (He) has two electrons. These two elements are the only ones in the top row of the periodic table because there are only two quantum states in the first energy level.

The second row

The next element, lithium (Li), has three electrons. Lithium begins the second row because the third electron goes into the second energy level. The second energy level has eight quantum states and there are eight elements in the second row of the periodic table, ending with neon. Neon (Ne) has 10 electrons, which exactly fill all the quantum states in the first and second levels.

The third row

Sodium (Na) has 11 electrons, and starts the third row because the eleventh electron goes into the third energy level. The third row ends with the noble gas, argon, which has 18 electrons. Eighteen electrons completely fills the third energy level.

Energy levels correspond to bonding properties

If you compare the energy level diagram with the periodic table, you find that all the noble gases have completely filled energy levels. All the elements which tend to form negative ions (F, Cl, Br) have one electron *less* than a full energy level. All the alkali metals that tend to form positive ions (Li, Na, K) have one electron *more* than a full energy level. This is a strong clue that the energy levels have a *lot* to do with the chemical properties of the elements.

Rows of the periodic table correspond to filling of the energy levels

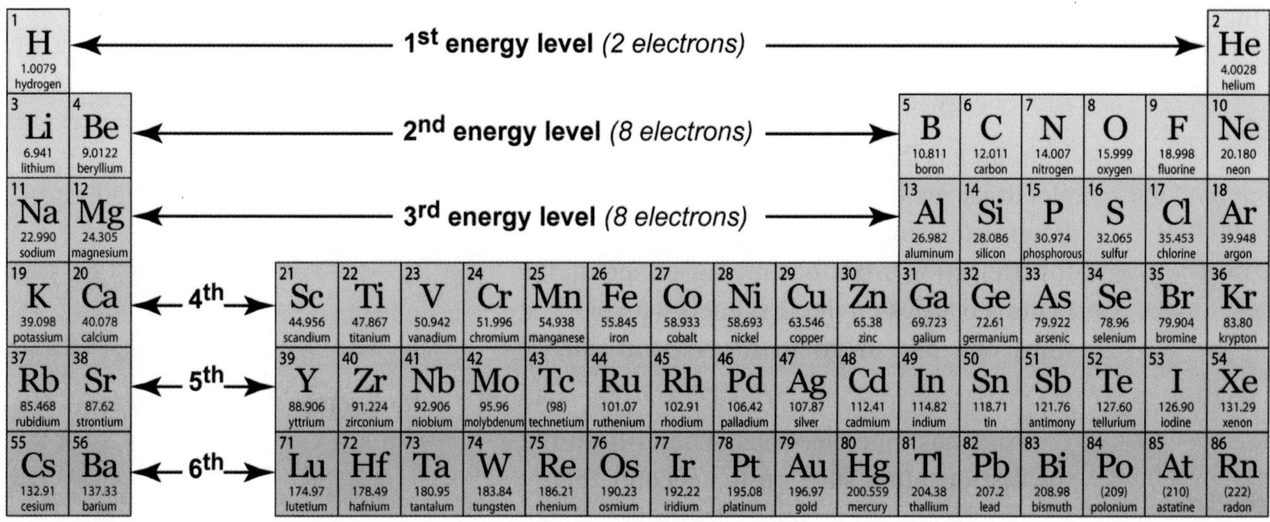

5.3 Electron Configurations

Organization of the energy levels

All the elements share a common organization for how the quantum states are grouped into energy levels. Because each element has a different nuclear charge, the actual energies of each level are unique to each element. However, the overall pattern is the same for all the elements, and determines the organization of the periodic table.

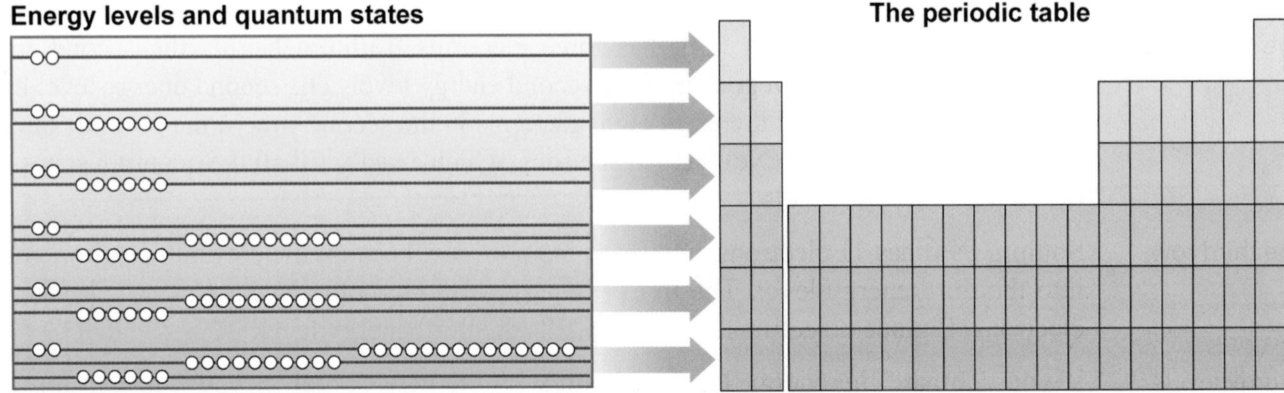

Electrons fill up lowest energy orbitals first

Every proton in the nucleus of an atom will attract an electron. Each of those electrons must exist in one of the quantum states in the diagram. Like a ball rolling downhill, each electron settles into the lowest unoccupied quantum state.

Electrons settle into the lowest unfilled quantum states

How the first few elements fill the energy levels

The first and lowest energy level holds two electrons. Hydrogen has one electron so it belongs in the first energy level. Helium has two electrons, and they completely fill the first energy level.

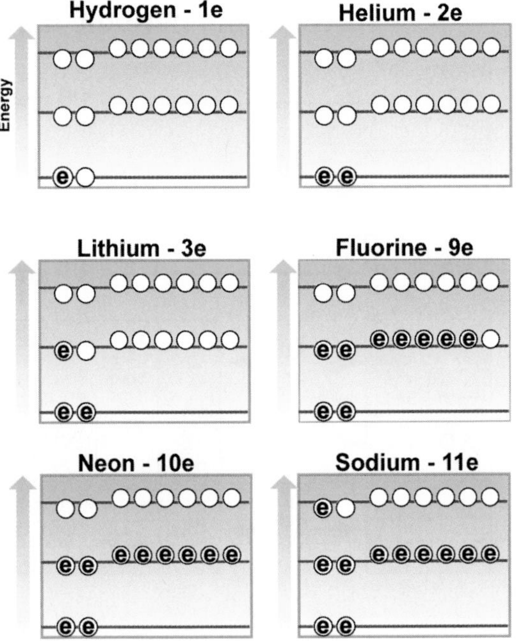

Beginning the second row

Lithium has three electrons. Two of lithium's electrons go on the first energy level. The Pauli exclusion principle forbids lithium's third electron from occupying either of the occupied states on the first energy level. The third one has to go to the second energy level.

Ending the second row

Fluorine has nine electrons. They fill all but the last quantum state on the second energy level. Neon has ten electrons which completely fill the first and second energy levels.

Electron configuration

The principal quantum number

The **principal quantum number** is a number that provides a mathematical organization to the quantum states. It is also called just quantum number and it is related to the energy level but it is NOT the same as the energy level.

Quantum #	Orbitals				Total electrons
1	s 2 electrons				2
2	s 2 electrons	p 6 electrons			8
3	s 2 electrons	p 6 electrons	d 10 electrons		18
4	s 2 electrons	p 6 electrons	d 10 electrons	f 14 electrons	32

Electron configuration

The **electron configuration** is a shorthand way to describe exactly how the electrons in any atom are distributed among the orbitals. The first number is the principal quantum number. The letter identifies the orbital and the superscript is the number of electrons in that orbital. The table below gives the electron configuration for several elements.

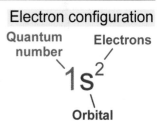

Electron configuration

Quantum number — Electrons

$1s^2$

Orbital

Orbitals and quantum numbers

Quantum number 1 includes only an s orbital. Each s orbital can hold two electrons. Quantum number 2 includes s and p orbitals. Each s orbital holds two electrons and each p orbital holds 6 electrons for a total of 8 electrons.

Element	# of electrons	Electron Configuration	Description of electron location
Hydrogen	1	$1s^1$	One electron at the 1s level.
Helium	2	$1s^2$	Two electrons at the 1s level
Lithium	3	$1s^2 2s^1$	Two electrons at the 1s level and one electron at the 2s level.
Beryllium	4	$1s^2 2s^2$	Two electrons at the 1s level and two electrons at the 2s level.
Boron	5	$1s^2 2s^2 2p^1$	Two electrons at the 1s level, two electrons at the 2s level, and one electron at the 2p level.
Fluorine	9	$1s^2 2s^2 2p^5$	Two electrons at the 1s level, two electrons at the 2s level, and five electrons at the 2p level.

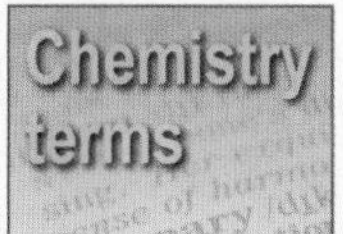

electron configuration - a description of which orbitals contain electrons for a particular atom.

principal quantum number - a number that specifies the quantum state and is related to the energy level of the electron.

Finding the electron configuration

The complete list of orbitals and the order in which they fill for all of the currently discovered elements is:

1s, 2s, 2p, 3s, 3p, 4s, 3d, 4p, 5s, 4d, 5p, 6s, 4f, 5d, 6p, 7s, 5f, 6d, 7p

The table is arranged according to the structure of the atoms and orbitals

This order tells a chemist exactly how the electrons are structured in an atom. It may seem somewhat random, but using the periodic table as your guide, there is a way to remember the orbital filling order. This is because the structure of the periodic table is actually based on the structure of the atoms. By understanding the connection between the orbitals and the periodic table, we learn about both the structure of the atom and its connection to chemical and physical properties. First, lets look at how the periodic table would be laid out if we didn't pull out the block of elements known as the rare earth elements and place them below the table.

Periodic Table Showing Which Orbital Typically Contains the Highest Energy Electrons for that Element

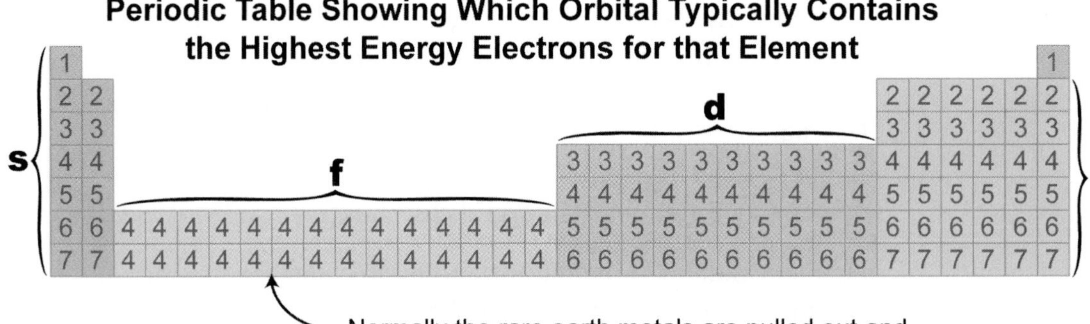

s
f
d
l
Normally the rare earth metals are pulled out and placed below the rest of the table to make the table more compact. This is only done to save space.

Finding the electron configurations

To find an electron configuration, start with the number of electrons (the atomic number). Use the chart below to find the largest number of electrons that is still less than the configuration you are trying to find. Subtract that number from the number of electrons you have and the remainder is the superscript on the orbital that is unfilled. For noble gases the chart will give you the electron configuration exactly.

Electron configuration chart

$1s^2$,	$2s^2$,	$2p^6$,	$3s^2$,	$3p^6$,	$4s^2$,	$3d^{10}$,	$4p^6$,	$5s^2$,	$4d^{10}$,	$5p^6$
2	4	10	12	18	20	30	36	38	48	54
Total number of electrons for full orbitals										

Write the electron configuration for silicon (atomic # 14)

Solve: *There are 14 electrons. The chart shows that 12 electrons fill up to* $3s^2$*. Therefore, the remaining 2 electrons must go into a 3p orbital making the electron configuration* $1s^2$, $2s^2$, $2p^6$, $3s^2$, $3p^2$.

5.4 Light and Spectroscopy

Visible light

Light is a form of electromagnetic energy that comes mainly from electrons in atoms. If you are reading this book under a fluorescent light then you are seeing the energy levels in the atom right now! In fact, we "see" because the light energy that enters our eyes is absorbed by electrons in molecules at the back of our eyes. When these molecules absorb light energy the cells that contain them send electrical signals to your brain and you "see".

Only visible light interacts with molecules in the eye.

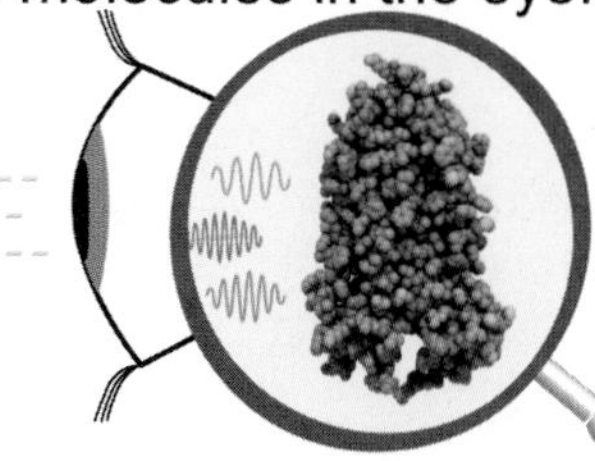

The spectrum of visible light

Energy and color

The different colors of light come from the energy of each photon. Within the range of colors that humans can see, the lowest energy is red. The highest visible energy is blue-violet. All the colors form the visible *spectrum*. A **spectrum** is a representation of the different energies present in light. Since energy depends on frequency and wavelength, the colors of light also depend on frequency and wavelength. The spectrum often specifies wavelength on the x-axis. One nanometer (nm) is 10^{-9} meters.

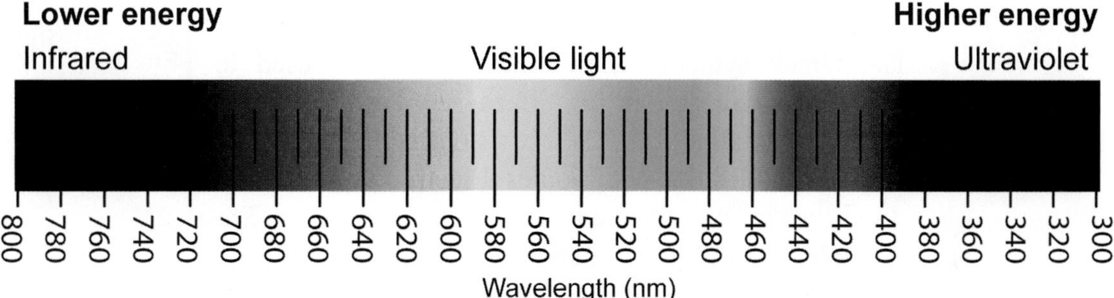

White light is a mixture of colors

The "white" light from a lamp is actually a mix of many different colors and energies. White light from the sun is not truly white. If you use a prism, you can split white light into a spectrum of colors. A device that splits light into its component spectrum is called a **spectrometer**. Spectrometers provided one of the first and best clues to unraveling the mystery of the structure of the atom.

Analyzing Starlight With a Prism
(one of the first spectrometers)

Chemistry terms

spectrum - a representation that analyzes a sample of light into its component energies or colors, can be a picture, graph or table of data.

spectrometer - a device that measures the spectrum of light.

The electromagnetic spectrum

There are many types of "light"

The light that we can see (visible light) is really only a small subset of a much larger spectrum of electromagnetic energy. The full **electromagnetic spectrum** includes "light" of lower energy (radio and microwaves) and "light" of higher energy (ultraviolet and x-rays). Many scientists use the word "light" loosely to mean the entire electromagnetic spectrum.

- Radio waves
- Microwaves
- Infrared
- Visible light
- Ultra violet
- X-rays
- Gamma rays

low energy
high energy
microwaves for cell phone communication
infrared lasers read CDs
radio waves
visible light

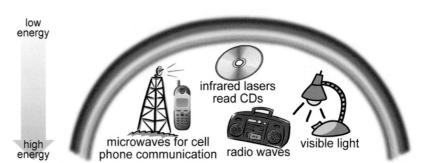

Energy and frequency

The energy of a photon depends on its frequency and wavelength accoring to the Planck relationship $E = h\nu$. The Greek symbol, ν, pronounced "nu" is used to represent frequency in Hz. Because photon energies tend to be small (10^{-18} J) scientists define the **electron volt** (eV) as 1.602×10^{-19} joules. Electron volts are also about the size of typical energy changes in an atom. For example, the difference between the second and third energy level in hydrogen is 1.89 eV (3.03×10^{-19}J).

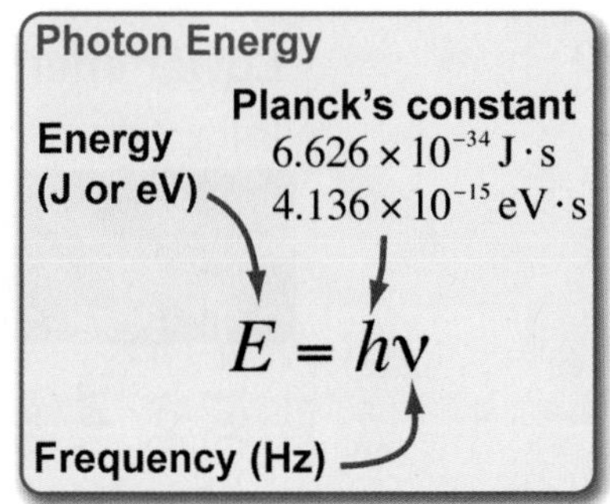

Energy, Frequency, and Wavelength of Various Form of Electromagnetic Radiation (EMR)

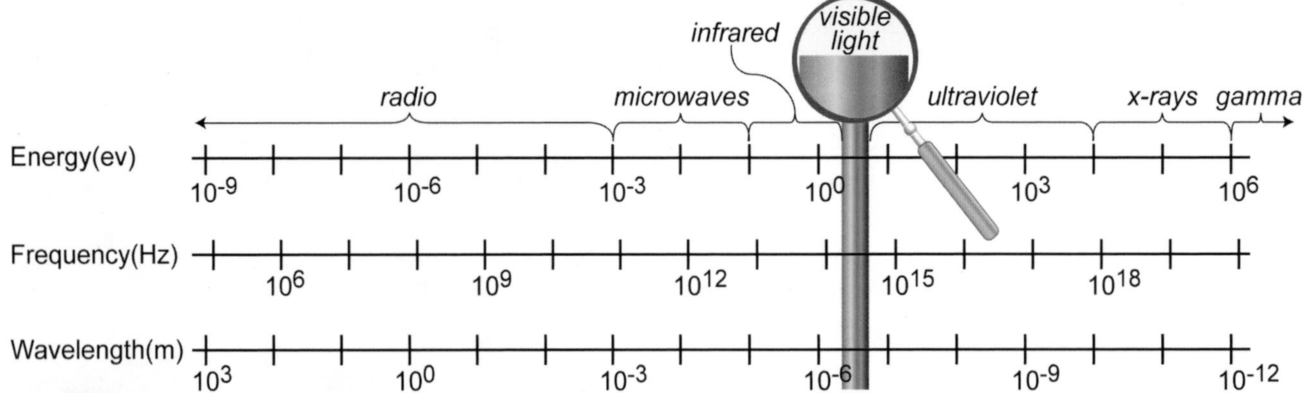

Chemistry terms

electromagnetic spectrum - the complete range of electromagnetic waves, including visible light.

electron volt - a unit of energy equal to 1.602×10^{-19} joules

The speed of light

The speed of a wave is its frequency times its wavelength

Photons and other waves move one wavelength with each oscillation. Since the number of oscillations per second is the *frequency*, and one wavelength is the distance the wave advances, the *speed* of a wave is its frequency multiplied by its wavelength. This relationship allows us to calculate the frequency if we know the wavelength and vice versa.

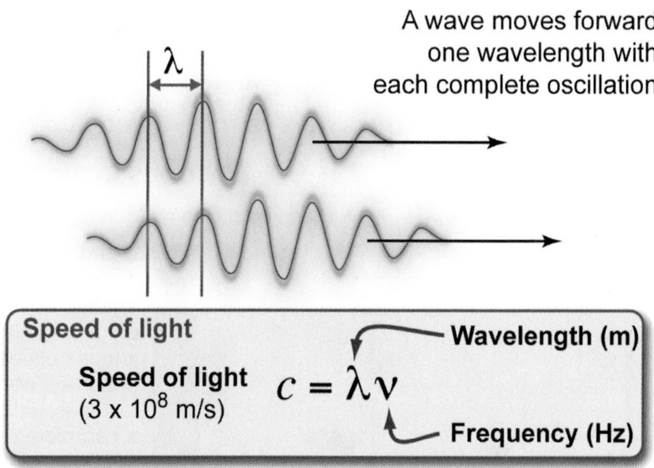

The speed of light

Photons move very fast, 3×10^8 m/s, or 186,000 miles per second! This is the ultimate speed limit in our universe. Nothing can move faster than light. The **speed of light** is so important it has its own special symbol, a lower case letter "*c*".

c is a constant

The speed of light in vacuum is a *constant*. That means it is the same for all frequencies and wavelengths. In chemistry, the speed of light is often most useful for converting between frequency and wavelength. If you know the wavelength you can calculate the frequency from the speed of light.

The wavelength of red laser light is 652 nm. What is the frequency? How much energy does a photon of this light have in eV?

Asked: *frequency and energy*

Given: $\lambda = 652 \times 10^{-9}$ m.

Relationships: $c=\lambda\nu$, $E=h\nu$

Solve:

$$c = \lambda\nu \quad \text{therefore} \quad \nu = \frac{c}{\lambda} = \frac{3 \times 10^8\,\cancel{m}/s}{652 \times 10^{-9}\,\cancel{m}} = \frac{4.6 \times 10^{14}}{\text{sec}} = 4.6 \times 10^{14}\,\text{Hz}$$

$$E = h\nu = (4.136 \times 10^{-15}\ \text{eV}\cdot\cancel{s})(4.6 \times 10^{14}\ \cancel{Hz}) = 1.9\ \text{eV}$$

Answer:

The frequency is 4.6×10^{14} Hz and the energy is 1.9 eV.

Chemistry terms

speed of light, c - a constant speed at which all electromagnetic radiation travels through vacuum, including visible light. The speed of light in vacuum is 3×10^8 m/s.

Interactions between light and matter

Experimenting with excited electrons.

Light from an incandescent light bulb shows a continuous spread of colors. This tells us the atoms that produced the light can absorb and release any amount of energy. Light from hydrogen does something completely different. Instead of a continuous rainbow, *we see a few very specific colors, and darkness in between.*

Electrons Losing Energy in the Visible Light Range

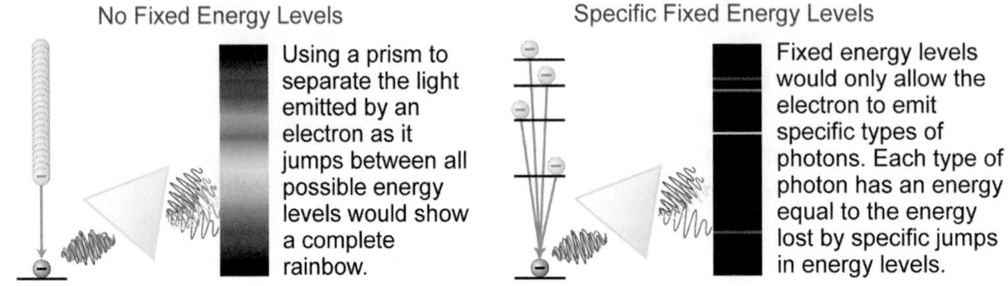

No Fixed Energy Levels
Using a prism to separate the light emitted by an electron as it jumps between all possible energy levels would show a complete rainbow.
Specific Fixed Energy Levels
Fixed energy levels would only allow the electron to emit specific types of photons. Each type of photon has an energy equal to the energy lost by specific jumps in energy levels.

The spectra tells us there are energy levels

The spectrum from hydrogen tells us that hydrogen atoms can only absorb and emit light of very specific energies. It is like viewing the inner workings of a hydrogen atom. The fact that the spectrum shows discrete colors tells us directly that electrons in the hydrogen atom can only have discrete energies.

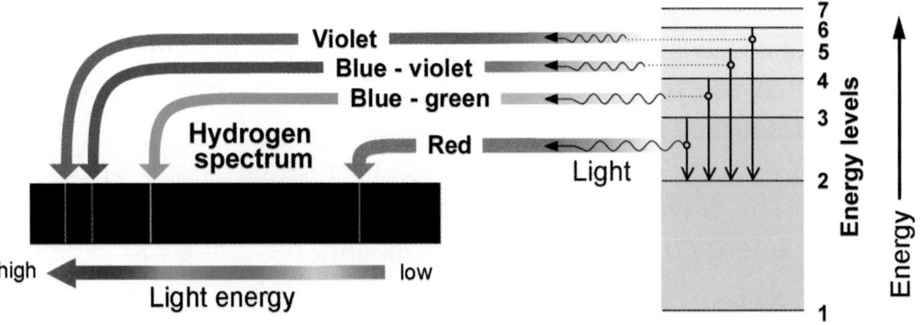

Interactions between matter and light

As light moves through matter, any electrons in atoms oscillate in response to the wave. On the atomic scale, two kinds of interaction can occur: absorption or nothing. If a photon is absorbed, another equal-energy photon may be re-emitted. The scattering of light is actually a two step process of absorption and re-emission

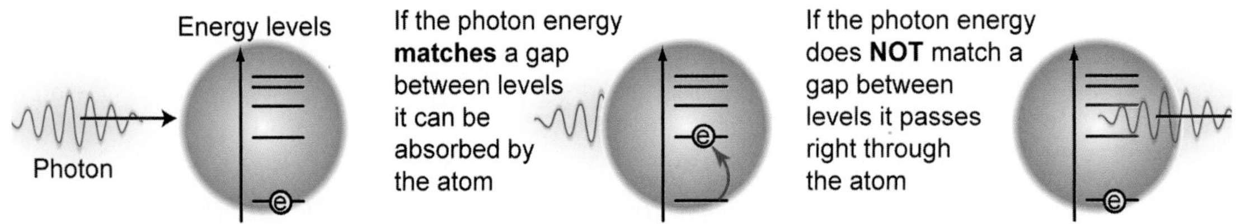

Energy levels
Photon
If the photon energy **matches** a gap between levels it can be absorbed by the atom
If the photon energy does **NOT** match a gap between levels it passes right through the atom

Electrons can get excited

Which of the two interactions occurs depends on how the energy of the photon compares to the energy levels in the atom. If the photon energy matches a difference between the atom's energy levels, the photon may be absorbed. If not, the photon typically passes right through.

Spectroscopy

Different elements have different energy levels

Quantum theory tells us that any electron confined to a small space, like an atom, results in energy levels. The specific energy levels depend on the strength of the force between the nucleus and the electrons. This depends on the number of protons in the nucleus, and on the number of other electrons in the atom that might be "shielding" the attractive force from the nucleus. For this reason *the energy levels are different and unique for each element.*

Each element has unique energy levels

Every element and compound has a unique spectra

If the energy levels of electrons are different for different elements, then the light emitted must also be different. In fact, each element emits a characteristic spectrum. Chemists refer to the emission spectrum as the "fingerprint" of an element. Chemical laboratories routinely identify elements and compounds by their spectra.

Emission and absorption spectra

Atoms both emit *and absorb* light at the energies corresponding to their energy levels. If white light is passed through a sample of matter, some light will be absorbed by the atoms in the sample. Not all light will be absorbed, however. Only colors corresponding to specific energy levels are strongly absorbed, resulting in dark lines in a continuous spectra. This is called an absorption spectra. Can you see the similarities between the absorption spectra and emission spectra?

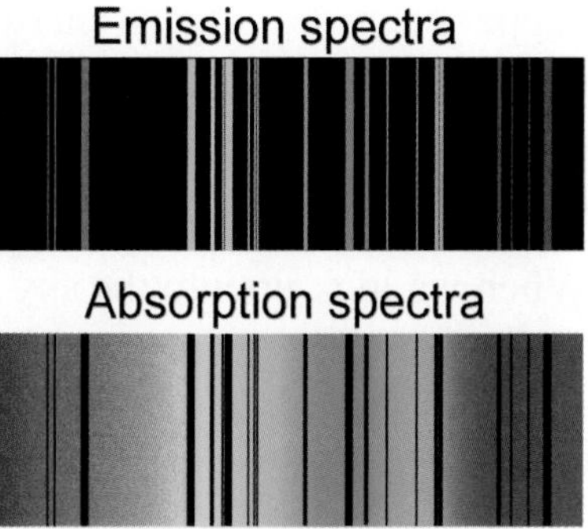

Spectroscopy

Using spectra to analyze substances is called **spectroscopy**. Spectroscopy can tell you what elements produced the light being observed. Spectroscopy is a tool for knowing what distant stars are made of, identifying unknown compounds at a crime scene, and even identifying forgeries. Right now, satellites are searching for water on Mars and astronomers are studying the composition of distant stars and galaxies by using spectroscopy. Even the makeup of our own atmosphere and global scale environmental research is done via satellites using spectroscopy.

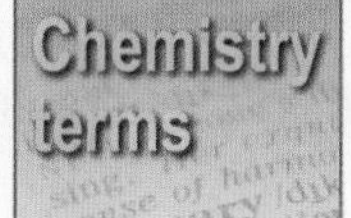

spectroscopy - the science of analyzing matter using electromagnetic emission or absorption spectra.

Everyone loves the festive display of fireworks. Fireworks have long been associated with special occasions and annual holidays like the fourth of July. The brilliant display of colors and the fantastic explosions attract crowds from miles around. Fireworks similar to the ones we use today, where first discovered in China during the early 1800's. The use of fireworks first began after the discovery of gunpowder. Gunpowder is an explosive mixture of sulfur (S), charcoal, (primarily carbon), and potassium nitrate (KNO_3), also known as saltpeter. The gunpowder mixture was used to heat chemicals that made the flames and sparks turn different colors.

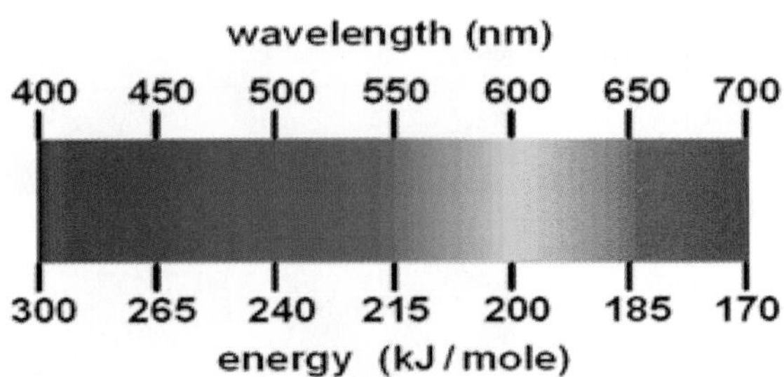

What do fireworks have to do with chemistry? Certain chemicals release light in the visible spectrum when they are heated. The chemicals that do this are positive ions, or cations, of ionic compounds (or salts). For example the copper ion (Cu^{2+}) in $CuCl_2(s)$ gives off a green color when heated in a bunsen burner flame. Which wavelength corresponds to green?

Thinking back to what you have learned you might remember that the heat from the flame causes the copper electrons to gain energy and become "excited". This excitation is fleeting, and when the electrons return back to their original energy levels, they release the "excess" energy in the form of light. In this example, Cu^{2+} releases light around 550 nm and we can see it. The amount of energy determines the color of light emitted, and the amount of energy emitted is characteristic of the element.

When we look up into the sky and see flashes of colors, what chemicals are responsible for these colors?

Color	Wavelength	Compounds
Purple	430 nm	Mixture of $SrCl_2$(red) + CuCl(blue)
Blue	500 nm	CuCl, copper compounds
Green	550 nm	$BaCl_2$, barium compounds
Yellow	600 nm	NaCl, sodium salts
Red	650-700 nm	$SrCO_3$, strontium salts, $LiCO_3$ lithium salts

Chemists have learned what salts to add to achieve just the right color. To make fireworks these metallic salts are added to a "star." A star is a small clay lump or cube that contains binders, which hold everything together, and other chemicals that help to make the colors bright. The appearance of a firework is determined by the stars, which are handmade and packed with care inside the cardboard shell. The stars are ignited by a time-delay fuse.

Quite a lot happens when a firework is lit! First, the power to lift each firework is provided by the gunpowder, or black powder as it is sometimes called. The exothermic reaction provides the thrust to lift the firework into the air. The firework compartment is designed to trap the gases produced inside the shell until the force is enough to shoot the firework upward. The firework actually has two fuses. One goes to the gunpowder, and this one burns quickly. The second one, which is a time delay fuse, goes up toward the stars.

Fireworks actually require some careful choreography! The lift charge of gun powder is supposed to run out at exactly the same time the second fuse ignites the first compartment of stars. The colors become visible and loud "bangs" are heard simultaneously. Sometimes, if the timing is off fireworks detonate too early near the ground or too late as they are beginning to fall back down. These are the variables that make fireworks dangerous to people who are not trained to handle explosives.

The old saying "what goes up must come down" is an issue with fireworks and the environment. When a firework combusts it leaves behind particulate matter and chemical by-products that find their way into the nearby soil and water. Of primary concern is the perchlorate anion (ClO_4^-). Inside the stars where the metallic salts are contained, chemicals that produce a hotter, more exothermic reaction are required to energize the colorful metal ions. Here chemicals such as potassium chlorate ($KClO_3$) and potassium perchlorate ($KClO_4$) are used. These chlorate (ClO_3^-) compounds are more dangerous to handle than perchlorate (ClO_4^-) compounds. The fact that chlorate has one fewer oxygen in its structure causes this instability. Fireworks containing chlorate compounds can even be detonated by simply dropping them! These have caused many deadly explosions. Because of this we now use perchlorate or ClO_4^- which is far more stable than (ClO_3^-). However, perchlorates are toxic to the environment and have been shown to be a human health hazard. Perchlorate (ClO_4^-), is very water soluble and finds its way into the water supplies very easily. When consumed in drinking water, perchlorate is known to cause thyroid problems. In several lakes tested after fireworks displays, the perchlorate levels were up to 1,000 times the normal base level and researchers found it took up to 80 days for the levels to return to normal.

Now there are many alternatives being developed that will make fireworks safer for our environment. Researchers have found that by cutting down on the smoke given off by a firework, fewer chemicals are needed to get a strong visual effect, and they are also substituting other chemicals for the perchlorates.

Chapter 5 Review

Vocabulary

Match each word to the sentence where it best fits.

Section 5.1

element	law of definite proportions
law of conservation of mass	

1. The ________ tells you that every molecule of carbon dioxide has 1 atom of carbon to two atoms of oxygen, even the frozen carbon dioxide found on Mars.
2. If you have an ________ then you can't break it down into a simpler substance.
3. The ________ means that you must end with everything that you start with.

Section 5.2

cathode ray	x-ray
electron	radioactivity
phosphorescence	nuclear radiation

4. When Becquerel was experimenting with uranium he accidentally found ________.
5. When glow in the dark materials give off light, this is called ________.
6. The first subatomic particle to be discovered was negative with almost no charge, and became known as an ________.
7. A ________ can be deflected by a magnet or electric field because it is actually made from negatively charged particles.
8. Various kinds of ________ are given off by unstable atoms, and can penetrate materials to different depths.
9. An ________ is a form of light that can penetrate through paper (and other materials).

Section 5.3

electromagnetic radiation	ground state
wavelength	excited electron
frequency	spectroscopy

10. The ________ is also known as how often a wave moves up and down.
11. Using ________ astronomers can tell what distance stars are made of.
12. Photons of ________ can be radio wave, microwaves, infrared light, visible light, ultra violet light, x-ray, or gamma rays.
13. Electrons give off light as they travel back down to their ________.
14. A short ________ indicates a higher energy wave.
15. When it absorbs energy it becomes an ________.

Section 5.4

uncertainty principle	atomic mass
orbital	isotopes
atomic number	ion
mass number	

16. An ________ is an atom (or small molecule) that has an unequal amount of protons and electrons.
17. The ________ describes the number of protons and neutrons.
18. Calculating the ________ involves using a weighted average of the masses of the isotopes.
19. An ________ describes where you are likely to find an electron.
20. Because of the ________ we can only know the location or the momentum of an electron, but not both at the same time.
21. The ________ tells us what kind of element an atom is.
22. Every atoms of a particular element has a nucleus with the same number of protons. However, the number of neutrons in the nucleus can be different making different ________.

Conceptual Questions

Section 5.1

23. Draw a time line showing the development of atomic theory starting with Democritus and ending with Heisenberg. Be sure to mention the person, a date, and their contribution to our current knowledge of the atom.
24. What was the main goal of the alchemists and how did this differ from the chemists that followed them.
25. Give at least two experiments that support the idea that everything is made from atoms, and explain why.
26. Describe in your own words the law of conservation of mass.
27. Describe in your own words the law of definite proportions.
28. Which parts of Dalton's atomic theory are now known to be untrue?

Section 5.2

29. What was the first subatomic particle to be discovered?
30. Describe the connection between cathode rays and electrons.
31. Explain how J.J. Thomson determined that cathode rays are made from tiny, negatively charged particles.
32. What is the "plum pudding" model of the atom?
33. Becquerel discovered which of the following:
    a. phosphorescence
    b. x-rays
    c. radioactivity
    d. radio waves
34. What are the three main types of nuclear radiation and how are they different from each other?
35. Which is the only kind of nuclear radiation that is also an electromagnetic wave?
36. Describe the connection between an electron and a beta particle.
37. Explain why Rutherford's experiment proved that the plum pudding model of the atom was incorrect.
38. Given Rutherford's results, why did he determine the nucleus of an atom was small, dense, and positively charged?
39. From what particles are atoms constructed and where are they located in the atom?
40. Which subatomic particle has almost no mass, the proton, neutron, or electron?

Section 5.3

41. Compare nuclear radiation to electromagnetic radiation. How are they different and how are they similar?
42. Using the diagram below compare the wavelength, frequency and energy of the two waves:

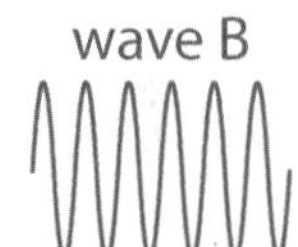

43. Which type of electromagnetic radiation carries more energy per photon, microwaves or visible light?
44. Which type of photon has more energy a red visible light photon or a blue visible light photon?
45. Explain why shining an intense red laser at a particular metal does nothing, while a very weak ultra violet light can cause electrons to be ejected from the atoms?
46. How does an electron absorb energy?
    a. It jumps to a higher energy level.
    b. It heats up.
    c. It moves closer to the nucleus.
    d. It jumps to a lower energy level.
47. How does an electron emit a photon of light?
    a. It jumps to a higher energy level.
    b. It heats up.
    c. It moves farther away from the nucleus.
    d. It jumps to a lower energy level.
48. Name three different ways you can excite an electron (give it more energy).
49. What is spectroscopy and how can it help us learn about distant stars, planets, and own atmosphere?
50. If you shine a light into a prism it:

a. creates new types of light so you see photons other than the color of the original light.
b. separates the photons that were mixed together in the original beam of light.
c. subtracts some of the photons that are in the original light.

51. If an electron could be in four different energy levels, how many different photons of electromagnetic energy could be given off by that electron, assuming it keeps absorbing and releasing energy so that it visits every possible energy level?
    a. two different photons
    b. four different photons
    c. six different photons
    d. eight different photons

Section 5.4

52. What is the main difference between the Bohr model and the current orbital model of the atom?
53. Why did Bohr think electrons were orbiting in fixed orbits with fixed energy levels?
54. Describe the difference between an orbit and an orbital.
55. Why do we need to use probably maps like the orbital picture above to show where the electron is? Why can't we say the electron orbits the nucleus like a planet?
56. What is the same for all oxygen atoms?
    a. the number of protons
    b. the number of neutrons
    c. the number of electrons
57. What is the number on the periodic table that tells you which kind of element an atom is?
58. If you have an atom of carbon-12 and an atom of carbon-14 then you have:
    a. two different ions.
    b. two different elements.
    c. two different isotopes.
59. Describe the difference between the mass number for an atom and the atomic mass of an element.
60. Give two examples of how things rubbing together cause an electric charge to build up.
61. Explain why rubbing almost anything together can cause an electric charge to build up.
62. If something is neutral, then which one of the following must be true:
    a. It is constructed only from neutrons.
    b. The positive and negative charges are exactly balanced.
    c. It has been discharged by touching something metallic.
    d. It is made from an equal number of protons and neutrons.
63. What must be true of the lithium atoms in a "lithium ion" battery, the kind that power most laptops.

Quantitative Problems

Section 5.1

64. An experiment is done measuring the respiration reaction common in most organisms, and the following measurements were taken:

Starting Materials	Ending Materials
sugar = 2.00 g	water = 1.20 g
oxygen = 2.13 g	carbon dioxide = 2.93 g

Does this experiment satisfy the law of conservation of mass?

65. An experiment was done so that the carbon and oxygen content of carbon dioxide found in three different samples could be compared. The samples came from Earth, a passing comet, and a meteor that originated from Mars. Below is the data:

Sample	carbon (g)	oxygen (g)
Earth	3.00	8.00
comet	0.15	0.40
Mars meteor	6.00	16.00

Does the data above support the law of definite proportions?

66. If the nucleus of an atom is about 1/10,000th the radius of an atom. What percentage of the volume

of the atom is empty space assuming the atom is a perfect sphere?

67. Make up some analogy that gets across how big an atom is compared to its nucleus. For example: If the nucleus were as big as a baseball (about 3.5 cm radius), the outer edge of the atom would be about 3.5 football fields away (about 350 meters or 35,000 cm).

Section 5.2

68. Below is a diagram showing which kinds of electromagnetic radiation are produced when an electron jumps from a higher energy level to a lower one. Why does it make sense that when an electron jumps from a higher level to level 3 that it emits infrared photons, but when it jumps to level 1, ultraviolet photons are emitted.

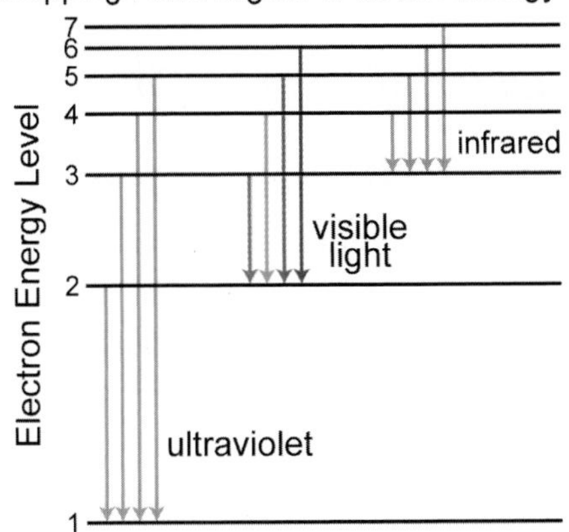

69. If you have an instrument that uses microwaves with a frequency of 10^{12}Hz, what is the energy in (ev) and the wavelength in (nm)?

70. An electron absorbs an amount of energy equivalent to 12.75eV, and three photons are emitted as the electron loses energy and returns to its ground state. The first photon has an energy of 0.66eV, and the second photon has an energy of 1.89ev. What is the energy of the last photon emitted by that electron?

Section 5.3

71. How many protons does a nitrogen atom have?
    a. 7
    b. 14
    c. 28
    d. It depends on the charge of the atom.

72. How many electrons does a nitrogen atom have?
    a. 7
    b. 14
    c. 28
    d. It depends on the charge of the atom.

73. Below is a table of several different isotopes. Specific isotopes are used in medical imaging and research. Fill in the table with the appropriate numbers.

Element	Isotope	Protons	Neutrons
potassium	$^{40}_{19}K$		
potassium	$^{39}_{19}K$		
oxygen	$^{15}_{8}O$		

74. Below is a table of ions or atoms. Fill in the table with the appropriate numbers.

Element	Isotope	Protons	Electrons
lithium	Li^{1+}		
argon	Ar		
nitrogen	N^{2+}		
nitrogen	N^{5-}		

75. Explain why you don't need the atomic symbol to determine the number of electrons.

76. Calculate the approximate atomic mass for sulfur given the following isotopes and abundances:

sulfur-32:	94.93%
sulfur-33:	0.76%
sulfur-34:	4.29%
sulfur-36:	0.02%

CHAPTER 6
Elements and the Periodic Table

Where did all the different elements come from?

How is the periodic table organized and how is it used?

Which elements are essential for life here on Earth?

Where did matter come from? The answer is found in astronomy! The universe contains billions of galaxies full of stars and planets. When they look with powerful telescopes, astronomers find that all the galaxies are rushing away from each other. The universe is expanding. An expanding universe means galaxies that are far apart today must have been closer together in the distant past. If we extrapolate back 14 billion we believe all matter in the entire universe was concentrated in a single point. That point "exploded" into the vast expanse of matter we see today in a cataclysmic event astronomers call the *Big Bang*.

Because of some complex physics, only four different elements were made in the big bang. Hydrogen accounted for 75% of all atoms. Helium accounted for almost all of the remaining 25%. with tiny amounts of lithium and beryllium. So, where did carbon, oxygen, iron, and the rest of the elements come from?

Stars release their prodigious energy by nuclear reactions that fuse hydrogen and helium into heavier elements, such as carbon and oxygen. When a star explodes in a *supernova* even heavier elements are created, such as iron, silver, and gold. The supernova blows all the heavy elements back into space, where they eventually condensed into a new star and planets: our own Solar System. The picture is shows the remnant of a supernova that occurred in 1054 AD. It was so bright that it was visible during the day for almost a month.

The Crab Nebula (result of a supernova)

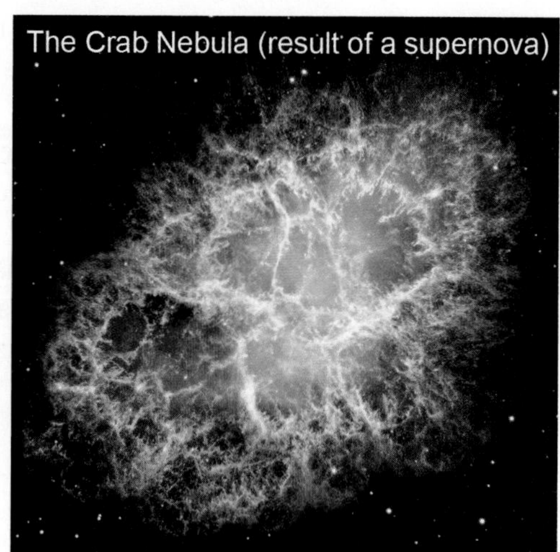

Credit: NASA, ESA and Allison Loll/Jeff Hester (Arizona State University).

Testing properties of elements

How does the periodic table help us to understand the properties of the elements? By making some simple observations we can begin to see patterns in the way elements are organized on the periodic table. Here will will experiment with four metals: calcium (Ca), magnesium (Mg), copper (Cu) and zinc (Zn), to see how they react with water.

To begin you will need four test tubes set up in a test tube rack. Fill each test tube with about 10.0 mL of water.

Place a small piece of each metal in the test tubes, using a metal spatula.

*Use caution as some test tubes may get quite hot!

Look carefully at the test tubes and record some observations over the next 3-5 minutes. After the reactions have stopped touch the test tubes to see which ones are warm.

You many have noticed that calcium and magnesium had the most vigorous reactions. Where are they on the periodic table? They are both in group 2 (or in the 2nd vertical row). Which one had the most impressive reaction? Look at where these elements are placed relative to each other. Could it be that the further down a group the more reactive the metal? If the metal is more reactive that means it caused the most observable change; for example showed the most bubbling and got warm.

Where are copper and zinc located? They are over in the middle of the periodic table, group 11(Cu) and group 12 (Zn). Neither of these seemed to react with water much, if at all. What could be different? These metals are located more toward the middle of the periodic table. Perhaps metals located on the left side react more with water.

Which metal would you select to cover the hull of a ship?

You will learn in this chapter that elements are grouped in vertical rows based on the number of outer electrons they have. Group 2 elements such as Ca and Mg react more with water than elements that have a larger number of electrons in their outer shell like Cu and Zn. Sodium, Na which is a metal in group one bursts into flames when it comes into contact with water! Perhaps you can go on-line and see an example of this reaction.

6.1 The Periodic Table

The elements can be divided into metals, non-metals, and metalloids.

There are about 118 known elements. Based on their physical and chemical properties they belong to one of three broad categories: metals, non-metals, and metalloids. Metals are malleable, shiny, and conduct electricity, and generally solid at room temperature. Examples are aluminum and copper. Non-metals do not conduct electricity and many are gasses or liquids at room temperature. Examples are nitrogen and oxygen. Metalloids have some properties similar to metals and some similar to non-metals. Silicon is an important metalloid.

The Periodic Table

Atomic Number → 6
C
Atomic Mass → 12.011
carbon

metal
metalloid
non-metal

group 1	group 2	group 3	group 4	group 5	group 6	group 7	group 8	group 9	group 10	group 11	group 12	group 13	group 14	group 15	group 16	group 17	group 18
1 H 1.0079 hydrogen																	2 He 4.0028 helium
3 Li 6.941 lithium	4 Be 9.0122 beryllium											5 B 10.811 boron	6 C 12.011 carbon	7 N 14.007 nitrogen	8 O 15.999 oxygen	9 F 18.998 fluorine	10 Ne 20.180 neon
11 Na 22.990 sodium	12 Mg 24.305 magnesium											13 Al 26.982 aluminum	14 Si 28.086 silicon	15 P 30.974 phosphorous	16 S 32.065 sulfur	17 Cl 35.453 chlorine	18 Ar 39.948 argon
19 K 39.098 potassium	20 Ca 40.078 calcium	21 Sc 44.956 scandium	22 Ti 47.867 titanium	23 V 50.942 vanadium	24 Cr 51.996 chromium	25 Mn 54.938 manganese	26 Fe 55.845 iron	27 Co 58.933 cobalt	28 Ni 58.693 nickel	29 Cu 63.546 copper	30 Zn 65.38 zinc	31 Ga 69.723 galium	32 Ge 72.61 germanium	33 As 79.922 arsenic	34 Se 78.96 selenium	35 Br 79.904 bromine	36 Kr 83.80 krypton
37 Rb 85.468 rubidium	38 Sr 87.62 strontium	39 Y 88.906 yttrium	40 Zr 91.224 zirconium	41 Nb 92.906 niobium	42 Mo 95.96 molybdenum	43 Tc (98) technetium	44 Ru 101.07 ruthenium	45 Rh 102.91 rhodium	46 Pd 106.42 palladium	47 Ag 107.87 silver	48 Cd 112.41 cadmium	49 In 114.82 indium	50 Sn 118.71 tin	51 Sb 121.76 antimony	52 Te 127.60 tellurium	53 I 126.90 iodine	54 Xe 131.29 xenon
55 Cs 132.91 cesium	56 Ba 137.33 barium	71 Lu 174.97 lutetium	72 Hf 178.49 hafnium	73 Ta 180.95 tantalum	74 W 183.84 tungsten	75 Re 186.21 rhenium	76 Os 190.23 osmium	77 Ir 192.22 iridium	78 Pt 195.08 platinum	79 Au 196.97 gold	80 Hg 200.559 mercury	81 Tl 204.38 thallium	82 Pb 207.2 lead	83 Bi 208.98 bismuth	84 Po (209) polonium	85 At (210) astatine	86 Rn (222) radon
87 Fr (223) francium	88 Ra (226) radium	103 Lr (262) lawrencium	104 Rf (267) rutherfordium	105 Db (268) dubnium	106 Sg (271) seaborgium	107 Bh (272) bohrium	108 Hs (270) hassium	109 Mt (276) meitnerium	110 Ds (281) darmstadtium	111 Rg (280) roentgenium	112 Uub (285) ununbium	113 Uut (284) ununtrium	114 Uuq (289) ununquadium	115 Uup (288) ununpentium	116 Uuh (293) ununhexium		118 Uuo (294) ununoctium

57 La 138.91 lanthanum	58 Ce 140.12 cerium	59 Pr 140.91 praseodymium	60 Nd 144.24 neodymium	61 Pm (145) promethium	62 Sm 150.36 samarium	63 Eu 151.96 eruopium	64 Gd 157.25 gadolinium	65 Tb 158.93 terbium	66 Dy 162.50 dysprosium	67 Ho 164.93 holmium	68 Er 167.26 erbium	69 Tm 168.93 thulium	70 Yb 173.06 ytterbium
89 Ac (227) actinium	90 Th 232.04 thorium	91 Pa 231.04 protactinium	92 U 238.03 uranium	93 Np (237) neptunium	94 Pu (244) plutonium	95 Am (243) americium	96 Cm (247) curium	97 Bk (247) berkelium	98 Cf (251) californium	99 Es (252) einsteinium	100 Fm (257) fermium	101 Md (258) mendelevium	102 No (259) nobelium

Some elements are "synthetic".

Technetium (#43), promethium (#61) and all of the elements above atomic number 92 do not have any stable isotopes. These elements are highly radioactive and were synthesized in a laboratory by bombarding the nuclei of existing atoms with high energy particles such as neutrons, or other extremely fast moving atomic nuclei.

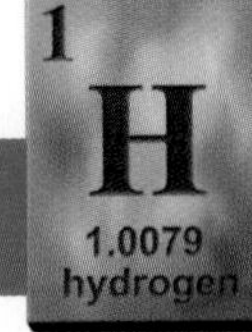

The periodic table and you

Most of you is hydrogen, carbon, oxygen, and nitrogen.

Take approximately 44,000,000,000,000,000,000,000,000,000 atoms of the right kind, bond some of them together, mix the resulting bunch of molecules and ions together, and you could end up with a typical person weighing about 75kg (or 165lbs). Of those atoms, over 99% of them would need to be only four different elements: hydrogen, carbon, oxygen, and nitrogen. Other elements are also essential for life, but they exist in much smaller numbers.

TABLE 6.1. Elemental composition of the human body

Element	% by mole
Hydrogen	63.0%
Oxygen	26.0%
Carbon	9.0%
Nitrogen	1.25%
Calcium	0.25%
Phosphorus	0.19%
Potassium	0.06%
Sulfur	0.06%
Sodium	0.04%
Chlorine	0.025%
Magnesium	0.013%
Iron	0.00004%
Iodine	0.000002%

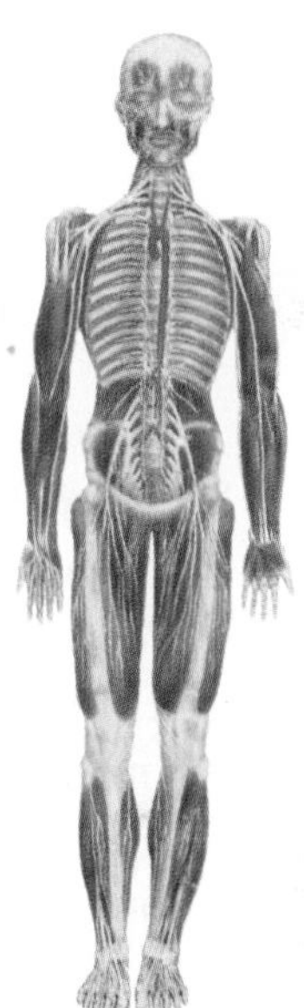

There would also be **trace amounts** of almost every element on the table, because "you are what you eat" (and what you breathe), absorbing everything from your environment.

You are made from star dust.

While hydrogen is the most common atom in your body, virtually all of the hydrogen in the universe was created in the Big Bang, so you have atoms in you that were present at the beginning of space and time. Even more incredible is the fact that almost all of the other atoms in your body must have been produced in the core of exploded stars that eventually provided the material from which the Earth was formed. You are literally made from star dust!

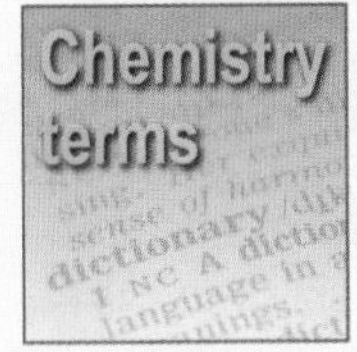

trace amount - refers to a very small quantity. For example, while radioactive and dangerous, uranium is naturally found in the environment including common soil, so it can be found in small amounts in your body as well. Normal sea water contains about 3 parts per billion of uranium. We evolved in this environment, so our bodies have mechanisms to repair the DNA damage caused by normal trace amounts of uranium.

Essential elements

Macronutrients and trace elements.

While traces of almost all naturally occurring elements exist in your body, there are a small number that are considered "essential" to life. To maintain a healthy balanced body there are some elements which you need to have a lot of called **macronutrients**. These are the ones that form the bulk of your body. There are also some elements which are beneficial in trace amounts, but toxic if you have too much.

Essential Elements

macronutrients
trace elements

H																	He
Li	Be											B	C	N	O	F	Ne
Na	Mg											Al	Si	P	S	Cl	Ar
K	Ca	Sc	Ti	V	Cr	Mn	Fe	Co	Ni	Cu	Zn	Ga	Ge	As	Se	Br	Kr
Rb	Sr	Y	Zr	Nb	Mo	Tc	Ru	Rh	Pd	Ag	Cd	In	Sn	Sb	Te	I	Xe
Cs	Ba	Lu	Hf	Ta	W	Re	Os	Ir	Pt	Au	Hg	Tl	Pb	Bi	Po	At	Rn
Fr	Ra	Lr	Rf	Db	Sg	Bh	Hs	Mt	Ds	Rg	Uub	Uut	Uuq	Uup	Uuh		Uuo

The molecules of life

To make fats, carbohydrates, DNA, and proteins, you need lots of hydrogen, carbon, nitrogen, oxygen, phosphorous, and sulfur. Calcium is needed to build your bones, and as dissolved ions it functions in many other ways. Sodium, magnesium, potassium, and chlorine, all exist as dissolved ions and work in many different ways, helping to transport other molecules into and out of cells, firing your nerves so you can see, smell, hear, and touch, and controlling the amount of water that enters and leaves your cells. These are only a few of the functions done by these other macronutrients.

Trace elements play special roles

The **trace elements** tend to have more specialized functions and are only present in extremely small amounts. You might be surprised to see chromium (Cr) on the list of essential elements. Some forms of this are know to cause cancer. However, in small amounts Cr^{+3} ions are necessary to enhance insulin function, helping to control blood sugar levels. Other trace elements work together with enzymes to help the enzyme perform whatever biological function is necessary. You would not be able to live without functioning enzymes.

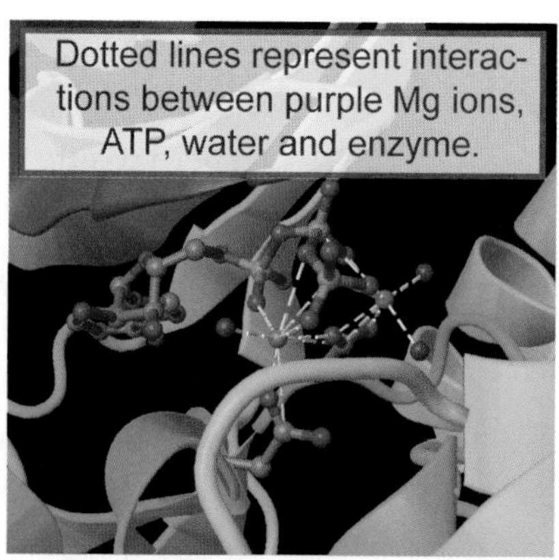
Dotted lines represent interactions between purple Mg ions, ATP, water and enzyme.

macronutrients - elements needed in large quantity by your body. Note: Often only Na, Mg, P, S, Cl, K, and Ca are listed here because they come from mineral sources.The other elements (C,H,N, and O) are found in organic matter.

trace elements - elements which are needed in very small quantities to maintain optimum health. Too much can be toxic. Not enough can be fatal or cause disease.

Periodic properties of the elements

Searching for order in the elements.

Back in 1869, when the science of chemistry was still young, a scientist and teacher of chemistry Dimitri Mendeleev, was trying to figure out if there was any kind of organization to the elements, some kind of pattern he could use to help organize them in a logical way. At that time little was known about atoms, just that elements had an atomic mass, and some other physical and chemical properties, so Mendeleev tried arranging the elements in order of atomic mass. Protons were not yet discovered, and atomic number was not known at that time.

Mendeleev

When he graphed the density of the elements vs. their atomic mass he saw a pattern.

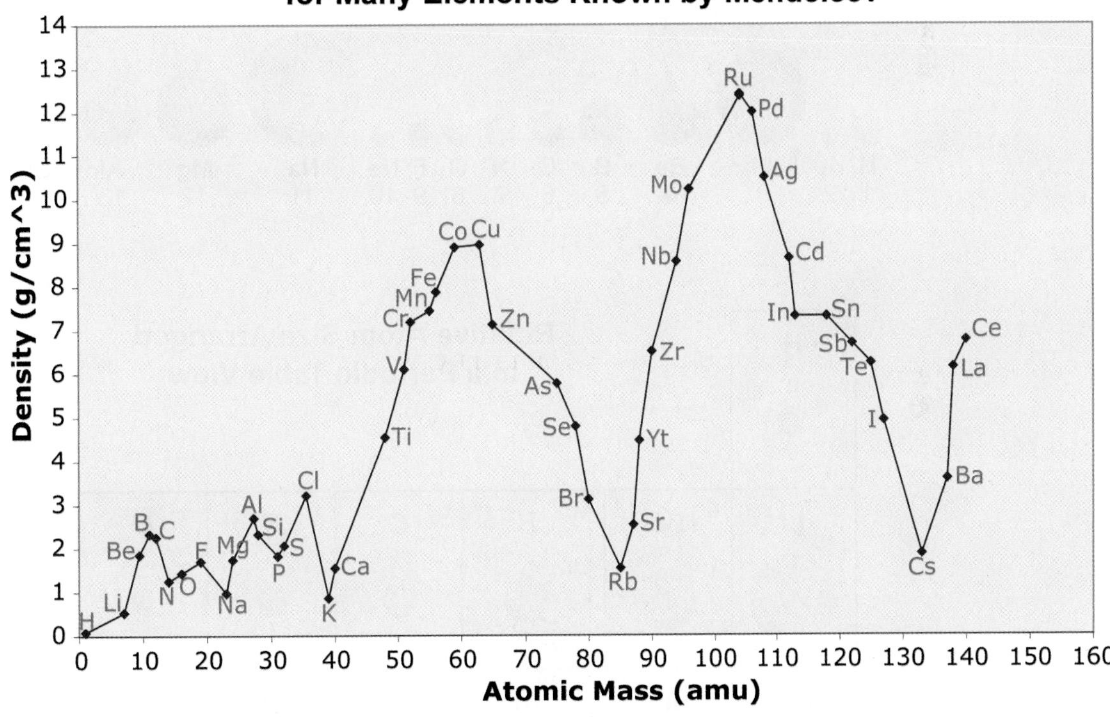

Density is a periodic property.

He observed that the density of the elements followed a **periodic** pattern, meaning that at regular intervals the pattern would repeat. Here he saw the density increase and then decrease in a repeating pattern.

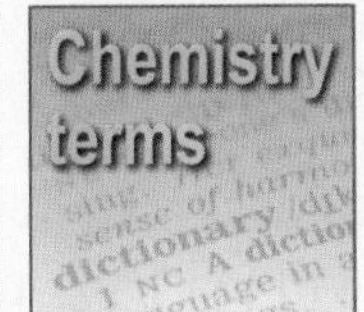

periodic - repeating at regular intervals. The periodic table is named for this because the rows are organized by repeated patterns found in both the atomic structure and the properties of the elements.

Atomic level periodic properties

Atomic level properties match the pattern of easily observable properties like density.

Mendeleev could have known nothing about the properties of individual atoms. Very little was known about atoms during his lifetime other than the belief by many chemists that atoms existed. The internal structure consisting of protons, neutrons, and electrons was not known. Only the bulk properties that could be measured like density, melting point, boiling point, and the ratios of how one element would combine with another were known. However, it would probably not surprise Mendeleev that atomic level properties would follow the same periodic patterns, closely matching the easily observable properties already discussed.

One simple example of a periodic atomic property is the size of the atoms. The image below shows the relative size of atoms from the first 18 elements on the periodic table.

Relative Size of Atomic Radii for the First 18 Elements

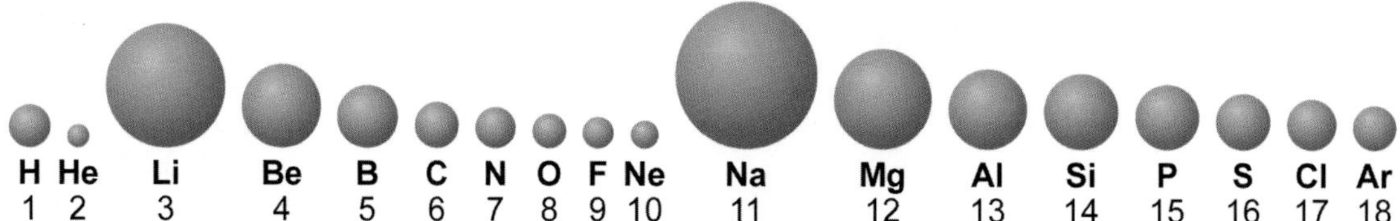

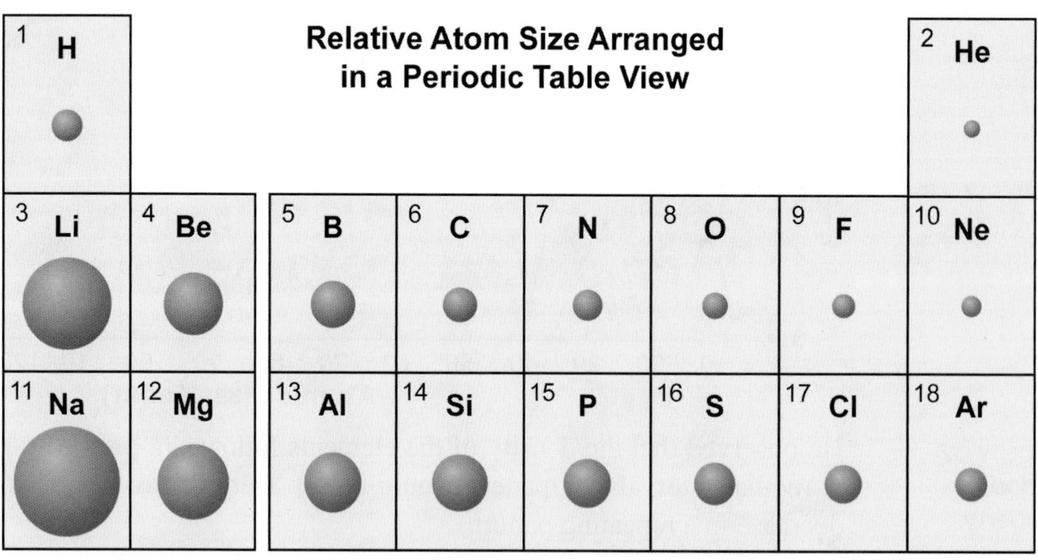

Atomic level properties explain bonding.

As you can see, every time the pattern repeats, it coincides with the beginning of a new row in the periodic table, just like density, and just like the reaction ratios with oxygen.

There are three important atomic level properties that we will consider, because they become very helpful in understanding the different kinds of bonding we will learn about in the next chapter: atomic radius, electronegativity, and ionization energy.

Atomic radius, electronegativity, and ionization energy

Atomic radius tells you the size of an atom.

Atomic radius describes the distance from the center of an atom to its "outer edge." The words "outer edge" are in quotes, because atoms don't have a sharply defined edge. The tiny nucleus is surrounded by a cloud of electrons. The space an electron occupies is defined by the orbitals occupied by its electrons. These orbitals describe where you are likely to find electrons, so the outer edge of an atom is defined by the place where the likelihood of finding its electrons gets very low.

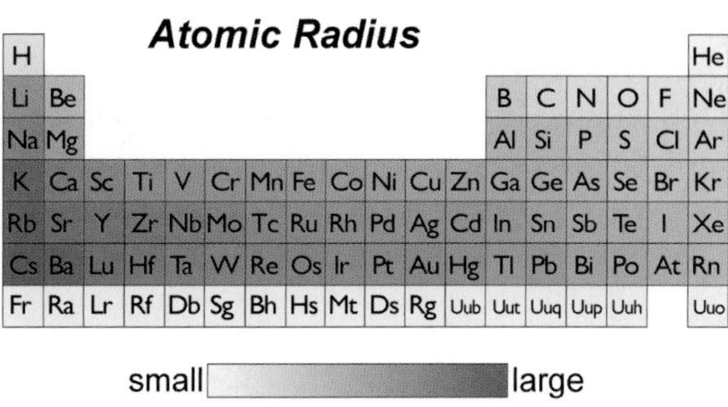

Electronegativity describes how well one atom grabs electrons from another.

Whenever atoms bond together they share electrons. In some cases they share the electrons equally or almost equally, causing a covalent bond to form. In other cases the "sharing" is so uneven that we consider one atom to have taken electrons from the other. This causes the formation of ions. **Electronegativity** is the measure of how well an atom can attract electrons from another atom to which it is bonded.

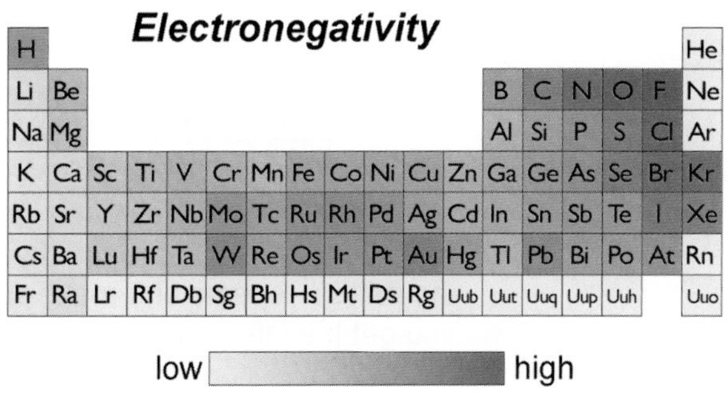

Ionization energy described how well an atom holds onto its own electrons.

Ionization energy is the energy it takes to remove an electron from an atom. Electronegativity measures how well an atom can take electrons from another bonded atom. Ionization energy measures how well an atom holds onto its own electrons. You don't need to have an atom bonded to another atom when considering ionization energy. If you hit an individual atom with a photon of light that carries enough energy, you can cause an electron to get so excited that it jumps completely off of the atom, forming an ion. The minimum energy needed to make this happen is called the ionization energy.

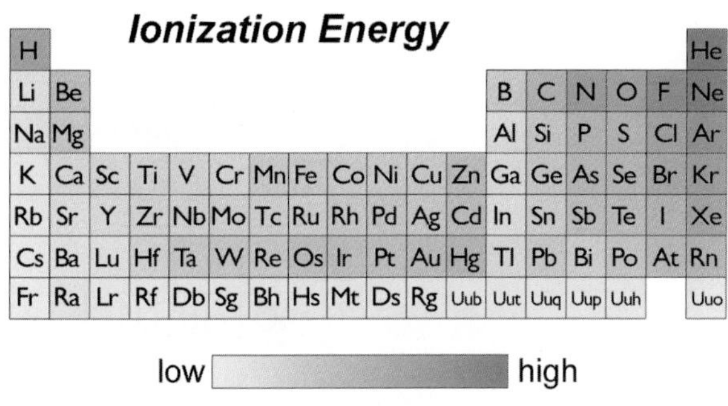

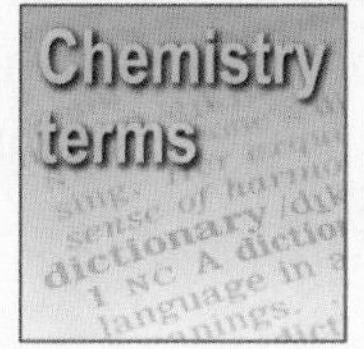

atomic radius - the distance from the center of an atom to its "outer edge."

electronegativity - the ability for an atom to attract another atom's electrons when bonded to that other atom.

ionization energy - the energy required to remove and electron from its atom.

The first periodic table

Chemical properties have a repeating pattern.

Density was a physical property of the elements which Mendeleev used in looking for patterns, but he also used chemical properties. For most of the elements, the ratio of how they react with either hydrogen or oxygen was also known. Again, the elements were arranged in order of increasing atomic mass as was done with density, but this time the oxide or hydride formula was written instead of the value for density. Another pattern emerged.

Oxides and Hydrides of First 35 Elements Known in 1869

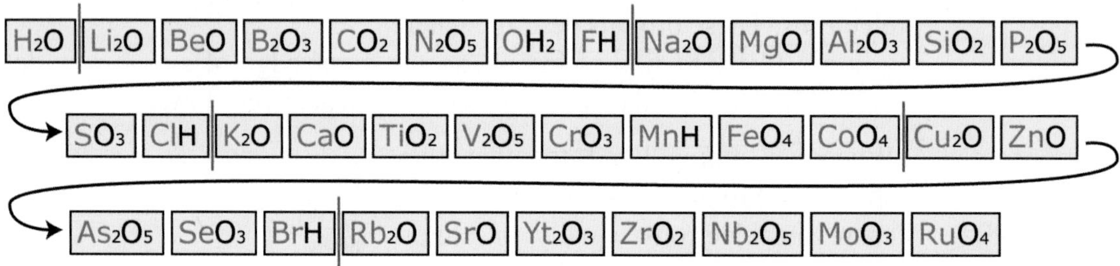

The green lines represent a place where the pattern of chemical combination with oxygen (or hydrogen) begins to repeat. If you take the same set of elements and start a new row each time the pattern repeats and make sure elements that react in the same way are in the same column you get the following.

Same Elements as Above but new Row Started For Each Green Line

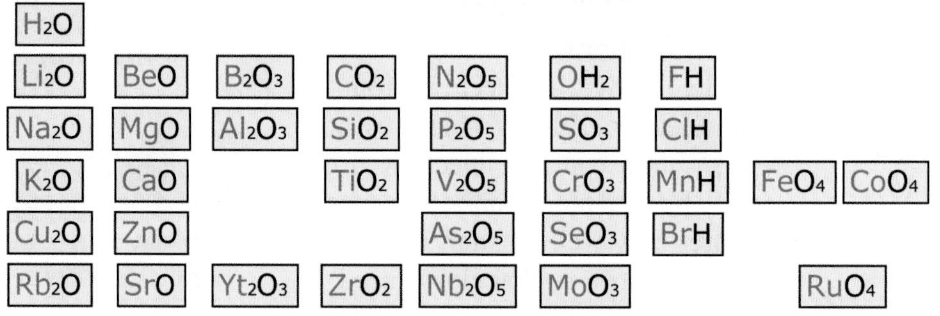

The first "periodic" table suggested by Mendeleev in 1869.

The first periodic table was created.

If you take a look back at the density graph, you will find another reason to start a new row with Li, Na, K, Cu, and Rb. All of them, except for Cu, are at a point when the density is lowest and the pattern of densities starts to repeat. So, the repeating physical property of density seems to match up well with the repeating chemical properties of the elements too. Mendeleev noticed this periodic pattern in both chemical and physical properties and he arranged the elements into a table like you see above, with columns representing elements with similar chemical properties.

The modern periodic table

The modern table is organized by atomic number

At the time of Mendeleev nothing was known about the internal structure of atoms. P protons were not yet discovered so the more logical ordering by atomic number was not possible. Today's table includes many more elements, and is ordered not by atomic mass, but by atomic number.

However, two things are still true of the periodic table, each column represents a group of elements with similar chemical properties, and each row (or period) marks the beginning of some repeated pattern of physical or chemical properties. While elements can be broadly categorized into metals, non-metals, and metalloids, an understanding that each column has similar chemical properties had lead to names for some of these element groups. Groups of particular interest are the Alkali Metals (group 1), the alkaline earth metals (group 2), the Halogens (group 17), and the Noble Gases (group 18).

Periodic Table Colored by Element Groups

metals
- alkali metals
- alkaline earth metals
- transition metals
- rare earth metals
- other metals

non- metals
- halogens
- noble gases
- other non-metals

metalloids
- metalloids

group 1	group 2	group 3	group 4	group 5	group 6	group 7	group 8	group 9	group 10	group 11	group 12	group 13	group 14	group 15	group 16	group 17	group 18
1 H 1.0079 hydrogen																	2 He 4.0028 helium
3 Li 6.941 lithium	4 Be 9.0122 beryllium											5 B 10.811 boron	6 C 12.011 carbon	7 N 14.007 nitrogen	8 O 15.999 oxygen	9 F 18.998 fluorine	10 Ne 20.180 neon
11 Na 22.990 sodium	12 Mg 24.305 magnesium											13 Al 26.982 aluminum	14 Si 28.086 silicon	15 P 30.974 phosphorous	16 S 32.065 sulfur	17 Cl 35.453 chlorine	18 Ar 39.948 argon
19 K 39.098 potassium	20 Ca 40.078 calcium	21 Sc 44.956 scandium	22 Ti 47.867 titanium	23 V 50.942 vanadium	24 Cr 51.996 chromium	25 Mn 54.938 manganese	26 Fe 55.845 iron	27 Co 58.933 cobalt	28 Ni 58.693 nickel	29 Cu 63.546 copper	30 Zn 65.38 zinc	31 Ga 69.723 galium	32 Ge 72.61 germanium	33 As 79.922 arsenic	34 Se 78.96 selenium	35 Br 79.904 bromine	36 Kr 83.80 krypton
37 Rb 85.468 rubidium	38 Sr 87.62 strontium	39 Y 88.906 yttrium	40 Zr 91.224 zirconium	41 Nb 92.906 niobium	42 Mo 95.96 molybdenum	43 Tc (98) technetium	44 Ru 101.07 ruthenium	45 Rh 102.91 rhodium	46 Pd 106.42 palladium	47 Ag 107.87 silver	48 Cd 112.41 cadmium	49 In 114.82 indium	50 Sn 118.71 tin	51 Sb 121.76 antimony	52 Te 127.60 tellurium	53 I 126.90 iodine	54 Xe 131.29 xenon
55 Cs 132.91 cesium	56 Ba 137.33 barium	71 Lu 174.97 lutetium	72 Hf 178.49 hafnium	73 Ta 180.95 tantalum	74 W 183.84 tungsten	75 Re 186.21 rhenium	76 Os 190.23 osmium	77 Ir 192.22 iridium	78 Pt 195.08 platinum	79 Au 196.97 gold	80 Hg 200.559 mercury	81 Tl 204.38 thallium	82 Pb 207.2 lead	83 Bi 208.98 bismuth	84 Po (209) polonium	85 At (210) astatine	86 Rn (222) radon
87 Fr (223) francium	88 Ra (226) radium	103 Lr (262) lawrencium	104 Rf (267) rutherfordium	105 Db (268) dubnium	106 Sg (271) seaborgium	107 Bh (272) bohrium	108 Hs (270) hassium	109 Mt (276) meitnerium	110 Ds (281) darmstadtium	111 Rg (280) roentgenium	112 Uub (285) ununbium	113 Uut (284) ununtrium	114 Uuq (289) ununquadium	115 Uup (288) ununpentium	116 Uuh (293) ununhexium		118 Uuo (294) ununoctium

57 La 138.91 lanthanum	58 Ce 140.12 cerium	59 Pr 140.91 praseodymium	60 Nd 144.24 neodymium	61 Pm (145) promethium	62 Sm 150.36 samarium	63 Eu 151.96 eruopium	64 Gd 157.25 gadolinium	65 Tb 158.93 terbium	66 Dy 162.50 dysprosium	67 Ho 164.93 holmium	68 Er 167.26 erbium	69 Tm 168.93 thulium	70 Yb 173.06 ytterbium
89 Ac (227) actinium	90 Th 232.04 thorium	91 Pa 231.04 protactinium	92 U 238.03 uranium	93 Np (237) neptunium	94 Pu (244) plutonium	95 Am (243) americium	96 Cm (247) curium	97 Bk (247) berkelium	98 Cf (251) californium	99 Es (252) einsteinium	100 Fm (257) fermium	101 Md (258) mendelevium	102 No (259) nobelium

Orbitals and atomic radius

The cloud of electrons in overlapping orbitals is what gives an atom its size.

One of the atomic level properties that we looked at previously was atomic radius, an indication of the size of an atom. The nucleus is tiny, so the size of the atom is really due the space occupied the by the cloud of electrons surrounding the nucleus. To the right is a view of a boron atom showing a unique color for each orbital containing electrons. the cloud of electrons created by the overlapping 1s, 2s, and 2p orbitals is what gives a boron atom its size.

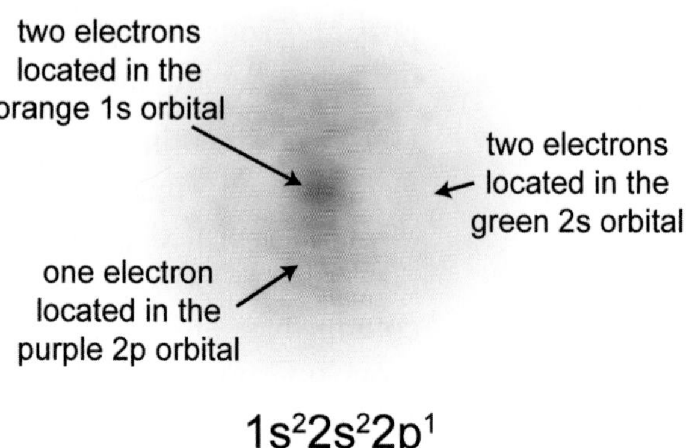

Here is another version of the table showing which orbitals are the last to be filled as the electrons fill up around an atom's nucleus.

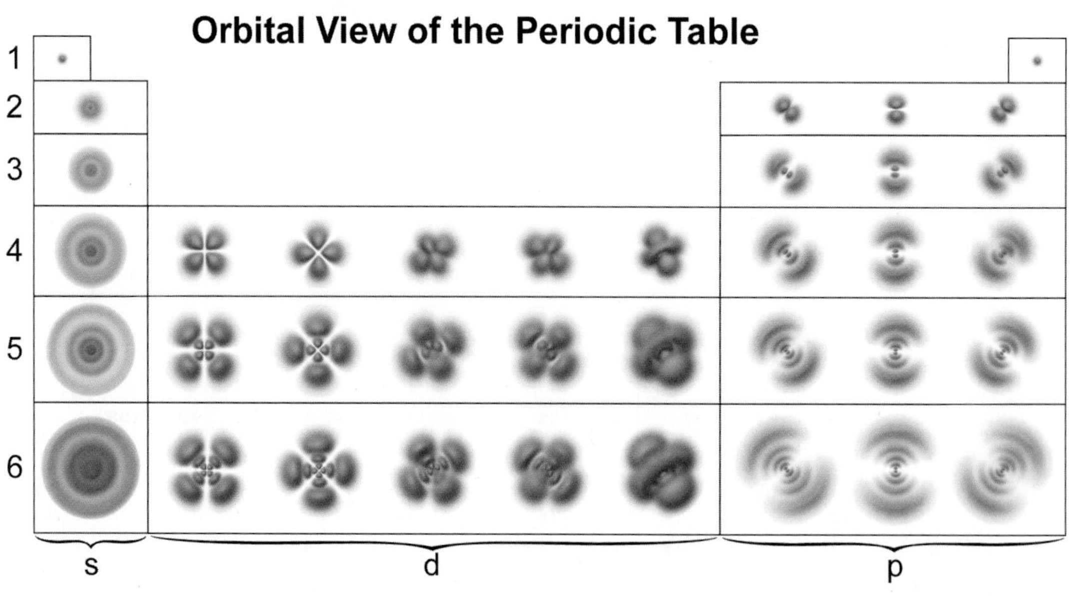

Atomic radius matches orbital size

Compare the table above with the chart to the right. Each time you start a new row the atom gets bigger, because you start filling a new primary energy level with bigger orbitals. As you go across a row, atoms get smaller, because the increased number of protons pulls electrons at the same primarily level closer to the nucleus.

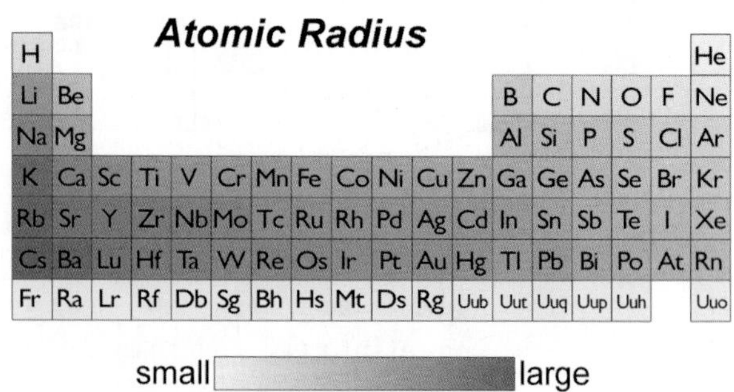

6.2 Properties of Groups of Elements

Column 1 forms the alkali metal family.

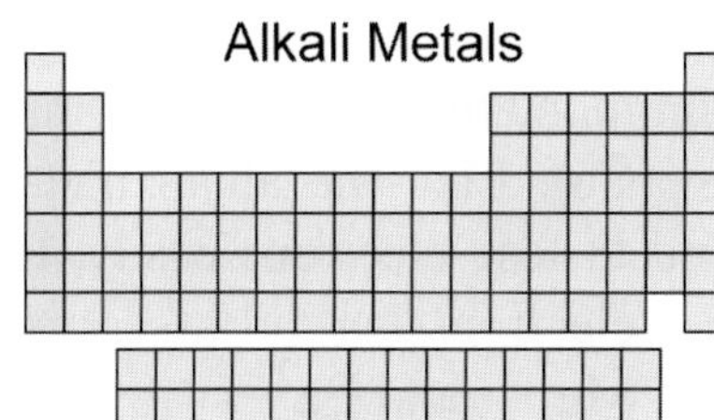

Group 1 is known as the alkali metals. Hydrogen is in this column, but due to the peculiar properties of hydrogen, it is not considered to be part of the alkali metals. The most abundant alkali metals include lithium (Li), sodium (Na), and potassium (K). Less common are rubidium (Rb), cesium (Cs), and francium (Fr).

Alkali metals form positive ions

The elements in group 1 all lose electrons easily. For example, in many compounds sodium becomes a positive ion (Na^+) by giving up an electron. Lithium and potassium also become positive ions (Li^+, K^+). Sodium never becomes a negative ion, neither does lithium or potassium. This characteristic comes directly from their electron structure. Can you see from the diagram what is common about the alkali metals?

Electron configuration of the alkali metals

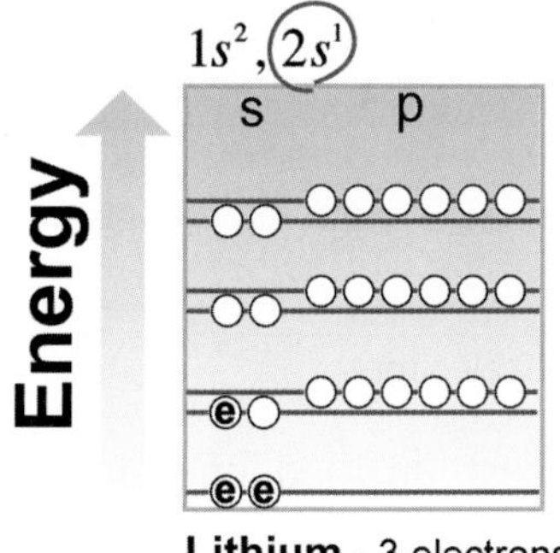

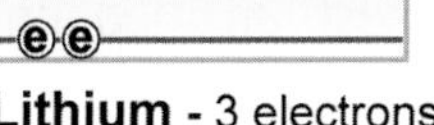

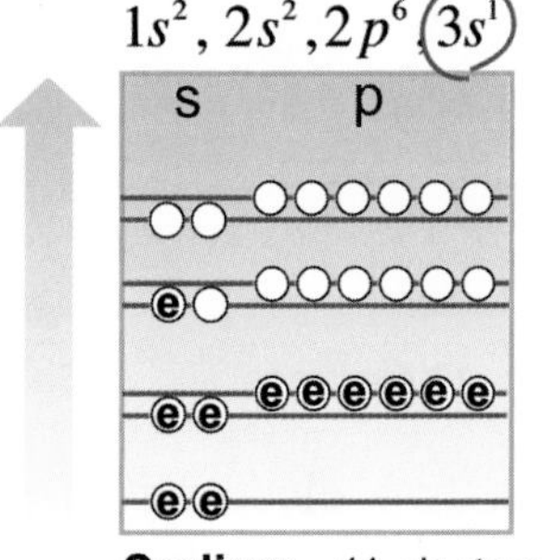

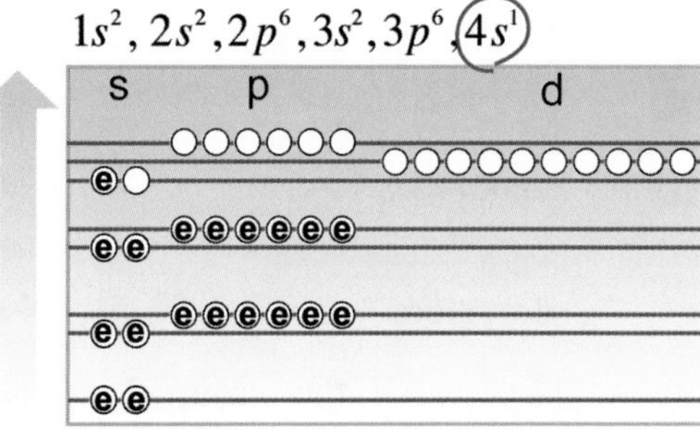

Electron structure

All the alkali metals have a single electron in the highest unfilled energy level. Lithium has a full first energy level and its third electron is alone on the second energy level. Sodium has the first and second energy levels full and its 11th electron is alone on the third level. Potassium has the 1st, 2nd, and 3rd energy levels full and its 19th electron is alone on the fourth level.

Oxides and chlorides

All of the alkali metals combine with oxygen in a 2:1 ratio. For example, lithium oxide is Li_2O. Sodium oxide is Na_2O and potassium oxide is K_2O. The alkali metals all combine with chlorine in a 1:1 ratio to make ionic salts. Examples are sodium chloride (NaCl), and potassium chloride (KCl) both of which are common in food.

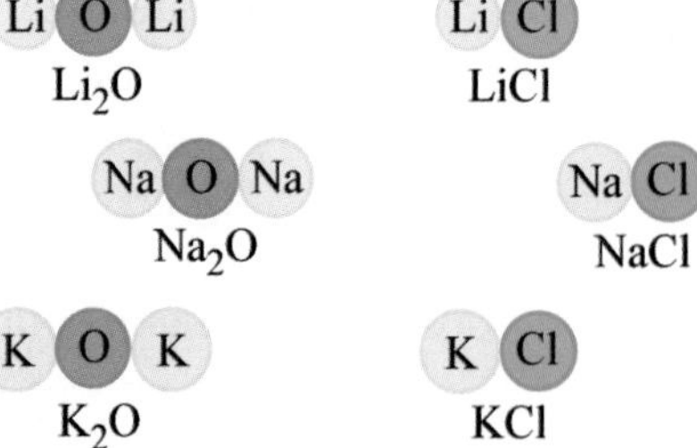

Properties of the alkali metals

The alkali metals are soft and react explosively with water. Many of the alkali metals have interesting biological functions. Their ions are used in nerve signalling, proper water retention, and maintaining correct blood chemistry levels.

The alkaline earth metals

The alkaline earth metals

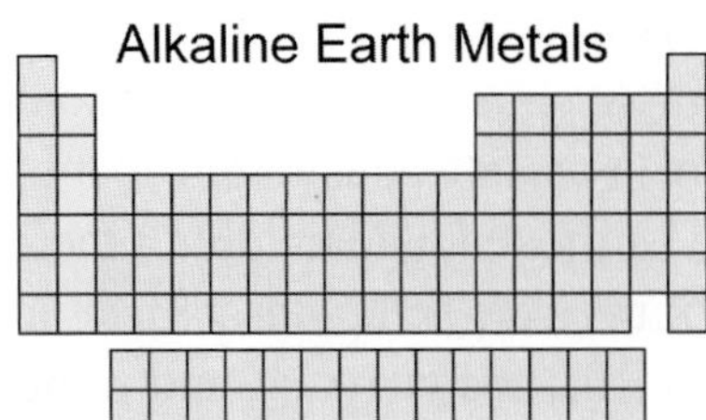

Group2 is known as the alkaline earth metals. The most common are beryllium (B), magnesium (Mg) and calcium (Ca). Like the alkali metals, these elements also tend to lose electrons easily. However, they tend to lose 2 electrons instead of 1, making their positive ions B^{2+}, Mg^{2+}_ and Ca^{2+}.

Electron structure

Looking at the electron structures of the group 2 elements provides a clue to why they make ions with a charge of 2+.

Electron configuration for the alkali earth metals (group 2)

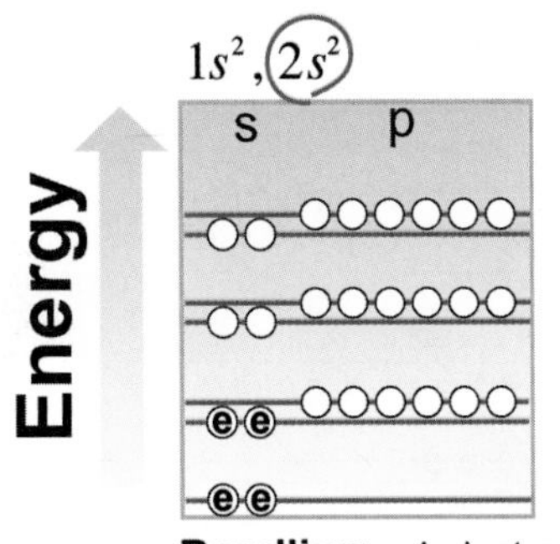

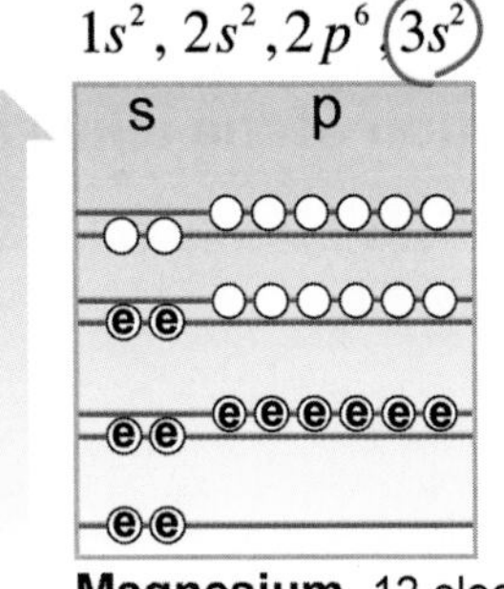

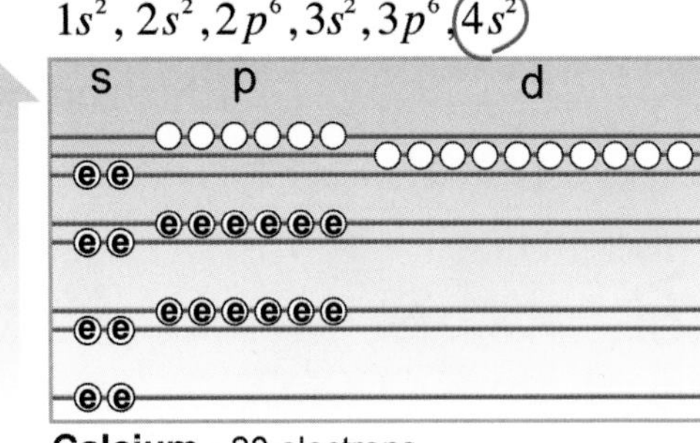

All the group 2 metals have a two electrons in the highest unfilled energy level. Beryllium has a full first energy level and its third and fourth electrons are on the second energy level. Magnesium has the first and second energy levels full and its 11th and 12th electrons are on the third level. Potassium has the 1st, 2nd, and 3rd energy levels full and its 19th and 20th electrons are on the fourth level.

Oxides and chlorides

The alkali earth metals combine with oxygen in a 1:1 ratio. For example, Beryllium oxide is BeO, magnesium oxide is MgO and calcium oxide is CaO. The alkali metals all combine with chlorine in a 1:2 ratio to make ionic salts. Examples are beryllium chloride ($BeCl_2$), and magnesium chloride ($MgCl_2$) and Calcium chloride ($CaCl_2$).

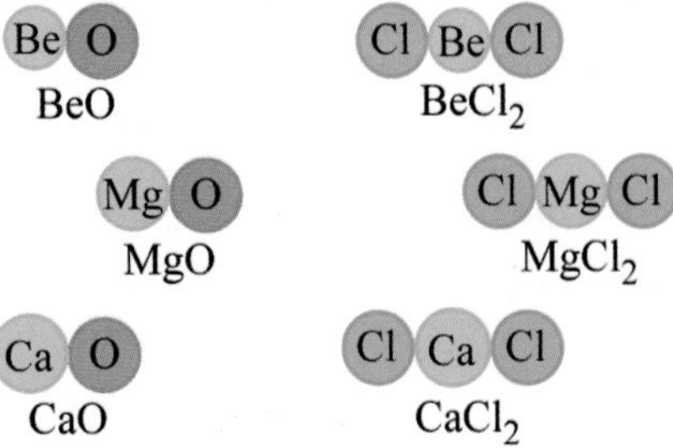

Group 2 metals in the body

Both magnesium and calcium are crucial for life. Calcium is one of the main structural components of bones and teeth. The disease *osteoporosis* is caused by lack of calcium in bones. Your nervous system also relies on calcium. A flood of calcium ions is the electrical impulse that triggers the contraction of your muscles. Magnesium ions are crucial to the chemistry of many enzymes in your body. Without sufficient magnesium, these enzymes would not be able to function efficiently, if at all.

The transition metals

About the transition metals

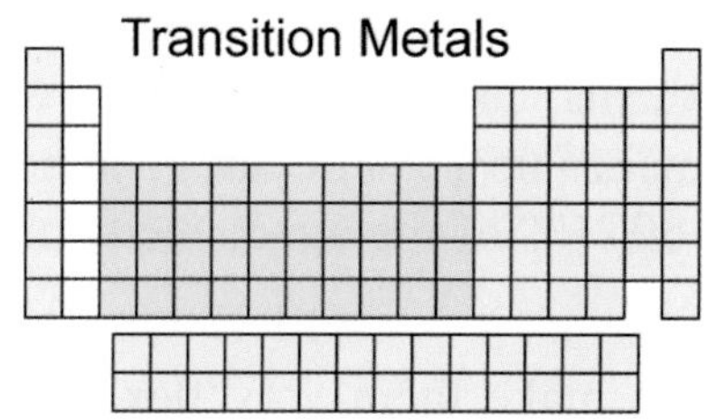

In the center of the periodic table are the transition metals. These include familiar "metals" like titanium (Ti), chromium (Cr), iron (Fe), nickel (Ni) and copper (Cu). All the transition metals are solid at room temperature except mercury (Hg). All are excellent conductors of electricity. Platinum is the densest metal with a density of 21.45 g/cm^3.

The transition metals

The transition metals start on the fourth row of the periodic table because they all have electrons in *partly filled d orbitals*. There are no d orbitals in the first, second, and third rows of the table. Iron (Fe) and silver (Ag) are characteristic of the group.

Electron configurations for the transition metals

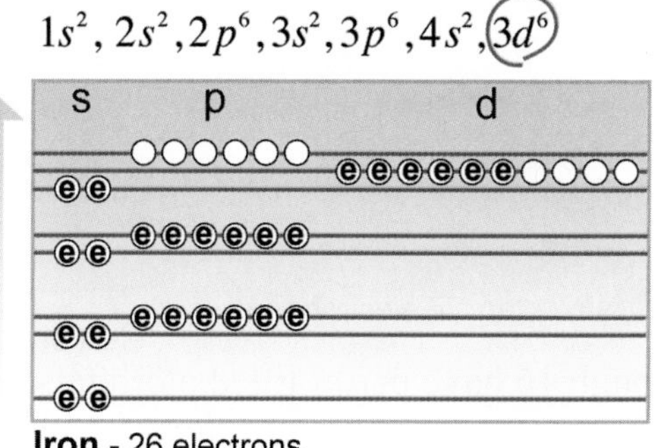

Iron - 26 electrons

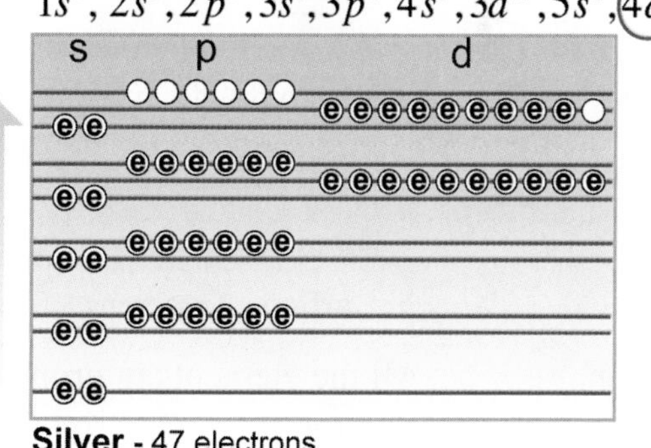

Silver - 47 electrons

How the orbitals fill

The transition metals illustrate a peculiar fact: the 3d orbitals have higher energy than the 4s orbitals! This is true even though the principal quantum number (3) of 3d is less than the 4 in "4s". Energy is the real, physical quantity that determines how the electrons act in atoms. The real energy levels correspond to the rows of the periodic table. The quantum number is an important mathematical construction, but is not the same as the energy level.

Bonding properties of transition metals

The bonding properties of the transition metals are complicated. For example, iron can form two different compounds with oxygen: FeO and Fe_2O_3. The second compound (Fe_2O_3) is what you know as *rust*. Silver bonds with oxygen in a ratio of 2 silver atoms to each oxygen atom (Ag_2O). The dark tarnish on fine silverware is silver oxide.

Rust is Fe_2O_3

Silver tarnish is Ag_2O

Carbon, oxygen, and nitrogen

These are important elements

Groups 13 through 16 contain some extremely important elements. Carbon (C) is the backbone of the chemistry of life. Proteins, fats, carbohydrates, and even DNA are carbon-based molecules. Nitrogen (N) makes up 78% of earth's atmosphere in the form of a diatomic gas (N_2). Oxygen makes of 21% of the atmosphere, also in the form of a diatomic gas (O_2).

Bonding properties

The electron structures of these elements make them very flexible in their chemistry. Oxygen and nitrogen tend to accept electrons, rather than donate them. Carbon can go either way, sometime accepting electrons and sometimes donating them.

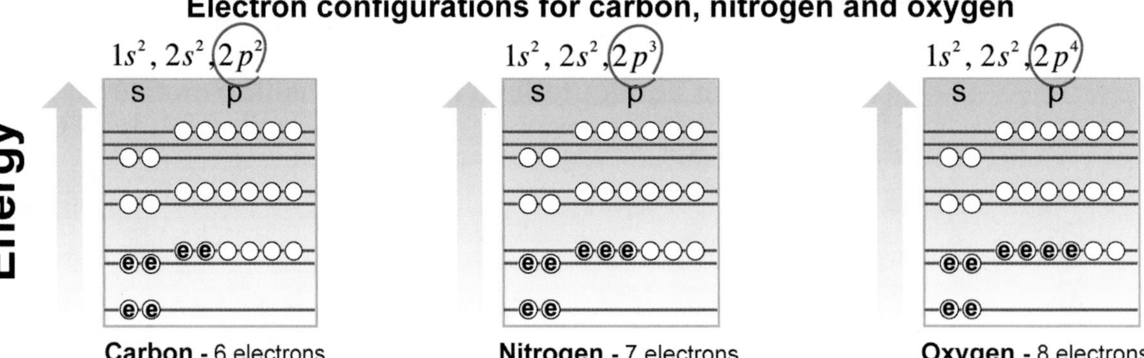

Electron configuration is partially filled p orbitals

All the elements in groups 13 through 16 have partly filled p orbitals. Boron has a single electron in the 2p orbital. Carbon, nitrogen, and oxygen have 2, 3, and 4 electrons in the 2p orbital. The heavier elements in groups 13-16 have similar electron configurations - partly filled p orbitals.

Example compounds

The chemistry of these three elements is extraordinarily rich. Carbon bonds to many elements, as well as itself. Carbon is so important that there are two entire chapters of this book devoted to the chemistry of carbon and its compounds (Chapters 17 and 18). Diamond, methane gas, and sugar are carbon compounds. Nitrogen and oxygen likewise form bonds with many other elements and with themselves. Ammonia and laughing gas (nitrous oxide) are nitrogen compounds.

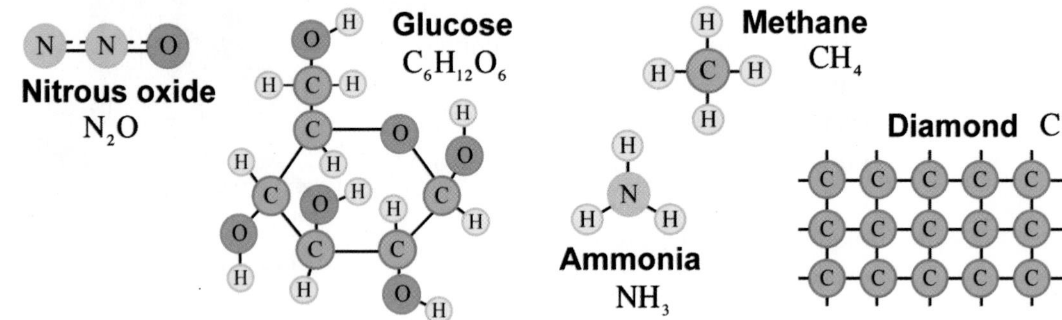

The halogens

Column 17 forms the halogen family.

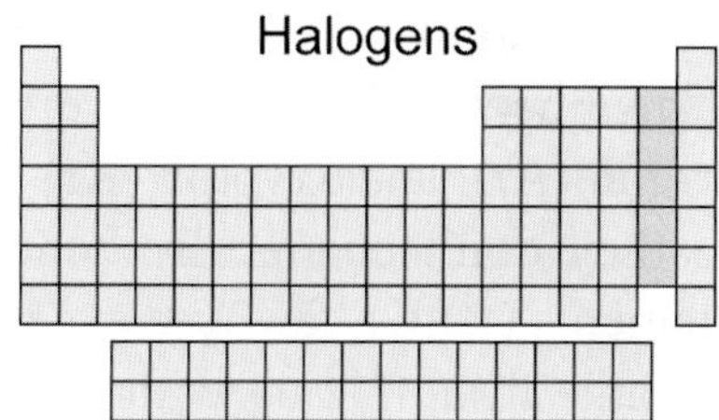

Group 17 is known as the halogens. This group includes fluorine (F), chlorine (Cl), and bromine (Br). The halogens tend to be colored gases at room temperature, or if not gases at room temperature, they easily form gases when heated slightly. The pure form of these gaseous elements is not healthy to inhale. Chlorine gas was once used as a poison gas. However, chlorine in its ionic form is essential for life.

Halogens form negative ions

Unlike metals, the halogens tend to grab one electron instead of giving any up. For this reason they all form singly charged negative ions. Examples are chloride (Cl^-) and fluoride (F^-). Looking at the electron configurations is a clue to why. Notice that each of the halogens has only one open quantum state in the p orbital. Atoms with this configuration have a high affinity for electrons to fill up that last open space and reach a full energy level.

Electron configurations of the halogens

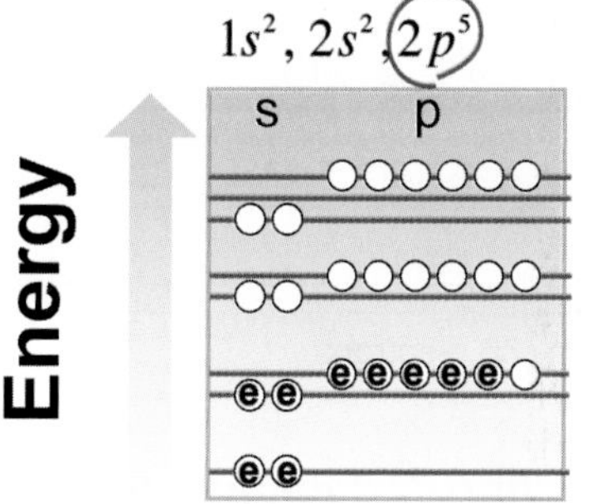

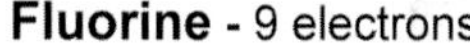

Fluorine - 9 electrons

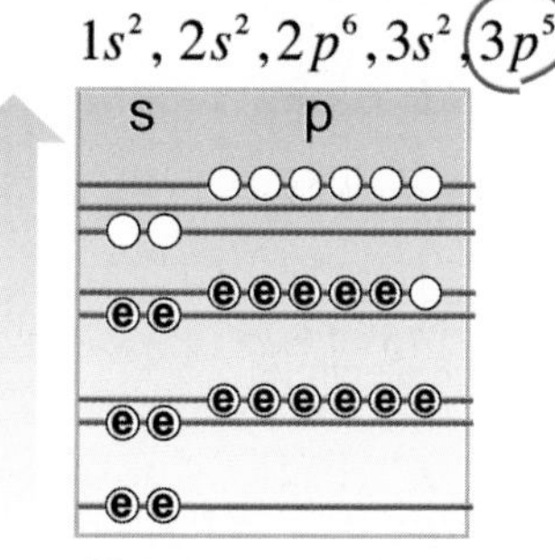

Chlorine - 17 electrons

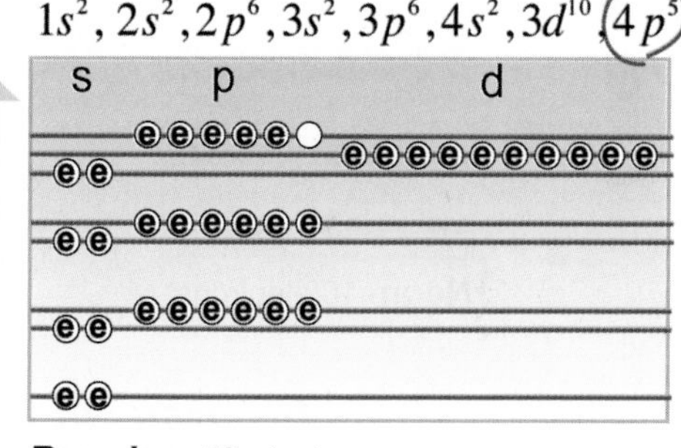

Bromine- 35 electrons

Pure halogens are toxic but compounds are not

In their pure form, these elements form diatomic molecules such as Cl_2 and F_2. If pure, the halogens are so reactive that they are highly toxic to many organisms. For example, chlorine is used in drinking water and swimming pools to kill bacteria and other harmful microorganisms. Pure chlorine is a greenish, foul-smelling gas that does not occur in nature even though chlorine compounds are very common. When combined with a metal, the resulting compound is generally an ionic salt, such as sodium chloride (NaCl).

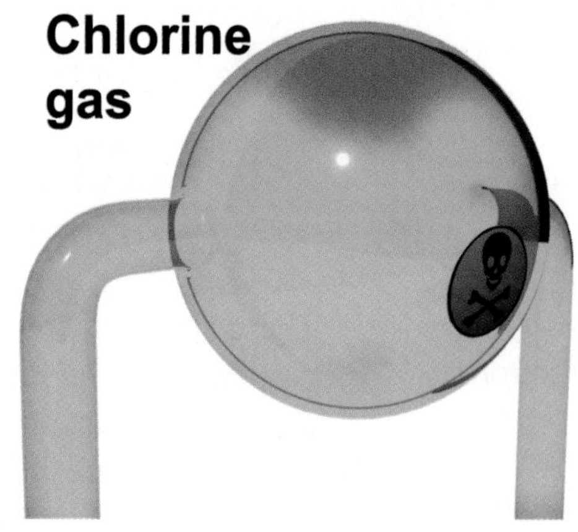

The noble gasses

Column 18 forms the noble gas family.

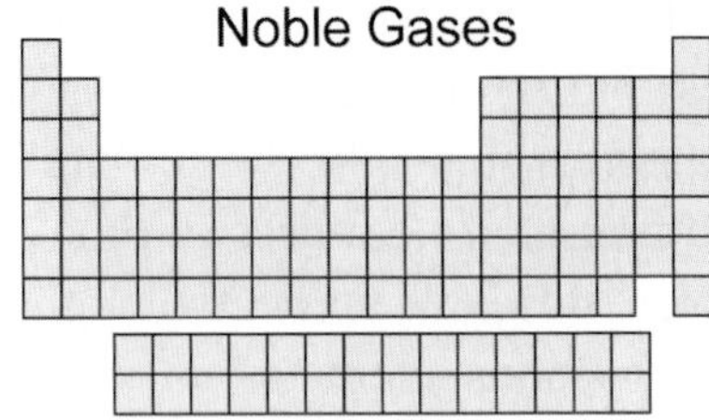

Group 18 is known as the noble gases. The three lightest members of this group are helium (He), neon (Ne), Argon (Ar), and krypton (Kr). They are "noble" because they don't chemically bond with *any* of the other elements (with a few rare exceptions). The name "noble" came from mixing with the "lower" elements, like royalty not socializing with the peasants!

Why the noble gasses are important

The noble gasses are extremely important because understanding why they *don't* form chemical bonds is the key to understanding why other elements *do* form chemical bonds. Look careful at the electron configuration of neon, argon, and krypton. What is common to all three?

Electron configurations of the noble gasses

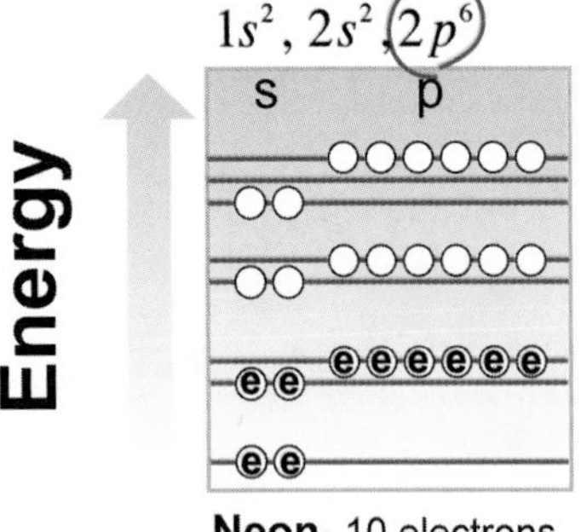

Neon- 10 electrons

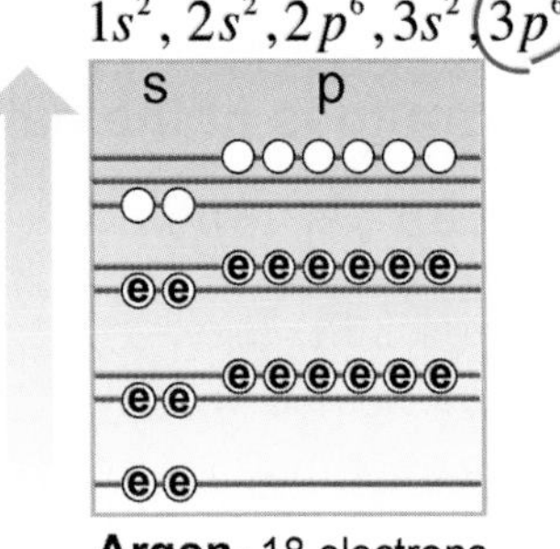

Argon- 18 electrons

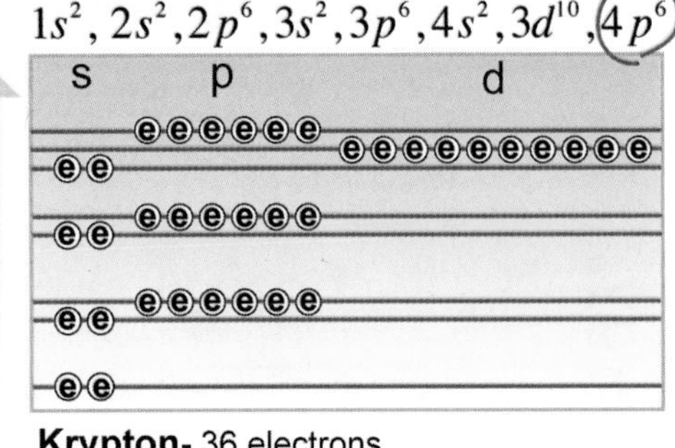

Krypton- 36 electrons

Electrons in noble gasses are ALL in completely filled energy levels

When chemical bonds may or may not occur

Elements which have completely filled energy levels *do not make chemical bonds*. The converse is also true. Elements that have unfilled energy levels *DO* make chemical bonds. This observation leads to two very important ideas.

ONLY electrons in an unfilled energy level are available to make bonds

Electrons in completely filled energy levels do NOT make bonds

The outermost electrons determine chemical properties

These observations are the key to explaining the periodic table, and also many other properties of the elements. Remember that we said earlier: electrons determine all the properties of an element except its mass. Now we can refine that statement because not all electrons are important to chemical behavior. Only the outermost electrons in the highest energy level interact with other atoms. These outermost, or highest energy electrons are the ones that determine chemical behavior.

Why compounds form

The "nobility" of the noble gasses

The noble gasses are "noble" because they have only completely filled or completely empty energy levels. This electron structure makes the noble gasses chemically inert and we find them as pure, single atoms in the environment. Every other element has a partially filled energy level. These other elements are found bonded to each other (as O_2) or to other elements in compounds (NaCl or H_2O).

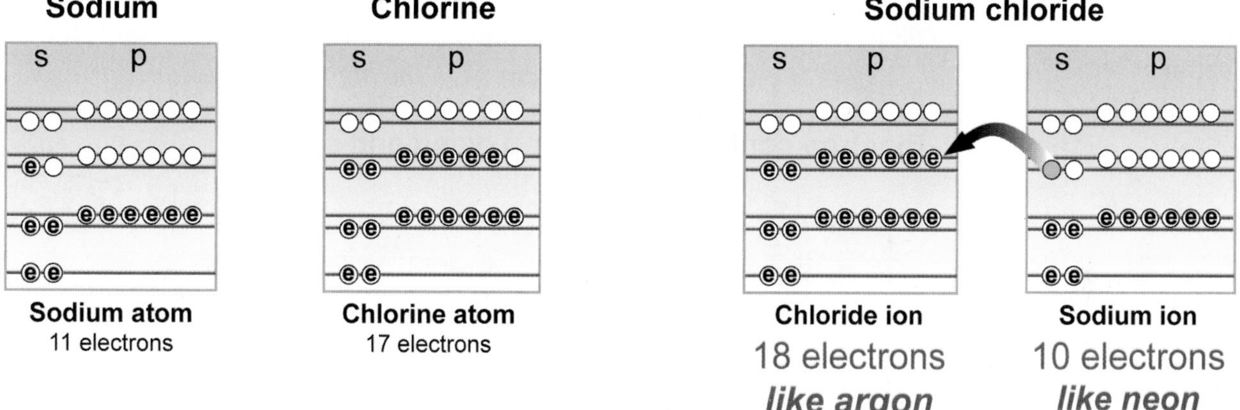

Why sodium chloride forms

Consider the electron structures of sodium and chlorine side-by-side. The pure elements have partially filled energy levels. However, by exchanging electrons to form the ionic compound sodium chloride, *both sodium and chloride ions achieve a noble gas electron structure!*

Compounds give each participating atom a noble gas electron structure

Electron structures are the "why" of chemical bonds

Electron structures are important because they explain the "why" of chemical bonding. The noble gasses have the most stable, lowest energy configuration for electrons in an atom. Noble gasses don't form bonds because they have nothing to gain. Other elements do form bonds and they do it so they too can have noble gas electron configurations. The diagram below shows how one oxygen atom gets a noble gas electron structure by sharing electrons with two hydrogen atoms (explaining the 2 in H_2O).

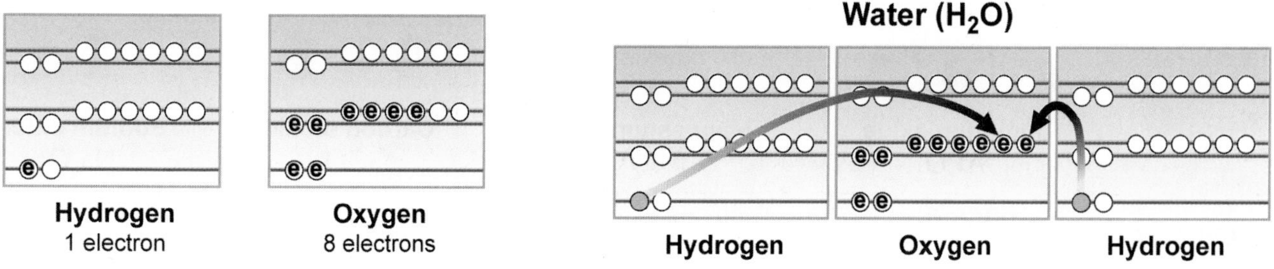

6.3 Valence

The meaning of valence

In the last section we learned that only the electrons in the highest unfilled energy level form chemical bonds. Think of the electrons in any completely filled energy levels as "permanently grounded!" This greatly reduces the problem of understanding chemical bonding. We don't need to care about electrons in filled energy levels. We need only concern ourself with the electrons in the highest unfilled level.

Sulfur has 6 valence electrons

For example, Sulfur has 16 electrons. Ten of these are in the completely filled first and second energy levels. That leaves 6 electrons in the unfilled third energy level. *Sulfur's chemical properties come from the 6 electrons in the highest unfilled energy level.*

Electron configurations for oxygen and sulfur

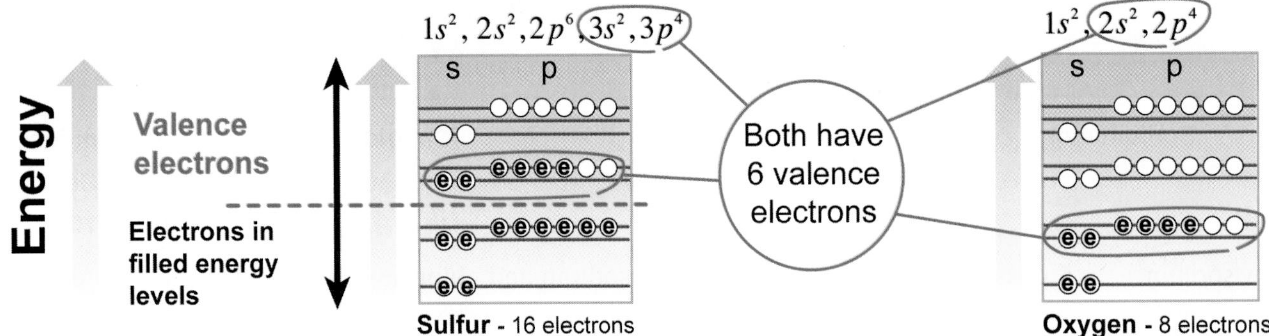

Oxygen has 6 valence electrons

Oxygen has 8 electrons. Two of these are in the completely filled first energy level. That leaves 6 electrons in the partly filled second energy level. All of oxygen's chemical properties are due to having 6 electrons in the highest unfilled energy level.

Valence

Oxygen and sulfur have the same number of valence electrons. **Valence electrons** are the special name given to electrons in the highest energy level. Because oxygen and sulfur have similar valence electrons, they form similar chemical compounds.

Oxygen and sulfur form similar chemical compounds

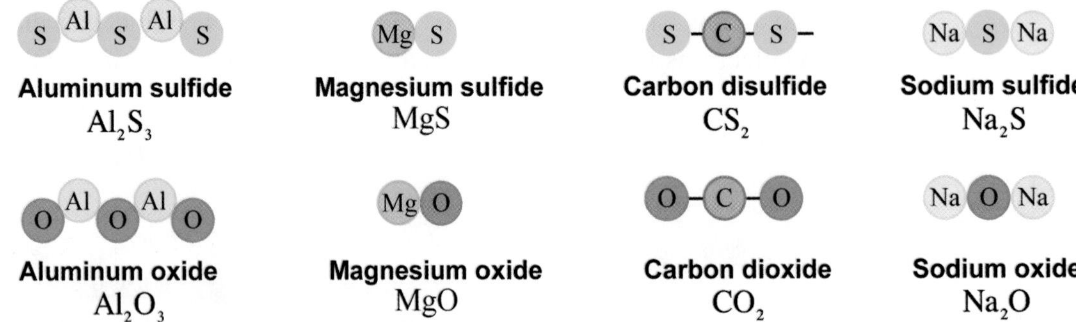

Chemistry terms

valence electrons - electrons in highest unfilled energy level. These are the electrons that make chemical bonds.

Determining valence electrons

Identifying valence electrons

The electrons that are in the highest energy level are the valence electrons. To determine how many there are you start with an element's electron configuration.

1. Write down the electron configuration
2. Count how many electrons are in the highest s and p orbitals. You should get a number between 1 and 8.
3. These are the valence electrons .

Counting Valence Electrons

$$Cl = 1s^2 2s^2 2p^6 3s^2 3p^5$$

2 + 5 = 7 valence electrons

level 3 is the highest principle energy level

$$Ga = 1s^2 2s^2 2p^6 3s^2 3p^6 4s^2 3d^{10} 4p^1$$

2 + 1 = 3 valence electrons

level 4 is the highest principle energy level

d orbitals do not count

For elements *other than the transition metals*, electrons in d orbitals do not count as valence electrons. For a complicated reason of quantum theory, the s and p orbitals become mixed when atoms bond with each other. This has the technical name *hybridization* During bonding the separate s and p orbitals become a hybrid s-p orbital.

How many valence electrons does tin (Sn) have?

Given: *Valence electrons are the electrons found in the highest principle energy level orbitals.*

Relationships: *Tin is in the 5th period (or row) of the periodic table so the highest principle energy level electrons will be in level five.*

Solve: *The electron configuration of tin is:*

$$Sn = 1s^2 2s^2 2p^6 3s^2 3p^6 4s^2 3d^{10} 4p^6 5s^2 4d^{10} 5p^2$$

or via the shorter notation

$$Sn = [Kr] 5s^2 4d^{10} 5p^2$$

There are a total of 4 electrons in level five s and p orbitals, so tin has four valence electrons.

The main group elements

The main group elements Valence is such an important concept that many chemists prefer to label the periodic table groups according to valence electrons. In this scheme groups 1A through 8A are called the *main group elements*. The transition metals are placed in a separate category according to how the d-orbital electrons form bonds (groups 1B - 8B).

Main group elements

Transition metals (valence electrons vary)

1	2	Valence electrons										3	4	5	6	7	8
1A	2A	3B	4B	5B	6B	7B	8B	8B	8B	1B	2B	3A	4A	5A	6A	7A	8A
1 H 1.0079 hydrogen																	2 He 4.0028 helium
3 Li 6.941 lithium	4 Be 9.0122 beryllium											5 B 10.811 boron	6 C 12.011 carbon	7 N 14.007 nitrogen	8 O 15.999 oxygen	9 F 18.998 fluorine	10 Ne 20.180 neon
11 Na 22.990 sodium	12 Mg 24.305 magnesium											13 Al 26.982 aluminum	14 Si 28.086 silicon	15 P 30.974 phosphorous	16 S 32.065 sulfur	17 Cl 35.453 chlorine	18 Ar 39.948 argon
19 K 39.098 potassium	20 Ca 40.078 calcium	21 Sc 44.956 scandium	22 Ti 47.867 titanium	23 V 50.942 vanadium	24 Cr 51.996 chromium	25 Mn 54.938 manganese	26 Fe 55.845 iron	27 Co 58.933 cobalt	28 Ni 58.693 nickel	29 Cu 63.546 copper	30 Zn 65.38 zinc	31 Ga 69.723 galium	32 Ge 72.61 germanium	33 As 79.922 arsenic	34 Se 78.96 selenium	35 Br 79.904 bromine	36 Kr 83.80 krypton
37 Rb 85.468 rubidium	38 Sr 87.62 strontium	39 Y 88.906 yttrium	40 Zr 91.224 zirconium	41 Nb 92.906 niobium	42 Mo 95.96 molybdenum	43 Tc (98) technetium	44 Ru 101.07 ruthenium	45 Rh 102.91 rhodium	46 Pd 106.42 palladium	47 Ag 107.87 silver	48 Cd 112.41 cadmium	49 In 114.82 indium	50 Sn 118.71 tin	51 Sb 121.76 antimony	52 Te 127.60 tellurium	53 I 126.90 iodine	54 Xe 131.29 xenon
55 Cs 132.91 cesium	56 Ba 137.33 barium	71 Lu 174.97 lutetium	72 Hf 178.49 hafnium	73 Ta 180.95 tantalum	74 W 183.84 tungsten	75 Re 186.21 rhenium	76 Os 190.23 osmium	77 Ir 192.22 iridium	78 Pt 195.08 platinum	79 Au 196.97 gold	80 Hg 200.559 mercury	81 Tl 204.38 thallium	82 Pb 207.2 lead	83 Bi 208.98 bismuth	84 Po (209) polonium	85 At (210) astatine	86 Rn (222) radon

Valence of main group elements All of the elements in the same column (group) in the main group elements have the same number of valence electrons. For example, carbon, silicon, germanium, tin, and lead have four valence electrons.

Valence of transition metals The number of valence electrons for the transition metals do not follow a simple pattern. Many act as if they have 2 valence electrons, like the group 2A elements. However, there are more exceptions than rules, and it depends strongly on what other elements are in a compound. For example, chromium makes bonds that involve 2, 3, 4, 5, or even 6 electrons.

How many valence electrons does magnesium (Mn) have?

Given: *Magnesium - atomic number is 12*

Relationships: *Mg is a group 2B element - all group 2B elements have 2 valence electrons.*

Answer: *Magnesium has 2 valence electrons.*

Introduction to Lewis dot notation

Lewis dots show valence electrons.

Because valence electrons are the most important electrons involved in bonding, a special graphic notation called *Lewis dot diagrams* was invented to represent them. A **Lewis dot diagrams** shows each valence electron as a dot surrounding the element symbol. There are up to 8 dots in Lewis dot diagram. Electrons in filled energy levels don't count because they do not participate in bonding.

Lewis Dot Diagram for Carbon

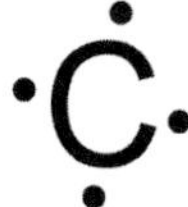

$C = 1s^2 2s^2 2p^2$
4 valence electrons

Draw individual dots before pairing them

What if there are more than four valence electrons? Where do they go? Once each side of the atom symbol has one valence electron, you start to pair them up. Drawing the dots this way, first drawing them individually, and then pairing them up when necessary, will become important when we learn about covalent bonding in the next chapter. Below is a part of the periodic table, showing you how many valence electrons each atom has.

Same Number of Valence Electrons Causes Similar Bonding Patterns

Lewis dot diagram → H; compound with oxygen or hydrogen → H_2O

H							He
H_2O							n/a
Li	Be	B	C	N	O	F	Ne
Li_2O	BeO	B_2O_3	CO_2	N_2O_5	H_2O	HF	n/a
Na	Mg	Al	Si	P	S	Cl	Ar
Na_2O	MgO	Al_2O_3	SiO_2	P_2O_5	H_2S	HCl	n/a
K	Ca	Ga	Ge	As	Se	Br	Kr
K_2O	CaO	Ga_2O_3	GeO_2	As_2O_5	H_2Se	HBr	n/a
Rb	Sr	In	Sn	Sb	Te	I	Xe
Rb_2O	SrO	In_2O_3	SnO_2	Sb_2O_5	H_2Te	HI	n/a

n/a ← indicates no reaction

Elements in the same group have the same number of valence electrons.

Atoms in the same family (or column) have the same number of valence electrons. They also react with oxygen in the same way, forming the same kind of chemical formula when bonding with oxygen. By understanding the underlying structure of the electrons, especially the valence electrons, we are now set up to better understand why atoms bond together the way they do. The next chapter will take a deeper look at bonding.

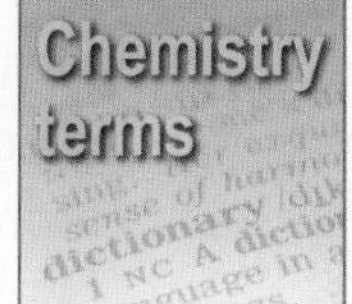

Lewis dot diagrams - a diagram showing one dot for each valence electron an atom has. These dots surround the element symbol for the atom.

More about valence electrons

Why valence electrons bond

Valence electrons participate in bonding because they are the most loosely bound electrons in an atom. The loosely bound electrons are the easiest to share. Electrons in completely filled energy levels are tightly bound to an atom. That explains why the noble gasses form no chemical bonds. There are no loosely bound electrons to bond with.

Valence electrons and Lewis dots

All the alkali metals have a single valence electron, like sodium. Their Lewis dot diagrams have a single dot representing the single electron available for bonding. All the alkali earth metals (group 2) have two valence electrons, like calcium. Elements in boron's group have three valence electrons. Carbon's group has four valence electrons. Nitrogen has five, oxygen has six and fluorine has seven. The second row ends with neon which has 8 valance electrons.

Lewis dot diagrams

Li	Be	B	C	N	O	F	Ne
1	2	3	4	5	6	7	8

Valence electrons

Filled d orbitals do not contribute valence electrons

For elements in the main groups, electrons in the d orbitals do not count as valence electrons. These elements have only completely filled d orbitals and the electrons in a completely filled d orbital act like they are in filled energy levels. That is why selenium (Se) has 8 valence electrons, putting it in the oxygen group. The electron configuration of selenium shows a completely filled d orbital.

Valence electrons for selenium

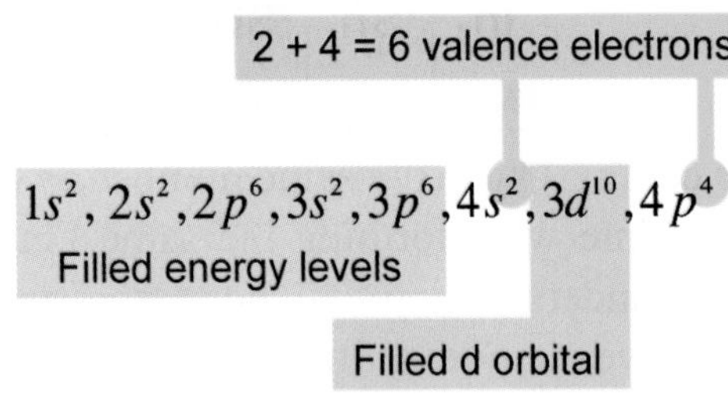

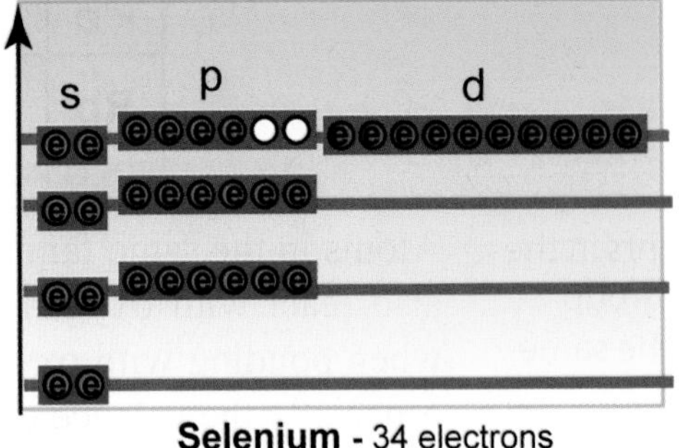

Selenium - 34 electrons

The transition metals are different

The situation is different with the transition metals. These elements have partially filled d orbitals and therefore transition metals may have more than 8 valence electrons. This is one reason that the bonding patterns among the transition metals are so complicated. How many valance electrons there are depends partly on which other element or elements are participating.

Using the periodic table

Who uses the periodic table?

So, why should we care about the periodic table and its guide to periodic properties? Chemists, biologists, material scientists, engineers and physicists use the periodic table every day to help design new drugs, create recyclable construction materials, utilize sources of renewable energy, and invent the next cool electronic device. This is just a tiny fraction of the applications in which the periodic table is a crucial guide.

A stronger type of glass.

Take, for example, the making of strong glass. Standard glass from which drinking bottles and window panes are made from contains the element sodium. If you've ever accidentally dropped a glass on the floor, you know that it breaks pretty easily.

Normal glass can be made stronger by replacing Na with K on its surface.

Common glass is primarily made from silicon and oxygen, but it contains some sodium, calcium, and aluminum as well. To make the glass stronger you could replace the sodium atoms on the surface with another atom that is a little bigger than sodium. However, the replacement atom must have similar chemical properties as sodium in order to interact with its neighboring atoms in a similar way and be incorporated into the glass. To have similar chemical properties we would want to choose an element from the same column (or family) as sodium, and the next largest atom would be potassium. If you were a chemist working for a glass maker, choosing potassium would be a good option.

A pottery glaze is made from a mixture of several materials. One of those materials is called a "flux", which changes the melting point of the glaze. Here are several fluxes that are commonly used: MgO, CaO, BaO, SrO. Why does it make sense that if one of these compounds is a flux, the others might make good fluxes too?

Asked: *Why does is it make sense that MgO, CaO, BaO, and SrO all act as fluxes in pottery glazes if any one of them do?*

Given: *A flux is a component of a pottery glaze. The compounds listed above are all commonly used as fluxes.*

Relationships: *Mg, Ca, Ba, and Sr are all from the same family of the periodic table - the alkaline earth metals.*

Solve: *Elements in the same column have similar chemical properties and follow either an increasing or decreasing trend in other properties. If they come from the same column, then they can behave the same way (with slight variations depending on their atomic level properties) in the glaze.*

Group 1 and 2 Metal Ions and our Bodies

Alkali metal ions and alkaline earth metal ions have some very important functions in your body. Among the most biologically important metal ions are Na^+, K^+, Li^+ (Group I) and Ca^{2+} and Mg^{+2} (Group II). Animals and humans could not survive with out the proper amounts of these ions in their blood and tissues.

Biologically Important Group 1 and 2 ions

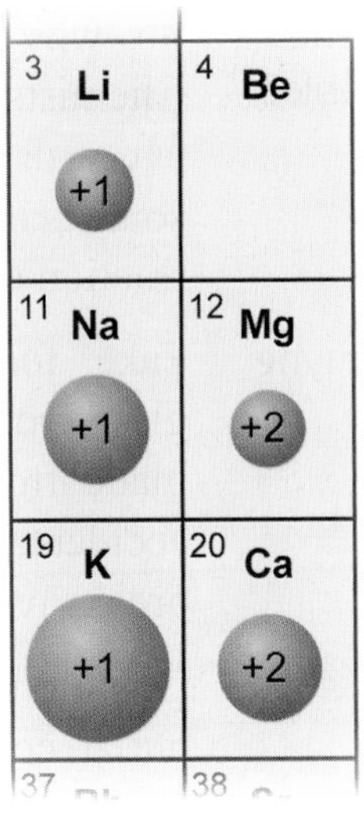

The biological importance of the metal ions is a marked contrast to their ho-hum behavior in the chemistry lab. For most of us a chemical reaction is interesting when we notice a dramatic chemical change, such as a color change, an explosion, or the release of light. However, the group 1 and 2 metal ions are soluble in aqueous solutions and usually play the role of "spectator ions" in most chemical reactions. However, these dissolved ions do very interesting things in the living body that we cannot see directly. They regulate nerve impulses, heart muscle contractions, and proper water retention inside our cells, just to name a few of their important roles.

Calcium makes a good example. When Ca^{2+} ions are released inside your heart muscle this stimulates contraction of the heart muscle, causing it to squeeze or push your blood. To relax the heart muscle the calcium ions are reabsorbed (taken in) by the nearby tissues. The presence or absence of the calcium ion, Ca^{2+} causes contraction or relaxation of the heart!

A nerve impulse is stimulated by an increase in the amount of sodium ion, Na^+, inside the neuron or nerve cell. When a nerve is "fired" or stimulated it sends a message to a part of our body. In this case the sodium ion allows a message to travel to the proper place in our bodies. Nerve cells are designed to send the "message" from one cell to another by absorbing and releasing sodium ion. The sodium ion is absorbed into the cell and this stimulates it to "fire". The next nerve cell or neuron is then stimulated. In this way, the message travels from one cell to the next following a domino effect.

Sodium/Potassium Ion Pump

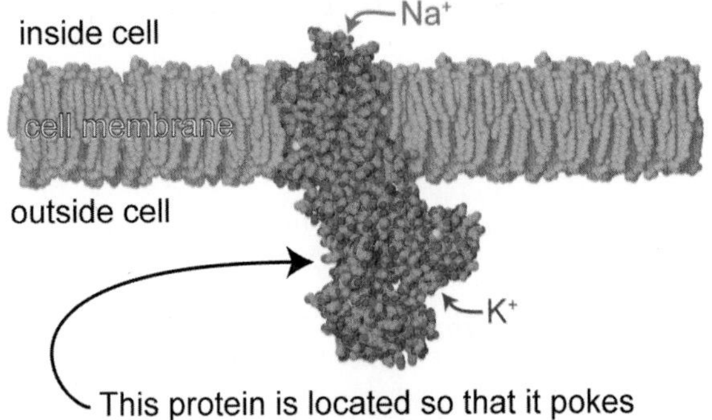

This protein is located so that it pokes through the cell membrane. It uses the chemical energy in ATP to pump Na^+ out of the cell and K^+ into it.

Electrolytes contain metal ions such as potassium, K^+, sodium, Na^+, and calcium, Ca^{2+}. Sports drinks usually provide electrolytes, and some vitamin waters contain sugars and electrolytes. These metal ions such as Na^+ and K^+ are responsible for regulating how much water is absorbed by our tissues. These ions are major components of blood plasma and inter cellular fluids. Scientists have long been studying the amounts of these metal ions inside our body tissues.

It has been discovered that other group one ions have effects on the body. Some of these effects have been discovered through circumstance. Lithium, Li^+, for instance has no known function in the human body. However, in 1817 salts containing lithium were thought to have mystical healing power. The legendary "fountain of youth" was a spring that provided magical healing and restored youth to the elderly. This spring was thought to contain lithium ions! In 1927, Mr. C.L. Grigg began to sell a lithium containing lemon-lime soft drink, which later became known as Seven-Up.

A natural spring in Florida was thought to be the Fountain of Youth by the Spanish explorer Ponce de Leon.

Concerns from the Federal Drug Administration (FDA) caused lithium ions to be removed from the drink in the 1950's. Right around this same time it was discovered that lithium ions had a remarkable therapeutic effect on mental illness. Manic depressants and people suffering from bi-polar disorder were prescribed lithium drugs to help regulate their mood swings. Lithium when prescribed as a medication is able to help people function more successfully in their daily lives, because it "evens-out" extreme mood swings. Today lithium is given in the form of lithium carbonate, Li_2CO_3. These drugs are helpful to about 70% of the people experiencing manic depression.

Interestingly, scientists are still not sure exactly how the lithium ion helps these patients, but they think it has something to do with it's similarities to the sodium ion, Na^+. It is hypothesized that the smaller size of the lithium ion may alter the function of some neurotransmitters, and this may account for its therapeutic effect. One theory is that Li^+ is transported across the cell membrane instead of Na^+ and this decreases the amount of Na^+ entering the cell, or neuron. In this way the excitability associated with the release of some neurotransmitters decreases. Periodicity shows us that the lithium ion is smaller in size than the sodium ion, and this has led to some ideas about how it affects the human body.

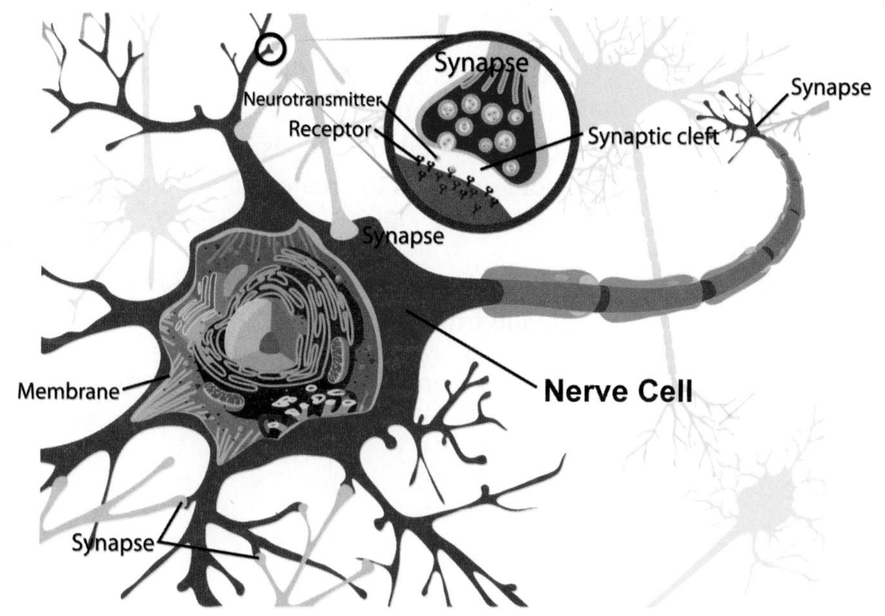

Lithium ions my help to regulate neurotransmitters, chemicals that allow nerve cells to communicate with each other. See the enlarged synapse, which is the place where one nerve cell "touches" another.

Chapter 6 Review.

Vocabulary

Match each word to the sentence where it best fits.

Section 6.1

supernova	macronutrients
trace amounts	trace element
electronegativity	periodic
ionization energy	atomic radius

1. You'll need a significant amount of _____________ if you want to stay healthy.
2. In a _____________ most of the elements in your body were created.
3. While necessary for life, too much of any one _____________ can be toxic.
4. Almost every element on the periodic table can be found in our bodies in ________.
5. The changing of the seasons is an example of a _____________ shift in weather patterns.
6. There is a phenomena called the photoelectric effect in which shining a light with high energy photons on a metal can cause the metal to eject electrons. Those photons must have energy equal to or greater than the ____________ of the atoms of that metal.
7. If I want to find an element which will strip electrons from another element during a chemical reaction, then I want one with a high _______________.
8. The _____________ of an atom is a result of the size of the largest orbital occupied by its electrons.

Section 6.3

electron configuration	Lewis dot diagrams
valence electrons	

9. By knowing the ________, it gives us deeper knowledge of the structure of the atom.
10. You find ________ at the highest principle energy level of an atom.
11. When trying to figure out a chemical formula, it is helpful to use ________, a visual representation of the outermost electrons.

Conceptual Questions

Section 6.1

12. What were the two most abundant elements created in the Big Bang?
13. If only two elements were created in the Big Bang (with trace amounts of two other elements), where did the rest of the elements on the periodic table come from?
14. Describe the distribution of metals, non-metals, and metalloids on the periodic table.
15. All elements higher than atomic number 92 are not found mineral ores, because they don't have any stable isotopes. Without a supernova here on Earth to create heavier elements, how were these elements discovered?
16. Explain what is meant by the phrase: "You are made from star dust."
17. Which element in your body was most likely created early in the birth of our universe?
    a. hydrogen
    b. oxygen
    c. carbon
    d. nitrogen

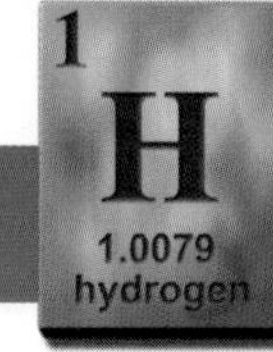

18. Describe what is meant by an "essential element."

19. If you don't eat a well rounded diet, it may be necessary for you to take mineral supplements often found in multi-vitamins. These can help provide you with, necessary trace elements that may be missing from your diet. Explain why it would not be good to take too many multi-vitamins at one time.

20. Name two functions performed by essential elements in your body.

Section 6.2

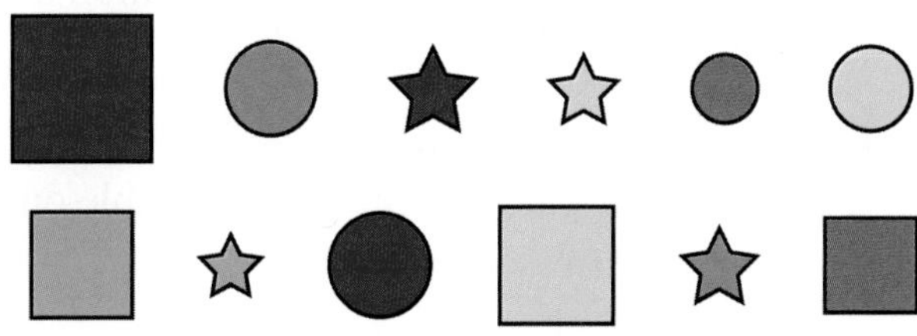

21. Given the collection of objects above, come up with at least one table that organizes them according to their properties of shape, size, and color - a periodic table of objects.

22. There are three missing objects that would complete your table. Describe the properties of size, shape, and color for these missing objects.

23. How did density and reaction patterns help Mendeleev to create the first periodic table?

24. When Mendeleev first started looking for patterns in elements, he arranged the elements first in order of increasing atomic mass. How is this different from how we would order elements today?

25. What makes a group of elements a "family?"

26. One of the most dangerous radioactive isotopes to get into your body would be some radioactive version of strontium. Strontium tends to be incorporated into your bones and teeth, instead of passing through your system quickly. Why does it make sense that strontium would be captured by your bones and teeth given the family of elements strontium belongs to?

27. Chlorine gas was once used as a chemical weapon. Using the periodic table suggest another gaseous element that may also be dangerous and explain why you chose this element.

28. Explain the difference between electronegativity and ionization energy.

29. Compare and contrast the periodic properties of atomic radius, electronegativity, and ionization energy. What is similar and what is different in how these properties map to elements on the periodic table?

Section 6.3

30. Which sub-atomic particle most directly affects the properties of an atom or molecule?
    a. proton
    b. neutron
    c. electron

31. Explain the reasoning behind your choice for the question above.

32. If the primary energy levels for electrons are represented by numbers such as 1, 2, 3, 4, etc., what letters are used to describe the energy sublevels (or orbitals) within these primary levels?

33. What are two differences between a 1s and 2s orbital? How about a 2p and 3p orbital?

34. In what order will the electrons fill up the orbitals?
    a. They will fill the lowest energy ones first.
    b. They will fill the highest energy ones first.
    c. The orbitals are filled in a random order.
35. Fill in the table below to show which type of orbital is being filled by the highest energy electron in each of the large block of elements.

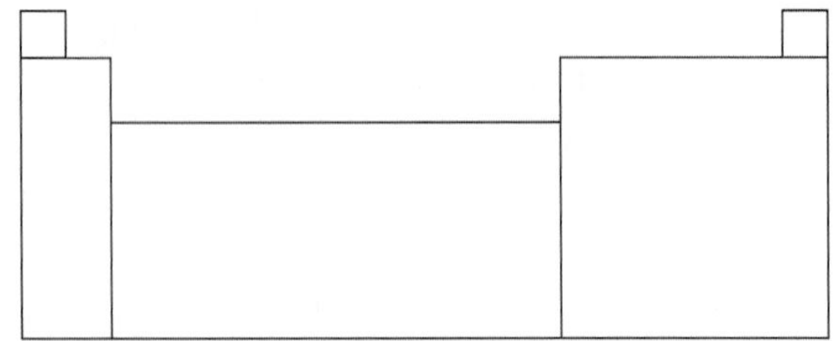

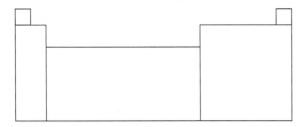

36. Of all the electrons surrounding the nucleus, why are the valence electrons considered to be the most important?
37. What is the best predictor of an element's chemical properties?
    a. Atomic radius.
    b. Density.
    c. Number of valence electrons.
    d. Its full electron configuration.
38. What is the relationship between the size of an orbital and the energy level of electrons found in that orbital?
39. Carbon can form the following two compounds: carbon dioxide (CO_2) and methane (CH_4). This means that:
    a. The number of valence electrons on carbon atoms can vary.
    b. The number of valence electrons of hydrogen and oxygen must be different.
    c. Carbon always has the same number of valence electrons.
    d. Both b and c are true.

Quantitative Problems

Section 6.1

40. If a carbon-12 atom fused with a helium-4 atom, what would the resulting new atom be?
41. Of the 4.40 x 10^{28} atoms in a typical person, 3.96 x 10^{27} of them are nitrogen. What percent of a typical person's atoms are nitrogen?
42. Of the 4.40 x 10^{28} atoms in a typical person, 8.80 x 10^{20} of them are iodine. Why is iodine considered to be found in "trace amounts" when the typical person has 880 billion billion atoms of iodine in his or her body?
43. Place the following elements in the appropriate place on the periodic table outline shown below.
    a. Hydrogen
    b. Helium,
    c. Potassium,
    d. Krypton,
    e. Copper
    f. Boron,
    g. Calcium,
    h. Phosphorus
    i. Iron,

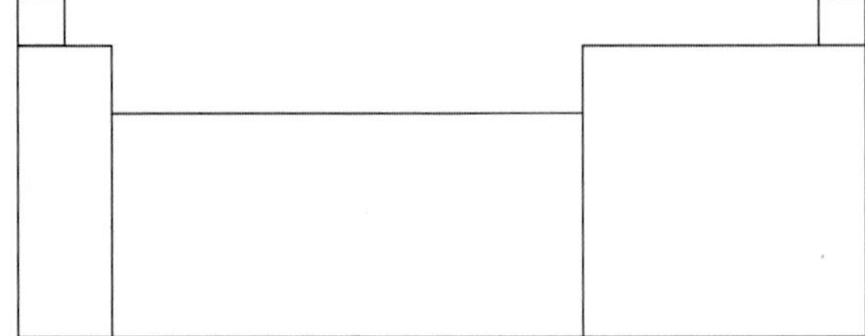

44. Which of the following atoms has the biggest radius?
    a. H or He
    b. Li or Be
    c. He or Ne
    d. Na or Mg
    e. O or S

Section 6.2

45. How many electrons can fit in an orbital?

46. Write the electron configuration for the following elements.
    a. Nitrogen (N),
    b. Neon (Ne),
    c. Sodium (Na),
    d. Boron (B),
    e. Iron (Fe),
    f. Chlorine (Cl).

47. Give electron configurations both the full and short version for each of the following elements:
    a. Silicon (Si),
    b. Molybdenum (Mo),
    c. Cesium (Cs),
    d. Uranium (U).

48. What are the similarities and the differences in the electron configurations of the following pair of atoms?
    a. Ne and Kr
    b. F and Cl
    c. Li and Na
    d. O and N
    e. C and Si

49. Given the graph below, what would you estimate would be the melting point of titanium (Ti)?

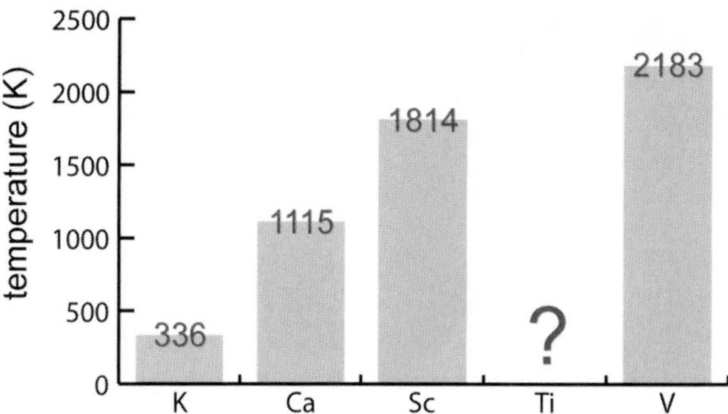

Section 6.3

50. How many electrons can be at primary energy level 3? Explain why.

51. How many valence electrons does each of the following elements have?
    a. Kr,
    b. F,
    c. Na,
    d. Al,
    e. Br,
    f. Zn.

52. Draw a Lewis dot structure for each of the following elements:
    a. N,
    b. Ne,
    c. Kr,
    d. Na,
    e. O
    f. Be,
    g. Xe,
    h. F,
    i. Li

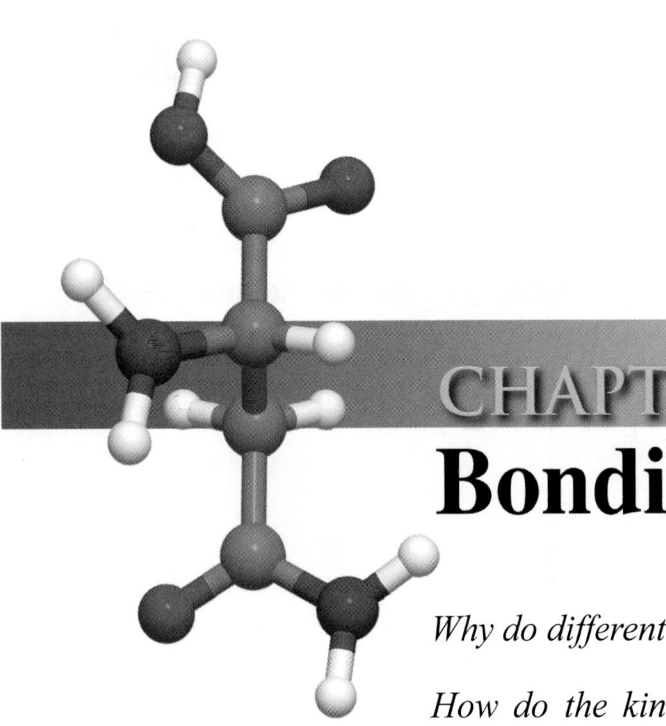

CHAPTER 7

Bonding

Why do different kinds of bonds form?

How do the kind of bonds formed between atoms give us the entire variety of materials we see all around us?

How can we use the periodic table to predict what will bond?

The elements are not what determines the variety and physical properties of matter. With the exception of the noble gasses, the elements are rarely found in pure form on Earth. *Compounds* are the matter we experience. How the elements bond together in compounds is what determines the properties of matter that we observe.

We have learned that sodium chloride (table salt) is essential for life. In solution, NaCl dissociates forms the sodium Na^{+} and chlorine Cl^{-} ions. Both ions are necessary for good health, however, the pure elements sodium and chlorine are both deadly. Pure sodium is a soft metal that reacts violently with water, producing flames and explosions. Pure chlorine is a yellowish gas that is highly toxic. There are many examples like sodium chloride where one compound is toxic yet the same elements arranged as another compound are essential nutrients!

Explosive Sodium Metal + Toxic Chlorine Gas

Edible Table Salt

This chapter is about how the elements form chemical bonds with each other. Ionic compounds, like salt, dissociate into ions in solution. Covalent compounds, like sugar, remain whole molecules even in solution. Metals such as copper and iron are neither ionic nor molecular, but something in between. Solid metals conduct electricity because their electrons are not fixed to a particular atom. Pure ionic and covalent compounds are usually insulators, and do not conduct electricity.

Iron (Fe); magnetic. Iron Sulfide (FeS); non-magnetic

What do you know about iron and sulfur? Both are present in your body and are essential nutrients. Sulfur is a yellow solid and its crystals are prized by mineral collectors. The most common form of sulfur in a chemistry lab is a finely ground yellow powder. Biologically, sulfur is in the amino acids *cysteine* and *methionine*, which are part of many proteins. Sulfur is also associated with many foul-smelling compounds, including the odor of garlic, and the smell of a skunk!

Elemental sulfur crystals

Powdered laboratory sulfur

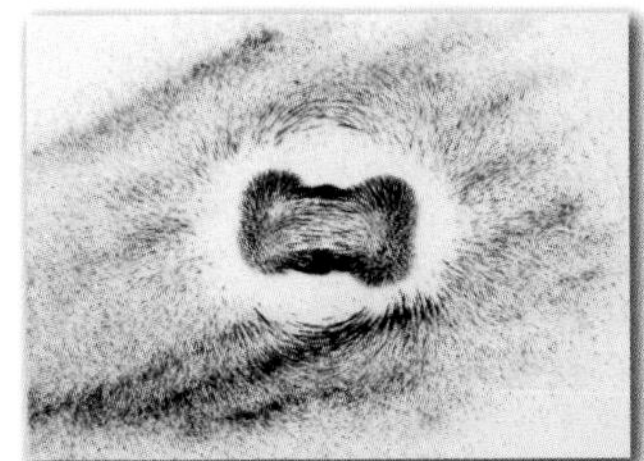
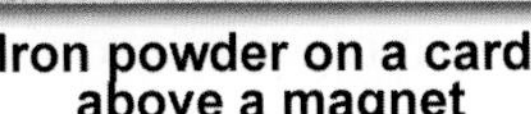
Iron powder on a card above a magnet

Iron powder

Besides being the main ingredient of steel, Iron is also the active element in *hemoglobin*, the chemical that transfers oxygen in your blood. As pure elements, iron is a gray metal. The photograph shows iron powder that has been finely ground. One of the interesting properties of iron is that it is magnetic. Iron powder sprinkled on an index card responds to the lines of force created by a magnet held below.

Mixing iron and sulfur powder makes a yellow-grey powder that can easily be separated with a magnet. However, if you heat the two substances past a certain temperature, a chemical reaction occurs. As heat is added, at first, the powders melt and seem to react. However, a magnet shows this is still a mixture of iron and sulfur. The sulfur has melted and dispersed the iron powder.

Near the "red-hot" temperature of steel the iron and sulfur DO form chemical bonds to produce the compound *iron sulfide* (FeS). Iron sulfide is a grey-brown molecular solid that is found in several naturally occurring minerals. Iron sulfide formed this way is not magnetic! The chemical bond between iron and sulfur atoms, as shown on the picture to the right, changes the magnetic properties of the substance. Magnetism, like all physical properties, depends much more on the specific compound than on the component elements.

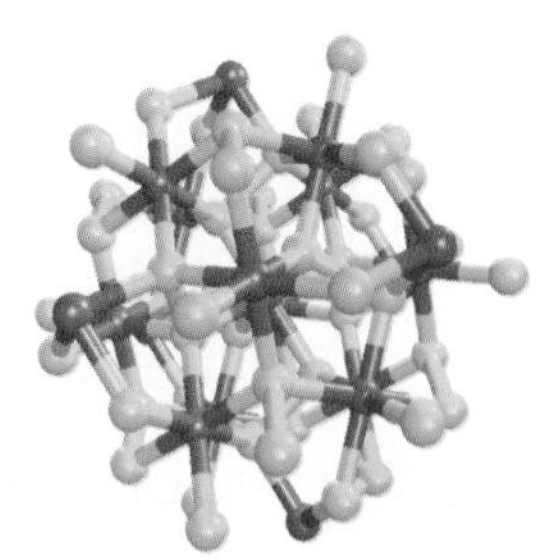

7.1 What is a Chemical Bond?

Chemical Bonds and electro-magnetic force

In your molecule building kit, a chemical bond is a "stick" joining atoms together. Diagrams show bonds as lines or sticks. Of course, in reality there are no "sticks!" Sticks and lines are used to represent the bond between atoms so you can "see" them on a macroscopic scale. Inside the atoms, the attraction between positive and negative electric charge is what really does the job. In Chapters 4 and 5, we talked about the *electromagnetic force*. In this chapter we will see that chemical bonds are not just covalent, or just ionic. There is a sliding scale between covalent and ionic, so many bonds have a little of both characteristics.

Bonds shown as "sticks" represent the binding force between atoms.

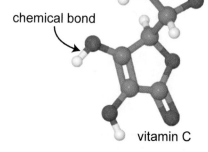

chemical bond

vitamin C

Orbital shape and charge.

Chapter 5 described the electron cloud, the three-dimensional region of space around an atom where the electrons are. The electron cloud is held together by the attraction between the negative electrons and the positive nucleus. However, the electron cloud of an atom responds to *any* electric charge, not just the attraction of that same atom's nucleus. If positive and negative charges are brought near an atom, the negative charge repels electrons and attracts the nucleus. The positive charge attracts electrons and repels the nucleus. The result is a distortion in the shape of the atom.

A Normal Helium Atom

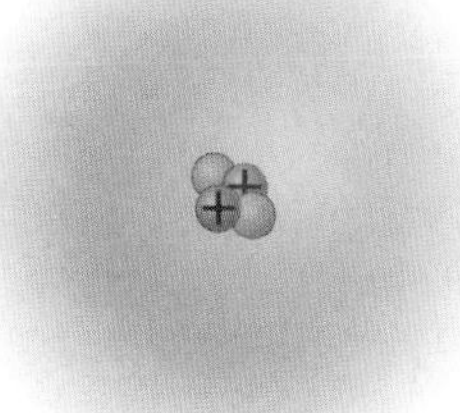

Electron distribution distorted by nearby charges

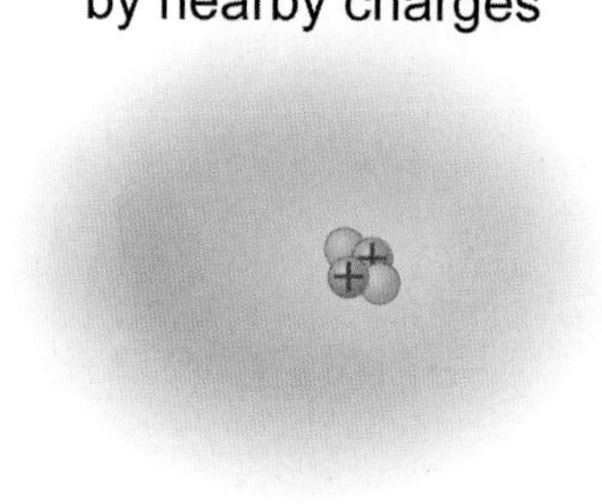

Polarization

The diagram (above left) shows a normal hydrogen atom. The diagram on the right shows the distortion caused by external positive and negative charges. The atom's negative electrons are attracted to the positive plate and its positive nucleus is attracted to the negative plate. This distortion is called *polarization*. A **polarized** atom has an uneven distribution of positive and negative charge. Polarization occurs whenever anything creates a charge outside the atom, including another atom. Polarization is how many chemical bonds form.

Chemistry terms

polarization - uneven distribution of positive and negative charge.

Forming bonds

Overlapping orbitals form a bond

The diagram on the right shows what happens as two hydrogen atoms approach each other. The electrons in each atom are attracted both to their own nucleus, and also to the nucleus of the other atom.

Orbitals of approaching hydrogen atoms feel each other's nuclei.

When atoms are close they polarize each other

As atoms begin to get close, the attraction of the "other" nucleus slightly polarizes each atom. The result of the polarization is an *attractive force* between the atoms. However, if the atoms get too close, the electrons are also repelled by the electrons of the other atom. The more the electron clouds overlap, the higher the repulsive forces between electrons.

There is a certain distance apart when the attractive forces equal the repulsive forces. If this "equilibrium distance" is close enough, then an electron will be transferred or shared and a chemical bond forms.

Two bond types

When two atoms approach each other close enough to form a bond there are two possible things that can occur. One possibility is equal sharing of electrons. Another possibility is for one atom to take electrons from the other.

Covalent bonds

Covalent bonds occur when electrons are shared between atoms. Take two hydrogen atoms as an example (see the image to the right). As the two atoms approach each other, each nucleus begins to tug on the other atom's electrons. Eventually, the orbitals of the two atoms overlap, forming a new three dimensional pattern that we call a *molecular orbital*. Notice that the highest concentration of negative charge is now between the two positive nuclei. This is what holds the atoms together in a covalent bond, the attraction between positive nuclei and negative shared electrons.

A Covalent Bond: positive nuclei attracted to negative electrons being shared between them.

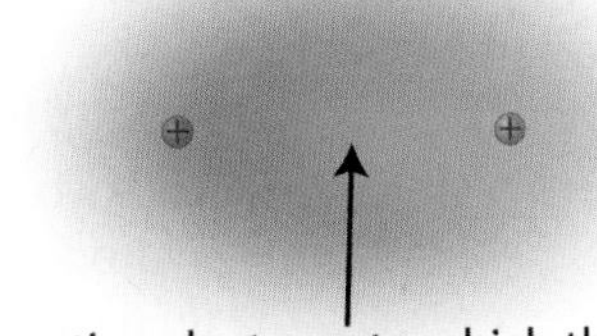

negative electrons to which the positive nuclei are attracted

Ionic bonds

The opposite extreme is when one atom takes electrons from the other. Ionic bonds happen when one or more electrons are transferred between atoms, causing the one that lost electrons to become positive, and the one that gained electrons to become negative. Now the two atoms have opposite charges, so they are attracted to each other, again through the electromagnetic force.

An Ionic Bond: the attraction between + and - ions.

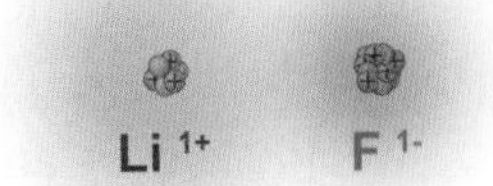

protons = 3 electrons = 2

protons = 9 electrons = 10

Showing bonds in models and diagrams

Model building limitations

Models give you a good intuitive feel for the composition, bonding patterns and overall shape of molecules. However, the models are approximations of the real thing. As with all approximations, there are limitations to how "real" any model can be to the quantum world. Size is the obvious approximation, but "solidity" and "static" are even more important and less obvious. Real atoms and molecules do not have hard surfaces and are in *constant, vigorous motion.*

Overlapping orbitals for a molecular "surface."

Atoms are not solid balls. They are made of a tiny nucleus surrounded by a cloud of electrons. When covalent bonds occur there is an overlapping of the electron clouds which forms a surface most closely resembling the "surface" view in the diagram on the right. This surface represents the boundary within which you are most likely to find all the electrons in the entire molecule.

Molecular Surfaces

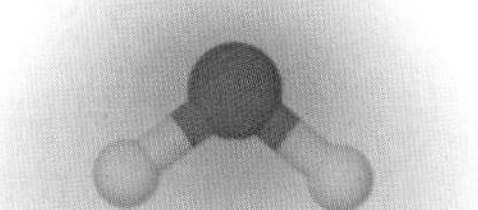

electron view combined with ball and stick view

molecular "surface" view

Using water as a simple example we can show this surface view in a couple of ways, showing a typical orbital view, or showing a surface that encloses an area where the electrons are likely to be found.

Surface color shows charge

It helps to add information that shows how evenly or unevenly the electrons are being shared. If we color the surface by charge, then we can see to what degree electrons are shared evenly or unevenly. The diagram below uses blue to represent negative, white to represent neutral, and red to represent positive.

Molecular Surfaces Colored by Charge

− +

The two atoms of hydrogen that make up a hydrogen molecule share electrons equally, so there is no surface charge.

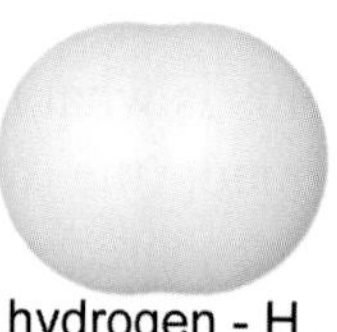

hydrogen - H_2

Electrons are unevenly shared between oxygen and hydrogen causing a surface charge to occur on water molecules.

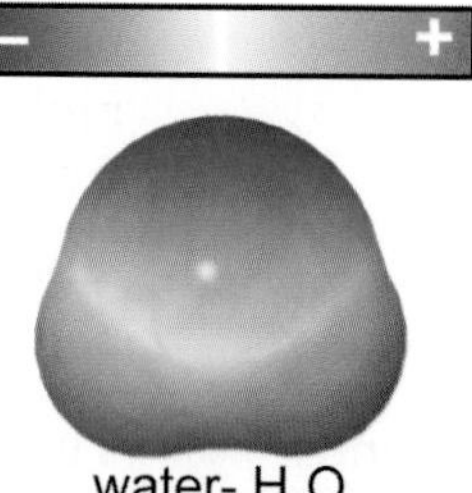

water- H_2O

Water is a polar molecule

You can see that a diatomic hydrogen molecule (H_2) is neutral all over, or *unpolarized.* A water molecule is quite different. The electrons are attracted more to the oxygen atom than the hydrogen atoms. That makes the oxygen end of the molecule more negative (blue) and the hydrogen end more positive (red). Unlike hydrogen molecules, water molecules are strongly polarized, they have a charge separation.

Predicting bond type

Periodic properties and bond type.

Since the type of bond is related to how the electrons are shared (or transferred), two periodic properties we introduced in Chapter 6 are particularly important. **Electronegativity** describes how strongly an atom attracts electrons from another atom. The higher the electronegativity, the more strongly the element attracts electrons. **Ionization energy** describes how much energy it takes to *remove* an electron from an atom. High ionization energy means it takes a lot of energy to remove an electron. Ionization energy is a measure of how well an atom can hold onto its own electrons.

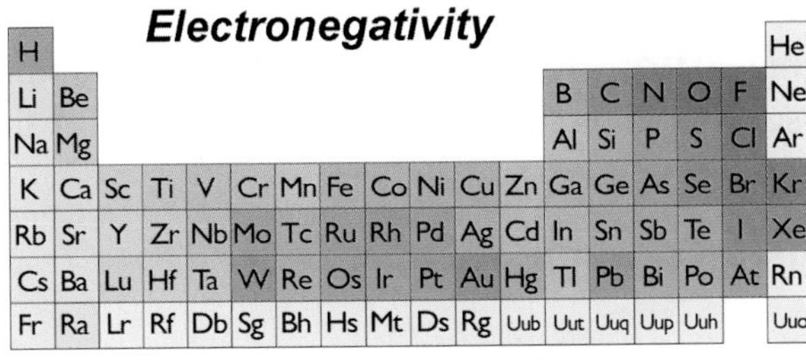

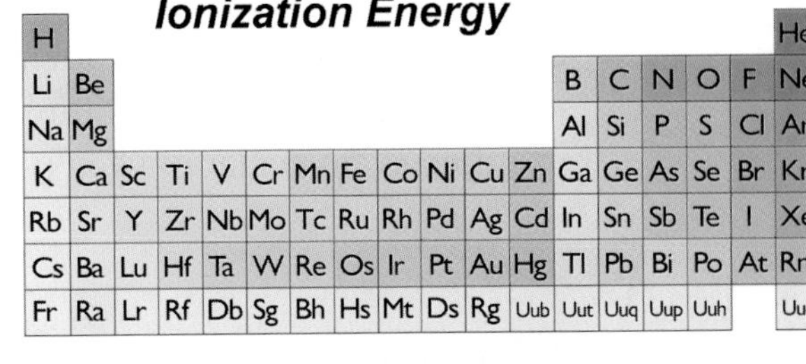

low high

low high

Bond types

The type of chemical bond (if any) between any two elements depends on their relative differences in electronegativity and ionization energy. There are three possible scenarios.

1. Both atoms have similar electronegativity and ionization energy;
2. There is a moderate difference. One atom has medium electronegativity and ionization energy, and the other has high (or low) electronegativity and ionization energy;
3. There is a large difference. One atom has high electronegativity and ionization energy, and one has low electronegativity and ionization energy.

The chart below shows how the different kinds of chemical bonds correspond to the differences in electronegativity and ionization energy.

	Non-polar Covalent Bond	Polar Covalent Bond	Ionic Bond
Atom 1	high electronegativity	high electronegativity	high electronegativity
Atom 2	high electronegativity	medium electronegativity	low electronegativity
Difference in Electronegativity	very little difference	a moderate difference	a large difference
Electron Sharing	equal or nearly equal sharing	uneven sharing of electrons	Transfer of electrons

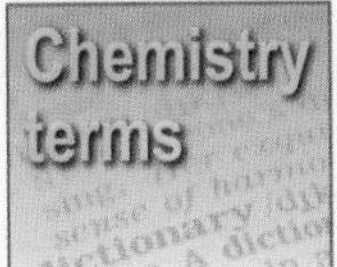

electronegativity - a value between 0 and 4 that describes the relative "pull" of an element for electrons from other atoms. High numbers mean stronger attraction for electrons.

ionization energy - the energy required to completely remove an electron

Covalent bonds

Definition of covalent bond

Covalent bonds occur when electrons are shared between atoms. Sometimes the sharing is perfectly equal, and sometimes one atom has a slight advantage over the other in the contest for electrons.

Non-polar covalent bonds

If you bring two atoms together that have high ionization energy and high electronegativity, then they will pull strongly on each other's electrons, but they also hold onto their own electrons well. These tend to be the non-metals found in the upper right part of the periodic table, as well as hydrogen.

Since their electronegativities and ionization energies are very similar, they will share electrons evenly between them, forming a **non-polar covalent bond**. Non-polar means there is no (or very little) charge separation on the surface of the molecule. A good example is molecular oxygen (O_2)

A Non-polar Covalent Bond

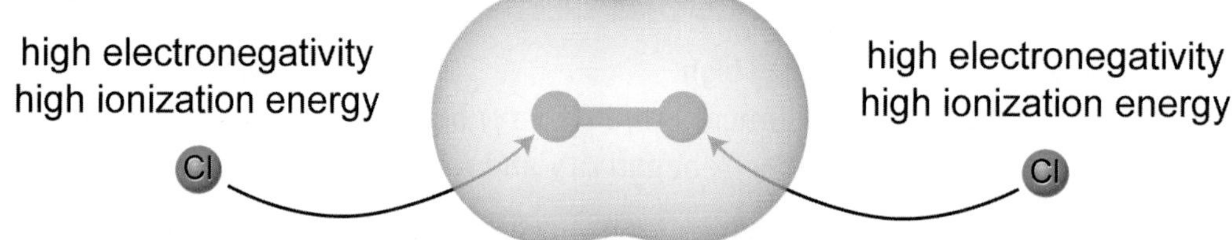

Polar covalent bonds

When one atom has an electronegativity that is moderately different from the other, then the tug of war for electrons between the two atoms will be unequal. The unequal sharing of electrons makes a **polar covalent bond**. Polar means there are one or more regions of the molecule that have an average positive or negative charge. A good example is water. Water is a polar covalent molecule.

A Polar Covalent Bond

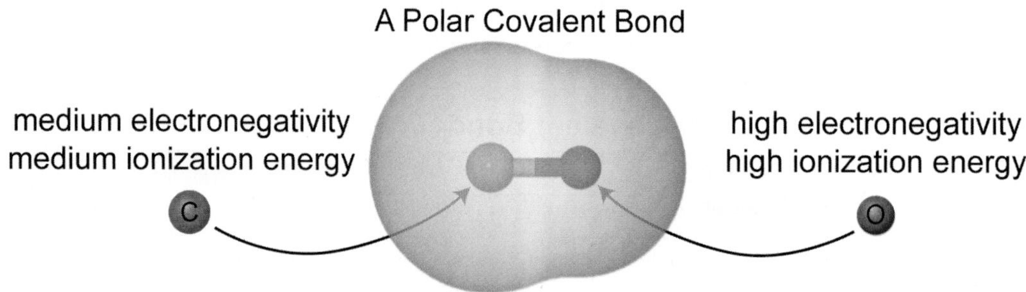

non-polar covalent bond - a bond formed between two atoms in which electrons are shared equally or almost equally between the two atoms.

polar covalent bond - a bond formed between two atoms in which electrons are unequally shared.

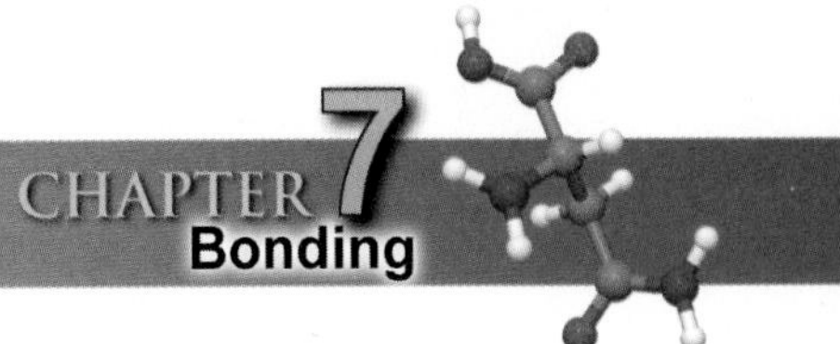

Ionic and metallic bonds

Transfer of electrons

If two atoms have electronegativities that are very different, electron "sharing" is so uneven that one or more electrons are essentially *transferred* from one atom to the other. This kind of bond often forms between a non-metal (upper-right periodic table), and a metal. For example, the bonds in calcium chloride ($CaCl_2$) would be transferred electrons because chlorine is a non-metal and calcium is a metal.

Ionic bonds

When an electron is transferred, both atoms involved become charged ions. The atom losing a electron becomes a positive ion and the atom gaining an electron becomes a negative ion. The strong attraction that between the positive and negative ions is called an **ionic bond**.

An Ionic Bond

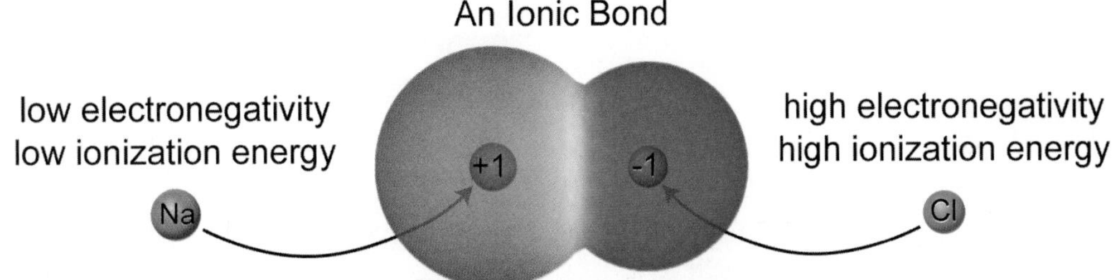

Ionic substances form crystals

Ionic bonds connect atoms to all neighbors, not just a single neighbor, as in a molecule. This is because a positive ion attracts all nearby negative ions and vice versa. For example in sodium chloride each positive sodium ion attracts all neighboring chlorine atoms. This is very different from what happens in a covalent bond. In a covalent bond, one atom is bonded to only one other atom and the pair stay together as a unit. In an ionic bond, there is no one-to-one pairing of atoms. Each ion is bonded to all of its oppositely charged neighbors.

An Ionic Crystal

Metallic bond

A metal is sort of like an ionic substance and sort of like a covalently bonded substance. Metals tend to have both a low electronegativity and low ionization energy, so they don't attract each other's electrons well, and they don't hold onto their own electrons well. The result is a bunch of atoms that share their electrons with every other atom forming a **metallic bond**. Electrons are shared like in covalent bonds, but no two atoms are specifically bonded together as in ionic compounds.

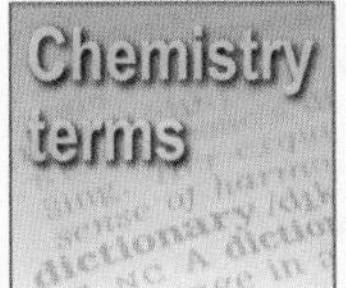

ionic bond - an attraction between oppositely charged ions. This attraction occurs with all nearby ions of opposite charge.

metallic bond - an attraction between metal atoms that loosely involves many electrons.

Electronegativity

Values of electro-negativity

The periodic table below shows the values of electronegativity for the elements. The values range from a low of 0.7 for francium (Fr) and a high of 3.98 for fluorine (F). The alkali metals are all low since they all have a single valence electron and they try to donate it. The halogens are highest since they need one electron to reach the magic number of eight electrons to make a full energy level.

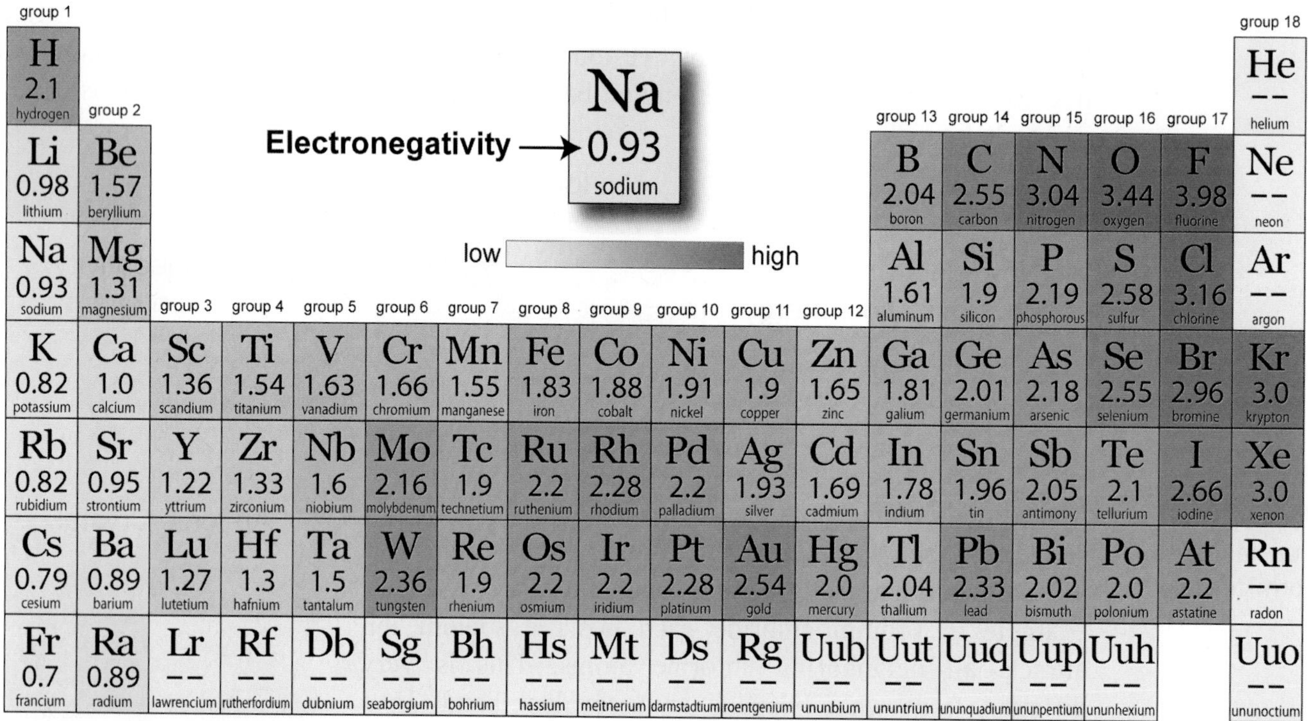

Electro-negativity and bond type

Whether a bond is covalent or ionic mostly depends on the difference in electronegativity between the two bonded atoms. If the difference is low, then you get a non-polar covalent bond. If it's high, then you get an ionic bond. In reality, there is a sliding scale of bond type ranging between these two extremes. The smallest possible difference in electro negativities is zero, while the largest difference is 3.28 (3.98 - 0.7 = 3.28). The chart below shows what kind of bond forms over the range of possible differences.

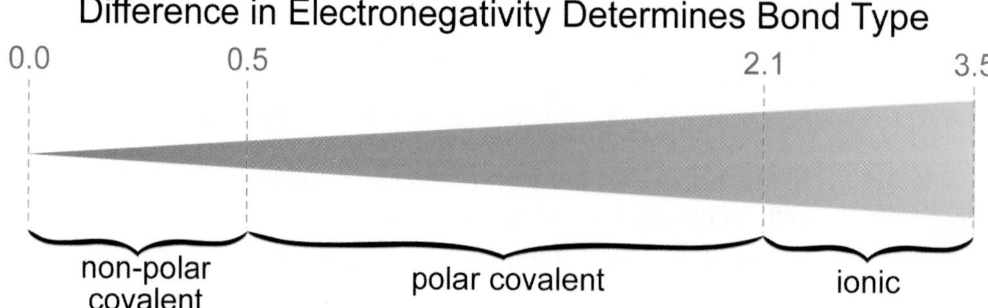

Differences in electronegativity determine bond type

Classifying chemical bonds

While there really is no sharp dividing line, most chemists classify bond types into the three categories: non-polar covalent, polar covalent, and ionic. If you take the electronegativities of the two bonded atoms and subtract them, you get a number that tells you how to classify the bond. If the difference is less than 0.5, it is considered to be non-polar covalent. If it is between 0.5 and 2.1, then the bond is considered to be polar covalent. If the difference is greater than 2.1, then the bond is considered to be ionic.

Bond Type	Electronegativity Difference
non-polar covalent	less than 0.5
polar covalent	between 0.5 and 2.1
ionic	greater than 2.1

Two examples

Diatomic compounds make the best examples since there are only two atoms to consider. The diagram shows a charge-shaded molecular surface with the ball-and-stick inside. Diatomic iodine (I_2) is a non-polar covalent bond, since both atoms have the same electronegativity. Carbon monoxide is a polar covalent bond since the difference in electronegativity is 0.89. The oxygen atom has the higher electronegativity so the bonding electrons spend proportionally more time near the oxygen than the carbon nucleus. The shaded surface shows that the oxygen end of the molecule is more negative (blue) and the carbon end is more positive (red),

Non-polar Bond Calculation

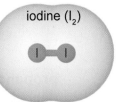

iodine (I_2)

Electronegativity of atoms:
I = 2.66
I = 2.66

Electronegativity difference:
I - I = 2.66 - 2.66 = **0**

Polar Bond Calculation

Electronegativity of atoms:
C = 2.55
O = 3.44

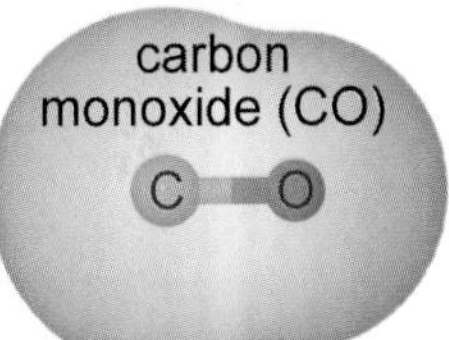

Electronegativity difference:
O - C = 3.44 - 2.55 = **0.89**

Water is made by bonding two hydrogens to an oxygen. Are the bonds non-polar covalent, polar covalent or ionic?

water

Asked: *What kind of bonds are used to make water?*

Given: *The electronegativity of oxygen is 3.44.*
The electronegativity of hydrogen is 2.1.

Relationships: *A difference in electronegativities < 0.5 is non-polar covalent, between 0.5 and 2.1 is polar covalent, and >2.1 is ionic.*

Solve: *3.44 - 2.1 = 1.34*

Answer: *The bond formed between hydrogen and oxygen is polar covalent.*

Polar bonds may make a polar molecule (or may not!)

Molecules may contain both polar and non-polar bonds

Most molecules are made up of more than two atoms and have more than one bond. Molecules that have multiple bonds could include all polar bonds, all non-polar bonds, or some mixture of the two. When chemists talk about substances being polar or non-polar, they are talking about the *whole molecule*, not a single bond.

A Phospholipid Molecule
(found in cell membranes)

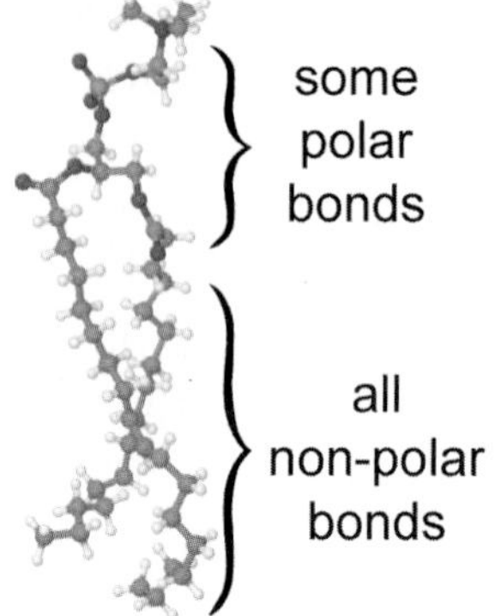

Polar bonds versus polar molecules

In a molecule, when electrons are pulled more towards one atom than another, it causes an uneven distribution of the electrons. Since the electrons are negative, this results in an uneven distribution of charge. Polar molecules have at least one area where the electrons are not distributed evenly. Non-polar molecules have an even distribution of electrons over the entire molecule.

Polar Bonds in a Molecule Make the Molecule Polar

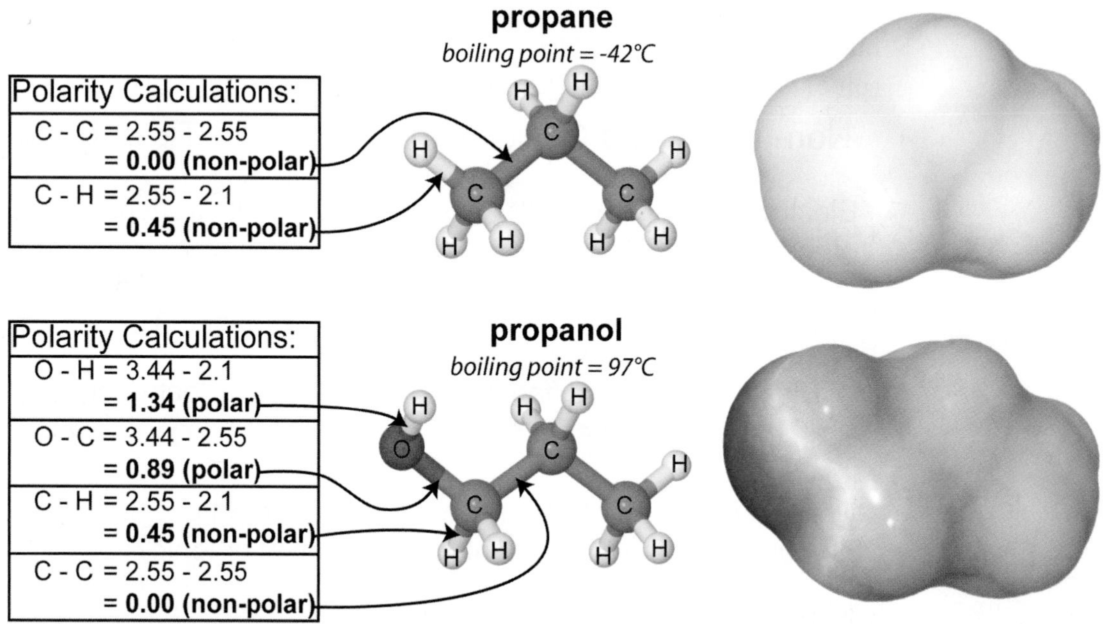

Polarity and physical properties

Understanding the polar or non-polar nature of molecules is crucial for understanding the physical properties of substances. For example, on the previous page you calculated that both bonds in water are polar covalent, so that makes water a polar molecule. It is the polar nature of water that makes life on this planet possible. If you want to invent chocolate that will melt in your mouth, but not in your hand, then you need to understand the polar or non-polar nature of your ingredients!

7.2 Valence Electrons and Bonding Patterns

Only valence electrons form bonds

When a chemical bond forms some valence electrons are either shared or transferred between atoms. Only the *unpaired* valence electrons in an atom participate in chemical bonds. for complex reasons, the 5th, 6th, and 7th valence electrons pair up and reduce the number of electrons available for bonding. The diagram below shows the main group elements along with their paired and unpaired valence electrons.

The Number of Valence Electrons Affects Bond Number and Ion Charge

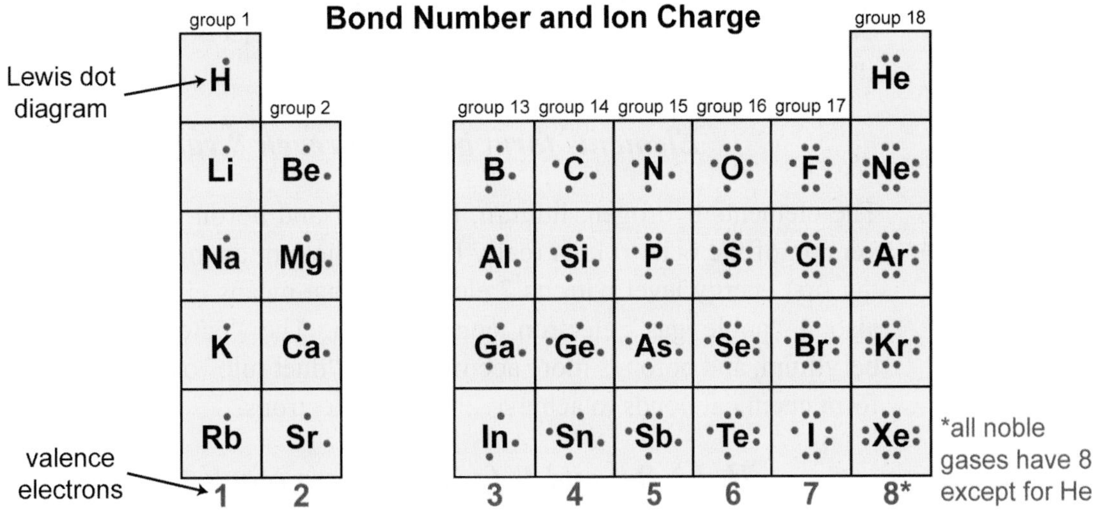

Lewis dot diagram

group 1	group 2	group 13	group 14	group 15	group 16	group 17	group 18
H							He
Li	Be	B	C	N	O	F	Ne
Na	Mg	Al	Si	P	S	Cl	Ar
K	Ca	Ga	Ge	As	Se	Br	Kr
Rb	Sr	In	Sn	Sb	Te	I	Xe
1	2	3	4	5	6	7	8*

valence electrons

*all noble gases have 8 except for He

Each unpaired valence electron can form ONE covalent bond

In a molecular compound, each unpaired valence electron can form one covalent chemical bond. For example, both nitrogen (N) and phosphorous (P) atoms each have three unpaired valence electrons. In molecular compounds, these elements both form three covalent bonds.

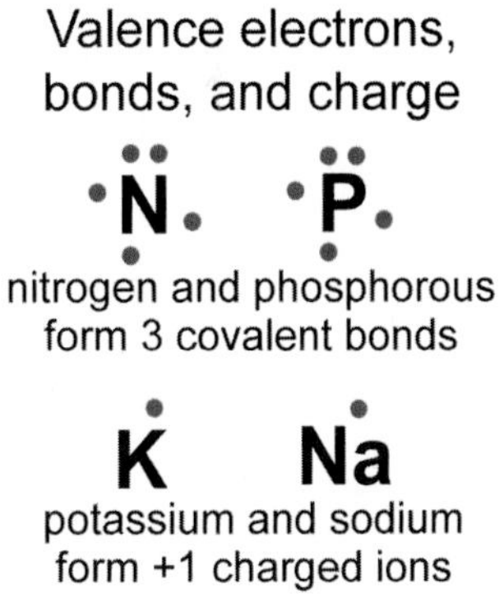

Valence electrons, bonds, and charge

N P

nitrogen and phosphorous form 3 covalent bonds

K Na

potassium and sodium form +1 charged ions

In a molecular compound, each unpaired valence electron forms one covalent bond

Ion charge and valence electrons

When forming a positive ion, each valence electron can contribute 1 unit of positive charge to the ion. This occurs when an atom loses its valence electron(s) to another atom, leaving a charge of +1 for every valence electron lost. For example, sodium (Na) and potassium (K) have a single valence electron and form singly charged ions (+1). Magnesium (Mg) and calcium (Ca) have two valence electrons each. These elements form ions with a charge of +2.

In positive ions, each valence electron can contribute 1 unit of positive charge

The octet rule

eight valence electrons = chemical stability

In Chapter 6 you learned that the noble gasses have eight valence electrons (except for He which has two). These elements form no bonds because they are already at the most stable configuration of electrons. From the noble gasses we infer that elements with eight valence electrons are chemically stable. The observed behavior of the rest of the elements confirms that elements form chemical bonds to achieve the "magic" configuration of 8 valence electrons. This observation is known as the *octet rule*. The **octet rule** states that elements transfer or share electrons in chemical bonds to reach a stable configuration of 8 valence electrons.

8 Valence e⁻ are stable

:Ne: neon forms no bonds or ions

Elements form bonds to reach 8 valence electrons

2 valence electrons are most stable for H, Li, Be, and B

The elements hydrogen, lithium, beryllium, and boron have so few electrons that their version of the octet rule is really based on helium as the closest noble gas. Helium fills the first energy level with its 2 electrons. That means either 0 or 2 valence electrons are also a "noble gas" electron configuration. The "octet rule" for hydrogen, lithium, beryllium, and boron is more accurately the "duet rule" or "the rule of 2." These elements form chemical bonds to achieve 2 valence electrons.

H, Li, Be, and B form bonds to reach 2 valence electrons

Shared electrons are counted by both atoms

In a covalent bond, a shared electron gets counted as a valence electron by *both atoms*. For example, molecular hydrogen (H_2) shares electrons to get 2 valence electrons - like helium. In water (H_2O) each of the two hydrogen atoms shares 1 electron with oxygen giving each hydrogen 2 valence electrons and oxygen 8 valence electrons.

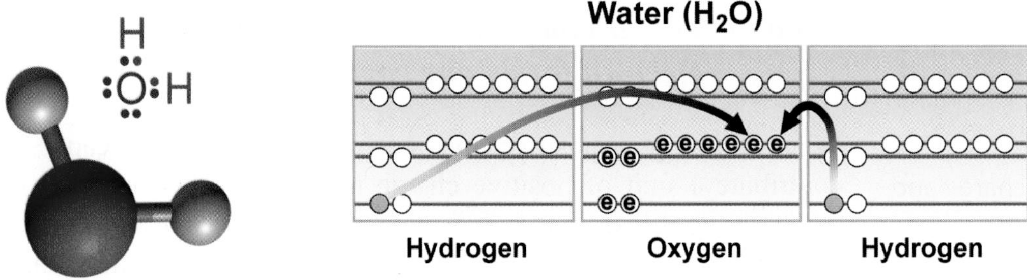

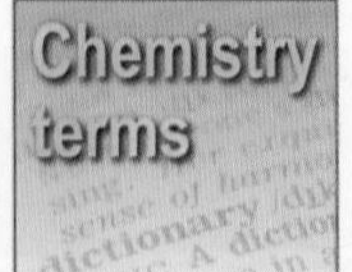

octet rule - elements transfer or share electrons in chemical bonds to reach a stable configuration of 8 valence electrons. The light elements H, Li, Be, and B have helium as the closest noble gas so the preferred state is 2 valence electrons instead of 8.

Valence electrons and ion formation

ionic charge and the periodic table

There are two ways in which elements satisfy the octet rule: by sharing electrons in a covalent bonds, or by transferring electrons in an ionic bond. When the difference in electronegativity is larger than 2.1, the chemical bond is ionic. Most elements form some ionic compounds. The table below shows the most common charges for ions of the main group elements.

Valence Electrons in Combination With Properties of Ionization Energy and Electronegativity Determine Specific Ionic Charge

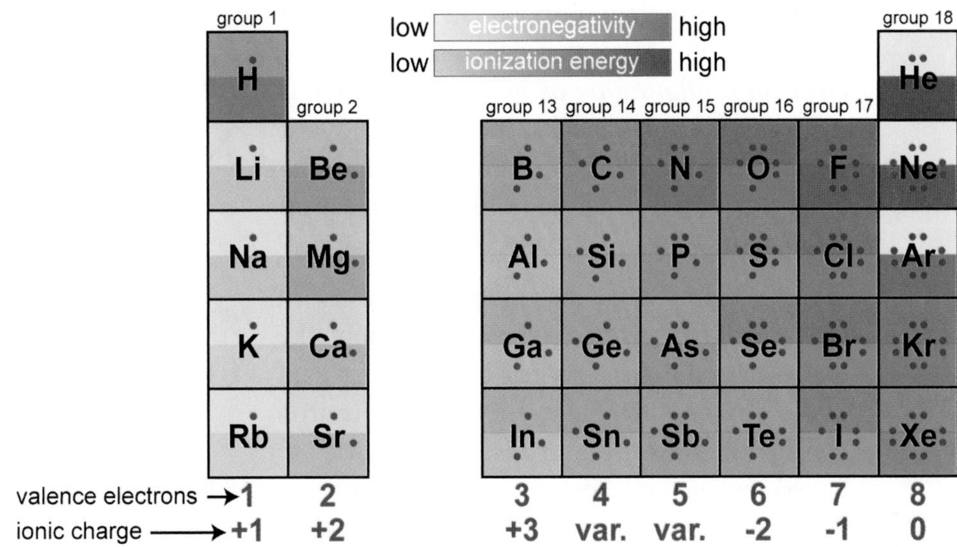

Ion patterns

You should see several patterns:

1. Elements on the left side tend to form positively charged ions, and elements on the right side form negatively charged ions;
2. Elements in the middle sometimes form positive and sometimes form negative ions and are marked as "variable" charge;
3. The alkali metals (group 1) form +1 charged ions because they have one valence electron;
4. The alkali earth metals (group 2) form +2 charged ions because they have two valence electrons;
5. Boron-like elements (group 13) form +3 charged ions because they have three valence electrons;
6. Oxygen-like elements (group 16) form -2 charged ions. These elements have 6 valence electrons. Their easiest path to the octet rule is to gain 2 electrons;
7. The halogens (group 17) form -1 charged ions. These elements have 7 valence electrons. Their easiest path to the octet rule is to gain one electron.

Electron configuration of ions

Why sodium forms a +1 ion

Sodium is an alkali metal with one valence electron. Sodium usually loses its one valence electron to become Na^{+} in order to satisfy the octet rule and have an electron configuration like that of a noble gas. Theoretically, sodium could also gain 7 electrons to become Na^{-7}, but losing one is so much more likely that we do not ever observe Na^{-7} ions in nature. The electron configuration of Na, Na^{+}, and Ne are shown below. Note that Na^{+} *has the same electron configuration as neon*. Sodium tends to form Na^{+} ions because this is the lowest energy path by which sodium can satisfy the octet rule and reach a noble gas electron configuration.

Neutral sodium

$$Na = 1s^2, 2s^2, 2p^6, 3s^1$$

Lose 1 electron

Sodium ion Na^{+}

$$Na^{+} = 1s^2, 2s^2, 2p^6$$

Neon

$$Ne = 1s^2, 2s^2, 2p^6$$

Why oxygen forms a -2 ion

Oxygen atoms have six valence electrons. In order to have an electron configuration like that of a noble gas, oxygen could either gain two electrons to have the same electron configuration as neon, or lose six electrons to have the same electron configuration as helium. Considering that oxygen has high ionization energy, it is not likely to lose six electrons. Also, oxygen has high electronegativity, so it has the ability to grab electrons from other atoms. For oxygen it makes sense that it will gain two electron rather than lose six.

Neutral oxygen

$$O = 1s^2, 2s^2, 2p^4$$

Gain 2 electrons

Oxygen ion O^{2-}

$$O^{2-} = 1s^2, 2s^2, 2p^6$$

Neon

$$Ne = 1s^2, 2s^2, 2p^6$$

Write the electron configuration for a magnesium ion (Mg^{2+})

Asked: *electron configuration of Mg^{2+}*

Given: *Mg, atomic #12, charge of +2.*

Relationships: *The electron configuration of magnesium is $1s^2$, $2s^2$, $2p^6$, $3s^2$*

Solve: *Mg must lose 2 electrons to become Mg^{2+}. Therefore it loses the pair of $3s^2$ electrons.*

Answer: *The electron configuration of Mg^{2+} is $1s^2$, $2s^2$, $2p^6$, which is identical to neon.*

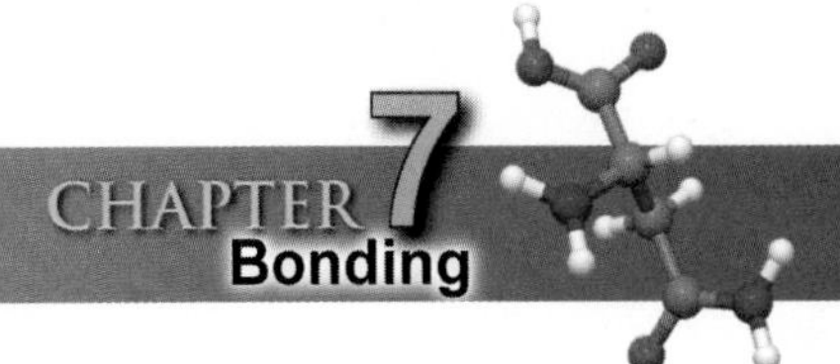

Simple ionic formulas

Ionic crystals

Ionic substances typically form a crystal, which is a large group of oppositely charged ions arranged in a regular pattern. The calcium chloride ($CaCl_2$) crystal in the diagram is a good example. Calcium chloride is often used to melt ice on roads because it is better for the environment than sodium chloride (also used to melt ice). Ionic crystals are neutral even though they are formed through the attractions of trillions of charged ions. You can have any number of ions in the crystal as long as the positive charges exactly balance the negative charges.

Calcium Chloride - $CaCl_2$
two chloride ions for each calcium

Why calcium chloride has the formula $CaCl_2$

To determine the formula of an ionic compound, you need to balance the positive and negative charges. For example, calcium makes a Ca^{2+} ion. Chlorine makes a Cl^- ion. Each calcium atom loses two electrons and each chloride ion gains only one. This means the compound requires two chloride ions to have the same amount of negative charge as one calcium ion. This makes the ratio of calcium to chlorine 1:2, and the formula is therefore $CaCl_2$.

Ca^{2+} ← each calcium ion comes with a +2 charge

Cl^{1-} ← each chloride ion comes with a -1 charge

Ca^{2+} Cl^{1-}
Cl^{1-}

+2 -2
total charge from ions

→ $CaCl_2$
forumula shows a 1:2 ratio of calcium to chlorine

What is the correct formula for calcium oxide, a compound used in making paper, pottery, and adjusting the pH of soils?

Asked: *What is the formula for the ionic compound calcium oxide?*

Given: *Calcium oxide is made from calcium and oxygen ions. Calcium forms +2 ions and oxygen forms -2 ions.*

Relationships: *Ca^{+2} and O^{-2} must combine in a ratio that will balance out the positive and negative charges.*

Solve: *The charge on one Ca^{+2} ion will balance out with the charge on one O^{-2} ion. Therefore the ratio is 1:1 and the formula is CaO.*

Covalent bonds

Covalent bond formation

In covalent bonds, electrons are shared between atoms, not transferred. The number of covalent bonds is equal to the number of *unpaired* valence electrons. The diagram below shows the number of covalent bonds by the main group elements.

Valence Electrons in Combination With Properties of Ionization Energy and Electronegativity Determine Number of Covalent Bonds Formed

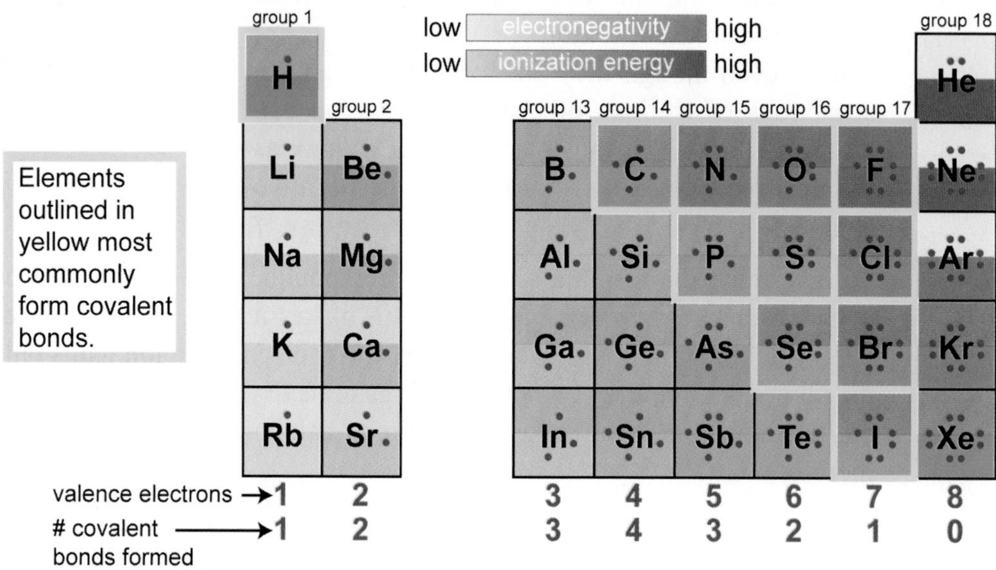

Covalent bond patterns

1. Only non-metals and hydrogen are commonly found as covalently bonded parts of molecules.
2. Carbon-like elements (group 14) form four covalent bonds.
3. Nitrogen-like elements (group 15) form three covalent bonds.
4. Oxygen-like elements in (group 16) form two covalent bonds.
5. Halogens (group 17) form one covalent bond.
6. The number of covalent bonds is equal to the number of unpaired valence electrons.

Paired and unpaired electrons

Going from left to right across the periodic table, notice that nitrogen has three unpaired electrons and *one pair*. The 8 electrons in s and p orbitals act like 8 strangers filling up four bench seats on a bus. Everyone prefers their own seat so up to four, each person sits solo. The fifth person must pair up with someone. The same is true of electrons. The fifth valence electron pairs up and is *no longer available for bonding*! For reasons of quantum mechanics, only unpaired electrons form bonds.

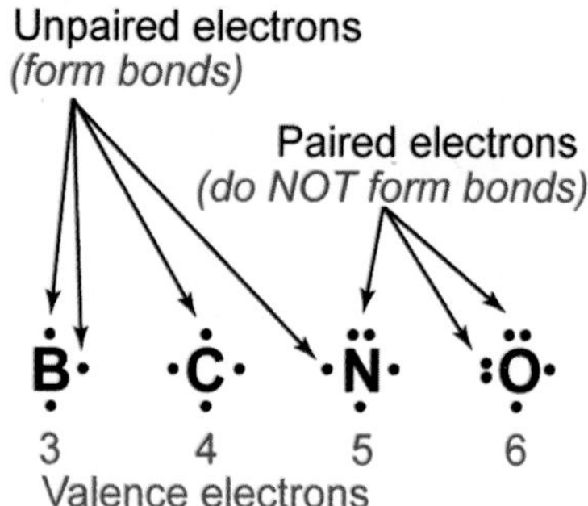

Atoms form bonds with every unpaired electron

Every unpaired electron forms a bond

Take a look at the molecule of vitamin C to the right. This molecule is made from several carbon, oxygen, and hydrogen atoms. Take a close look at each carbon atom. You will see that every carbon atom has four covalent bonds to other atoms. The oxygen atoms all from two covalent bonds, and the hydrogen atoms all form one covalent bond.

Vitamin C - All atoms of the same element form the same number of bonds.

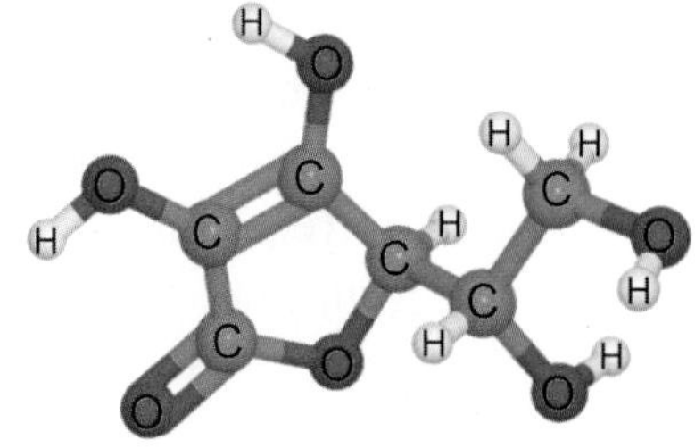

Covalent bonds and noble gases

Carbon has four valence electrons and they are all unpaired. Therefore carbon forms four chemical bonds. Oxygen has six valence electrons. Four of these electrons are paired and two are unpaired. Oxygen always forms two covalent bonds because it has two unpaired electrons. Hydrogen has one valence electron, and its closest noble gas is helium with two valence electrons. Hydrogen needs one more electron to be like helium. It gets this by forming one covalent bond with another atom.

bonds formed to have same number of valence electrons as noble gas

H = $1s^1$ → 1 bond → He = $1s^2$

C = $1s^2 2s^2 2p^2$ → 4 bonds → Ne = $1s^2 2s^2 2p^6$

O = $1s^2 2s^2 2p^4$ → 2 bonds → Ne = $1s^2 2s^2 2p^6$

Acetic Acid (Vinegar)

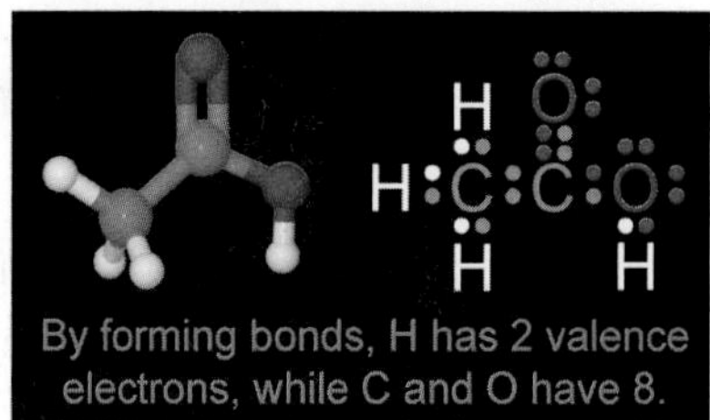

By forming bonds, H has 2 valence electrons, while C and O have 8.

Unpaired electrons create very reactive atoms or molecules

One quick way to tell how many covalent bonds an atom will form is to look at its Lewis dot structure. Atoms will form one covalent bond for each unpaired valence electron. Atoms or molecules that have unpaired electrons are highly reactive and are known as **free radicals**. These are the kind of molecules which can be responsible for aging and diseases like cancer. Sometimes free radicals are created as part of our own natural metabolism, and sometimes they are caused by outside sources like ultraviolet radiation from the sun. **Antioxidants** are considered an important part of a person's diet because of their role in preventing free radicals from reacting with and damaging DNA.

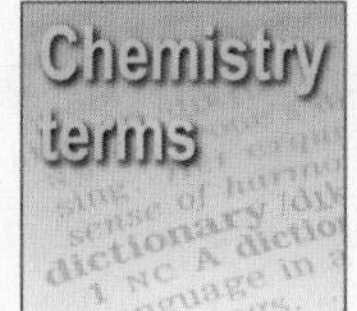

free radical - a molecule or atom that is highly reactive due to its having one or more unpaired valence electrons.

antioxidant - a molecule that reacts easily with free radicals. Some sources of antioxidants include brightly colored fruits and vegetables, vitamin E, and chocolate.

7.3 Molecular Geometry and Lewis Dot Structures

Molecular formulas and bonding

Everyone knows that the formula for water is H_2O. This tells us two hydrogens and one oxygen bond together to form a molecule of water. However, it doesn't tell us *how* they are bonded together. Are the hydrogen atoms attached to each other? You could imagine both possibilities shown in the diagram on the right. One is correct and one is not. Lewis structures are a tool for figuring out the correct way to connect atoms into molecules.

Two ways to bond two hydrogens and one oxygen

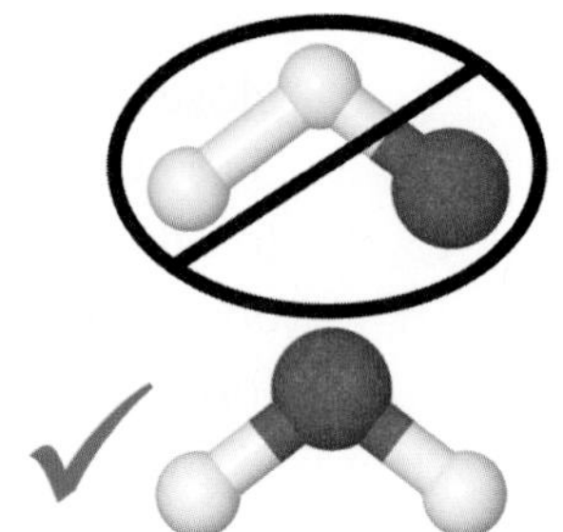

Lewis dot diagrams

If you use the Lewis dot diagrams, you can predict that both hydrogen atoms bond to a central oxygen. You can also predict the formula of H_2O. To assemble a molecule from individual atoms, you share one unpaired electron from one atom with an unpaired electron from another atom. Now both electrons "belong" to each atom, and a covalent bond is formed. Let's see how this would be done to form a water molecule.

Forming Water by Using Lewis Dot Diagrams

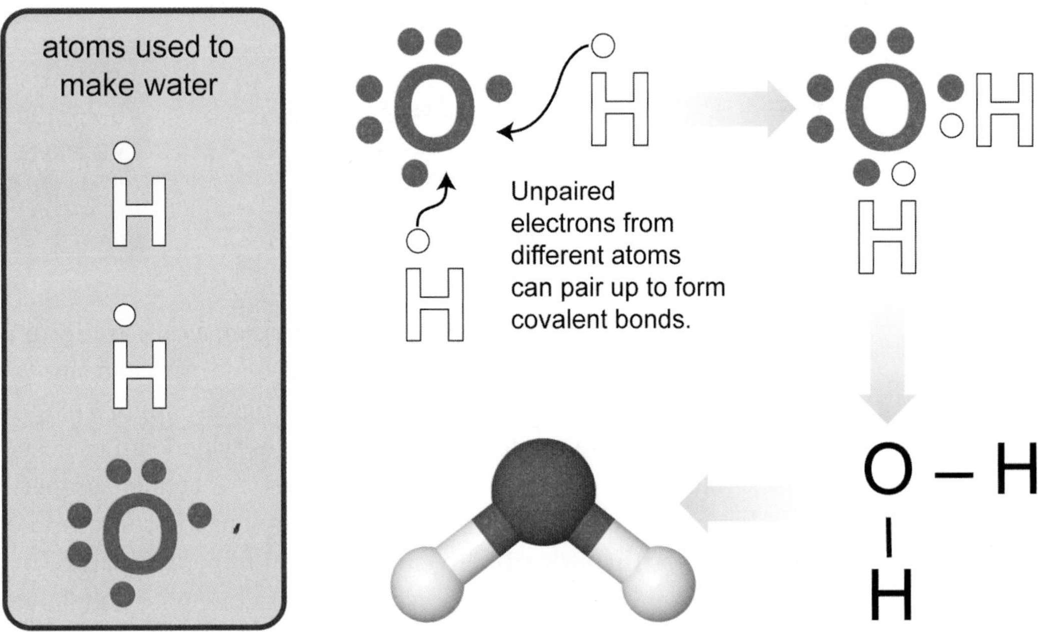

Solving the "molecule" puzzle

You can think of the Lewis structures for individual atoms like puzzle pieces that can be put together to assemble molecules. The goal is to end up with each atom having no unpaired electrons, and surrounded by eight valence electrons (unless the atom is hydrogen, in which case two paired electrons should be next to each hydrogen atom).

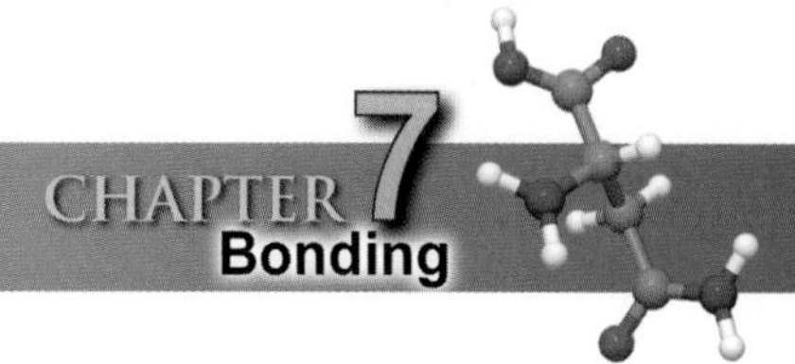

Isomers

Same molecular formula, different properties?

Sometimes there are two or more different but correct ways to fit the atomic puzzle pieces together. This gives you different molecules with the same chemical formula. For example, C_2H_6O can form ethanol, the alcohol that is found in drinks like beer, wine, and liquor. The same C_2H_6O it can also be *dimethyl ether*, often used in spray cans as a propellant. These two molecules have the same chemical formula but are very different substances with different properties. When you have multiple molecules that can be represented by the same chemical formula, the molecules are called **isomers**. The diagram shows two isomers of C_2H_6O.

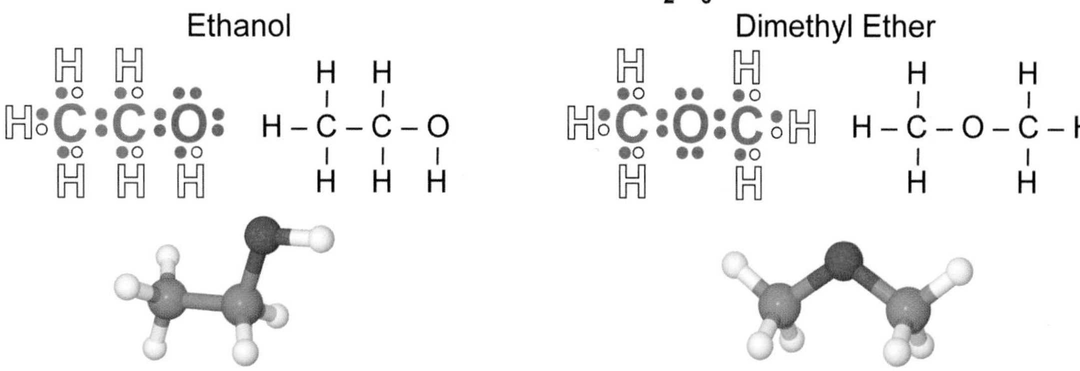

Give three isomers for the formula C_3H_8O. Show the Lewis dot diagram and the structural formula for each molecule.

Asked: *What are the Lewis dot diagrams and structural formulas for the three molecules represented by the formula C_3H_8O?*

Given: *Carbon has 4 unpaired electrons, hydrogen has 1, and oxygen has 2. 3 carbons, 8 hydrogens, and 1 oxygen form each molecule.*

Relationships: *The atoms will bond together such that all unpaired electrons will be paired up with electrons from other atoms.*

Solve:

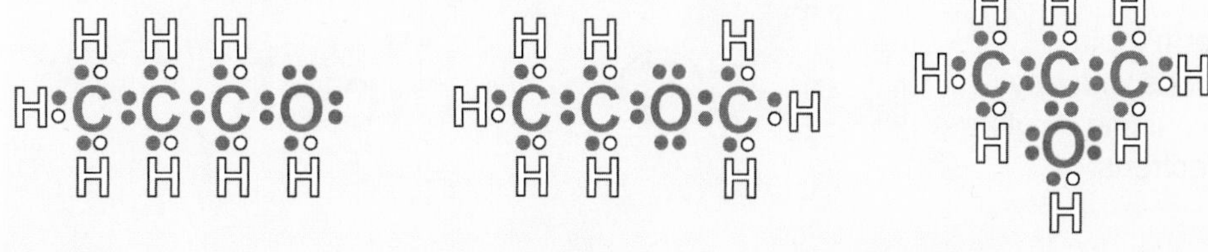

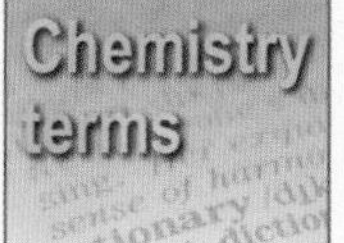

isomer - one version of how atoms can come together to form a molecule. This term is only used when a chemical formula could represent more than one molecule. By bonding the same group of atoms together in different ways you form isomers.

Double and triple bonds

Double and triple bonds

Sharing a pair of electrons (one from each atom) is called a *single bond.* It is possible for atoms to share two or even three pairs of electrons. Sharing two pairs of electrons is a double bond and sharing three is a triple bond. Carbon, nitrogen, and oxygen commonly form double and triple bonds.

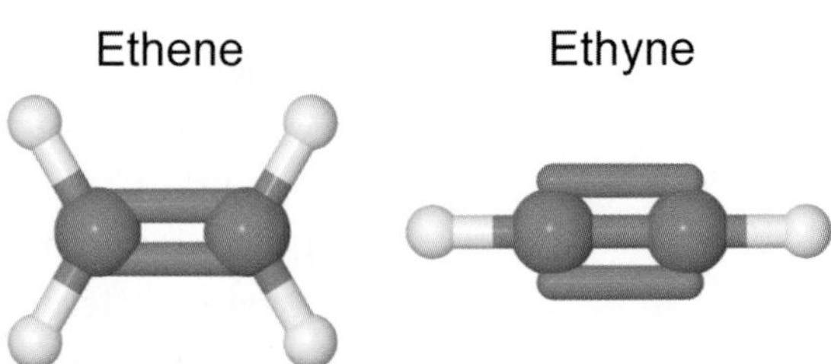

Order of bond formation

The process for writing Lewis dot structures for molecules with double and/or triple bonds is very similar to what you have done already. First you single bond all the atoms together. Once you do this you will find that there are still some unpaired electrons left over. If those unpaired electrons are on two atoms that are already sharing electrons, then you can bring those two unpaired electrons into the same region where the first pair of shared electrons is already drawn. **This can only be done with unpaired electrons on two atoms that are already sharing electrons.** If, after single bonding all of the atoms together, your unpaired electrons are not on two atoms that are already bonded, then you need to change how you originally put the atoms together to form single bonds.

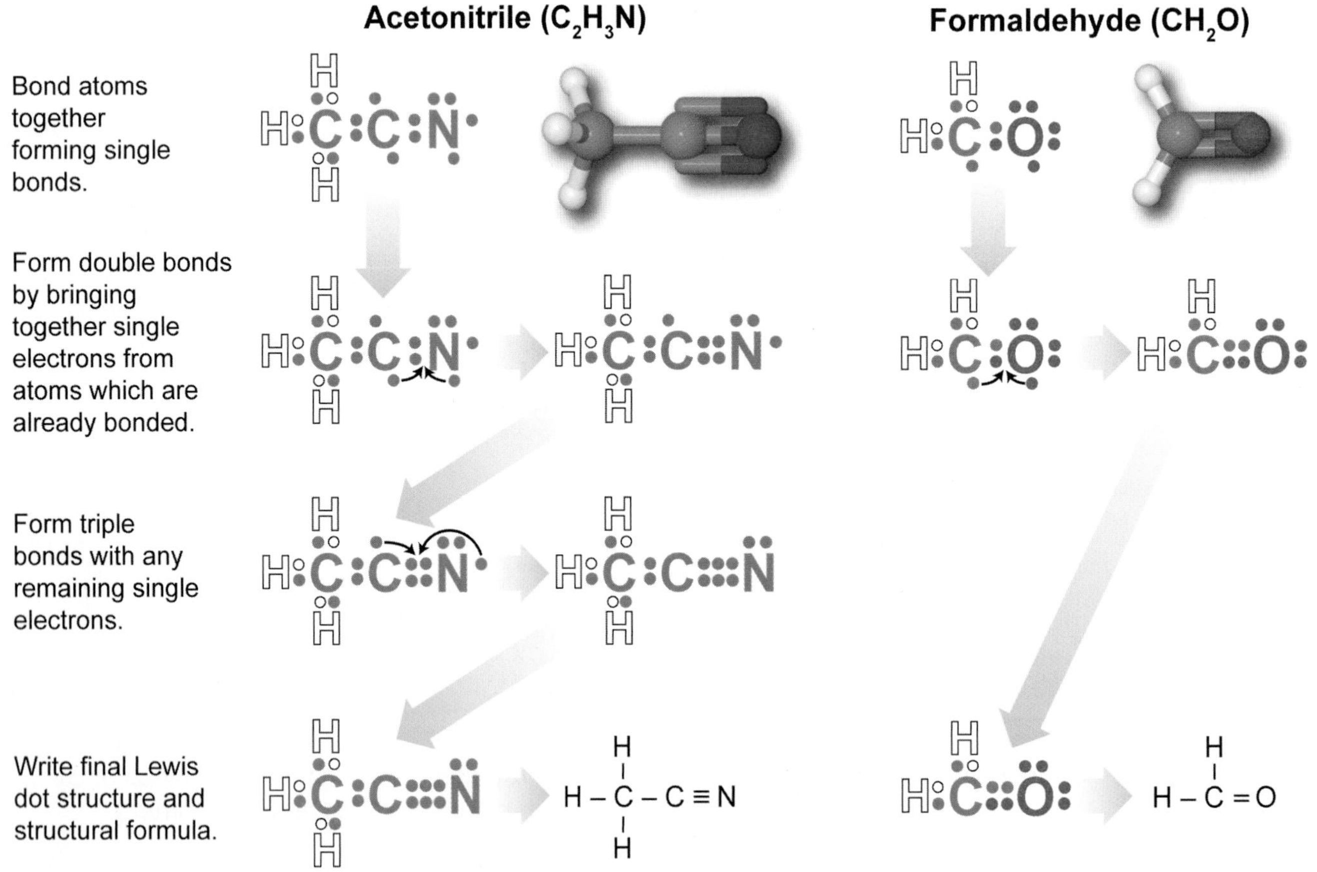

Acetonitrile (C_2H_3N)

Formaldehyde (CH_2O)

Bond atoms together forming single bonds.

Form double bonds by bringing together single electrons from atoms which are already bonded.

Form triple bonds with any remaining single electrons.

Write final Lewis dot structure and structural formula.

H–C(H)(H)–C≡N

H–C(H)=O

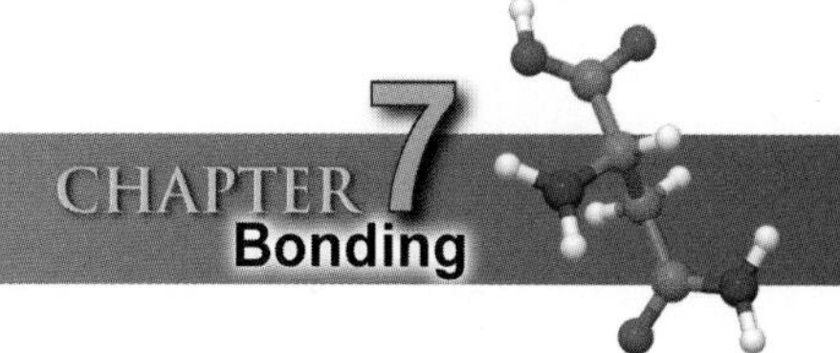

Molecular geometry

Lewis dot structures are 2D, but real molecules are 3D

Lewis dot structures can all be drawn on a flat two-dimensional piece of paper. However, molecules are three dimensional objects. In fact, the precise 3-D shape of a molecule is very important, in determining its physical and chemical properties, such as boiling point and viscosity. For complex molecules, like those typically found in biological systems, shape can make the difference between a drug that treats cancer or one that does nothing at all.

A biological molecule involved in regulating the breakdown of sugar

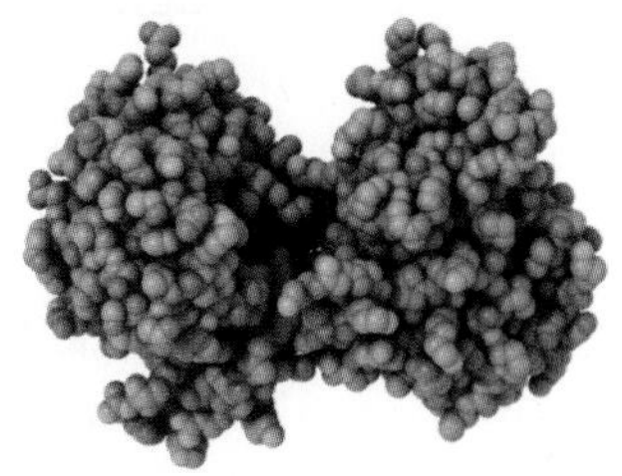

VSEPR

Using Lewis dot structures, we can predict the shapes of simple molecules and gain insights on shapes of larger ones. Using a theory called Valence Shell Electron Pair Repulsion (**VSEPR**), you can make predictions about the angles that attached atoms will form. The first three letters of VSEPR stand for "Valence Shell Electron." That means we are talking about the valence electrons, the ones represented by Lewis dots. The last two letters stand for "Pair Repulsion." Paired electrons are not shared in chemical bonds but they *do* affect the shape of a molecule. Paired electrons repel each other and also repel shared electrons that *are* involved in a chemical bond.

unshared pair of electrons

O H H

shared pair of electrons

Bonds and electron density

Because similar charges repel each other, electrons that are being shared in bonds *and* unshared pairs of electrons will push away from each other. This repulsion is part of what causes a molecule to form a particular shape. Each bond or unshared pair of electrons represents a **region of electron density** around an atom. Depending on how many regions of electron density exist, different shapes will be formed.

Examples of Central Carbon Atom With 2,3, and 4 Regions of Electron Density

two regions

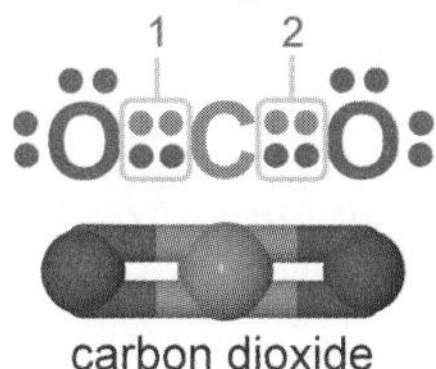

carbon dioxide

three regions

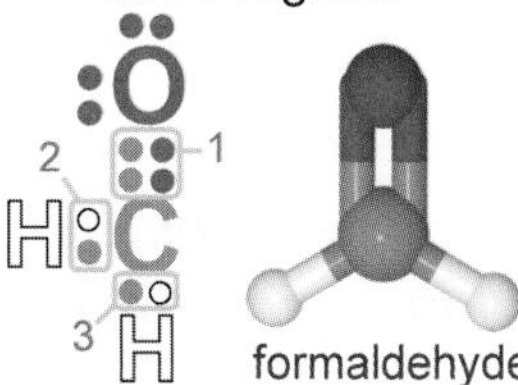

formaldehyde

four regions

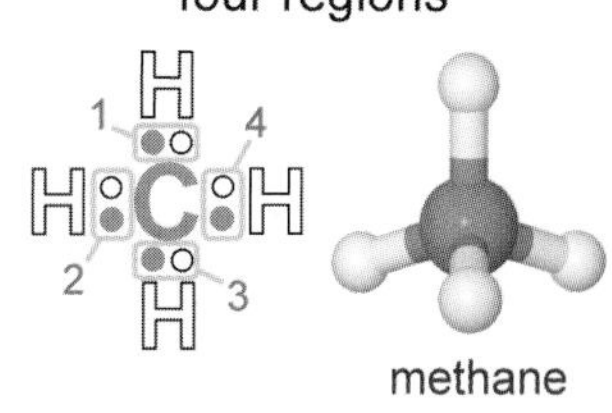

methane

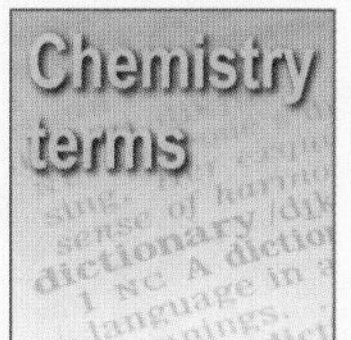

VSEPR - an acronym that stands for Valence Shell Electron Pair Repulsion, a theory that states the shapes of molecules are dictated, in part, by the repulsion of the shared electrons and the unshared pairs of electrons.

region of electron density - an area represented by shared or unshared electrons around an atom.

Two areas of electron density

Linear bond shape: two points form a line

If you rub a balloon against your hair, the balloon will grab some electrons from your hair, making the balloon negative. If you had two negatively charged balloons and brought them close together they would repel each other because they are both have the same sign of charge. They push away from each other in opposite directions until they are as far away as possible from each other.

Two charged balloons repel in opposite directions.

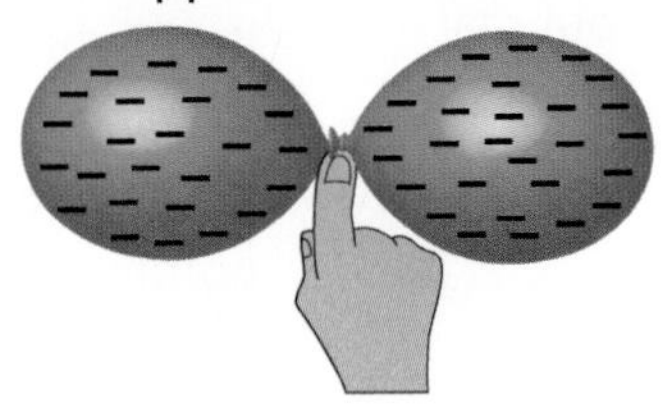

Repulsion and bond shape

The same thing happens when there are two regions of electron density around an atom. The electrons in both regions push away to be as far apart as possible. For example, consider acetylene (C_2H_2). Around each carbon there are only two areas where electrons are located, the ones being shared in the triple bond between the carbons, and the ones being shared in the single bond between the carbon and hydrogen atoms. These electrons repel each other until they are as far away as possible from each other. This causes the molecule to be *linear* around each of the carbon atoms.

Two Areas of Electron Density Repel to Form Linear Shapes

These two regions of electron density repel each other forming a 180° angle.

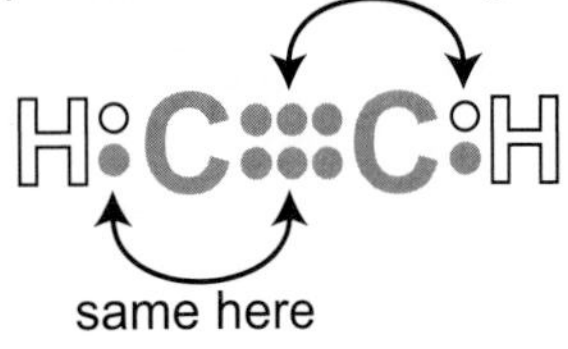

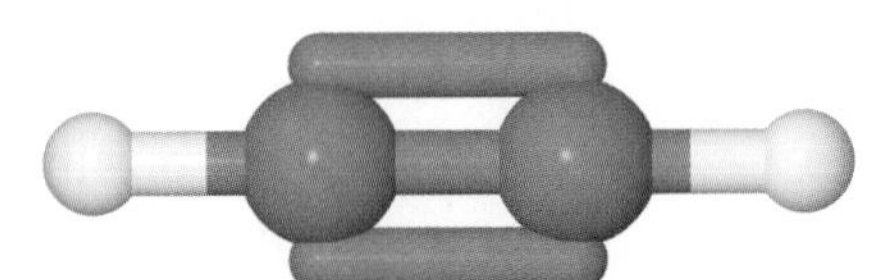

The two 180° angles formed around each carbon make the entire molecule straight.

There are two isomers for the formula C_3H_4. Show the Lewis dot diagram for each molecule, and indicate which atoms are at the center of a linear part of the molecules.

Asked: *What are the linear parts of each isomer of C_3H_4?*

Given: *There are two different isomers. Part of each molecule will be linear. The molecules are made from three carbons and four hydrogens.*

Relationships: *Each atom that has two regions of electron density around it will form a linear part of the molecule.*

Solve:

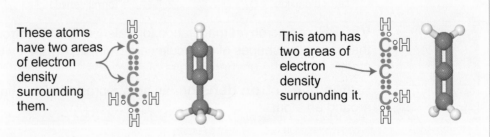

These atoms have two areas of electron density surrounding them.

This atom has two areas of electron density surrounding it.

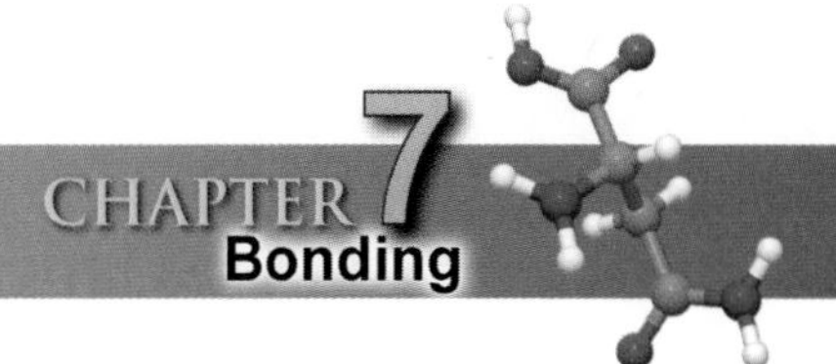

Three areas of electron density

Three points form a plane

Now imagine that you have three negatively charged balloons. Again they will repel each other and move as far away from each other as they possibly can. In this case the balloons will point into the corners of a triangle, forming a 120° angle between the balloons.

Three charged balloons repel into the corners of a triangle.

Trigonal planar

When there are three regions of electron density around an atom, the electrons in all three regions push away until they are as far apart as possible. A good example is ethene (C_2H_4). Around each carbon are three areas of electron density. One area is being shared in the double bond between carbons. The other two are shared in the single bonds between the carbon and each of two hydrogen atoms. When three regions of electron density repel each other, a **trigonal planar** shape is formed.

Three Areas of Electron Density Repel to Form Trigonal Planar Shapes

These three regions of electrons density around each carbon repel, forming 120° angles between the three atoms bonded to each carbon atom.

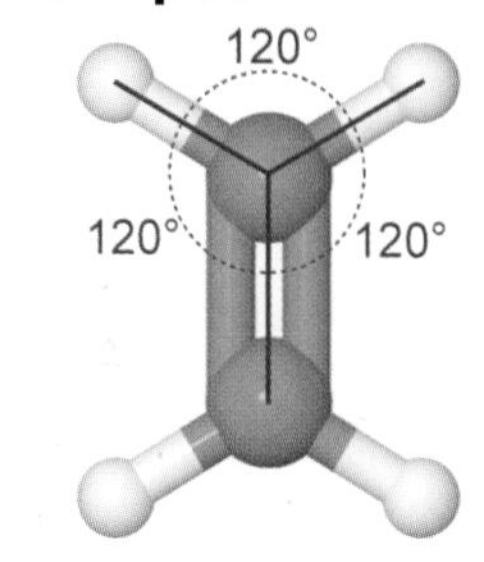

Acetic acid when mixed with water is commonly known as vinegar, and has the formula $C_2H_4O_2$. The correct isomer has both oxygens bonded to the same carbon. Draw the Lewis dot structure for this isomer and indicate where the molecule will be trigonal planar.

Asked: *What are the trigonal planar parts of acetic acid?*

Given: *The formula for acetic acid is $C_2H_4O_2$ and both oxygens are bonded to the same carbon.*

Relationships: *Each atom that has three regions of electron density around it will form a trigonal planar part of the molecule.*

Solve:

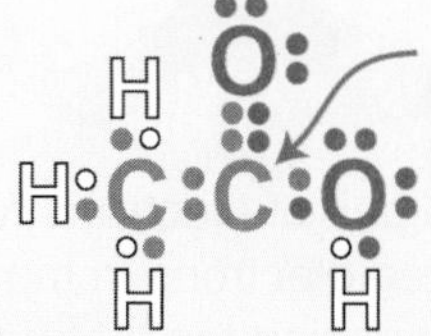

This carbon has three regions of electron density around it, so the molecule will form a trigonal planar shape around this atom.

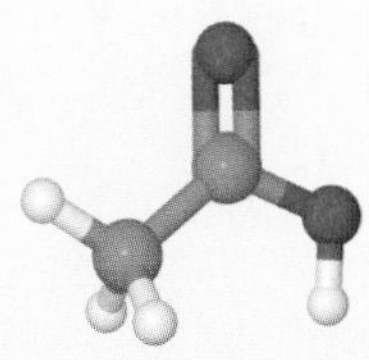

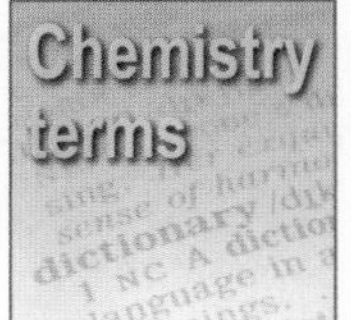

trigonal planar - a shape formed when a central atom has three regions of electron density surrounding it. Typically, this means a central atom has three other atoms bonded to it and no unshared electrons, so the three atoms that are bonded to the central atom all lie in a plane and point to the corners of a triangle. The angle between the atoms will be 120 degrees.

Four areas of electron density

Tetrahedral geometry: 3D

You might think that the shape formed by four charged balloons would look something like an "x" or a "+" with all four balloons as far apart from each other as they can get in a plane. However, there is a way to orient the balloons so that they are even farther apart – have the balloons point into the corners of a *tetrahedron*, a pyramid shape formed from four triangles. If the four balloons lie in a plane the angle between each balloon is 90°, but pointing the balloons into the four corners of a tetrahedron increases the angle to 109.5°, moving them even farther apart.

Four charged balloons repel into the corners of a tetrahedron making an angle of 109.5° between them, larger than the 90° angle that could form if they were all in a plane.

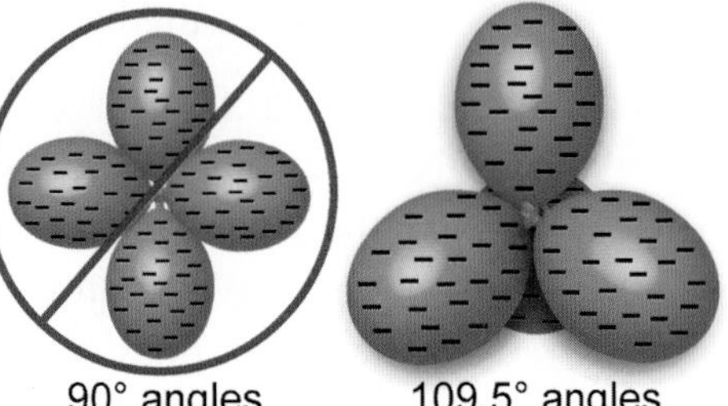
90° angles
109.5° angles

Lone pairs

In all of the Lewis dot structures we have seen so far, all of the electrons have been shared between atoms in covalent bonds. When looking at the 3D version of the molecule, each bond represented one or more pairs of shared electrons. However, atoms like nitrogen and oxygen have some electrons that are already paired, even before any bonding has occurred. These electrons, known as **lone pairs,** are "invisible" in the ball and stick view of a molecule, but their presence affects the geometry of the molecule just as much (if not more) than the electrons being shared in bonds.

Lone Pairs of Electrons Repel Just Like Shared Pairs

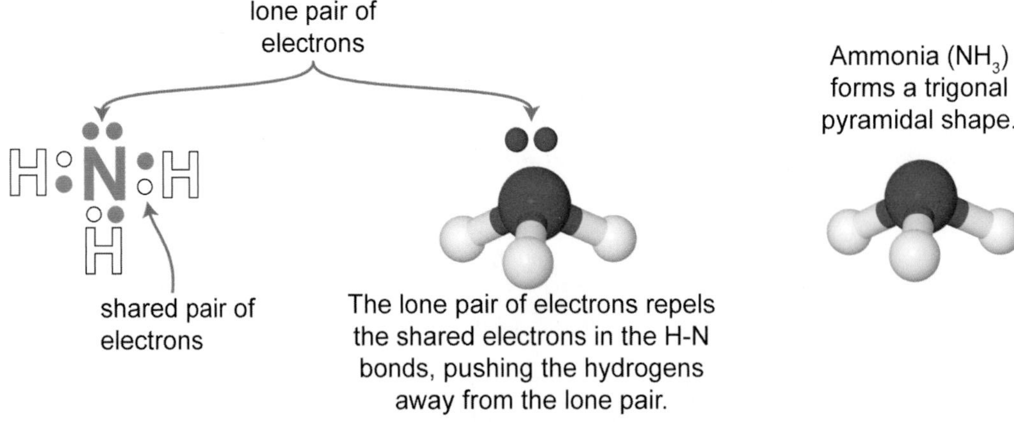
lone pair of electrons
H:N:H
H
shared pair of electrons
The lone pair of electrons repels the shared electrons in the H-N bonds, pushing the hydrogens away from the lone pair.
Ammonia (NH_3) forms a trigonal pyramidal shape.

Chemistry terms

lone pairs - electrons that are paired up in a Lewis dot diagram, but are not shared between atoms in a covalent bond.

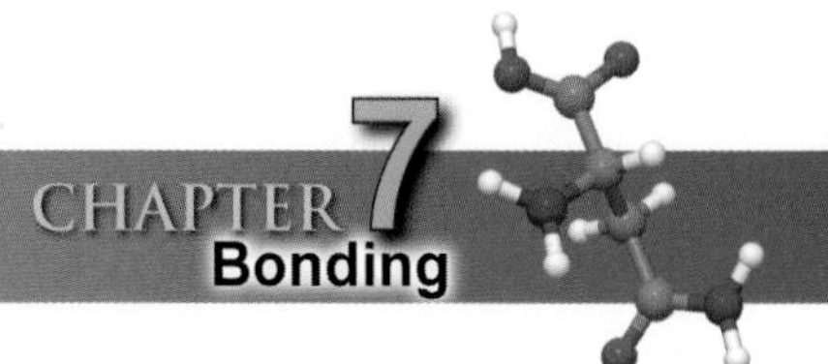

Multiple shapes due to four areas of electron density

Different geometric shapes

There are three shapes that are all based on having electron density in four places around an atom: tetrahedral, trigonal pyramidal, and bent. If there are four single bonds to a central atom, then the four bonded atoms mimic the balloons and form a **tetrahedral** shape. If there are three atoms bonded to a central atom that also has one unshared pair of electrons, then the unshared pair acts just like a shared pair, repelling the other electrons which are being shared in the three bonds. However, because one corner of the tetrahedron is "missing," we call this **trigonal pyramidal**. If there are two atoms bonded to a central atom that has two pairs of unshared electrons, then they all repel as in previous examples, except the three atoms form a **bent** shape.

Different Geometries Formed By Atoms with Four Regions of Electron Density

Tetrahedral

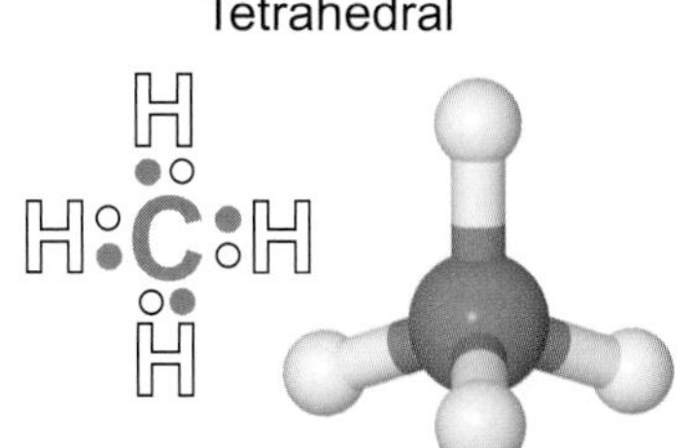

Trigonal Pyramidal

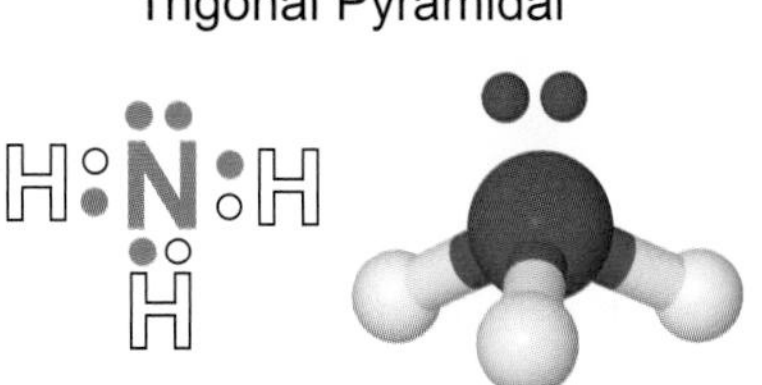

Bent

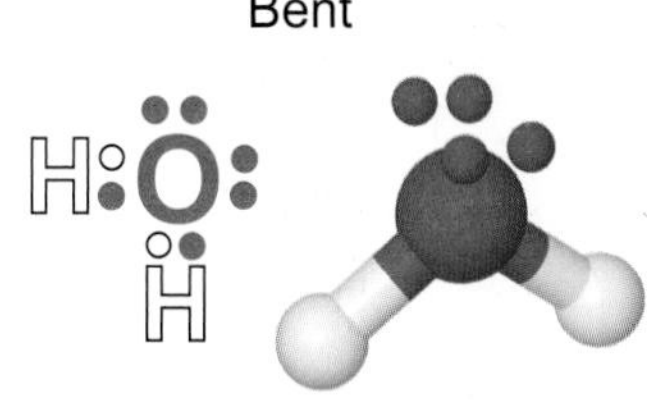

What shapes are formed within the isomer of C_4H_5NO, which has a triple bond connecting nitrogen?

Solve:

These two carbons have 3 regions of electron density, so a **trigonal planar** shape is formed around them.

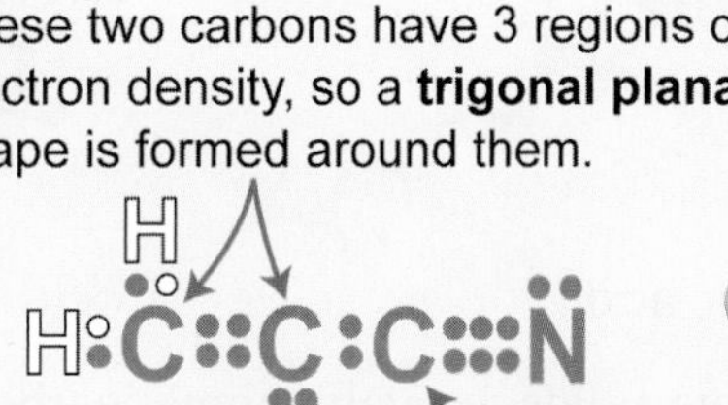

This carbon has 4 regions of electron density and is bonded to 4 atoms, so a **tetrahedral** shape is formed.

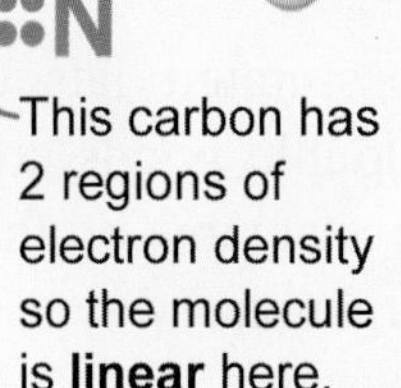

This carbon has 2 regions of electron density so the molecule is **linear** here.

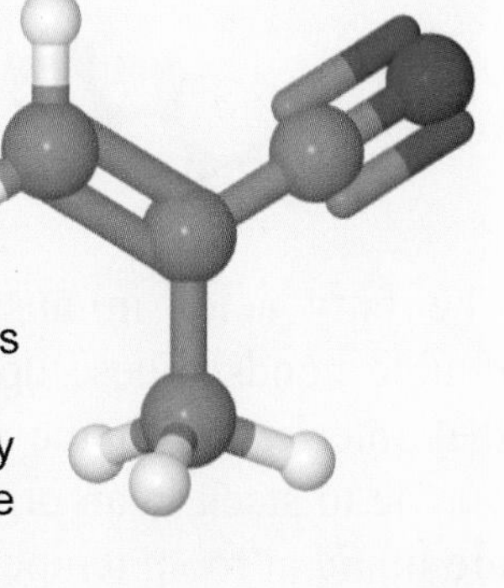

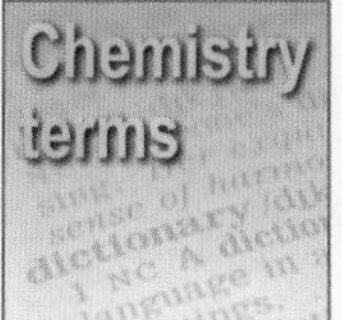

tetrahedral - a shape formed when a central atom is bonded to four other atoms that all point into the corners of a tetrahedron.

trigonal pyramidal - a shape formed when a central atom has one unshared pair of electrons and is bonded to three other atoms.

bent - a shape formed when a central atom has two unshared pairs of electrons and is bonded to two other atoms.

Why shapes and charge matter: Trans-fat and your health

You have probably heard how some fats are good and some are bad for you. Terms like *saturated*, *unsaturated*, *trans fats*, or *partially hydrogenated* can all be found on nutritional labels. These labels are really describing the different structures of fat molecules. Fats and oils are made by both plants and animals as a means for storing chemical energy. Each molecule is composed of a glycerol backbone containing 3 carbons and each carbon is attached to a long chain fatty acid. Chemists often refer to fats and oils as triglycerides. Tri- meaning three, and glyceride meaning glycerol backbone. The triglyceride resembles a three pronged fork, with each prong representing the long chain fatty acids.

A chemical reaction between fatty acids and glycerol results in a triglyceride, a typical "fat" or "oil" molecule found in living things.

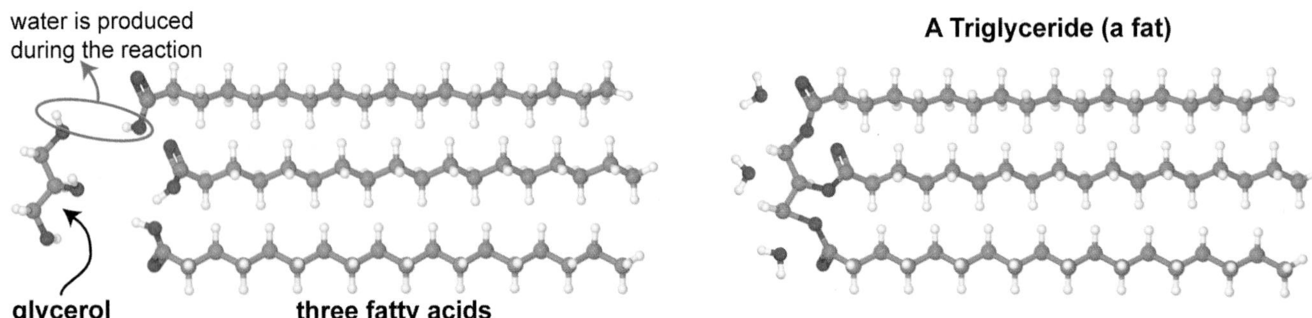

The term "saturated" means each carbon in a fatty acid chain is bonded to the largest number of hydrogen atoms possible. Saturated fats, like butter, are solid at room temperatures, because their long straight chain fatty acids contain only single carbon-to-carbon bonds. This allows them to stack very neatly together, like the logs in the walls of a log cabin.

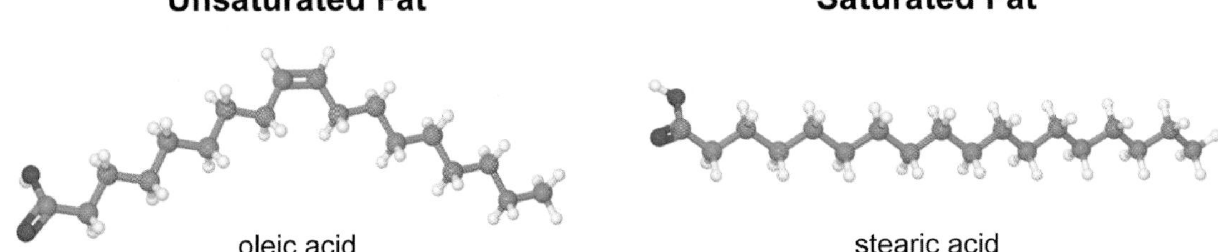

The fatty acids in unsaturated fats contain some carbon-carbon double bonds. These double bonds create "kinks" along the chain that interfere with the stacking of these fat molecules. They are unable to stack in an organized way. Unsaturated fats, like olive oil, are liquid at room temperature.

Unsaturated fats tend to be liquid at room temperature.

Saturated fats can have a negative impact on cardiovascular health. Unsaturated fats, while still high in calories, are beneficial, helping to reduce the bad cholesterol and increase the good type of cholesterol. Originally, scientists thought the story of good fat vs. bad fat ended with saturated vs. unsaturated. However, "trans" fats, unnatural fats made during the "partial hydrogenation" process, are as bad or even worse than saturated fats. The difference in this example is the way the carbons are bonded together around the double bond.

A Trans (unsaturated) Fat

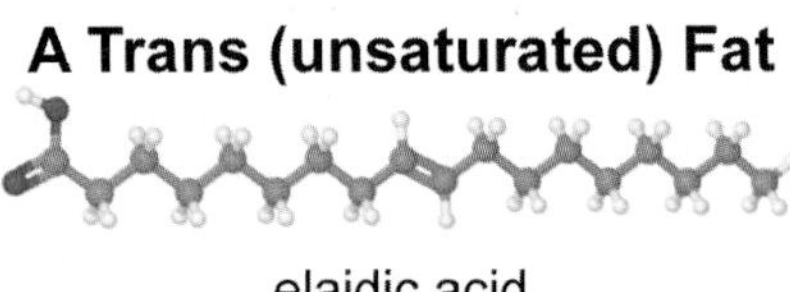

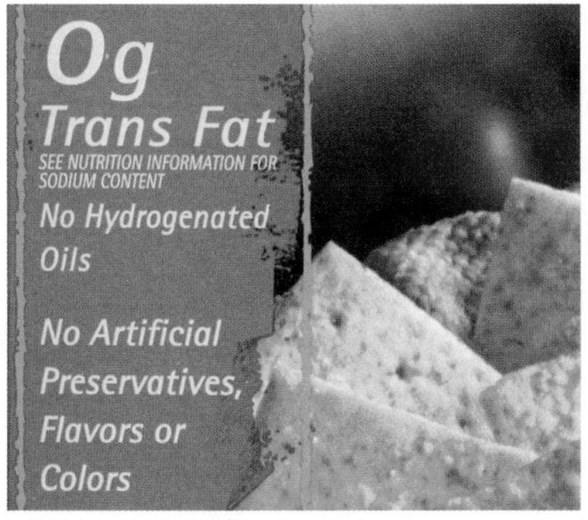

If you walk down the grocery store aisles, you will see items being advertised as containing zero grams of "Trans Fat". Some food labels also say that they are "heart healthy." Companies are making sure consumers notice their efforts to reduce or eliminate the harmful trans fats. So what is a "trans" fat, and why are they considered bad for us?

There are two very interesting facts about naturally occurring unsaturated fats. Naturally occurring unsaturated fats form "cis" double bonds. "Cis" in Latin literally means "on the same side". This refers to the fact that both hydrogen atoms are on the same side of the double bond. Secondly, "cis" fats react more readily with oxygen in the air and result in a shorter shelf life for a food product. When "cis" fats react with oxygen they break apart, forming smaller chains that do not taste good and smell bad. Manufacturers discovered that when they used the healthier naturally occurring fats to make their food products, it resulted in a much shorter shelf life. So manufacturers searched for a way to increase the shelf life of items made with unsaturated fat. The method of "partial" hydrogenation was discovered in the early 1900's, and this process allowed for a longer shelf life for food products. Crisco$^{(TM)}$ was the first product of this new technology. Proctor and Gamble$^{(TM)}$ first marketed Crisco$^{(TM}$ in 1911.

Part of a Cis-fat Molecule

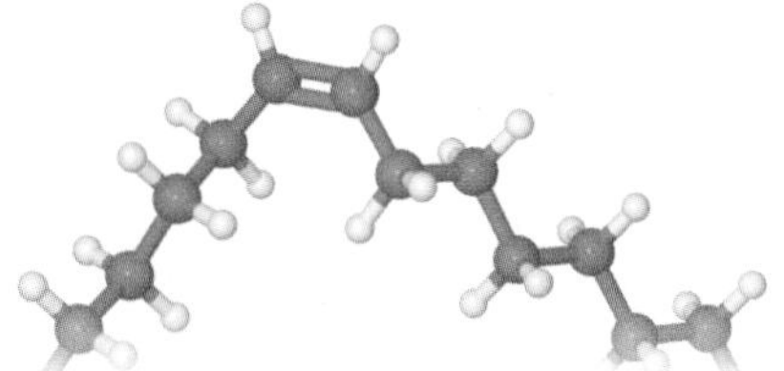

Part of a Trans-fat Molecule

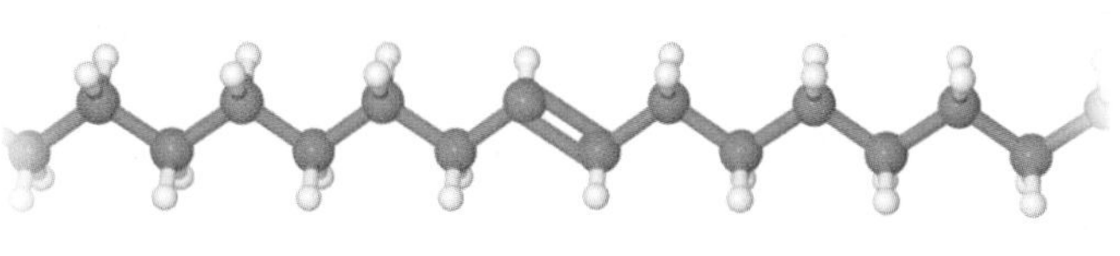

"Trans" in Latin means "across", which indicates the hydrogen atoms are on opposite sides of the double bond. When a naturally occurring "cis" fat is heated at a high pressure in the presence of hydrogen gas (H_2) and a metal catalyst, such as nickel, one hydrogen atom is added to each carbon joined by the double bond. This results in the carbon becoming single bonded as it is in a saturated fat. During the partial hydrogenation process only some of the double bonds are transformed to single bonds. The remaining double bonds are converted from "cis" to "trans". This is how a trans fat is made. Trans fats are less likely to react with oxygen than the naturally occurring "cis" fats, and this made them last longer in food products, increasing shelf life.

Originally, it was thought that these fats were healthier for us than saturated fats. Margarine was advertised as a heart healthy alternative to butter! Doctors recommended margarine for patients with heart disease or high cholesterol up until the late 1980's. Current research has shown that trans fats cause an increase in blood levels of low density lipoproteins, LDL's. These LDL's are known as the "bad" cholesterol, because as it travels in the blood it can deposit some cholesterol in the arteries. It is also important to note that just because a label says "zero grams of Trans fats" does not mean it is healthy to consume large quantities. Try to avoid eating these unhealthy fats. Instead, get as much as possible of your daily fat calories from "good oils" such as olive oil.

Chapter 7 Review.

Vocabulary

Match each word to the sentence where it best fits.

Section 7.1

electrostatic force	polarize

1. The ________________ is what holds atoms together in bonds.
2. When the electrons are shared unevenly between two atoms they cause the molecule to ________________ in that area.

Section 7.2

non-polar covalent bond	polar covalent bond ionic bond

3. A ________ forms when electrons are shared unevenly between atoms.
4. If one or more electrons are transferred between atoms causing them to become positively and negatively charged, then an ________ forms.
5. If two atoms have the same or very similar electronegativity the bond between them will be a ________.

Section 7.3

free radical	antioxidant

6. A ________ is highly reactive due to having one unpaired electron, and can do damage to other necessary molecules in your cells.
7. One reason dark colored fruits and vegetables are considered healthy is due to their high content of ________ molecules.

Section 7.4 and 7.5

isomer VSEPR region of electron density trigonal planar	lone pair tetrahedral trigonal pyramidal bent

8. If an atom has a ________ then some valence electrons are not shared in covalent bonds.
9. Using ________ theory you can predict the shapes of simple molecules.
10. Each bond and each pair of electrons not shared in a bond around an atom is considered to be a ________.
11. If you can make more then one molecule from the same set of atoms then each molecule made is called an ________.
12. The shape typically formed around an oxygen atom is ________.
13. There are two shapes that involve four atoms. The one that has all the atoms lying flat is ________, while the one where they can't be all placed in the same flat surface is ________.
14. If you had four things that are trying to repel as far away from each other as possible, they will point into the corners of a ________ shape.

Conceptual Questions

Section 7.1

15. Why don't all atoms repel each other if the outer part of the atom consists of negative electrons?
16. Describe why sharing electrons causes atoms to bond together.
17. In a ball and stick view of a molecule, what do the "sticks" represent?
18. Describe the main difference between a covalent and an ionic bond.
19. The surface view of a molecule makes it look like the molecule is "solid." What does this surface really represent?

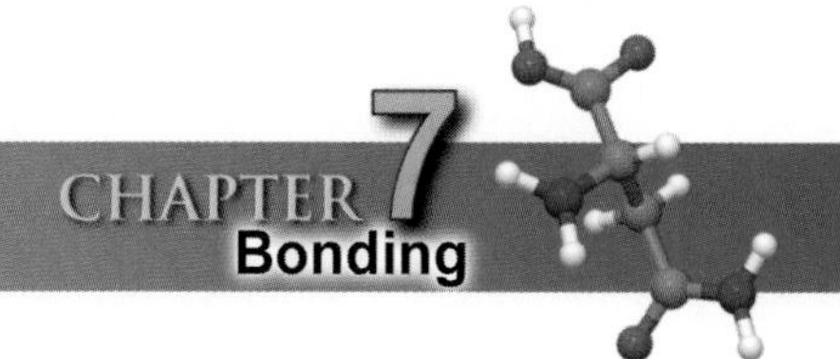

20. Describe how the red, white, and blue colors applied to some surface views of molecule tells you about the charge of the molecule near that part of its surface.

Section 7.2

21. If a covalent bond is formed when electrons are shared between two atoms, then what must be true of the electronegativity of both atoms.
    a. The electronegativity must be very different for each atom.
    b. The electronegativity must be the exactly the same for each atom.
    c. The electronegativity must be similar, but not necessarily the same.
22. Explain the reasoning behind your answer to the previous question and be sure to talk about the effect an atom's electronegativity has on electrons.
23. Describe the difference in how electrons are shared in a non-polar covalent bond versus a polar covalent bond.
24. Which of the molecules below is bonded together with a polar covalent bond, and how do you know?

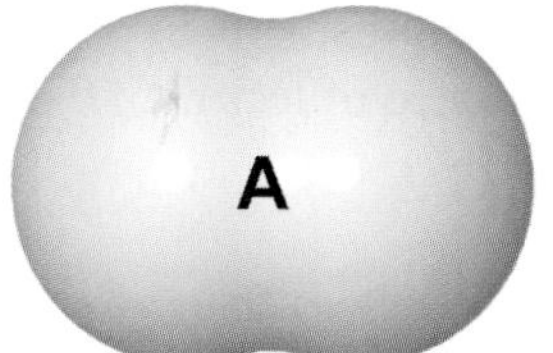

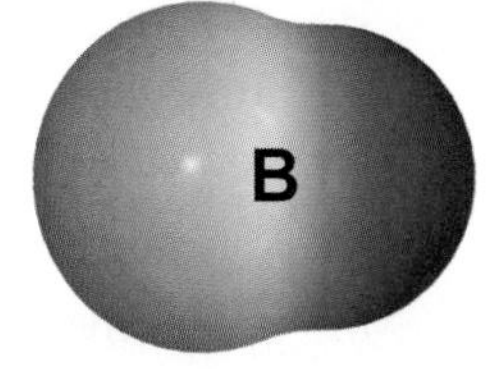

25. Which part of molecule "B" in the previous question is partially negative?
26. Why does having one atom with low ionization energy and low electronegativity paired with and atom that has high ionization energy and high electronegativity result in the formation of ions and an ionic bond?
27. Why do most atomic level images of ionic substances show many atoms forming a crystal, but images of molecular substances usually just show a specific set of bonded atoms?
28. In a chunk of metal all of the atoms are the same, so they all have the same ionization energy and electronegativity. They are both low. Why don't we call the type of bonds formed between the atoms in a metal covalent bonds?
29. Most molecules are made from more than two atoms bonded together, so there are more than just a single bond in a typical molecule. If one of the bonds is a polar covalent bond, but the rest of the bonds are non-polar covalent bonds:
    a. The molecule is considered to be non-polar.
    b. The molecule is considered to be polar.
    c. The molecule has a polar part or may be entirely polar.

Section 7.3

30. Why to atoms of elements in the same group on the periodic table tend to form the same charge if they become an ion or form the same number of covalent bonds when bonded in this way to other atoms?
31. Which valence electrons are the ones involved in ionization and covalent bonding?
    a. The unpaired electrons from the Lewis dot diagram of the neutral unbonded atom.
    b. The paired electrons from the Lewis dot diagram of the neutral unbonded atom.
32. Using the properties of ionization and electronegativity, explain why atoms of elements from the left side of the periodic table tend to form positive charges while atoms from elements on the right side of the table form negative charges?

Section 7.4

33. How many different isomers do you see below?

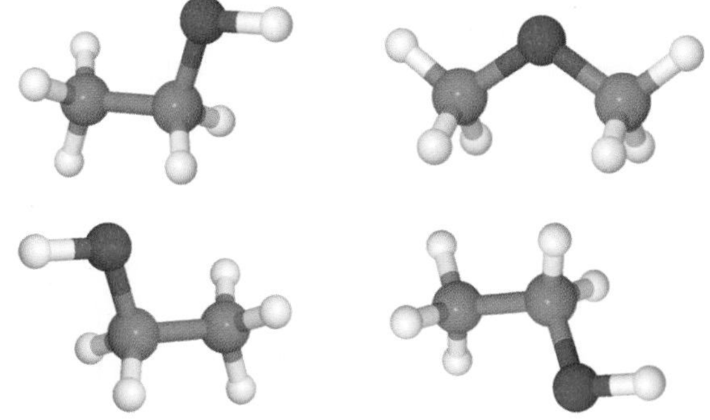

Chapter 7 Review.

34. Are these two molecules isomers of each other or not? Explain.

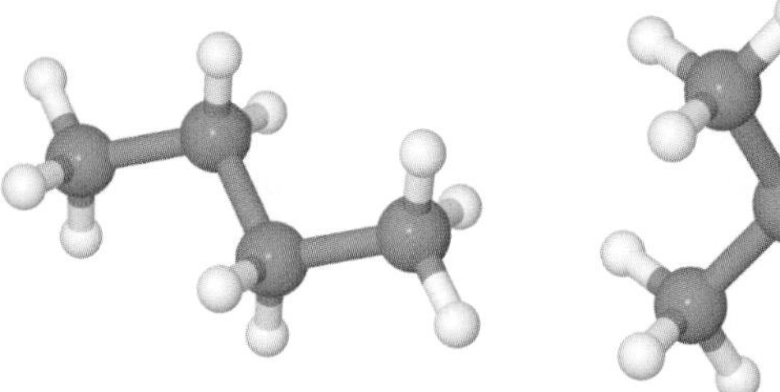

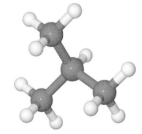

35. What is the number of valence electrons surrounding a stable atom in a Lewis dot configuration?

36. Formaldehyde has the formula CH_2O, and has one double-bond to allow all atoms to form stable valence structures. Explain why this initial attempt to come up with the correct Lewis dot structure will not work.

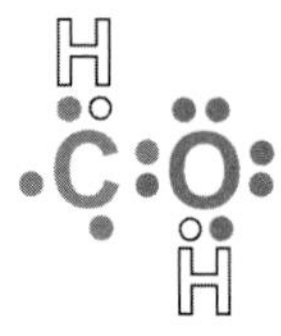

Section 7.5

37. Describe in your own words what Valence Shell Electron Repulsion Theory is and what is is used for.

38. A region of electron density is:
    a. The electrons shared in a bond.
    b. The electrons that form the inner core of the atom.
    c. The valence electrons that are not shared and are called "lone pairs."
    d. Both A and C

39. Which number of regions of electron density will give you which molecular shapes? Match the correct number with the correct shape.

Regions of Electron Density	Shape
2	tetrahedral
3	linear
4	trigonal planar

40. Describe the difference between trigonal planar and trigonal pyramidal.

41. What are the three different possible shapes formed around an atom that is surrounded by four regions of electron density?

Quantitative Problems

Section 7.1

42. Polar covalent bonds form when electrons are shared unevenly. Because electrons are not fully transferred, this means they form partially negative and partially positive charges on the atoms. What color indicates a partial positive and what color indicates a partial negative charge on parts of a molecule?

Section 7.2

43. Calculate the type of bond that will form between the following pairs of atoms:
    a. carbon and hydrogen
    b. oxygen and hydrogen
    c. lithium and bromine
    d. carbon and oxygen
    e. fluorine and fluorine

44. Bonds formed between atoms with an electronegativity difference of less than 0.5 are considered to be:
    a. non-polar covalent
    b. polar covalent
    c. ionic

45. Bonds formed between atoms with an electronegativity difference between 0.5 and 2.1 are considered to be:
    a. non-polar covalent
    b. polar covalent
    c. ionic

46. Bonds formed between atoms with an electronegativity difference greater than 2.1 are considered to be:
    a. non-polar covalent
    b. polar covalent
    c. ionic

47. Most of the molecules in gasoline have the formula C_8H_{18}. Is gas a polar or non-polar molecule?

48. Some amino acids have polar side chains and some do not.
    a. Is the side chain for tyrosine polar or not?
    b. If polar are all of the bonds polar or just some of them?
    c. If just some of them are polar which bonds are polar and which ones are not?

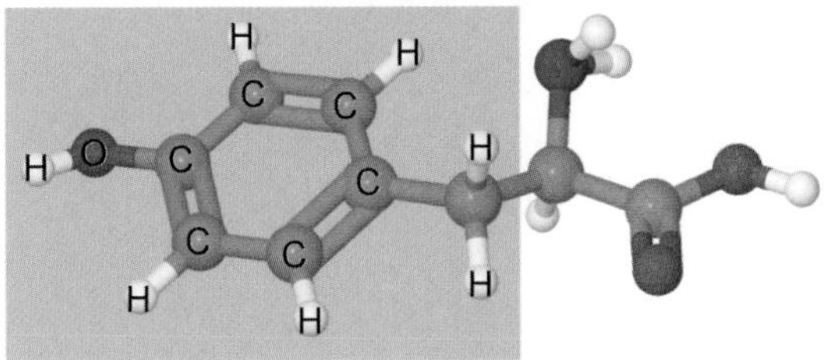

tyrosine side chain in green box

Section 7.3

49. Describe the relationship between the electron configuration of the noble gases and the charge formed by specific ions. Give at least two examples and use electron configurations to support your answer.

50. Describe the relationship between the electron configuration of the noble gases and the number of covalent bonds formed by specific atoms. Give at least two examples and use electron configurations to support your answer.

51. What is the most stable number of valence electrons to have and how do you know?

52. What ratio of ions will be formed in the following substances, so that the positively charged ions will cancel out the negatively charged ions?
    a. Sodium Sulfide (Na and S)
    b. Magnesium Chloride (Mg and Cl)
    c. Aluminum Oxide (Al and O)

Section 7.4

53. Write the Lewis dot structure and structural formula for PH_3.

54. Write the Lewis dot structure and structural formula for C_2H_6.

55. Write the Lewis dot structure and structural formula for CBr_4.

56. Write the Lewis dot structure and structural formula for CH_5N.

57. Write the Lewis dot structure and structural formula for P_2H_2.

58. Write the Lewis dot structure and structural formula for SiH_3N.

59. Write the Lewis dot structure and structural formula for HCP.

60. Write the Lewis dot structure and structural formula for N_2.

61. Write as many isomers of the Lewis dot structure and structural formula for CH_3ON as you can.

62. What is the Lewis dot structure and structural formula for erythrose:

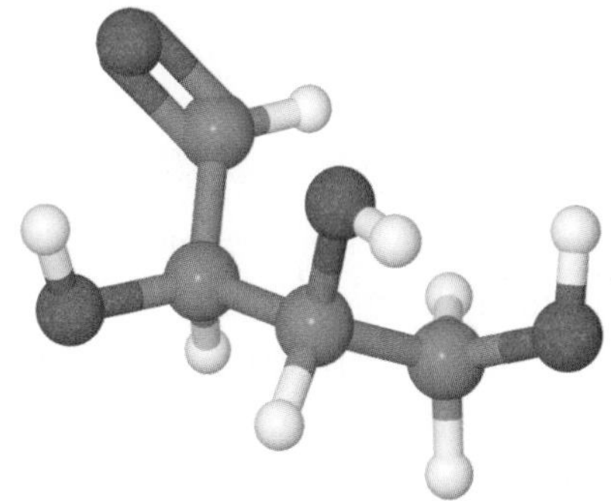

Section 7.5

63. For each of the previous problems (54,55,56, and 59) describe how many regions of electron density exist for the carbon atoms in those molecules.

64. If part of a molecule is linear, what is the angle formed between adjacent bonds?

65. If part of a molecule is trigonal planar, what is the angle formed between adjacent bonds?

66. If part of a molecule is tetrahedral, what is the angle formed between adjacent bonds?

CHAPTER 8 Compounds and Molecules

Why are diamonds so hard?

If sugar and salt look so similar, why do they have such different melting points and other properties?

How do water bugs "walk" on water?

A good, healthy diet consists of a range of foods. You can't survive on only one thing, because there is no single fruit, vegetable, grain or meat that contains every kind of chemical our bodies need to survive. Eating provides us with four basic things: calories, protein, vitamins and minerals.

Calories are not an actual substance, they are a measure of the energy we get from eating three types of molecular substances: carbohydrates, fats and proteins. Most of our energy typically comes from carbohydrates and fats.

We also get energy from eating proteins, but more importantly, proteins get broken down into smaller molecules called amino acids, which we then reassemble into different proteins. The DNA in our body is there for one reason: to provide instructions for building proteins from amino acids. Meat tends to be the food with the highest concentration of protein, but all living things make proteins, so vegetarians can also get the amino acids they need by eating the right combinations of plants.

Multi-vitamins contain ionic and molecular substances

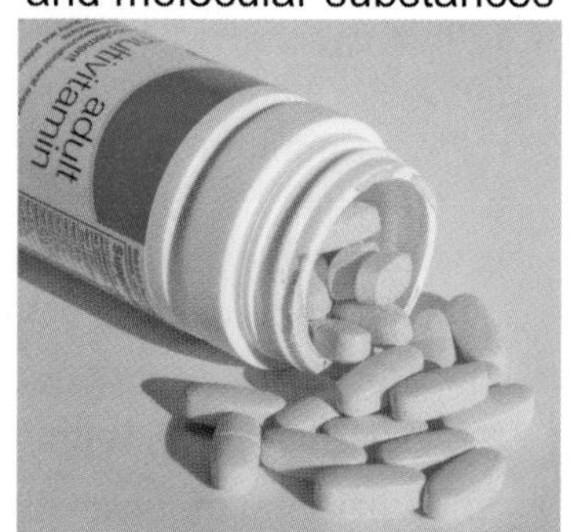

magnesium oxide

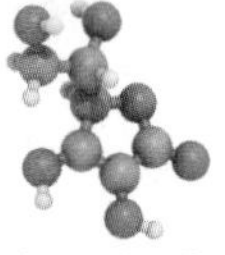

vitamin C

The other two necessary "foods" often get grouped together: vitamins and minerals. Many breakfast cereals have these added to them, and many people take a multi-vitamin that actually provides many minerals as well. Both vitamins and minerals are needed in small quantitates by your body, but going without any one of them for long periods of time can be very bad for your health. Vitamins are actually molecular substances, while minerals generally come in ionic form. Most of what we eat is molecular, but minerals and table salt (also considered a mineral) are important ionic components of a healthy diet.

Classifying solids

There are four main categories of solid substances: ionic, covalent network, molecular, and metallic. Chemicals that belong to these different categories share some common physical properties. We will test sand(SiO_2), salt(NaCl), sulfur(S), and copper(Cu). You are already familiar with several of the things we will be looking for. For example we will compare a soft powder to a hard and granular substance.

To begin make a table with the name of each of the four substances across the top. Next, list the properties down the side: including texture, solubility in water, malleability(ability to bend), and conductivity.

1. Carefully observe each of the substances below. Record anything interesting about their appearance.

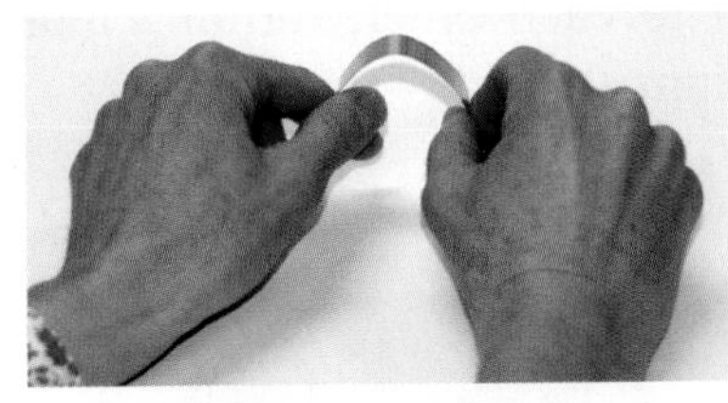

2. Try each of the following tests on the solids.

a) Can they bend? Are they malleable?

b) How do they feel? Soft, smooth, hard, granular?

c) Place the conductivity apparatus in contact with each substance. Which substance(s) conduct electricity as a solid?

d) Place a small amount of each of them in water, and mix. What happens?

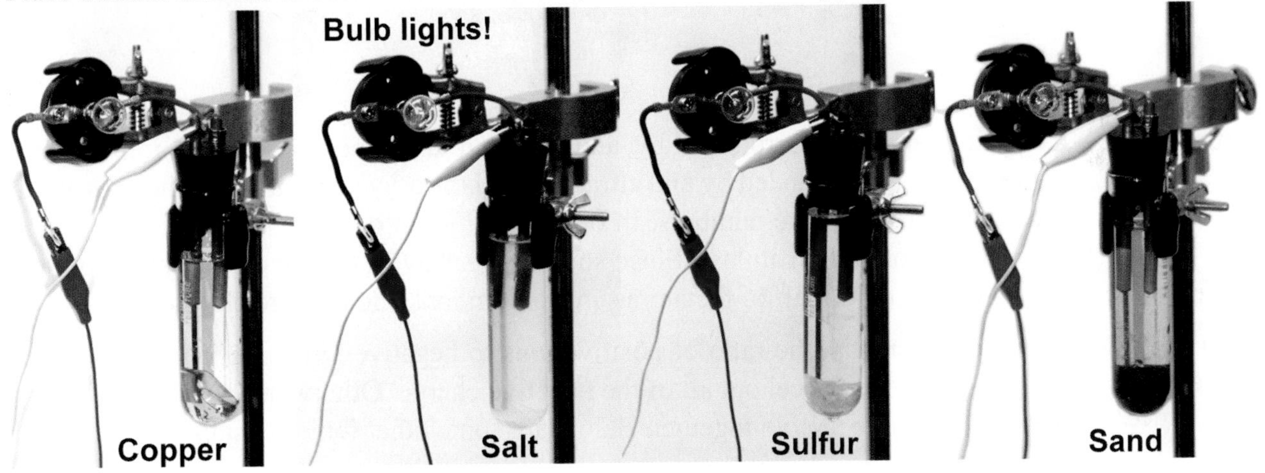

e) Repeat the conductivity test on the substances in water from part (d). Record your observations about each substance.

You have just gathered data about all of the different classes of solids that we will study in this chapter.

The characteristics of texture, conductivity, and malleability all represent different types of attractive forces and bonding patterns.

This information will be a useful guide as you learn about these solids. In this chapter, we will learn how bonding affects a solids chemical properties.

8.1 Ionic compounds

Properties of ionic substances

Unlike molecular compounds, ionic substances tend to have very similar properties. They are typically:

- hard and brittle
- solid at room temperature
- have very high melting points
- conduct electricity if heated to a liquid state
- conduct electricity if dissolved in water or some other solvent

Ionic properties and atomic level structures

The properties we observe can be traced back to the underlying properties and atomic level structure of the ionic substances. An ionic substance is formed when a positively charged ion (due to loss of one or more electrons) is attracted to a negatively charged ion (due to gaining one or more electrons). Ions will attract to each other, forming ionic bonds with all neighboring atoms. Let's take a look at sodium chloride more closely:

Structure of sodium chloride - NaCl

Each chloride ion is ionically bonded to six nearby sodium ions.

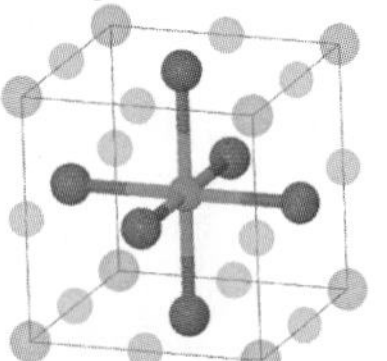

Each sodium ion is ionically bonded to six nearby chloride ions.

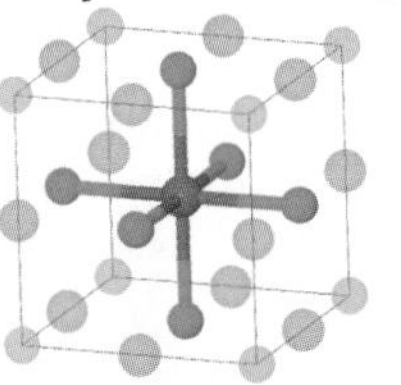

Ion attraction

Every sodium ion is surrounded by, and attracted to, six nearby chloride ions, and every chloride ions is surrounded by and attracted to six nearby sodium ions. The ions attract to each other in very large numbers. In one typical grain of table salt there are about 10^{18} atoms (a billion billion atoms). Since sodium ions have a charge of +1, and chloride ions have a charge of -1, half of the atoms are sodium ions and half are chloride ions.

Positive and negative charges cancel out

In any ionic crystal, the ratio of positive ions to negative ions must allow for all of the positive charge to cancel out all of the negative charge. Otherwise, the ions in the crystal would not be able to hold together. Below are some other ionic substances:

lead(II) sulfide - PbS

aluminum oxide - Al_2O_3

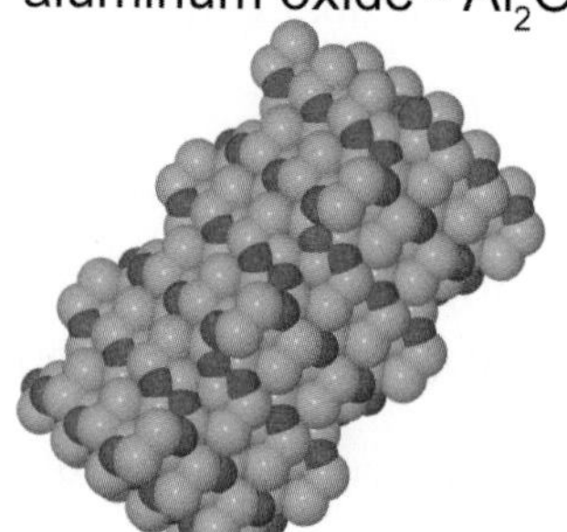

copper(II) sulfate - $CuSO_4$

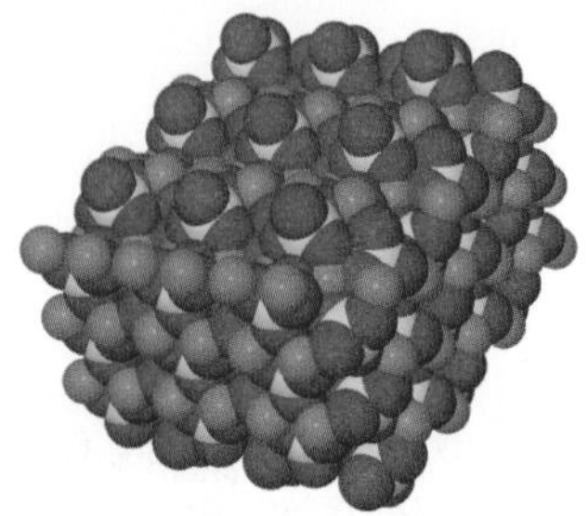

Connecting ionic structure to properties

Bonding patterns of ionic crystals

Now let's consider how ionic structure brings about their properties. Three of the properties are all related: hardness, melting point and state of matter.

Ionic substances are hard, have very high melting points, and so they are solid at room temperature. All of this can be explained by the bonding patterns found in ionic crystals. Each ion is attracted to and bonded with all of its neighbors, so the ions have an interconnected network of bonds holding the entire crystal together. This gives the overall crystal strength, making it hard. This also makes it difficult to melt. To be in a liquid state, the ions have to have enough kinetic energy (have a high enough temperature) to continually break free of the attractions allowing them to flow by each other.

Network of ionic bonds makes ionic compounds hard, and have high melting points.

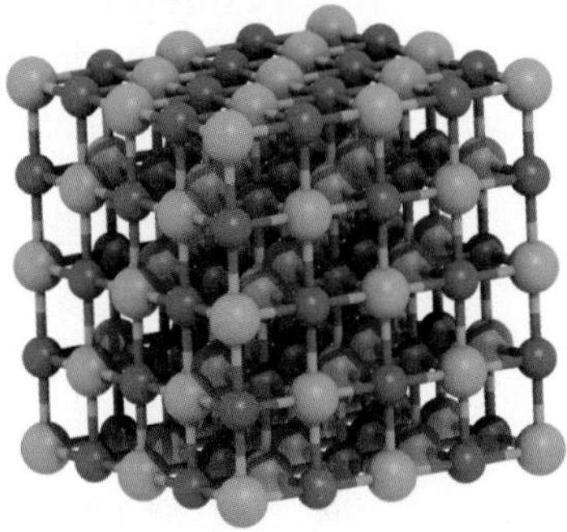

Ionic substances are brittle

Ionic substances are **brittle** because putting pressure along one edge can cause the ions to shift place so that the positive and negative ions from one layer are not properly aligned with the oppositely charged ions in the next layer. If this happens there will be a repulsion between the layers and the crystal will break.

When aligned the ions attract well, but add enough pressure and the crystal will break.

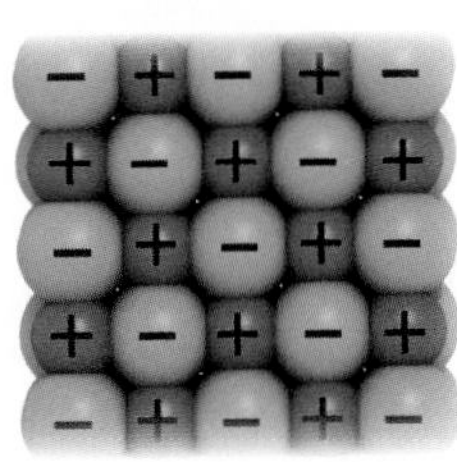

Apply force and the ions repel breaking the crystal.

Electric current

An **electric current** is the movement of many electric charges in a particular direction. When melted or dissolved, the ions are free to move around, making it is possible to conduct electricity.

Ions dissolved in water move in random directions by colliding with water molecules.

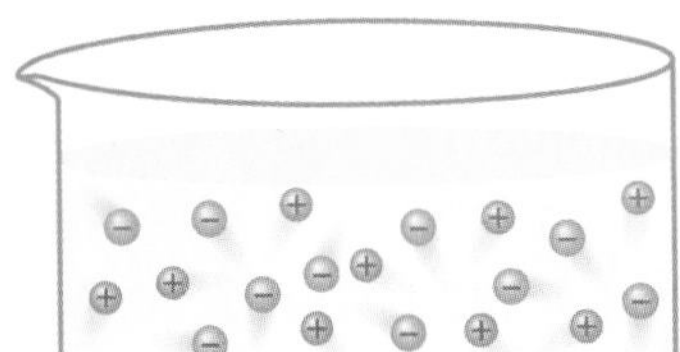

Overall ion movement is toward the electrodes. Directed moving charge IS an electrical current.

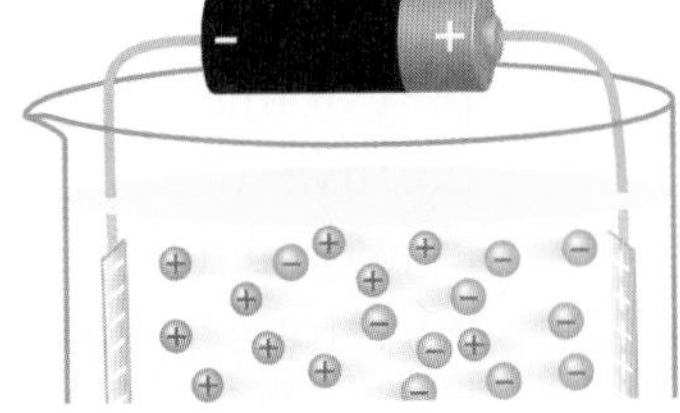

Chemistry terms

brittle - hard, but with the possibility of breaking or shattering relatively easily.

electric current - the directed movement of electric charges, either ions or electrons.

Polyatomic ions

An ion can also be a small molecule with a charge

Until now all of our examples of ionic compounds have been made from ions that are single atoms with either a positive or negative charge. However, an ion can also be a small molecule with a charge. Remember, something becomes charged when there is an unequal amount of protons and electrons, so for a molecule to become charged you need the total number of protons and electrons in the entire molecule to be unequal. One example of a **polyatomic ion** (an ion made from more than one atom), is carbonate with the formula CO_3^{2-}. One carbon is bonded to three oxygens and the entire molecule has a -2 charge.

Structure of the carbonate (CO_3^{2-}) ion.

CO_3 molecule needs two extra electrons for all atoms to have 8 valence electrons

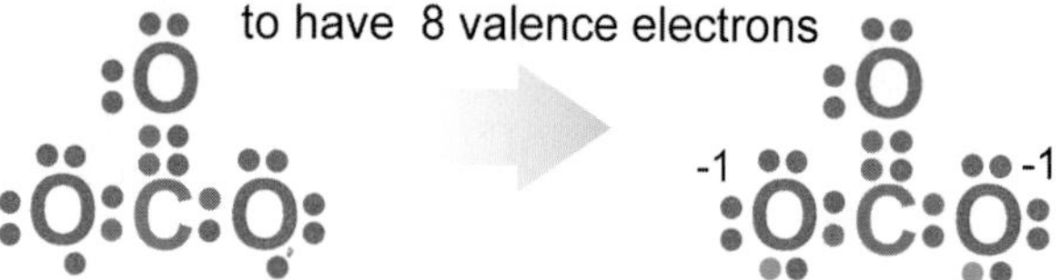

two oxygens now have a -1 charge giving the whole molecule a -2 charge

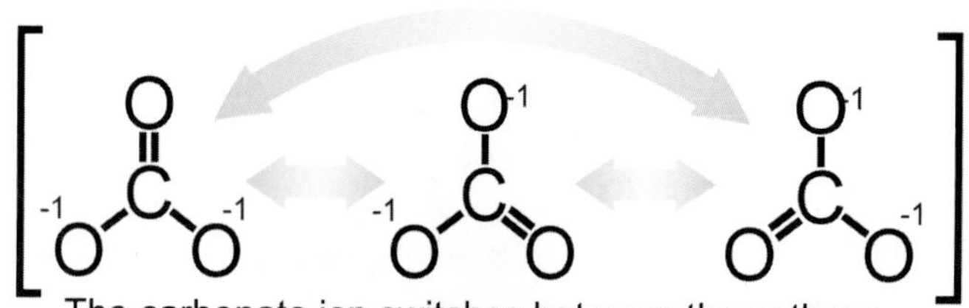

The carbonate ion switches between these three versions of the molecule.

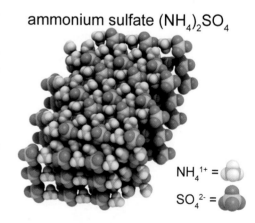

Bonding patterns are similar in ionic substances with polyatomic ions

Ionic crystals with polyatomic ions are very similar to the ones we have seen with single atom ions. They pack together in a regular pattern with each positive ion attracting to all of its negative neighbors, and each negative ion attracting to all of its positive neighbors. Most polyatomic ions are negative, but there are a few positive ones. To the right is an example of ammonium sulfate which is made from two polyatomic ions: ammonium (NH_4^{1+}) and sulfate (SO_4^{2-}). It has the formula $(NH_4)_2SO_4$.

ammonium sulfate $(NH_4)_2SO_4$

NH_4^{1+} =

SO_4^{2-} =

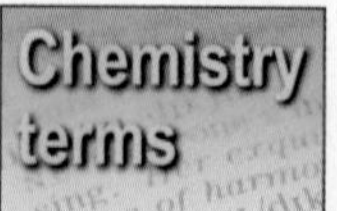

polyatomic ion - a small molecule with an overall positive or negative charge.

Writing formulas for ionic compounds

Ionic crystals come in many sizes

A grain of table salt is a small piece of an ionic substance, but as we learned earlier, that grain of salt is made from approximately 1 billion billion ions. Not all grains of salt are the same size. If you carefully grow a salt crystal under the right conditions, there is no limit to how big the "grain" could be. In an underground cavern in Germany there are single "grains" that measure over one meter in length on a side. The image to the right shows a large single crystal of table salt from the Fersman Mineralogical Museum in Moscow.

Ionic formulas tell you the ratio of ions in the compound.

The formula for table salt is NaCl, but what does that mean if you can have a single crystal that has enormous variations in size? For ionic compounds, the chemical formula tells us the ratio of positive to negative ions in the entire substance. The ratio must be just right so that the total positive charge from all of the positive ions will equal the total negative charge from all of the negative ions.

Balancing Positive and Negative Charge in Ionic Formula Writing

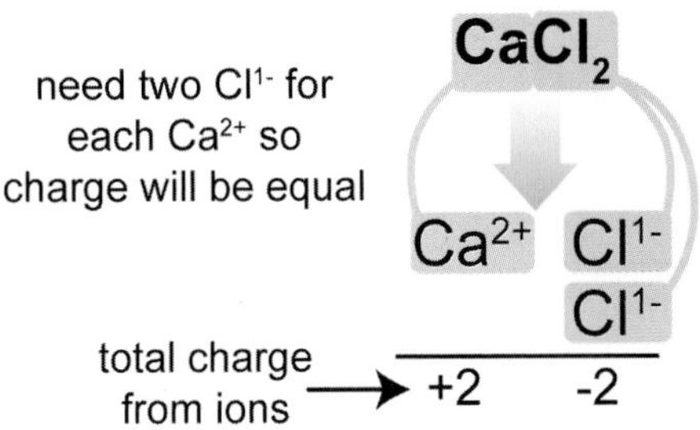

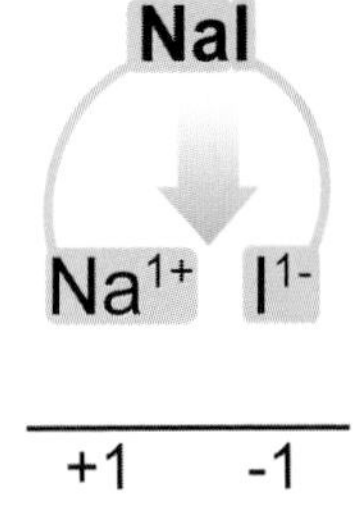

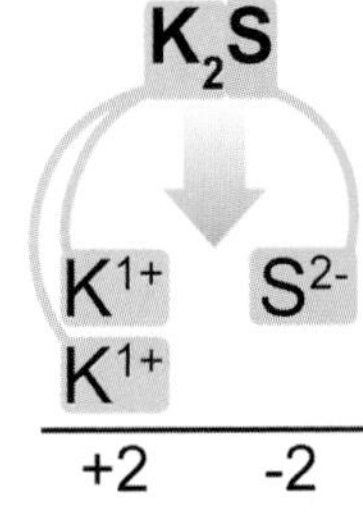

Write the correct formula for the compounds which will form using the following ion pairs (Mg^{2+} and Cl^{1-}; Na^{1+} and S^{2-}; Al^{3+} and O^{2-}).

Asked: *What is the correct ionic formula for three different ionic substances?*

Given: *The charges on all of the ion pairs; see above.*

Relationships: *The positive ions will combine with the negative ions in a ratio so that the positive and negative charges from the ions will be equal.*

Solve: *For each Mg^{2+} we need two Cl^{1-} (+2 -1-1=0) and the formula is $MgCl_2$*

For each S^{2-} we need two Na^{1+} (-2 +1+1=0) and the formula is Na_2S

To balance charges between Al^{3+} and O^{2-} we need to multiply Al by two and O by three (2(+3)+3(-2)=0). The formula is Al_2O_3

Writing formulas with polyatomic ions

formulas with polyatomic ions

Writing ionic formulas using polyatomic ions is basically the same as writing them with simple single atom ions. The strategy is the same. You need to write a formula that has the correct ratio of positive to negative ions. The only difference is how you show that you want more than one polyatomic ion. Since an ion like nitrite (NO_2^{1-}) is a single unit – a small molecule with a negative charge – we treat that ion like it is a single atom. To write a formula that says you need two or three NO_2^{1-} ions you need to put parentheses around the ion and write it like this: three nitrites = $(NO_2^{1-})_3$.

Polyatomic ions represent a single unit with a charge just like single atom ions.

Cl^{-1} NO_2^{-1}

Both ions are individual particles carrying charge.

Ionic Formula Writing with Polyatomic Ions

The chlorate ion (ClO_3^{1-}) is made by covalently bonding chlorine to three oxygens and adding an extra electron. To say we want more than one chlorate, parentheses must be used.

ClO_3^{1-} =

Using Ca^{2+} and ClO_3^{1-} to make calcium chlorate:

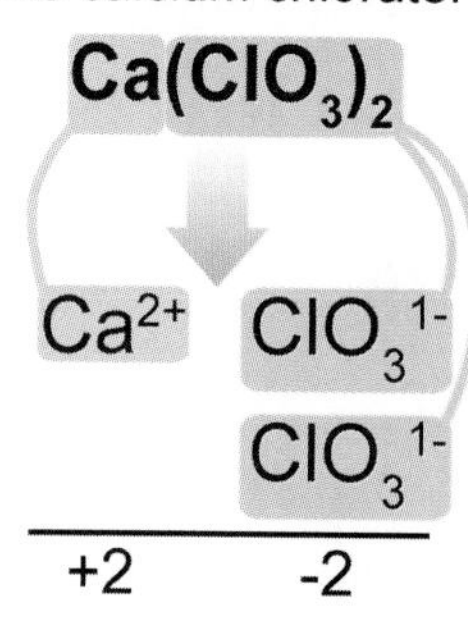

Using Al^{3+} and OH^{1-} to make aluminum hydroxide:

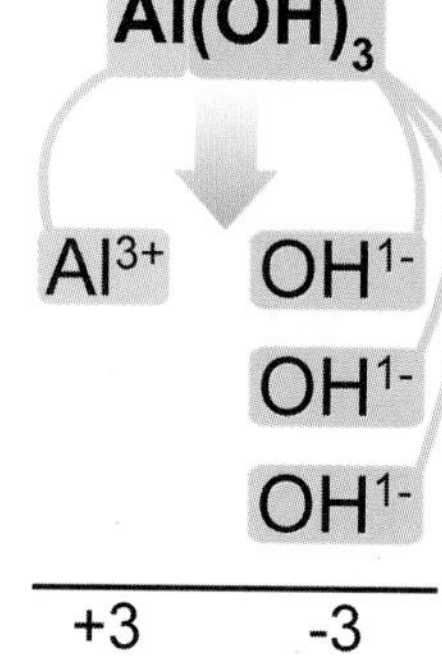

Note: The hydroxide ion (OH^{1-}) is made by bonding one atom of hydrogen to one atom of oxygen, so it is polyatomic even though the ion has no subscripts.

Write the correct formula for the compounds which will form using the following ion pairs; (Mg^{2+} and SO_4^{2-}; Ca^{2+} and PO_4^{3-}; NH_4^{1+} and S^{2-}).

Asked: *What is the correct ionic formula for three different ionic substances?*

Given: *The charges on all of the ion pairs. See above.*

Relationships: *The positive ions will combine with the negative ions in a ratio so that the positive and negative charge from the ions will be equal.*

Solve: *For each Mg^{2+}, need one SO_4^{2-} (+2 -2=0) and the formula is $MgSO_4$*

To balance charges between Ca^{2+} and PO_4^{3-}, multiply Ca by three and PO_4 by two: (3(+2)+2(-3)=0). The formula is $Ca_3(PO_4)_2$

For each S^{2-}, need two NH_4^{1+} (-2 +1+1=0) and the formula is $(NH_4)_2S$

Writing names for ionic compounds

Naming binary ionic compounds

When naming an ionic compound, you just write the names of the ions as they appear in the formula. For simple binary ionic compounds (ionic formulas from two single atom ions), like NaCl, the first part of the name is just the name of the element, and the second part of the name is a modified version of the name of the element. Typically, you drop the standard ending and add -ide.

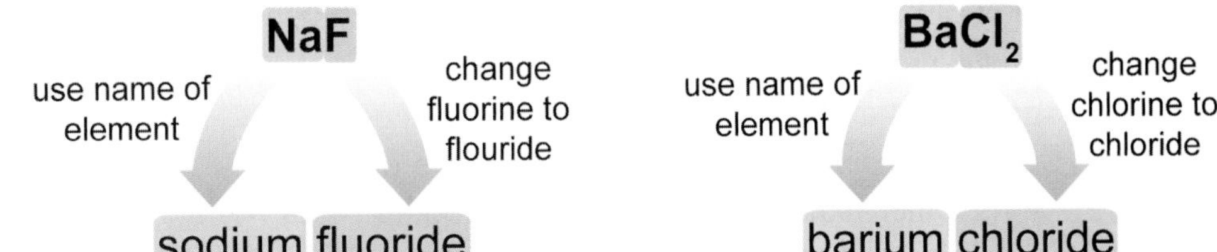

Names for polyatomic ions are found on the common ion table

For polyatomic ions you will need to get the name from the table below. Otherwise, it is done the same way as it is for binary ionic compounds. Write the name of the ions in the order that they appear in the formula.

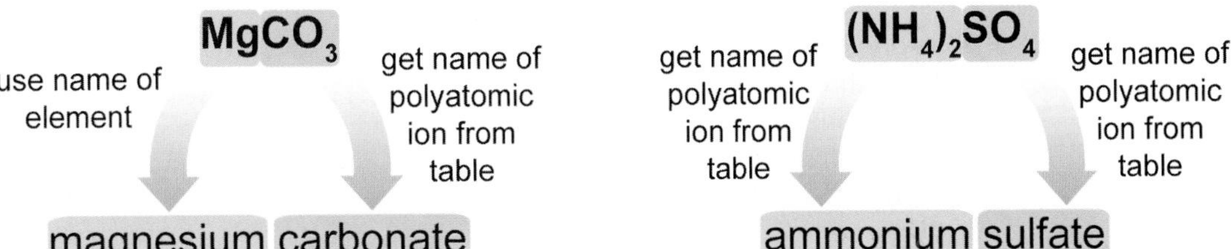

TABLE 8.1. Common Ions

positive ions		negative ions	
Aluminum Al^{3+}	Lead(II) Pb^{2+}	Acetate $C_2H_3O_2^{1-}$	Hydrogen sulfate (bisulfate) HSO_4^{1-}
Ammonium NH_4^{1+}	Lead(IV) Pb^{4+}	Chloride Cl^{1-}	Hydroxide OH^{1-}
Barium Ba^{2+}	Magnesium Mg^{2+}	Carbonate CO_3^{2-}	Nitrate NO_3^{1-}
Copper(I) Cu^{1+}	Mercury(I) Hg_2^{1+}	Chromate CrO_4^{2-}	Nitrite NO_2^{1-}
Copper(II) Cu^{2+}	Mercury(II) Hg^{2+}	Chlorate ClO_3^{1-}	Oxide O^{2-}
Calcium Ca^{2+}	Potassium K^{1+}	Chlorite ClO_2^{1-}	Peroxide O_2^{2-}
Chromium(II) Cr^{2+}	Silver Ag^{1+}	Cyanide CN^{1-}	Phosphate PO_4^{3-}
Chromium(III) Cr^{3+}	Sodium Na^{1+}	Dichromate $Cr_2O_7^{2-}$	Sulfate SO_4^{2-}
Hydrogen* H^{1+}	Tin(II) Sn^{2+}	Fluoride F^{1-}	Sulfite SO_3^{2-}
Iron(II) Fe^{2+}	Tin(IV) Sn^{4+}	Hydrogen carbonate (bicarbonate) HCO_3^{1-}	Sulfide S_{2-}
Iron(III) Fe^{3+}	Zinc Zn^{2+}		

*Hydrogen ions rarely exist by themselves. Often they combine with water to form hydronium ions: H_3O^{1+}.

Ionic naming with transition metals

Roman numerals in the ion name

On the previous table you will see several names for positive ions that are followed by a Roman numeral. This number describes the charge on that ion. Most transition metals form multiple charges, so to know which charge the ion should have when naming the compound, you need to specify it with a Roman numeral.

A roman numeral indicates the charge on a positive ion.

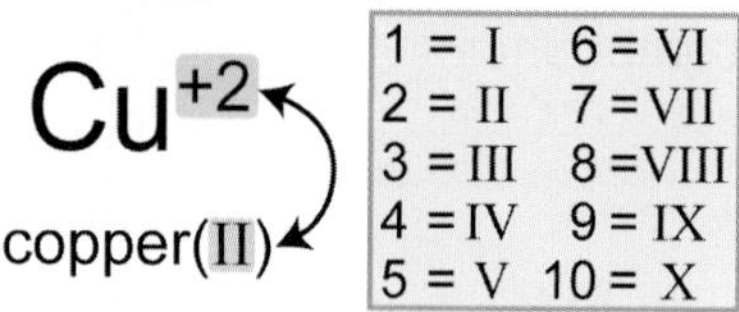

The common ion table

Since the transition metals and those metals in groups 14 and 15 can form varying charges, you need to check the common ion table before you write the name of an ionic compound that includes one of those elements. Most transition metals do, but there are two common exceptions: silver and zinc. Silver almost always forms a +1 ion and zinc a +2 ion, so you don't need Roman numerals in names of silver or zinc compounds.

Need to check if Roman numeral is necessary for elements in highlighted part of the periodic table.

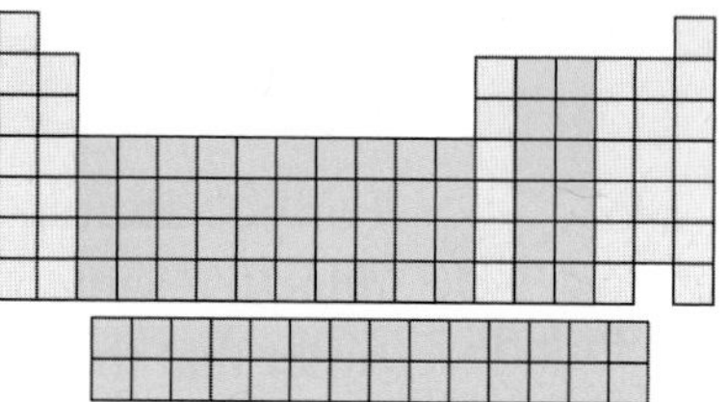

Pb is one of the elements to check for multiple charges. The ion table shows common ions for lead are Pb^{+2} and Pb^{+4}, so you need to use the negative ion to determine the charge on the postive one:

$Pb^{?+}$ Cl^{1-}
Cl^{1-}

total charge from ions → +2 -2

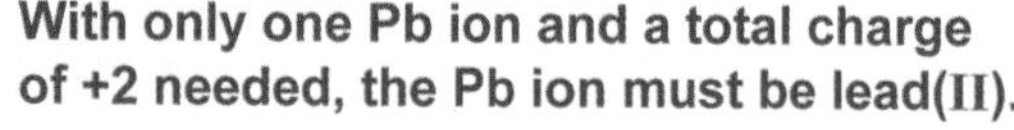

With only one Pb ion and a total charge of +2 needed, the Pb ion must be lead(II).

$PbCl_2$ → lead(II) chloride

$ZnSO_4$ → Zinc sulfate

Zinc needs to be checked, but the ion table shows it only forms a +2 charged ion, so no Roman numeral is needed.

Write the name for each of the following formulas: $CaCl_2$, $Zn(NO_3)_2$, $Fe_3(PO_4)_2$:

Asked: *Name the formulas above.*

Given: *Three formulas, and a common ion table.*

Relationships: *The name of the formula is constructed from the name of the ions. If the negative ion is a single atom, modify the name of the element to end in "-ide". If the positive ion might form multiple charges, make sure to use a Roman numeral in the name.*

Solve: $CaCl_2$ = *calcium chloride*

$Zn(NO_3)_2$ = *zinc nitrate*

$Fe_3(PO_4)_2$ = *iron(II) phosphate*

8.2 Molecular Compounds

Properties of molecular substances

Properties of molecular compounds vary widely compared to those of ionic compounds:

- Some are hard and brittle, while others are flexible, or soft and mushy.
- They could be solid, liquid or gas at room temperature.
- Their boiling points vary from -253°C to over 4800°C.
- Most do not conduct electricity well regardless of their state of matter.
- When dissolved in a solvent, they don't typically conduct electricity.

A molecule is the building block of a molecular substance

The properties of molecular substances depend on two things: the structure of the individual molecule, including the types of covalent bonds formed within the molecule, and the attractions between molecules. Take water for example. The smallest piece of water you can have is a single water molecule. However, when we think of water we imagine a drop, or a cup, or an ocean. One cup of water is made from approximately eight million billion billion (8×10^{24}) water molecules. They bounce off of each other, constantly in motion, forming no particular crystalline organization when in liquid form. As ice, water can form an organized crystal structure, and in that form it has the property of being hard and brittle like an ionic compound. The great variety of structures possible with molecular substances is what gives them their great variation in properties.

An individual molecule of water.

About eight million billion billion (10^{24}) water molecules make up this cup of water.

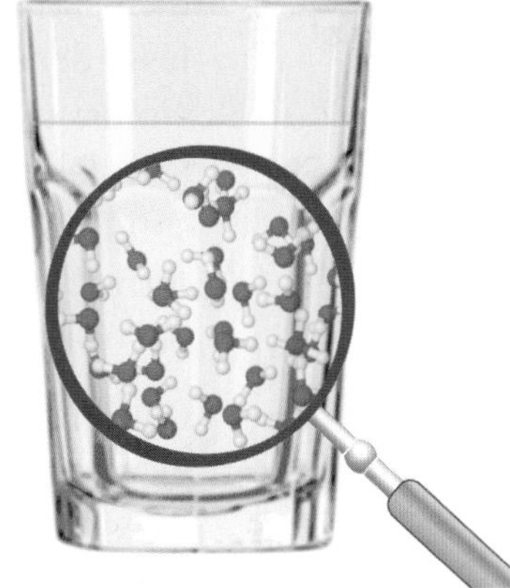

Molecules can be classified into different categories

First we will look at the different types of molecular building blocks. Let's study how their interactions affect the overall properties of the substances made of those molecules. Molecules are typically small, medium or large. Within the "large" category there are polymers and network covalent substances.

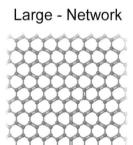

Small	Medium	Large - Polymer	Large - Network
caffeine - found in tea, coffee, and soda	wax - one of the main hydrocarbons in beeswax	protein - molecule on the right shows a thick line tracing the linear chain, with red and blue marking either end	graphene - layers of this make graphite

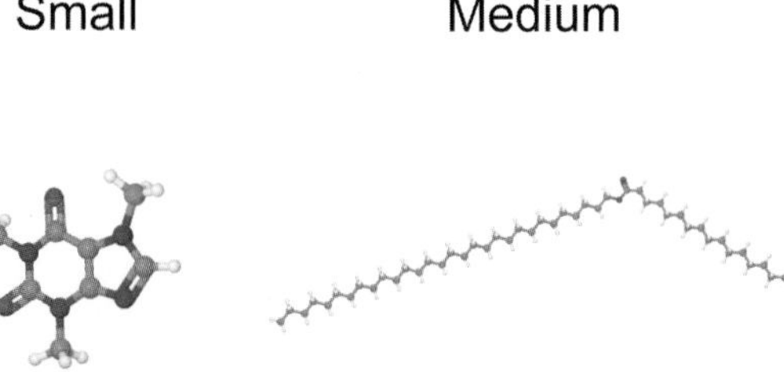

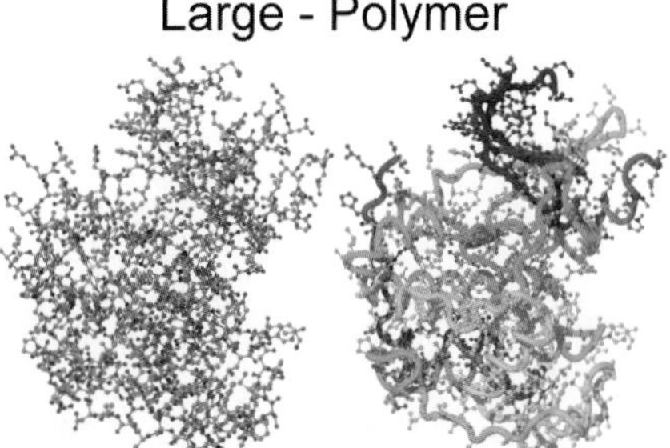

Small molecules

Most small molecules are liquids or gases at room temperature

Small molecules are those that are constructed from about a dozen atoms or less. There is no specific number of atoms that puts a molecule in the "small" category, but substances made from small molecules tend to have some similar properties. They are almost always gases or liquids at room temperature and include some of the most important molecules related to the functioning of our entire ecosystem. Often they are over the counter drugs, such as acetaminophen, which in its pure form is a liquid.

Acetaminophen

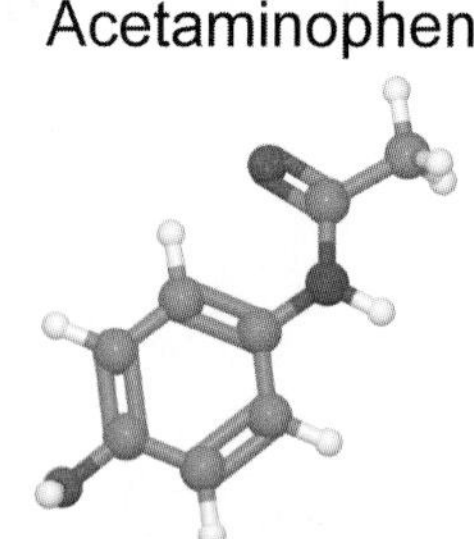

The lower density of ice, compared to liquid water, made the evolution of human life possible

Water is a good example. A single polar water molecule is made from only three atoms, but this molecular substance is essential for life. It is how the rest of the molecules in our body organize themselves. When it freezes, the solid form is less dense than the liquid form, something almost unique to water. If ice were more dense, then lakes, ponds, and even oceans could freeze solid in cold weather, killing all higher life forms, and preventing the evolution of anything beyond the simplest of creatures.

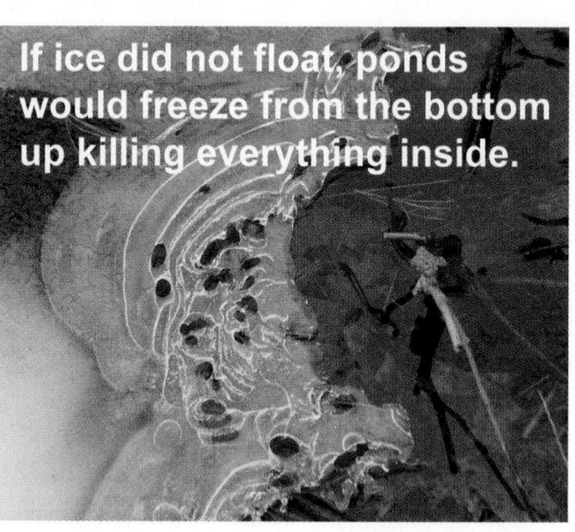

If ice did not float, ponds would freeze from the bottom up killing everything inside.

Most common gases are small molecules

Carbon dioxide, and oxygen fall into the small molecule category as well. These are all gases at room temperature, and it is not by chance that they are all non-polar. Carbon dioxide is also a greenhouse gas, but it is also crucial for the lives of plants, which use the carbon to build up their physical structure, and as a temporary chemical storage for the light energy absorbed by their leaves. Typically it is the polarity of the molecule that makes the difference between it being liquid or gas at room temperature.

Some other small molecules you may be familiar with are ethene, which is a non-polar gas given off by ripening fruit, and ethanol which is almost the same size but polar and is liquid at room temperature. When fruits and other sugar-containing material ferment, this substance is produced, making what is commonly called alcohol.

Ethene

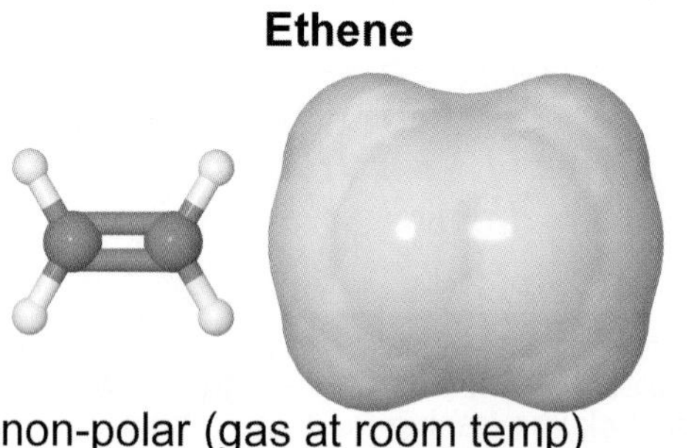

non-polar (gas at room temp)
boiling point = -103.7°C

Ethanol

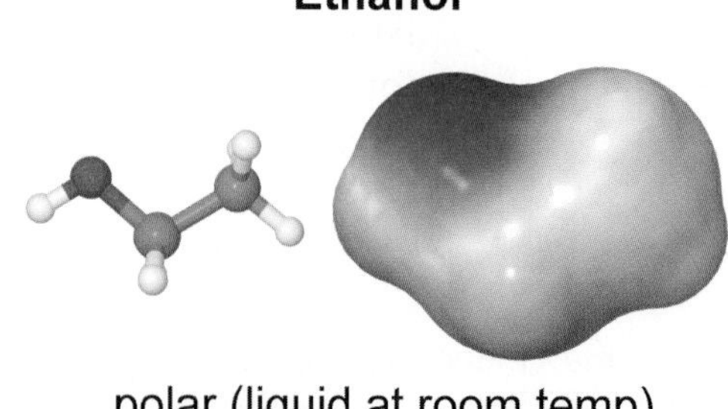

polar (liquid at room temp)
boiling point = 78.4°C

Medium-sized molecules

Liquid or soft solids

Medium sized molecules are those that are typically formed from more atoms than small molecules, but not much more than 100 atoms. As with small molecules, there is a significant variety of molecule types that fall into the "medium" category, but they tend to share some properties. Typically, they are liquids or soft solids at room temperature, and often fall into one of two official chemical categories: lipids or long chain hydrocarbons.

Candlewax is a mixture of medium-sized hydrocarbons.

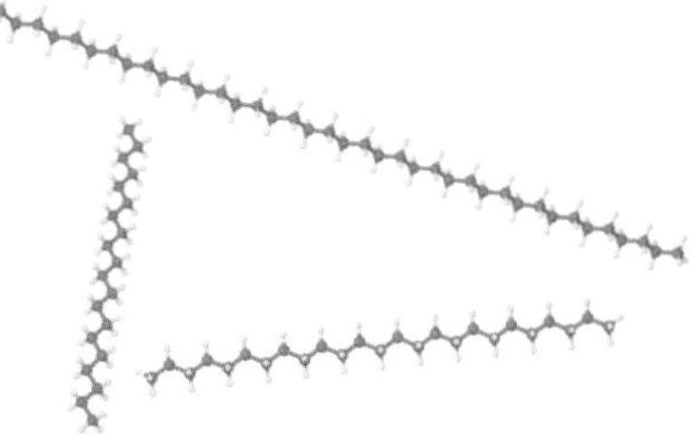

Lipids

Lipids are typically fats or steroids, molecules primarily made from carbon and hydrogen, but may include small numbers of other elements. Depending on the size and shape of a fat molecule, the substance made from those molecules might be liquid or solid at room temperature, for example vegetable oil or butter. Lipids are typically non-polar and insoluble in water.

Linoleic acid, a major component of vegetable oil.

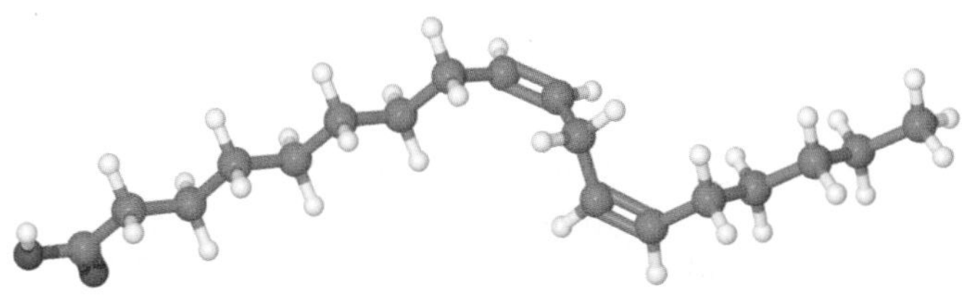

Steroids are another type of lipid

Steroids also fall into the lipid category, but have a larger percentage of non-hydrogen and carbon atoms. Most have four rings of carbon atoms, and could function as hormones, drugs, vitamins or poisons. Anabolic steroids, which are used as performance enhancing drugs, all mimic the shape and function of testosterone, a natural steroid.

Testosterone

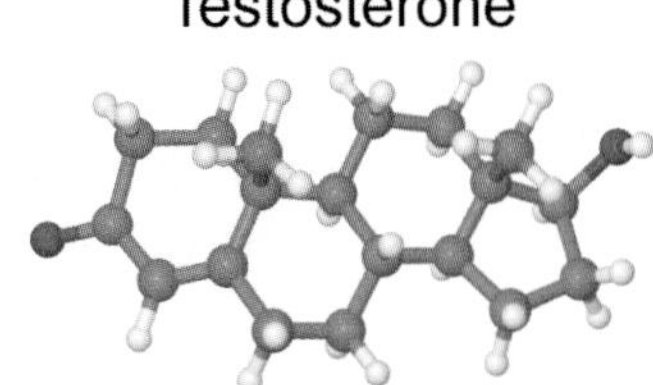

Hydrocarbons

Hydrocarbons are molecules formed purely from long chains of hydrogen and carbon atoms. They are very similar to fats, except they lack the typical oxygen atoms that are part of the fatty acids produced in living organisms. Medium-sized hydrocarbons tend to form soft solids like petroleum jelly, and ones made from even larger numbers of atoms tend to form harder substances like wax.

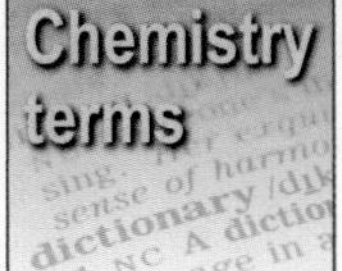

lipid - a molecule that typically falls into the category of fat or steroid.

steroid - a molecule with four carbon rings that is biologically active as either a hormone, vitamin, drug or poison.

hydrocarbon - a molecule made almost entirely from carbon and hydrogen atoms.

Polymers made from a single monomer

Polymers and monomers

The prefix "poly" means "many," which is appropriate for the kinds of molecules we call polymers, because **polymers** are made from many smaller molecules. The smaller molecules that covalently bond together to make polymers are called **monomers**. All of the plastic in the world is made of various kinds of polymers. The recycling numbers on bottles and packages indicate which kind plastic/polymer the object is made from. Number 3, for example, is polyvinyl chloride (PVC), which is made into pipes and siding. Many vinyl chloride monomers bond together to form long chains, making individual polyvinyl chloride molecules that consist of hundreds or thousands of atoms.

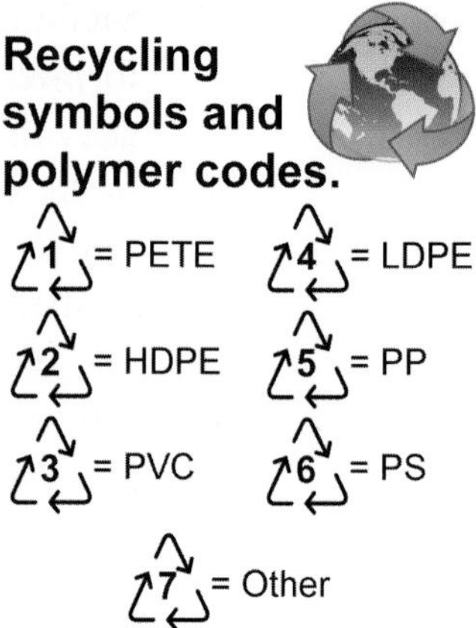

A Single Monomer of Vinyl Chloride

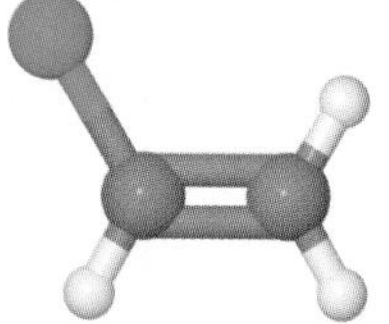

Polyvinyl Chloride (PVC) Made by Bonding Many Vinyl Chloride Monomers Together

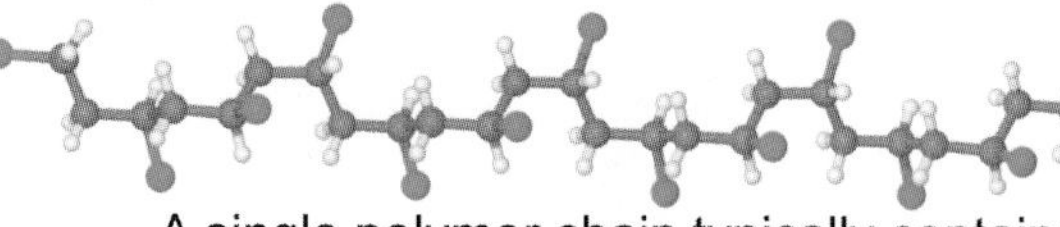

A single polymer chain typically contains hundreds to thousands of atoms.

Polymers can be synthetic

Plastics are synthetic polymers, typically made in labs and factories. However, many natural polymers exist, made by living organisms. The simple sugar glucose is a key monomer for making two different natural polymers: starch and cellulose. By bonding the glucose monomers together in a slightly different way, you either get something we can digest, starch, which is found in many foods like bread, pasta and potatoes, or something we can't digest, cellulose which is what wood, paper and most plant material is made from.

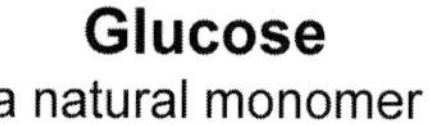

Glucose

a natural monomer

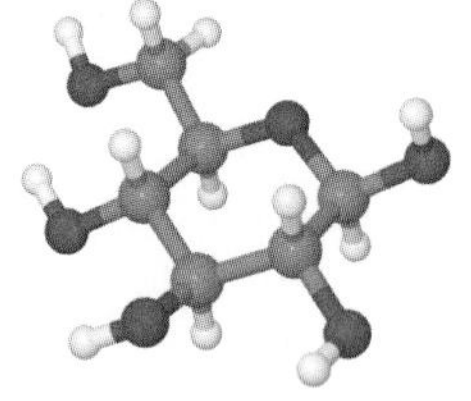

Starch Fragment

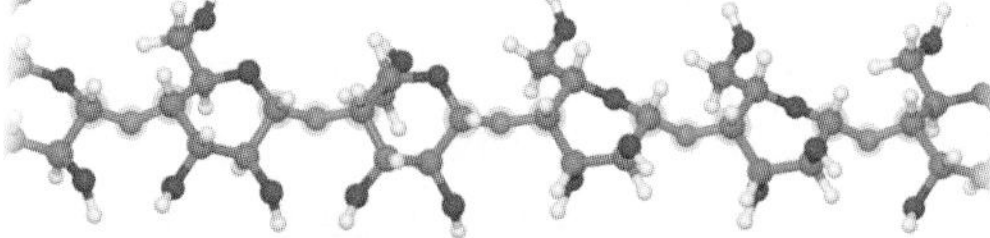

Cellulose Fragment

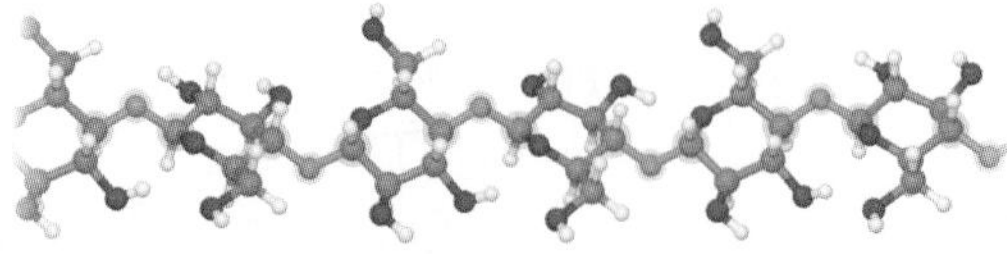

Both are polymers of glucose, but the glucose monomer is bonded together slightly differently in starch vs. cellulose. Note the pattern of "purple" oxygen atoms where the monomers are bonded together.

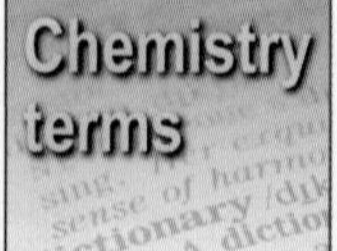

polymer - a molecule made from bonding many small molecules together.

monomer - a small molecule which is a building block of larger molecules called polymers.

Polymers made from multiple monomers

A single polymer can be made from more than one type of monomer

Each of the polymers we saw on the previous page were all made from a single type of monomer. Sometimes more than one monomer is used. DNA is a polymer that uses four different monomers that biologists call "nucleotides." Proteins are constructed from up to 20 different monomers called amino acids. To distinguish between these two types of polymers, we use the term **homopolymer** to refer to polymers made from only one type of monomer, while **copolymer** describes polymers made from multiple types of monomers.

DNA fragment colored by base and backbone

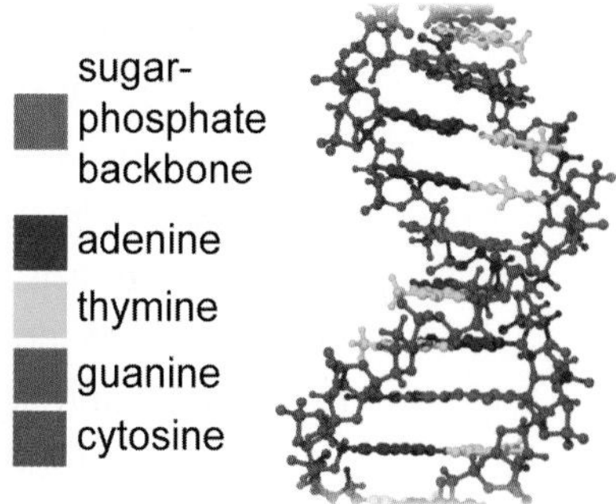

The four monomers used to make strands of DNA
(colored to help see them in the polymer above)

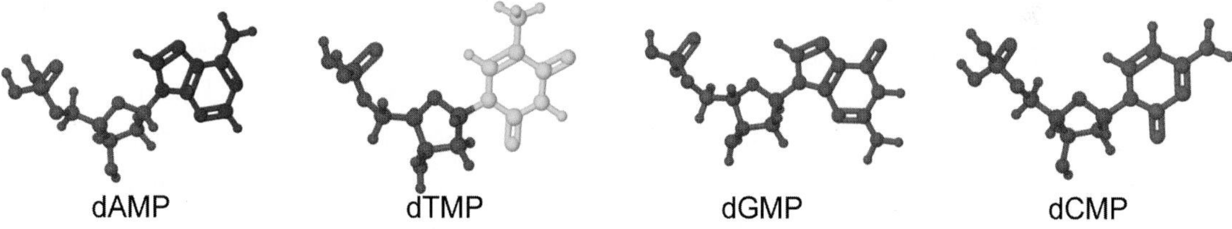

DNA: the code for building proteins

DNA is the molecule that provides the code necessary to build proteins. The proteins are built by adding one monomer (amino acid) at a time. The order of the amino acids in the protein polymer is based upon the order of the monomers (bases) in the DNA. Once the full protein is made, various parts of the molecule attract to itself, causing it to form into a particular shape, giving the protein a specific function.

A Folded Protein Molecule

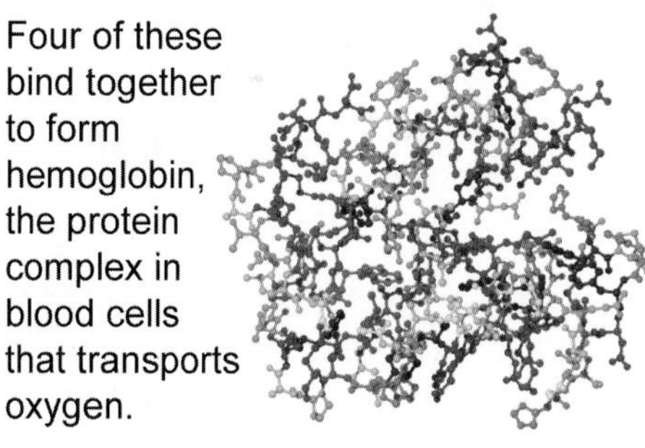

Four of these bind together to form hemoglobin, the protein complex in blood cells that transports oxygen.

An unfolded protein fragment

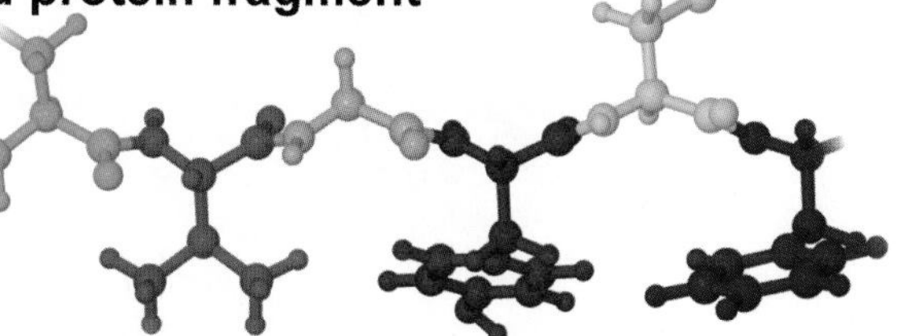

Proteins are made from up to 20 different amino acids. Here they are colored by amino acid, so you can see each monomer clearly.

Chemistry terms

homopolymer - a molecule made from repeatedly bonding the same monomer together.

copolymer - a molecule made from repeatedly bonding more than one type of monomer together.

Network covalent molecules

Network covalent

Usually, when we think of molecules we think of things so small that we can't even see them with a normal microscope. However, when you see sand, it is likely that at least some of the individual sand grains are actually made from a single individual molecule. Sand is usually a mixture of many different minerals, but the most common one is quartz, which is made from silicon and oxygen. Each silicon is covalently bonded to four neighboring oxygen atoms, and each oxygen atom is bonded to two neighboring silicon atoms. Since every atom in quartz is covalently bonded together, the entire grain can be considered a single molecule. Molecules in which large numbers of atoms are covalently bonded in an interlocking way are called **network covalent**.

Silicon Dioxide

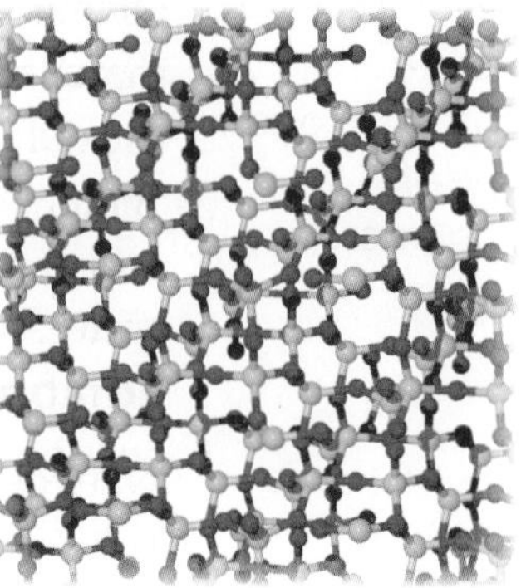

Carbon

Carbon is special because it can form many different kinds of network covalent substances, some of which have amazing properties and were among the first substances to bring about the era of nanotechnology. Below are several forms of carbon. Diamond and graphite are commonly found in nature, while buckyballs and nanotubes are primarily synthesized in labs. Buckyballs do occur in nature, but in only small amounts.

Carbon Forms Many Types of Network Covalent Substances

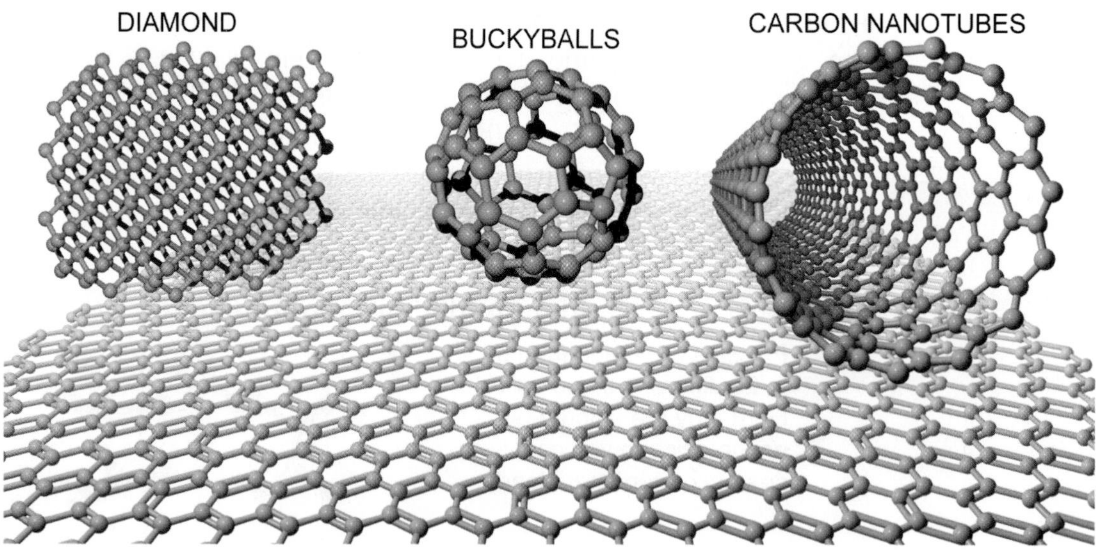

Chemistry terms

network covalent - a type of large molecule, usually made from hundreds to billions of atoms, in which each atom is covalently bonded to multiple neighboring atoms, forming a web of connections.

Chemical formulas of covalent compounds

Molecular substances have a smallest piece

With ionic compounds, the formula indicates the ratio of ions in the compound. This makes sense, because there is no specific boundary between each "piece" of the ionic crystal. In other words, there is no specific group of ions within the larger structure of the substance that we would isolate for any special reason. In molecular substances, like water, or ethanol, there is a "smallest piece" that is significant - a single molecule of that substance.

A sample of ionic or network covalent substance is a single bonded unit of material.

Empirical formulas

Network covalent substances are treated more like ionic substances in which the ratio of elements is used. This is because the large number of covalently bonded atoms forms a single unit, just like the large number of bonded ions do in ionic substances. For example, the formula for quartz is SiO_2, indicating that in a grain of typical quartz-based sand, the ratio of silicon to oxygen will be 1:2. A formula that describes the simplest ratio of elements in a substance is called an **empirical formula**.

A molecular (non-network) substance is made from weakly attracted molecules.

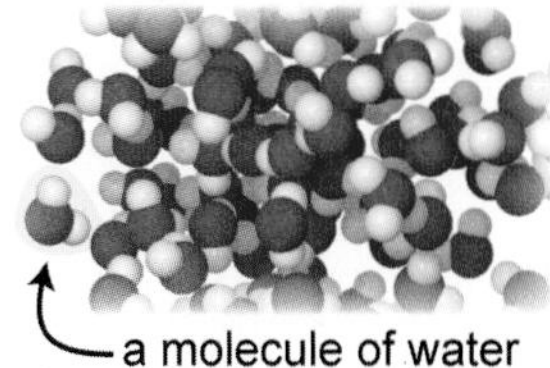

Molecular formulas

Chemical formulas for most molecular substances indicate the exact type and number of each atom in a single molecule of that substance. This is called a **molecular formula**. The elements in a molecular formula are ordered alphabetically, unlike in ionic compounds where the positive ion is listed first. Sometimes the empirical formula is the same as the molecular formula, as is the case for water (H_2O), but sometimes the molecular formula is not the empirical formula, as is the case for benzene (C_6H_6).

Benzene

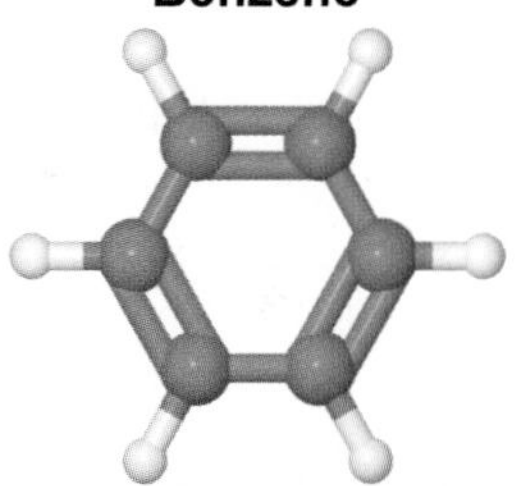

empirical formula: CH
molecular formula: C_6H_6

Write the molecular formula for each of the molecules below:

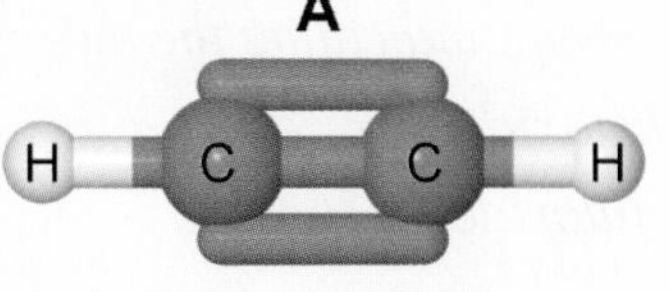

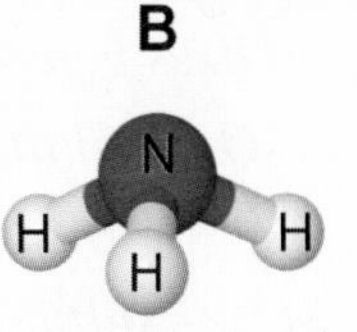

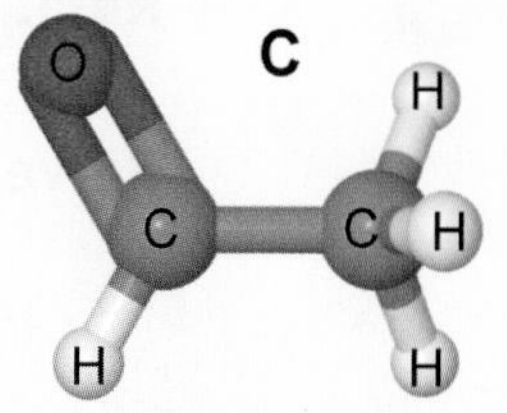

Asked: *Use the molecular structures to determine the molecular formula.*

Relationships: *The formula should reflect the exact number and type of atoms in an individual molecule of the substance.*

Solve: *a)* C_2H_2 *b)* NH_3 *c)* C_2H_5O

Naming for simple binary molecular compounds

Organic molecules

Molecular substances vary so much in their basic structures that different naming systems are used depending upon the molecule. **Organic molecules** (those made primarily of carbon, hydrogen, oxygen, and the halides) have an entire, very complex naming system dedicated to that class of substances. In this chapter we will look at naming simple binary molecular substances, those made from only two elements.

Naming organic molecules can be complex.

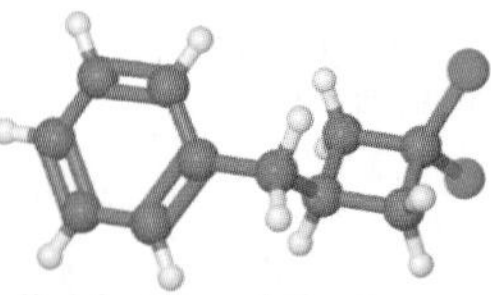

2-benzyl-1,1-dichlorocyclobutane

Binary molecular compounds have simpler names.

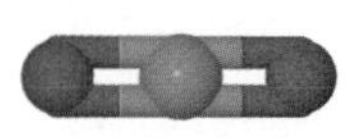

carbon dioxide

Naming simple organic molecules

Naming these simple inorganic molecules is much easier than with ionic substances, because there is a basic two-step procedure for creating the name:

1. Write down the name of the compound as if it were a simple binary ionic compound.
2. Add prefixes to each name which tell you how many of that atom are in the molecule (Note: If there is only one atom of the first element in the molecule you can leave out the "mono" prefix from the name.).

Prefix definitions
- mono = 1
- di = 2
- tri = 3
- tetra = 4
- penta = 5
- hexa = 6
- hepta = 7
- octa = 8
- nona = 9
- deca = 10

Two phosphorous atoms with three sulfur atoms. P_2S_3 → diphosphorous **tri**sulfide

Water has another name. H_2O → dihydrogen **mon**oxide

If there is only one of the first element drop the mono. CO → carbon **mon**oxide

Write the name for each of the following formulas: N_2O_4, S_2F_{10}, SO_3

Asked: *Name the formulas above.*

Given: *Three formulas, and a table of prefix names.*

Relationships: *The name of the formula is constructed from the simple ionic name with prefixes before each element, indicating the number of each atom.*

Solve: N_2O_4 = *dinitrogen tetraoxide*
S_2F_{10} = *disulfur decafluoride*
SO_3 = *sulfur trioxide*

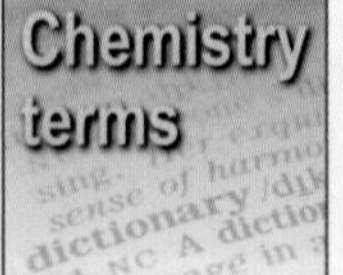

empirical formula - the simplest ratio of atoms in a substance.

molecular formula - the exact number and types of atoms in a molecule.

organic molecule - a molecule primarily made from carbon and hydrogen, but often with some oxygen, nitrogen, one of the halides, or some other non-metal atoms.

8.3 Intermolecular Forces

Why do molecules stick together in liquids and solids?

Ionic substances stay together through the attraction between positive and negative ions. Network covalent substances stay together because all the molecules or atoms are interconnected. What about water? What force attracts water molecules to other water molecules in a liquid or solid?

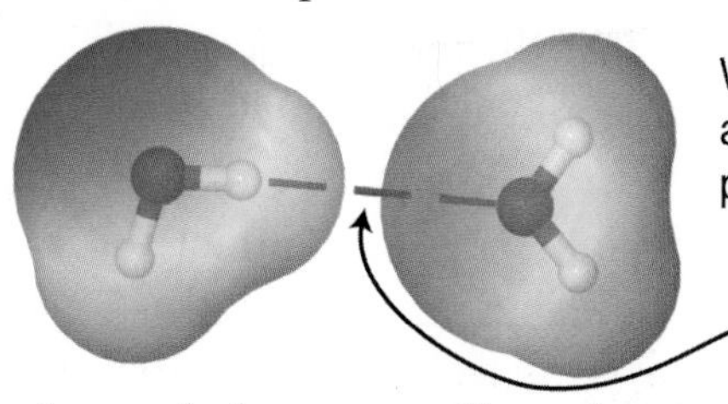

Water is polar, so the partially positive hydrogen atom from one water molecule attracts to the partially negative oxygen atom on another molecule.

intermolecular attraction

Intermolecular and van der Waals attractions

We learned in an earlier chapter that all atoms and molecules are in continual motion, so why don't the individual molecules, which are moving at extremely high velocity at room temperature, bounce off of each other forming a gas? Why are there liquid and solid molecular substances? The answer must be that the molecules stick to each other. These attractions between molecules are called **intermolecular attractions**, and are sometimes referred to as **van der Waals attractions**. These attractions are much weaker than ionic or covalent bonds which tend to be 100 to 1000 times stronger than intermolecular attractions.

A Tiny Drop of Water

Dipole-dipole and London dispersion attractions

There are two broad categories of intermolecular attractions: dipole-dipole attractions and London dispersion attractions. All molecules attract to each other using the London dispersion type attraction. However, for polar molecules the dipole-dipole attraction tends to be the more important of the intermolecular forces due to higher strength of the dipole-dipole attraction. There is one kind of dipole-dipole attraction that is significantly stronger and that has been given a special name - hydrogen bonding. We will look at each of these kinds of attractions in more detail on the following pages.

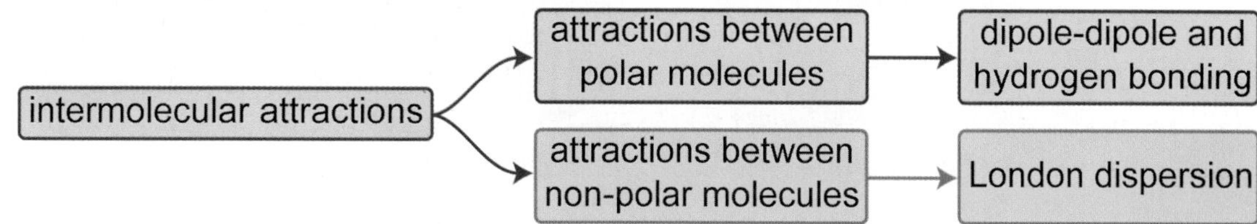

Chemistry terms

intermolecular attractions - the attractions between molecules.

van der Waals attractions - another term used to describe the attractions between molecules. Most sources consider these identical to the broad term "intermolecular attractions". However, some people only associate the term van der Waals attractions with the London dispersion type of intermolecular attraction.

Dipole-dipole attractions

Some molecules are polar and form dipole-dipole attractions

Some molecules are polar because they have some polar covalent bonds as part of their structure. This means that at least one part of the molecule will be partially positive and another part partially negative. This allows for the attractions between the oppositely charged parts of different molecules. The attraction between two polar molecules is call a **dipole-dipole attraction**. In the case of very long polymers like proteins, different parts of the same molecule can attract to itself, stabilizing the overall structure of a single molecule.

Polar Formaldehyde

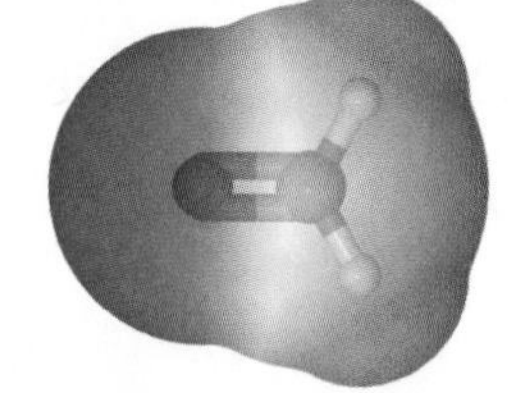

The polar covalent C=O bond makes the entire molecule polar.

Formaldehyde is a substance used as a preservative, and is a polar molecule. Because the molecule is polar, dipole-dipole attractions will form between the molecules.

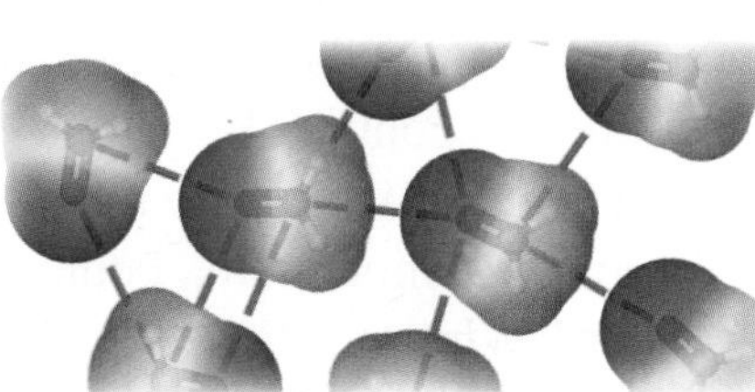

The green dotted lines represent dipole-dipole attractions between the molecules, which cause formaldehyde to condense into a liquid at room temperature.

The more polar a molecule is, the stronger it will attract to other molecules

The strength of the attraction between polar molecules can be measured by their boiling points. The higher the boiling point, the stronger the individual molecules cluster together, preventing them from separating and forming a gas. Below you can see that more polar molecules have higher boiling points. Usually, when we talk about a molecule being more polar than another, we mean that it has more polar covalent bonds. Oxygen has a high electronegativity, so by adding oxygen to a molecule, you typically form polar covalent bonds and cause a molecule to become polar (or more polar). Anytime atoms with high electronegativity are part of a molecule you usually have a polar molecule.

Higher Polarity Molecules Attract More Strongly

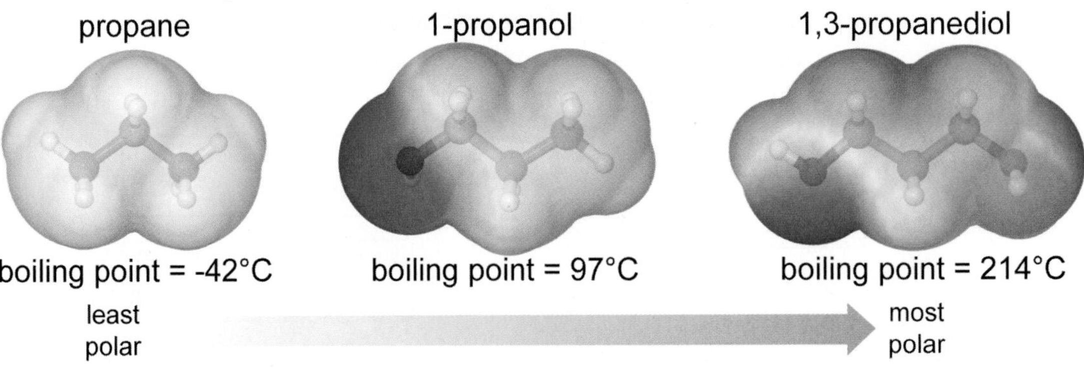

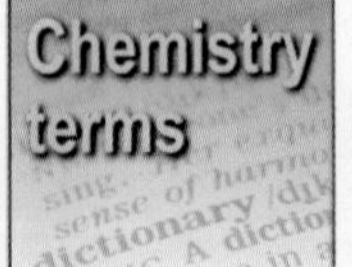

dipole-dipole attraction - the attractions between the positive part of one polar molecule and the negative part of another polar molecule.

Hydrogen bonding and the special role of water

Hydrogen bonding

In many cases, a polar covalent bond is formed between hydrogen and nitrogen or oxygen. Since there is a moderate difference in electronegativity between hydrogen and either nitrogen or oxygen, you get a polar covalent bond with hydrogen becoming partially positive. When two molecules form a dipole-dipole attraction between hydrogen on one molecule and nitrogen or oxygen on another molecule, the attraction is stronger than other dipole-dipole attractions, so we give it a special name: **hydrogen bonding**.

Hydrogen Bonds (dashed lines) in Liquid Water

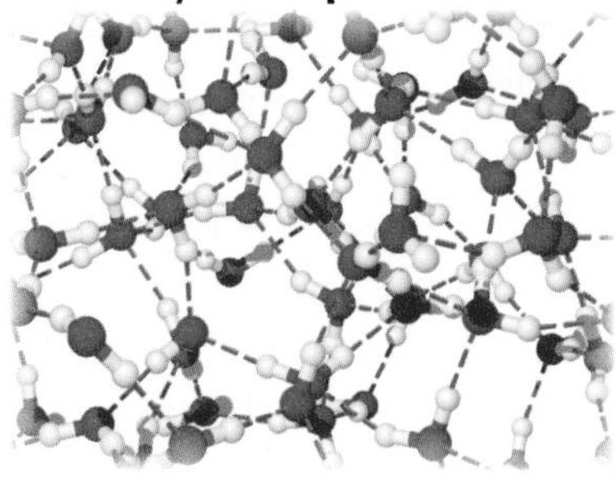

Water has special properties

Water is composed of only oxygen and hydrogen, so hydrogen bonds form between water molecules and between water and other polar molecules with similar kinds of bonds.The unique physical and chemical properties of water explain why it is so important. For example, when water freezes, it becomes LESS dense! Most materials are *more* dense as solids than as liquids. Ice is less dense than liquid water because hydrogen bonds force water molecules to align in a crystal structure where molecules are farther apart than they are in a liquid.

Iceburg photo curtesy of: NOAA

Surface tension

Another feature of water and other polar liquids is **surface tension**. As water molecules pull together they tend to form spherical drops. That's why water drops are round. At the surface of liquid water the molecules pull together, forming a "skin." This is how water striders are able to walk across the water instead of sinking into it.

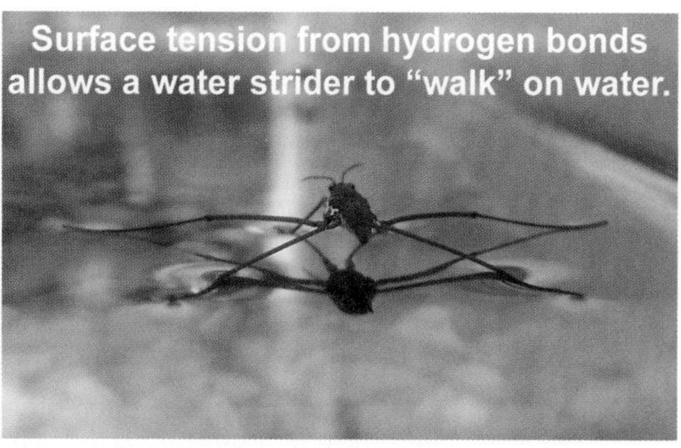

Surface tension from hydrogen bonds allows a water strider to "walk" on water.

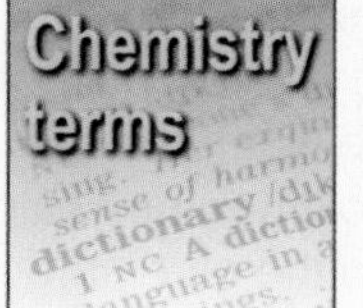

hydrogen bonding - the attractions between the partially positive hydrogen from one polar molecule to the partially negative oxygen or nitrogen on another polar molecule.

surface tension - a force that acts to pull a liquid surface into the smallest possible area, for example, pulling a droplet of water into a sphere.

Other examples of hydrogen bonding

DNA is held together with hydrogen bonds

When we refer to DNA, we usually mean the spiralling polymer that encodes all the genetic information needed by an organism to function and reproduce. In fact, DNA is really two molecules spiraling around and attracted to each other. In order to read the code, the two strands of DNA must be separated from each other and then go back together. It is hydrogen bonding which holds the DNA strands together, but not so strongly that they can't be separated and have the genetic code do its job.

DNA Uses Hydrogen Bonds to Hold the Two Strands Together

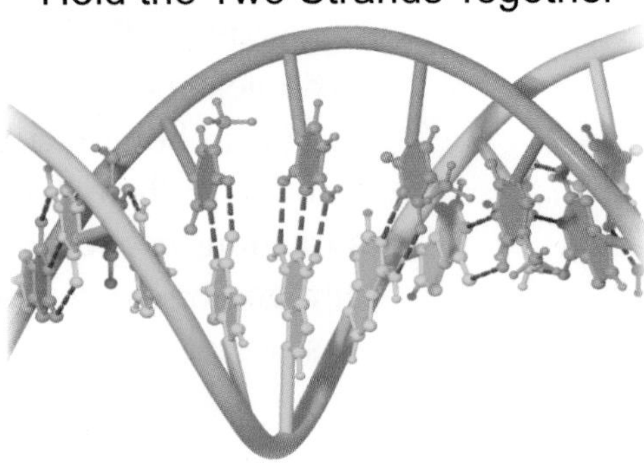

Hydrogen bonds influence protein shapes

In proteins, the shape of the molecule is one of its most important features, giving the molecule a specific function. Many forces help to shape the molecule into its final form, and hydrogen bonding plays a key role. Two common structures that are found in proteins are the alpha helix and the beta sheet. Both of these structures are stabilized by forming hydrogen bonds across different parts of the same molecule.

Ball and Stick View of Small Protein

With Backbone Trace to See Folding Pattern

Protein Structure Stabilized with H-bonds

PVA

Paper glue is a mixture of the polymer polyvinyl acetate (PVA) and water. When water is present the glue is a liquid, allowing the long polymer molecules to flow by each other. As the water evaporates, the long PVA molecules attract each other directly forming many hydrogen bonds and locking the molecules in place, forming a solid.

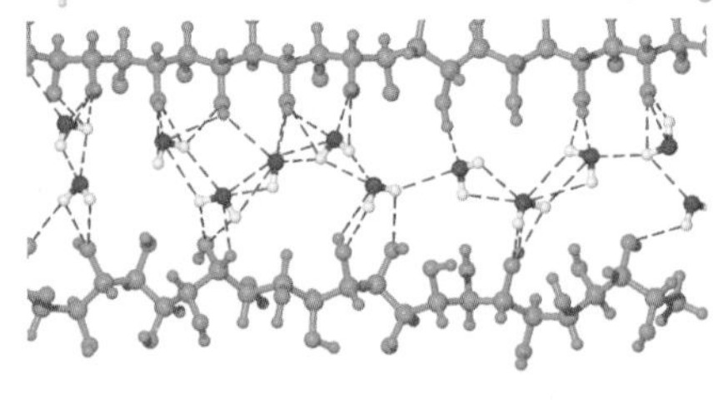

Normal "wet" glue. Polymer molecules "lubricated" by water.

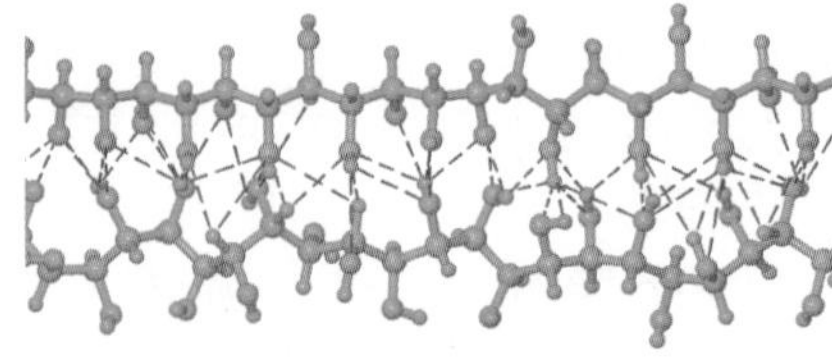

As glue dries, many more H-bonds form between the polymer molecules so the glue hardens

London dispersion attraction

Non-polar molecules can change

Even non-polar molecules cluster together into liquids and solids, so there must be an attraction even between them even if it is usually much weaker than the dipole-dipole type attractions. In fact, when two non-polar molecules come very close to each other they can become very slightly polar and stabilize themselves. This polarization is temporary and easily disrupted by the motions of molecules, yet it is enough to allow them to condense together. This attraction between non-polar molecules is called **London dispersion attraction**.

Isolated hydrogen molecules are non-polar.

A temporary, very slight polarity can be induced when non-polar molecules are close enough.

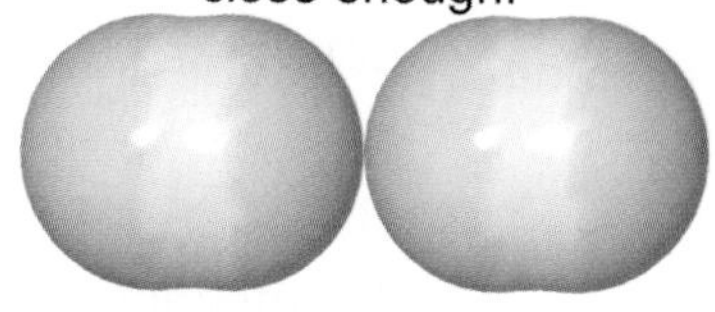

Large surface area molecules

However, if a molecule is large enough, then there is more surface area over which it can form these temporary attractions, so even non-polar molecules can be solids at room temperature. However, they tend to be soft solids like wax, or butter.

Molecules with Larger Surface Area Attract More Strongly

propane	butane	pentane
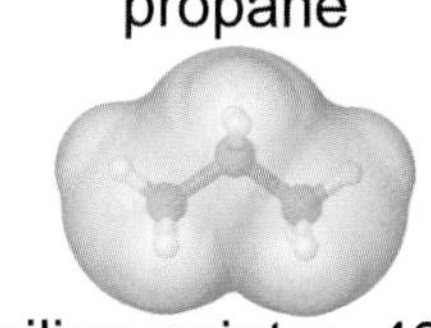		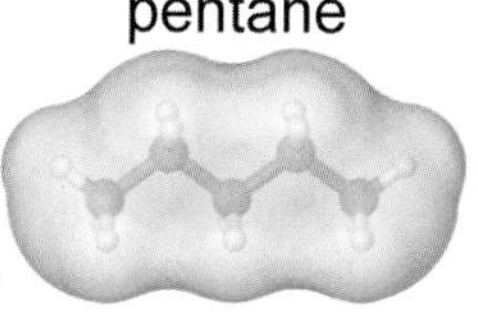
boiling point = -42°C	boiling point = 0°C	boiling point = 36°C

least surface area → most surface area

Getting close counts

The key to stronger overall attractions between non-polar molecules is not just pure size though. All intermolecular attractions are only felt when very close, so molecules that can pack together more tightly also have an advantage.

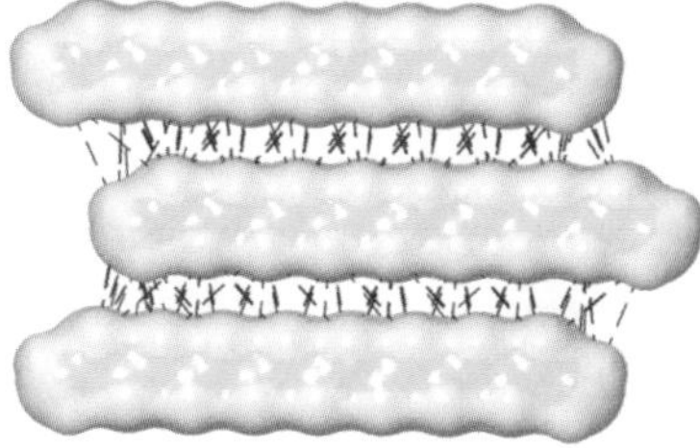

The molecules on the left and right both have about the same surface area. However, the bent shape of the molecules on the right prevents them from forming as strong a London dispersion attraction as the molecules on the left.

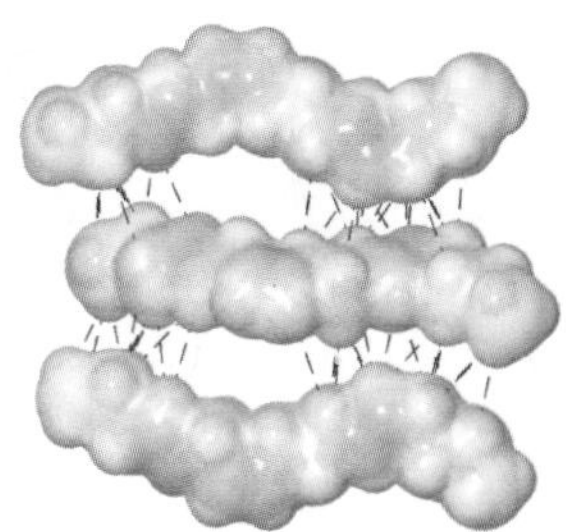

Chemistry terms

London dispersion attraction - the attraction that occurs between non-polar molecules due to temporary slight polarizations that occur when the normally equal distribution of electrons is shifted.

8.4 Formula Masses

Chemical forensics

By measuring the different amount of elements used to make a particular substance, we can determine its formula and often identify the specific substance. The field of chemical forensics is focused on determining the identity of unknown substances. Often, unidentified substances are collected at crime scenes, but they might also be found at archeological digs, in a biology lab, to verify the production of new materials, or other sources.

A scientist working for the National Water Quality Assessment Program

Molar mass and molecular formulas.

To be able to understand the data produced from various chemical analyses we need to clearly understand what a chemical formula tells us. You learned the basics of this in chapter 2, but we'll look at some more advanced formulas here. For many calculations you need to know the molar mass of a substance. To calculate this, you need to be able to use the formula to determine how many moles of each atom type there are in a mole of that substance. See examples below:

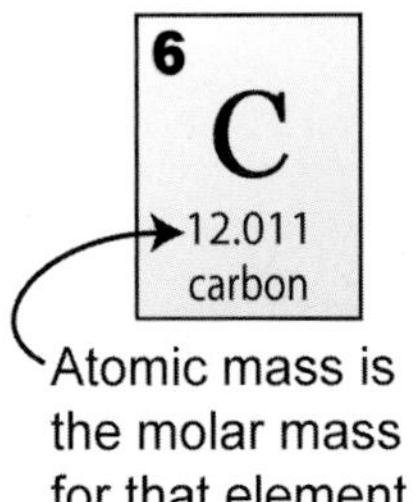

Atomic mass is the molar mass for that element.

Calculating Molar Mass

Subscripts apply only to the element or group they follow.

If the subscript follows a parenthesis, then multiply everything inside the parentheses by the subscript.

H_2SO_4

H = 2 x 1.0079 = 2.02
S = 1 x 32.07 = 32.07
O = 4 x 15.999 = 64.00
molar mass → 98.08 $\frac{g}{mole}$

$(NH_4)_3PO_4$

N = 3 x 14.007 = 42.02
H = 12 x 1.0079 = 12.09
P = 1 x 30.974 = 30.97
O = 4 x 16.00 = 64.00
molar mass → 149.08 $\frac{g}{mole}$

What is the molar mass of aluminum carbonate?

Asked: *The molar mass.*

Given: *The name of the formula.*

Relationships: *The name of the formula can be used to determine the ratios of elements in the formula. Then the periodic table can be used to get the molar mass of each of the elements used.*

Solve: *Aluminum carbonate is ionic, so the formula is* $Al_2(CO_3)_3$.

Al = 2x26.98 = 53.96 g/mol
C = 3x12.011 = 36.03 g/mol
O = 9x15.999 = 144.00 g/mol

molar mass = 233.99 g/mol

Percent composition

Mass percent composition

One way to identify compounds is by the percentage by mass of each element. This is like a fingerprint for a particular compound. To calculate the percentage of each element you need to know the total mass of your compound and the mass of each individual element.

NaCl

% by atom number
Na = 50% Cl = 50%

% by mass
Na = 39% Cl = 61%

Calculating a percent

$$\text{\% by mass} = \frac{\text{mass of element in the compound}}{\text{total mass of the compound}} \cdot 100$$

Calculate percent composition

Calculating % using a formula:

1. Calculate the molar mass.
2. Calculate the percent of each element in a mole of the substance.

Calculating % with a measured sample:

1. Add up the individual mass measurements to get the total mass.
2. Calculate the percentage as above.

$(NH_4)_3PO_4$

Ca = 3 x 40.078 = 120.24
P = 1 x 30.974 = 30.97
O = 4 x 15.999 = 64.00
molar mass → 149.12 $\frac{g}{mol}$

$$\%Ca = \frac{120.24}{149.12} \cdot 100 = 38.76\%$$

$$\%P = \frac{30.97}{149.12} \cdot 100 = 19.97\%$$

$$\%O = \frac{64.00}{149.12} \cdot 100 = 41.27\%$$

You find a white powder which can be broken down into 1.05 g of carbon, 0.16 g of hydrogen and 1.29 g of oxygen. Is this powder common table sugar - sucrose ($C_{12}H_{22}O_{11}$)?

Asked: *Is the unknown powder sugar?*

Given: *The mass of each element in the powder and the formula for sugar.*

Relationships: *The % of each element should match the percent composition of sugar if the power is a sample of that substance.*

Solve: *Calculate the % composition of sugar and the unknown compound.*

sucrose

C = 12 × 12.01 = 144.13 g/mole
H = 22 × 1.0079 = 22.17 g/mole
O = 11 × 15.999 = 175.99 g/mole
molar mass = 342.29 g/mole

$$\%C = \frac{144.13}{342.29} \bullet 100 = 42.1\%$$

$$\%H = \frac{22.17}{342.29} \bullet 100 = 6.5\%$$

$$\%O = \frac{175.99}{342.29} \bullet 100 = 51.4\%$$

unknown

C = 1.05 g
H = 0.16 g
O = 1.29 g
total = 2.50 g

$$\%C = \frac{1.05}{2.50} \bullet 100 = 42.0\%$$

$$\%H = \frac{0.16}{2.50} \bullet 100 = 6.4\%$$

$$\%O = \frac{1.29}{2.50} \bullet 100 = 51.6\%$$

YES!

The percent composition is very close, so the unknown is probably sugar

Calculating empirical formulas

The simplest ratio of atoms

An empirical formula tells you the simplest ratio of elements in a compound. All ionic formulas are empirical, and many molecular formulas are empirical as well. Table salt is a one-to-one ratio of sodium to chloride. In a typical grain of salt there are about a billion billion ions. What should the formula be? Should we write $Na_{1000000000000000000}Cl_{1000000000000000000}$? What about the salt crystals that are bigger or smaller? The simplest way is to write the empirical formula, showing the most basic ratio - NaCl.

The empirical formula is the one with the simplest ratio.

Not empirical. All show the same ratio, but not in its simplest form.	Na_2Cl_2 Na_5Cl_5 $Na_{12}Cl_{12}$
The empirical formula for sodium chloride is a simple 1:1 ratio.	**NaCl**

Finding the mole ratio of atoms

To calculate the empirical formula for a substance you need to know the ratio of atoms. In one mole of NaCl there will be one mole of Na^{+1} ions and one mole of Cl^{-1} ions. The way chemists count atoms (or moles of atoms) is by measuring their mass, so you will typically start with a mass which needs to be converted to moles, to find the mole ratio.

Calculating an Empirical Formula

Convert grams to moles:
Assume you have a sample that is 0.504g hydrogen and 4.00g oxygen.

$$H = 0.504g \cdot \frac{1mol}{1.01g} = 0.500mol$$
$$O = 4.00g \cdot \frac{1mol}{16.00g} = 0.250mol$$

Simplify the mole ratio:
Divide each answer by the smallest number of moles.

$$H = \frac{0.500}{0.250} = 2$$
$$O = \frac{0.250}{0.250} = 1$$

Write the formula:

$$H_2O$$

You find a jar of powder labeled copper oxide, but there is a problem because you know it should have a Roman numeral in the name because copper can be Cu^{+1} or Cu^{+2}. After taking a small sample you find that it is made from 2.96g of copper and 0.37 g of oxygen? What is the correct name and formula for the ionic compound in that jar?

Asked: *What is the correct name and formula for the compound in the jar?*

Given: *The mass of each element in a small sample of the powder.*

Relationships: *The mole ratio of elements can be simplified to give an empirical formula.*

Solve: *Convert the grams to moles and then simplify the ratio.*

convert grams to moles

$$Cu = 2.96g \cdot \frac{1\ mole}{63.55g} = 0.0466mole$$
$$O = 0.37g \cdot \frac{1\ mole}{15.999\,g} = 0.023mole$$

simplify mole ratio

$$\frac{0.0466\ mole}{0.023\ mole} = 2.0$$
$$\frac{0.023\ mole}{0.023\ mole} = 1.0$$

The mole ratio is 2 moles of Cu to 1 mole of O, so the formula is Cu_2O, and the name is copper(I) oxide.

Calculating molecular formulas

Sugar and salt crystals.

Like a grain of salt, a grain of sugar contains billions and billions of atoms. However, there is a significant difference in the structure of a sugar crystal compared to a salt crystal. In the ionic salt crystal every atom is bonded with strong chemical bonds to all of its neighboring atoms. In the sugar crystal there are small clusters of atoms strongly bonded to form a molecule, and then these molecules are weakly attracted to each other through intermolecular attractions to form the overall crystal.

A molecule of glucose has six carbon atoms, twelve oxygen atoms, and six hydrogen atoms.

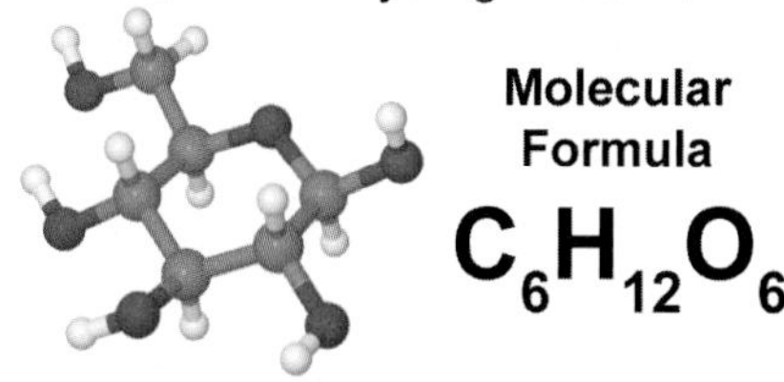

The importance of the molecular formula

For the sugar glucose, the cluster of covalently bonded atoms (the glucose molecule) is made from 6 carbons, 12 hydrogens, and 6 oxygens. That means the smallest possible piece of glucose is a molecule with the formula $C_6H_{12}O_6$. Take a look at that ratio of 6:12:6, and you will see that by dividing everything by 6 you can simplify the ratio to 1:6:1, giving the empirical formula CH_2O. For covalently bonded compounds it is the *molecular* formula which is crucial, because CH_2O is the molecular formula for formaldehyde, not glucose. For formaldehyde the empirical formula is the same as the molecular formula, a situation which is often true even for molecular compounds.

The molecular formula is a multiple of the empirical formula

In order to find the molecular formula, you need one more piece of information than you needed for finding the empirical formula - the molar mass of the substance. The molecular formula will always be equal to, or a multiple of, the empirical formula.

Empirical Formula		Molecular Formula
CH_2O	x6	$C_6H_{12}O_6$
$30.03\frac{g}{mol}$	x6	$180.16\frac{g}{mol}$

Given the following empirical formulas and molar masses, determine the molecular formula for each compound: CH:78.11 g/mol and H_2O:18.02 g/mol.

Asked: *What is the molecular formula for each compound above?*

Given: *The empirical formula and the molar mass of the compound.*

Relationships: *The ratio of molar mass of the empirical formula to the molar mass of the compound will give a multiplier for applying to the empirical formula to get the molecular formula.*

Solve: *Divide the molar mass of substance by the molar mass of the empirical formula, and multiply empirical formula by the result.*

calculate empirical molar mass	determine multiple	calculate molecular formula
$CH = 12.011 + 1.0079 = 13.018 g/mole$	$\frac{78.11}{13.02} = 6$	$6 \times CH = C_6H_6$
$H_2O = 2(1.0079) + 15.999 = 18.015 g/mole$	$\frac{18.02}{18.02} = 1$	$1 \times H_2O = H_2O$

Nanotechnology: Food, Cosmetics and Medicine

Nanotechnology is a very important and exciting new area in science. The hope with nanotechnology is that we can build molecules atom by atom! We can't do this yet, but it is quite possible that scientists will be able to soon. Nanotechnology has broad applications in all of the core areas of science.

A nanometer (nm) is 10^{-9} meters, and there are one billion nanometers in one meter. To gain perspective on this small size, a human hair is about 50,000 nm in diameter, a DNA double helix is about 2 nm in diameter, and an average carbon-to-carbon bond length is about 0.15 nm. The term "nano" technology usually refers to structures having dimensions around 100 nm or smaller. Although it is difficult, some modern technology has made it possible for chemists to develop materials with these small dimensions.

Eyelash magnified 400 times

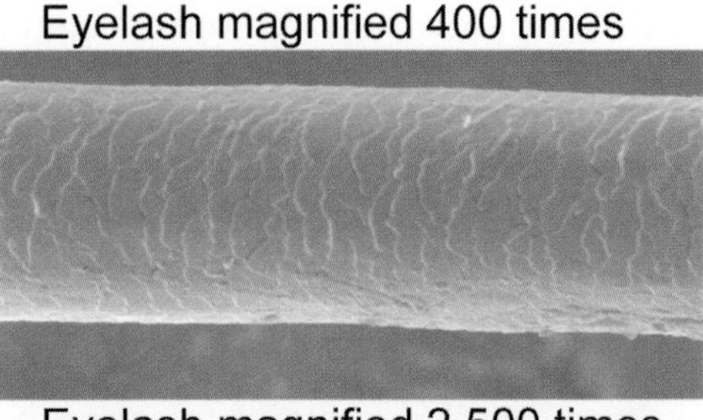

Eyelash magnified 2,500 times

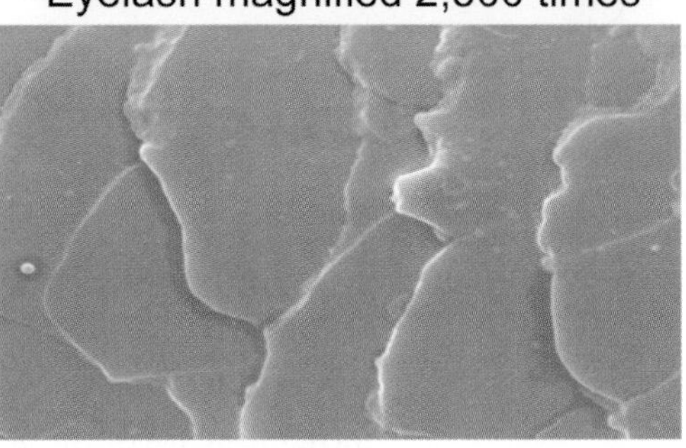

One of the biggest hurdles for scientists in this field is developing equipment that will work effectively on such a small scale. Much of our current technology is not sensitive enough to handle individual atoms with precision. The field of nanotechnology is very broad and diverse. More than 800 consumer products are already on the market that contain nanotechnology. In most cases consumers are unaware that their products contain nanoparticles.

What is a nanomaterial? Nanomaterials are generally of two types: fullerenes, such as buckyballs and carbon nanotubes, and inorganic nanoparticles. These substances have many properties of interest to scientists including heat resistance, superconductivity, strong mechanical strength and special electrical properties. Nanoparticles are made of metals, metal oxides, superconductor materials, and can have catalytic properties. It is very important to note that these materials have very different properties at the nanoscale level than they do in larger everyday amounts. Metals that are nonreactive in normal quantities become very reactive at the nanoscale level. For instance, gold can act as a catalyst in nanoparticle form, but in larger quantities it is nonreactive.

DNA Nano Robot

Nanotechnology also holds significant promise for medicine. Scientists are researching ways to use these small particles to deliver drugs to specific tissues, and to develop nanorobots that will help repair cells after damage from surgery or disease. Nanosized imaging tools are also being developed that can be placed inside cells and used to monitor cellular changes. Nanoparticles are used as contrast agents to enhance current imaging techniques such as MRI's and ultrasound. By inserting small particles that fluoresce, small areas of tissue can be seen more reliably. For cancer patients, nanoparticles can help to detect tumor growth early-on, and allow for more effective treatment. Small nanoparticles can penetrate and travel to hard to reach places in the body where they can be absorbed. Researchers link antigens or RNA strands that interact and identify the specific target cell surface. In this way, medicine can reach specific tissues and last longer so that it can work more effectively.

Scientists hope to build molecular-sized biological machines or nanorobots to deliver medication and to repair damaged tissues. Molecular self-assembly is a very important concept in nanotechnology. Since it is difficult to "build" small structures, scientists hope to make the component parts and allow the molecules to form the desired structures on their own based on intermolecular force attractions such as: Van der Waals forces, hydrogen bonding and hydrophobic interactions. One application of this approach would be to build regions of nucleic acids, and allow them to form sections of DNA or RNA that carry out specific and known functions in the human body. These regions could be used to produce needed proteins and enzymes.

Molecules may be arranged to construct elaborate mechanical devices

Nano Machines may be used to assist targeted drug delivery

The food industry is adding nanosized particles to their foods to improve their taste, nutritional value, flavor and consistency. For example, a chocolate shake with "nanoclusters" gives a richer chocolate flavor without the need to add extra sugar. Certain teas or "nanoteas" offer increased selenium absorption by the body. In this case, a person could drink tea instead of taking a supplement. For children there are nutritional supplement drinks on the market that contain iron nanoparticles.

Cosmetic companies are using nanotechnology to add small oil droplets to make the application of their face creams smoother. Most face creams contain nanosized mineral sunscreen agents such as zinc oxide or titanium dioxide. The small sized mineral particles cause sunscreens to be more transparent instead of white. Other common chemicals that are added to face creams such as Retinol A and anti-aging chemicals are more easily absorbed by the skin when they are in tiny nanocapusles.

Currently, scientists are working hard to gain the ability to precisely control every atom and nanoparticle so that they can better control the properties and behavior of the nanomaterials they are making.

Chapter 8 Review

Vocabulary

Match each word to the sentence where it best fits.

Section 8.1

brittle	polyatomic ion
electric current	

1. Something that is ________ can be strong but still break easily.
2. If you have a ________ you have a small molecule with a charge.
3. When charge moves in a directed way you have an ________.

Section 8.2

lipids	copolymers
steroids	network covalent
hydrocarbon	empirical formula
polymers	molecular formula
monomer	organic molecule
homopolymer	

4. A ________ will tell you exactly how many atoms there are in an individual molecule.
5. Most of our fuels are one kind of ________ or another, because the carbon and hydrogen they contain have a large amount of chemical energy.
6. Molecules that are made from repeatedly bonding together smaller molecules are broadly called ________.
7. ________ are bonded together to form polymers.
8. Polypropylene is only made from one monomer, so it is a ________.
9. For a molecule to be an ________ it must be made primarily from carbon and hydrogen with the possibility of some other non-metals also part of the molecule.
10. An ionic formula is always an ________.
11. Some ________ are natural and are necessary for correct biological functioning, while others are sometimes taken as drugs.
12. Some of the highest calorie foods contain ________.
13. Protein and DNA are molecules that are examples of ________.
14. The type of molecular substances that form very large interconnected molecules are called ________.

Section 8.3

intermolecular attractions	hydrogen bonding
van der Waals attractions	surface tension
dipole-dipole attraction	London dispersion attraction

15. ________ is a slightly stronger version of dipole-dipole attraction, but it is not as strong as a covalent bond.
16. Two broad terms that describe why molecules condense into liquids and solids are ________ and ________.
17. When two polar molecules attract to each other you have an example of ________.
18. You can "float" a paper clip on water if you very carefully lay it flat on top of the water. This is due to what is known as ________.
19. ________ is particularly important for non-polar molecules.

Conceptual Questions

Section 8.1

20. What is one property shared by most ionic substances?

21. Why do ionic substances not conduct electricity in their solid form, but will if they are dissolved or in their liquid form?

22. True or false. The positive ion in an ionic compound must have the same charge as the negative ion. Explain.

23. Describe the structure of an ionic crystal.

24. In what ways is an ionic crystal like a molecule and in what ways is it different?

25. Describe why you sometimes need parentheses around a polyatomic ion when used in a formula. Give one example of when you need parentheses and one where you don't.

26. Why must the total positive and total negative charge in an ionic compound be equal?

27. Why do some ionic names need a Roman numeral, and some don't. Give an example.

Section 8.2

28. Describe a major difference between the properties of ionic compounds and molecular compounds.

29. What are some common properties of substances made from small molecules?

30. If two small molecules are about the same size, but one is polar, which one will have the higher boiling point?

31. Gasoline is a hydrocarbon. What does that mean?

32. What is a common property of substances made from medium-sized molecules?

33. If you wanted to avoid lipids, what foods should you not eat?

34. Describe the relationship between a monomer and a polymer.

35. How many different monomers are used to make proteins?

36. What do we call a monomer that is used to make a protein?

37. Which kinds of molecular substances most resemble ionic compounds and in what ways are they similar?

38. Describe the difference between an empirical formula and a molecular formula.

Section 8.3

39. What are the major categories of intermolecular attractions?

40. Is a hydrogen bond more like a dipole-dipole attraction or more like a London dispersion attraction? Explain.

41. Dipole-dipole attractions occur between which types of molecules:
    a. network covalent
    b. polar
    c. non-polar

42. Explain why water has a boiling point that is so much higher than most molecules that are so small.

43. List three special properties of water, not including its high boiling point.

44. Describe the difference between London dispersion attraction and dipole-dipole attraction.

45. How does the size of a molecule play a role in the strength of its intermolecular attractions?

46. How does the shape of a molecule play a role in the strength of its intermolecular attractions?

47. How does the polarity of a molecule play a role in the strength of its intermolecular attractions?

Section 8.4

48. What does an empirical formula tell you? Give an example.

49. Give an example of when a molecular formula is the same as its empirical formula.

50. Give an example when the molecular formula is different from the empirical formula.

Chapter 8 Review.

51. Describe how the percent composition helps in determining the molecular formula of an unknown substance.

Quantitative Problems

Section 8.1

52. What does the formula $BaCl_2$ tell you about the substance barium chloride?

53. Write the formula for each of the following names:
    a. lithium bromide
    b. magnesium fluoride
    c. calcium sulfide
    d. aluminum iodide
    e. aluminum sulfide

54. Write the formula for each of the following names:
    f. chromium(III) nitrate
    g. copper(I) oxide
    h. copper(II) oxide
    i. lithium phosphate
    j. iron(III) sulfate

55. Write the formula for each of the following names:
    a. barium hydroxide
    b. sodium sulfide
    c. tin(IV) carbonate
    d. cobalt(II) chlorate

56. Write the name for each of the following formulas:
    a. K_2O
    b. BeI_2
    c. $SrCl_2$
    d. $PbCl_4$

57. Write the name for each of the following formulas:
    a. FeO
    b. Fe_2O_3
    c. $ZnBr_2$
    d. $AgNO_3$
    e. K_2CO_3

58. Write the name for each of the following formulas:
    a. $NaHCO_3$
    b. $Sn(NO_3)_2$
    c. $(NH_4)_2SO_4$
    d. $Al(C_2H_3O_2)_3$
    e. $CuSO_4$

Section 8.2

59. Write the formula for each of the following names:
    a. dinitrogen monoxide
    b. dichlorine hexoxide
    c. iodine tribromide

60. Write the formula for each of the following names
    a. iodine heptafluoride
    b. phosphorous trichloride
    c. chlorine monoiodide
    d. dinitrogen tetrafluoride

61. Write the names for each of the following formulas:
    a. SF_2
    b. P_4O_{10}
    c. P_2Cl_4

62. Write the names for each of the following formulas:
    d. Cl_2O_7
    e. PF_5
    f. S_2F_{10}

Section 8.3

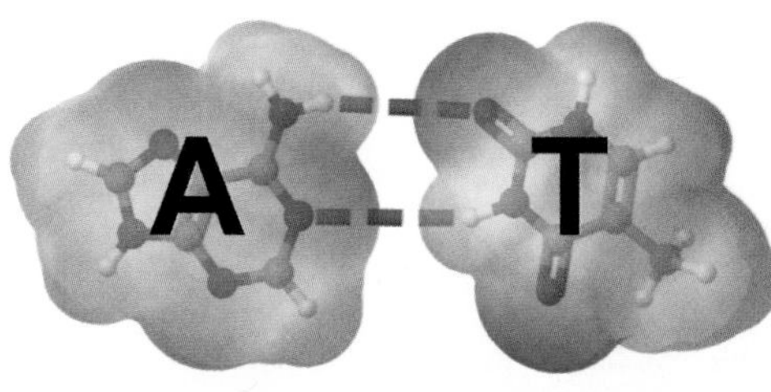

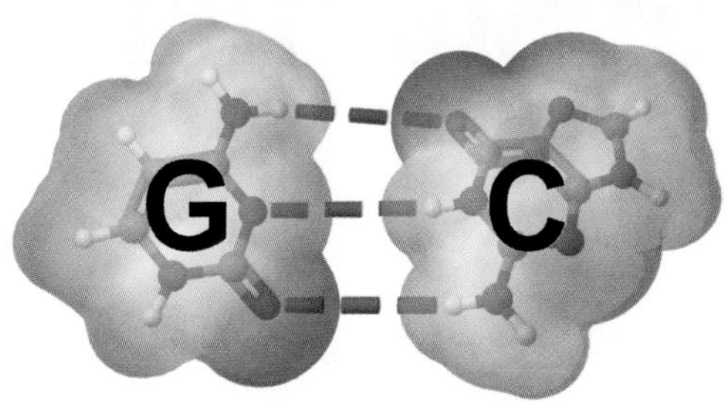

63. There are four DNA bases that are part of the nucleotide monomers that make up the DNA molecule. When two DNA strands wrap around each other they only stick together if there is an "A" on one strand across from a "T" on the other or a "G" on one strand across from a "C" on the other. Which pair do you think sticks together better and why?

64. Why does it make sense that "A" and "T" pair up and "G" and "C" pair up?

Section 8.4

65. Calculate the molar mass for the following substances:
    a. H_2O
    b. $NaNO_3$
    c. $(NH_4)_2S$
    d. $Ca_3(PO_4)_2$

66. What is the percent composition by mass for each of the substances in the previous question?

67. If you add up all the % of each element in a compound you should get what number?

68. Calculate how many molecules are in the following
    a. 6.5 g of H_2O
    b. 19.3 g of C_8H_{10}

69. Give the empirical formula for each of the molecular formulas:
    a. N_2O_4
    b. P_2O_5
    c. P_4H_{10}
    d. H_2O_2
    e. P_3Cl_5

70. Give three possible molecular formulas for the empirical formula CH_2O.

71. If you had a sample of a substance that had 20.23g of aluminum and 79.77g of chlorine, what would the empirical formula be for this substance?

72. What is the empirical formula for a substance with the following % composition:
    40% carbon
    6.7% hydrogen
    53.3% oxygen

73. You have 16.0g of some compound and you perform an experiment to remove all of the oxygen. What is left is 11.2g of iron. What is the empirical formula?

74. After burning 1.5g of scandium in oxygen, a new compound is formed that has a mass of 3.1g. What is the empirical formula.

75. The empirical formula for cyclohexane is CH_2, and its molar mass is 84.18g/mol. What is the molecular formula?

76. What is the molecular formula for a compound with a molar mass of 34.02g/mol and empirical formula HO?

77. A sample of an unknown compound was determined to be made from 8.56g of carbon and 1.44g of hydrogen. The molar mass of the compound was found to be 28.03g/mol. What is the molecular formula?

Water and Solutions

Why is water so important?
What are the chemical properties of water?
How do we describe the things that are mixed into water?
Is steel different from iron, or 14k gold different from pure gold?

We may not realize it, but we drink solutions every day. Coffee is a solution, and so is apple juice, soda and mineral water. In fact the fluid that makes up our blood is a solution too.

In the environment, sea water is a solution of salt, water and other dissolved minerals. Solutions are important because they are all around us and within us. In fact, almost all of the chemical reactions that occur in our body only work in solutions.

A solution is a homogeneous mixture of two or more substances. Molecules in a solution are mixed so well that there are no "clumps" of one type of molecule. For example, think about sugar dissolved in water to make a solution. In a true solution the molecules of sugar are completely separated from other molecules of sugar.

True Solution
(**no** molecular clumps)

Not a Solution
(molecular clumps)

Most of the solutions we talk about in this chapter will be solutions in water; however, solutions can be solids or gases too! Air is a solution of gases, mostly oxygen and nitrogen. Steel is also a solution, a solid solution of iron and carbon. Fourteen carat gold is a solution of gold and silver. The number "14" in 14k gold means 14 out of every 24 grams are gold (and 10 out of 24 grams are silver).

When is a solution like a conducting wire?

Dissolve 10 grams of salt or sugar in water and what do you get? Both look like clear water, but one tastes sweet and the other tastes salty. Other properties than taste have also changed. Sugar and salt are different types of compounds (ionic and molecular) and they have quite different properties in solution. Both are critical to living things, and many popular sports drinks have both.

Here is an interesting experiment. Start with two solutions that are both made with 10 grams dissolved in 100 mL of water. One solution is sugar and the other is salt.

An electrical voltage from a battery is connect so the electricity has to pass through the solution. The results are quite different. Electricity passes right through the salt solution as if it were a conducting metal wire. The sugar solution stops electricity from flowing. A sugar solution is an insulator and the bulb stays dim.

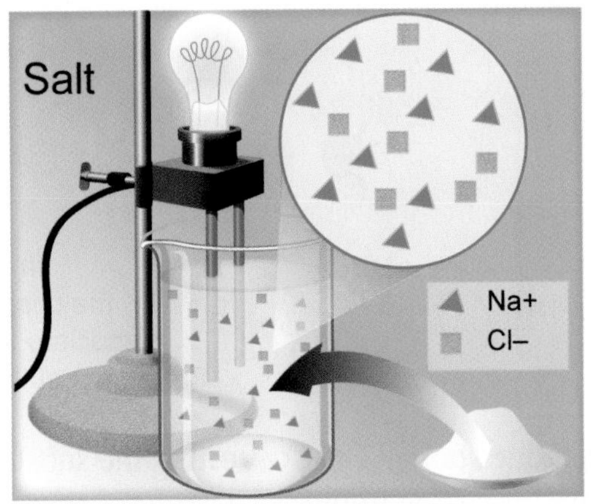

Both the sugar and the salt dissolve completely in water! They are both white granular solids.

Why is the result different for two substances which appear to be so much the same?

By itself, pure water is an electrical insulator. The ability to conduct electricity is created by particles in solution. Dissolved salts make conductive solutions because they dissociate into charged ions. Copper sulfate solution is also a good conductor for the same reason. The blue color comes from dissolved copper ions (Cu^{2+}). Dissolved molecular substances make insulating solutions.

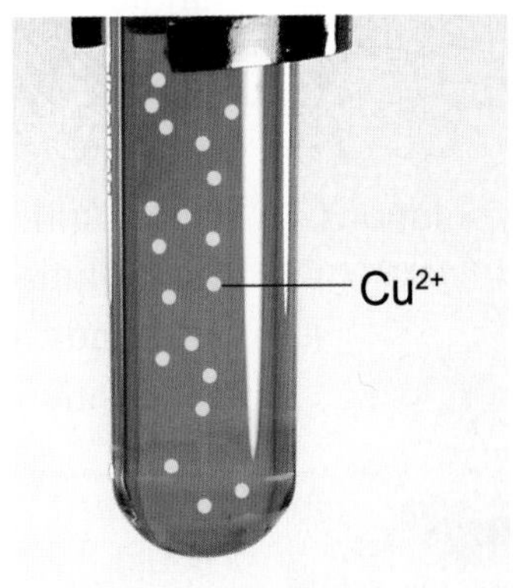

This fact could save your life some day! Your body's own electrical control signals travel in conductive salt solutions within and between your cells. The presence of dissolved salts in body fluids explains why wet skin is a more effective electrical conductor than dry skin. In fact, wet skin is more than 1,000 times as conductive as dry skin. That is why your hands should always be dry when working around electrical tools or plugging in appliances.

9.1 Solutes, Solvents and Water

Definition

In Chapter 2 we talked about solutions and you have worked with them extensively in the laboratory. This chapter will look at solutions more closely and develop some of their properties. A true solution is *homogenous on the molecular level.* That means there are no clumps bigger than a molecule.

An example solution

Sixteen grams of copper sulfate ($CuSO_4$) in 100 mL makes a deep blue solution when every particle is dissolved. This solution is 1 molar because there is 1 mole of copper sulfate per liter of solution.

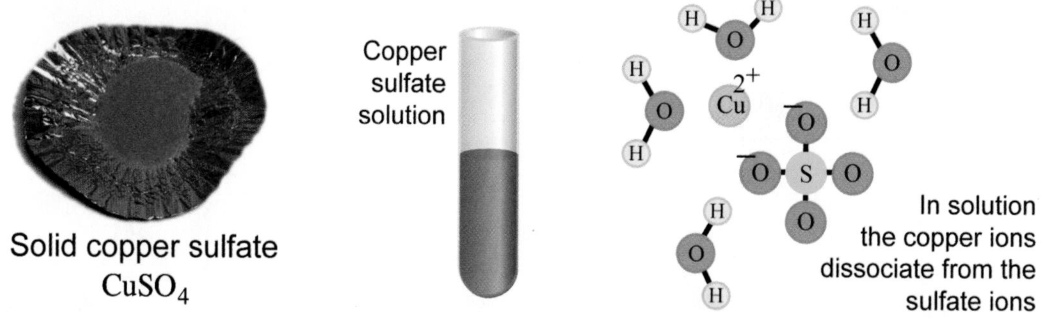

Solvent and solute

A solution always has a *solvent* and at least one *solute*. The **solvent** is the substance that makes up the biggest percentage of the mixture, or is liquid. The copper sulfate solution has water as the solvent. The **solutes** are the other substances in the solution. Copper sulfate is the solute in the example.

Dissolving

When the solute particles are evenly distributed throughout the solvent, we say that the solute has **dissolved**. The copper sulfate starts as a solid powder. Water is added and the mixture is carefully stirred until all the solid powder has dissolved. Once the copper sulfate has dissolved, the solution is transparent again.

All true solutions are transparent

Why solutions are transparent

True solutions are transparent because all the particles are single molecules. Milk is not a true solution because each of the tiny fat particles in milk are thousands, or tens of thousands, of molecules. Particles this size are large enough to block light and scatter it to the side. The scattering of light is the reason milk is opaque.

solvent - the substance that makes up the biggest percentage of the mixture, or is liquid.

solute - any substance in a solution other than the solvent.

dissolved - when molecules of solute are completely separated from each other and dispersed into a solution.

The water molecule

Water is critical to life

Your body is about 60% water by weight. In 1 hour of exercise, you may lose as much as a half-gallon of water sweating and breathing! You also lose small amounts of dissolved salts. If the lost water and dissolved salts are not replaced, your body stops working. The unique physical and chemical properties of water explain why it is so important.

Water is a polar molecule

The water molecule is **polar**. That means the electric charge is unevenly distributed in a water molecule. The two shared electrons from the hydrogen atoms are attracted to the oxygen atom. This makes the oxygen side of the molecule slightly negative and the hydrogen side slightly positive. The diagram on the right shows the surface of the molecule shaded red where it is more positive and blue where it is more negative.

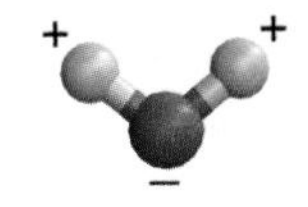

Water molecule
ball-and-stick model

Water molecule
molecular surface

Hydrogen bonding

Water molecules tend to stick to each because of the polarity. Positive hydrogen atoms from one water molecule are attracted to the negative oxygen atoms in another molecule. This attraction is not as strong as a chemical bond. BUT, it is strong enough that we give it its own name: **hydrogen bonding**. Hydrogen bonding explains some curious properties of water.

Why ice floats

When water freezes, it becomes LESS dense! This is unusual, most materials are *more* dense as solids than as liquids. Ice is less dense than liquid water because hydrogen bonds force water molecules to align in a crystal structure where molecules are farther apart than they are in a liquid.

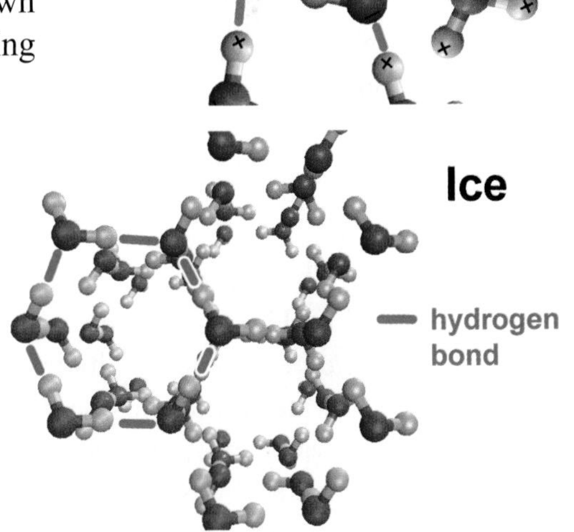

Chemistry terms

polar - a molecule is polar when there is a charge separation that makes one side (or end) of the molecule more positive or negative than the other side (or end).

hydrogen bond - an intermolecular bond that forms between a hydrogen atom in one molecule and the negatively charged portion of another molecule (or another part of the same molecule).

The importance of hydrogen bonding

Comparing water and methane

Consider two molecules with similar molecular masses. Methane (CH_4) has a molecular mass of 16 g/mole and water (H_2O) has a molecular mass of 18 g/mole. Methane boils at -161°C, so at room temperature, methane is a gas. Water has a similar mass, but boils at an astounding +100°C! This is far hotter than almost any other low molecular-weight hydrogen compound. What is the reason? The answer has to do with hydrogen bonding and the *shape* of the two molecules.

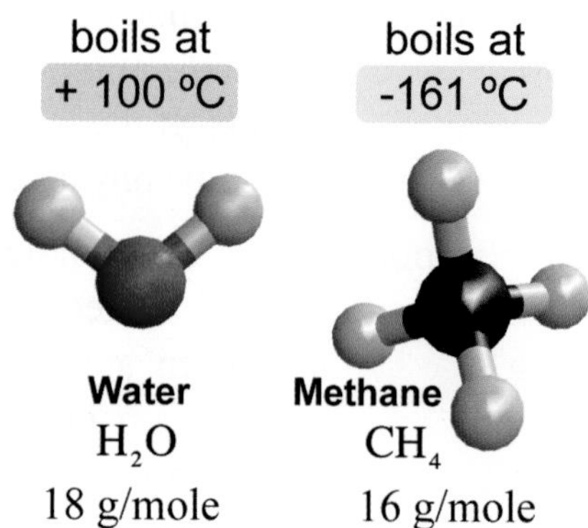

Hydrogen bonding causes water's high boiling point

Due to its symmetrical shape, a methane molecule is non-polar! Hydrogen bonding does not occur between neighboring methane molecules. Without hydrogen bonding, it is easier for a methane molecule to escape its neighbors and become a gas. By comparison, water has strong hydrogen bonding between neighboring molecules. The hydrogen bonding is the reason for water's unusually high boiling point. The high boiling point of water is of critical importance to life on earth. Life needs *liquid* water. If water was a gas at room temperature, it would be impossible for life to exist as we know it.

Making solid steel float!

Gently place a pin on the surface of water and the surface of alcohol. The pin sinks immediately in the alcohol. The pin *floats* on the water! Pins are made of steel, which is dense enough to *sink* in water. Why can a pin float on water but not on alcohol?

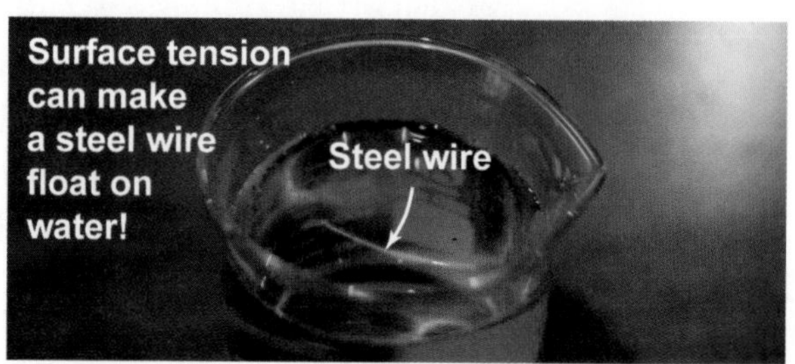

Surface tension can make a steel wire float on water!
Steel wire

Surface tension

The answer is a property called **surface tension**. Surface tension is a force that pulls molecules together along a liquid surface. The high surface tension of water is a direct consequence of hydrogen bonding. The pin floats on water because it has to separate water molecules on the surface to break through and sink. The hydrogen bonds between water molecules are held tightly enough that a gently placed pin will float. The fact that water drops pull themselves into spheres is also caused by surface tension. Alcohol has a lower surface tension, because it has fewer hydrogen bonds than water and the pin easily drops through the surface to sink.

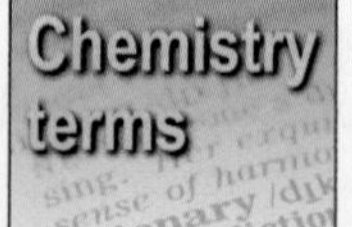

surface tension - a force created by intermolecular attraction in liquids, such as hydrogen bonding in water. Surface tension acts to pull a liquid surface into the smallest possible area, for example, pulling a droplet of water into a sphere.

Water as a solvent

The universal solvent

Water is often called the "universal solvent." While water doesn't dissolve everything, it does dissolve many substances such as salts and sugars. Water is a good solvent for two reasons:

1. A water molecule is small and light;
2. Water molecules are polar.

How water dissolves salt

The polar molecules of water are especially good at dissolving ionic compounds like salt (NaCl). When salt is mixed with water, the polar water molecules are attracted to sodium and chlorine ions in a solid salt crystal. When they hit, they transfer some energy which causes the ions in the salt crystal to separate.

The negative side of the water molecules are attracted to the Na+ ions and the positive ends are attracted to the Cl- ions. Once a Na^+ or Cl^- ion has broken off the crystal, it instantly attracts the oppositely charged sides of nearby water molecules. This process is called **hydration**. The hydrated sodium and chlorine ions stay in solution because of their accompanying polar water molecules.

Water dissolves many molecular compounds

Glucose is a covalent molecule, but it dissolves easily in water. The glucose molecule has areas of positive and negative charge (red and blue shading). When glucose is mixed with water, the charged areas attract many water molecules. Molecules of glucose are not dissociated into ions, but remain together (and neutral) in solution.

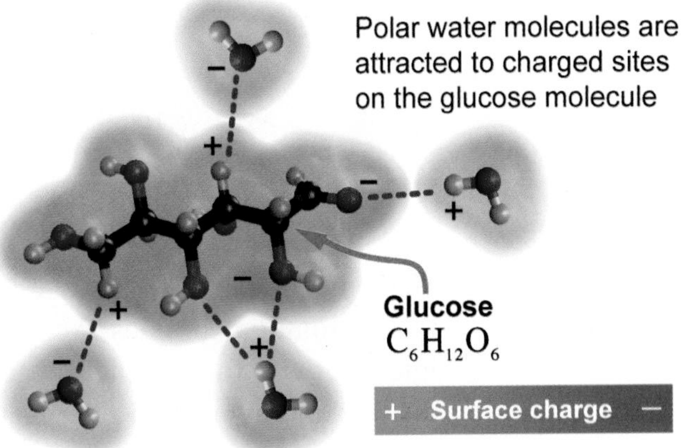
Polar water molecules are attracted to charged sites on the glucose molecule
Glucose
$C_6H_{12}O_6$
+ Surface charge −

Chemistry terms

Hydration - the process of molecules, with any charge separation, to collect water molecules around them. While not "chemically bonded", a hydrated molecule does hold fairly tightly to its "private collection" of water molecules.

Distilled water vs. tap water

Distilled water is used to make solutions

When making solutions with water in the chemistry lab we often use **distilled water**. Distilled water is water that has been purified so that it contains no dissolved substances. The distillation process heats water to it's boiling point of 100°C. The steam that forms is collected in a separate clean container. Any ions and solid contaminants are left behind as "residue" in the original container. Distilled water is sometimes incorrectly referred to as deionized water however, this is incorrect. **Deionized water** is water that has had its ions removed using a specific filtration process. Deionized water is sometimes used instead of distilled water, because it is cheaper to produce and, for most purposes, yields sufficiently pure water.

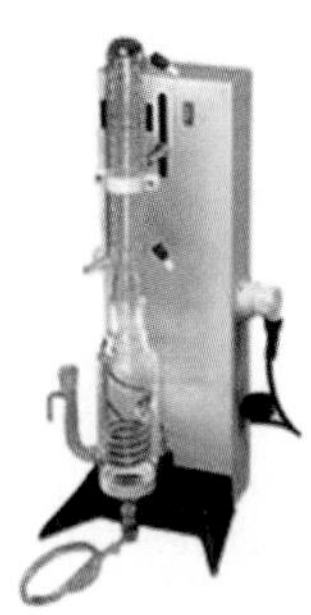

In contrast, **tap water** often contains dissolved salts and minerals from the ground water supply and even from the pipes that carry it from the reservoir. Many minerals, such as Fe^{2+}, Ca^{2+} and Mg^{2+}, that may be present in water are important for our health. Other sources of ions in tap water come from water supplies that are treated with chloride ions (Cl^-) for purification purposes. Chloride ions kill bacteria that may be present in the water supply. Fluoride (F^- ion) is also added at 4 mg/L to help with dental health.

Evidence?

How do we know that tap water contains ions and distilled water does not?

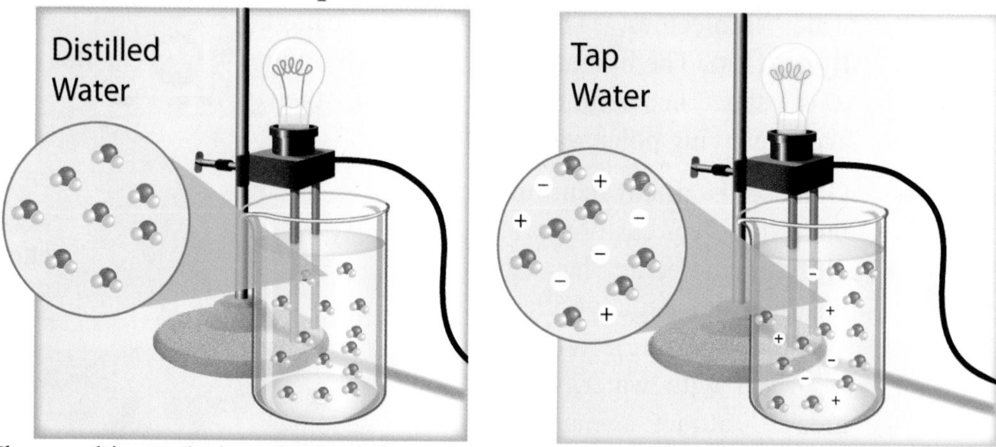

Why do we use distilled water?

When making solutions in the laboratory we need to know there are no other chemicals in the solution, besides the ones that we add. Contaminating ions and minerals can interfere with the chemical reaction we are trying to perform. For this reason we use distilled or deionized water when we make our solutions in the chemistry laboratory.

Distilled water - water purified by heating water to steam and condensing the steam into a clean or sterilized container.

Deionized water - water without ions, purified using filtration methods.

Tap water - drinking water from the faucet or "tap."

Reactions in aqueous solutions

Aqueous means dissolved in water

A solution with water as the solvent is called an **aqueous** solution. Aqueous solutions are so important that chemists consider being dissolved in water to be almost a fourth state of matter! In writing reactions, we use the symbols *(s)*, *(l)*, *(g)* and *(aq)* to show what state of matter the reactants and products are in.

(s)	indicates a solid
(l)	indicates a liquid
(g)	indicates a gas
(aq)	indicates a substance dissolved in water

A good example is the familiar baking soda and vinegar reaction. All four states of matter appear in this reaction.

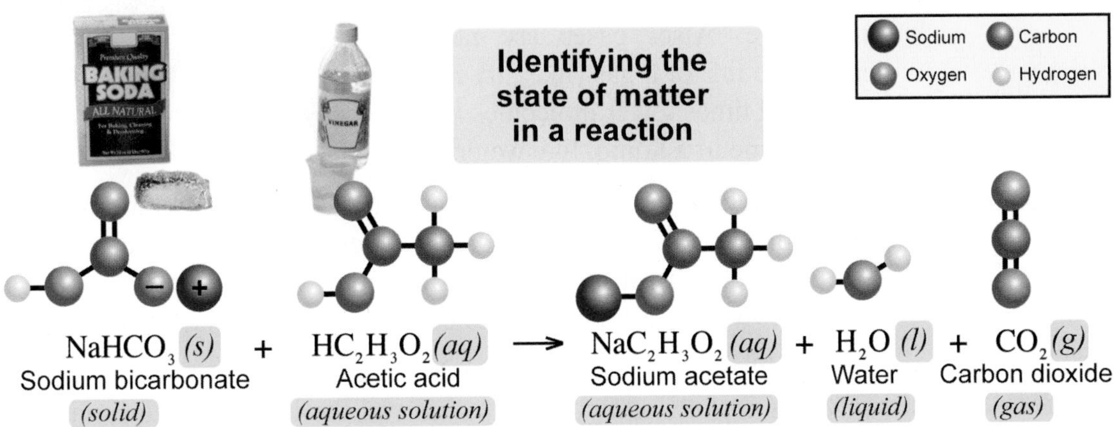

Reading the symbols (s), (g), (l) and (aq)

The *(s)* tells us the baking soda is a solid in the reactants and the *(aq)* tells us the acetic acid (vinegar) is an aqueous solution. On the product side of the reaction, the first *(aq)* tells us the sodium acetate is dissolved in the water that was part of the vinegar. The water *(l)* is a liquid. The carbon dioxide *(g)* is a gas. The gas is what makes the bubbles. If you are not careful, any product that is a gas quickly escapes, taking its mass with it! If the CO_2 is allowed to escape, we would expect to measure less mass in the products of this reaction compared to the reactants. You can learn a lot about a reaction by knowing how to read these special symbols!

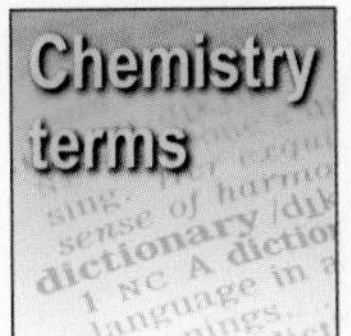

Aqueous solution - An aqueous solution is any solution where the solvent is water. Molecules, atoms, or ions in aqueous solutions are identified by the symbol (aq) after a chemical formula. For example CO_3^{2-}(aq) is how you would write a carbonate ion (CO_3^{2-}) that is dissolved in water.

The importance of aqueous solutions

Reactions in solids

In order for a chemical reaction to occur, molecules must *touch* each other. This seems obvious, but think about what it means. In a solid, molecules cannot easily move around. Chemical reactions DO occur in solids, but they are slow, like changing limestone into marble under heat and pressure in the earth. This takes thousands of years.

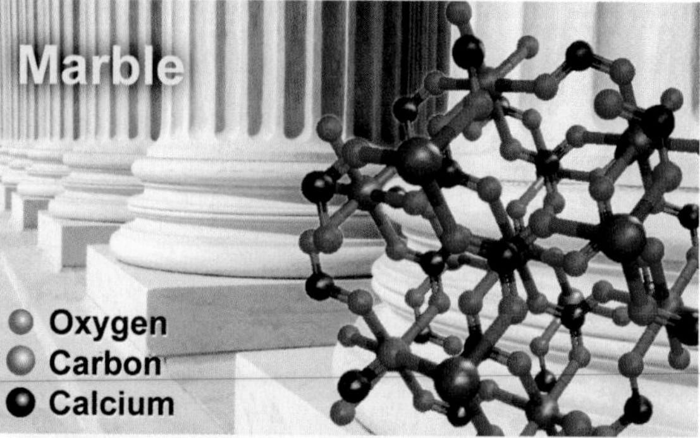

Reactions in gases

In a gas, molecules move around quickly. Reactions (like combustion) can happen quickly in a gas, but there are very few chemicals that are gases at room temperature. Also, gases are not very dense. One mL of a gas (like air) contains a thousand times fewer molecules than the same volume of a liquid, like water.

Fire is a chemical reaction that depends on oxygen gas

Reactions in liquids

Liquids combine the advantages of both high density and mobility. In a liquid there are lots of molecules, close together, and they can move around and reach each other. Relatively few chemicals are liquid at room temperature either. Sugar is solid, salt is solid, and virtually all proteins and carbohydrates are also solids at room temperature.

Why life requires water

Life involves reactions of thousands of complex chemicals. *This can only happen in aqueous solutions!* Only in a solution can the molecules in living things circulate and reach each other to have reactions. For example, a sequence of reactions called the Krebs Cycle occurs in virtually all living things on Earth. In one step of this reaction, malic acid reacts with a chemical called NAD+ to produce oxaloacetic acid and NADH. This is part of how energy is extracted from glucose in every cell of your body.

NAD+ *(aq)* + Malic acid *(aq)* → Oxaloacetic acid *(aq)* + NADH *(aq)*

Life involves many complex chemical reactions in aqueous solutions

Other solvents

Non-polar solvents

Water may be the most important solvent to living things, but lots of other solvents exist, and you will probably use them from time to time in your life. For example, artists thin oil-based paints with odorless mineral spirits. Mineral spirits is the common name for hexane (C_6H_{14}), a linear molecule derived from oil. Hexane is non-polar, and good at dissolving other non-polar solutes.

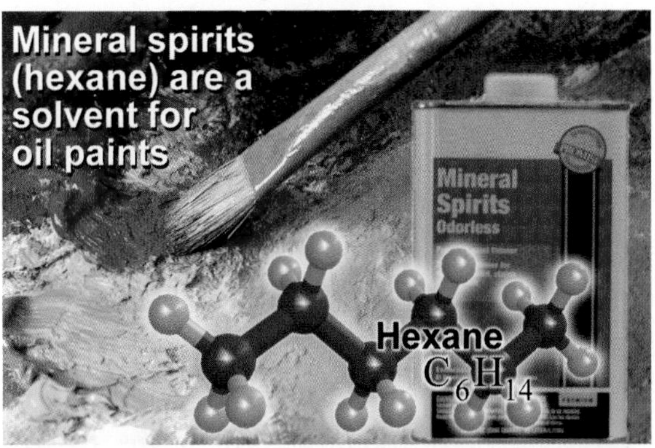

Why oil does not dissolve in water

Oil does not dissolve in water because water is a polar molecule and oil molecules are nonpolar. However, oil dissolves easily in mineral spirits. In general, polar solvents dissolve polar solutes, like water dissolving salt. Non-polar solvents dissolve non-polar solutes, such as mineral spirits dissolving oil.

In general, like dissolves like
Polar solvents dissolve polar solutes
Non-polar solvents dissolve non-polar solutes

"Solvent" means different things in everyday use

In every-day conversation the word "solvent" usually means organic liquids such as mineral spirits, acetone (nail polish remover), turpentine, alcohol, and similar chemicals. Many of these chemicals are flammable, and noxious to breathe. In chemistry, the term "solvent" means something that dissolves something else. The common meaning and the chemistry meaning are similar, *so don't get confused*! Water is a solvent in chemistry, the most important one to living things.

TABLE 9.1. Some common solvents other than water

Name	Formula	Typical uses	Hazards
Mineral spirits	C_6H_{14}	paint thinner	flammable, toxic
Methanol	CH_3OH	solvent, fuel	flammable, toxic
Acetone	CH_3COCH_3	nail polish remover	flammable, toxic, carcinogen
Tetrachloroethylene	C_2Cl_4	dry cleaning chemical	flammable, toxic, carcinogen

9.2 Concentration and Solubility

Concentration and dilute solutions

The **concentration** of a solution describes how much of each solute there is compared to the total solution. A solution is **dilute** when there is very little solute. A solution is **concentrated** when there is a lot of dissolved solute compared to solvent. The distinction is hard to determine, since even a concentrated solution of acid contains less solute than solvent.

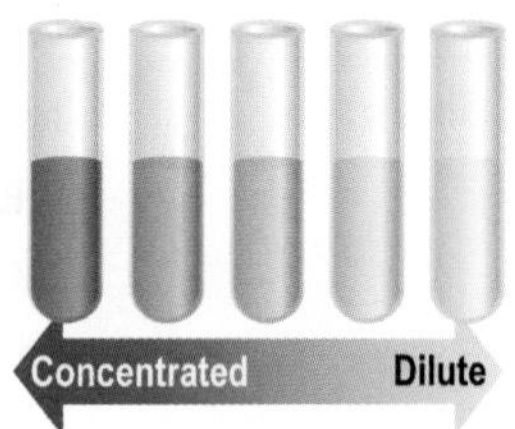

What is solubility?

Solubility means the amount of solute (if any) that can be dissolved in a quantity of solvent. The most accurate solubility data is given as a mass percent, which is the mass of solute per 100 grams of solution. In the laboratory it is sometimes easier to work with solubility in terms of grams per 100 mL. Solubility is always given at a specific temperature since temperature strongly affects solubility.

TABLE 9.2. Solubility of common solids in water at 20°C

Name	Solubility		Name	Solubility	
	% mass	g/100 mL		% mass	g/100 mL
Salt (NaCl)	26.4	32	Marble ($CaCO_3$)	0	0
Sugar (suc. $C_{12}H_{22}O_{11}$)	67.1	88	Copper sulfate ($CuSO_4$)	16.7	20
Baking soda ($NaHCO_3$)	8.73	9.2	Epsom salts ($MgSO_4$)	25.1	32
Zinc sulfate ($ZnSO_4$)	35	45	Glucose ($C_6H_{12}O_6$)	45.0	81

Comparing mass % to g/mL

Notice that the values in g/mL are always larger than the mass percents. This is because the density of a solution varies with concentration. Adding more solute raises the density. This happens for two reasons.

1. Solute molecules fit in between solvent molecules.
2. Solutes may themselves be denser than water.

Adding one gram of solute always increases the mass of the solution by one gram. However, adding one cm^3 of solute does NOT increase the volume by one cm^3 because solute molecules squeeze between water molecules. The volume of a solution does increase, but not as fast as the mass does, and that is why the density changes.

concentration - describes the amount of each solute compared to the total solution.

dilute - a solution with relatively little solute compared to solvent.

concentrated - a solution with a lot of solute compared to solvent.

solubility - the amount of a solute that will dissolve in a particular solvent at a particular temperature and pressure.

Concentration

Potassium concentration in your body

Potassium is critical to many of your body's functions, such as transmitting nerve signals. In a healthy person, potassium is dissolved in blood at a concentration of 140 to 200 milligrams (mg) per liter. A blood potassium concentration of less than 130 mg/L causes muscle weakness and heart rhythm instability (*hypokalemia*). A concentration of more than 215 mg/L may also result in heart instability (*hyperkalemia*). Since potassium is lost through excretion and sweat, you need to eat foods containing 4 - 5 grams of potassium (4,000 - 5,000 mg) each day. Fruits and vegetables are the best dietary sources.

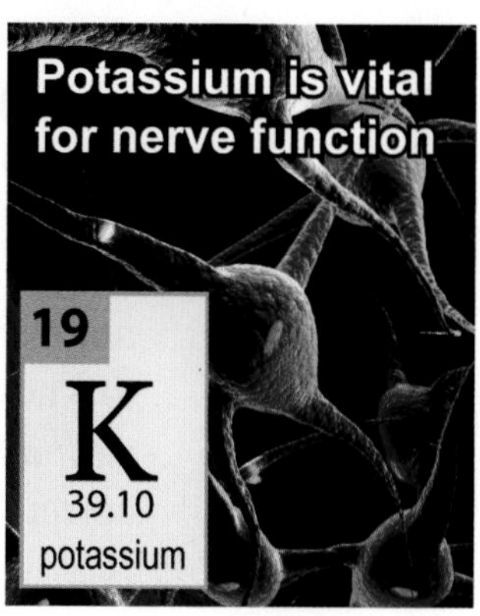

Different ways to measure concentration

Three common ways to express concentration are:

- As a percent, using mass of solute per liter of solution (g/L or mg/L)
- As a percent using mass of solute ÷ by total mass of solution
- In molarity, which is moles of solute per liter of solution (M)

Three ways to calculate concentration

grams / liter

$$\text{concentration}_{g/L} = \frac{\text{mass of solute}}{\text{liters of solution}}$$

% mass

$$\text{concentration}_{\%} = \frac{\text{mass of solute}}{\text{mass of solution}} \times 100\%$$

Molarity

$$\text{concentration}_{\text{Molarity}} = \frac{\text{moles of solute}}{\text{liters of solution}}$$

Very low concentrations

Environmental scientists often use a variation of percent concentration. Parts per million (ppm), parts per billion (ppb), and parts per trillion (ppt) are commonly used to describe very small concentrations of solutes in the environment. These terms are measures of the ratio (by mass) of one material in a much larger amount of another material. For example, a pinch (gram) of salt in 10 tons of corn chips is about 1 g salt per billion grams chips, or a concentration of 1 ppb.

Suppose you dissolve 10.0 grams of sugar in 90.0 grams of water. What is the mass percent concentration of sugar in the solution?

Asked: *What is the mass percent concentration?*

Given: *10 grams of solute (sugar) and 90 grams solvent (water)*

Relationships:

$$\text{concentration} = \frac{\text{mass of solute}}{\text{total mass of solution}} \times 100\%$$

Solve:

$$\text{concentration} = \frac{10 \text{ grams sugar}}{(10 + 90) \text{ grams of solution}} \times 100\% = 10\% \text{ sugar}$$

Molarity: a review

Molarity is the most common unit of concentration

Since we will using aqueous solutions so often in chemistry. It is important to review what molarity means as well as how to calculate it. Solution concentrations inside living organisms must be carefully maintained. Dehydration is a common side effect of serious illness. Dehydration, or loss of water, can be very dangerous because it affects the concentration of ions inside our cells and tissues. Recall that **molarity** of a solution is the number of moles of solute per liter of solution. Molarity is useful because chemists need to know the number of reacting molecules. For this reason, *chemists make solutions with a known molarity*. This allows them to make the appropriate concentration, with the correct number of particles for the reaction.

How to calculate molarity:

Calculate molarity

To find molarity, you will need to know how many moles are dissolved per liter of solution. Lets find the molarity of a salt solution containing 6.0 g of NaCl in 100 mL of water. The problem is solved in three steps:

1. Calculate the formula mass of NaCl so you know the mass of one mole;
2. Use the formula mass to figure out how many moles there are in 6.0 grams;
3. Calculate molarity by dividing the number of moles by the volume of solution in litters (L).

Calculate the molarity of a salt solution made by adding 6.0 g of NaCl to 100mL of distilled water.

Asked: *Molarity of solution*

Given: *Volume of solution= 100.0 mL , Mass of solute (NaCl) = 6.0 g*

Relationships: $M = \frac{\text{moles}}{\text{L}}$ *Formula mass (NaCl)= 22.99+ 35.45 = 58.44 g/mole*

1000 mL = 1.0 L therefore 100mL = 0.10 L

Solve:

$$\text{moles NaCl} = 6.0\cancel{\text{g NaCl}} \times \frac{1\ \text{mole NaCl}}{58.44\ \cancel{\text{g NaCl}}} = 0.103\ \text{mole NaCl}$$

$$M = \frac{0.103\ \text{moles}}{0.100\ \text{L}} = 1.03\ \text{M}$$

Answer: *1.03 M solution of NaCl*

molarity - the number of moles of solute per liter of solution.

Saturation and equilibrium

Saturation

Suppose you add 250 grams of sugar to 100 mL of water at 30°C. What happens? According to the table, 219 grams will dissolve in the water. *The rest will remain solid.* That means you will be left with 31 grams of solid sugar at the bottom of your solution. A solution is **saturated** if it contains as much solute as the solvent can dissolve. The table tells us that 219 grams per 100 mL is the saturation concentration of sugar at 30°C. Any solute added in excess of the solubility *does not dissolve.*

Solubility of Sugar in Water

Temp (°C)	Solubility g/100mL	Temp (°C)	Solubility g/100mL
0	177	50	259
10	189	60	284
20	204	70	318
30	219	80	360
40	238	90	410

Equilibrium

A solution is in **aqueous equilibrium** when the concentration does not change over time. If you cover a saturated sugar solution and let it sit, the concentration of dissolved sugar will stay the same. However, the individual molecules that are dissolved do NOT just stay dissolved! Dissolving is really a two-way process.

Dissolving and "undissolving"

Consider a molecule of dissolved sugar that happens to touch some undissolved sugar. That molecule might bounce off and stay dissolved. It might also stick and come out of the solution to become "undissolved" (solid) again. When a solid solute is placed in a solvent, both dissolving and "undissolving" are going on at the same time! If the solution is NOT saturated, there is more dissolving than "undissolving". The amount of solid solute decreases and the amount of dissolved solute increases.

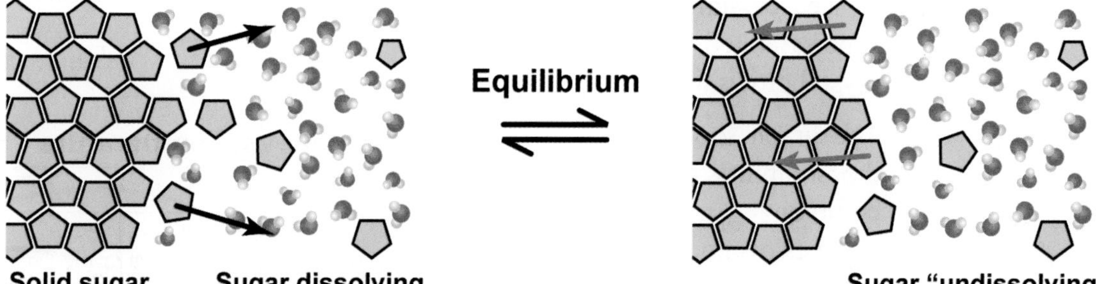

Why saturation happens

Of course, the more solute there is in solution, the higher the chance that a solute molecule will come out of solution and become solid again. Saturation happens when the rate of molecules becoming "undissolved" exactly balances the rate of dissolving.

saturated - when a solution has dissolved all the solute it can possibly hold. Saturation occurs when the amount of dissolved solute gets high enough that the rate of "undissolving" matches the rate of dissolving.

aqueous equilibrium - when the amount of dissolved solutes remains constant over time.

Solubility and temperature

Solubility of solids increases with temperature

Higher temperature normally increases the solubility of solids in liquids. This is mostly because warmer solvent molecules have more energy with which to hit a solid and knock off molecules of solute. A warm solid is also easier to dissolve because the molecules in the solid have more energy, and are easier to separate.

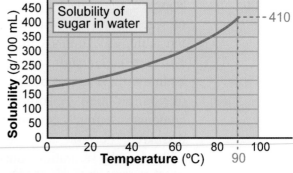

Solubility of sugar in water

Solubility (g/100 mL)

Temperature (°C)

410

90

Making a supersaturated solution

Suppose you make a saturated solution of sugar in water at 90°C. According to the graph, you can dissolve 410 grams of sugar in 100 mL of water! That's more than four times as much sugar as water (by mass).

Now let the solution cool down to 20°C. What happens? At 20°C only 204g of sugar can stay in solution. The rest must come out of solution. This is how rock candy is created! The solution is supersaturated as it cools. A seed crystal of sugar is put in the supersaturated solution and the seed crystal grows as sugar solidifies.

Dissolve to make a saturated solution

Cool down

Excess sugar crystallizes into rock candy!

Solubility curves

The graph shows how the solubility of some common substances changes over a range of temperature. Note that the solubility changes a lot for some solutes, including copper sulfate ($CuSO_4$), but very little for others, such as salt (NaCl).

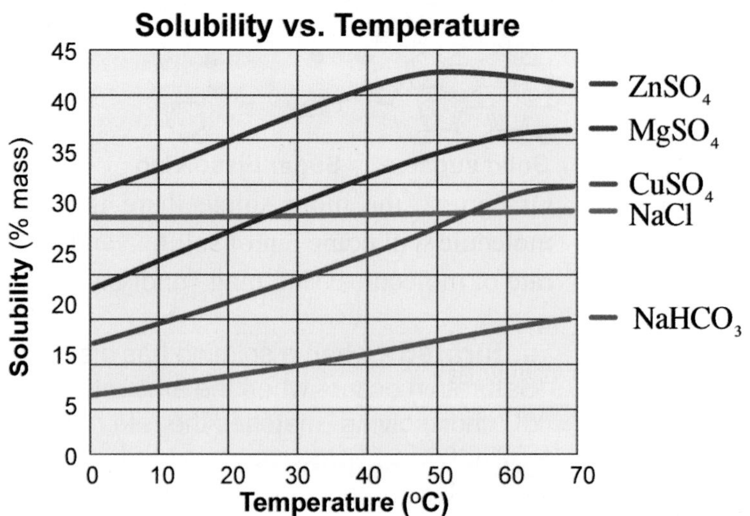

Dissolving rate

Dissolving happens faster when solutes are ground into powder

Dissolving can only happen at the surface between solvent and solute. Most things that are meant to be dissolved, like salt and sugar, are ground up to a powder to increase the surface area. Increased surface area speeds dissolving because more solute is exposed to the solvent.

Substances are often ground up into powder to make them dissolve faster

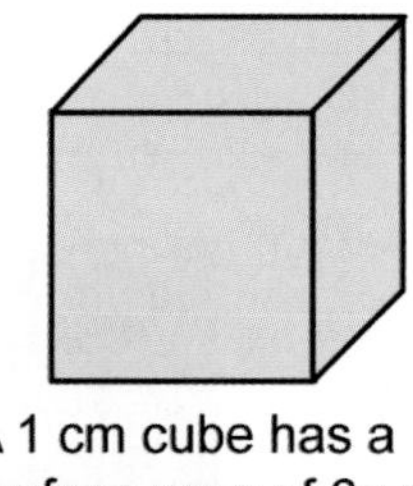

A 1 cm cube has a surface area of 6 cm^2

The same volume has a surface area of 9 cm^2 when divided up into smaller cubes.

Powders dissolve faster but not more

Making sugar or salt into a powder does not change the solubility of either compound! It just makes all the dissolving that is going to happen, happen *faster.* If you add more powdered sugar to water than will dissolve, the excess will stay powder (solid) in a sludge at the bottom of the beaker.

Temperature increases dissolving rate

Increasing the temperature of a solution will not change the solubility of sodium chloride much. Its solubility at 20°C is 26.4% by mass and at 70°C it is only a little over 27%. However, salt dissolves much faster in hot water than cold water. The same amount of salt will go into solution, but increasing the temperature makes it dissolve faster.

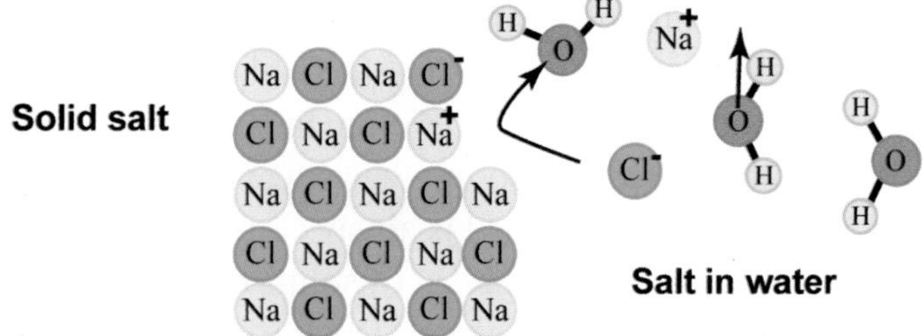

Dissolving is a collision process and slow (cold) molecules are not as effective as fast (hot) molecules

The molecular process of dissolving

Dissolving rate increases with temperature because dissolving is a physical process at the molecular level. Molecules of water literally slam into crystallized (solid) sodium and chloride ions, knocking them out of their solid form and into solution. Higher temperature means each water molecule has more energy to transfer when it strikes the salt crystal. More energy means more effective dislodging of ions from the crystal.

The solubility of gases in liquids

Gas dissolves in water

Gases can also dissolve in liquids. When you drink carbonated soda, the fizz comes from dissolved carbon dioxide gas (CO_2). The table below lists the solubility of CO_2 as 1.74 grams per kilogram of water at room temperature and atmospheric pressure (1 atm).

TABLE 9.3. Solubility of common gases in water at 25°C

Name	Solubility (g/L)	Name	Solubility (g/L)
Oxygen (O_2)	0.04	Methane (CH_4)	0
Nitrogen (N_2)	0.02	Hydrogen (H_2)	0
Carbon dioxide (CO_2)	1.74	Helium (He)	

Solubility of gases in liquids decreases with temperature

The solubility of gases in liquids *decreases* with temperature. The table shows that 0.04 grams of oxygen dissolves in a kilogram of water. Dissolved oxygen keeps fish and other underwater animals alive. When water gets warmer, the dissolved oxygen content goes down. Many fish seek deeper, cooler water during the summer. The fish swim deep because there is more dissolved oxygen in the cooler water.

Deeper water is cooler and contains more dissolved oxygen

Solubility of gases in liquids increases with pressure

The solubility of gases in liquids increases with pressure. Soda is fizzy because the carbon dioxide was dissolved in the liquid at high pressure. When you pop the tab on a can of soda, you release the pressure. The solution immediately becomes **supersaturated**, causing the CO_2 to bubble out of the water and fizz. Supersaturation means the solution has more dissolved solute than it can hold. A solution that is supersaturated is always unstable. The excess solute usually drops out of solution, like the fizz bubbling out of soda water. Rock candy is made by super saturating a sugar solution causing the sugar to “undissolve” back into sugar crystals.

Seltzer water is a supersaturated solution of CO2 in water

Chemistry terms

supersaturation - when a solution contains more dissolved solute than it can hold. Supersaturated solutions are always unstable and the excess solute becomes “undissolved”, often rapidly.

Preparing a solution

The importance of preparing solutions

Many laboratory experiments require a specific molarity (M) solution. Preparing solutions for an experiment is an important, frequently used laboratory skill. A solution of known molarity allows us to determine how many particles (ions, or molecules) are present quite accurately just using a graduated cylinder.

How do you begin?

Basic Steps for Preparing a Solution of Known Molarity (M).

1. Determine the formula mass of the solute being used.
2. Use the formula mass of the solute to determine the grams of solute needed. To determine the grams required, you multiply the moles of the solid by the formula mass.
3. Weigh the grams of solute on the balance.
4. Add the solute to a volumetric flask or graduated cylinder. For many purposes a graduated cylinder is sufficient, but chemists use a volumetric flask to ensure the best accuracy
5. Fill the flask about two-thirds of the way up with distilled water.

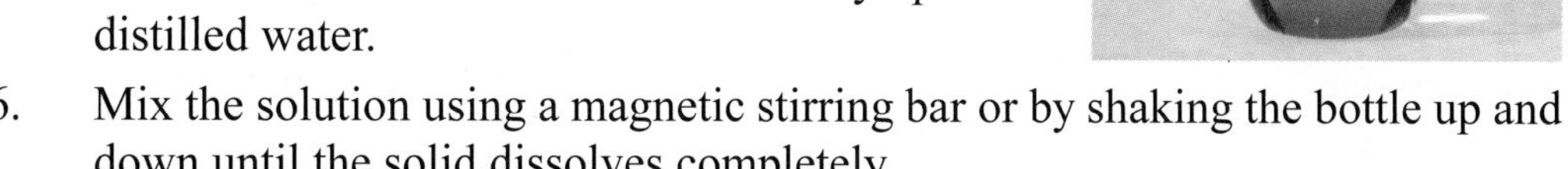

6. Mix the solution using a magnetic stirring bar or by shaking the bottle up and down until the solid dissolves completely.
7. Fill the volumetric flask or the graduated cylinder up to the correct volume mark. Sometimes it is necessary to use an eye dropper to get the meniscus right on the line. Remix the solution.

Let's make 500.0 mL of a 1.0 M $CaCl_2$ solution.

Asked: *Make a 1.0M solution of* $CaCl_2$

Given: *500 mL* $CaCl_2$ *or half a liter*

Relationships: $M = \frac{\text{moles}}{\text{L}}$ *Molar mass of* $CaCl_2 = 40.078 + (2 \times 35.43)$

$= 110.98$ g/mole

Solve: moles $CaCl_2$ = 0.5 mole $CaCl_2 \times 110.98$ g/mole = 55.49 g $CaCl_2$

Weigh out 55.49 g of $CaCl_2$ *and place it in the 500 mL graduated cylinder or volumetric flask. Add distilled water about two thirds of the way, and mix. Once the solid is dissolved, fill the water up to the 500 mL mark and remix.*

Answer: *Add 55.49 g* $CaCl_2$ *to a volumetric flask and fill to the 500 mL mark.*

9.3 Properties of Solutions

Reaction rate and concentration

Caves are formed when the carbonic acid in rainwater dissolves limestone in rock over long periods of time. Suppose you place a sample of solid limestone in a 1M solution of carbonic acid, and another piece in a 2M solution of carbonic acid. Will the reactions be the same? How might they be different?

Which acid will dissolve the limestone fastest?

Chemically, the same reaction occurs in both solutions, BUT, the 2M solution will make the reaction happen *twice as fast* as the 1M solution. That is because there are twice as many carbonic acid molecules per mL in the 2M solution. Twice as many molecules means twice the chance of molecules meeting to have a reaction.

Higher concentration generally means a faster reaction rate

Concentration and medicine

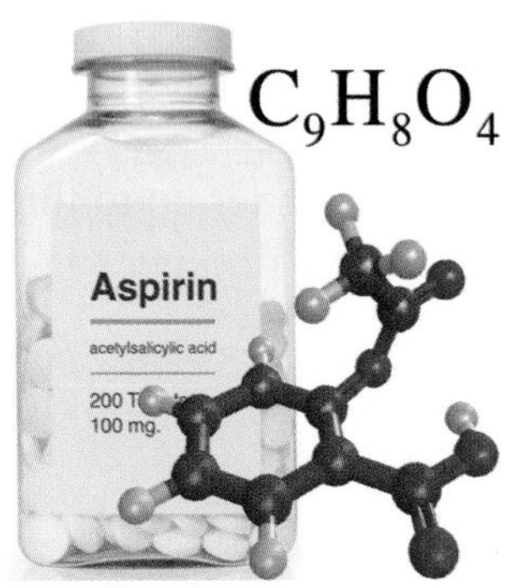

The effect of concentration on reaction rate is extremely important in medicine. When you take a medicine, such as aspirin, the molecules in the medicine (acetylsalicylic acid) become dissolved in your blood. Aspirin blocks pain, reduces swelling, and reduces blood clotting; however, if the concentration is too high, aspirin can cause internal bleeding and other problems. Medically, this is called an *overdose* and it can even be fatal. An overdose is a concentration which produces a reaction rate that is dangerous to the body.

Higher temperature generally means a faster reaction rate

Reaction rate is affected by temperature

Reaction rates usually go up with temperature because reaction molecules have more energy. Therefore, less energy has to be added to break the bonds and start the reaction. This effect is stronger in solutions because higher temperature increases the motion of solute molecules. Faster molecules mean more collisions between molecules, and therefore a greater chance that molecules meet in a reaction.

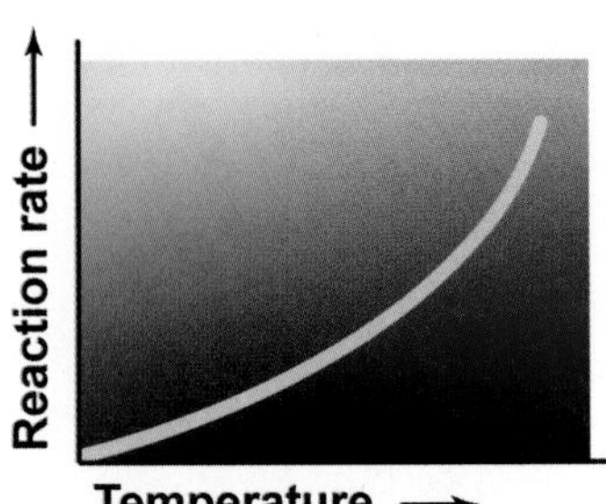

Energy and solutions

Energy and the heat of solution

When a solute dissolves, interparticulate bonds are broken. Depending upon the type of solute being dissolved ions, are hydrated by water molecules and hydrogen bonds may form between solute and solvent molecules. Since energy is involved whenever bonds form or break, *the overall process of dissolving may use or release energy.* The energy that is used or released in dissolving is called the **heat of solution**.

Negative heat of solution

A solute that *loses* energy when it dissolves has a *negative* heat of solution. The negative sign tells you energy is released into the solution. The released energy would make a thermometer placed in it rise in temperature. A good example is calcium chloride ($CaCl_2$). When calcium chloride dissolves in water, each mole releases 82.8 kJ of energy!

$$CaCl_2(s) + H_2O(l) \rightarrow Ca^{2+}(aq) + O^{-}(aq) + 82.8 \text{ kJ}$$

When dissolving is exothermic

A negative heat of solution means dissolving $CaCl_2$ is an *exothermic* process. Exothermic means it "gives off energy".

Heat packs use solutes with negative heat of solution

Dissolving calcium chloride in water makes the solution HOT! That's because the heat lost by the calcium chloride as it dissolves is *gained* by the solution. A *heat pack* contains water and calcium chloride in a thin plastic vial. Breaking the plastic vial allows the calcium chloride to dissolve and the solution heats up! Skiers often use heat packs to warm cold fingers and toes.

The heat comes from calcium chloride dissolving

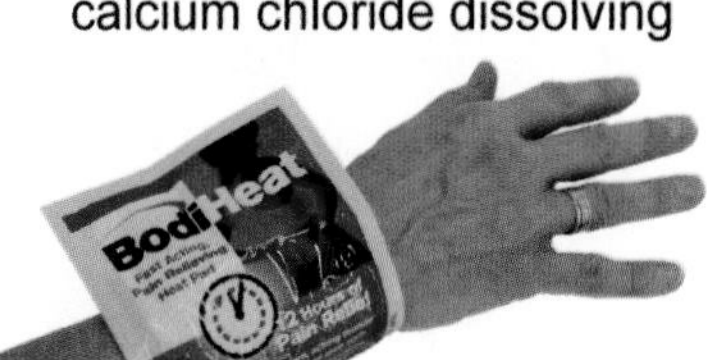

Positive heat of solution

Ammonium nitrate (NH_4NO_3) has a positive heat of solution of +25.7 kJ/mole. That means one mole of ammonium nitrate absorbs 25.7 kJ of energy when it dissolves. A thermometer placed in this solution would decrease in temperature. The solution becomes *colder.*

$$NH_4NO_3(s) + 25.7kJ \xrightarrow{H_2O} NH_4^{-}(aq) + NO_3^{-}(aq)$$

Cold packs use solutes with positive heat of solution

A cold pack contains a vial of ammonium nitrate in a bag of water. Breaking the vial allows the ammonium nitrate to dissolve. The solution gets so cold that frost often forms on the surface of the cold pack! Athletes use cold packs to quickly cool injuries.

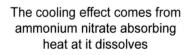

The cooling effect comes from ammonium nitrate absorbing heat at it dissolves

heat of solution - the energy absorbed or released when a solute dissolves in a particular solvent.

Measuring heat of solution: Coffee cup calorimetry

Experiments with heat and solutions

The amount of heat released or absorbed by a solution can be measured using a simple *coffee cup calorimeter.* A coffee cup calorimeter is a double styrofoam cup that prevents the reaction in solution from losing heat to the environment. You may use this technique in the lab to investigate three types of chemical changes:

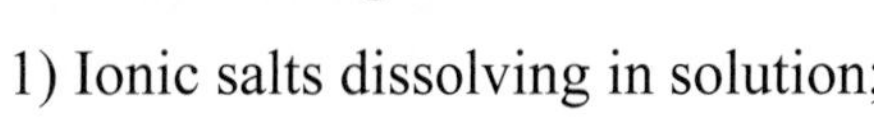

1) Ionic salts dissolving in solution;

2) Neutralization reactions between an acid and a base;

3) Oxidation-reduction reactions between a metal and an acid.

Enthalpy

The heat energy involved in a chemical reaction is given the special name **enthalpy**. Enthalpy is measured in joules per mole (J/mole) or kilojoules per mole (kJ/mole). The symbol for enthalpy is ΔH, and this is written next to the chemical equation. The enthalpy *change* tells us how much heat is released or absorbed during the chemical change.

ΔH for endothermic reactions

For instance, if the enthalpy is positive, we write the ΔH with a positive sign indicating the reactants *absorb* heat when forming products.

$$NH_4NO_3(s) + H_2O(l) \rightarrow NH_4^+(aq) + NO_3^-(aq) \quad \Delta H = +25.7 \text{ kJ}$$

In a solution this absorbed heat results in a decrease in temperature. The temperature decreases because the reaction absorbs thermal energy from the solution. You measure the temperature change of the solution with a temperature probe immediately before and after just after mixing the reactants.

ΔH for endothermic reactions

When an acid and a base are mixed, heat is released by the chemical change (exothermic). The solution becomes warmer. The change in enthalpy (ΔH) is *negative* because the reaction gives off energy to the surroundings.

$$HCl(aq) + NaOH(aq) \rightarrow NaCl(aq) + H_2O(l) \quad \Delta H = -56 \text{ kJ/mole}$$

This reaction takes place under the relatively constant pressure of the atmosphere, because the coffee cup calorimeter is not sealed. The reaction takes place "in" the solution inside the cups, so if the solution gains heat and the temperature rises then we can assume that the reaction lost heat, based on the First Law. For the short time that we are measuring the heat change in a reaction such as this one, we usually make the assumption that no heat is absorbed by the styrofoam cups.

Chemistry terms

enthalpy - ΔH, the heat energy of a chemical change measured in joules per mole (J/mole) or kilojoules per mole (kJ/mole).

Solution calorimetry

Assumptions

In the laboratory we know that over a long period of time heat would eventually be lost to the cups and surroundings because styrofoam is not a "perfect" insulator. We know this from our real life experiences with hot beverages. However, for many experiments the temperature changes over the course of a few minutes, so the actual heat loss can be neglected. For dilute solutions, we can also assume the specific heat of water, because the solute concentration is so low it does not affect the specific heat significantly.

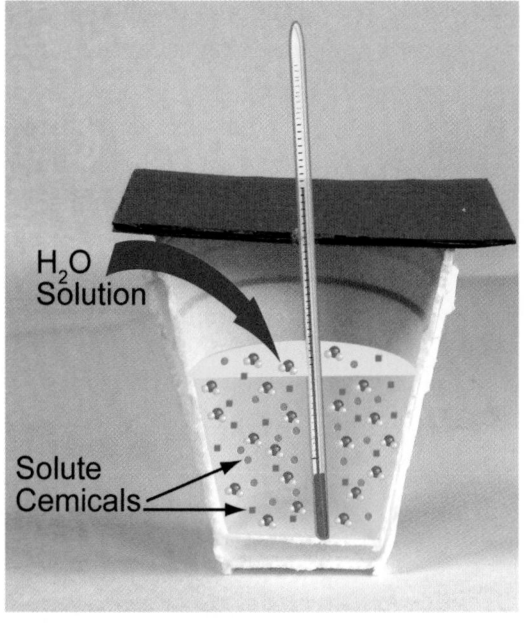

Determining heat lost or gained by the solution

To find out how much heat is lost or gained by the solution, we can use the equation for calculating energy used earlier in chapter three. The reacting chemicals are in the solution however, the solvent (water) is NOT part of the chemical change. Using the law of energy conservation we know heat must be conserved. Here, if the reaction is exothermic, the solution temperature will rise because the chemicals released heat.

Heat lost must equal heat gained: Net change is zero

$$\Delta H_{solution} = (\text{grams of solution}) \times (\text{specific heat solution}) \times \Delta T$$

"water"

$$\text{Where } \Delta H_{solution} = -\Delta H_{rxn}$$

When using this equation it is important to remember that you measure the heat change of the solution by massing your reacting mixture and by measuring the temperature change. You assume the conditions of temperature and pressure stay the same during the time it takes you to perform these measurements. You are then able to use your data to calculate the heat of the reaction, which takes place between the chemicals in the solution. The following pages present a few examples using some realistic data similar to what you will collect in the laboratory.

Coffee cup calorimetry calculations

The first example uses an acid-base reaction. The acid is HCl and the base is NaOH. The reaction is exothermic and heat is released.

When a student mixes 40.0 mL of 1.0 M NaOH and 40.0 mL of 1.0M HCl in a coffee cup calorimeter, the final temperature of the mixture rises from 22.0°C to 27.0°C. Calculate the enthalpy change for the reaction of hydrochloric acid (HCl) with sodium hydroxide (NaOH). Assume that the coffee cup calorimeter loses negligible heat, that the density of the solution is that of pure water 1.0 g/mL, and that the specific heat of the solution is the same as that of pure water.

$$NaOH(aq) + HCl(aq) \rightarrow NaCl(aq) + H_2O(l) \quad \Delta H = ?$$

Asked: *Amount of heat change (ΔH) for NaOH and HCl*

Given: *Volume of solutions = 40.0 mL + 40.0 mL, molarity of solutions 1.0M, and $\Delta T = 5°C$*

Relationships: $q_{solution} = m(solution) \times Cp(solution) \times \Delta T$

Solve: *First note the temperature increased so the reaction released energy to the solution, this means the reaction is exothermic and will have a negative ΔH.*
Total volume of solution is 40.0 mL+ 40.0 mL = 80.0 mL
Total mass of the solution is 80.0 g using the density of water=1.0 g/mL

$q_{solution} = 80.0\ g \times 4.18\ J/g\ °C \times (27.0°C - 22.0°C)$

$q_{solution} = +1672\ J$ *the positive sign indicates heat is absorbed*

$q_{solution} = +1.67\ kJ$, *Therefore* $q_{rxn} = -1.67\ kJ$ *we reverse the sign as heat gained by the solution is lost by the reaction.*
To find heat on a per mole basis we use the molarity times the volume in liters to calculate moles ; 1.0M × 0.040 L = 0.040 moles of both reactants (NaOH and HCl equimolar amounts).

Lastly $\dfrac{-1.67\ kJ}{0.040\ mole} = -41.8\ kJ/mole = \Delta H$

Since the temperature increased, heat was released from the reaction, making the ΔH negative.

Answer: $\Delta H = -41.8\ kJ/mole$

Any time an acid and a base are mixed, heat will be evolved. The amount of heat depends upon the amount of moles of acid and base that are present in the solution. You will experience firsthand how the solution warms up when doing some neutralization experiments in the laboratory.

Coffee cup calorimetry calculations

This example involves dissolving a salt or an ionic compound in water. In this case, heat is absorbed when the ions separate in water.

In one experiment a student mixed 100.0mL of water with 1.50 g of ammonium nitrate ($NH_4NO_3(s)$) in a "coffee cup" calorimeter. The initial temperature of the water was 23°C. After the ammonium nitrate was added, the solution was stirred and the final temperature of the mixed solution was 20.6°C. Calculate the heat change (enthalpy) for the dissolving of $NH_4NO_3(s)$, in kilojoules per mole of solid dissolved. Assume the specific heat of the solution is the same as that of pure water.

$$NH_4NO_3(s) + H_2O(l) \rightarrow NH_4^+(aq) + NO_3^-(aq) \quad \Delta H = ?$$

Asked: *Amount of heat change (ΔH) when dissolving ammonium nitrate. This is equal to the* $-q_{solution}$

Given: *Mass of water = 100.0 mL, mass of solid,* $NH_4NO_3(s)$ = 1.50g

Relationships: *Molar mass of* NH_4NO_3 = (14.007x2) +(15.999x3)+(1.0079 x 4) =80.04 g/mole. $q_{solution} = m(solution) \times Cp(solution) \times \Delta T$

Solve: *First note the temperature decreased so the dissolving process absorbed energy from the solution, this means the change is endothermic and will have a positive ΔH.*

density of water = 1.0 g/mL for 100.0 mL = 100.0 g

total mass of the solution is 100.0 + 1.50 g = 101.5 g

$q_{solution}$ = 101.5 g x 4.18 J/g.°C x (23°C - 20.6°C)

$q_{solution}$ = - 1018.2 J *the negative sign indicates heat is lost.*

$q_{solution}$= - 1.02 kJ,

Therefore ΔH = 1.02 kJ; we reverse the sign as heat lost by the solution is gained by the reaction.

Now we calcualte the number of moles in 1.5 g of NH_4NO_3

$$1.50\text{ g } NH_4NO_3 \times \frac{1\text{ mole } NH_4NO_3}{80.04\text{ g } NH_4NO_3} = 0.0187\text{ mole } NH_4NO_3(s)$$

Lastly $\frac{1.02\ kJ}{0.0187\ mole} = +54.5\ kJ/mole = \Delta H$

Since the temperature decreased, heat from the surroundings was absorbed by the solution making the q rxn positive.

Answer: *ΔH = + 54.5 kJ/mole*

If you recall, ammonium nitrate is the chemical used in cold packs! No wonder it feels cool on our skin, it is absorbing +54 kJ/mole of heat from the warmth of our skin.

Density, freezing and boiling

Mass and volume of solutions

When you add 25 grams of salt to 100 mL of water you DON'T get 125 mL of solution. That's because the sodium ions (Na^+) and chlorine ions (Cl^-) fit partially in between water molecules. The volume of solution is a little greater than 100 mL, but not as much as the added volume of solid salt. The mass of solution, however, is 125 grams. The matter has to go somewhere and 25g of salt + 100g of water makes 125 g of solution.

20 g salt + 80 mL water = 87 mL solution!

Volumes of solute and solvent do not add up to the volume of solution

Density of solutions

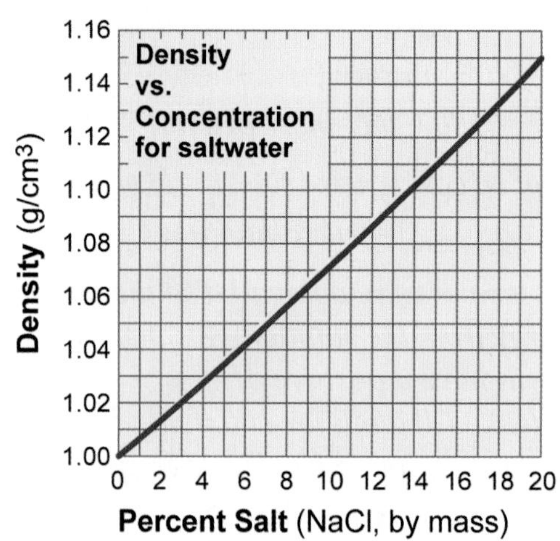

If the volume stays about the same and the mass increases, the *density* of a solution increases. Salt water is more dense than fresh water. Almost any solid solute increases the density of the solution above the density of the pure solvent. This explains why a can of diet soda floats and a can of sugared soda sinks. Diet soda has only a dilute concentration of sweetener. Sugared soda may have 40 g of sugar in a single can!

Density of gas/liquid and liquid/liquid solutions

If gases or liquids are dissolved in a liquid solvent, the density can either increase or decrease depending on the solute. Alcohol is less dense than water, so a solution of alcohol and water is less dense than pure water. Carbon dioxide gas is also less dense than water, so soda water is less dense than pure water. However, the amount of CO_2 dissolved in soda water is not very large, so the density change is very small.

Solutes usually lower the freezing point of a solution

The presence of solute molecules usually lowers the freezing point of a solution compared to the pure solvent. The freezing point goes down by about 1.86°C for every mole of solute particles. For example, sea water is about 3.5% salt. Since salt dissociates into Na^+ and Cl^- ions, each mole of dissolved salt adds 2 moles of particles. Pure water freezes at 0°C. The dissolved salt causes seawater to freeze at -2.2°C.

Solutes usually raise the boiling point of a solution

Adding solutes increases the boiling point of a solution (compared to the pure solvent). For water, each mole of solute particles increases the boiling point by about 0.5°C. That means seawater boils at 100.6°C instead of 100°C.

Colligative properties

Colligative properties

Why does ice melt when we sprinkle salt on it? Understanding how salt interacts with the ice at the molecular level will give us insight into this phenomenon. We know ice freezes at 0°C (or 32°F) but why does it become a liquid when we add salt to it at the same outdoor temperature? To understand this, we need to study **colligative properties**. Colligative properties are the properties of a solution that depend upon the number of dissolved particles collectively, but not on the nature or type of particle. One common colligative property is freezing point depression.

Lowering the freezing point

For dilute solutions there is a direct relationship between the increase in the number of solute particles and a decrease in the freezing point of a solution. Why does this occur? First let's consider a pure solvent such as water. Pure water freezes at 0°C. When water freezes and goes from a liquid to a solid, the molecules are held more tightly. Remember, the particles can vibrate but not slide past one another in the solid phase. The process of solid formation causes a decrease in randomness in the molecules, because they are no longer "free" to move and slide past one another. Ice is a more ordered structure than water. Solid formation from a liquid is always associated with a decrease in randomness, or **entropy, S**. When solute particles are introduced into a liquid their mixing introduces a greater randomness to the solution. You could think of it as more variation of particle type, which increases the disorder of the system. These dissolved particles get "in the way" of the molecules trying to solidify and this makes it harder for an ordered lattice to form this slows down the freezing process.

Entropy

Molality

To calculate the change in the freezing point due to a particular solute that is added, it is necessary to use a new unit of concentration called **molality, m**. Molality is the number of moles of solute per kilogram of solvent. This is *different from molarity, M* because the mass of solvent does not vary with temperature. When solutions are undergoing temperature changes, the volume of the solvent can change, using molality corrects for this.

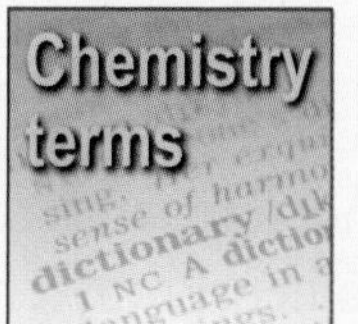

colligative property - physical property of a solution that depends only on the number of dissolved solute particles not on the type (or nature) of the particle itself. Freezing point depression is an example of a colligative property.

entropy, S - is a measure of the disorder or randomness of a system.

molality, m - unit of concentration used when temperature varies, moles of solute per kilogram of solvent.

Colligative properties

Calculation of molality

Below is the equation that allows us to calculate the molality, m of a solution. Remember that the density of water is 1.0 g/mL, so 1.0 gram = 1.0 mL for this solvent.

$$m = \frac{\text{moles solute}}{\text{kg of solvent}}$$

molality

One practical application of colligative properties is the use of antifreeze in our car's radiator. The radiator fluid must be able to circulate, as a liquid, around the engine in order to maintain the proper engine temperature. By adding ethylene glycol ($C_2H_6O_2$) a nonvolatile molecular compound to water and mixing, a solution is created which resists freezing at cold temperatures inside our car radiator. Why does this happen? It happens because the solution now, has an increased number of ethylene glycol solute particles, which interfere with waters ability to form an ordered solid structure.

Calculating freezing point depression

To calculate the change in freezing point we need to use the equation:

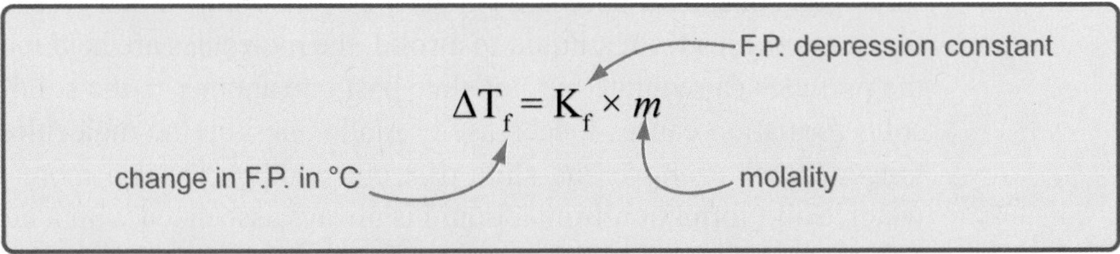

Here is an example, using antifreeze, that explains how you can apply the formula and calculate the freezing point depression:

Calculate the freezing point of a 1.8 m aqueous solution of antifreeze which contains ethylene glycol ($C_2H_6O_2$) as the solute.

Asked: *Calculate the freezing point of a 1.8 m solution of ethylene glycol*

Given: *Molality is 1.8 m ; $K_f = 1.86^{o}C/m$ (for water the solvent)*

Relationships: $\Delta T_f = K_f \times m$

Solve: $\Delta T_f = K_f \times m = 1.86^{o}\ \frac{C}{m} \times 1.8\ m = 3.35\ ^{o}C$

Freezing point of antifreeze solution = $0^{o}C - 3.35^{o}C = -3.35^{o}C$

Answer: *The freezing point is lowered by $3.35^{o}C$.*

By adding sufficient amounts of antifreeze, the solution can withstand temperatures below -25°C, and this allows our vehicles to start in the cold winter weather. Propylene glycol ($C_3H_8O_2$) is a typical de-icer used on aircraft, and it works in a similar way. This substance is sprayed on the airplane to decrease the ice-crystal formation.

Electrolyte solutions

Electrolytes conduct electricity

When salts dissolve in water they dissociate (or split apart) into their counter part ions. Aqueous solutions containing dissolved ions are able to conduct electricity. The dissolved ions are capable of transferring an electric charge through a solution. The term **electrolyte** refers to a solute that is able to conduct electricity in an aqueous solution. For example, you know that you are not supposed to blow dry your hair when you are in the tub! This is because your skin contains salt that dissolves in the bath water, the water itself contains some mineral ions, and our blood contains ions.

Single ion dissolving

Some salts dissolve in a one-to-one ratio producing the same concentration of individual ions as the original solution.

$$NaCl(aq) \rightarrow Na^+(aq) + Cl^-(aq)$$

$$1.0\ M \qquad 1.0\ M \qquad 1.0\ M$$

Multiple ion dissolving

Other salts dissolve and yield a larger number of ions and therefore a higher concentration.

$$MgCl_2(aq) \rightarrow Mg^{2+}(aq) + 2Cl^-(aq)$$

$$1.0\ M \qquad 1.0\ M \qquad 2.0M$$

On the label of any solution that says 1.0 M NaCl or 1.0 M $MgCl_2$, it is more correct to think of the ions as split apart. Imagining the ions split apart is representative of how they actually are in the aqueous solution.

Greater number of particles causes greater effect

With colligative properties the greater the number of particles in solution the greater the effect. In this case a salt with three component ions vs. a salt with two would have a greater freezing point depression.

TABLE 9.4. Solute Concentration of Dilute Solutions

Solute	# Particles	0.10 m	0.01 m	0.0010 m
NaCl (Na^+, Cl^-)	2 ions	1.87	1.94	1.97
$CaCl_2$ (Ca^{2+}, $2Cl^-$)	3 ions	2.13	2.63	2.89

Chemistry terms

electrolyte - solute capable of conducting electricity when dissolved in an aqueous solution.

Evolving from the Ocean

For years scientists have speculated that all life forms originated in the ocean waters. Since life first began, water's role as a solvent was important in bringing chemicals together. When we search for life on other planets scientists search for evidence of water, because we know it must be present for life to exist. Most of our earth is covered by vast oceans. It is in these oceans that scientists find clues to early life forms and information about how living things evolved and adapted to their environments. The chemical composition of the ocean plays a major role in what elements were available for living things to use. At the dawn of time, during the Precambrian era, evidence suggests that life existed only in the ocean.

How could living organisms form from ocean water?

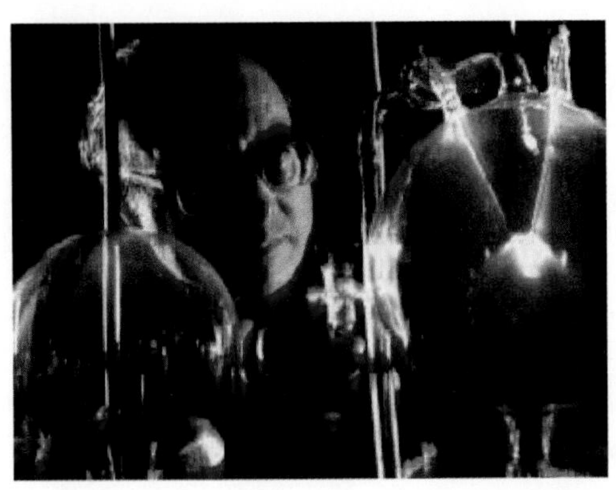

Water's ability to hydrate charged ions, and to dissolve minerals and gases, made it a mixing pot for elements of life. Ocean water contained nutrients and organic compounds that were formed during energetic reactions. The famous Miller and Urey experiment showed that organic molecules such as amino acids, sugars, and lipids could be formed from simple molecules like water, with methane, ammonia and hydrogen gases. Miller and Urey used electrodes to simulate the energy thought to be provided by lightning. Lightning was an available source of energy sufficient to cause molecules to break apart and reform. There is recent evidence that also suggests meteorite impact could have supplied the energy for chemical reactions in the oceans. Experiments suggest that meteorites could have supplied both the energy and the carbon to synthesize key molecules used as building blocks for life. When energy is sent through a solution containing molecules chemical bonds are broken and reformed. It is hypothesized that some of these new molecules came together and formed unicellular organisms. Water's polarity helped of molecules get together and form cellular-like structures. Blue-green algae or cyanobacteria were among the first forms of life.

Blood and Ocean Water.

Strong evidence for life evolving from the ocean comes from the observation that blood and ocean water contain similar chemical substances! Blood can be separated into two portions: a liquid layer called plasma and a solid layer containing cells. Blood plasma is about 92% water and has a concentration of salt and other ions that is amazingly similar to ocean water. Could it be that as life evolved it developed cells that were adapted to the chemical environment of sea water? Blood bathes our tissues inside our bodies. It is responsible for maintaining a very delicate chemical balance inside and outside of our tissues.

Today the ocean's total ionic concentration is several times that of blood, but in the ancient oceans it was thought to be more dilute. Like our blood, the concentration of ions in the ocean is remarkably constant. The ocean is always undergoing localized changes in concentration, just like our blood, yet the concentration of both fluids remains constant overall. Scientists feel that the only way to account for this is through thorough mixing of the ocean water.

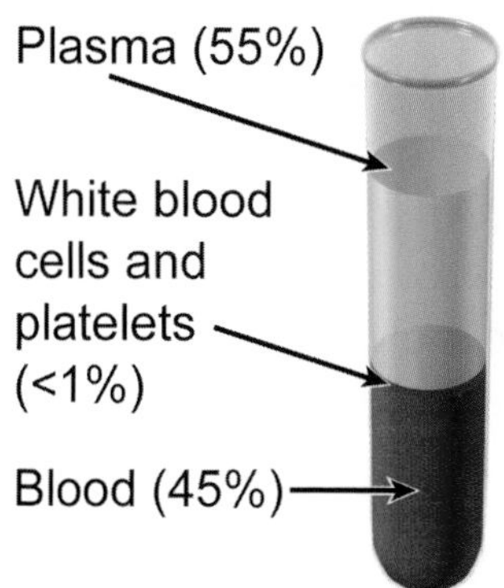

Chemistry of Ancient Oceans.

To understand the chemistry of early oceans, scientists have analyzed samples of sedimentary rock dating back to the first half of the earth's history. They found that early oceans had large amounts of dissolved iron, in the form of Fe(II) or Fe^{2+} ion. Virtually every living organism we know of requires iron for its life functions. Scientists wonder if this is because organisms evolved in an iron-rich environment? Today's oceans contain very little iron and organisms have had to develop clever ways to capture it. Once photosynthesis began to make oxygen present in the atmosphere and in the water, it combined with the dissolved iron in the oceans to form layers of rust. The rust then settled to the bottom of the ocean. The reaction of $2Fe^{2+}(aq) + O_2(g) \rightarrow Fe_2O_3(s)$ caused the iron to form solid deposits seen in the geological record dating back 1.8 million years ago.

Sulfur became abundant in the oceans as sulfate (SO_4^{2-}) and hydrogen sulfide (H_2S). Sulfur is another very important element found in living organisms. Disulfide bridges or covalent sulfur bonds help to stabilize many molecules such as DNA double helices. Deep on the ocean floor, where there is no sunlight available for photosynthesis, bacteria get energy by oxidizing hydrogen sulfide (H_2S). These bacteria form the base of the ecosystem food chain on the ocean floor. The H_2S is made available from volcanic gases that erupt from geothermal vents in the ocean floor. These vents release mineral rich "smoke." Overall, the elemental composition of the ocean has changed dramatically over time.

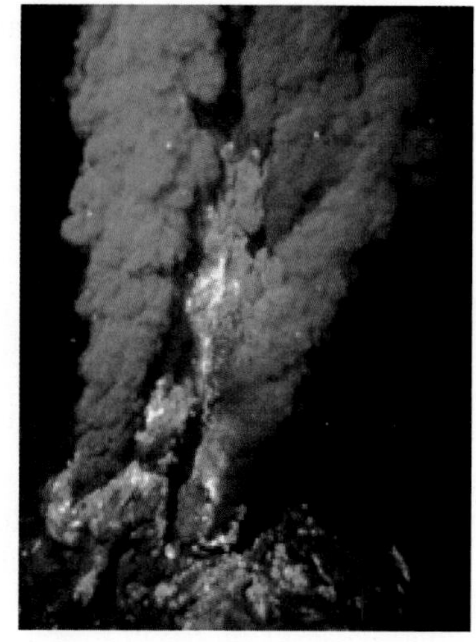

By studying which elements were present in the early oceans and comparing that to which elements are present today, scientists are learning about the key elements that may have influenced selection pressures of early life forms.

Chapter 9 Review

Vocabulary

Match each word to the sentence where it best fits.

Section 9.1

dissolved	solute
solvent	polar
hydrogen bond	surface tension
hydration	distilled water
tap water	deionized water
aqueous	

1. The biggest percentage of a solution is made up of the _______________. Often this is a liquid.
2. A molecule with a charge separation is called _____________.
3. A substance is said to be ______________ when it completely separates and disperses into the solution.
4. The _________ is the substance dissolved in the solvent.
5. A property named _____________ describes when intermolecular forces pull a liquid surface into the smallest possible area.
6. A _____________ is a type of intermolecular attraction formed when a hydrogen atom bonded to an N, O or F atom is attracted to a negative charge on another molecule or even the same molecule.
7. The process known as ____________ describes the surrounding of water molecules around a charged ion or polar molecule.
8. Water that comes from the sink is called ______ water, and conducts electricity.
9. An ______________ solution is one that has water as its solvent.
10. Water that is condensed after boiling to separate solute particles from the water is called ________________.
11. Water that does not contain ions is called _________________.

Section 9.2

concentration	dilute
solubility	concentrated
insoluble	Molarity
aqueous equilibrium	saturated
supersaturation	

12. A solution that contains more dissolved solute that it can "hold" is called __________.
13. An ________________ has been achieved when the amount of dissolved solute becomes constant over time.
14. A solution is ___________ when it is holding the maximum amount of solute that it can hold.
15. A solution that has very little solute dissolved in it is said to be ______________.
16. If a solution is called ____________ then it has a large amount of dissolved solute.
17. When the concentration of a solution is measured as moles solute divided by liters of solution the concentration unit is called ___________.
18. When a substance will not dissolve it is said to be ______________.
19. The amount of solute that will dissolve in a particular solvent is called its __________. (at a given temperature and pressure)

Section 9.3

suspension	heat of solution
enthalpy	colloid
entropy	colligative property
electrolyte	molality

20. An ___________ is a solute capable of conducting electricity when it is dissolved.
21. When a liquid mixture can be separated by filtration it is called a ___________.
22. A uniform mixture that contains particles that are too small to filter and do not settle is called a ______________.

23. The ______________________ is the energy released or absorbed when a solute dissolves in a solution.

24. ____________ is the concentration unit that does not vary with temperature. It is measured as moles solute per kilogram of solvent.

25. The heat energy of a chemical change is called ________________, or ΔH.

26. The ___________ of a system is a measure of the disorder or randomness.

27. A ___________________ is a property that depends on the number of solute particles in solution and not the type of solute particle.

Conceptual Questions

Section 9.1

28. Suppose you make a solution of ice tea using a mix, what are the solute and solvent. Explain.

29. Describe what happens to the ice tea mix when you mix it.

30. Give an example of a common solid that dissolves in water. Can you explain how it does this?

31. Draw a picture of a water molecule. Label the positive ends and the negative ends of the molecule.

32. Now draw 3 water molecules and show how they form hydrogen bonds to each other. Label the hydrogen bonds with a different color pen.

33. Describe why hydrogen bonding is important to life on earth.

34. Look up hydrogen bonding on the internet. Write down the definition.

35. Give an example of surface tension that you have witnessed in your everyday life. If you have not seen this in your everyday life ask some friends.

36. When we are told to "stay hydrated" during exercise, what does that actually mean? Explain.

37. Draw a picture of a potassium ion, K^+ surrounded by water molecules. Show how the water molecules orient themselves around potassium.

38. Sometimes we use distilled water to iron or to soak our contact lenses in. Why don't we use tap water?

39. Describe the difference between distilled water and deionized water.

40. Give 3 reasons why aqueous solutions are important.

41. Write out a chemical equation that represents salt, NaCl(s) dissolving in water. On the products side of your arrow show the appropriate state of matter for the ions.

42. Write the chemical formula for mineral spirits. Would this substance be soluble in water?

Section 9.2

43. If you were to mix water and chalk, would the chalk dissolve? Explain. Give an example of something that would dissolve (for comparison).

44. Describe how you would make a concentrated solution of lemonade.

45. There are several ways to calculate concentration. List 3 of them.

46. Pretend you are sitting at the kitchen table with a glass of water and a sugar bowl:
    a. You take a teaspoon of sugar and add it to the water. What happens?
    b. You add two more teaspoons of sugar to the water and stir. What happens now?
    c. Slowly you keep adding small amounts of sugar, what will eventually happen?

47. Refer to the previous question - What would happen if your glass of water was hot? Would this affect what you observe? Explain.

48. Can gases "dissolve" in water? If so, give an example.

49. Explain why the fish swim deeper during the summer months?

50. Describe how you would prepare a 1.0 M solution of copper(II) chloride, $CuCl_2$. List each step and explain as necessary.

Chapter 9 Review.

51. What would be the difference between a 1.0 M solution of potassium chloride, KCl and a 6.0M solution of potassium chloride, KCl?

52. Explain why it is important to be very accurate when preparing a solution of known concentration?

53. Do you think it would be easy to measure a drinking water supply that contained 1 ppb of lead? Is it important to be able to measure such small amounts accurately?

54. Explain what would happen if you dropped a sugar cube in a cup of water and did not stir. How would this compare to the same amount of granular sugar added to water?

55. What is the density of water? How does this compare with the density of ice? Briefly explain.

Section 9.3

56. If you added a piece of chalk, $CaCO_3(s)$ to a solution of 1.0 M carbonic acid, $H_2CO_3(aq)$ and another piece of chalk to a 3.0 M solution of carbonic acid, would you expect there to be a difference in what you see? Explain. Describe what you think would be different at the microscopic or molecular level.

57. Describe why sugar dissolves better in hot tea than in cold ice tea. Be sure to discuss the water molecules involved.

58. The ocean does not "freeze-over" as easily as a lake or pond. What does an ocean have in it that could describe this behavior? Explain.

59. When we add y 2.0 grams of a solute to a solution, the mass of the solution always increases by 2.0 grams as well, but if we add 2.0 mL or 2.0 cm^3 the volume does not increase by the same amount. Why not? Briefly explain.

60. How does the density of a solution change with the addition of a solute? Explain.

61. How is the Tyndall effect used to identify a colloid from a suspension? Explain.

62. Sometimes there is a change in temperature in a solution when as solute is added to it. For example the temperature can either increase or decrease when a salt is dissolved in water. Give two practical applications that take advantage of this phenomena.

63. In a case where a salt dissolves and the solution absorbs heat, is the heat written on the products side of the arrow or the reactants side? Explain.

64. Why are styrofoam coffee cups good to use when we are investigating chemical changes in solution?

65. When a chemical change takes place in a solution the water molecules either absorb heat or give heat to the chemical solute. What are these two processes called?

66. Based on the law of conservation of energy, we can say that $\Delta Hsoltn = - \Delta Hrxn$. Explain why.

67. In your own words describe what entropy means.

68. Why does spreading salt on icy roads make the salt melt, when the temperature is still at the freezing point for ice? Explain to the best of your ability.

69. Compare molality(m) to molarity(M).

70. What is the formula for calculating molality?

71. Why is it important to use molality when applying the principles of colligative properties?

72. Why can electrolyte solutions conduct electricity?

73. Which solution would be a better conductor, a 1.0 M solution of NaCl or a 1.0 M solution of $CaCl_2$. Explain your answer.

74. Which substance listed below would you expect to give the lowest freezing point when added to water? Assume that the same mass of each solute is added to 1.0 L of water. Explain your choice.
a) NaCl b) Sugar, $C_{12}H_{22}O_{11}$ c) $MgCl_2$ d) KI

75. Look at table 9.4 and compare the 0.1 m solution of $CaCl_2$ to the 0.001 m solution of $CaCl_2$. Why do you think the molarity goes up in a more dilute solution?

Quantitative Problems

Section 9.2

76. If you dissolve 12.0 g of sugar in 120.0 g of water.

a) Calculate the concentration, in percent by mass, of sugar in the solution.
b) If the density of water is 1.0 g/mL, calculate the concentration in grams /liter.
c) Calculate the molarity of this solution.

77. You need to prepare a solution with 20.0 g of $MgCl_2$ in 500 mL of water.
    a. Calculate the percent by mass of $MgCl_2$ in the solution.
    b. Calculate the concentration in grams/L, assume density of water is 1.0g/m.
    c. Calculate the molarity of this solution.

78. What is the molarity of a solution that contains 8.5 g of NaCl in 250mL of water?

79. If there is 30 g of $CaCO_3$ dissolved in a 2.0 L of solution what is the percent by mass of $CaCO_3$ in the solution (assume density of water is 1.0 g/ mL)?

80. Calculate the number of moles of potassium iodide (KI) needed to make one liter of a 0.75 M solution.

81. When a student mixes 50.0 mL of 1.0 M NaOH and 50.0 mL of 1.0M HCl in a coffee cup calorimeter, the final temperature of the mixture rises from 21.0°C to 25.0°C. Calculate the enthalpy change for the reaction of hydrochloric acid (HCl) with sodium hydroxide (NaOH). Assume that the coffee cup calorimeter loses negligible heat, that the density of the solution is that of pure water 1.0 g/mL , and assume the specific heat of the solution is the same as that of pure water. $NaOH(aq) + HCl(aq) \rightarrow NaCl(aq) + H_2O(l)$ $\Delta H = ?$

82. In one experiment a student mixed 100.0mL of water with 2.60 g of ammonium nitrate, $NH_4NO_3(s)$, in a coffee cup calorimeter. The initial temperature of the water was 24.5°C. After the ammonium nitrate was added the solution was stirred and the final temperature of the mixed solution was 20.2°C. Calculate the heat change (enthalpy) for the dissolving of $NH_4NO_3(s)$, in kilojoules per mole of solid dissolved. Assume the specific heat of the solution is the same as that of pure water.
$NH_4NO_3(s) + H_2O(l) \rightarrow NH_4^+(aq) + NO_3^-(aq)$
$\Delta H = ?$

83. In a coffee cup calorimetry experiment 1.75 g of Zn(s) is added to 100.0 mL of 1.0M HCl initially at 21.0°C. The temperature rises to 30.1°C. What is the heat of the reaction per mole of Zn? Assume that the specific heat of the acid is the same as that of pure water.
$Zn(s) + 2HCl(aq) \rightarrow ZnCl_2(aq) + H_2(g)$ $\Delta H = ?$

84. Calculate the molality of the following solutions:
    a. A solution containing 2.0 g of NaCl in 500.0g of water.
    b. A solution containing 6.5 g of $CaBr_2$ in 1200 g of water.
    c. A solution prepared by adding 100 g of sugar ($C_{12}H_{22}O_{11}$) to 2.0 L of water.

85. Calculate the freezing point of a 2.20 m aqueous solution of antifreeze which contains ethylene glycol ($C_2H_6O_2$) as the solute. K_f= 1.86°C/m for water.

86. Adding 1.50 g of benzene, C_6H_6, to 80.70 g of cyclohexane (C_6H_{12}) causes a decrease in the freezing point of the cylcohexane from 7.0 to 3.8°C. What is the Kf (freezing point depression constant) for cyclohexane?

87. Determine the freezing point of a solution containing 0.750 kg of chloroform, $CHCl_3$ and 48.0 g of eucalyptol ($C_{10}H_{18}O$) a compound found in eucalyptus trees. The K_f for $CHCl_3$ is 4.68°C/m and the normal freezing point of $CHCl_3$ is -63.5 °C.

CHAPTER 10 Chemical Reactions

How do we represent a chemical reaction?

How do we classify chemical reactions?

How do we evaluate the effectiveness and the economics of chemical reactions?

Chemical reactions occur all around us. Chemical reactions also occur inside our bodies. In fact the human body is the ultimate chemical factory. Digestion of food is a chemical reactions. The nutrients from the food we eat are then transported to muscle and organ cells. where other chemical reactions extract energy and make life possible.

The atmosphere of our planet is 78% nitrogen (N_2) and 21% oxygen (O_2). However, it was not always this way. 4 billion years ago we believe Earth's atmosphere contained a lot more CO_2 and almost no oxygen. All of the oxygen we breath in the atmosphere today was created by plants! Over millions of years the photosynthesis reaction gradually changed the entire atmosphere of the entire planet by turning carbon dioxide into oxygen and organic matter. In the same way that nature uses the energy from the sun to capture carbon, human technology also uses chemical reactions to make new compounds from existing fundamental elements. From your kitchen to the pharmaceutical laboratories where medicines are produced, to the petroleum refining factories where gasoline and plastics are produced, chemical reactions are taking place that make something tasty, healthy or useful.

Rusting generates heat

Rusting is a process that we are very familiar with. From cars to bridges, we have all seen a shiny piece of iron become covered with the brown flaky substance we know as *rust*. The process that turns iron metal to reddish rust is the result of a very common chemical reaction. It is the reaction of iron with oxygen. This counter-top experiment is a quick way to make rust, and also to investigate the chemical reaction that occurs. All chemical reactions involve energy, and rusting is no exception. By making the reaction go quickly, you can see the energy change with a thermometer. To do the experiment you need:

1. Two thermometers.
2. A piece of #000 steel wool.
3. A small rubber band for wrapping around the thermometer as indicated in the picture.

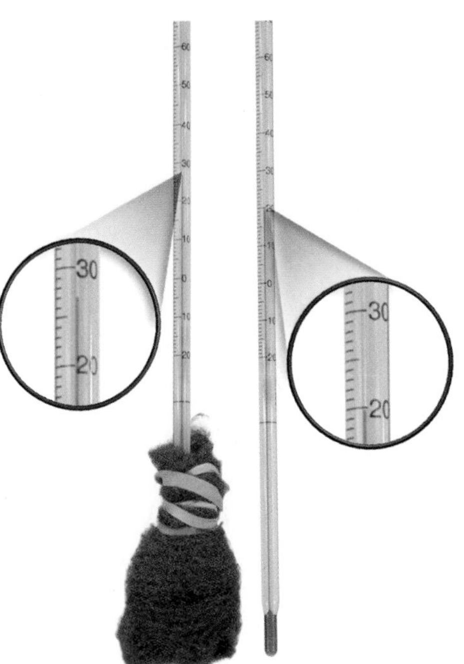

Steel wool always has some oil on it from the manufacturing process. This oil must be removed before the iron in the steel can be exposed directly to moist air. Washing the steel wool with dish washing soap usually does the trick.

Step 1: Wrap one of the thermometers with moist steel wool

Step 2: Place the two thermometers next to each other as shown on the picture so that you can compare their readings easily.

After some time you should notice that the temperature of the thermometer wrapped with the steel wool starts to increase while the other thermometer stays constant.

Why does the temperature increase?

Hypothesis #1: There is something going on in the steel wool that releases energy.

Hypothesis #2: There must be another source of heat for the temperature to increase.

The "bare" thermometer is not in contact with any heat source that does not also affect the thermometer with the steel wool equally. That rules out hypothesis #2. So, where does the energy come from? The best explanation is that energy must come from whatever is going on inside the steel wool. In fact it comes from the chemical reaction between iron (Fe) and oxygen (O_2) that results in the rusting of Fe. The rust is called iron oxide and has the chemical formula Fe_2O_3. We can write a simple expression that summarize our observations as follows:

Iron reacts with oxygen to produce rust and energy.

This is a very good start for a deeper investigation into the subject of chemical reactions because: 1) it is a simple reaction that occurs all around us and 2) it shows directly how energy is involved and energy is involved in all chemical reactions.

10.1 Chemical Equations

Developing the language of chemistry

Think of the element symbols as the *alphabet* of chemistry and the chemical formulas as the *words* of chemistry. Chemical equations are the *sentences* of chemistry and they tell the story of the material world. All the complex changes we see in the material world are the result of chemical reactions. The "big" story of the world is the inter-relationships between all the countless chemical reactions that occur. To understand that story, even in part, we need the language of chemical equations, which really are the sentences of chemistry.

The Language of Chemistry

Element symbols H, O, Na, Fe	Alphabet of chemistry
Compound formulas H_2O, CO_2, NaCl	Words of chemistry
Chemical equations $2H_2 + O_2 \rightarrow 2H_2O$	Sentences of chemistry

Chemical reactions and chemical change

A **chemical reaction** is any process of chemical change. Chemical reactions convert one or more substances into new substances. In the process atoms are not created or destroyed. They are simply rearranged to form new substances. In general we represent a chemical reaction by a **chemical equation** as follows.

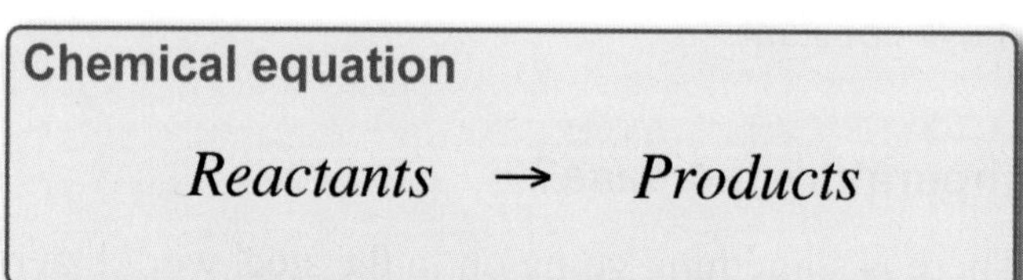

Chemical equation

Reactants → *Products*

Chemical equations and chemical reactions

The **reactants** are the starting materials used in the reaction. The **products** are the substances that are produced by the reaction. The reactants are always written on the left side of the arrow and the products are written on the right side of the equation. The arrow indicates the direction of the chemical reaction, pointing from the reactants to products. It means that the reactants "give" or "yield" or "react to form" the products.

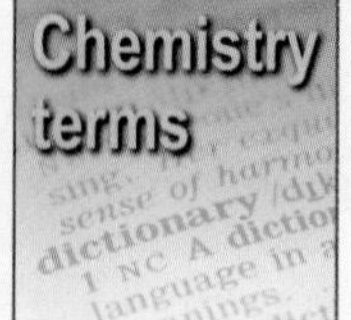

chemical reaction - is the process of chemical change.

chemical equation - A chemical equation is an expression that describes the changes that happen in a chemical reaction.

reactants - the starting materials or substances in a chemical reaction.

products - the materials or substances resulting from a chemical reaction.

Constructing a chemical equation

An example chemical reaction

As an example let's construct the chemical equation for one of the overall reactions that takes place in a fire. In this reaction, oxygen (O_2) from the atmosphere reacts with carbon (C) from wood, to produce carbon dioxide gas (CO_2). In this reaction, O_2 and C are the *reactants* since they are "used up" by the reaction. Carbon dioxide (CO_2), is the *product*. In words, the reaction can be written

Carbon reacts with oxygen to produce carbon dioxide.

We can write the expression in a more compact way using the arrow.

carbon + oxygen → carbon dioxide

Chemical equation with symbols

However, the use of the chemical formulas for each substance involved in the reaction is the most accurate, and least ambiguous way to describe the reaction. For our reaction we write the general relationship with the equation.

$$C + O_2 \rightarrow CO_2$$

Interpreting a chemical equation

The diagram below summarizes the interpretation of the chemical equation

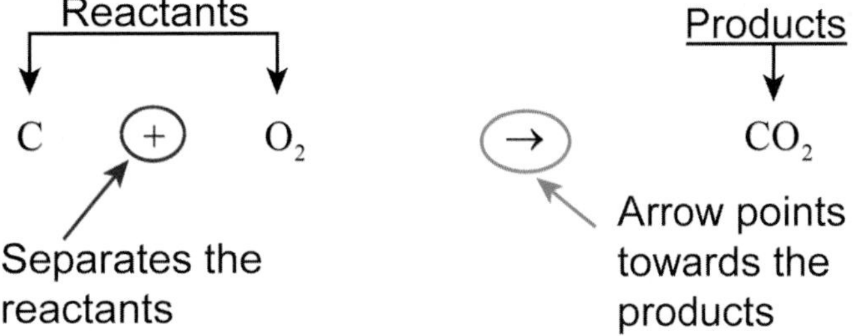

Why we use chemical formulas

Using the chemical formulas to write the chemical equation gives us a very compact way to express chemical reactions. It also gives a clear picture of the types of atoms that are involved in the reaction and how they are bonded together in both the products and the reactants.

Energy may appear on either side of a chemical equation

Does the chemical equation $C + O_2 \rightarrow CO_2$ give a complete representation of the burning of wood? When we look at a wood burning fire we see that fire gives off light and we also feel that fires gives off heat. So, if the chemical equation $C + O_2 \rightarrow CO_2$ does represent the phenomenon of wood burning then *where is the heat* and *where is the light?* Later in the chapter we will add energy to the chemical equation and explicitly recognize the energy as either a reactant (absorbed) or a product (given off).

Conservation of mass

Chemical reactions conserve mass

One of the fundamental properties of chemistry is that of the conservation of mass. In fact this property is so important that scientists call it the **law of conservation of mass**. The law states that the total mass of the reactants is equal to the total mass of the products. In other words, mass is neither created or destroyed during a chemical reaction.

The number of atoms in a chemical reaction does not change.

The fundamental building block of chemistry and chemical reactions is the *atom*. Conservation of mass really means that a chemical reaction does not change the total number and type of atoms. A chemical reaction rearranges the atoms in the reactants into a new arrangement *of the same atoms* in the products.

Law of Conservation of Mass

Each and every atom in the ***reactants*** must also appear in the ***products***.
Each and every atom in the ***products*** must also appear in the ***reactants***.

Mass of products = Mass of reactants

Burning hydrogen in oxygen

For example, consider burning hydrogen in oxygen. This is a reaction that is observed experimentally and gives water as the product. It is also the reaction that fuels the space shuttle. The white "smoke" that comes out of the shuttle engines is water vapor which is the product of the reaction.

Beginning a chemical equation

To start developing a chemical equation we write down the reaction between hydrogen and oxygen with the correct chemical formulas for participating substances.

Hydrogen	+	Oxygen	→	Water
H_2	+	O_2	→	H_2O
H–H	+	O=O	→	H–O–H

An unbalanced equation

This chemical equation correctly shows that water can be created by combining hydrogen and oxygen, but it does NOT satisfy the law of conservation of mass. There are *two* oxygen atoms in the reactants and only *one* in the products!

Chemistry terms

law of conservation of mass - the total mass of reactants (starting materials) and the total mass of products (materials produced by the reaction) is the same.

Reading chemical equations

Why an unbalanced equation is inaccurate

One problem with the equation $H_2 + O_2 \rightarrow H_2O$ is that if you mixed one mole of oxygen and one mole of hydrogen you would NOT get one mole of water molecules. You would actually get one mole of water molecules and a 0.5 moles of leftover oxygen molecules. You get leftover oxygen molecules because there are more oxygen atoms in the reactants than there are in the products. The chart below summarizes the reaction by counting the type of each atom on the reactant and product side.

"balanced" means the same number of atoms appear on both sides

The number of *hydrogen* atoms is the same in both the reactants and the products. We therefore can say that hydrogen atoms are *balanced.*

	Number of Atoms	
	Reactants	Products
Hydrogen	2	2
Oxygen	2	1

missing oxygen

The unbalanced chemical equation

The chemical equation $H_2 + O_2 \rightarrow H_2O$ is an **unbalanced chemical equation**. The unbalanced equation tells you what substances are involved in the reaction but it does *not* tell you how *much* of each are involved. In order to make it balanced, we need to adjust the number of hydrogen molecules, oxygen molecules, and water molecules until there are the same type and number of each atom on both sides of the chemical equation.

Coefficients

The *number* of molecules in a chemical equation is given by the coefficient in front of the chemical formula. For example, the following might appear in a chemical reaction to describe 2 hydrogen molecules, 1 oxygen molecule or 3 water molecules.

Coefficient how many molecules

$2H_2$

Subscript how many atoms in 1 molecule

		Molecules	Atoms
2 hydrogen molecules	$2H_2$	H-H H-H	H H H H
1 oxygen molecule	O_2	O=O	O O
3 water molecules	$3H_2O$	H-O-H H-O-H H-O-H	H H H H H H O O O

Coefficients versus subscripts

The subscript tells you how many atoms are in the molecule. For example the subscript 2 in H_2 tells you there are 2 hydrogen atoms in a hydrogen molecule. The coefficient tells you how many molecules there are in the reaction. If "$2H_2$" appears in a chemical equation you know there are two molecules of hydrogen that include a total of 4 atoms of hydrogen (2 molecules × 2 atoms per molecule).

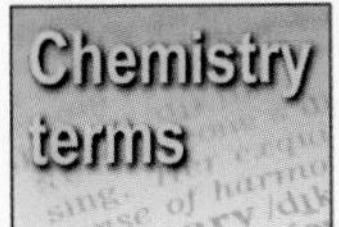

unbalanced chemical equation - a chemical equation that does not satisfy the law of conservation of mass. The number of each type of atom on the reactant side of the equation does not equal the same number for each atom on the product side.

A balanced chemical equation

The balanced equation

Multiplying the hydrogen molecule by a factor of 2 and the water molecule by a factor of 2 we obtain a chemical equation that satisfies mass conservation.

The balanced chemical equation

$$2H_2 + O_2 \rightarrow 2H_2O$$

	Reactants	Products
Molecules	H-H, H-H, O=O	H-O-H, H-O-H
Atoms		
	4 hydrogen atoms (H)(H)(H)(H)	4 hydrogen atoms (H)(H)(H)(H)
	2 oxygen atoms (O)(O)	2 oxygen atoms (O)(O)

The same number of each type of atom appear in both products and reactants

balanced equation conserves mass

This chemical equation is **balanced**.

There are two rules for balancing chemical equations.

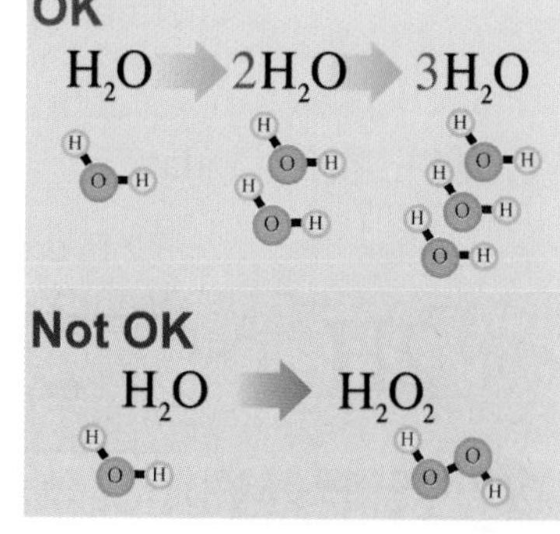

1. DO NOT CHANGE the subscripts in the chemical formulas because this would change the substances involved in the reaction, in essence making a *different* reaction. For example if we change H_2O to H_2O_2 we have changed water to hydrogen peroxide.
2. DO CHANGE the coefficients to adjust the number of each molecule in the reaction

The balanced equation correctly tells you amounts

The balanced equation tells us that when we combine 2 molecules of hydrogen with one molecule of oxygen we can obtain 2 molecules of water. This is the complete recipe for making water. It correctly represents the amount of each substance in the reaction and it satisfies mass conservation because the same number of each type of atom appear on the reactant and product sides of the chemical equation.

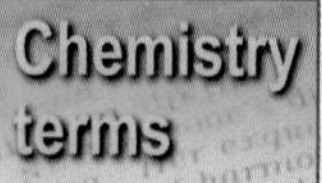

balanced chemical equation - a chemical equation that satisfies the law of conservation of mass.

Balancing chemical equations

Why balance chemical equations?

The goal of balancing a chemical equation is to get the numbers of each type of atom to be the same on both the reactant and product sides of the chemical equation. This is really just counting atoms on either side then adjusting the coefficients until the counts match. At this point the process of balancing a chemical equation is trial and error. You guess and try different coefficients until it works out. For many reactions this approach works fine. However, for reactions that involve a number of reactants and products a more structured approach is needed and will develop it in the next few pages.

Coefficients of one

When a single molecule is in a chemical equation the coefficient of 1 not written explicitly. For example, O_2 in a chemical equation means 1 *molecule* of oxygen (which contains 2 atoms).

Coefficients apply to whole molecules

The coefficient of 2 in front of a molecule applies to the entire molecule. For example, $2H_2O$ means that we have two water molecules. Since each water molecule has 2 hydrogen atoms, the total number of hydrogen atoms is 4.

Balance pure elements last

It is always easiest to balance pure elements last. For example if O_2 appears in a chemical equation, you can change the coefficient to make $2O_2$ or $3O_2$ and change only the number of oxygen atoms. Making CH_4 into $2CH_4$ changes both the number of carbon atoms *and* the number of hydrogen atoms.

Balance one element at a time

Another good strategy is to start with one element, often carbon. Then check oxygen, and so on, checking and balancing one element at a time. However, every time you change a coefficient, you should make sure you have not unbalanced an element that you already balanced!

The glucose ($C_6H_{12}O_6$) contained in biomass is used as a biofuel to produce ethanol (C_2H_6O) and carbon dioxide (CO_2). Write the balanced equation for this reaction.

Given: *The unbalanced chemical equation* $C_6H_{12}O_6 \rightarrow C_2H_6O + CO_2$

Asked: *Find the coefficients in order to balance the chemical equation.*

Relationships: *The same number of each type of atom must appear on each side.*

Solve: *All atoms involved in the reaction are unbalanced.*
There are 6 C atoms in the reactants and 3 C atoms in the products
There are 12 H atoms in the reactants and 6 H atoms in the products
There are 6 O atoms in the reactants and 3 O atoms in the products
From this accounting we notice that the number of atoms in the products is half the number of the reactants.
By multiplying the products by a factor of 2 we obtain a balanced chemical equation of the reaction.

Answer: *The balanced chemical equation is* $C_6H_{12}O_6 \rightarrow 2\ C_2H_6O + 2\ CO_2$

10.2 Methods for Balancing Chemical Equations

The respiration reaction

A chemical reaction between blood sugar (glucose, $C_6H_{12}O_6$) and oxygen provides the energy that your body needs. The reaction, called *cellular respiration*, produces carbon dioxide (CO_2) and water (H_2O). Respiration also generates heat.

Cellular respiration - unbalanced chemical equation

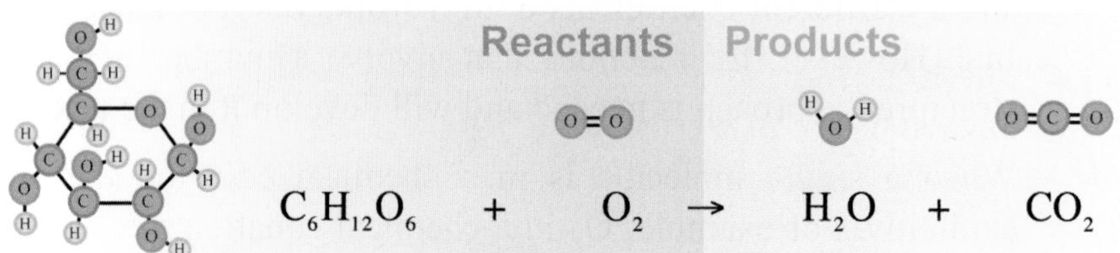

In this equation we have three different elements to consider: C, O and H. These elements make up the four substances involved in the reaction.

A general rule

The general rule for developing an effective way to balance chemical equations is to start with the elements that occurs in the fewest substances and end with the element that is involved in the most substances.

Start with one element

For our example both hydrogen and carbon are involved in only one reactant and one product substance. We could start with either carbon or hydrogen. Let's start with hydrogen. The number of hydrogen atoms in the products is balanced by using a coefficient of 6 in front of the H_2O product.

$$C_6H_{12}O_6 + O_2 \rightarrow CO_2 + 6\,H_2O$$

Balanced H

Do the next element

Next carbon, which is involved in two substances is balanced by inserting a coefficient of 6 in front of the CO_2 product

$$C_6H_{12}O_6 + O_2 \rightarrow 6\,CO_2 + 6\,H_2O$$

Balanced C

Balance pure elements last

The last element to be balanced is oxygen. We have left oxygen for last since it is involved in the largest number of compounds in the reaction. Also, another reason for leaving it last is that it exists as a pure substance in the reaction.

Let's do some counting of the oxygen atoms as they appear in the above equation.

Total number of oxygen atoms in the reactants:	6+2=8
Total number of oxygen atoms in the products:	12+6=18

Balancing pure elements and checking

$$C_6H_{12}O_6 \quad + \quad O_2 \rightarrow 6H_2O \quad + \quad 6CO_2$$

6 + 2 = 8 oxygen atoms | 6 + 12 = 18 oxygen atoms

The last step in balancing the equation

The reaction above is balanced for all but oxygen. As written there are 10 more oxygen atoms in the products than the reactants. In order to make up the difference we must add 10 additional oxygen atoms, or 5 oxygen molecules (O_2), to the reactants.

The five additional molecules plus the existing oxygen molecule give us a total of 6 oxygen molecules in the products.

$$C_6H_{12}O_6 \quad + \quad 6\,O_2 \rightarrow 6\,CO_2 \quad + \quad 6\,H_2O$$

Check the atom counts

The next step is to check to see that the equation is indeed balanced. We do this by counting the number of atoms for each element on either side of the equation. The equation is balanced when the number of atoms for each element are balanced.

$$C_6H_{12}O_6 \quad + \quad 6O_2 \rightarrow 6H_2O \quad + \quad 6CO_2$$

Reactants	Products
6 carbon	6 carbon
12 hydrogen	12 hydrogen
18 oxygen	18 oxygen

Check that coefficients are not multiples

The last step is to check that the coefficients are the smallest whole numbers possible. For example, the same balanced reaction could also be written as:

$$2\,C_6H_{12}O_6 \quad + \quad 12\,O_2 \rightarrow 12\,CO_2 \quad + \quad 12\,H_2O$$

Here, we have doubled the amount of reactants and products. This is still a correct chemical equation. It is however preferred and established by convention, to write the coefficients as the smallest whole numbers.

The steps for balancing chemical equations

In summary here are the steps for balancing chemical equations.

Step 1. Write down the unbalanced chemical equation

Step 2. Identify the element that occur in only one compound on both sides of the equation and balance it first. If more than one element satisfies this condition you may balance them in any order.

Step 3. Continue with the rest of the elements. If a free element is present in the reaction it is balanced last.

Step 4. Check each element to make certain that the equation is balanced.

Step 5. Make sure the coefficient are the smallest possible whole numbers

Methane (CH_4) reacts with oxygen (O_2) to produce carbon dioxide (CO_2) and water (H_2O). Write the balanced equation for this reaction.

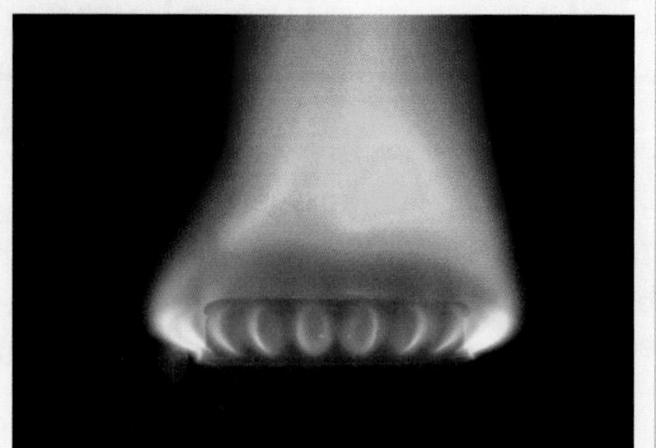

Given: *The unbalanced chemical equation is:* $CH_4 + O_2 \rightarrow CO_2 + H_2O$

Solve: *We solve the problem by following the 5 steps*

Step 1. *Let's start with carbon (C)*
Reactants have 1 C atom
Products have 1 C atom
Carbon is balanced already

Step 2. *Let's now proceed by balancing H*
Reactants have 4 H atoms
Products have 2 H atoms
Need to add 2 H atoms to the products to balance H
$CH_4 + O_2 \rightarrow CO_2 + 2H_2O$

Step 3. *Next we balance oxygen*
Reactants have 2 O atoms
Products have 4 O atoms
Need to add 2 O atoms to the reactants to balance O
$CH_4 + 2O_2 \rightarrow CO_2 + 2H_2O$

Step 4. *Check the number of atoms for each element on both sides of the reaction*

Reactants	Products
1 C atom	1 C atom
4 O atoms	4 O atoms
4 H atoms	4 H atoms

Step 5. *Coefficients in front of the various compounds are the smallest whole numbers possible.*

Answer: *The balanced equation for the chemical reaction is*
$CH_4 + 2O_2 \rightarrow CO_2 + 2H_2O$

10.3 Types of Chemical Reactions

chemical equations follow certain patterns

The number of possible chemical reactions is very large. However, there is a limited number of structural patterns that every chemical reaction may fall into. By learning these patters we will be able to better understand these chemical reactions.

These patterns can be best seen by considering the following four classifications of chemical reactions:

1. Combination reactions (Synthesis reactions)
2. Decomposition reactions
3. Single replacement reactions
4. Double replacement reactions (metathesis reactions)

We will now describe the structure of each one of these classes or reactions.

Combination reactions (Synthesis reactions)

A reaction in which two substances combine to form a third substance is called a combination or synthesis reaction. The general form of a combination reaction is

Synthesis reaction

Two compounds combine to make a third compound

$$A + B \rightarrow AB$$

Synthesis of compounds from elements

This symbolic form shows substances A and B combining to form a new substance AB. The substances A and B can be two compounds, two elements or an element and a compound. The formation of water from hydrogen and oxygen is a synthesis reaction that combines two pure elements to form a compound.

$$2H_2(g) + O_2(g) \rightarrow 2H_2O\ (g)$$

Synthesis of compounds from compounds

Another example of a synthesis reaction combines magnesium oxide (MgO) with water to make magnesium hydroxide ($Mg(OH)_2$), also known an *milk of magnesia.* In this example, two compounds combine to form a third compound.

$$MgO(s) + H_2O(l) \rightarrow Mg(OH)_2(s)$$

Synthesis of magnesium hydroxide from magnesium oxide and water

Milk of magnesia, a common medicine is a suspension of $Mg(OH)_2$ in water

Synthesis reactions *combine* different substances therefore the number of products is less than the number of reactants.

Decomposition and single displacement reaction

Decomposition is the opposite of synthesis

A decomposition reaction is the opposite of a synthesis reaction. In a decomposition reaction a single substance breaks apart (decomposes) to form two or more new substances. The new substances may be elements or compounds.

Decomposition reaction

One compound breaks apart into two or more compounds or elements

AB → A + B

Decomposition of limestone

Calcium oxide (CaO), also called lime, has many uses including water treatment, glass manufacturing, food preservation and cement manufacturing. Lime is produced from the decomposition of calcium carbonate ($CaCO_3$) in limestone. The reaction occurs when limestone is heated to 825°C.

$$CaCO_3(s) \rightarrow CaO(s) + CO_2(g)$$

Another important reaction is the decomposition of water which happens when electrical current passes through it.

$$2H_2O(l) \rightarrow 2H_2(g) + O_2(g)$$

Decomposition usually requires energy

Most decomposition reactions require the addition of some type of external energy to proceed. The energy can be in the form of heat or electrical current as in the previous examples. Energy can also be in the form of light as in the case of the destruction of ozone (O_3) in the upper atmosphere.

$$O_3 + \text{light} \rightarrow O_2 + O$$

Single displacement reactions

A single displacement reaction happens when an element reacts with a compound replacing an element in it. The general form of a displacement reaction is

Single displacement

Two compounds swap a single element or polyatomic ion

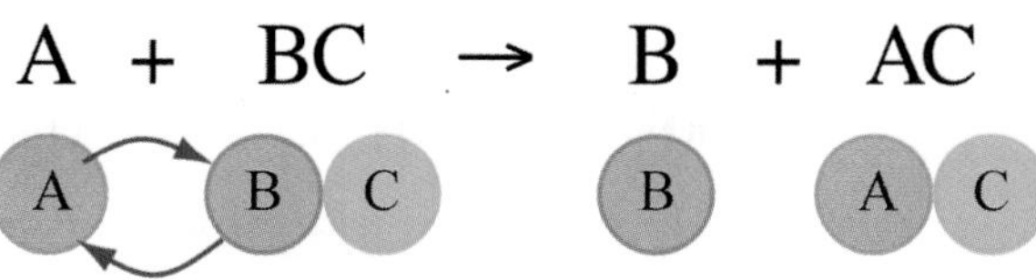

Examples of displacement reactions

The element A replaces the B part of the compound BD. It could also replace the D part of the compound BD. A single displacement reaction occurs when zinc (Zn) or iron (Fe) are immersed in a solution of copper sulfate ($CuSO_4$)

$$Zn(s) + CuSO_4(aq) \rightarrow Cu(s) + ZnSO_4(aq)$$

$$Fe(s) + CuSO_4(aq) \rightarrow Cu(s) + FeSO_4(aq)$$

Double replacement reactions (precipitate reactions)

Ions "trade" partners

Precipitate reactions often involve the exchange of parts of the reactants. The general form of this type of reaction is

$$AB + DX \rightarrow AX + DB$$

Ions "trade partners" in double displacement

Here we see that parts B and X , which represent the negative ions, of the compounds AB and DX exchange places giving us the new compounds AX and DB. Basically the positive and negative ions "trade partners."

Double replacement or precipitate reactions are one very common reaction type that take place in aqueous solutions. Recall that a precipitate is the formation of an insoluble solid which forms when certain aqueous solutions are mixed. The precipitate separates from the solution and can be easily observed.

Precipitate reaction

A good example is the reaction between lead nitrate ($Pb(NO_3)_2$) and potassium iodide (KI). The pictures show a 0.1 M solution of lead nitrate, and a 0.1 M solution of potassium iodide. Look carefully at their colors. Notice that they are both essentially clear and colorless. When $Pb(NO_3)_2$ is added notice the brilliant yellow color! The yellow comes from an insoluble precipitate (PbI_2) which is a product of the reaction.

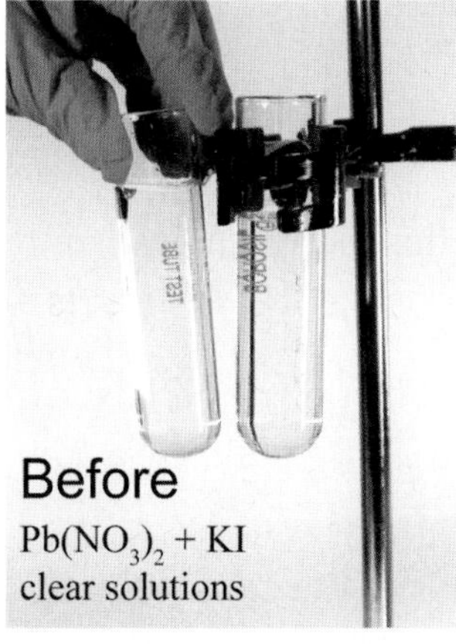

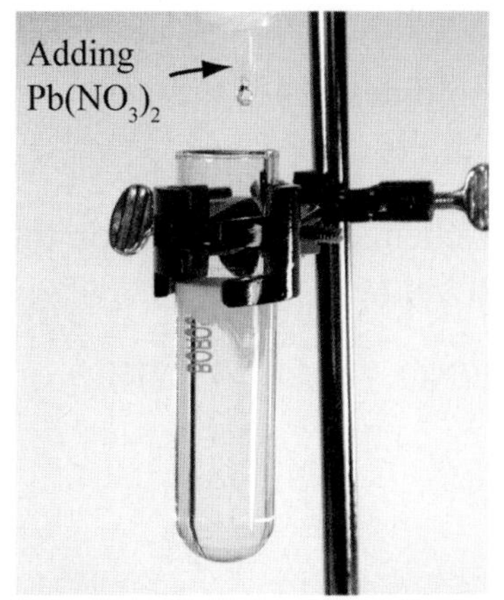

The chemical equation for the reaction

The chemical reaction shows that the lead ion displaces the potassium ion in KI. The potassium ion in turn displaces the lead ion in $Pb(NO_3)_2$.

$$Pb(NO_3)_2(aq) + 2KI(aq) \rightarrow PbI_2(s) + 2KNO_3(aq)$$

Precipitates and solubility

The nitrate anion

Before we look why a precipitate forms, consider the polyatomic nitrate ion (NO_3^-). The nitrate ion acts as a "unit" so the nitrogen atom and three oxygen atoms remain covalently bonded together. The negative charge is the charge of the whole nitrate ion. A helpful comparison is to treat nitrate ion as if it were a monatomic ion Cl^-.

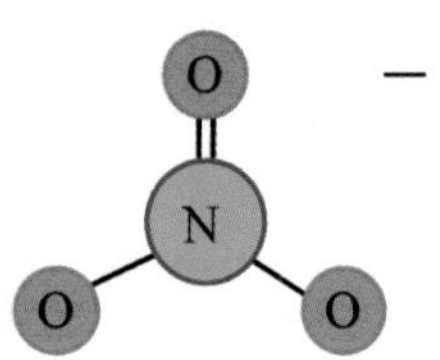

When a precipitate forms

A precipitate forms when one of the products of a reaction is *insoluble* in water. To be able to predict which ionic compound is the solid precipitate we need refer to a set of *solubility rules*. The rules summarize which compounds are soluble and which are not.

TABLE 10.1. Solubility Rules for Common Ionic Compounds in Water

Soluble Compounds	Insoluble Compounds (except when with group I metal ions and NH_4^+)
Group I metal ions: Li^+, Na^+, K^+, Rb^+, Cs^+	Carbonates, CO_3^{2-}
Ammonium: NH_4^+	Hydroxides, OH^-, (ex Ba^{2+})
$C_2H_3O_2^-$	Chlorides of Cu, Pb, Ag, and Hg
Nitrates, NO_3^-	Bromides of Cu, Pb, Ag, and Hg
SO_4^{2-}	Iodides of Cu, Pb, Ag, and Hg
Chlorides, Bromides, and Iodides except with Cu, Pb, Ag, and Hg	Sulfides, S^{2-},
	Phosphates, PO_4^{3-}

Using the solubility table

The balanced chemical equation for the reaction between lead nitrate and potassium iodide is given below.

$$Pb(NO_3)_2(aq) + 2KI(aq) \rightarrow PbI_2(s) + 2KNO_3(aq)$$

Solid (precipitate)

K^+ K^+ NO_3^-
Dissolved ions

Note from the table that *nitrates* are typically soluble. Therefore we expect potassium nitrate to be soluble, and it is. The table also says that iodides are generally soluble except for compounds with copper (Cu), lead (Pb), silver (Ag), and mercury (Hg). Potassium iodide is soluble. Lead iodide is insoluble, and forms a precipitate.

Precipitate reaction uses and analysis

"Hard" water

Precipitate reactions are important to our environment and drinking water supplies. Minerals such as Mg^{2+} and Ca^{2+} leach into our water supplies and cause a phenomena known as water "hardness". You have experienced water hardness if you ever been in the shower and the soap did not lather well. "Hard" water leaves your skin feeling sort of sticky and filmy, because soaps and oils do not come off as easily. Iron, Fe^{2+} and manganese, Mn^{2+} also cause some other problems, such as staining of clothing and increased bacterial growth.

Water softening

To condition (or soften) the water in areas where there are lots of dissolved minerals, water treatment plants add special ions that precipitate out some of the Mg^{2+} and Ca^{2+} ions. Hydroxide, OH^- and carbonates, CO_3^{2-} are added to the water supply to precipitate out large amounts of these contaminating ions. For example Mg^{2+} + OH^- forms $Mg(OH)_2$ (s) and this solid settles to the bottom and is easily removed by filtering.

Mg^{2+} and Ca^{2+} ions in solution

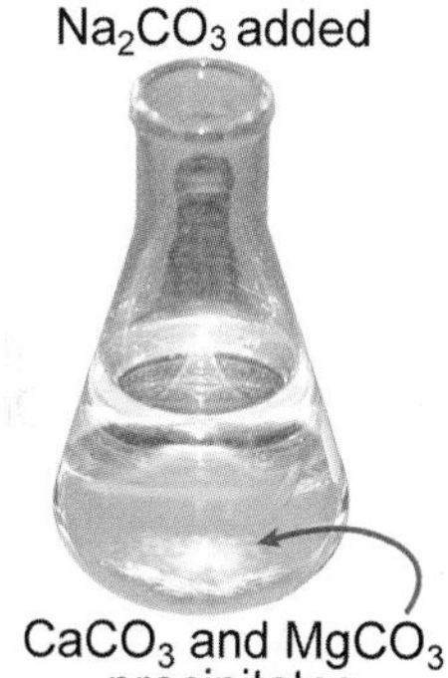

Na_2CO_3 added

$CaCO_3$ and $MgCO_3$ precipitates

Gas forming reactions

Some double replacement reactions involve the production of a gas. For example when magnesium carbonate, $MgCO_3$, and hydrochloric acid, HCl, are mixed:

$$MgCO_3(s) + 2HCl(aq) \rightarrow MgCl_2(aq) + H_2O(l) + CO_2(g)$$

In this reaction, Mg^{2+} and H^+ ions exchange places resulting in magnesium chloride, $MgCl_2$ and H_2O and carbon dioxide, CO_2 is evolved.

Predict what happens when we mix a solution of silver nitrate ($AgNO_3$) with a solution of sodium chloride (NaCl). Follow the steps and write the complete balanced reaction.

Asked: *What is the reaction that represents the mixing of $AgNO_3$ and NaCl?*

Given: *formulas for both reactant solutions*

Relationships: *Solubility Rules tell us that group I metal ions (here Na+) and nitrates are soluble, but chlorides of silver are insoluble.*

Solve: *$AgNO_3(aq) + NaCl(aq) \rightarrow AgCl(s) + NaNO_3(aq)$*
According to the solubility table, sodium nitrate is soluble and dissociates to $Na^+(aq)$ and $NO_3^-(aq)$.
Silver chloride is insoluble and forms a precipitate.

Discussion: *We know the precipitate is AgCl because it does not have a group I metal ion or nitrate. All ions are balanced.*

Polymerization reactions

Polymers are long chains of repeated units

Many important molecules in both technology and biology are *polymers*. The word is a clue to how a polymer is made. The prefix "poly" means "many" and the Greek word "polymeres" means "of many parts". A **polymer** is a molecule made of many identical parts called *monomers*. A good example of a naturally occurring polymer is *starch*. Starch is a long chain molecule, built from repeated addition of sugar molecules.

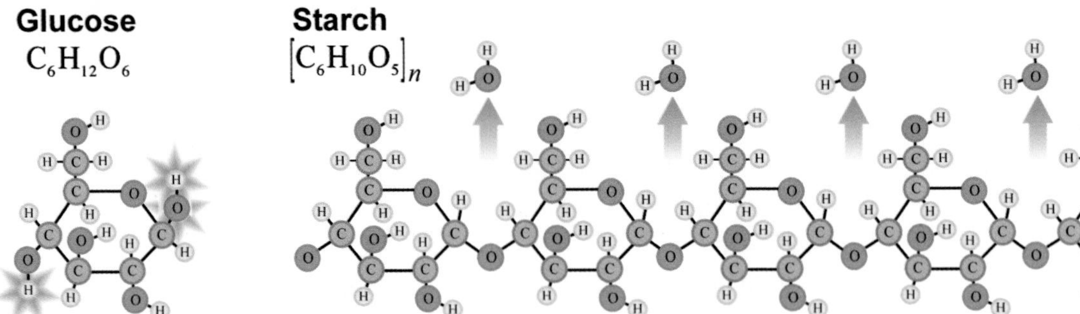

Polymerization

Photosynthesis in plants produces glucose. However, glucose is also rapidly used to release energy, even by plants. To save chemical energy for long term use, plants convert glucose into starch by *polymerization*. **Polymerization** is a repeated addition reaction that assembles a polymer from smaller molecules. Anyone who likes fresh corn has tasted the difference between sugar and starch. Fresh picked ripe corn has a high sugar content. Over a period of a day or two, enzyme reactions in the corn convert the sugar to starch. Starch does not taste sweet and the polymerization reaction is responsible for the "old corn" taste.

Dehydration synthesis

The polymerization reaction that produces starch is called *dehydration synthesis*. Look carefully and notice that glucose has the chemical formula $C_6H_{12}O_6$ while the monomer for starch is $C_6H_{10}O_5$. The difference is one water molecule. To *dehydrate* means to "take away water" and to synthesize means "to put together." Dehydration synthesis builds up the starch molecule by removing an oxygen and two hydrogens from each glucose molecule that gets added to the chain. The bond between adjacent monomers is through the oxygen.

Starch is a staple food

Starch is the staple carbohydrate of many animal diets, including humans! Bread, pasta, and all grains are mostly starches. The body reverses the synthesis process and digests starch back into glucose.

polymer - a molecule built up from many repeating units of a smaller molecular fragment.

polymerization - a reaction that assembles a polymer through repeated additions of smaller molecular fragments.

10.4 Chemical Reactions and Energy

chemical bonds and energy

Every chemical reaction involves the breaking and the formation of chemical bonds. In order to break chemical bonds energy must be provided. When chemical bonds are formed energy is released. Therefore every chemical reaction either releases or absorbs energy. The amount of energy involved depends on the particular reaction. From the burning of fossil fuels to the melting of ice on the street we rely on the energy released or absorbed by chemical reactions.

hot packs release energy cold packs absorb energy

Cold and hot packs have become the best tools for athletic trainers. When an athlete is injured, the trainer reaches in the first aid kit, grabs a small plastic bag, punches it and applies it on the injured muscle. Within seconds the bag is ice cold. How does this happen? At another time the trainer may reach in the first aid kit and get a small bag which in an instant becomes very hot and can be used to treat an injury. Where does this heat come form? Well, both the hot and the cold packs rely on chemical reactions to do their job. Inside the hot pack a chemical reaction releases energy and inside the cold pack another chemical reaction absorbs energy.

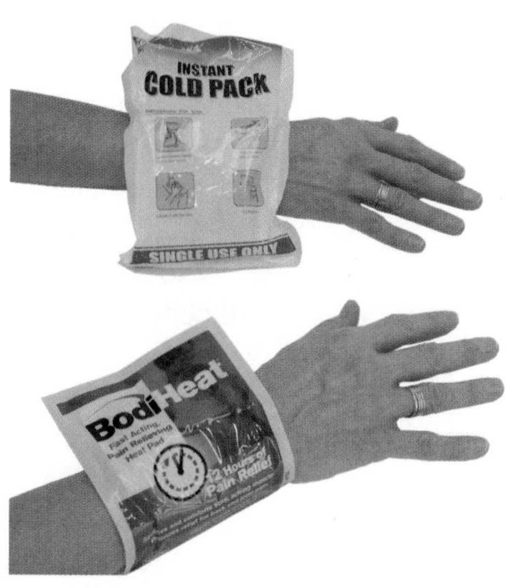

exothermic: release of energy. endothermic: absorption of energy

When a chemical reaction releases energy is called **exothermic reaction**. When a chemical reaction absorbs energy is called **endothermic reaction**.

The amount of energy that is released or absorbed is denoted by the symbol ΔH which is called the **enthalpy** of the reaction. Enthalpy, which means to put heat into, is another word for energy and it has the units of Joule. A chemical reaction is presented completely by writing the enthalpy information on the right of the chemical equation.

$$\text{Reactants} \rightarrow \text{Products} \qquad \Delta H$$

ΔH < 0 exothermic

For exothermic reactions ΔH *is a negative number indicating that energy is released by the reaction.*

ΔH > 0 endothermic

For endothermic reactions ΔH *is a positive number indicating that energy is absorbed by the reaction.*

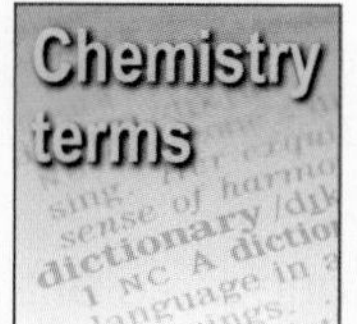

exothermic reaction - a reaction that releases energy. ΔH<0

endothermic reaction - a reaction that absorbs energy. ΔH>0

enthalpy - is related to the amount of energy that a chemical reaction releases to the environment or absorbs from the environment.

Thermochemical equations

When carbon is burned with oxygen it produces carbon dioxide and releases energy in the form of heat and light. Since the reaction releases energy it is an exothermic reaction.

enthalpy is a measured quantity

For this reaction the enthalpy change is measured to be -393.5 kJ, where kJ (kilojoule) is 1000 Joules. The complete reaction is written as follows

$$C(s) + O_2(g) \rightarrow CO_2(g) \qquad \Delta H = -393.5 \text{ kJ}$$

This is called a **thermochemical equation** and it includes the chemical equation and the information about the enthalpy change on the right. The negative sign for ΔH means that the reaction is exothermic. As expected the enthalpy change is related to the amount or material involved in the reaction. For example, the more carbon we burn the more CO_2 is produced and the more energy is released. The enthalpy change refers to a certain amount of a substance involved in the reaction. In our case the denoted enthalpy change refers to the formation of 1 mole of CO_2. If we burn 2 moles of C the enthalpy change will be twice as great

$$2C(s) + 2O_2(g) \rightarrow 2CO_2(g) \qquad \Delta H = -787 \text{ kJ}$$

enthalpy of formation is used to calculate reaction enthalpy

The enthalpy change is a measured quantity and it is obtained experimentally for the various reactions. Since there are millions of possible reactions it is impossible to list all of them with their enthalpy values. Chemists have measured and cataloged the standard **enthalpy of formation** for many common substances. The enthalpy of formation corresponds to the enthalpy change during the formation of one mole of a substance. These values are given at standard conditions (25 °C and 1 atmosphere pressure). Using these values they are then able to calculate the enthalpy of most reactions. Some enthalpy values for the formation of some common substances are shown on Table 10.2.

TABLE 10.2. Enthalpies of Formation

Substance	ΔH_f(kJ/mole)	Substance	ΔH_f (kJ/mole)
CO_2	-393.5	NO_2	33.2
CO	-110.5	O_3	142.7
$CuSO_4$	-771.4	C (diamond)	1.9
H_2O (l)	-285.5	C_6H_6	49.0
Fe_2O_3	-824.2	O_2	0

Chemistry terms

thermochemical equation - the equation that gives the chemical reaction and the energy information of the reaction.

enthalpy of formation - the enthalpy change during the formation of one mole of a substance.

Exothermic and endothermic reactions

Reverse reactions

The enthalpy change of the reverse reaction is the negative of the enthalpy change of the forward process. For our CO_2 example reaction we have

$$CO_2(g) \rightarrow C(s) + O_2(g) \qquad \Delta H = 393.5 \text{ kJ}$$

The enthalpy of formation for each substance involved in a reaction (ΔH_f) is related to the reaction enthalpy (ΔH) by the equation

$$\Delta H = \Delta H_f\,(\text{products}) - \Delta H_f\,(\text{reactants})$$

This is a mathematical expression of energy conservation and may be applied in order to calculate an unknown reaction enthalpy or an unknown enthalpy of formation.

our bodies are powered by exothermic reactions

A chemical reaction that takes place continuously inside our bodies results from the combination of glucose ($C_6H_{12}O_6$) with oxygen. This is an exothermic reaction and it is given by the thermochemical equation

$$C_6H_{12}O_6\,(s) + 6O_2\,(g) \rightarrow 6CO_2\,(g) + 6H_2O\,(g) \qquad \Delta H = -2808 \text{ kJ}$$

The negative sign for ΔH indicates that the reaction releases energy and thus it is exothermic. This reaction tells us that when one mole of glucose ($C_6H_{12}O_6$), 180 g, is combined with 6 moles of oxygen it releases 2808 kJ of energy. This energy is used by our bodies to help us grow and move. It is the energy that makes our life possible.

The complete combustion of glucose ($C_6H_{12}O_6$) releases 2,808kJ of energy per mole of glucose. Calculate the enthalpy of formation of glucose.

Given: *The enthalpy of combustion of one mole of glucose is 2,808 kJ.*
The combustion reaction
$C_6H_{12}O_6\,(s) + 6O_2\,(g) \rightarrow 6CO_2\,(g) + 6H_2O\,(g)\ \Delta H = -2,808\ kJ$
From Table 10.2 we read: The enthalpy of formation of O_2 is 0.
The enthalpy of formation of CO_2, $\Delta H_f\,(CO_2)$, is -393.5 kJ
The enthalpy of formation of H_2O, $\Delta H_f\,(H_2O)$ is -285.5 kJ

Relationships: $\Delta H = \Delta H_f\,(products) - \Delta H_f\,(reactants)$

Solve: *We write down the reaction for the formation of glucose*
$6CO_2\,(g) + 6H_2O\,(g) \rightarrow C_6H_{12}O_6\,(s) + 6O_2\,(g)\quad \Delta H = +2,808\ kJ$
From the enthalpy relation equation we have:
$\Delta H = \Delta H_f(glucose) + 6\,\Delta H_f\,(O_2) - 6\,\Delta H_f\,(CO_2) - 6\,\Delta H_f\,(H_2O)$
$2,808 = \Delta H_f(glucose) + 0 - 6\,(-393.5) - 6\,(-285.5)$ *which gives*
$\Delta H_f(glucose) = -1,266\ kJ$

Answer: *The enthalpy of formation for glucose is -1,266 kJ per mole.*
or -7.0 kJ per gram since there are 180 g/mole

Calculating the enthalpy change of a reaction

rusting of iron releases energy

The rusting of iron, as we saw in the experiment at the beginning of this chapter, results from a chemical reaction that is exothermic. When iron (Fe) reacts with oxygen (O_2) it produces iron oxide (Fe_2O_3), rust, and it generates heat. For each mole of iron oxide the enthalpy change is -824.2 kJ. This reaction is written as

$$2\ Fe + 3/2\ O_2 \rightarrow Fe_2O_3 \qquad \Delta H = -824.2\ kJ$$

Coefficients multiply enthalpies of formation

We have written the above thermochemical equation for the formation of one mole of Fe_2O_3. This is the reason for the use of the fractional coefficient in front of O_2. If we were to write the coefficients using the smaller whole number possible the thermochemical reaction would be

$$4\ Fe + 3\ O_2 \rightarrow 2\ Fe_2O_3 \qquad \Delta H = -1648.4\ kJ$$

The energy released per mole of Fe_2O_3 is still the same (-824.2 kJ) but now since the chemical equation has a coefficient of 2 in front of iron oxide we must multiple the noted enthalpy change by 2.

Endothermic reactions

Have you noticed that when you chew certain types of gum your mouth feels cooler? You actually feel the result of an endothermic reaction taking place in your mouth. Ingredients in gum called polyols dissolve in the saliva resulting in an endothermic reaction. The reaction absorbs energy from your mouth which as a result feels cooler. A common gum ingredient is xylitol ($C_5H_{10}O_4$) which when dissolved in water or saliva absorbs about 22.3 kJ/mole or about 167 J/g. This is considerable energy resulting in a cool and refreshing feeling in your mouth.

photosynthesis stores energy

The glucose that our bodies burn was generated by plants during **photosynthesis**. Photosynthesis reaction captures the energy of the sun and it is the fundamental energy storage reaction. The photosynthesis reaction takes place inside plants and combines water and carbon dioxide to make glucose and oxygen. The thermochemical reaction of photosynthesis is

$$6CO_2\ (g) + 6H_2O\ (g) \rightarrow C_6H_{12}O_6\ (s) + 6O_2\ (g) \quad \Delta H = +2808\ kJ$$

photosynthesis is an endothermic reaction

Photosynthesis is an endothermic chemical reaction as indicated by the positive sign for ΔH. The energy required to make this reaction happen is provided by sunlight. Photosynthesis is thus the ultimate energy storage reaction. It stores the energy from the sun and it produces oxygen which are in turn used for sustaining life on earth. Photosynthesis has been capturing sunlight and storing it in compounds that contain carbon and hydrogen, the hydrocarbons.

Chemistry terms

photosynthesis - the chemical reaction that combines CO_2 and H_2O to form glucose $C_6H_{12}O_6$ and oxygen. The reaction is endothermic and it driven by sunlight.

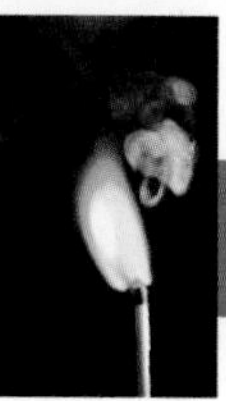

Energy profile of chemical reactions

Why exothermic reactions do not start spontaneously

The synthesis reaction of carbon dioxide is an exothermic reaction releasing a substantial amount of energy.

$$C(s) + O_2(g) \rightarrow CO_2(g) \qquad \Delta H = -393.5 \text{ kJ}$$

However, carbon does not spontaneously catch fire! In order to burn a piece of carbon it requires energy to *start* the process. It is not enough that we simply put carbon and oxygen together. Some initial energy is needed to break the bonds between oxygen atoms in O_2 before they can re-form with carbon atoms to make CO_2.

A spontaneous reaction

Another exothermic reaction that releases considerable energy is the reaction of sodium (Na) with water (H_2O).

$$Na(s) + H_2O \rightarrow NaOH(aq) + 1/2H_2\ (g) \qquad \Delta H=-140.9 \text{ kJ}$$

This reaction IS spontaneous. When we drop a piece of sodium metal in water the reaction starts immediately with an impressive release of energy.

Both the carbon-oxygen and sodium-water reactions are exothermic but one of them happens spontaneously while the other needs to be initiated.

The reason for the difference

The answer has to do with chemical bonds and the energy that is required in order to break them and to form them. This is best shown with the energy diagram of the reaction. In an energy diagram we show the bond energies of the products and the reactants as well as the energy path that they follow during the reaction. This presentation gives us an energy profile for the reaction. On the vertical axis of the energy diagram we give the total bond energy of reactants and products. The horizontal axis is used to show the time progression of the reaction from reactants to products. The energy profile may have a barrier represented as a "hump" in the diagram. This energy barrier corresponds to the energy input needed for the reaction to proceed. Once the reaction happens the net energy of the reaction is the difference between the initial and the final energy levels. For an exothermic reaction the energy level of the products is less than the energy level of the reactants. The difference in the energy is the amount that is released from the reaction.

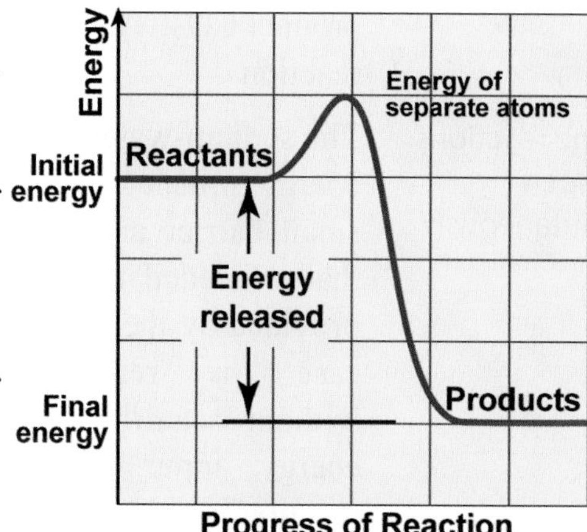

The energy barrier

some reactions need external energy to proceed

The presence of an energy barrier in the carbon-oxygen reaction means that we must provide external energy for the reaction to proceed.

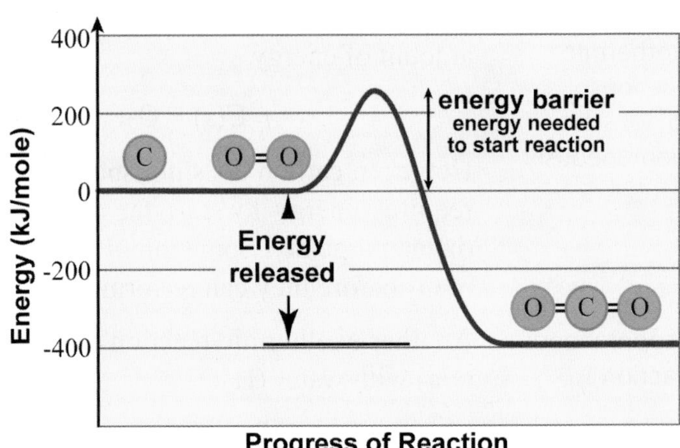

energy barrier

The external energy is needed in order to overcome the energy barrier. Only the molecules that have enough energy to go over the **energy barrier** can react.energy of a reaction is stored in chemical bonds

Without getting into the fine details of molecular bonds we can say that all the energy information about this reaction is included in the bonds of molecular oxygen and carbon dioxide.

In order for the reaction to proceed the first thing that has to happen is to break the bonds of molecular oxygen. This requires a certain amount of energy input, indicated by the barrier in the energy profile of the reaction.

The "energy content" in the bonds of C and O_2 is higher than the energy content of the product CO_2. The difference in the two energy levels corresponds to the enthalpy of the reaction.

some reactions happen spontaneously

The sodium-water reaction has an energy profile that has a very small barrier and all the energy that is needed to overcome it is provided by the reactants. In this case the reaction proceeds without requiring any external energy input. This type of reaction is called **spontaneous** which means that it can happen spontaneously by simple bringing the two reactants together.

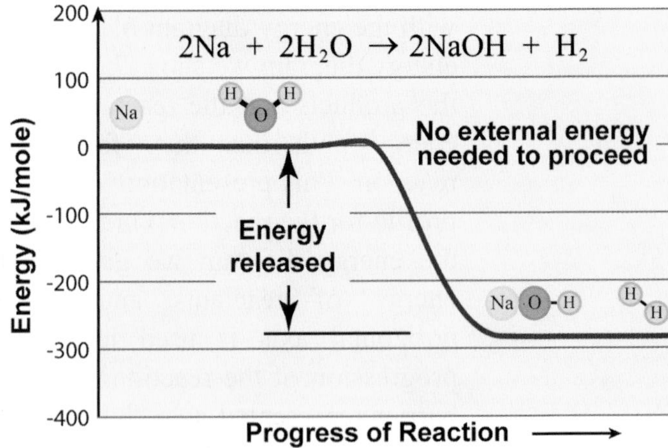

Chemistry terms

energy barrier - The energy needed to initiate a chemical reaction.

spontaneous reaction - A chemical reaction that happens without the need for external energy input.

Hess's law: A method for calculating enthalpy change.

enthalpy does not depend on the path taken

enthalpy is a state function

Enthalpy, like energy, is conserved and it depends only on the initial and the final states of the process that generate it. In chemistry we say that enthalpy is a state function to denote this property. Since the enthalpy depends only on the initial and the final states, the value of the enthalpy change depends only on the initial and the final states. For a chemical reaction the initial state is related to the reactants and the final state is related to the products. As a reaction proceeds from reactants to products, the actual path that we take in calculating the enthalpy change does not matter. If we go directly from Reactants to Products following path 1 as indicated in the schematic the enthalpy change is ΔH. The change is also ΔH even if we follow path 2 going from Reactants to A then to B and finally to Products. Note that the enthalpy going from Reactants to A is negative, while the enthalpy going from A to B is positive. This is a general rule called Hess's law.

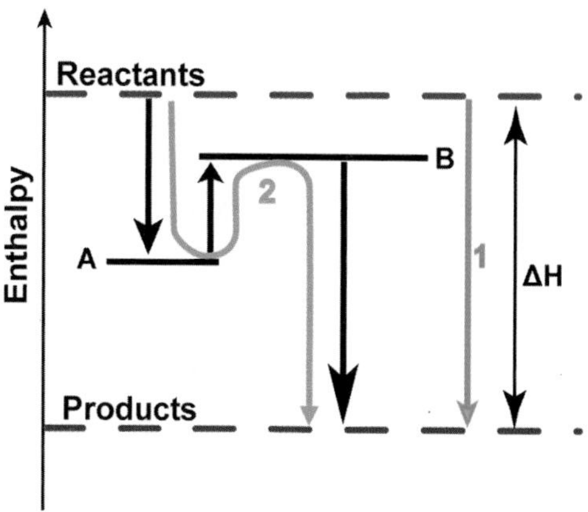

Hess's law

The overall enthalpy of a reaction is the sum of the reaction enthalpies of the various steps into which a reaction can be divided.

Given that the enthalpy of combustion of graphite (C_{gr}) and diamond (C_d) are -393.5 kJ/mole and -395.4 kJ/mole respectively, calculate the formation enthalpy of diamond from graphite.

Given: *The entahlpies of combustion and the compounds of interest.*

Asked: *We are asked to calculate the enthalpy of the reaction:*

$C_{gr}(s) \rightarrow C_d(s) \quad \Delta H = ?$

Relationships: *The basic relationship is Hess's law and the combustion reactions*

$C_{gr}(s) + O_2(g) \rightarrow CO_2(g) \quad \Delta H_1 = -393.5 kJ/mole$

$C_d(s) + O_2(g) \rightarrow CO_2(g) \quad \Delta H_2 = -395.4\ kJ/mole$

Solve: *We will apply Hess's law by considering a path that will start with Cgr(s) and give Cd(s). We can create such a path by considering the two reactions*

$C_{gr}(s) + O_2(g) \rightarrow CO_2(g) \quad \Delta H_1 = -393.5 kJ/mole$

$CO_2(g) \rightarrow C_d(s) + O_2(g) \quad \Delta H_2 = +395.4\ kJ/mole$

Note the change of sign for ΔH_2.

The first reaction produces CO_2 and uses O_2 and the second does the exact opposite. The sum of the two reactions gives

$C_{gr}(s) \rightarrow C_d(s) \quad \Delta H = (395.4 - 393.5)\ kJ/mole = +1.9\ kJ/mole$

Sustainable chemistry

Much of the technological world is created through *chemical engineering*. **Chemical engineering** is the profession of using chemistry to solve human problems. For example, plastics and medicines are produced by chemical engineers. So are fabrics, paints, dyes, metals, food additives, soaps, make-up, fuels, and thousands of products you use every day.

Goals of green chemistry

The goal of *green* chemistry, also known as *sustainable chemistry*, is to reduce or eliminate any harmful impact of human chemical technology on our environment and our bodies. Green chemistry is a very important idea because human civilization cannot simply give up using substances such as plastic and metal. What we can do is:

1. Redesign the chemistry of products so they are biodegradable, nontoxic, based on renewable or recycled resources or otherwise have minimal environmental or health impact.
2. Redesign manufacturing processes so harmful chemicals or excess waste are reduced or eliminated and overall energy input is reduced.

Goals of green chemistry

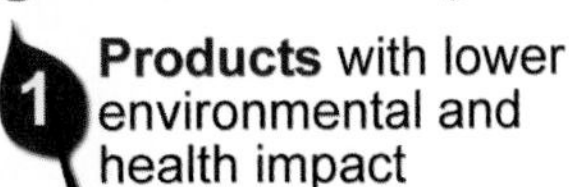

1 **Products** with lower environmental and health impact

2 **Processes** with lower environmental and health impact

Dry cleaning and PERC

Have your ever owned clothes that say "dry clean only?" What is "dry cleaning" and why is it necessary for some clothes? The answer is that some fabrics such as certain wools are damaged by washing with water and soap. In the mid 1800's a French dye-maker noticed that his tablecloth looked cleaner after spilling kerosene on it. He came up with the idea of washing garments in organic solvents such as kerosene instead of water, even gasoline was used! Today, the process is called "dry cleaning" and by 1990 the U.S. was using 140 million kg of tetrachloroethylene, or "PERC" to wash clothing. Long term exposure to PERC has been linked to kidney and liver damage and other health problems. PERC is a major contaminant in ground water and many states have asked for it to be banned.

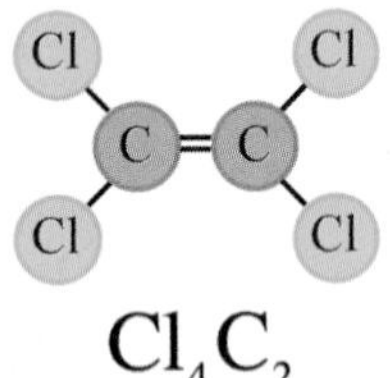

Cl_4C_2

Tetrachloroethylene
(or perchloroethylene abbr. PERC)

Liquid CO_2 is a green alternative to dry cleaning

Is there a better way? In 1997 a green chemistry researcher named Joseph DeSimone won a presidential award for developing special soaps that work with liquid carbon dioxide instead of PERC. With the special soaps, liquid CO_2 works as a cleaning fluid and does not damage either clothes or groundwater. Washing machines that use liquid CO_2 instead of PERC are a *product innovation* that reduces harmful chemical impacts on our world. You can participate in this innovation by using alternatives to dry cleaning or buying clothes that can be washed in water.

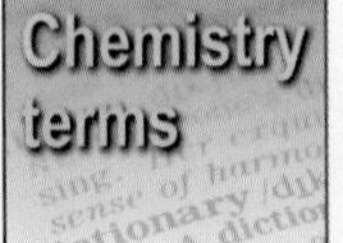

chemical engineering - The application of chemistry to solving human problems, such as creating materials, fuels, medicines, or processes.

green chemistry - The practice of chemical science in a manner that is safe, sustainable and produces minimal waste.

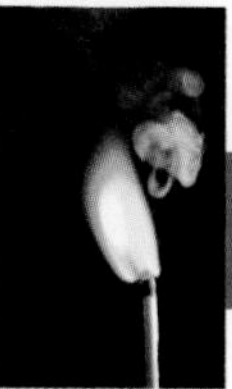

Biodegradable plastics

Plastics are widely used

In the last decade, the worldwide production of plastics totaled more than 300 million tons per year. Of this total, less than 8% were recycled. A significant part of the remainder becomes garbage every year. Plastic is predominantly hydrogen and carbon, which can be re-used by plants and animals. However, some plastic garbage lasts hundreds or even thousands of years before breaking down into substances that can be reused by the environment. How can this problem be solved?

Polymers are long chains of repeated units

Plastics are a special kind polymer. Like all polymers, plastic molecules are long chains of identical monomers. A good example is *polyethylene.* A single polyethylene molecule might include 1,000 or more repeated ethylene monomers. The chemical formula for polyethylene is C_nH_{2n} where n can be any number from a few thousand to ten thousand or more. Individual chain lengths vary from a few thousand to ten thousand or more carbon atoms! This relatively soft plastic is used in sandwich bags, milk containers, toys, and a wide range.

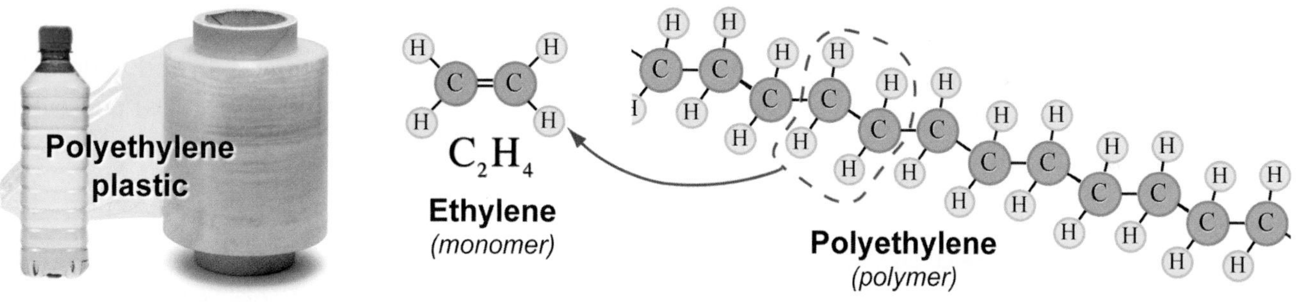

Polymers can last a long time

While polyethylene can be recycled, more than 90% ends up in landfills instead. Polyethylene takes hundreds of years in a landfill to break down into short fragments which can be digested by microorganisms.

Biodegradable polymers

Biodegradable plastics can decompose quickly in the natural environment. Microorganisms in soil extract the carbon to make biologically useful molecules such as glucose. However, a long chain of 1,000 carbon atoms cannot be "eaten" by microorganisms because the molecule is simply too large.

How biodegradable polymers work

A biodegradable polyethylene includes a special chemical called a *catalyst.* The catalyst is like a chemical "self destruct" system that slowly breaks the carbon-carbon bonds in the polyethylene backbone after weeks or months. This breaks up single molecules into fragments a few hundred carbon atoms long. These shorter chains can be digested by soil microbes. The final products of the process are carbon dioxide and water.

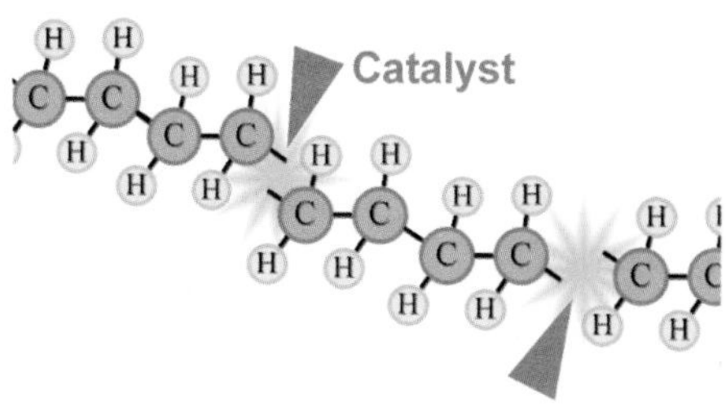

Chemical manufacturing

Chemical manufacturing

What does it mean to *manufacture* chemical products, such as plastics, fertilizers, dyes, and medicines? We cannot make the atoms or elements. Instead, chemical manufacturing typically starts with reactants that are available raw materials such as air, water, and natural minerals or hydrocarbons. A sequence of chemical reactions converts the reactants into the desired products. Unfortunately, many of the steps create waste products that are not used and must be disposed of.

A traditional manufacturing process

A good example of traditional chemical manufacturing is the LeBlanc process for making soda ash (Na_2CO_3). Soda ash is used in dying fabrics and in making glass.

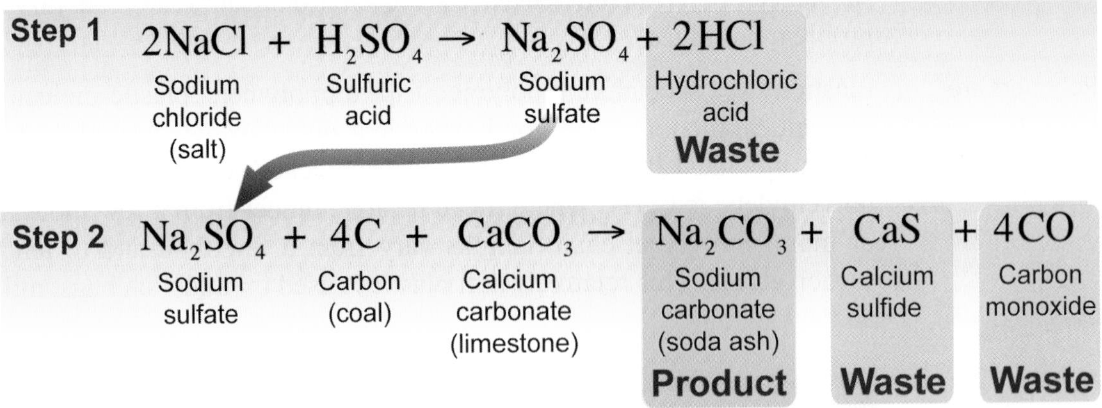

Analyzing the process

In the first step, salt and sulfuric acid react to produce sodium sulfate and hydrochloric acid. The second step combines the sodium sulfate with carbon from coal and calcium carbonate from limestone to produce sodium carbonate, the desired product. Notice that the reaction produces calcium sulfide and carbon monoxide as waste products.

Desired product		Waste					
1 mole Na_2CO_3		2 moles HCl + 1 mole CaS + 4 moles CO					
84.0 g	**+**	**73.0 g**	**+**	**72.1 g**	**+**	**112.0 g**	**= 341.1 g**
$\frac{84.0\text{ g}}{341.1\text{ g}} = 24.6\%$		$\frac{73.3\text{ g} + 72.1\text{ g} + 112.0\text{ g}}{341.1\text{ g}} = 75.4\%$					

This process generates 75% waste!

When you analyze the process, you see that it produces one mole (84 g) of soda ash for every 2 moles of hydrochloric acid (73 g), 1 mole of calcium sulfide (72.1 g) and 4 moles of carbon monoxide (112 g). Overall, only 24.6% of the mass of the products is soda ash! The rest of the "products" of the reaction are waste and were once dumped into the environment.

Reuse can raise efficiency

One way to make a chemical process more environmentally friendly is to recycle or reuse the "waste products" in other processes. Early on, chemical companies realized that the hydrochloric acid given off by the LeBlanc process was a valuable resource in itself so this "by-product" was sold for other uses. Counting HCl as another useful product raises the over efficiency of the Leblanc process to 46%.

Atom economy

The concept of atom economy

A major goal of green chemistry is to reduce waste by designing chemical manufacturing processes to be more efficient. In an ideal chemical process, all the atoms of the reactants become useful products and none are wasted. Applying the law of mass conservation to this goal provides a useful measure of how well a chemical process uses materials. This is called the *percent atom economy* and it is given by the ratio of the mass of *useful* products divided by the total mass of products. An ideal process would have 100% atom economy, meaning all the mass of the reactants was converted into useful products.

$$\text{Atom Economy} = \frac{\text{Mass of Desired Product}}{\text{Total Mass of Product}} \times 100$$

The Solvay process

The Solvay process invented by Ernest Solvay is a far more efficient way to produce soda ash. The Solvay process starts with salt (NaCl), water, ammonia (NH_3) and limestone ($CaCO_3$). There are four reactions that eventually produce 1 mole of sodium carbonate for every mole of calcium carbonate and salt used. The process also produces 1 mole of calcium chloride which is used to melt snow on roads.

The Solvay process for producing Na_2CO_3

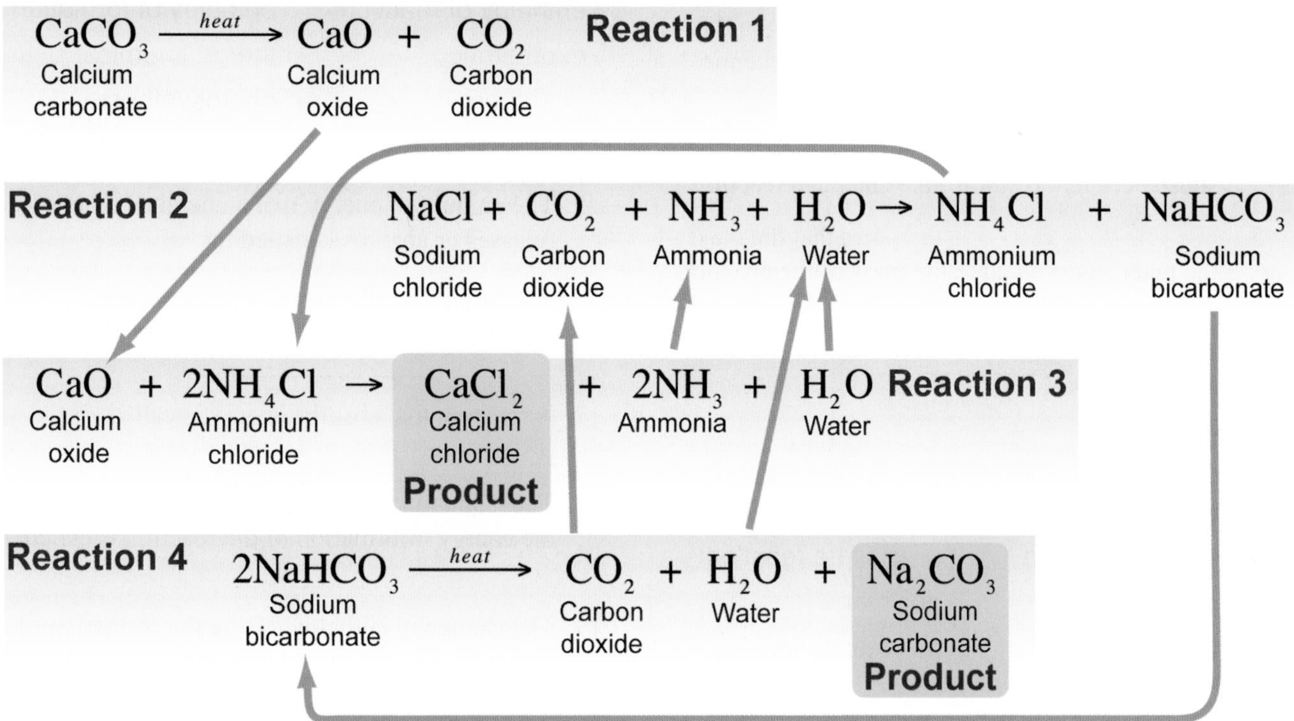

The Solvay process has an atom economy of 100%

Notice how perfectly engineered the Solvay process is! Every product feeds back into the reaction as a reactant in another step *except* the two desired end products! For example, the ammonia used in Reaction 2 is produced in Reaction 3. The carbon dioxide used in reaction 2 is produced by Reactions 1 and 4. The atom economy of the Solvay process is 100% and this has become a model for other chemical manufacturing processes.

Chapter 10 Review.

Vocabulary

Match each word to the sentence where it best fits.

Section 10.1

chemical reaction	mass conservation
reactants	balanced equation
chemical equation	unbalanced equation
products	

1. The process that converts one or more substances into new substances is called __________________.
2. The starting materials of a reaction are called ______________.
3. A __________________ is an expression that describes the changes that happen in a chemical reaction.
4. We call ____________ the materials or substances that result from a chemical reaction.
5. ____________________ states that the mass of the reactants is equal to the mass of the products.
6. A chemical equation that obeys mass conservation is called a ________________.

Section 10.3

combination reaction	displacement reaction
synthesis reaction	precipitate reaction
decomposition	polymer
	polymerization

7. A Combination reaction is also called a __________________.
8. The reaction AB + DX → AX + DB is called ________________________.
9. In a __________________ a single substance decomposes to form two or more new substances.
10. The reaction A + BD → B + AD is called single __________________________.
11. The reaction in which two substances combine to form a new substance is called __________________.
12. The reactions that form an insoluble solid when various aqueous solutions are mixed are called _________________________.
13. A molecule that is built up from many repeating units of a smaller molecule is called a __________________.
14. The reaction that assembles a polymer is called _________________________.

Section 10.4

enthalpy of reaction	enthalpy of formation
exothermic	thermochemical
endothermic	photosynthesis
energy barrier	spontaneous reaction

15. The amount of energy that a chemical reaction releases or absorbs is called the _________________________.
16. A reaction that releases energy is called ______________________.
17. A reaction that absorbs energy is called ______________________.
18. The equation that gives the chemical reaction and the energy information of the reaction is called __________________________.
19. The change in enthalpy during the formation of one mole of a substance is called ______________________.
20. The reaction that takes place in plants and combines carbon dioxide and water and gives glucose and oxygen is called ____________.
21. A chemical reaction that happens without the need for external energy input is called ______________________.

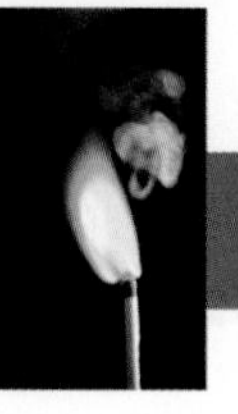

22. The energy needed to initiate a chemical reaction is called ____________________.

23. Spontaneous reactions have a very small ____________________ .

Connection

green chemistry	sustainable chemistry
atom economy	hazardous substances
chemical engineering	

24. ________________minimizes or eliminates the generation of hazardous substances.

25. ____________________are compounds that are harmful to the environment.

26. Green chemistry is also called ____________________.

27. __________________gives us a way to correctly calculate the fraction of reactants that goes into the final product.

28. ____________________ is the application of chemistry to solve human problems.

Conceptual Questions

Section 10.1

29. Describe what is a chemical reaction and give three examples.

30. What is a chemical equation? Give an example and identify all relevant parts.

31. What is the differences between a chemical equation and a chemical reaction.

32. Which of the following chemical equations conserve mass?
    a. $N_2\ (g) + H_2\ (g) \rightarrow NH_3\ (g)$
    b. $H_2\ (g) + Cl_2\ (g) \rightarrow 2HCl(g)$
    c. $2N_2H_4(g) + N_2O_4(g) \rightarrow 3N_2(g) + 4H_2O(g)$

33. Describe in your own words the various components of the language of chemistry.

34. Explain what is meant by the statement: "The grammar of the chemical language is given by the science that describes the chemical processes".

35. Describe why water is the most important chemical compound for our life and the environment.

36. Which of the following chemical equations are balanced and why?
    a. $Fe + Cl_2 \rightarrow FeCl_3$
    b. $2NaN_3(s) \rightarrow 2Na\ (s) + 3N_2\ (g)$
    c. $NaOH(g) \rightarrow Na^+\ (aq) + OH^-\ (aq)$
    d. $NO(g) + CO(g) \rightarrow N_2(g) + CO_2(g)$

37. Explain what is missing from the chemical equation that describes the burning of carbon $C + O_2 \rightarrow CO_2$.

Section 10.2

38. Describe the procedure for balancing chemical equations. What is the fundamental law that a balanced reaction must satisfy?

Section 10.3

39. Classify the following chemical reactions as: combination, decomposition, single displacement, or double displacement reaction.
    a. $Pb\ (s) + 2\ AgNO_3(aq) \rightarrow Pb(NO_3)_2(aq) + 2\ Ag(s)$
    b. $NH_4NO_3 \rightarrow N_2O + 2H_2O$
    c. $2\ NaOH(aq) + CuCl(aq) \rightarrow 2\ NaCl(aq) + Cu(OH)_2(s)$
    d. $2Fe + 3Cl_2 \rightarrow 2FeCl_3$

40. Determine which of the following compounds are soluble and which are insoluble.
    a. $CaCl_2$
    b. $CaSO_4$
    c. $PbSO_4$
    d. $Pb(NO_3)_2$
    e. CuI_2
    f. K_2SO_4

g. Na_3PO_4
h. $CuCO_3$

41. Describe the polymerization process that produces starch.

Section 10.4

42. Explain the meaning of ΔH and the significance of the positive or negative sign that it may have?
43. Explain why it is important to know the enthalpy of a chemical reaction.
44. What is the relationship between ΔH and exothermic or endothermic reactions.
45. How is the enthalpy of formation, ΔH_f, related to the reaction enthalpy, ΔH?
46. Oil and natural gas are chemical compounds assembled from carbon, oxygen and hydrogen. These compounds were formed millions of years ago and have since been stored in the earth.
    a. Was the reaction that created these compounds an exothermic or an endothermic reaction?
    b. What is the significance of that reaction in the current state of our world?

Connection

47. What are the main goals of Green Chemistry?
48. Give two examples of products that you use that are produced by green chemistry methods.
49. Why is atom economy important when we make chemical compounds?
50. What are the main goals of green chemistry?
51. Organic farming is becoming very important as people realize the harmful effects of fertilizers and pesticides. How do you think is green chemistry related to organic farming? Could you give a specific example of organic farming that affects, or may have affected, your life?

Quantitative Problems

Section 10.1

52. When sugar ($C_{12}H_{22}O_{11}$) ferments it reacts with water to form aqueous ethyl alcohol (C_2H_5OH) and carbon dioxide. Write the general chemical equation for this reaction.

Section 10.2

53. When solid sodium in dropped into liquid water it reacts and it produces aqueous sodium hydroxide ($NaOH(aq)$) and hydrogen gas (H_2). Write the balanced equation of this reaction.
54. When propane (C_3H_8) reacts with oxygen it produces carbon dioxide and water vapor.
    a. Write the chemical equation
    b. Balance this chemical equation.
55. Acid rain is a major environmental pollutant. Acid rain is water that contains aqueous sulfuric acid ($H_2SO_4\ (aq)$). Acid rain is generated when sulfur dioxide (SO_2), which mainly comes from the combustion of coal, reacts with oxygen and liquid water.
    a. Write the general chemical equation that describes this reaction.
    b. Balance the chemical equation.
56. The process of treating iron ore for the production of iron metal is given by the unbalanced equation.
    $Fe_2O_3\ (s) + CO(g) \rightarrow Fe(s) + CO_2\ (g)$
    Balance this equation
57. The artificial sweetener aspartame ($C_{14}H_{18}N_2O_5$) combusts to produce carbon dioxide, nitrogen and liquid water. Write the balanced chemical equation for this reaction.
58. The catalytic converter in the exhaust system of a car removes nitrogen oxides and carbon monoxide.
    The unbalanced equation for the reaction is:
    $NO(g) + CO(g) \rightarrow N_2(g) + O_2(g)$
    Balance this chemical equation.

59. Balance the following chemical equation:
$NH_3(g) + CO_2(g) \rightarrow CO(NH_2)_2(s) + H_2O(g)$

60. The production of iron metal from iron oxide ore can be considered to take place in two steps.
First step:
$Fe_2O_3 + CO(g) \rightarrow Fe_3O_4 + CO_2$
Second step:
$Fe_3O_4 + CO(g) \rightarrow Fe(s) + CO_2$
Write the balanced equation for each step.

61. When ultraviolet radiation interacts with ozone (O_3) in the upper atmosphere it is absorbed by the ozone molecule.* In the process ozone decomposes into molecular oxygen. What other element or atom results from this reaction? Write the balanced chemical equation of this important reaction. *(When ozone diminishes it creates the so called ozone hole in the upper atmosphere. The ultaviolet radiation can't be absorbed and the radiation reaches the earth where it is very damaging to living organisms.)

62. Balance the following chemical equations:
    a. $O_2 + PCl_2 \rightarrow POCl_3$
    b. $C_4H_{10} + O_2 \rightarrow CO_2 + H_2O$
    c. $Ca + H_2O \rightarrow Ca(OH)_2 + H_2$
    d. $SiCl_4 + H_2O \rightarrow SiO_2 + HCl$

63. Balance the chemical equations:
    a. $NO_2 + H_2O \rightarrow HNO_3 + NO$
    b. $CaCl_2 + Na_2CO_3 \rightarrow CaCO_3 + NaCl$
    c. $N_2O_5 \rightarrow NO_2 + O_2$
    d. $CH_3OH + O_2 \rightarrow CO_2 + H_2O$

Section 10.3

64. Classify each of the following reactions as a synthesis, decomposition, single-displacement, or double-displacement.
    a. $2Pb(s) + 2AgNO_3(aq) \rightarrow PbNO_3(aq) + 2Ag(s)$
    b. $2Na(s) + O_2(g) \rightarrow Na_2O_2(s)$

65. Classify each of the following reactions by its type:
    a. $2KNO_3(aq) \rightarrow 2KNO_2(aq) + O_2(g)$
    b. $NH_4Cl(s) \rightarrow NH_3(g) + HCl(g)$
    c. $Na_2O(s) + H_2O \rightarrow 2NaOH(aq)$

Section 10.4

66. From the enthalpy of formation data given in the text calculate the enthalpy of the reaction.
$4C(s) + 6H_2O(g) + O_2(g) \rightarrow 2C_2H_5OH(l)$
Calculate the enthalpy of formation of 1 mole of liquid ethanol (C_2H_5OH).

67. Let's assume that coal is made only of carbon (you know it is not really true but it is a very good approximation). When carbon burns the thermochemical reaction is:
$C(s) + O_2(g) \rightarrow CO_2(g)$ $\Delta H = -394$ kJ
If we burn 1 kg of coal, how many grams of water can we heat from 10°C to 40°C?

68. The thermochemical equation for the combustion of ethanol is
$C_2H_5OH(l) + 3O_2(g) \rightarrow 2CO_2(g) + 3H_2O(l)$
$\Delta H = -1368$ kJ.
How much energy is generated when we burn 1 kg of ethanol?

69. From the thermochemical equation
$C_3H_8 + 5O_2(g) \rightarrow 3CO_2(g) + 4H_2O(l)$
$\Delta H = -2220$kJ
and the data for the enthalpy of formation for O_2, $H_2O(l)$ and CO_2, calculate the enthalpy of formation for the compound C_3H_8.

70. The tank of a home barbeque grill contains 10 kg of propane (C_3H_8). The propane burns with the exothermic reaction
$C_3H_8(g) + O_2(g) \rightarrow 3CO_2(g) + 4H_2O(g)$
$\Delta H = -2043$ kJ.
How much energy in kJ is contained in the tank?

71. From the information given in the previous problem and from other reactions covered in this chapter of the book calculate the enthalpy of the reaction:
$3C(s) + 4H_2(g) \rightarrow C_3H_8(g)$

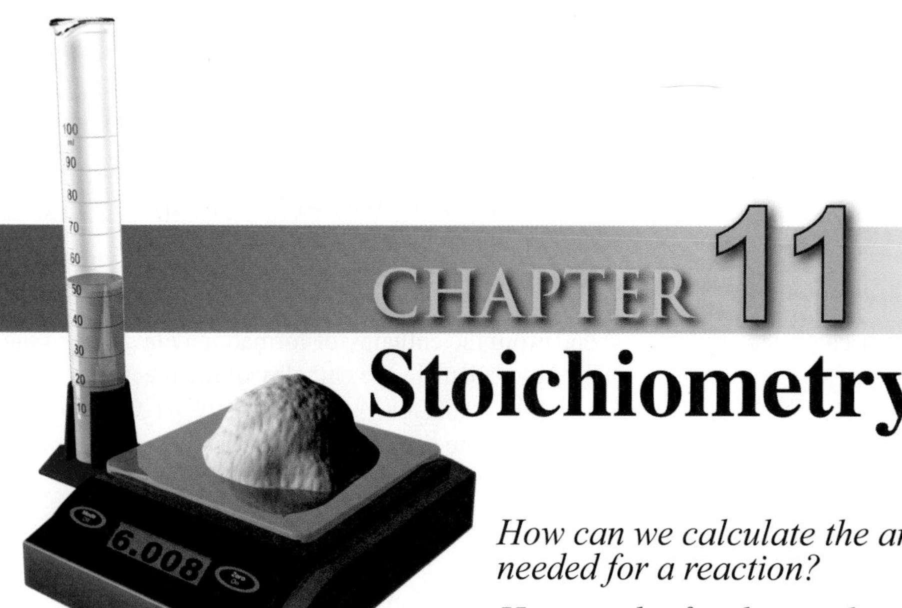

CHAPTER 11 Stoichiometry

How can we calculate the amounts of each chemical needed for a reaction?

How much of a chemical will be produced?

Why is it important to know the amounts of reactants and products?

Do chemists follow a "recipe" when mixing chemicals?

Many interesting substances such as Gortex(TM), Claratin(TM), Nutra-sweetTM), and Polar Fleece(TM) were developed initially in the chemistry lab. To make these substances a chemist needed to decide what chemicals to mix together and in what amounts. This is similar to cooking a special dish. The chef first needs to know how many people will be eating the meal and then he needs to decide how much of each ingredient is required for the correct amount of food. With chemicals and with food we want to make a close estimate of how much of each ingredient is required. Predicting the necessary amount allows us to decrease the waste of unused chemicals, or food, and it helps to keep the cost of production low.

Suppose you worked as a chef for a large restaurant. How would you figure out how to make enough food to serve all of your customers? Do you need to allow for a certain amount of waste or unused food in order to be efficient? Sometimes when you are baking, things do not always come out like you would expect, right? Are there any surprises? When you cook a particular type of food over and over, it becomes easier to flavor it and adjust the recipe to your particular taste. Chemists adjust ingredients to enhance certain chemical properties, and to increase production. In the following pages you will see how chemists are able to predict the amounts of chemicals. Industrial chemists work hard to make sure the correct amount of a chemical, such as aspirin, is produced, and that it costs the company and the environment as little as possible.

Spliting water

In this experiment we will actually "see" the relative amounts of products formed in a chemical reaction. Although we cannot "see" chemical changes taking place at the molecular level with our eyes, we can observe the end result. In this experiment, where hydrogen and oxygen gases are collected, we can *witness the relative mole amounts* as given by the balanced equation. Some molecules can be separated by using an electrical current, this process is known as electrolysis. Here we use the process of electrolysis to separate water into its elements hydrogen and oxygen.

Why are the amounts of the gases produced not the same?

This experiment you will see that as water decomposes into it's elements there are two hydrogens for every one oxygen. This makes sense when we look at the composition of the water molecule.

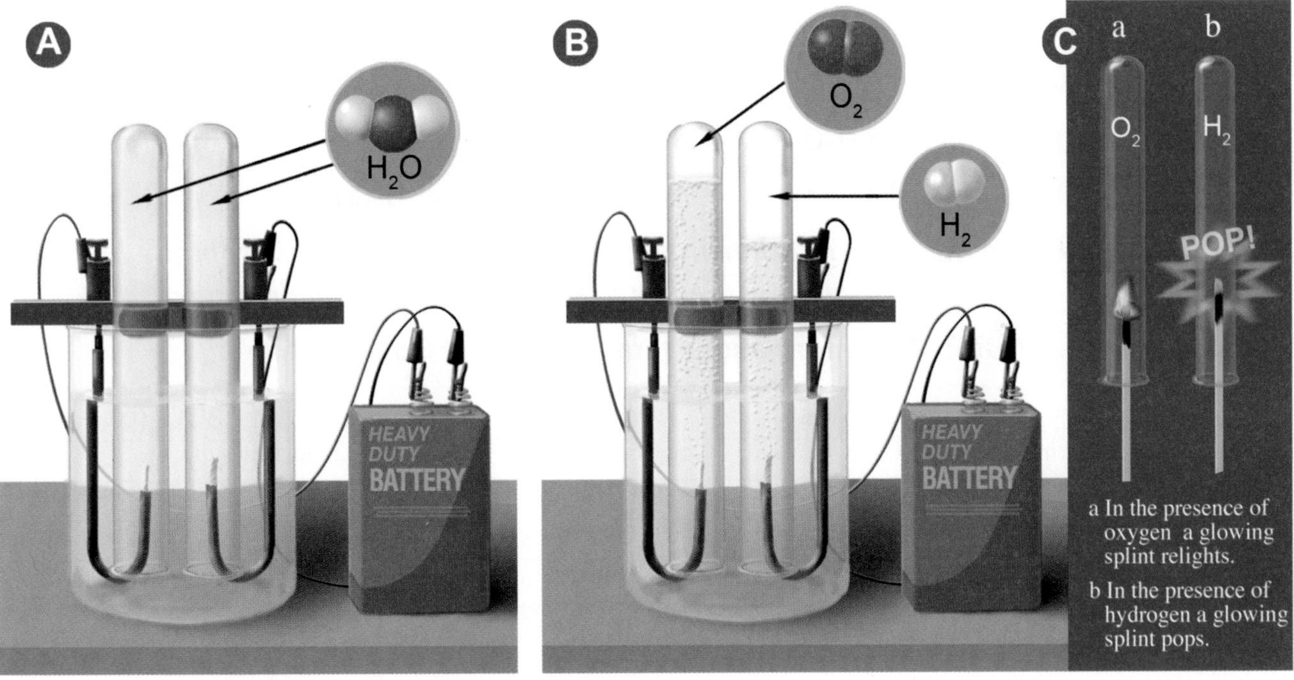

Did you notice that the volume of hydrogen produced is twice as much as the volume of oxygen?

$$2H_2O(l) \rightarrow \mathbf{2}H_2(g) + O_2(g)$$

From looking at the balanced equation we see that the products are hydrogen and oxygen gas. *The coefficient of 2 in front of the hydrogen gas (H_2) indicates that there is twice as much hydrogen produced as compared to oxygen.* Molecules are too small for us to see, but here because of the water volume in each test tube we are provided with concrete evidence that twice the hydrogen is produced for every volume of oxygen. This evidence supports the coefficients in the balanced equation.

11.1 Analyzing a Chemical Reaction

Interpreting the chemical equation

Stoichiometry is all about amounts! How much reactant is needed, and how much is used up? How much product can be formed? Stoichiometry allows us to answer these questions. To use stoichiometry we need to be able to correctly interpret a balanced chemical equation. A chemical equation tells us what substances are added together as reactants and what substances are produced as products. As you learned earlier we must always balance a chemical reaction because atoms can not be created or destroyed, just rearranged. To balance a chemical reaction we select the smallest coefficients that provide us with the same number of each type of atom on the reactants and products side of the equation. Using the reaction for water decomposing to its elements:

$$2H_2O\ (l) \rightarrow O_2(g) + 2\ H_2(g)$$

We can interpret the information from this equation in two ways.

Coefficients in a balanced equation tell us how much of each chemical

1) 2 molecules of $H_2O(l)$ yields 1 molecule of $O_2(g)$ and 2 molecules of $H_2(g)$.

2) 2 moles of $H_2O(l)$ yields 1 mole $O_2(g)$ and 2 moles $H_2(g)$

Using the mole relationship of reactants and products from the balanced equation is the key to understanding stoichiometry

TABLE 11.1. Information from a Balanced Equation

$2CO(g)$ +	$O_2(g)$ →	$2CO_2(g)$
2 molecules of CO	1 molecule of O_2	2 molecules of CO_2
2 dozen CO molecules	1 dozen O_2 molecules	2 dozen CO_2 molecules
2 moles of CO molecules	1 mole of O_2 molecules	2 moles of CO_2 molecules
$2(6.022 \times 10^{23})$CO molecules	6.022×10^{23} O_2 molecules	$2(6.022 \times 10^{23})$ CO_2 molecules

Why is the mole so important? Because it is impossible to weigh out an individual molecule or even 12 molecules! For this reason a chemist rarely carries out a reaction using individual molecules, instead the larger unit of the mole is chosen. The mole is the unit commonly used to determine the relative amounts of reactants and products. Since the mole represents a "quantity" of measure it is often referred to as the "chemists dozen". Do you remember how chemists measure out a quantity in moles? They convert the amount of moles to grams, using the molar mass of the substance. The amount in grams can then be weighed out on a balance. This very easy to do and very useful.

Chemistry terms

stoichiometry - STOY-KEE-AHM-EH-TREE - is the study of the amounts of substances involved in a chemical reaction. The amounts can be studied in moles or in mass relationships.

Practice relating moles to molecules

Explaining the meaning of a chemical equation in words often helps us to understand what it represents. Here we are going to practice interpreting balanced chemical equations, and looking at what the coefficients give us for information.

Propane is a fuel commonly used for gas grills and sometimes for heating homes. The combustion of propane is shown below:

$$C_3H_8(g) + 5O_2(g) \rightarrow 3CO_2(g) + 4H_2O(g)$$

Asked: *Explain the meaning of the equation in terms of the number of moles and molecules.*

Answer: *1 molecule of C_3H_8 reacts with 5 molecules of O_2 to yield 3 molecules of CO_2 and 4 molecules of H_2O*

Answer: *1 mole of C_3H_8 reacts with 5 moles of O_2 to yield 3 moles of CO_2 and 4 moles of H_2O*

Moles and Molecules are NOT conserved

It is important to notice that the total number of moles of reactants, 1 mole C_3H_8 plus 5 moles of O_2, do not add up to the total number of moles of products, 3 moles CO_2 plus 4 moles H_2O. Six does not equal seven! It is fine that the moles are not conserved.

Why is it OK that the total number of moles is not conserved on the reactant and products side?

Essentially because the atoms ARE conserved! The atoms were rearranged and they now make up new compounds. The atoms are the building blocks of molecules and they are still present in the same amount..

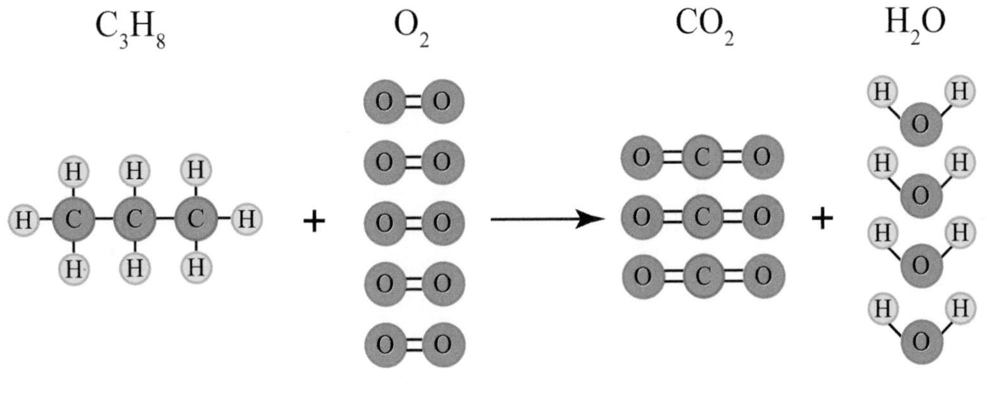

C_3H_8 O_2 CO_2 H_2O

Mole to mole relationships

Comparing amounts of ingredients

The balanced equation gives us important information about how much of each reactant is required to form products. This is helpful because we need to know the relative amounts of chemicals in order to make the "chemical recipe".

Here is an example:

Cake mix + 3 eggs + 1/3 cup oil + 1 cup water yields 1 batch cup cakes.

This information tells us what the ingredients are, how much you need of each ingredient, and how much product will be formed using the required ingredients.

Only compare chemicals in moles

When we use the balanced equation to determine amounts of chemicals we are comparing amounts in moles. This method allows us to understand how elements combine. For example:

1 mole of $CO(g)$ + 2 moles of $H_2(g)$ yields 1 mole of $CH_3OH(l)$

You can only compare the amounts of chemicals in moles!

This is because chemicals combine in simple molar proportions. When we work in the unit of the mole all substances relative amounts are equivalent.

Atoms have different masses so we can never compare amounts in grams, because grams are never equivalent for different molecules

We cannot compare in grams

For example, we can compare 12 apples to 12 oranges and we see the amounts are equivalent, but we know the apple and the orange are different fruits and have different masses. So we compare the amounts of each fruit and we can say twelve apples to is equivalent to twelve oranges. Using the above chemical example, $CO(g)$ and hydrogen gas, $H_2(g)$ are different chemicals, so their masses are different, but we can determine what amounts combine using the mole, because the amounts in moles represent the same quantity.

Mole to mole relationships

Lets consider the chemical reaction shown below:

$$CO(g) + 2H_2(g) \longrightarrow CH_3OH(l)$$

1 mole 2 moles 1 mole

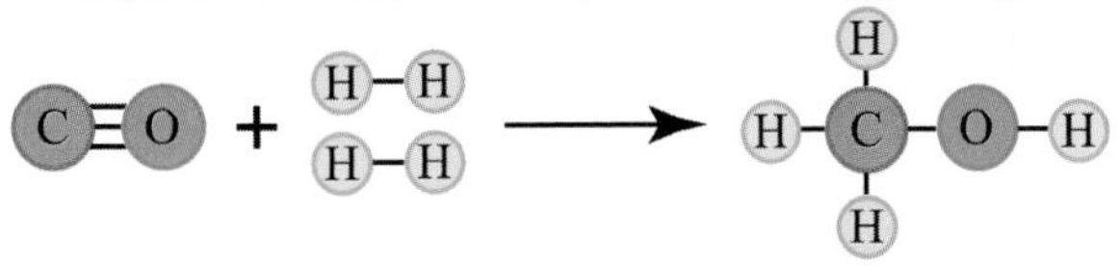

The mole amounts given by the balanced equation are called **stoichiometric equivalent** amounts. The mole amounts of each substance are proportionate to one another in a balanced equation. This means they combine in the amounts given by the coefficients of the balanced equation.

As shown above: 1 mole $CO(g)$ = 2 mole $H_2(g)$ =1 mole $CH_3OH(l)$,

Based on this equation the recipe calls for 1 mole of $CO(g)$ and 2 moles of $H_2(g)$ to make the product methanol, CH_3OH.

The only way to know how to mix chemicals together is to use the balanced chemical reaction

Equivalent amounts are also used in cooking

The concept of relative amounts are used in cooking also. For example, if you are baking cupcakes and the recipe calls for 1 package of mix, 1/4 cup of oil and 3 eggs to make one batch of cupcakes, then you could say that 1 package of mix is stoichiometrically equivalent to 1/4 cup of oil and 3 eggs. The fact that they are "equivalent" means for this particular recipe they are in a proportionate amounts. When you mix all the ingredients you get one batch. For some other recipe it is likely that the ratio of mix to oil and eggs is quite different.

Different chemical reactions require different amounts of chemicals

Proportionate amounts in chemistry is an important concept, because we need to know how to mix our ingredients and how much will be formed when we do mix them.

Chemistry terms

stoichiometric equivalent - the mole amounts of each substance in a balanced chemical equation are proportionate.

Determining mole relationships

Finding Mole Amounts

Glucose, $C_6H_{12}O_6$, is the primary sugar used by our bodies for energy. In yeast, glucose undergoes a process called fermentation. This process produces ethanol, the common form of alcohol. When making blackberry wine the sugar in the berries ferments according to the following chemical equation :

$$C_6H_{12}O_6(aq) \rightarrow 2\ C_2H_5OH(aq) + 2\ CO_2(g)$$

This equation says that 1 mole of $C_6H_{12}O_6$ yields 2 moles of C_2H_5OH plus 2 moles of CO_2. What if we have 3 moles of glucose? If 3 moles of $C_6H_{12}O_6$ decomposes how many moles of products do we obtain?

Rebalance the chemical equation

One way to approach this is to multiply the entire equation by 3.

$$3(C_6H_{12}O_6(aq) \longrightarrow 2\ C_2H_5OH(aq) + 2\ CO_2(g))$$
$$3\ C_6H_{12}O_6(aq) \longrightarrow 6\ C_2H_5OH(aq) + 6\ CO_2(g)$$

This gives us 3 moles of $C_6H_{12}O_6$ yields 6 moles of C_2H_5OH plus 6 moles of CO_2. This tells us we get six moles of ethanol and carbon dioxide as products, which answers our question.

Non-integer amounts of moles

What if we have 7.5 moles of glucose, $C_6H_{12}O_6$, that decomposes? How many moles of products will be formed then? Here we can multiply the equation by 7.5 to obtain the answer.

$$7.5(C_6H_{12}O_6(aq) \longrightarrow 2\ C_2H_5OH(aq) + 2\ CO_2(g))$$
$$7.5\ C_6H_{12}O_6(aq) \longrightarrow 15\ C_2H_5OH(aq) + 15\ CO_2(g)$$

This gives us 7.5 moles of $C_6H_{12}O_6$ yields 15 moles of C_2H_5OH plus 15 moles of CO_2.

This process of rebalancing the chemical equation to obtain the number of moles always works, but it can get tedious. If we want to know the amount of only one product there is no need to calculate each of them. In the next section we will work with a more convenient method that uses mole ratios. This method allows us to calculate the amount of one substance relative to another. These ratios are used as conversion factors that help us to determine the amounts of chemicals needed as ingredients, or produced as product.

The mole ratio

Finding the mole ratio

The **mole ratio** comes from the coefficients in front of the reactants and products in the balanced chemical equation. Given the following reaction:

$$CO(g) + 2H_2(g) \longrightarrow CH_3OH(l)$$

1 mole 2 moles 1 mole

We can set up mole ratio's that allow us to compare between reactants, or between reactants and products. Comparing the reactants we see that 1 mole CO is proportionate to 2 mole H_2 and we can show this as a ratio:

$$\frac{1 \text{ mole CO}}{2 \text{ moles } H_2} \approx \frac{2 \text{ moles } H_2}{1 \text{ mole CO}} \quad \text{Just Like} \quad \frac{1 \text{ dozen}}{12 \text{ eggs}} \approx \frac{12 \text{ eggs}}{1 \text{ dozen}}$$

mole ratio of reactants

In these ratios the quantities are the same but the comparison is reversed

Suppose you want to know *how much hydrogen gas, H_2* you need to react with 4 moles of carbon monoxide, CO? To answer this question we can multiply by the mole ratio. We select the ratio that allows the proper units to cancel, and leaves the desired unit in the numerator (top).

$$\cancel{4 \text{ moles CO}} \times \frac{2 \text{ moles } H_2}{\cancel{1 \text{ mole CO}}} = 8 \text{ moles of } H_2 \text{ are needed}$$

(Top)

If we want to know how much CO will react with 3.5 mole of H_2, we select the ratio with CO in the numerator (or top).

$$\cancel{3.5 \text{ moles } H_2} \times \frac{1 \text{ mole CO}}{\cancel{2 \text{ moles } H_2}} = 1.75 \text{ moles of CO are needed}$$

(Top)

The important point is that the mole ratio is consistent. No matter what the starting amounts are these chemicals will always combine in the same ratio. We can use this ratio to determine how much of each chemical is consumed or produced.

Chemistry terms

mole ratio - A ratio comparison between substances in a balanced equation. The ratio is obtained from the coefficients in the balanced equation. The ratio allows for the conversion of one substance to another substance by using molar equivalent amounts.

Using the mole ratio

Comparing reactant to product

Not only can we compare reactants to reactants, we can compare reactants to products. For instance using the same balanced equation: $CO(g) + 2H_2(g) \rightarrow CH_3OH(l)$, the ratio between CO(g) and $CH_3OH(l)$ is one to one, but the ratio of H_2 to $CH_3OH(l)$ is two to one.

Comparing the reactant carbon monoxide, CO to the product methanol, CH_3OH we get:

$$\frac{1 \text{ mole CO}}{1 \text{ mole } CH_3OH} \approx \frac{1 \text{ mole } CH_3OH}{1 \text{ mole CO}}$$

Starting with product to find reactant

Another question might be: if the reaction produces 5 moles of methanol, CH_3OH, how many moles of H_2 were consumed? Here we start with the amount of product and work backward to the reactant.

To select the correct mole ratio we look at the numerator and select the one with H_2 in the top, because we are being asked how many moles of H_2.

$$5 \text{ moles } CH_3OH \times \frac{2 \text{ moles } H_2}{1 \text{ mole } CH_3OH} = 10 \text{ moles of } H_2 \text{ are consumed}$$

Mole ratios are used as conversion factors that allow us to convert from one substance to another in a balanced chemical reaction . *The important part is to select the ratio that allows the proper units to cancel.*

Lets try this again to practice selecting the correct mole ratio.

A mixture of aluminum metal and chlorine gas reacts to form the compound aluminum chloride. How many moles of aluminum chloride ($AlCl_3$) will form when you react 5 moles of chlorine gas with excess aluminum metal?

$$2Al(s) + 3Cl_2(g) \rightarrow 2AlCl_3(s)$$

Asked: *How many moles of $AlCl_3$ are produced?*

Given: *5 moles of Cl_2 are reacting*

Relationships: *Mole ratio: 3 moles Cl_2 = 2 moles $AlCl_3$*

Solve: *Set up the mole ratios as proportions and select the one with $AlCl_3$ on the top because that is what you are solving for.*

Solve: $5 \cancel{\text{ moles } Cl_2} \times \frac{2 \text{ moles } AlCl_3}{3 \cancel{\text{ moles } Cl_2}} = 3.3 \text{ moles } AlCl_3$

Answer: *3.3 moles of $AlCl_3$ are produced.*

Practicing with mole ratios

Notice the molar coefficients in the balanced equation below :

$$CH_4(g) + 2\,O_2(g) \longrightarrow CO_2(g) + 2\,H_2O(l)$$

Practice example

Using the steps outlined below, we will determine how many moles of oxygen, O_2(g) are required to react with 4 moles of methane, CH_4(g).

Steps for Determining the Amount of Moles

Step 1	Step 2	Step 3
Determine which mole ratio between O_2 and CH_4 to use	Multiply by the appropriate mole ratio	Clearly state your answer
$\frac{1\text{mole } CH_4}{2\text{moles } O_2} = \frac{2\text{ moles } O_2}{1\text{ mole } CH_4}$	$4\,\cancel{\text{moles } CH_4} \times \frac{2\text{ moles } O_2}{\cancel{1\text{mole } CH_4}}$ $= 8\text{ moles } O_2$	It requires 8 moles of O_2 to react with 4 moles of CH_4

Our second example involves ammonia, which is the principal nitrogen fertilizer. It is prepared by the reaction between hydrogen and nitrogen.

$$3\,H_2(g) + N_2(g) \longrightarrow 2\,NH_3(g)$$

Practice example

How many moles of ammonia, NH_3(g), can be prepared using 6.5 moles of H_2(g). Assume that there is excess nitrogen.

Steps for Determining the Amount of Moles

Step 1	Step 2	Step 3
Determine which mole ratio between H_2 and NH_3 to use	Multiply by the appropriate mole ratio	Clearly state your answer
$\frac{2\text{moles } NH_3}{3\text{moles } H_2} = \frac{3\text{ moles } H_2}{2\text{ moles } NH_3}$	$6.5\,\cancel{\text{moles } H_2} \times \frac{2\text{ moles } NH_3}{3\,\cancel{\text{moles } H_2}}$ $= 4.33\text{ moles } NH_3$	4.33 moles of NH_3 is formed from 6.5 moles of H_2

Gram to gram conversions

Our balances measure in grams

We have no scale that measures the mole! Students, chemists and people that work in the laboratory measure in grams, but substances combine in mole proportions. So in a very practical sense we must convert grams to moles so we can see how the chemicals will combine, we then convert back to grams, so that we can easily measure the amount. In the flow chart below, the mass of substance A represents the grams of reactant or product, that you are starting with, and substance D represents the amount of reactant used up or product formed.

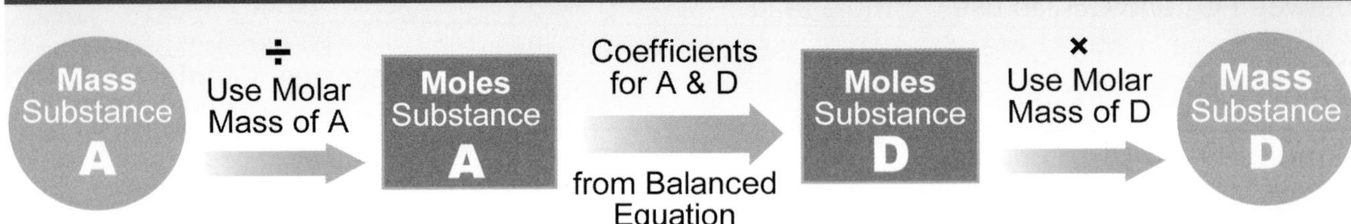

You can't get there from here!

This flow chart shows us that you CANNOT get from grams of substance A directly to grams of substance D!

You must go through the unit of the MOLE

Steps for Using the Flow Chart to calculate grams from grams given:

1. First convert to moles of substance A by dividing by the molar mass of substance A (Remember this is found by using the periodic table).
2. Use the mole ratio from the balanced equation to convert from moles of substance A to moles of Substance D.
3. Convert moles of substance D back to grams by multiplying by the molar mass of substance D (From the periodic table).

Why is it important to know how much product is formed?

Chemists need to know how much product will be formed by a particular reaction for many reasons. Sometimes the products formed are dangerous and not planing ahead could be harmful to the people working and to the environment. For industrial production, companies have to produce enough of a chemical to fill an order.

Gram to gram Calculations

To understand how the flow chart steps work we will consider the decomposition of calcium carbonate, $CaCO_3(s)$, which is also known as limestone.

$$CaCO_3(s) \rightarrow CaO(s) + CO_2(g)$$

Calcium carbonate is present in rocks found in many locations around the world. It is also found in nature as the primary component of egg shells and most sea shells. Calcium carbonate decomposes, or breaks apart when it is heated. How much carbon dioxide will be released when $CaCO_3(s)$ is heated? Lets find out.

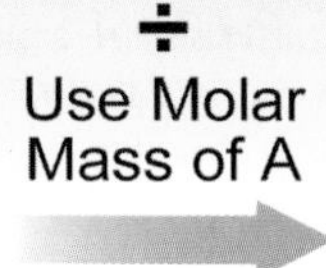

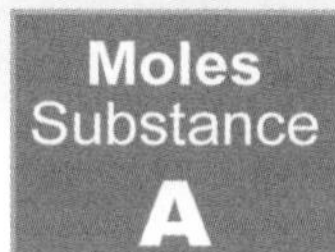

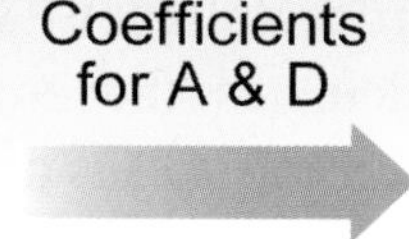

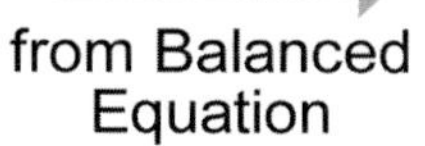

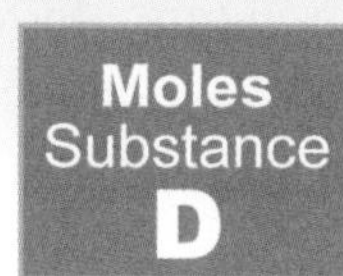

If 45.0 grams of calcium carbonate, $CaCO_3$ decomposes, how many grams of carbon dioxide, $CO_2(g)$ are produced? $CaCO_3(s) \rightarrow CaO(s) + CO_2(g)$

Asked: *How many grams of CO_2 are produced?*

Given: *45.0 g of $CaCO_3$ reacting*

Relationships: *Molar mass of $CaCO_3 = 40.078 + 12.011 + (15.999 \times 3)$*

$= 100.09 \frac{\text{g}}{\text{mole}}$ *Mole ratio : 1 mole $CaCO_3$ = 1mole CO_2*

Molar mass of $CO_2 = 12.011 + (15.999 \times 2) = 44.01 \frac{\text{g}}{\text{mole}}$

Solve: *Refer to flow chart and follow the steps (1-3).*

Step 1) $45 \text{ g } CaCO_3 \times \frac{1\text{mole } CaCO_3}{100.09 \text{ g } CaCO_3} = 0.450 \text{ moles } CaCO_3$

Step 2) $0.450 \text{ moles } CaCO_3 \times \frac{1 \text{ mole } CO_2}{1 \text{ mole } CaCO_3} = 0.450 \text{ mole } CO_2$

Step 3) $0.450 \text{ mole } CO_2 \times \frac{44.01 \text{ g } CO_2}{1 \text{ mole } CO_2} = 19.8 \text{ g } CO_2$

Answer: *19.8 grams of CO_2 are produced.*

Gram to gram calculations

Astronauts use lithium hydroxide canisters to absorb exhaled carbon dioxide during their space missions. These canisters are critical in keeping carbon dioxide at safe levels for the astronauts to breath.

During the Apollo 13 space mission that began on April 11, 1970 there was an electrical explosion that caused a decrease in the oxygen supply. Astronauts had to move to a different cabin where the LiOH canisters were fresh and could support the crew members. LiOH canisters are changed roughly every twelve hours. In order to know when to replace them someone calculates the amount of CO_2 they can absorb. Lets see how this is done.

Photo courtesy of NASA images.

Astronaut Christopher Ferguson changes a lithium hydroxide canister in Space Shuttle Atlantis.

Solid lithium hydroxide is used in space vehicles to remove exhaled carbon dioxide from the air inside the cabin. Lithium hydroxide reacts with the carbon dioxide to form solid lithium carbonate and water. How many grams of CO_2 can be absorbed by 100.0g of LiOH(s)?

$$2LiOH(s) + CO_2(g) \rightarrow Li_2CO_3(s) + H_2O(l)$$

Asked: *How many grams of CO_2 will be absorbed?*

Given: *100.0 g of LiOH reacting*

Relationships: *Molar mass of LiOH = 6.941 + 15.999+ 1.0079 = 23.95* $\frac{g}{mole}$

Mole ratio : 2 moles LiOH = 1 mole CO_2

Molar mass of CO_2 = 12.011 + (15.999×2)= 44.01 $\frac{g}{mole}$

Solve:

$$100.00\ g\ \cancel{LiOH} \times \frac{1\ mole\ LiOH}{23.95\ \cancel{g\ LiOH}} = 4.18\ moles\ LiOH$$

$$4.18\ \cancel{moles\ LiOH} \times \frac{1\ mole\ CO_2}{2\ \cancel{mole\ LiOH}} = 2.09\ moles\ CO_2$$

$$2.09\ \cancel{mole\ CO_2} \times \frac{44.01\ g\ CO_2}{1\ \cancel{mole\ CO_2}} = 91.98\ g\ CO_2$$

Answer: *91.98 grams of CO_2 are absorbed*

11.2 Percent Yield and Concentration

What is percent yield?

When you pop a bag of popcorn you expect the bag to fill up as the corn expands, but

amount you get to eat!

$$\text{Percent yield} = \frac{\text{amount of corn popped}}{\text{amount of kernels in the bag}} \times 100$$

you know that there will be un-popped kernels. In other words, we know that potentially all of the kernels could pop or open up, but we know from experience that this is not what actually happens. Although the popping of popcorn is not a chemical reaction it is a good analogy for the **percentage yield** of a chemical reaction.

Calculation

$$\frac{82}{100} \times 100 = 82\%$$

Often with chemical reactions you do not obtain the exact amount of product you predicted with your stoichiometric calculation. This is not because you did the calculation wrong! It is simply because in real life things are often not perfect, for a variety of reasons. In the laboratory when we perform an experiment, it is rare that we obtain 100% of the desired product(s).

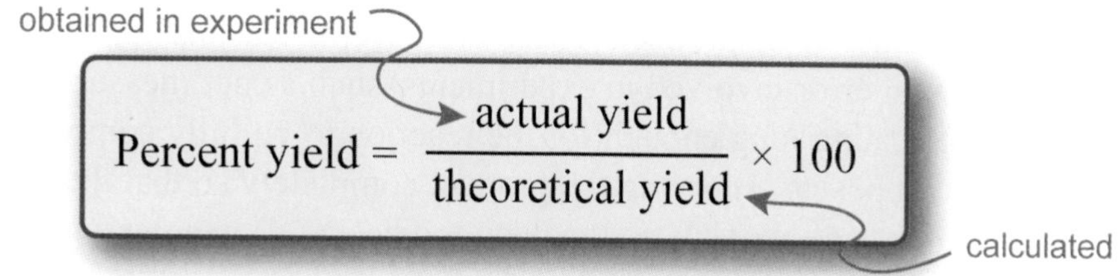

$$\text{Percent yield} = \frac{\text{actual yield}}{\text{theoretical yield}} \times 100$$

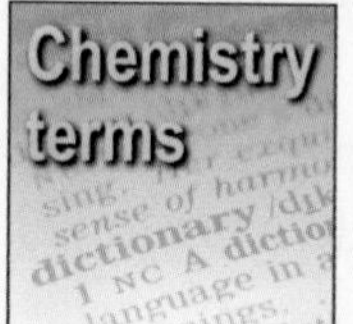

percent yield - the ratio of the amount of product actually obtained by experiment (actual yield) as compared to the amount of product calculated theoretically (theoretical yield) multiplied by 100.

actual yield - the amount obtained in lab by actual experiment.

theoretical yield - the calculated amount produced if everything reacts perfectly.

Percent yield

In the Lab

As an example of how percent yield can be obtained in the laboratory lets investigate a common reaction. When baking soda is heated it decomposes according to the following equation: $2NaHCO_3(s) \rightarrow Na_2CO_3(s) + H_2O(l) + CO_2(g)$. We can weigh the baking soda before heating and weigh the solid product Na_2CO_3 after heating. Stoichiometry allows us to calculate the amount of product we should obtain from our initial mass of baking soda, $NaHCO_3$. We can then compare our actual mass of the solid product Na_2CO_3 with the calculated mass and see how successful we were with our experiment.

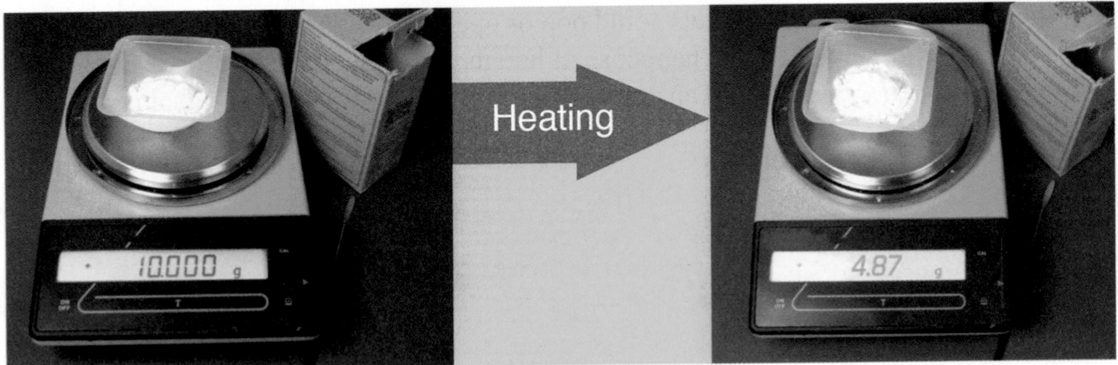

TABLE 11.2. Data $2NaHCO_3(s) \rightarrow Na_2CO_3(s) + H_2O(l) + CO_2(g)$

Initial Mass of $NaHCO_3$	Final Mass of Na_2CO_3
10.0 g	4.87 g

Question: "Can you think of reasons why the final mass of the remaining solid, Na_2CO_3 may not be accurate?

Reaction yields are often not 100%

When we mass the solid at the end of the experiment we generally obtain less than the full amount of product. There are several reasons for this: 1) generally there is some human error involved in experiments–such as not measuring exact amounts carefully; 2) not heating long enough for the reaction to be fully complete, so not all of the $NaHCO_3$ decomposes; 3) drying the product completely so that the water liquid does not increase the mass of Na_2CO_3 and make it too heavy; 4) weighing the product before it is cool.

Chemicals do not always fully react to form product due to impurities in the samples, small surface area, or insufficient reaction time. Chemical reactions can also produce side reactions that can cause the desired product to turn into something else, less desirable. Chemists are often faced with the challenge of "stopping" a chemical reaction so that it ends with the desired product. Lastly, some product is often lost during the isolation and purification steps.

Calculating percent yield

Applying the formula for percent yield

Lets use the data obtained from heating 10.0 grams of baking soda (on the previous page) to calculate the percentage yield.

First we need to calculate the theoretical yield, using our Flow Chart According the balanced equation:

$$2NaHCO_3(s) \rightarrow Na_2CO_3(s) + H_2O(l) + CO_2(g)$$

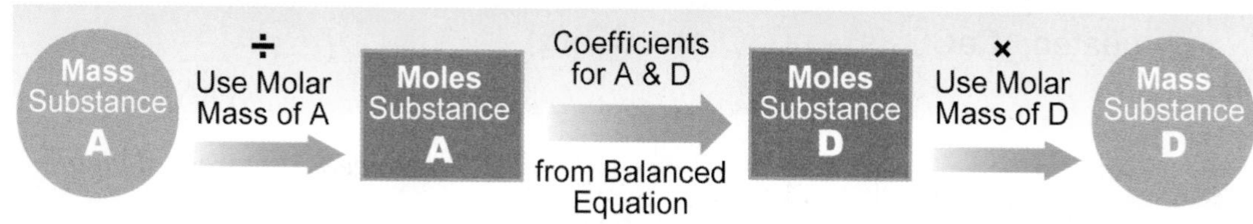

Step 1: Convert 10.0 g of $NaHCO_3$ to moles.

Molar mass of $NaHCO_3$ = 22.99 + 1.0079 + 12.011 + (15.999 x 3) = 84.01g/mole

$$10.0\ g\ \cancel{NaHCO_3} \times \frac{1\ mole\ NaHCO_3}{84.01\ \cancel{g\ NaHCO_3}} = 0.119\ moles\ NaHCO_3$$

Step 2: Convert moles $NaHCO_3$ to moles of Na_2CO_3

$$0.119\ \cancel{moles\ NaHCO_3} \times \frac{Na_2CO_3}{2\ \cancel{mole\ NaHCO_3}} = 0.0595\ moles$$

Theoretical yield is the calculated yield

Step 3: Convert moles Na_2CO_3 to grams.

Molar mass of Na_2CO_3 = (22.99 x 2) + 12.01 + (15.999 x 3) = 105.99 g/mole

$$0.0595\ \cancel{moles\ Na_2CO_3} \times \frac{105.99\ g\ Na_2CO_3}{1\ \cancel{mole\ Na_2CO_3}} = 6.31\ g\ Na_2CO_3$$

We just determined our theoretical yield to be 6.31 g Na_2CO_3. From the experiment a mass of 4.87 g was obtained. Now we can calculate the percentage yield.

Actual yield is obtained by experiment

$$\text{Percent Yield} = \frac{4.87\ g\ (\text{actual yield})}{6.31\ g\ (\text{theoretical yield})} \times 100 = 77.2\ \%$$

Practice calculating percent yield

Another example

To calculate the percent yield you need to know the theoretical yield and the actual yield. Recall that the theoretical yield is calculated based on the amounts of reactants used and the mole relationship obtained from the balanced equation. The actual yield can only be obtained by experiment, so it will either be given to you in the problem or you determine it in the lab.

1.Calcium carbonate, which is often found in seashells, decomposes when heated. $CaCO_3(s) \rightarrow CaO(s) + CO_2(g)$

If 30.5 g of $CaCO_3$ is heated, calculate the theoretical yield of CaO produced.

Asked: *Calculate the amount of CaO produced. (theoretical yield)*

Given: *30.5 g of $CaCO_3$ reacts*

Relationships: *mole ratio = 1 mole $CaCO_3$ = 1 mole CaO*
molar mass of $CaCO_3$ = 40.078 + 12.011+ (15.999 × 3) =100.09 g/mole. Molar mass of CaO=40.078+15.999=56.08 g/mole

Solve:

$$30.5\ \cancel{\text{g CaCO}_3} \times \frac{1\ \text{mole CaCO}_3}{100.09\ \cancel{\text{g CaCO}_3}} = 0.305\ \text{mole CaCO}_3$$

$$0.305\ \cancel{\text{mole CaCO}_3} \times \frac{1\ \text{mole CaO}}{1\ \cancel{\text{mole CaCO}_3}} = 0.305\ \text{mole CaO}$$

$$0.305\ \cancel{\text{mole CaO}} \times \frac{56.08\ \text{g CaO}}{1\ \cancel{\text{mole CaO}}} = 17.10\ \text{g CaO}$$

Answer: *The theoretical yield is 17.10 g of CaO*

Now we use the theoretical yield to calculate the percentage yield.

2. Calculate the percentage yield if 12.5 g of CaO is actually obtained.

Asked: *Calculate the percentage yield of CaO.*

Given: *The actual yield of 12.5 g CaO and the answer in part 1.*

Relationships: *Formula for percent yield*

Solve:

$$\text{Percent yield} = \frac{12.5\ \text{g (actual)}}{17.10\ \text{g (theoretical)}} \times 100 = 73.1\%$$

Answer: *The percent yield is 73.1%*

Discussion: *The percent yield tells the experimentalist how good their isolation techniques were.*

Stoichiometry with solutions

Calculating amounts from reactions in solution

Chemical reactions often take place between solids and liquids and between two liquids. Here we will look at a solid zinc metal reacting with hydrochloric acid.

$$Zn(s) + 2HCl(aq) \rightarrow H_2(g) + ZnCl_2(aq)$$

To be able to calculate the amount of a substance that reacts in an aqueous solution, it is helpful to review our units of concentration. Recall that our concentration unit is molarity, M, which tells us the number of moles in one liter of solution, M = mole/L.

A sample of Zinc metal (Zn(s)) reacts with 50.0 mL of a 3.0 M solution of hydrochloric acid (HCl). How many grams of hydrogen gas ($H_2(g)$) will be produced? Assume zinc metal is in excess.

$$Zn(s) + 2HCl(aq) \rightarrow ZnCl_2(s) + H_2(g)$$

Asked: *How many grams of H_2 will be produced?*

Given: *50.0 mL of 3.0 M of HCl reacting with excess zinc*

Relationships: $M = \frac{\text{mole}}{\text{L}}$, *so mole HCl = 3.0 M × 0.050 L = 0.150 moles HCl*

Excess Zn means more than enough, so the exact amount is not important.

Mole Ratio: 2 moles HCl = 1 mole H_2

Molar mass of H_2 = 1.0079 × 2= 2.02 $\frac{\text{g}}{\text{mole}}$

Solve:

$$0.150 \cancel{\text{mole HCl}} \times \frac{1 \text{ mole } H_2}{2 \cancel{\text{mole HCl}}} = 0.075 \text{ mole } H_2$$

$$0.075 \cancel{\text{mole } H_2} \times \frac{2.02 \text{ g } H_2}{1 \cancel{\text{mole } H_2}} = 0.15 \text{ g } H_2$$

Answer: *0.15 grams of H_2 are produced.*

As you can see above you must always find moles of the reactant. If the substance is a solid you calculate moles by dividing by the molar mass. If the substance is dissolved in an aqueous solution, you must find moles using your molarity equation. In this case molarity (M) x volume (L) will give you the amount of moles. So when determining moles you need to be aware of the physical states of the chemicals you are using.

Stoichiometry with solutions

Mass percentage

One of the most common ways to express the concentration of a solution is to use percent by mass of the compound in solution. For instance vinegar contains about 5% by mass acetic acid. Acetic acid comes from apples and is responsible for the sour taste of vinegar. So if vinegar is 5% acetic acid by mass that means it contains about 5 grams of acetic acid for every 100 g of solution. You can use the formula:

$$\text{Mass \% of compound} = \frac{\text{mass of compound}}{\text{total mass of solution}} \times 100$$

A solution is made by dissolving 15.6 g of glucose ($C_6H_{12}O_6$) in 0.100 kg of water. Calculate the mass percent of glucose in the solution.

Asked: *Determine the mass percent of the solute glucose,* $C_6H_{12}O_6$.

Given: *15.6 g of* $C_6H_{12}O_6$ *in 0.100 kg of water*

Relationships: *Mass% = (mass of glucose / mass of solution) x 100*

Solve: $\text{Mass \% } C_6H_{12}O_6 = \frac{15.6\text{ g } C_6H_{12}O_6}{(15.6\text{ g} + 100\text{ g})} \times 100 = 13.5\ \%$

Answer: *13.5 % of glucose,* $C_6H_{12}O_6$

Example where you determine the amount of solute.

Commercial vinegar is reported to be 5% acetic acid by mass. How many grams of acetic acid are in 120 mL of commercial vinegar? (assume the density of vinegar is the same as pure water = 1.0 g/mL.

Asked: *How many grams of acetic acid,* $C_2H_4O_2$ *are in 120 mL of vinegar?*

Given: *120 ml of vinegar and 5% acetic acid by mass*

Relationships: *120 ml = 120 g given a density of 1.0 g/mL*
Mass% = (mass of glucose / mass of solution) x 100

Solve: *5 %* $C_2H_4O_2$ */ 100 = 0.05*
0.05 × 120 g of vinegar = 6.0 g of acetic acid, $C_2H_4O_2$ *are in 120 mL of vinegar*

Answer: *6.0 g of acetic acid,* $C_2H_4O_2$

11.3 Limiting Reactants

What is a "limiting" reactant?

When carrying out reactions in the laboratory it is common to have one reactant that is used up completely to make products and the other has some "left over". The reactant that is used up completely is called the **limiting reactant**. It's name is appropriate because it "limits" the amount of product that can be formed. The concept is simple, when you run out of an ingredient (or reactant) you can no longer continue to make product. The reactant that is left over is called your **excess reactant**.

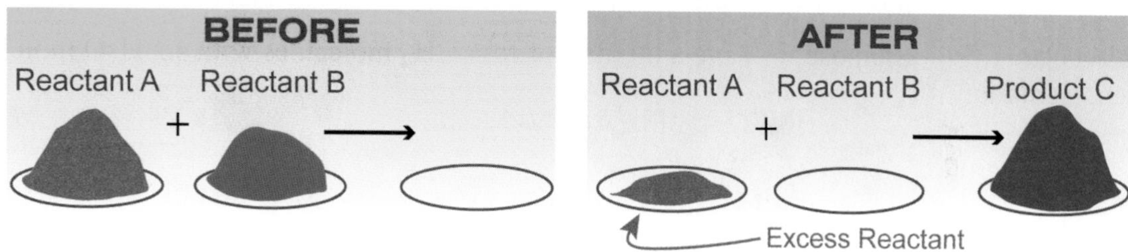

How do you recognize a limiting reactant?

Lets use an analogy of making sandwiches as an example of a limiting reactant. Suppose you need to make three ham and cheese sandwiches.

First we need our recipe! For each sandwich you will need:

2 slices of bread + 2 slices of ham + 1 slice of cheese = 1 ham and cheese sandwich

For three sandwiches you need 3 times that amount. So you go to the refrigerator and the cupboard, and you find a loaf of bread, a package of ham and some cheese. You begin to make sandwiches and discover there are only 2 slices of cheese left. In this case the cheese is your limiting reactant and you can only make 2 sandwiches.

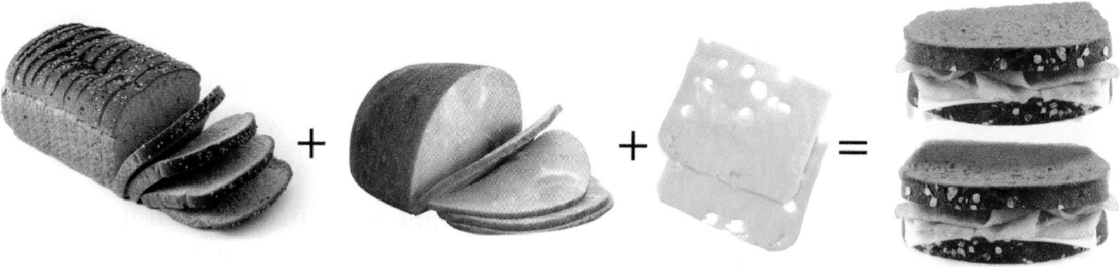

The chemical present in the least amount limits a chemical reaction in the same way.

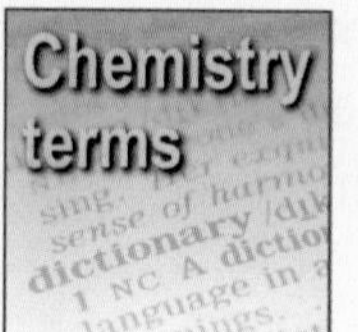

limiting reactant - the reactant present in the least amount. The reactant that "runs out" first.

excess reactant - the reactant that is remaining after the reaction is complete. The reactant that is not completely consumed.

Limiting reactants and chemical equations

Chemical example

For a chemical example, we will use the following equation:

$$CH_4(g) + H_2O(l) \rightarrow 3\ H_2(g) + CO(g)$$

Reacting methane, CH_4 with water, H_2O produces hydrogen gas that is then used to make fertilizer, carbon monoxide is also formed as a product of the reaction. From looking at the balanced equation we can see that if we react 1 mole of methane and 1 mole of water then we will (in theory) use up each reactant completely.

Molecular Representation

Suppose we have a mixture of four CH_4 molecules with six H_2O molecules:.

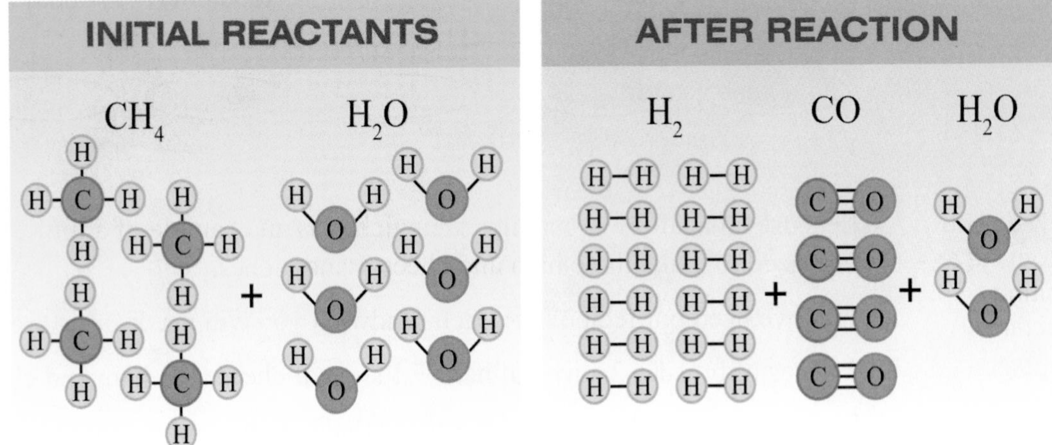

When looking at the reactants can you tell which one is limiting?

You can if you compare the molecules shown in the picture with the mole ratio from the balanced equation! When this is done we find that methane, CH_4 is limiting because there are only four molecules of methane. The reactants are in a 1 to 1 mole ratio so equal amounts are needed and there is more water! Notice in the final picture there are **NO** CH_4 molecules remaining.

Can you tell which reactant is in excess?

Can you tell how much excess reactant is left over? Again it's a one to one relationship and there are 2 molecules of water still remaining with the products after the reaction.

How many molecules of $H_2(g)$ form? You can find this out by counting.

Look at the mole ratio, between the limiting reactant, CH_4 and the products H_2 and CO. Does it make sense that 12 H_2's and 4 CO's are formed? If you are not sure compare these amounts carefully to the balanced equation above, and you will see that 4 CH_4's indeed yields 12 H_2's and 4 CO's. Remember you must use your limiting reactant to determine molecules of product.

Determining the limiting reactant

When we are given a problem in which both reactants are available in specified amounts, we must first determine which one is limiting in order to calculate the amount of product that can be formed. For an example lets use the reaction below :

$$Fe_2O_3(s) + 2\ Al(s) \longrightarrow 2\ Fe(s) + Al_2O_3(s)$$

Here we will use a strategy of comparing what we HAVE (the amounts given) to what we NEED (the amounts required to fully react each amount of reactant. What we have is analogous to what is in the cupboard available for us to use. What we need is what we "wish" we had available in the cupboard.

Never assume the smaller mass is limiting!

If we react 150 g of Fe_2O_3 and 60 g of Al, which one will be used up first?

Step 1: Convert our masses to moles

Molar mass of Fe_2O_3 = 159.7 g ; Molar mass of Al = 26.98 g

$$150\ \cancel{g\ Fe_2O_3} \times \frac{1\ \text{mole}\ Fe_2O_3}{159.7\ \cancel{g\ Fe_2O_3}} = 0.94\ \text{moles}\ Fe_2O_3\ \text{(HAVE)}$$

$$60.0\ \cancel{g\ Al} \times \frac{1\ \text{mole Al}}{26.98\ \cancel{g\ Al}} = 2.22\ \text{moles Al (HAVE)}$$

It is important to notice that the smaller mass gave us a larger amount of moles. We can never compare amounts in grams!

Step 2: Use mole ratio's to see which reactant will limit the reaction.

Finding the limiting reactant

$$0.94\ \cancel{\text{moles}\ Fe_2O_3} \times \frac{2\ \text{moles Al}}{1\ \cancel{\text{mole}\ Fe_2O_3}} = 1.88\ \text{moles Al (NEED) to react with all}\ Fe_2O_3$$

$$2.22\ \cancel{\text{moles Al}} \times \frac{1\ \text{mole}\ Fe_2O_3}{2\ \cancel{\text{mol Al}}} = 1.11\ \text{mole}\ Fe_2O_3\text{(NEED) to react with all the Al}$$

From these calculations we can see that Fe_2O_3 is our limiting reactant because we need more of it than we have available. We also see that Al is in excess.

Using the limiting reactant

Typically we want to know how much product can be formed in a particular reaction. We carry out reactions for the purpose of making a desired product. So to take this calculation a bit further lets determine how much iron, Fe(s) can be produced. To do this we add two final steps.

We must use the limiting reactant to determine the amount produced

Step 3: Use the limiting reactant to calculate the amount of product formed. Use the mole ratio between the limiting reactant and the desired product (here Fe).

$$0.94\ \cancel{\text{moles } Fe_2O_3}\text{(Have)} \times \frac{2 \text{ mole Fe}}{1\ \cancel{\text{mole } Fe_2O_3}} = 1.88 \text{ mole Fe}$$

Step 4: Convert the moles of product to grams using the molar mass.

$$1.88\ \cancel{\text{mole Fe}} \times \frac{55.85 \text{ g Fe}}{1\ \cancel{\text{mole Fe}}} = 105 \text{ g Fe Formed}$$

So in the end 105 grams of Fe are formed by reacting 150 g of Fe_2O_3 and 60 g of Al. At the end of the reaction some Al solid is left over.

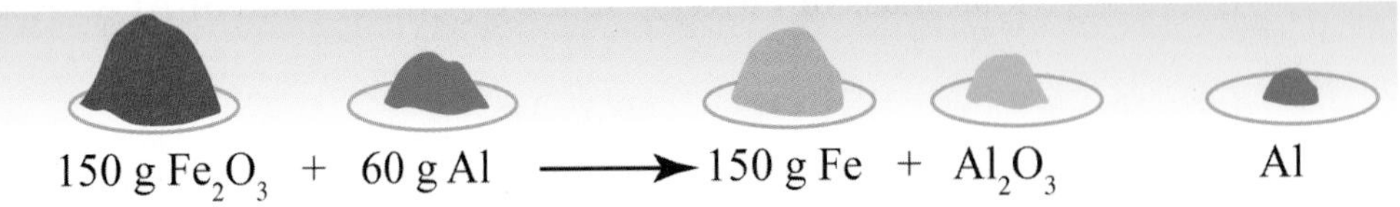

Alternative method

Another method for determining the amount produced based on the limiting reactant, involves doing two calculations one for each reactant forming product. In this case the amount of iron, Fe, formed for EACH reactant amount is calculated. Basically you do two calculations one for each reactant forming product.

$$150 \text{ g } \cancel{Fe_2O_3} \times \frac{1 \text{ mole } Fe_2O_3}{159.7\ \cancel{\text{g } Fe_2O_3}} = 0.94 \text{ moles } Fe_2O_3 \times \frac{2 \text{ moles Fe}}{1 \text{ mole } Fe_2O_3} = 1.88 \text{ mole Fe}$$

$$1.88\ \cancel{\text{mole Fe}} \times \frac{55.85 \text{ g Fe}}{1\ \cancel{\text{mole Fe}}} = 105\text{g Fe formed}$$

Mass produced is smaller because it is based on the limiting reactant.

$$60\ \cancel{\text{g Al}} \times \frac{1 \text{ mole Al}}{26.98\ \cancel{\text{g Al}}} = 2.22\ \cancel{\text{mole Al}} \times \frac{2 \text{ mole Fe}}{2\ \cancel{\text{mole Al}}} = 2.22\ \cancel{\text{mole Fe}} \times \frac{55.85 \text{ g Fe}}{1\ \cancel{\text{mole Fe}}} = 124 \text{ g Fe formed}$$

Mass produced is larger because it was based on the excess reactant.

Once you have the amounts of product formed from each reactant you can *select the reactant that made the least product*. Here the least amount was made using Fe_2O_3.

Summary of process

In this section we will outline and emphasize the steps you need to follow when doing a limiting reactant calculation. Sometimes organizing the steps in a table is useful.

Steps for Determining the Limiting Reactant		
Step 1	**Step 2**	**Step 3**
Convert both reactant masses to moles.	Multiply by the mole ratio from the balanced equation to find how much reactant is needed to use up all of the other reactant.	Compare the amounts of reactants. What you HAVE available to what you NEED.
This gives you the amount you HAVE available to use.	This gives you the amount you NEED to consume all of the reactant.	If what you NEED > HAVE then this is the Limiting Reactant.

Essentially you have to convert your given amounts to moles, and then to see which one will get used up first you need to look at the mole ratio between your reactants.

Almost all chemical reactions are carried out with one reactant in excess. So the one that is consumed needs to be identified.

At other times a flow chart that clearly lays out the sequence is helpful as a guide.

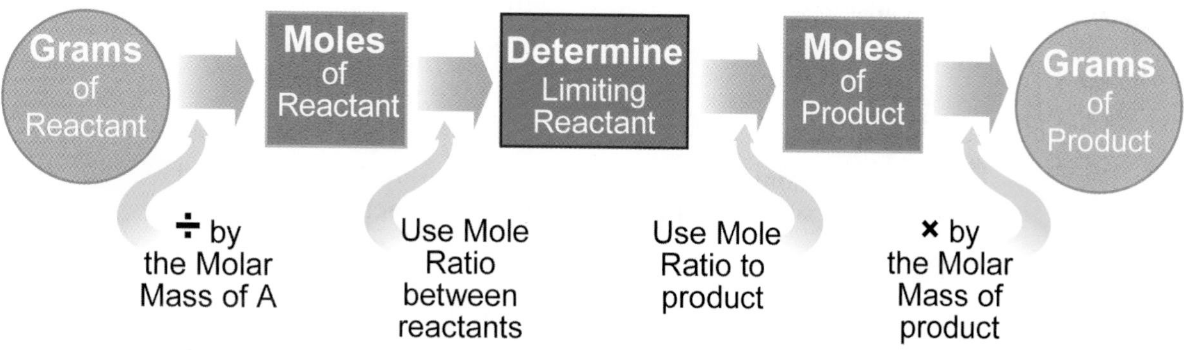

It may be helpful for you to sketch out this flow chart as you approach and try to solve a limiting reactant problem. Write out your information in each box and follow the steps as you go.

Practice determining the limiting reactant

Making Ammonia, NH_3

The Haber-Bosch process allowed for the production of ammonia on a large industrial scale. Before this process was developed ammonia was not able to be produced in large quantities. The production of ammonia was very important during World War I because ammonium nitrate was used to make explosives. Now it is important in the manufacture of fertilizer. Today ammonia fertilizers are responsible for sustaining approximately one third of the earths population. To practice calculations with limiting reactants we will use this reaction.

The Haber-Bosch process for the synthesis of ammonia uses the following chemical reaction:

$$N_2 + 3\,H_2(g) \rightarrow 2\,NH_3(g)$$

In one reaction 100 g of nitrogen gas, N_2 reacts with 10 g of hydrogen gas, H_2. Which reactant will limit the amount of ammonia that can be produced?

First we will use our flow chart method.

Step 1 Convert grams to moles

$$100.0\cancel{g\,N_2} \times \frac{1\text{ mole } N_2}{28.02\cancel{g\,N_2}} = 3.57\text{ mol } N_2\text{ (Have)}$$

$$10.0\cancel{g\,H_2} \times \frac{1\text{ mole } H_2}{2.02\cancel{g\,H_2}} = 4.95\text{ mol } H_2\text{ (Have)}$$

Step 2 Use mole ratio to determine amount of moles needed to react.

$$3.57\cancel{\text{ moles } N_2} \times \frac{3\text{ moles } H_2}{1\cancel{\text{ mole } N_2}} = 10.71\text{ moles of } H_2\text{ is needed}$$

Step 3 Identify the Limiting Reactant

Comparing 4.95 mole H_2 Have < 10.71 mole Need we see that H_2 is our Limiting Reactant!

Step 4 Multiply by mole ratio

$$4.95\cancel{\text{ moles } H_2} \times \frac{2\text{ moles } NH_3}{3\cancel{\text{ moles } H_2}} = 3.30\text{ moles of } NH_3$$

Step 5 Multiply by the molar mass

$$3.30\cancel{\text{ moles } NH_3} \times \frac{17.03\text{ g } NH_3}{1\cancel{\text{ mole}}} = 56.2\text{g } NH_3\text{ produced}$$

Now lets try this calculation using our steps as a formal solved problem. Both methods work well.

1. The Haber-Bosch process for the synthesis of ammonia uses the following chemical reaction:

$$N_2 + 3\ H_2(g) \rightarrow 2\ NH_3(g)$$

In one reaction 100 g of nitrogen gas (N_2) reacts with 10 g of hydrogen gas (H_2). Which reactant will limit the amount of ammonia that can be produced?

Asked: *Which reactant is the limiting reactant?*

Given: *100 g of N_2 and 10 g of H_2*

Relationships: *Write down the molar mass for each element.*
Molar mass of N_2: $14.007 \times 2 = 28.01$
Molar mass of H_2 : $1.0078 \times 2 = 2.02$
Mole ratio : 1 mole N_2 = 3 mole H_2

Solve: *Step 1: Convert grams to moles*

$$100.\cancel{0\ g\ N_2} \times \frac{1\ \text{mole}\ N_2}{\cancel{28.01\ g\ N_2}} = 3.57\ \text{mole}\ N_2\text{(Have)}$$

$$10.0\ \cancel{g\ H_2} \times \frac{1\ \text{mole}\ H_2}{2.02\ \cancel{g\ N_2}} = 4.95\ \text{mole}\ H_2\text{(Have)}$$

Step 2: Use mole ratio to determine what amount of moles are needed to react.

$$3.57\ \cancel{\text{mole}}\ N_2 \times \frac{3\ \text{mole}\ H_2}{\cancel{1\ \text{mole}\ N_2}} = 10.71\ \text{moles of}\ H_2\ \text{is needed}$$

Step 3: Comparing this to what we have, 4.95 mole H_2 (above), we see that H_2 is our Limiting Reactant!

OR *using the other reactant*

$$4.95\ \cancel{\text{mole}\ H_2} \times \frac{1\ \text{mole}\ N_2}{3\ \cancel{\text{mole}\ H_2}} = 1.65\ \text{moles}\ N_2\ \text{is needed}$$

Using this reactant and comparing this to what we have 3.57 mole N_2 we can see that N_2 is in excess.

Discussion: *You can determine the limiting reactant using either reactant once it is converted to moles.*
There is no need to do both calculations in step 2 as shown. Either one gives you the answer!

Keep in mind the reason we solve for the limiting reactant is to see how much product can be formed. On the next page we will use the limiting reactant to determine the amount of product, NH_3 formed.

Below we will use the limiting reactant to determine the amount of product formed.

2. Now we can calculate the amount of product based on the correct limiting reactant. How much ammonia, NH_3 can be produced from this reaction?

Asked: *How much product, NH_3 can be produced?*

Given: *H_2 is the limiting reactant*

Relationships: *There are $10.0\ g\ H_2 \times 1\ mole\ H_2 / 2.02\ g\ H_2 = 4.95\ mole\ H_2$*

Molar mass of NH_3 : $14.007 + (1.0079 \times 3) = 17.03\ \frac{g}{mole}$

Mole ratio : $3\ mole\ H_2 = 2\ moles\ NH_3$

Solve: *Use moles of limiting reactant to calculate moles of NH_3 produced*

Step 4: multiply by mole ratio

$$4.95\ \cancel{\text{mole H}_2} \times \frac{2\ \text{mole NH}_3}{3\ \cancel{\text{mole H}_2}} = 3.30\ \text{mole NH}_3$$

Step 5: multiply by the molar mass

$$3.30\ \cancel{\text{mole NH}_3} \times \frac{17.03\ \text{g NH}_3}{1\ \cancel{\text{mole NH}_3}} = 56.2\ \text{g NH}_3$$

Answer: *56.2 g of NH_3 are produced.*

As mentioned earlier ammonia is used to make fertilizer. Because of this it indirectly provides us with some of our nutritional needs. It is estimated that 150 million tons of fertilizer was produced globally in 2006. That is a lot of fertilizer! The production of ammonia using the Haber process requires a significant amount of energy and can be costly. Underdeveloped countries often do not have the advantage of using commercial fertilizer to grow their crops.

11.4 Solving Stoichiometric Problems

Why is it important to master this skill?

Stoichiometry gives us a sense of how chemical recipes work. For example: how do companies produce enough aspirin to have in all the drug stores? In much the same way they produce enough chocolate chip cookies for all the grocery stores! Learning about how much of a chemical is produced helps us to be more informed citizens. When we watch the news, read the paper, talk with our friends and family, discussions about pollution and the environment inevitably arise. If we understand how to calculate amounts, we have a better understanding about what these amounts represent. This also helps us to understand the significance of pollution amounts, and the impact of contamination by petroleum products.

Key Ideas

Chemicals combine in stoichiometric proportions. These proportions come from the balanced chemical equation. The mole allows us to compare between chemicals.

Lets continue to practice how to calculate amounts from a balanced chemical equation.

Here are some helpful tips that will make it easier for you to master these types of calculations:

1. Copy the flow chart for each question. Write down the chemical reaction and the amounts so that you can visually see what you are starting with and what steps are required in the calculation.
2. Write out in words, what you are doing next to each step.
3. Notice that you repeat the same sequence of steps each time.
4. Explain the meaning of your answer. What did you find? Try not to think of your answer as just a number, but a quantity that represents something meaningful
5. Remember to practice one or two examples without your notes, or the text. The process always looks easier when you are reading through an example.

Refer to the flow chart for the steps involved in the following calculations.

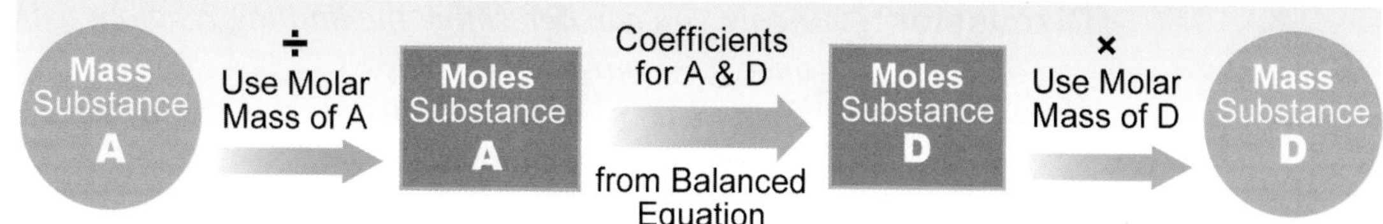

Calculating the limiting reactant

Here we will practice using a limiting reactant to calculate the amount of product formed. Refer back to the steps in the flow chart that guide us through determining the limiting reactant, as needed.

1. Lithium metal reacts directly with nitrogen gas to produce lithium nitride ($Li_3N(s)$). Lithium is the only group 1 metal that is capable of reacting directly with nitrogen gas.

$$6Li(s) + N_2(g) \rightarrow 2\ Li_3N(s)$$

When 48.0 g of lithium reacts with 46.5 g of nitrogen gas (N_2) which reactant is the limiting reactant?

Asked: *Which reactant is limiting?*

Given: *48.0 g Li and 46.5 g of N_2*

Relationships: *Molar mass of Li = 6.941 g/mole*
Molar mass of N_2 = 14.007 x 2 = 28.01 g/mole
Mole ratio = 6 mole Li = 1 mole N_2

Solve: *Step 1:* $48.0\ \text{g Li} \times \frac{1\ \text{mole Li}}{6.941\ \text{g Li}} = 6.92\ \text{mole Li (Have)}$

$46.5\ \text{g}\ N_2 \times \frac{1\ \text{mole}\ N_2}{28.01\ \text{g}\ N_2} = 1.66\ \text{mol}\ N_2\ \text{(Have)}$

Step 2: Use the mole ratio to determine what amount of moles are needed to react.

$6.92\ \text{mole Li} \times \frac{1\ \text{mole}\ N_2}{6\ \text{mol Li}} = 1.15\ \text{mole}\ N_2\ \text{(Needed)}$

Step 3: Comparing this 1.15 mole N_2 NEED < 1.66 mole N_2 HAVE we see that N_2 is our excess Reactant!

OR

$1.66\ \text{mole}\ N_2 \times \frac{6\ \text{mole Li}}{1\ \text{mole}\ N_2} = 9.96\ \text{mole Li (Needed)}$

Comparing this 9.96 moles Li NEED > 6.92 mole Li HAVE, so lithium is our limiting reactant.

Discussion: *Again you can determine the limiting reactant using either reactant once it is converted to moles.*
There is no need to do both calculations in step 2 as shown. Either one gives you the answer!

Calculating theoretical yield and percent yield

Now in part 2, of the solved problem below, we will calculate the mass of lithium nitride that can form using our limiting reactant, remember this is also called the theoretical yield of lithium nitride. The following two calculations also apply to the same balanced equation: $6Li(s) + N_2(g) \rightarrow 2\ Li_3N(s)$ as used on the previous page.

2. How much lithium nitride (LiN_3) can be produced from this reaction?

Asked: *How much product (LiN_3) can be produced?*

Given: *Li is the limiting reactant (from part 1 on previous page)*

Relationships: *Molar mass of Li_3N = (6.941 × 3) + 14.007 = 34.83 g/mole*
Mole ratio = 6 mole Li = 2 moles Li_3N

Solve: *Step 1: Use moles of limiting reactant to calculate moles of LiN_3 produced*

$$6.92\ \cancel{\text{mole Li}} \times \frac{2\ \text{mole Li}_3\text{N}}{6\ \cancel{\text{mole Li}}} = 2.31\ \text{mole Li}_3\text{N}$$

Step 2: Convert moles to grams using the molar mass

$$2.31\ \cancel{\text{mole Li}_3\text{N}} \times \frac{34.83\ \text{g Li}_3\text{N}}{1\ \cancel{\text{mole Li}_3\text{N}}} = 80.46\ \text{g Li}_3\text{N}\ \ \text{Produced}$$

Answer: *80.46 g of Li_3N are produced*

Now we will use the theoretical yield obtained above to calculate the percent yield of the reaction. Remember the reason "theoretical yield" is theoretical is because it is what you WOULD get in a "perfect" situation. Chemical reactions are not "perfect." So below we calculate percentage yield because we are given the actual amount produced. The actual yield can only be obtained in the laboratory.

3. Calculate the percentage yield of an experiment that actually produced 62.5 g of Li_3N .

Asked: *Calculate the percentage yield of Li_3N.*

Given: *The actual yield of 62.5 g Li_3N and the theoretical yield calculated above as 80.46 g of Li_3N.*

Relationships: *Formula for percent yield*

Solve:

$$\text{Percent yield} = \frac{62.5\ \text{g (actual)}}{80.46\ \text{g (theorectical)}} \times 100 = 77.7\%$$

Answer: *The percent yield is 77.7%*

Calculating the amount of excess reactant remaining

Often when a limiting reactant calculation is required it is necessary to calculate the amount of **excess reactant** that is not consumed in the chemical reaction. Calculating the amount of excess reactant adds a simple step in our calculation. We subtract the amount we HAVE from the amount we NEED of the excess reactant and then convert to grams by multiplying by the molar mass.

In a certain experiment 3.50 g of $NH_3(g)$ reacts with 3.85 g of $O_2(g)$.

$$4NH_3(g) + 5O_2(g) \rightarrow 4NO(g) + 6H_2O(g)$$

a) Which reactant is the limiting reactant? b) How much of the excess reactant remains after the limiting reactant is completely consumed?

Asked: *Which reactant is limiting? How much excess reactant remains?*

Relationships: *Molar Mass of* $NH_3 = (14.007 + 3(1.0079) = 17.03 \frac{g}{mole}$

Molar Mass of $O_2 = 15.999 \times 2 = 32.0 \frac{g}{mole}$

Mole Ratio: $4 \text{ mole } NH_3 = 5 \text{ mole } O_2$

Solve: *a)Step 1:* $3.50 \cancel{\text{g } NH_3} \times \frac{1 \text{ mole } NH_3}{17.03 \cancel{\text{g } NH_3}} = 0.206 \text{ mole } NH_3 \text{ (Have)}$

$$3.85 \cancel{\text{g } O_2} \times \frac{1 \text{ mole } O_2}{32.0 \cancel{\text{g } O_2}} = 0.120 \text{ mole } O_2 \text{ (Have)}$$

Step 2: $0.206 \cancel{\text{mol } NH_3} \times \frac{5 \text{ mole } O_2}{4 \cancel{\text{mole } NH_3}} = 0.26 \text{ mole } O_2 \text{ (Need)}$

$$0.120 \cancel{\text{mole } O_2} \times \frac{4 \text{ mole } NH_3}{5 \cancel{\text{mole } O_2}} = 0.096 \text{ mole } NH_3 \text{ (Need)}$$

Step 3: Compare Have and Need amounts for each reactant
For NH_3 *Have (0.205 mole) > Need (0.096 mole)*
For O_2 *Have (0.102 mole) < Need (0.26 mole)*
Here we can see that O_2 *is our Limiting Reactant*
b) To find excess reactant simply subtract Have - Need for NH_3
$0.205 \text{ mole } NH_3 \text{ (have)} - 0.096 \text{ mole } NH_3 \text{ (need)} = 0.109 \text{ mole}$
$0.109 \text{ mole } NH_3 \text{ remaining x } 17.03 \frac{g}{mole} = 1.86 \text{ g } NH_3 \text{ in excess}$

Answer: O_2 *is the limiting reactant and there are 1.86 g of* NH_3 *in excess.*

Chemistry terms

excess reactant - the reactant left over after the reaction is complete. The amount of reactant the did not react to form product.

Determining the amounts of solid formed from solutions

Stoichiometry with solutions

One of the reaction types you learned about in chapter 10 was a precipitate reaction. Chemists use precipitate reactions to test for pollutants in water supplies.

Here you can see two solutions form a dark brown solid precipitate:

$$CuSO_4(aq) + Na_2S(aq) \rightarrow CuS\ (s) + Na_2SO_4\ (aq)$$

You can probably guess that if the solutions are more concentrated they will form a larger amount of solid. This simply means the higher the molarities are, the more moles that react!

The Environmental Protection Agency (EPA) tests for particular ions by taking advantage of the fact that they form solids when mixed with the right solution. In the following example, Cu^{2+} ion is tested.

The concentration of Cu^{2+} ions (found as $CuSO_4(aq)$) in the water discharged from an industrial plant is found by adding an excess of sodium sulfide (Na_2S) solution to 1.0 L of the contaminated water. What is the concentration of $CuSO_4$ in the water if 0.021g of CuS(s) is formed? The reaction is:

$$CuSO_4(aq) + Na_2S(aq) \rightarrow Na_2SO_4(aq) + CuS(s)$$

Asked: *What is the concentration of the Cu^{2+} ion (from $CuSO_4(aq)$)?*

Given: *1.0 L of solution tested, and 0.021 g of CuS solid is formed.*

Relationships: *molar mass of CuS = 63.55 + 32.06 = 95.61 g/mole*
mole ratio = 1 mole CuS(s) = 1 mole $CuSO_4$ (aq)

Solve:

$$0.021\ \cancel{g}\ \text{CuS} \times \frac{1\ \text{mole CuS}}{95.\cancel{61\ g\ CuS}} = 2.20 \times 10^{-4}\ \text{mol CuS}$$

$$2.20 \times 10^{-4}\ \cancel{\text{mole CuS}} \times \frac{1\ \text{mole } CuSO_4}{1\ \cancel{\text{mole CuS}}} = 2.20 \times 10^{-4}\ \text{mole } CuSO_4$$

$$\frac{2.20 \times 10^{-4}\ \text{mole } CuSO_4}{1.0\ \text{L solution}} = 2.20 \times 10^{-4}\ \text{M}$$

Answer: *The concentration of $CuSO_4$ is 2.20×10^{-4} M.*

Discussion: If the legal limit is 5.0×10^{-4} M of Cu^{2+} ion, is the industrial plant following the environmental guidelines?
Yes because *2.20×10^{-4} M* is less than the legal limit.

Stoichiometry and nature

The primary elements here on earth are recycled over time throughout the different regions of the earth. For example carbon is available in different molecules in the atmosphere, land and ocean. It is important that the "balance" of elements is sustainable and able to support the living systems here on earth in the future. Today we hear about sustainability in many different contexts. Stoichiometry has quite a lot to do with how chemicals in the form of molecules and elements achieve a balance in our environment. We have established that chemicals make up our food, the air we breathe, the water we drink, and the buildings we work and live in. These are the essential things humans need to survive. What would happen if we did not have enough chemicals to make all of these things we need in order to live? Think about it! Stoichiometry looks at amounts of chemicals required for a chemical reaction, and chemical reactions make things. Knowing how much of a chemical is used or released is very important, if we want to understand the overall picture.

To be "sustainable" humans need to consume the earths resources at a rate that can be replaced by nature. There is much evidence out there to indicate that we are living in a manner that is unsustainable. Let's take the buring of fossil fuels as one example. As we burn fossil fuels there are specific stoichiometric ratios involved. Coal is made up of carbon, sulfur, hydrogen, oxygen and nitrogen. There are many different "types" of coal, however they are all primarily composed of the element carbon, C.

Coal ($C_xS_yH_z$)

$$\text{Coal}(C_xS_yH_z)(s) + O_2(g) \rightarrow CO_2(g) + H_2O\ (l) + SO_2(g)$$

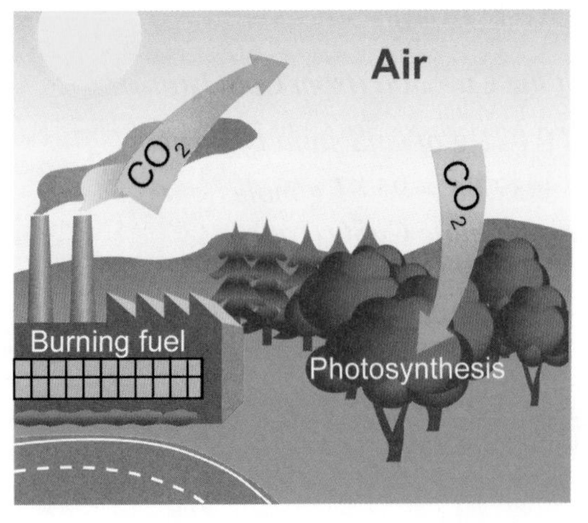

In this combustion reaction the mole ratio between the carbon and the carbon dioxide is 1 to 1. If we consider the fact that the carbon and other elements in the coal itself came from plant remains, we can begin to think about how elements cycle through earth processes. We can use stoichiometry to roughly calculate how much carbon is released in to the air when burning coal. We can also determine how much of this carbon will be returned to the earth when the plants use it for photosynthesis. Maybe in another million years the carbon in the glucose molecules will be in the form of coal again!

Stoichiometry can help us figure this out. If a coal powered electrical plant burns 1 ton of coal, and the coal itself is approximately 60% carbon. We can calculate the mass of carbon that will enter the atmosphere.
Lets see: 1 ton = 2240 lbs x .60 = about 1,344 lbs of carbon will enter the atmosphere as CO_2, because one mole of CO_2 forms for every mole of carbon in the coal. Here we are simplifying the process a little and assuming that there is complete combustion of the coal.

The amount of carbon entering the atmosphere as carbon dioxide gas is increasing the green house effect, which in turn is affecting the global climate. One method that is being used to remove some of the carbon dioxide is to capture the gas, concentrate it, and send it back below the earths surface. This is called carbon sequestering. In this way some carbon will be removed from our atmosphere and stored underground once again.

Of course some of the carbon dioxide is recycled naturally! How does this happen? There are two primary ways this is accomplished. Through the chemical reaction called photosynthesis and the dissolving of $CO_2(g)$ by the ocean water. Both of these chemical processes involve chemical reactions to which stoichiometry can be applied. Nature uses stoichiometry, because life processes such as photosynthesis and respiration are balanced chemical equations!

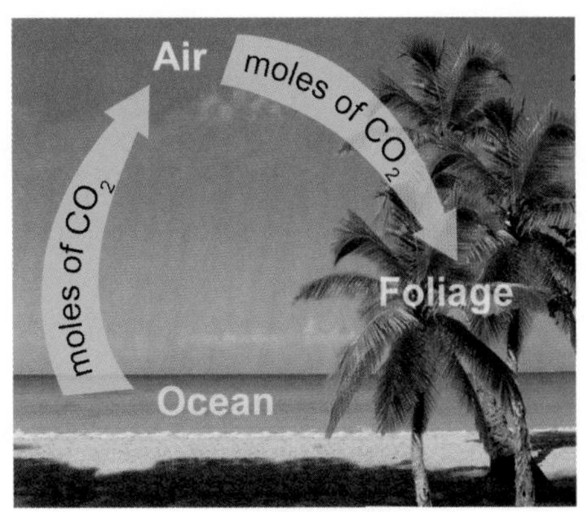

First plants use approximately one quarter of the CO_2 available in the air. So of the 1,344 lbs of carbon approximately 1,344 lbs x 0.25 = 336 lbs of CO_2 are cycled through photosynthesis

$$6CO_2(g) + 6H_2O(l) \rightarrow C_6H_{12}O_6(s) + 6O_2(g)$$

Here 6 moles of CO_2 are used to make 1 mole $C_6H_{12}O_6$.

Once photosynthesis is carried out carbon is now in the form of glucose. Glucose can enter the food chain when plants are consumed by animals. The animals metabolize the glucose and release CO_2 into the air, once again, through the process of respiration.

Secondly, another one quarter of our atmospheric carbon dioxide dissolves in the ocean by direct air and water exchange. This process is represented by the equation:

$$CO_2(g) + H_2O(l) \rightarrow H^+(aq) + HCO_3^-(aq)$$

According to this equation for every 1 mole of CO_2 dissolved 1 mole of HCO_3^- is formed.

In essence we use stoichiometry to help us determine how much carbon enters the atmosphere, and we take into account how much can be absorbed by natural processes. To do this we need an accurate understanding of the chemical reactions involved in these changes.

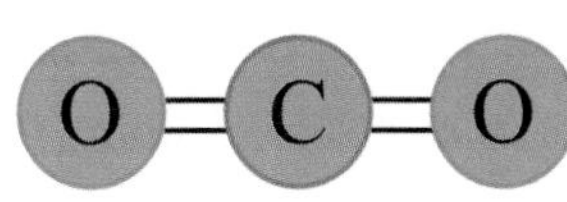

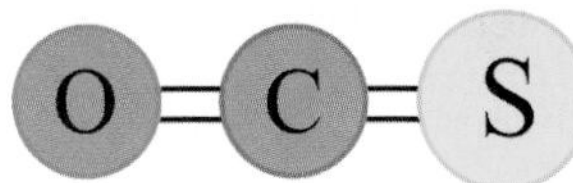

Scientists have been trying to discover more accurate methods for determining how much CO_2 is absorbed by photosynthesis and in November of 2008 a Professor at the University of California was able to measure the amount of CO_2 absorbed much more accurately, because he discovered that carbonyl sulfide (COS) used in a process that runs parallel (at the same time) as photosynthesis is absorbed at the same rate as CO_2. Notice how similar molecular structure for COS is to CO_2. It is easier to measure COS because plants do not release it the way they do CO_2. With these better methods of measurement we can apply stoichiometry more accurately. Because of this research we will now be able to include more accurate information about how photosynthesis will be able to help us battle climate change.

Chapter 11 Review

Vocabulary

Match each word to the sentence where it best fits.

Section 11.1

stoichiometry	stoichiometric-equivalent
mole ratio	

1. When analyzing a chemical equation we can say the mole amounts given in the balanced equation are ____________ amounts.

2. To study the amounts of substances used in chemical reactions we use ____________.

3. Using the coefficients from the balanced equation to make a ______________ allows us to compare between the amounts of chemicals required or produced in a chemical reaction.

Section 11.2

actual yield	percent yield
theoretical yield	

4. The ________________ is based on a calculation that assumes the reaction yields 100% of the product.

5. When we experiment in the laboratory we obtain the ________________ which is generally less than the theoretical yield.

6. We can calculate our _______________ which tells us how much of the product was obtained from our experiment.

Section 11.3

limiting reactant	excess reactant

7. The substance that limits the amount of product that can form is called the ____________.

8. The ________________ is left over after the chemical reaction has formed products.

Conceptual Questions:

Section 11.1

9. Interpret the following equation in terms of molecules and moles:
$C_3H_8(g) + 5O_2(g) \rightarrow 3CO_2(g) + 4\ H_2O(g)$

10. Are moles conserved in a chemical reaction? Explain.

11. Explain what the unit of the "mole" represents. In your own words.

12. Give an example from your everyday life where you use a recipe to make something.

13. Use the example you gave above and explain how this is similar to using a balanced equation to see how chemicals combine.

14. Write out the meaning of the following chemical equation in words.
$2NaN_3(s) \rightarrow Na(s) + 3N_2(g)$

15. Show the both mole ratios for comparing the reactants in the following equation:
$CH_4(g) + 2O_2(g) \rightarrow CO_2(g) + 2H_2O(l)$

16. Using the following balanced equation :
$Ca(s) + 2H_2O\ (l) \rightarrow Ca(OH)_2 + H_2(g)\ (aq)$
    a. Give an example of a stoichiometric equivalent amount.
    b. What mole ratio would you use to calculate how many moles of hydrogen, H_2 could form from a given number of moles of calcium, Ca?

17. When using the flow chart to calculate grams from grams given why do we need to:
    a. convert to moles in step 1? Explain.
    b. use the mole ratio in step 2? Explain.

18. Why do chemists need to calculate the amount of product produced in a chemical reaction?

19. Using the following balanced equation:
$S + O_2 \rightarrow SO_2$
Explain why 1.0 g of S will not react exactly with 1.0 g of O_2 ?

20. Since we weigh and measure amounts of chemicals in grams, why can't we compare chemicals in grams?

21. If 4 moles of Ca(s) react with water according to the equation:
$Ca(s) + 2H_2O\ (l) \rightarrow Ca(OH)_2 + H_2(g)\ (aq)$
How many moles of hydrogen gas, $H_2(g)$ are produced? Explain.

Section 11.2

22. Why don't we get 100% of the product formed in a chemical reaction? Give a couple of reasons.

23. Explain how the popcorn example helps us to understand the percent yield of a chemical reaction.

24. How does a chemist determine the actual yield of a chemical reaction?

25. Give an everyday life example of percentage yield (not popcorn!).

26. Would you expect the actual yield to be larger or smaller than the theoretical yield? Explain.

27. When performing the experiment where baking soda, $NaHCO_3(s)$ is heated to cause it to decompose into $Na_2CO_3(s) + H_2O(l) + CO_2(g)$, your final mass is close to the theoretical yield. List a couple of likely error sources.

28. What is the unit of concentration?

29. Describe how the process of calculating moles of a solution is different than calculating moles of a solid.

Section 11.3

30. Give one new (or original) real life example of a limiting reactant.

31. Look at the picture below and refer to the following equation:
$CH_4(g) + H_2O(l) \rightarrow 3H_2(g) + CO(g)$

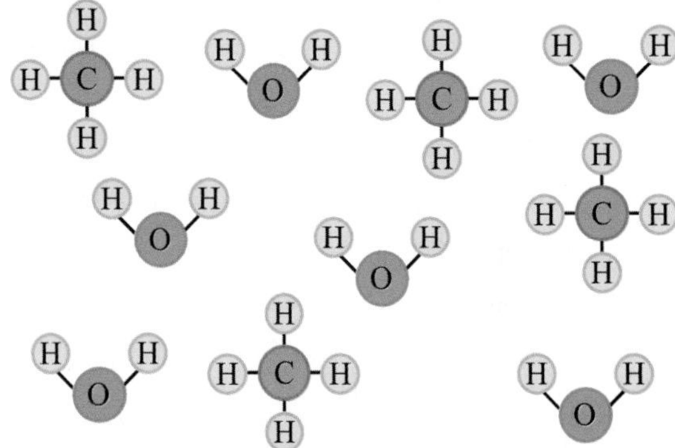

a. Which reactant is limiting? Explain.

b. Make a sketch of the molecules after the reaction.

32. Is is safe to assume the reactant present in the larger amount is the excess reactant?

33. Explain why knowing the limiting reactant is useful.

34. Given the following chemical reaction:
$6Li(s) + N_2(g) \rightarrow 2Li_3N(s)$
If you have 6 moles of Li(s) and 2 moles of $N_2(g)$ which reactant is limiting ? Explain.

35. Given the following chemical reaction:
$2Al(s) + 3Cl_2(g) \rightarrow 2AlCl_3(s)$
If you have 6 moles of each reactant, which one is limiting? Explain.

36. Why can't we use the excess reactant to determine the amount of product produced?

37. Sketch out the flow chart for determining the grams of product produced, when a limiting reactant is involved in your calculation.

Chapter 11 Review.

Quantitative Problems:

Section 11.1

38. Use the equation below to answer the following questions: Clearly show your work.
$4NH_3(g) + 3O_2(g) \rightarrow 2N_2(g) + 6H_2O(g)$
    a. How many moles of water form from 8 moles of ammonia, NH_3 reacting with excess oxygen?
    b. How many moles of nitrogen gas, N_2, form when 7 moles of oxygen, O_2 reacts with excess ammonia?
    c. If you have 3.5 moles of ammonia, NH_3, how many moles of oxygen will be consumed in making products?
    d. If 10 moles of water is formed how many moles of oxygen gas was required?
    e. If we need to produce 20 moles of water, how many moles of both ammonia and oxygen are required to produce it?

39. Nitrogen, N_2, reacts with iodine, I_2, according to the following equation:
$N_2\ (g) + 3I_2(s) \rightarrow 2NI_3(s)$
    a. Explain in words what the equation gives us for information.
    b. How many moles of $NI_3(s)$ can form when 127.0 g of $I_2(s)$ react with excess nitrogen gas?
    c. If you wish to prepare 2.0 g of $NI_3(s)$ how many moles of nitrogen are required?
    d. When 28.0 g of nitrogen, N_2, reacts with excess iodine, how many moles of nitrogen triiodide, NI_3 form ?

40. For each of the following equations listed below, calculate the moles of product that would be produced by reacting 4.0 moles of oxygen, $O_2(g)$ the reactant. Be sure to clearly show your mole ratio in your calculation.
    a. $4Fe(s) + 3O_2(g) \rightarrow Fe_2O_3(s)$
    b. $2\ Mg(s) + O_2(g) \rightarrow 2MgO(s)$
    c. $2Cu(s) + O_2(g) \rightarrow 2CuO(s)$
    d. $4Ag(s) + O_2(g) \rightarrow 2Ag_2O(s)$

41. Calculate the number of moles of each substance from each of the following masses given.
    a. 15.0 g of carbon dioxide, CO_2.
    b. 30.0 g of copper, Cu.
    c. 10.0 g of neon, Ne.
    d. 20.0 g of HCl.
    e. 32.00 g of $Cu(OH)_2$
    f. 50.0 g of $Al_2(SO_4)_3$

42. Given the following reaction:
$N_2(g) + O_2(g) \rightarrow 2\ NO(g)$
    a. How many moles of NO can form when 40 .0 g of N_2 react with excess oxygen?
    b. How many grams of NO can form, using the amounts in part a.

43. Using the unbalanced equation below answer the following questions:
$S(s) + H_2SO_4(aq) \rightarrow SO_2(g) + H_2O(l)$
    a. Balance the equation.
    b. Explain in words what the balanced coefficients tell us.
    c. Calculate the number of moles of $SO_2(g)$ that can be formed from 100.0 g of sulfur, S(s) and excess sulfuric acid, H_2SO_4.
    d. How many grams of SO_2 form given the amounts in part c) above?
    e. How many grams of water form when 100.0 g of sulfur, S reacts with excess sulfuric acid, H_2SO_4?
    f. If 64.0 g of sulfur dioxide, SO_2 form how many grams of sulfuric acid was required?

44. Silver tarnishes and turns a dull gray color when exposed to the oxygen in the air. The reaction for the tarnishing of silver is :
$Ag(s) + O_2(g) \rightarrow Ag_2O(s)$

a. Balance the equation.

b. If you react 8.0 g of silver with excess oxygen how many grams of silver oxide can form?

45. Iron oxide or rust, Fe_2O_3 reacts with aluminum metal, Al to form iron, Fe and aluminum oxide, Al_2O_3. The reaction is shown below :

$Fe_2O_3(s) + Al(s) \rightarrow Fe(l) + Al_2O_3(s)$

a. Balance the equation.

b. When 25.0 g of Aluminum reacts with excess iron oxide, how many grams of iron will be produced?

Section 11.2

46. A student performed a precipitate experiment based on the following equation:
$Ba(NO_3)_2(aq) + Na_2SO_4(aq) \rightarrow BaSO_4(s) + 2NaNO_3(aq)$
Before the experiment they calculated the theoretical yield of barium sulfate, $BaSO_4$, to be 2.58 g. After the experiment when they filtered and dried their precipitate they obtained 1.98 g of $BaSO_4$. Calculate the students percent yield.

47. Based on your prelaboratory calculations your experiment should yield 4.36 g of magnesium oxide. However, when you perform the experiment and weigh the MgO produced you find that you obtain 3.98 g MgO. What is your percent yield ?

48. When phosphorous reacts with chlorine it forms phosphorus trichloride, PCl_3.
$P_4(s) + 6Cl_2(g) \rightarrow 4PCl_3(l)$

a. How much PCl_3 can be formed by reacting 6.00 g of P_4 with excess chlorine?

b. If a 19.50 g sample of PCl_3 was obtained from the reaction, calculate the percent yield.

c. In another experiment, the theoretical yield was calculated to be 35.20 g PCl_3 liquid. What was the actual yield for this experiment, if it produced a 75.20% yield.

49. Limestone decomposes with heat, following the equation:
$CaCO_3(s) \rightarrow CaO(s) + CO_2(g)$
If 8.80 g of CO_2 is produced from the heating of 28.0g of $CaCO_3$. What is the percent yield of this reaction.

50. For each of the following calculate the number of moles present in the solution.

a. 50.0 mL of 1.33 M NaOH

b. 25.0 mL of 6.0 M NaCl

c. 10 mL of 0.10 M AgNO3

d. 10.0 g of NaCl in 250 mL of distilled water.

51. A sample of magnesium metal, Mg, reacts with 45.0 mL of 3.0M solution of hydrochloric acid, HCl.
$Mg(s) + 2HCl(aq) \rightarrow MgCl_2(aq) + H_2(g)$

a. How many moles of HCl are reacting?

b. Assume the Mg is in excess. How many moles of H_2 gas are produced?

c. How many grams of H_2 gas are produced?

52. A student performed a precipitate experiment based on the following equation:
$Cu(NO_3)_2(aq) + Na_2CO_3(aq) \rightarrow CuCO_3(s) + 2NaNO_3(aq)$

a. They mix 25.0 mL of 1.0 M $Cu(NO_3)_2$ with excess sodium carbonate, Na_2CO_3.

b. How many moles of $Cu(NO_3)_2$ are reacting?

c. How many grams of copper carbonate, $CuCO_3(s)$ will form ?

53. A student carried out a precipitate experiment based on the following equation:
$AgNO_3(aq) + NaCl(aq) \rightarrow AgCl(s) + NaNO_3(aq)$

a. They mix 10.0 mL of 3.0 M $AgNO_3$ with excess sodium chloride, NaCl.

b. How many moles of $AgNO_3$ are reacting?

c. How many grams of AgCl will form?

Chapter 11 Review.

54. A solution is made by dissolving 20.0 g of glucose, $C_6H_{12}O_6$ in 200.0 g of water. Calculate the mass percent of glucose in the solution.

55. Commercial hydrogen peroxide, H_2O_2, is advertised as 3% by mass. How many grams of hydrogen peroxide are in 200 mL of commercial solution. (assume the density of hydrogen peroxide is the same as for pure water.1.0 g/mL.

56. A certain brand name mouthwash contains 0.50 % alcohol by mass. How many grams of alcohol, CH_3OH, are in 100.0 mL of mouthwash?

57. Commercial vinegar is approximately 5% acetic acid by mass. How many grams of acetic acid, CH_3COOH, are in 25.00 mL of commercial vinegar?

Section 11.3

58. Iron, Fe(s) can be produced by a "thermite"reaction according to the following equation:
$Fe_2O_3(s) + 2\ Al(s) \rightarrow 2\ Fe(s) + Al_2O_3(s)$
    a. If 100.0 g of Fe_2O_3 reacts with 30.0 g of Al which one will be used up first?
    b. Explain your answer above.
    c. How much iron, Fe, in grams can be produced?
    d. Calculate the grams of excess reactant remaining after the reaction.

59. Lithium metal reacts with nitrogen gas to produce lithium nitride.
$6Li(s) + N_2(g) \rightarrow 2Li_3N(s)$
Use your flow chart to answer the following questions:
    a. If you have 20.0 g of Li and 15.0 g of N_2 which reactant will be used up first?
    b. Explain your answer above.
    c. How many grams of Li_3N can be produced?
    d. Calculate the amount of excess reactant remaining following the experiment.

60. Zinc metal reacts with copper(II) nitrate to produce copper solid.
$Zn(s) + Cu(NO_3)_2(g) \rightarrow Zn(NO_3)_2(aq) + Cu(s)$
Use your flow chart to answer the following questions:
    a. If you have 5.00 g of Zn and 25.0 mL of 1.00 M $Cu(NO_3)_2$ which reactant will be used up first?
    b. Explain your answer above.
    c. How many grams of Cu solid can be produced?
    d. Calculate the moles of excess reactant remaining.

61. Small bottles of propane gas are sold at hardware stores for propane torches. The combustion of propane is used for small campfires, heating ski wax, and soldering.
$C_3H_8(g) + 5O_2(g) \rightarrow 3CO_2(g) + 4H_2O(l)$
Use your flow chart to answer the following questions:
    a. If you have 1.0 Liter of propane gas, assume propane has a density of 0.82 g/L, how much oxygen will be required to burn it all?
    b. Explain your answer above.
    c. How many grams of H_2O are be produced?

62. For each of the following equations listed below, determine the liming reactant when 10.0 g of each reactant are used. Be sure to clearly support your answer with a calculation.
    a. $4Fe(s) + 3O_2(g) \rightarrow Fe_2O_3(s)$
    b. $2\ Mg(s) + O_2(g) \rightarrow 2MgO(s)$
    c. $2Cu(s) + O_2(g) \rightarrow 2CuO(s)$
    d. $4Ag(s) + O_2(g) \rightarrow 2Ag_2O(s)$

63. Using the reactions (a-d) above in number 62) calculate the mass of product formed in each reaction.

Section 11.4

64. Aluminium metal reacts with oxygen in the air to form a layer of corrosion, Al_2O_3. This layer protects the aluminum atoms underneath from further corrosion.
 $4\ Al(s) + 3O_2(g) \rightarrow 6Al_2O_3(s)$
   a. If 28.0 g of Al reacts with excess oxygen in the air, what mass of aluminum oxide is formed?
   b. How many moles of oxygen is consumed during this reaction?
   c. If 250 g of Al_2O_3 is formed how much Al reacted?

65. When gasoline or octane, C_8H_{18}, combusts the products are water and carbon dioxide gas.
 $2C_8H_{18}(g) + 25O_2(g) \rightarrow 16CO_2(g) + 18H_2O(l)$
   a. Calculate the mass of water formed from the combustion of 5.0 L of gasoline. The density of gasoline is 0.79 g/mL.
   b. What mass of carbon dioxide will be released into the air when the 5.0 L of octane is combusted?

66. Wine can spoil or become "vinegary" when ethanol, C_2H_5OH, is converted to acetic acid, CH_3COOH, by oxidation.
 $C_2H_5OH(aq) + O_2(g) \rightarrow CH_3COOH(aq) + H_2O(l)$
   a. Determine the limiting reactant when 5.00 g of ethanol and 2.0 g of oxygen are sealed in a wine bottle.
   b. Calculate the grams of acetic acid that will form.
   c. Calculate the grams of excess reactant remaining.

67. Make a sketch of what this chemical reaction looks like after the reaction takes place.
 $N_2(g) + 3H_2(g) \rightarrow 2NH_3(g)$

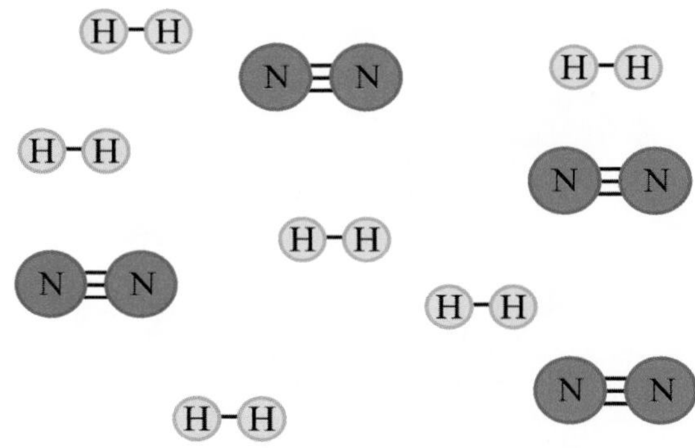

The blue molecules represent nitrogen and the white represent hydrogen. Put them together appropriately to make NH_3 as your product. Be sure to include any unreacted molecules as well as product formed in your sketch.

68. One step in the synthesis of rocket fuel is to make nitrogen dioxide, NO_2 from NO and O_2.
 $2NO(g) + O_2(g) \rightarrow 2\ NO_2(g)$
   a. If 20.0 g of NO is reacted with 10.0 g of $O_2(g)$, which reactant will limit the amount of NO_2 that can be formed?
   b. How many grams of NO2 will form?
   c. Calculate the amount, in grams, of the excess reactant remaining.

69. Using the following reaction:
 $Pb(s) + KNO_3(s) \rightarrow PbO(s) + KNO_2(s)$
 In the laboratory a yield of 20.2 g of potassium nitrite, KNO_2 was obtained, when 35.0 g of potassium nitrate, KNO_3 reacted with excess lead, Pb. Calculate the percentage yield of potassium nitrite, KNO_2.

CHAPTER 12
Reaction Rates and Equilibrium

What determines how rapidly chemicals will react?

Do chemical reactions "stop" after a period of time?

Can chemical reactions go "backwards?" Are they reversible?

Can you spot the chemical reactions?

Why are some chemical reactions fast, while others are slow? To consider this lets think about a few familiar everyday chemical reactions. What determines how quickly food spoils? For many third world countries this is a very important question. Chemistry has improved the methods for food preservation, and reduced the number of people suffering from starvation. We know from our own life experience that warm temperatures speed up the spoiling of food. We keep milk in the refrigerator, and we freeze our meat in order to make it last longer. In the poorer areas of the world, refrigeration is scarce and much too expensive to use, so they rely on chemical preservatives. These chemical preservatives slow down or interfere with the chemical reactions that cause food to spoil. In this chapter you will learn about the factors that affect the speed of a chemical reaction.

Our bodies are an excellent example of chemical equilibrium. The chemical systems in our bodies are in a delicate balance, called homeostasis, that is constantly maintained. One example of this is breathing. As you breathe you are continuously exchanging carbon dioxide and oxygen gases. We exhale carbon dioxide and inhale oxygen. During heavy exercise our breathing rate increases, which allows us to get rid of the excess carbon dioxide produced from our increased metabolic rate. The excess carbon dioxide produced needs to be eliminated in-order to restore your body's proper chemical balance.

Observe reactions over time

To study reaction rates or the speed at which a chemical process is happening it is helpful to have something that you can see or directly observe. Here we will use color change to determine how fast the chemical reaction is occurring. Many of us know that bleach removes stains from our laundry. Grass stains for instance are very difficult to remove with out it! Fabrics have natural and synthetic dyes that can be removed with bleach.

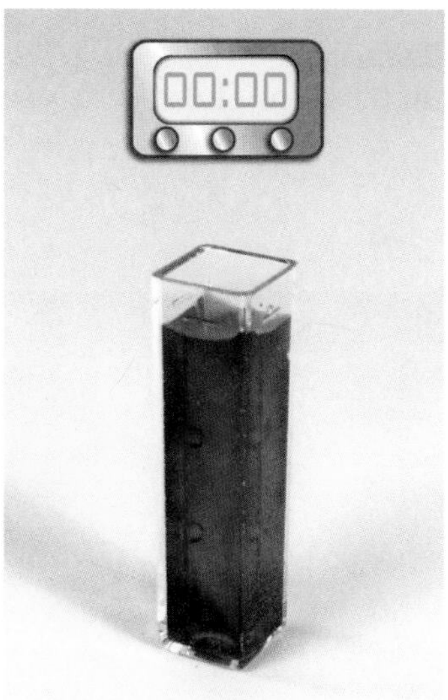

Reaction rates are determined by measuring the change in concentration over a period of time. Here the first blue solution is the most "concentrated," as you can tell from the deeper shade of blue. The initial solution was made using 100 mL of water and three drops of food coloring, it's concentration is about (3 drops dye/ 100 mL). As the chemical reaction between the dye and the bleach occurs we will see the solution become lighter.

In the first picture at time zero the bleach is just being added to the beaker containing the dye. You can see that the dye has not begun to fade. Two minutes after the bleach is added, the color is much lighter, indicating that the chemical change between the bleach and the dye has taken place. After a longer period of time (4minutes), or twice as long, we can see the solution in the cuvette is almost clear!

Bleach removes stains by breaking double bonds in the dye molecules, this alteration effects the dyes ability to absorb visible light. Therefore, no color can be seen and the stain is removed. The speed of this chemical process can be determined by noting the change in color and the change in time. Over a longer time period we can see that more of the color fades or disappears.

What effects how quickly the color fades?

The chemical formula for bleach is NaOCl and the formula for Blue dye #1 is $C_{37}H_{34}N_2Na_2O_9S_3$. Is shown here on the right.

12.1 Reaction Rates

Some reactions are fast while others are slow

Chemical reactions are the essence of chemistry. We know that reactions reorganize atoms into different molecules and compounds. In this section we are going to look at how fast the change occurs. Some reactions are fast, like a match burning. Some are slower, like milk spoiling when it is left out. The process of rusting occurs slowly over time when a metal object, such as a bicycle, is left out in the rain. The speeds of chemical reactions are affected by several factors.

Mg*(s)* strip in HCl*(aq)*

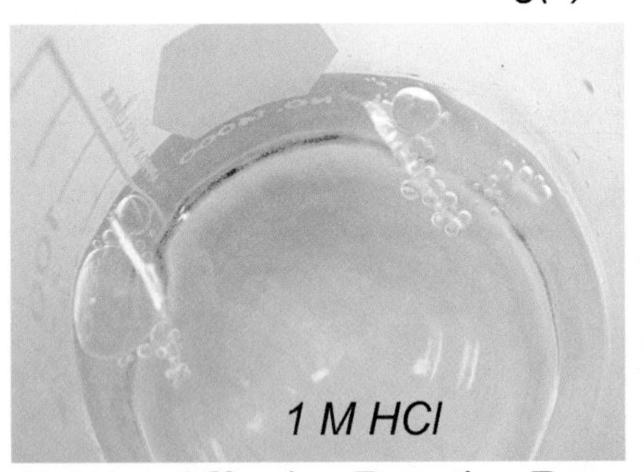

1 M HCl

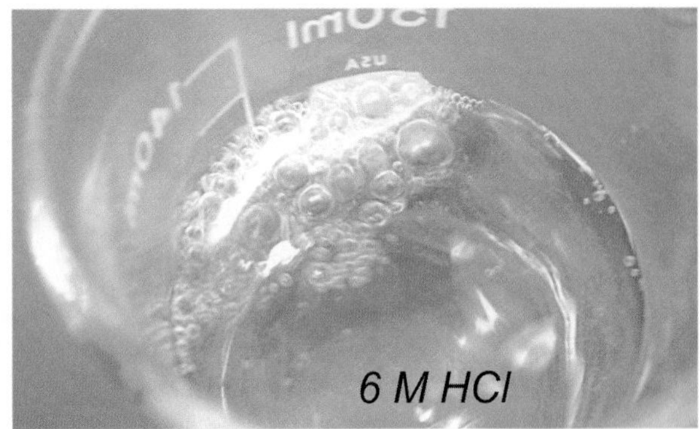

6 M HCl

Factors influencing reaction speed

Factors Affecting Reaction Rates:

1. The concentration of the reacting chemicals. Many chemical reactions happen more rapidly if the reactant concentrations are increased. The more reactants present the faster the reaction occurs.

2. The temperature at which the reaction is carried out. The higher the temperature the faster the reaction proceeds. The process of cooking speeds up the chemical reactions that take place in food.

3. Surface area - size of the reacting particles. Particularly, if the reactants are in solid form the reaction is affected by the surface area. The larger the surface area the faster the reaction will take place. The more atoms that come in contact with each other the more likely they are to react. A medicine will be absorbed more quickly by our bodies if we ingest it in the form of a powder (held in a dissolvable capsule), rather than in the form of a solid pill. This is because our stomach can dissolve a powder more quickly than a pill.

4. The addition of a catalyst. A catalyst is a substance that facilitates a chemical reaction without being used up itself. It helps the chemicals to react, but it does not break down or become chemically changed. In a catalytic converter, a transition metal such as rhodium acts as a catalyst, by providing a surface for molecules to attach to, which allows them to come into contact more easily. The rhodium metal itself does not participate in the chemical reaction, but it reduces the polluting gases that would be otherwise emitted in the exhaust.

The idea of rate

The concept of speed is similar to rate

When we drive in our cars we measure our speed by reading the speedometer. The speedometer tells us our speed in miles per hour. If we are traveling at a speed of 45 mph that means we will cover a distance of 45 miles in one hour. A thoroughbred race horse can maintain a speed of 45 mph for over a mile! In the Kentucky Derby which is 1.25 miles long, the average race horse covers this distance in approximately 2.00 minutes. The race time is measured with a stop watch.

To find the average speed traveled by a typical race horse:

$$\text{Speed (Rate)} = \frac{\text{Distance}}{\text{Time}} = \frac{1.25 \text{ miles}}{2.00 \text{ minutes}}$$

The speed of the horses varies during the race. Most race horses finish at a faster speed than they start at. Imagine as the horses begin the race, that they need a burst of energy to get started, the horses then maintain their speed until the home stretch, the final piece of the race, here the horses slowly accelerate as they approach the finish.

This example shows that the rate or speed can change during a horse race. The rate of a chemical reaction changes over time as well. The difference is that chemical reactions have a faster rate at the beginning and a slower rate at the end. Chemical reactions tend to slow down over time. We measure the rate of a chemical reaction by measuring a change in the amount of reactant or product per unit time.

$$\text{Rate} = \frac{\text{Change in Concentraton}}{\text{Change in time}}$$

Rate in chemistry

Reaction rate

Chemists measure the change in concentration, of a reactant or product, per unit time. The **reaction rate** is used to describe how reactions evolve over time.

Observing reaction progress

For example given a generic reaction where A $\rightarrow$ C, we can measure the rate at which the reaction proceeds by measuring the decrease in the concentration of reactant A over time, or by measuring the increase in concentration of product C over time. As the reaction proceeds the amount of reactant A decreases, because it is being converted to the product C.

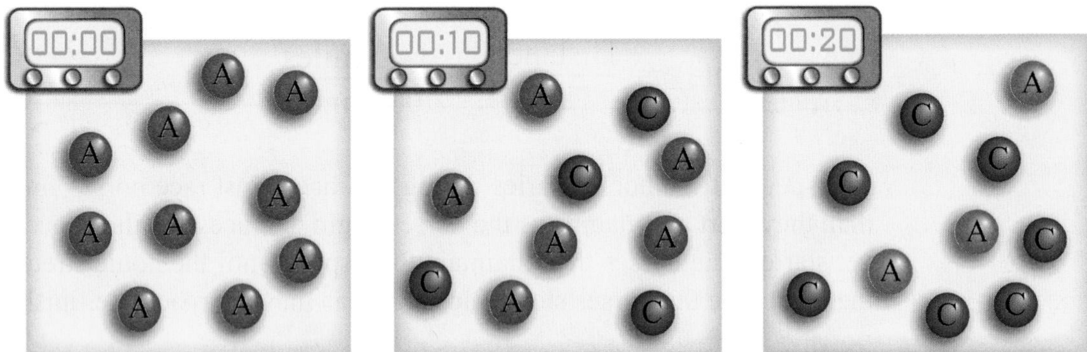

What do you notice about the reaction rate when you compare the three pictures?

Notice the time interval between each picture is constant, at 10 seconds, but the rate decreases. Why? Because fewer molecules of A are converting to C over time.

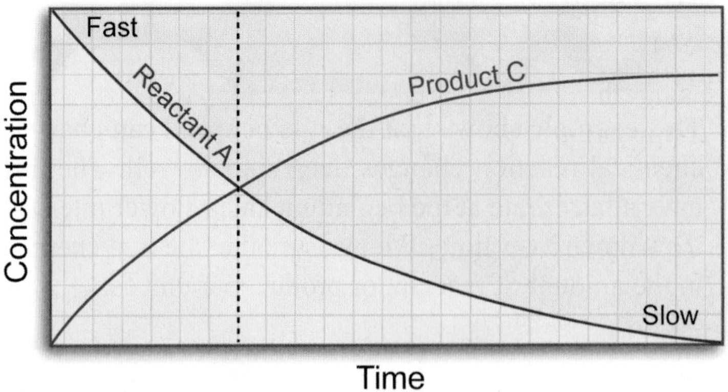

Chemistry terms

reaction rate - the speed at which a chemical reaction occurs, the change in concentration, of a reactant or product, per unit time.

Calculating average rate of reaction

The simplest way to calculate rate is to use the average rate equation. Average rate can be calculated between any two time periods during the chemical reaction.

Average rate equation

$$\text{Average Rate} = \frac{\text{Change in Concentraton}}{\text{Change in time}} = \frac{\Delta M \text{ (moles/L)}}{\Delta t \text{ (sec)}}$$

Delta means "change in"

TABLE 12.1. Data for Reaction A -> C (in a 1 Liter container)

Time (min)	Time (seconds)	Moles of A	Moles of C
0	0	1.00	0.00
10	600	0.74	0.26
20	1200	0.54	0.46
30	1800	0.40	0.60
40	2400	0.30	0.70

The average rate of appearance for the product C, can be found by:

$$\text{Average Rate C} = \frac{[\text{moles C (at t = 20) - moles of C (at t = 10)}]}{\text{20 min - 10 min}}$$

$$= \frac{\text{(0.46 mole C - 0.26 mole C)}}{\text{20 min - 10 min}} = \text{0.02 mole/min}$$

Using the information for the reactant A, we could calculate the average rate of disappearance for A. We would account for the disappearance by using a negative sign in front of the calculated rate, to indicate a decrease.

$$\text{Average Rate A} = \frac{\text{(0.54 mole A) - (0.74 mole A)}}{\text{20 min - 10 min}} = \text{- 0.02 mole/min}$$

Here the rate of appearance of C is equal to the rate of disappearance of A because they are in a one to one mole relationship in the equation: A → C.

The rate of reaction is not constant

If you calculate the rate between later time periods, you will see that the rate decreases. This indicates that the reaction goes slower and slower until there is no longer any measurable change in the concentration. This is because there is less and less reactant remaining, therefore less product will be formed per unit time.

Stoichiometry and rate

When using a balanced equation such as : $2A \rightarrow B + C$, where the mole ratios are not one to one, we use stoichiometry. You can correctly assume that the rate of appearance of B and C would be the same because they are present in the same, one to one, molar proportions. The rate of disappearance of A will be twice as much, because the balanced equation has a 2 in front of A.

Mole ratios and rate

The mole ratio from the balanced equation shows us the relationship between the rates of reactants and products. For example:

	$2HI(g)$	$\longrightarrow$	$H_2(g)$	+	$I_2(g)$
Rate:	$-2(6.0 \times 10^{-3}$ M/s)		$(6.0 \times 10^{-3}$ M/s)		$(6.0 \times 10^{-3}$ M/s)

if the rates of formation for I_2 and H_2 are 6.0×10^{-3}M/s, then the rate of HI will be $2(6.0\times10^{-3}$M/s), we then add a negative sign in front of the rate of HI because it is disappearing or decreasing.

The decomposition of N_2O_5 is shown below:

$$2N_2O_5(g) \rightarrow 4NO_2(g) + O_2(g)$$

The rate of decomposition of N_2O_5 was measured after 25 seconds and was found to be 5.60×10^{-6} M/s. What is the rate of appearance of NO_2?

Asked: *What is the rate of NO_2 production?*

Given: *Rate of decomposition of N_2O_5 is 5.6×10^{-6} M/s*

Relationships: *Mole ratio: 2 mole N_2O_5 = 4 mole of NO_2*

Solve: $$5.60 \times 10^{-6} \times \frac{4 \text{ mole } NO_2}{2 \text{ mole } N_2O_5} = 1.12 \times 10^{-5}$$

Answer: *The rate of appearance of NO_2 is 1.12×10^{-5} M/s.*

Discussion: *The mole ratio tells us that the rate of appearance of NO_2 is 2× the rate of N_2O_5 decomposition.*

It is very helpful to know how fast product is formed.

Using stoichiometry to calculate the rates of product formation, can help us to predict how fast the product will form, before we carry out the chemical reaction. In many cases this is helpful information. Sometimes reactions are so fast that they become explosive, for instance, when too much hydrogen gas is produced, and this can be dangerous. Other times we want to use a reaction that occurs rapidly, producing combustible gases for thrust, such as during a rocket launch.

The collision theory

Molecules must hit each other in order to react.

Chemical reactions take place at the particulate level, where molecules of reactant are colliding with each other. We can not directly observe this aspect of chemical change. As experimenters we can only observe the macroscopic changes. It makes sense that molecules need to hit one another in order to react. If they never collided how would atoms be exchanged? Chemists believe that there are certain criteria that must be met in order for a molecular collision to be "successful" and form product. Playing pool is a good analogy to understand these criteria or conditions.

First, if you wish to hit a pool ball into a pocket you need to line up the que ball with the ball you are attempting to hit. If the que ball is lined up properly, the angle at which it hits will cause the ball to roll in the direction of the pocket. What do you think some of the other factors are? Right! How hard you hit the que ball. If you do not hit it hard enough it will not make it to the pocket. You must hit the ball hard enough to get it in to the pocket, or else the two balls will just collide and roll away from each other.

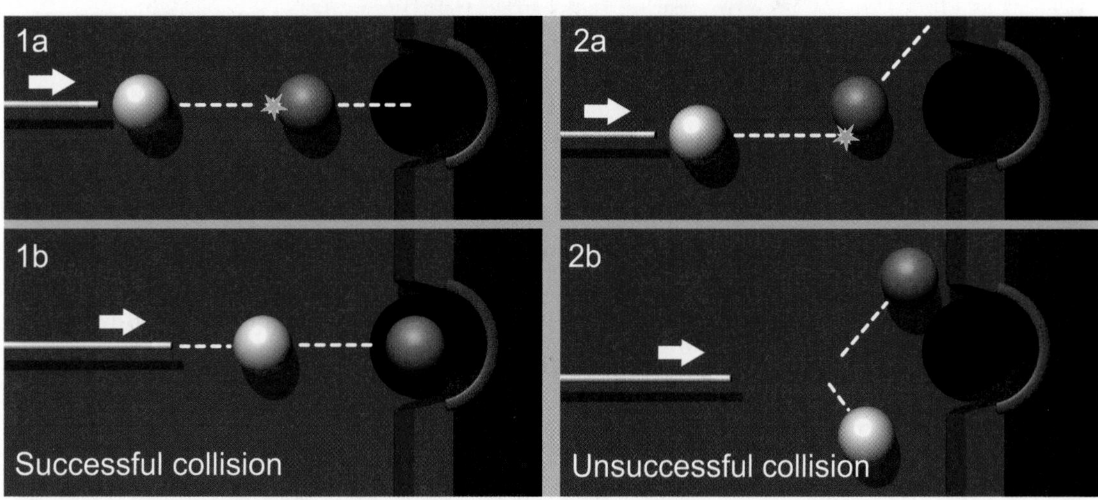

Activation energy

Transition state

Using the pool ball analogy as a comparison molecules have to hit each other properly as well. If the molecules hit each other with enough energy at the proper angle bonds can be broken, and this allows for bonds to be reformed. In order for atoms to be rearranged bonds need to be broken and reformed. The formation of a new compound would be analogous to the pool ball going into the pocket. The minimum amount of energy required for molecules to react is called the **activation energy, Ea**. When two molecules collide with sufficient activation energy a new high energy state can be achieved, called the **transition state**.

activation energy: minimum amount of energy required for molecules to react.

transition state: high energy state where bonds are being broken and reformed, also referred to as the activated complex.

Energy changes in a chemical reaction

A reaction profile

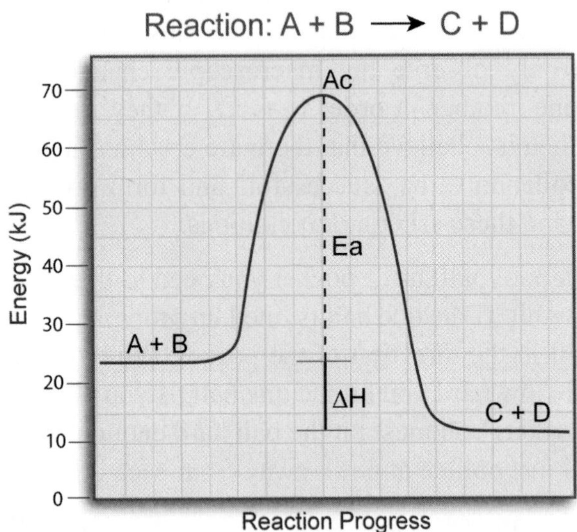

Energy changes always accompany chemical reactions. Here we use a graph called a **reaction profile** to describe what happens when molecules or particles collide.

The reaction profile helps us to visualize what happens when the reactant molecules collide with each other. The graph shows us that the reactants must have enough energy to get up the "hill." This is represented by the **activation energy**, **Ea**, labeled on the graph. Recall this is the minimum amount of energy required for the molecules to react. At the top of the hill, is the highest energy and this is place where the **activated complex**, **Ac**, forms. When the activated complex is formed it lasts for only a fraction of a second. Once the activated complex is formed, the colliding particles may form product or they may go back and reform reactants. The activated complex is extremely unstable because it forms at such a high energy. For this reason it cannot be isolated in the laboratory. From the high point on the graph where the activated complex exists there is an equal chance (probability) that products will be formed or that reactants will be reformed.

Activated complex

This means that even if the activated complex forms it may not lead to the formation of product!

This graph shows that the products are lower in energy than the reactants. In this case, energy is released and the reaction is exothermic, with a negative ΔH. If we subtract the final energy of the products C and D, approximately 12 kJ, from the reactants A and B, approximately, 23 kJ, we can see that the products are 11 kJ lower in energy. The fact that the products are lower in energy makes them more stable than the reactants. Molecules that are more stable are less reactive.

Chemistry terms

reaction profile : a graph showing the progress of a reaction with respect to the energy changes that occur during a collision.

activated complex: high energy state where bonds are being broken and reformed. Also referred to as the activated complex.

Analyzing the reaction profile

Now lets look at same reaction profile, but for the reverse chemical reaction.

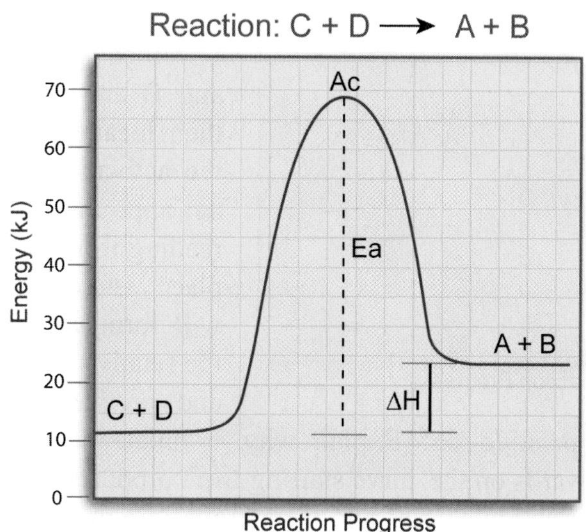

Here we see that the reactants, now C and D, are at a lower energy than the products, A and B. What does this tell us? Well lets assess the ΔH carefully here. The reactants are at about 12 kJ and the products are at about 23 kJ, so this means the products absorbed about 11 kJ of energy. This reaction profile shows us that the reaction is endothermic, with a positive ΔH, of 11 kJ.

We need to consider how this effects the energetics of the overall reaction! Look carefully at the hill. Is the hill higher now that we are looking at the reverse reaction?

Measuring the activation energy

To measure the activation energy, Ea we start at the reactants and measure the energy required to get to the top of the hill. Here we see that we start at 12kJ for the reactants and end at about 68 kJ at the top of the hill. The measured Ea here is approximately 56kJ. So indeed the hill is higher!

How can we relate this to the graph of the reverse reaction? If we take the Ea of the reverse reaction and add the ΔH to it we end up with the Ea of the forward direction that we see here! So remember that all chemical reactions are reversible. They often favor one direction over another. This explains why. The activation energy is smaller in one direction and when less energy is required the reaction is more likely to proceed in that direction. So when the "hill" is lower the reaction requires less energy to break bonds.

Exothermic reactions tend to be more common than endothermic reactions, because the hill is lower!

In nature exothermic reactions are more common. The energy they release is often used to make endothermic reactions that require more energy, go forward. This process is referred to as coupling reactions. Coupled reactions use the energy released from an exothermic reaction to make a harder endothermic reaction occur more easily.

Interpreting reaction profiles

Creating reaction profiles

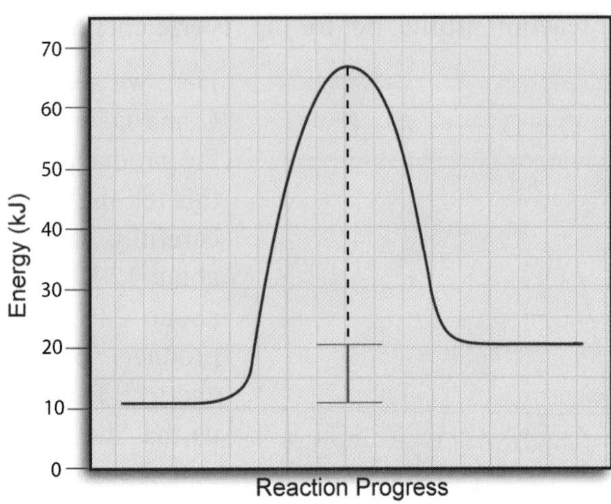

See if you can sketch this curve below and label the parts of the reaction profile.

Specifically, label the reactants, A and B and products, C and D, at their locations on the curve. Label the activation energy, Ea and give an approximate energy value by reading from the y-axis. Locate the place where the activated complex will form, and write Ac. Look at the relative energies of the reactants and products, do you think this graph has a negative or positive ΔH? Explain why. Estimate the value of the ΔH. What happens if you go backwards on the curve starting from products and going to reactants? Look carefully at the activation energy, and compare it with the activation energy of the forward reaction.

Lets look again at a molecular collision, and review the importance of molecular orientation.

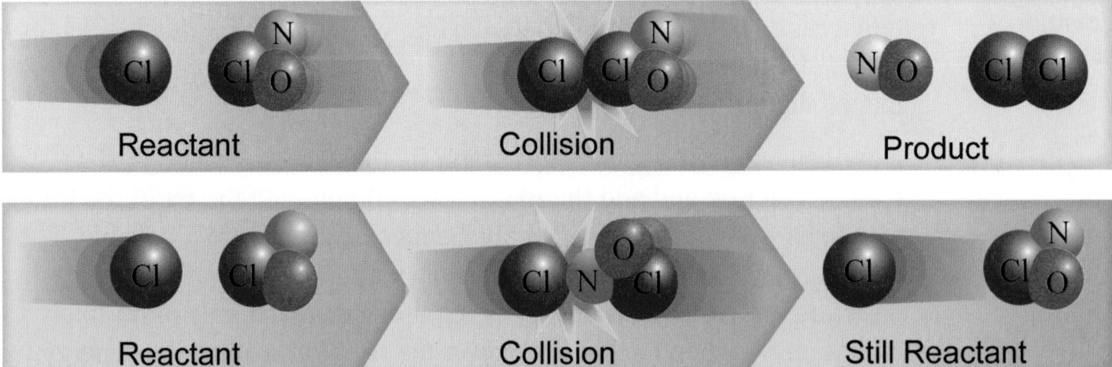

Molecules are always colliding

Molecules are colliding all the time! In the air for instance oxygen and hydrogen molecules are hitting each other constantly, but it is not always raining outside! Now we know that a collision does not always mean that a chemical reaction will take place.

Very few collisions result in the actual formation of product

Its a good thing too, because our bodies would be mush if every collision resulted in a reaction.

Collision theory and factors affecting rate

Higher concentration of reactant = more collisions

The collision theory explains why changing the concentration of the reactants changes the rate of a chemical reaction. Can you guess why? Look at the picture below.

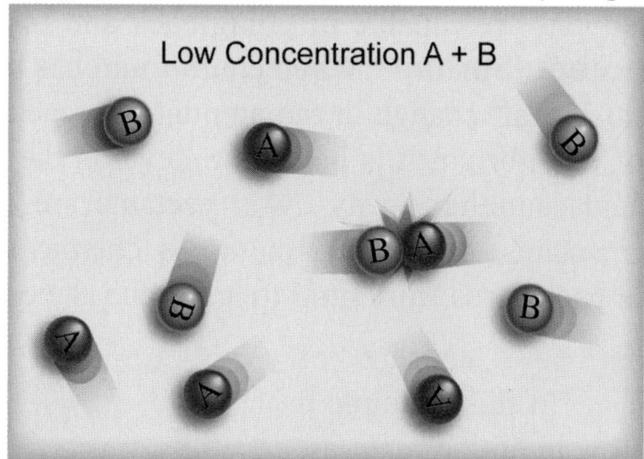

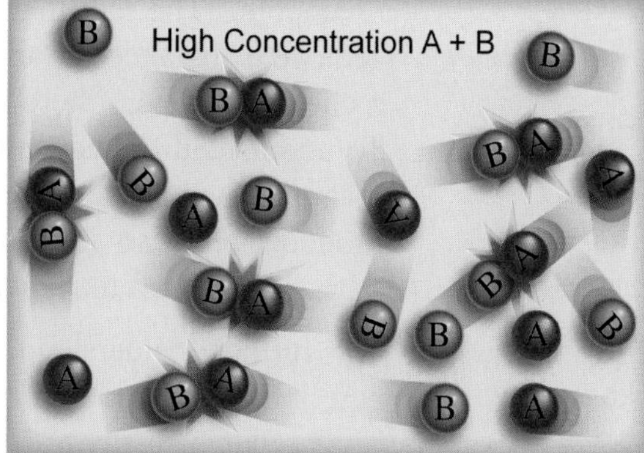

More particles leads to an increased number of collisions.

More collisions result when the concentration is higher, and this speeds up the reaction rate.

Increased temperature = more collisions

The collision theory also explains why increasing the temperature of a reaction mixture, generally increases the rate of the reaction. Temperature is a measure of the average kinetic energy of the reacting molecules. The higher the kinetic energy the higher the number of molecules that can collide with the necessary activation energy and get up the hill. If we raise the temperature of a reaction mixture a greater percentage of molecules will be moving faster and therefore they will have the energy, Ea, necessary to get up the hill.

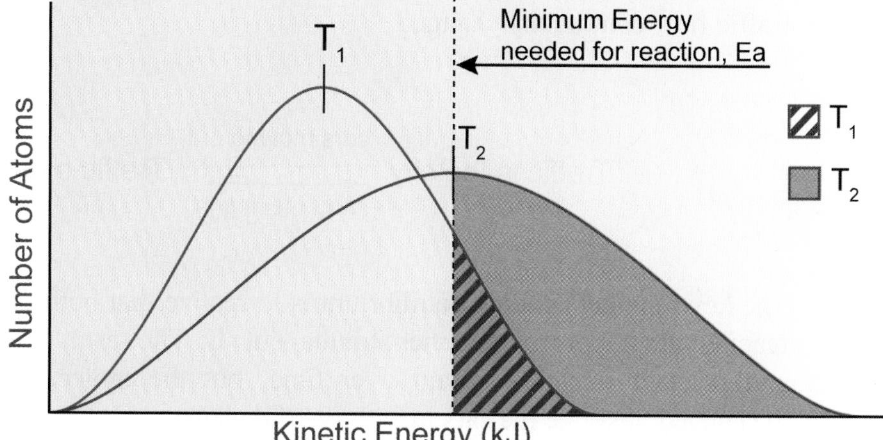

Increased surface area = more collisions

The affect of surface area can also be understood in terms of the collision theory. The larger the surface area the more particles available for collisions. For this reason solids are often ground up into a powder form, or dissolved in a solution, both of these methods allow for a separation of particles which allows them to react more readily.

12.2 Chemical Equilibrium

Chemical reactions are reversible

Up until now we have treated chemical reactions as though they only go in one direction, from reactants to products. In this section you will learn that chemical reactions are reversible. The fact that reactions are capable of going back and forth allows them to achieve a "balance", known as **equilibrium**. When equilibrium has been established in a chemical system there is no overall change in the amounts of reactants and products.

Equilibrium

To understand the concept of equilibrium it is helpful to develop the idea of chemicals forming products and reactants simultaneously. As reactants are forming products, products are also forming reactants, so we can picture a chemical reaction moving forward from left to right and backwards from right to left at the same time!

Equilibrium is like a two way bridge

To further understand this new concept of equilibrium, lets look a bridge. In this example a bridge with a steady traffic flow in both directions, represents a chemical reaction.

During a busy commute time, the traffic flowing in both directions on the bridge, is steady. The number of cars moving across the bridge into the city, and out of the city stays relatively constant.

When the number of cars moving into and out of the city, across the bridge, is constant we can say the system has established a balance. In contrast, a balance could not be established if the traffic flow was only in one direction, say into the city, or if the traffic flow continually changed.

$$\text{Traffic in the city} \underset{\text{cars moving in}}{\overset{\text{cars moving out}}{\rightleftharpoons}} \text{Traffic outside the city}$$

The key to understanding equilibrium is to realize that both the forward and the reverse reaction are happening together simultaneously. The result is that the *amount* of product and reactant remain constant over time, but the molecules are continuously being exchanged, just like the cars in and out of the city

Chemistry terms

equilibrium - a "balance" in a chemical system. At equilibrium the rate of the forward reaction is equal to the rate of the reverse reaction, and the concentration of reactants and products remain constant over time.

Physical equilibrium

Balance between evaporation and condensation

The evaporation of water in a closed container is a good example of physical equilibrium. It is called **physical equilibrium**, because it is an equilibrium between two phases of the same substance, and the changes that take place are physical changes. Here we will carefully consider what is happening to the water molecules as they move between the vapor and liquid phase. Lets take a glass jar, fill it about half way with water and then cover it. The flask will be kept at room temperature.

After we cover the jar what will happen? Over time, will anything change inside the flask?

We can show this cycle of water: $H_2O(l) \rightleftharpoons H_2O(g)$

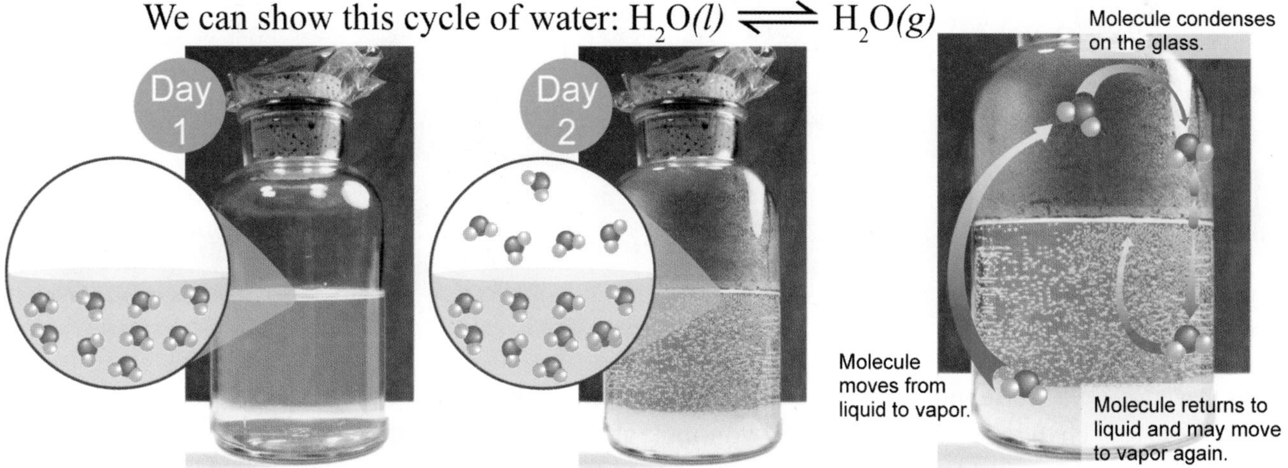

When the flask is first covered some molecules will be trapped as they evaporate and enter the gas phase. As a result some water vapor will build up over the water liquid in the flask. After a while some molecules in the vapor phase will begin to condense and drip back down to the liquid phase. When the number of molecules evaporating and the number of molecules condensing steadies-out and reaches a balance, a state of physical equilibrium will be achieved. Water molecules are always condensing and evaporating.

Change is always taking place at the microscopic (molecular) level

Water changes phases at equal rates

At the macroscopic level it may appear that nothing is happening, because the level of water in the flask remains the same. However, at the microscopic level molecules (particles) are continuously being exchanged between the liquid and the gas phase. Because the rate of the forward and reverse reactions are equal, there is not a noticeable change in the water level. Once equilibrium has been established, the water level appears to remain the same as long as the flask remains at a constant temperature.

Equilibrium is dynamic

Equilibrium systems are always changing.

One common misconception about equilibrium is that it is static and unchanging once it has been reached. This is because the system does not appear to change. To explain this it is helpful to consider a simple demonstration with beakers of water. Look a the picture of the beakers below. Notice they both have the same amount of water in them. If we remove 50mL of water from each beaker and we simultaneously pour it in to the other beaker the water levels in both beakers will still remain the same.

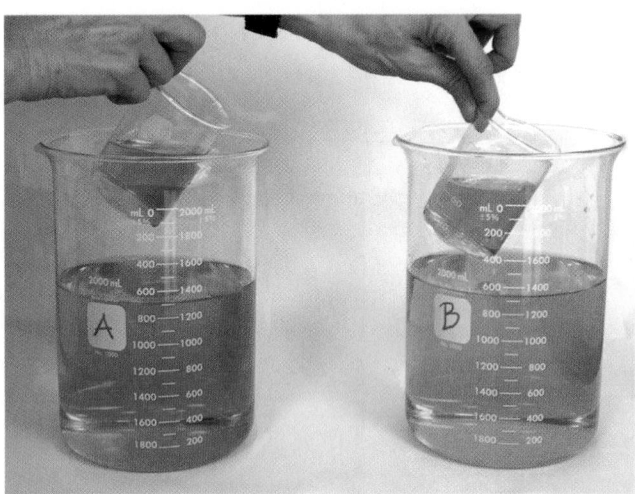

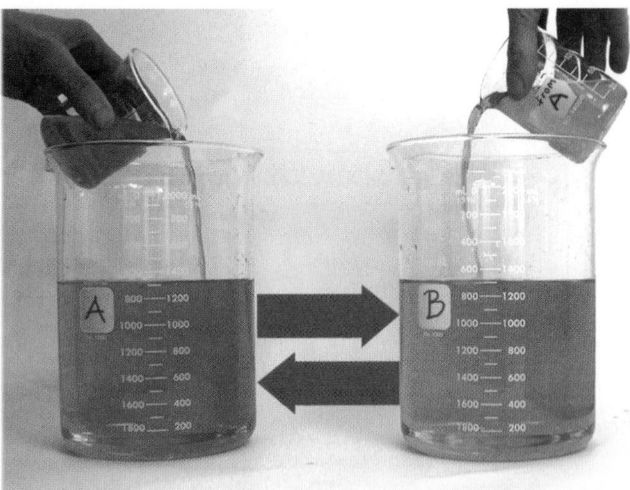

It looks like nothing is happening!

So while molecules of water moved from beaker A to beaker B, there was no noticeable change in either beaker. If you look at the beakers before and after the transfer of water you would not know that they were different. However we know that water molecules were transferred from beaker A to beaker B. In chemical reactions we are thinking about the rate of the forward reaction being equal to the rate of the reverse, this example mimics the equal rate by transferring equal amounts of water at the same time, so that the rate of change is equal over time. This is what fools us into thinking nothing happened. A chemical reaction has different chemicals that make up the forward and reverse reactions, however the rate at which the forward and reverse reaction take place is constant in equilibrium, so there is no overall observable change. Just like the beakers of water above.

It is important to emphasize that a balance is achieved between the two beakers. The amounts of water transferred are the same between the two beakers. In this example we started with beakers A and B having the same amounts of water in them. This would be similar to a chemical reaction where the amount of product and reactant are present in equal amounts. As you will see on the next page the equilibrium starting position and balance may not be the same. However, equilibrium still occurs.

Equilibrium balance

Balance is established no matter where the equilibrium system begins.

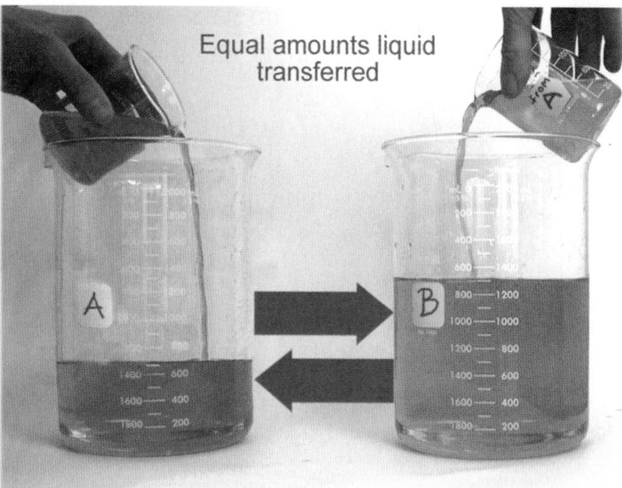

Here you can see that beakers A and B start with unequal amounts of liquid. However, the same amount of liquid is transferred between beakers A and B, which mimics the "rate" of the forward and reverse reaction being equal over time.

Equilibrium does not mean there must be equal amounts of reactants and products

Equilibrium simply means that a balance has been established. The "balance" is often different for different chemical reactions. Even though the beakers contain different amounts of reactant and product the water levels remain the same as long as the "rate" of transfer between the two is equal. Equal numbers of molecules are being exchanged between both beakers. The amount of reactant and product are not "equal" as shown by the unequal levels of liquid in the beakers.

Non-equilibrium

In a case of non equilibrium the amounts of water transferred would be *unequal* and over time *you would see a net change in the water levels*. The water levels would continue to become more unbalanced as "unequal" amounts of liquid are transferred. What would happen over time? We would notice that the water levels would noticeably change over time. One beaker would over-flow and one would become empty. In this case, there would be no balance achieved, because the rate of transfer is not equal between the beakers. Of course it is often impossible for us to see the molecules being exchanged back and forth, so to our eyes it appears as though nothing is happening. In the lab we rely on color changes and sampling amounts as they change over time. To measure the change over time requires some sensitive equipment, such as a spectrophotometer.

Chemical equilibrium

Chemicals establish equilibrium

In a **closed system** many chemical reactions do not form only products, but tend to have a mixture of reactants and products present. Recall that when a chemical reaction is in equilibrium we show a double arrow which indicates that both the forward and reverse reaction are taking place.

How can we observe the reversibility of a chemical reaction?

One method is to look for a color change inside the closed system. In this example the reactant $N_2O_4(g)$ is a clear and colorless gas, and the product $NO_2(g)$ is a red-brown gas. If we see a change in color it must be due to the change in amounts of the reactant and product. This is an example of chemical equilibrium, because the reaction shows a chemical change from reactant to product, not just a change in phase of the same substance.

$$N_2O_4(g) \rightleftharpoons 2NO_2(g)$$

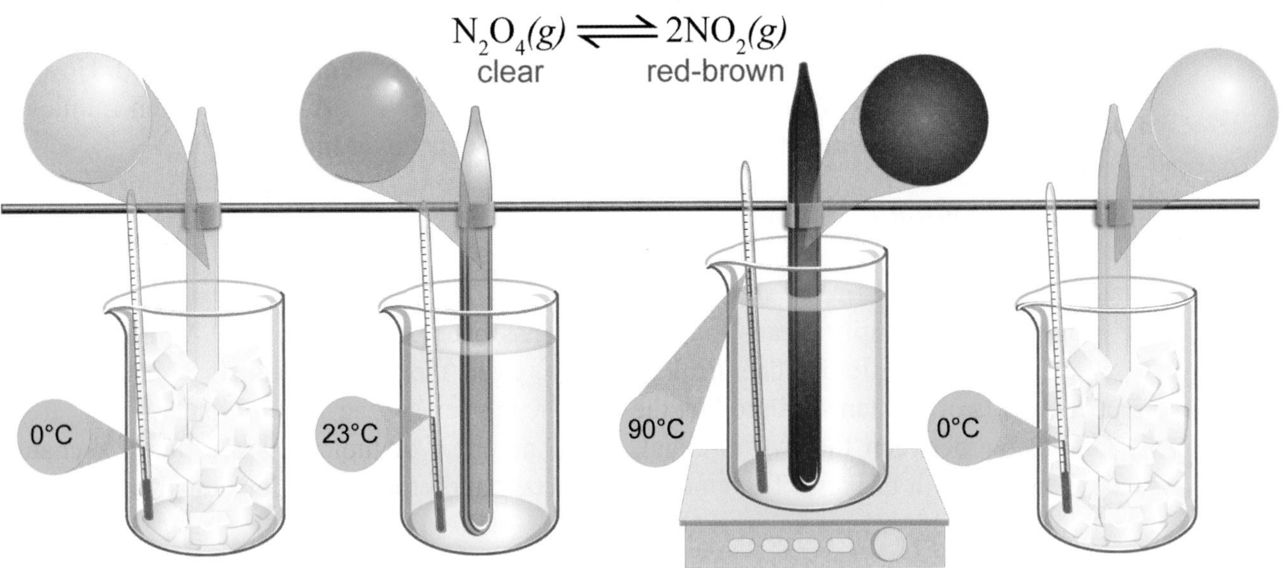

In each picture equilibrium has been established inside the tube, but the amount of product relative to reactant is different at different temperatures. Why the temperature causes the equilibrium balance to shift will be covered later in the chapter.

Chemistry terms

closed system - a system that is not open to the environment. Usually a chemical or a mixture of chemicals in a closed container.

Chemical equilibrium

Equilibrium in the air around us

This equilibrium of $N_2O_4(g) \rightleftharpoons 2\ NO_2(g)$ is present in the air around us. Nitrogen dioxide, $NO_2(g)$, is produced in automobile exhaust emissions. Depending upon where you live you are exposed to different amounts of the gases. Areas with high populations tend to have more people commuting to work, and therefore higher levels of NO_2 in the air. The NO_2 present in the air sets up a natural equilibrium balance with N_2O_4.

$NO_2(g)$ is considered a very dangerous air pollutant

Nitrogen dioxide, NO_2 is a poisonous gas that can cause serious trouble for people with breathing disorders such as asthma, or emphysema. When air quality is tested, and the pollutant levels are high, people are often warned to say indoors if they have any breathing difficulties. At hotter temperatures more NO_2 is present, and the air has more of the visible red-brown gas in it. Interestingly, because of NO_2's brown orange color it makes for lovely sunsets!

The Los Angeles skyline in smog.

Equilibrium can be established from either direction.

Equilibrium can be established by starting with reactant only or product only. If we start with pure $N_2O_4(g)$, which is frozen, in a closed flask, and we warm it to room temperature the forward reaction will create some $NO_2(g)$ and establish equilibrium. There needs to be no $NO_2(g)$ present initially in order for equilibrium to be established. In reverse, if we begin with just $NO_2(g)$ in the flask, the reaction will form reactant, N_2O_4, in order to establish equilibrium. In each case no matter what we start with equilibrium will be established.

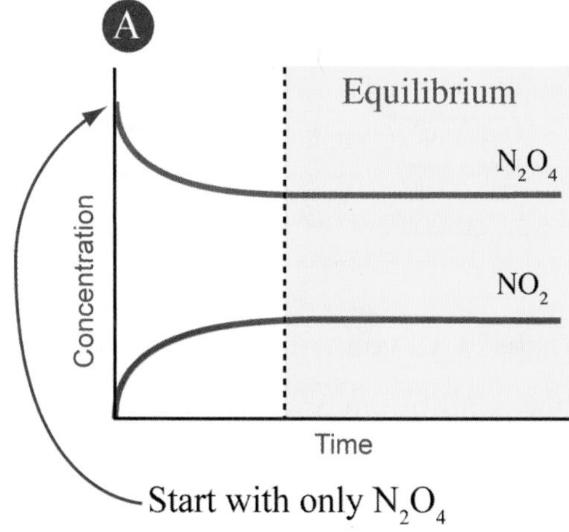

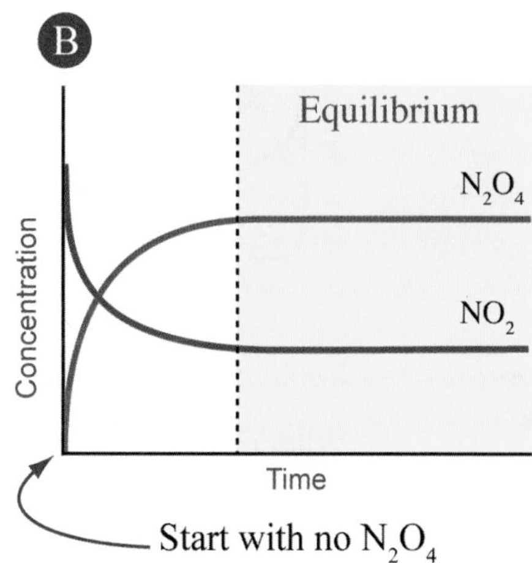

Equilibrium position

The balance that is achieved is different for different chemical reactions. Reversible chemical reactions can "favor" the direction of the products or reactants. It is quite rare that a reaction in equilibrium would have similar amounts of reactants and products.

Equilibrium does not mean equal amounts!

Equilibrium is achieved when the rate of the forward reaction is equal to the rate of the reverse reaction. However, this does not mean that the actual amounts of reactant and product are "equal" or the same. For any given chemical reaction we can determine the **equilibrium position** by measuring the amount of reactants and products present once equilibrium has been established.

In the N_2O_4/NO_2 system the reactant is "favored"

If you look carefully at the previous figure you will see that in each graph of the equilibrium system $N_2O_4(g) \rightleftharpoons 2NO_2(g)$ the concentration of reactant, N_2O_4 is higher than the concentration of product, NO_2. What does this tell us? Basically, that the reactants side of this equilibrium is "favored" under the conditions of 25°C and 1 atm. The fact that the N_2O_4 concentration is always greater, indicates that there is more of it present at equilibrium.

The relative equilibrium amounts of reactant and product present for a given reaction are dictated by the initial starting concentrations, and the temperature, and pressure of the system. For any given reaction it is possible that one direction of the reaction is favored over the other. To fully understand this behavior data is collected in the laboratory.

Below are 2 graphs of Concentration vs. Time of the $2SO_2(g) + O_2(g) \rightleftharpoons 2SO_3(g)$ equilibrium system. The first starts with all reactant, the second starts with just product. In each case over time equilibrium is established.

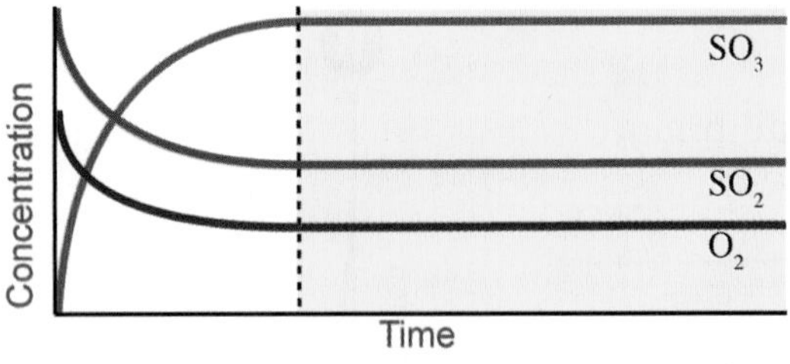

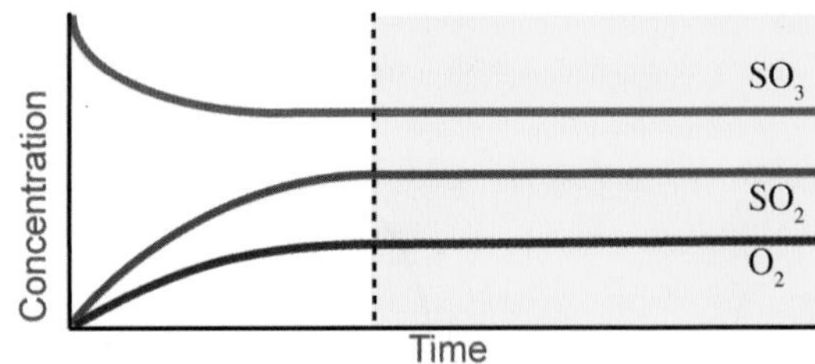

Analyzing graphs

Look carefully at the above graphs. Can you tell which side of the reaction is favored?

What does this tell you about the equilibrium position?

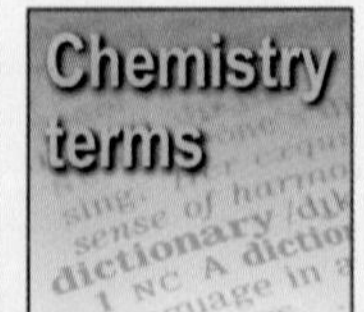

equilibrium position - the favored direction of a reversible reaction. Determined by each set of concentrations for the reactant(s) and product(s) at equilibrium.

Le Chatelier's principle

Chemists carefully study ways to influence equilibrium positions. In industry and in the environment it can be important to understand how to make a chemical reaction move more in one direction. What factors affect the direction an equilibrium reaction "favors"? How can we influence the reaction and make it go in the direction we desire? Four variables affect the direction of equilibrium: concentration, temperature, pressure and volume. In this section we will study how changes in conditions affect the equilibrium position.

Systems shift to "partially" offset the stress

A French chemist, Henri LeChatelier, proposed a good general rule that allows us to predict how an equilibrium reaction will respond to changes. **Le Chatelier's principle** states that when a "change" is placed on a system at equilibrium, the system will shift in a direction that partially offsets the "change". The change placed on an equilibrium system refers to changes in concentration, pressure, volume or temperature.

The "change" causes the equilibrium balance to be disturbed.

We avoid stress too!

Le Chatelier's principle is very applicable to our everyday lives! Just think about it. Most of us tend to avoid, or move away from things that cause us stress. Changes in our environment cause us to adjust our behavior. Chemical reactions behave the same way! Remember that equilibrium is a "balance" so anything that changes the balance would cause an adjustment

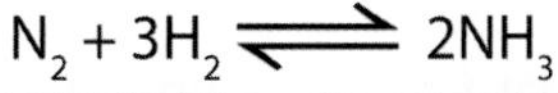

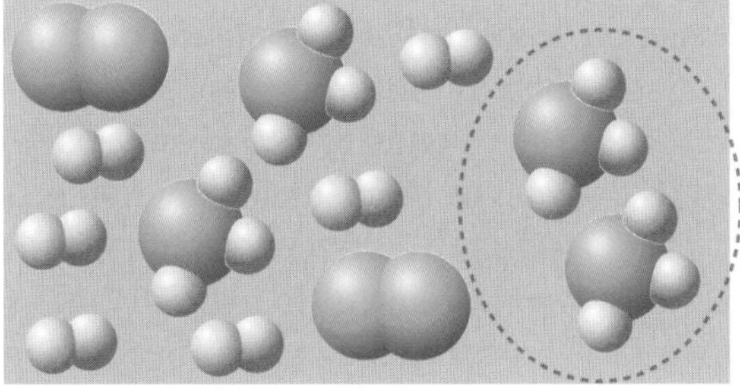

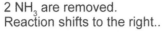

2 NH_3 are removed.
Reaction shifts to the right..

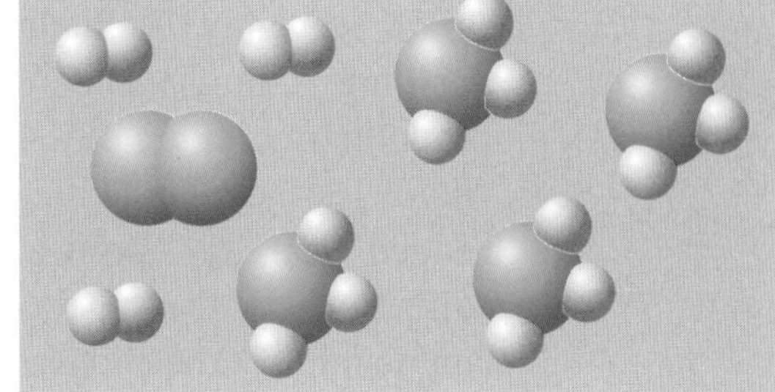

Some reactants are used to make more NH_3 to compensate for the removal.

Chemistry terms

Le Chateliers principle - states that when a "change" is placed on a system at equilibrium, the system will shift in a direction that partially offsets the "change". The change can be defined as a change in temperature, concentration, volume or pressure.

The effect of a change in temperature

Reversible reactions shift to absorb or release heat

Using the familiar equilibrium system for the N_2O_4/NO_2 system:

$$N_2O_4(g) + 58.0\ \text{kJ/mole} \rightleftharpoons 2NO_2(g)$$

clear red-brown

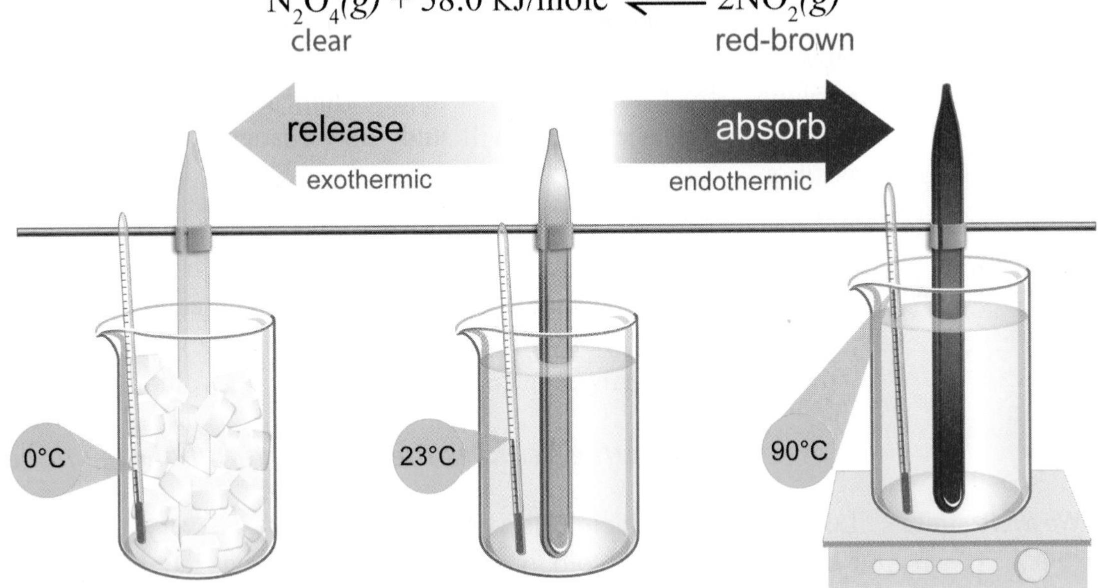

Here the reaction is absorbing heat, in the forward direction.

How does this system react to change in temperature?

1)An increase in temperature will cause the system to shift to decrease the temperature. Think about what you do if you are hot? You might remove your jacket or a layer of clothing, right? A chemical reaction "cools off" by taking in some of the added heat, and holding it in the chemical bonds, thereby absorbing some of the added heat. The endothermic side holds more heat in the bonds of the chemicals, the exothermic side holds less heat in the bonds of its chemicals. A chemical reaction gets rid of some of the added heat by "storing" it. The N_2O_4/NO_2 system is endothermic in the forward direction, so the reaction shifts toward products to partially off-set the increase in temperature. The picture above shows us that the system becomes a darker red-brown when placed in hot water.

2)A decrease in temperature will cause the system to want to release some heat to counter act the change. If the system is cooled down then it wants to compensate by heating up a little. How does the system release heat? It shifts to the exothermic side and releases some of the "stored" heat from the chemical bonds. Here the exothermic side is the reactants side: $N_2O_4(g) + 58.0\ \text{kJ/mole} \rightleftharpoons 2NO_2(g)$ When going from right to left using this chemical reaction heat is released and this partially compensates for the decrease in temperature. This reaction tells us that as 2 moles of $NO_2(g)$ form 1 mole of the reactant $N_2O_4(g)$, 58.0 kJ of heat is released.

Effect of change in concentration

Equilibrium shifts to counter act changes in concentration

Increasing or decreasing the concentration of the reactants or products, changes the amounts available for the overall reaction. If the amounts decrease the reaction will shift to increase them. If the amounts increase the reaction will shift to decrease them.

1) If the amount of SO_2 increases, the system will shift to the right, the products side to consume some of the added reactant.

$$2SO_2(g) + O_2(g) \rightleftharpoons 2SO_3(g)$$

Reactants Product

Shifts Right

In a reversible equilibrium reaction, forming product, means "using up" or consuming reactants

2) If the amount of SO_2 is decreased, the system will shift to the left, the reactants side to produce some of the lost reactant.

$$2SO_2(g) + O_2(g) \rightleftharpoons 2SO_3(g)$$

Reactants Product

Shifts Left

Forming reactants, means consuming some of the products

Just remember if there is an increase in concentration Le Chatelier's principle says that the system will shift to decrease some of the concentration. The reverse is also true.

Using the following equilibrium system: $PCl_5(g) \rightleftharpoons PCl_3(g) + Cl_2(g)$

Predict the direction the system will shift, as a result of an increase in the concentration of $Cl_2(g)$.

Asked: *Predict the direction the system will shift?*

Given: *Increase in* $Cl_2(g)$

Relationships: *Note that* $Cl_2(g)$ *is a product on the right side of the equation.*

Solve: *The system will shift toward the reactant to consume some of the added* $Cl_2(g)$.

Relationships: *Shifts left to produce some* PCl_5

Discussion: *Partially off-sets the increase in* $Cl_2(g)$

Effect of change in pressure or volume

Changes in pressure or volume only effect gaseous equilibrium systems. To understand how a change in pressure or volume can influence an equilibrium system we will first look a how pressure and volume are related. In the example below we will assume we have a container of fixed volume, with rigid walls that cannot expand or contract.

Gas particles can compress or expand when inside a piston

The picture below shows a piston with a gas trapped inside. In figure a, when the pressure on the gas is increased, the volume of the gas decreases. On the other hand in figure b, when the pressure on the gas decreases, the volume of the gas is increases and the gas expands.

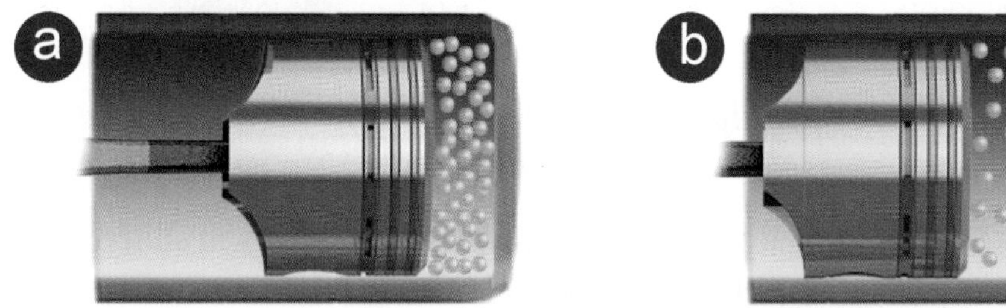

We can study the effects of pressure changes on the following equilibrium system:

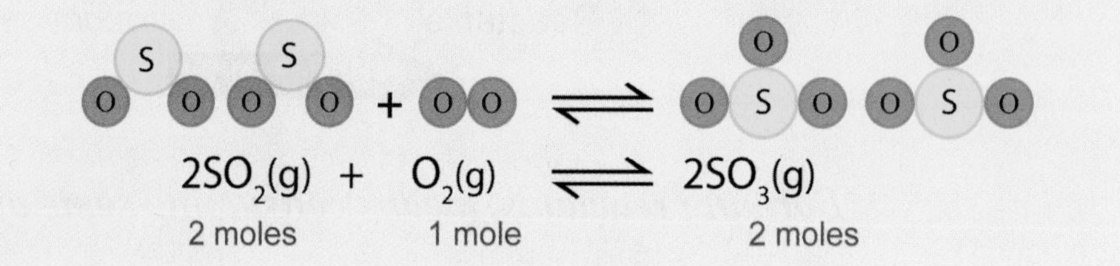

Consider the system to be in a container that has a movable piston, like the one shown above.

What happens when pressure changes?

1) Increased Pressure on the System: Le Chatelier's principle predicts the system will adjust to partially decrease the pressure. Chemical reactions do this by shifting to the side of the reaction with fewer moles. In this case that would be the products side. Two moles of SO_3 gas exerts less pressure than 3 moles of the reactant gases (2 moles SO_2 plus one mole $O_2(g)$). The fewer the gas particles the lower pressure. Think of each individual gas particle as a force that pushes, the more particles of a gas the higher the pressure.

2) Decreased Pressure on the System: Here the reaction shifts to the side with more moles to try to raise the pressure. The side with more moles in this case would be the left side or the reactants.

3) If the total moles of gas are equal on both sides of the gaseous equilibrium equation then there would be no shift, because neither direction would partially off set the change.

The equilibrium expression

ADVANCED AP

To gain a better understanding of equilibrium, we must study experimental data, that allows us to look at the relationship between the forward and the reverse reaction. How can we tell if a chemical reaction prefers to go one way or another? How can we approach finding out?

Which direction is "favored"?

Lets pretend there is a path that goes off in two directions. How could we tell if one direction is favored over the other? The easiest way would be to sit by the path and see which direction people go. We would then have some data to predict the favored direction. If several of our friends help and they record observations a few more times, the data could then be "pooled" and we could see if one direction is favored over the other. The problem with this analogy is that people have a choice, and they might make a different choice from day to day. However, chemicals have no choice, which direction they "favor" is dependent upon the reaction conditions, such as temperature. But we determine the favored direction in the same way. We observe and collect data.

Experimental observations led to the equilibrium expression

After many experiments, two Norwegian chemists proposed the following expression based on their observations. This was called the **law of mass action**, which is a general description of equilibrium. For any equilibrium reaction : $aA + bB \rightleftharpoons cC + dD$, a moles of reactant A and b moles of reactant B, react to yield c moles of product C and d moles of product D.

Using this generic reaction we can learn how to set up the **equilibrium expression**.

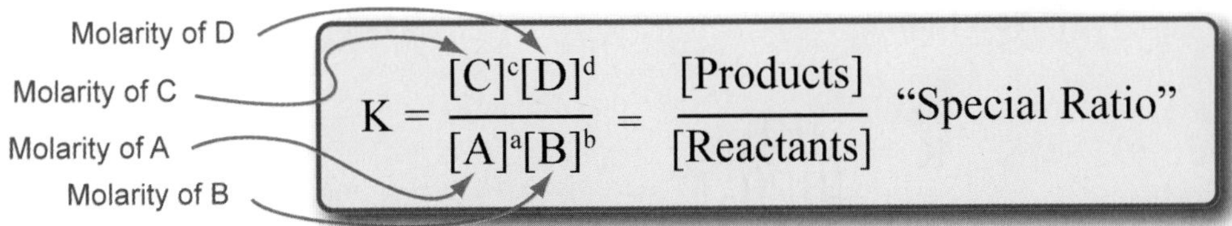

Equilibrium expression

Here our reactant and product amounts will be measured in units of concentration, which are molarity, M. The use of brackets around our reactants and products indicates that they are concentrations. Each concentration is raised to the power of the corresponding coefficient in the balanced equation. K is the equilibrium constant which represents the numeric value of the ratio. The **equilibrium constant** always has the same value, at a specified temperature, for a specified chemical reaction. Most importantly, equilibrium data repeatedly support this special ratio.

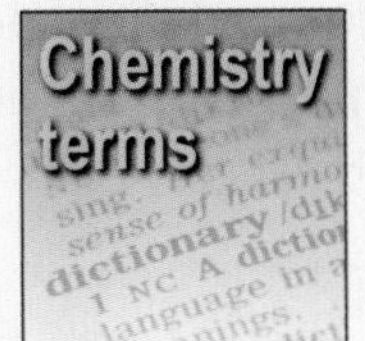

equilibrium expression - is a special ratio of product concentrations to the reactant concentrations. Each concentration is raised to the power of the corresponding coefficient in the balanced equation.

equilibrium constant - numeric value of the equilibrium expression.

law of mass action - general description of any equilibrium reaction.

The equilibrium constant

ADVANCED AP

The constant K is temperature dependent.

The equilibrium constant K is only affected by changes in temperature, otherwise it remains the same for a given reaction.

Lets write an equilibrium expression for the following reaction.

$$N_2O_4(g) \longleftrightarrow 2NO_2(g)$$

Reactants   Product

$$K = \frac{[NO_2]^2}{[N_2O_4]}$$

$$N_2O_4(g) \rightleftharpoons 2\,NO_2(g)$$

We can predict equilibrium concentrations using K

The importance of the equilibrium expression is that we can substitute known concentration values for reactants and products and calculate the constant, K. If we know the value of K, we can predict equilibrium concentrations, before doing the experiment!

<u>Practice Writing Equilibrium Expressions:</u>

Write the expression for: $H_2(g) + I_2(g) \rightleftharpoons 2HI(g)$

solution:

$$K = \frac{[HI]^2}{[HI][I_2]}$$

Write the expression for: $CH_4(g) + 2H_2S\,(g) \rightleftharpoons CS_2(g) + 4\,H_2(g)$

solution:

$$K = \frac{[CS_2][H_2]^4}{[CH_4][H_2S]^2}$$

Always remember that the products are raised to the power of their molar coefficients and go in the numerator, and the reactants are raised to the power of their molar coefficients too but go in the denominator. The number of moles of the reactants and products that appear after the equation is balanced make the equilibrium expression simple or more complex depending upon the number of moles of each substance.

Calculating equilibrium constants

For a specific reaction at a given temperature the equilibrium constant K will always be the same, or constant. Using data from many experiments we find that when we solve for K it is always the same.

Experimental Data

TABLE 12.2.

Results of Three Experiments for: $N_2(g) + 3H_2(g) \rightleftharpoons 2NH_3(g)$ @ 500°C				
	Equilibrium Concentrations			
Experiment	$[N_2]$	$[H_2]$	$[NH_3]$	$K = \frac{[NH_3]^2}{[N_2][H_2]^3}$
1	0.921	0.763	0.157	0.0603
2	0.399	1.197	0.203	0.0602
3	2.59	2.77	1.82	0.0602

Lets calculate K using the above data and see if it agrees with the last column value of 0.0602!

Solving for K

Applying our equilibrium expression to the data from experiment 1:

$$K = \frac{[NH_3]^2}{[N_2][H_2]^3} = \frac{[0.157]^2}{[0.921][0.763]^3} = 0.0603$$

Applying our equilibrium expression to the data from experiment 2:

$$K = \frac{[NH_3]^2}{[N_2][H_2]^3} = \frac{[0.203]^2}{[0.399][1.197]^3} = 0.0602$$

We can apply the same calculation to experiment 3 and indeed we do get the same value for K. This supports the fact that K is constant at the same temperature.

The very interesting thing to notice is that for each experiment the equilibrium concentrations are all VERY different!

However, when applied to our equilibrium expression we get the same value for K. Equilibrium concentrations may be different for each experiment, but the “special” ratio is the same in each case!

Using the equilibrium constant

Like you have just seen in the previous section, the equilibrium expression is used to determine a numeric value for the equilibrium constant. Chemists use the equilibrium constant to tell them how much a particular direction is favored by the equilibrium system, and to solve for equilibrium concentrations of reactant(s) or product(s).

The size of K predicts the favored direction

The numeric size of K, whether it is large or small, tells us to what extent the reaction goes to product. Using the simple reaction A(g) $\rightleftharpoons$ B(g). We can set up the expression as :

$$K = \frac{[B]}{[A]}$$

If K is very large ($K > 10^4$) then this means that [B] must be much larger than [A] and the products are favored for this reaction.

If K is very small ($K < 10^{-4}$) then this indicates that [A] must be much larger than [B] and the reactants are favored for this reaction.

Sometimes K is neither very large or very small it is somewhat in between, this indicates that there are appreciable amounts of BOTH reactant [A] and product [B].

Here are some equilibrium constants for some common reactions:

$2\,H_2(g) + O_2(g) \rightleftharpoons 2\,H_2OH(l)$ $K = 1.4 \times 10^{83}$ at 25°C	Large K favors product
$CaCo_3(s) \rightleftharpoons CaO(s) + Co_2(g)$ $K = 1.9 \times 10^{-23}$ at 25°C	Small K favors reactant

Solving for an unknown equilibrium concentration:

Calculate the equilibrium concentration of [HI] given the following information.

$H_2(g) + I_2(g) \rightleftharpoons 2HI(g)$. K = 50 at 450°C, the $[H_2]$ = 0.22 M and $[I_2]$= 0.22 M

Asked: *Calculate the concentration of HI at equilibrium.*

Given: *K = 50 at 450°C, and $[H_2] = [I_2] = 0.22$ M*

Relationships: *equilibrium expression :* $K = \frac{[HI]^2}{[H_2][I_2]}$

Solve: $K = 50 = \frac{[HI]^2}{[0.22][0.22]} = \frac{[HI]^2}{0.0484}$

so 50 x (0.0484) = 2.42 = $[HI]^2$

the square root of 2.42 = 1.56 M = [HI]

Answer: *The equilibrium concentration of [HI] = 1.56 M*

Discussion: *Here we used a known equilibrium constant (K) to solve for the unknown equilibrium concentration of product.*

12.3 Chemical Pathways

Chemical reactions take place through a series of steps.

So far we have always treated chemical reactions as though they go from reactant to product all in one step. Atoms in the reactants are rearranged to form product seemingly all at once. This is actually a big over simplification. Now we will consider the fact that each reaction occurs in a series of **elementary steps**. These elementary steps combine to form a "pathway" for a chemical reaction. The term "pathway" helps us to visualize the series of paths or steps that take place when reactants are converted to products. The series of steps is referred to as a **reaction mechanism.** A reaction mechanism is proposed based on experimental evidence, and it must be supported by what is observed in the laboratory. Sometimes there is more than one proposed reaction mechanism for a chemical reaction. Chemists cannot know for sure which mechanism, or pathway, is correct because as you learned in the first section 13.1 the high energy molecules formed during the reaction cannot be isolated, because they exist for a such a small fraction of time.

A good analogy to what a reaction mechanism represents, is following a map on a trip. You follow a series of roads that take you to your destination, the series of roads represents the reaction mechanism. An overall chemical reaction only shows the start of the trip and the end of the trip, but it is impossible to go directly to your destination. You have to follow the roads. The reaction mechanism shows the roads you take on your trip, as well as the start and finish.

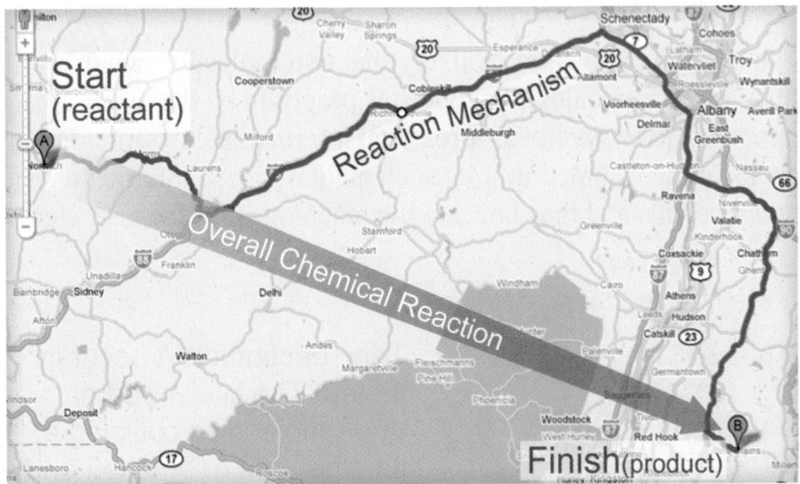

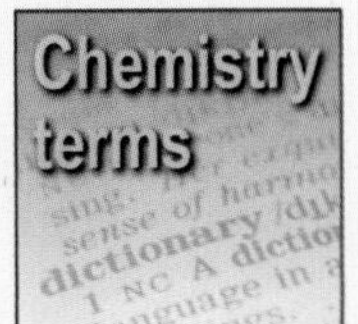

elementary steps - a series of simple reactions that represent the overall progress of the chemical reaction at the molecular level.

reaction mechanism- a proposed sequence of elementary steps that leads to product formation. Must be determined using experimental evidence.

Two step reaction mechanism

Lets look at an example of a reaction mechanism with two elementary steps. Here we will use the reaction between hydrogen, $H_2(g)$ and iodine monochloride, $ICl(g)$.

Overall reaction: $H_2(g) + 2\ ICl(g) \rightarrow I_2(g) + 2\ HCl(g)$

The proposed reaction mechanism that is supported by experimental evidence is :

The elementary steps must add to the overall reaction.

1) elementary step: $H_2 + ICl \rightarrow HI + HCl$

2) elementary step: $HI + ICl \rightarrow I_2 + HCl$

Overall reaction: $H_2 + HI + 2\ ICl \rightarrow I_2 + HI + 2\ HCl$

The sum of the elementary steps must yield the overall balanced equation. In the first elementary step, one H_2 molecule collides with one ICl molecule. This step is followed by the collision between HI and the second ICl molecule.

Note that it is much more likely that two molecules would collide favorably in each step than three molecules all at once, as proposed by the overall reaction. HI is referred to as a reaction **intermediate**. A reaction intermediate is formed in one of the early elementary steps and consumed in a later elementary step. Intermediates are higher in energy than the reactants and they become the reactant(s) for the next elementary step.

This reaction can't occur in 1 step! The likelyhood of three molecules colliding at just the right orientation and speed is just about zero.

Thinking back to the collision theory it is unlikely that all 3 reactant molecules collide with sufficient energy, and at the proper orientation to form products!

Chemistry terms

intermediate - chemical species that is formed during the elementary steps, but is not present in the overall balanced equation.

Reaction progress curves

Most reactions have more than one elementary step.

How reactions proceed can be graphed in the form of a reaction profile. The simple reaction profile that you learned about earlier is only for one elementary step. Here we will look at the reaction progress for a reaction with two elementary steps leading to product.

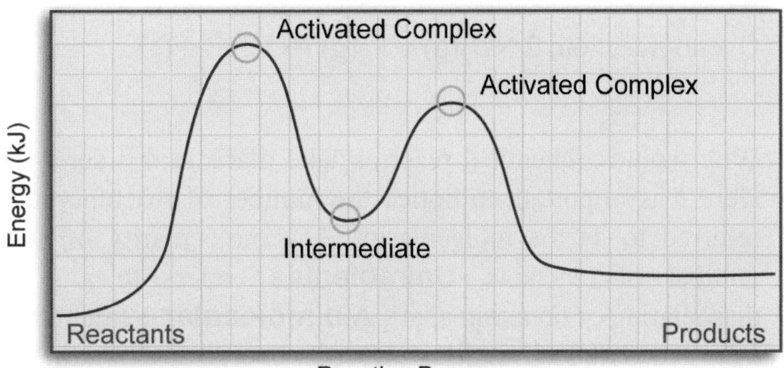

This graph shows the formation of reaction intermediate (label in picture). You can pinpoint the place on the curve where the intermediate is present by looking for a dip (or valley) in the curve that is higher in energy than the reactants. The reaction intermediate becomes a reactant in the following step. The overall graph has two peaks, which refer to the energies of the two activated complexes, one for each elementary step.

Analyzing Graphs:

Look at the graph below to help you answer the following questions.

How many elementary steps are there in this overall reaction? How many intermediates are formed?

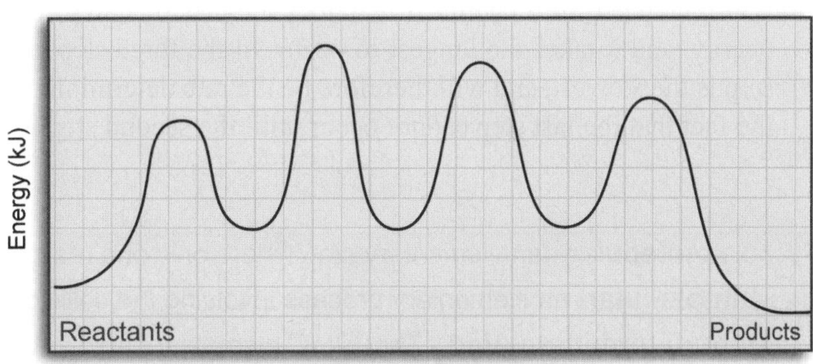

Reaction mechanisms

A more complex mechanism involves 3 steps.

As you just saw in the graph some reactions have more than just two elementary steps. Here is an example of a proposed mechanism with three steps:

elementary step 1: $NO_2(g) + F_2(g) \rightleftharpoons NO_2F_2(g)$ (fast)

elementary step 2: $NO_2F_2(g) \rightarrow NO_2F(g) + F(g)$ (slow)

elementary step 3: $F(g) + NO_2(g) \rightarrow NO_2F(g)$ (fast)

Overall reaction: $2\ NO_2(g) + F_2(g) \rightarrow 2\ NO_2F(g)$

Why would a chemical reaction take place in this series of complicated looking steps? Here it is important to notice the number of reactants in each elementary step. This is referred to as the molecularity. The molecularity of an elementary step tells us the number of reactants. **Unimolecular** refers to a process with a single reactant dissociating or breaking apart. A **bimolecular** elementary process involves the collision of just two molecules. Remember an overall reaction that involves three molecules, here two NO_2's and one F_2 reacting, is highly unlikely! So the complicated steps actually simplify the overall process. Each step involves either one or two reactants coming together, which is more probable, based on the collision theory.

All mechanisms must be determined by experiment.

The other very important aspect of a proposed reaction mechanism is that it must correspond to the experimentally observed rate law. The rate law is an equation that predicts the relationship between the concentration of reactants and the speed of the chemical reaction. From the predicted reaction mechanism it is possible to predict the "slow step" or the **rate determining step**. This can be identified as the elementary process that takes the longest, and therefore acts as a bottle-neck, in the overall reaction mechanism. A good analogy for the rate determining step is a toll booth on a road trip. You may be able to travel at 55 mph on the highway, but you must slow down and wait in line to get through the toll. Once you are through the toll you can once again resume your 55 mph speed. In a chemical reaction, there is one step that has the highest activation energy, and it takes the longest to occur. In the three step mechanism above, the second step is the slowest, and will therefore be the rate determining step. Remember to consider the fact that the last step cannot occur until the second step is complete.

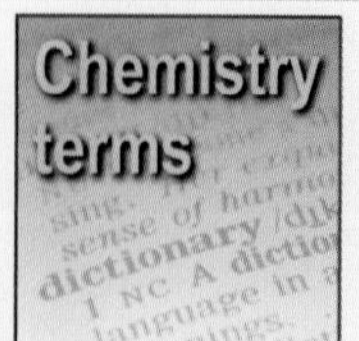

unimolecular - an elementary step where only one reactant is involved.

bimolecular- an elementary process involving the collision of two molecules.

rate determining step - The "slow" elementary step in the reaction mechanism determines the overall rate (or speed) of the chemical reaction.

Biological pathway

Your body carries out complex chemical reactions as a series of steps, as do many microorganisms. Most biological pathways involve many enzymes and are the subject of biochemistry. Here we will study the fermentation process as a simple example of how living organisms use a "pathway" to carry out an overall chemical reaction.

Fermentation looks like a simple chemical reaction!

Yeasts and a few other microorganisms use alcoholic fermentation to produce energy.

When starting with a six carbon sugar such as glucose, the overall chemical process is:

$$C_6H_{12}O_6 \rightarrow 2\,C_2H_5OH + 2\,CO_2 + \text{Energy (2 ATP's)}$$

In words: glucose(sugar) → ethanol + carbon dioxide + energy

During the process of fermentation the overall sugar molecule is modified during several chemical reactions which are facilitated by specific enzymes.

Breakdown of the glucose molecule via fermentation requires 12 enzymes!

In the first step of this chemical pathway the sugar, glucose, is broken into two pyruvic acid CH_3COCO_2H, molecules.

Step 1 splits the glucose molecule

$C_6H_{12}O_6 \rightarrow 2\,CH_3COCO_2H$ This step requires 10 different enzymes!

This one step is really a series of elementary steps where the substrate glucose is oxidized and the electrons are given to an electron acceptor NAD^+ to form NADH. The cells use the NADH as a form of energy for their metabolic processes.

Step 2 produces ethanol and carbon dioxide

The second step in this pathway, where pyruvic acid is broken down into ethanol and carbon dioxide yields NAD+ which is recycled back to the first step.

$2\,CH_3COCO_2H \rightarrow 2\,C_2H_5OH + 2\,CO_2$ This step requires 2 enzymes.

12.4 Catalysts

Catalysts facilitate chemical reactions

Catalysis is one of the most important areas of chemical research. A **catalyst** is a substance capable of speeding up a chemical reaction, without being "used-up" during the reaction. The fact that catalysts are unchanged by a chemical reaction, makes them able to be reused over and over. In our experiences with chemical reactions so far reactants are consumed when they make product. A catalyst is NOT a reactant. You could think of catalysts as facilitators of chemical reactions.

Catalysts work by providing a new pathway with a lower activation energy.

Reaction: A + B ⟶ C + D

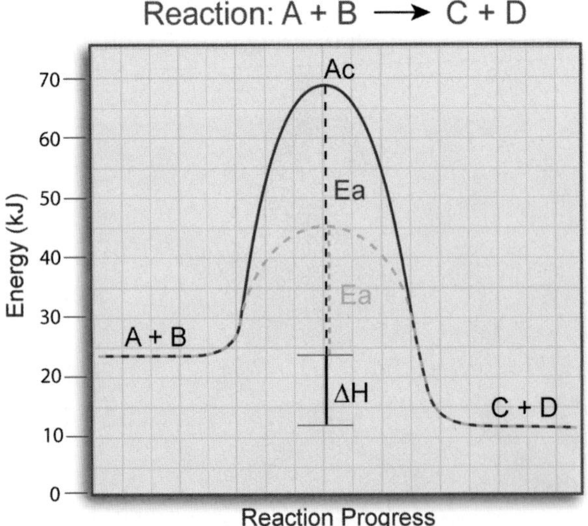

In the graph above, the reaction A + B → C + D has an uncatalyzed Ea (activation energy) of 250 kJ. The red line shows the Ea of the catalyzed reaction is approximately 125 kJ. The activation energy of the catalyzed reaction is half that of the uncatalyzed reaction. This is amazing! Because the activation energy is lower, more molecules now have sufficient energy to react, causing the reaction to happen at a faster rate overall.

We are alive right now, because our bodies contain thousands of catalysts, which speed up complicated reactions that would be too slow at normal body temperatures.

The types of catalysts that work in our bodies are called **enzymes**.

Chemistry terms

catalyst - a substance that speeds up the rate of a chemical reaction, by providing a pathway with a lower activation energy.

enzymes- biological catalysts, responsible for speeding up the chemical reactions inside our bodies.

Enzymes

One enzyme that is very important in our bodies is called carbonic anhydrase. Carbonic anhydrase is responsible for the rapid conversion of carbon dioxide and water to acid, H^+ and bicarbonate, HCO_3^-.

Carbonic anhydrase catalyzes the reaction of CO_2 with water

$$H_2O(l) + CO_2(g) \rightarrow H^+ + HCO_3^-$$

Carbonic anhydrase helps to remove carbon dioxide, $CO_2(g)$ from your tissues, by dissolving it in the form of HCO_3^- ion, which can be carried to the lungs by our blood. Once it reaches our lungs CO_2 converts back to it's gaseous form and is exhaled. Without carbonic anhydrase we would not be able to effectively remove the necessary carbon dioxide from our tissues.

If you take a sip of really cold soda you will feel a tingling on your tongue. This sensation occurs because your saliva contains carbonic anhydrase, which creates some acid, H^+ and when acid comes into contact with the nerve endings on your tongue it makes them tingle.

Bromelain is an enzyme found in pineapples.

Another natural enzyme, bromelain, is contained in certain fruits. Pineapple, contains the bromelain enzyme, which helps to digest or "break-down" proteins. At the molecular level, bromelain breaks the bonds which hold the protein's structure together. We can witness the effect that the enzyme has on proteins, by using a simple experiment with gelatin and pineapple. Gelatin is a protein made from collagen, which comes mainly from the bones and connective tissue of animals. The experiment is shown in the pictures below. Picture a) shows the control (without the pineapple enzyme) and it "gels" properly. In picture b) fresh pineapple was mixed with the gelatin. Here the gelatin will not "gel" or become firm. The protein in gelatin cannot hold together in the presence of the bromelain enzyme.

Holds it's shape.

Doesn't hold it's shape with pineapple.

Environmental effects of catalysts

Ozone O_3, absorbs harmful UV rays

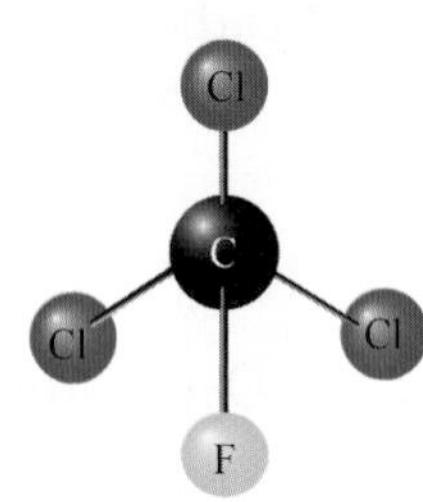

Freon –11

The ozone layer in the upper atmosphere protects us from the sun's harmful ultraviolet radiation. Chemicals called chlorofluorocarbons, or CFC's, have been shown to interfere with ozone formation. Chlorofluorocarbons were widely used in air conditioning systems, as refrigerants, in aerosols, and in the manufacture of styrofoam(TM) products.

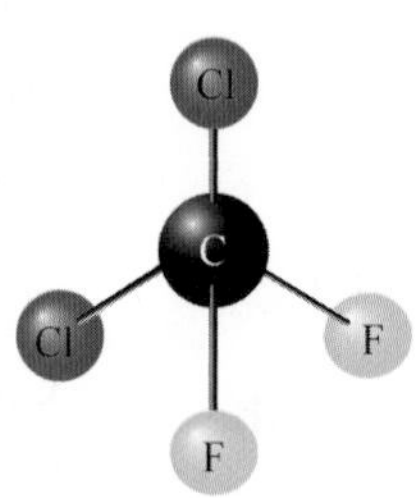

Freon –12

Cl atoms act as catalysts in the destruction of ozone, O_3.

These CFC's are relatively inert, or unreactive, near the earths surface, but as they float up into the atmosphere the sun's energy breaks off a chlorine atom, $Cl^{\bullet}$, which is a free chlorine radical with an unpaired electron. This chlorine atom acts as a catalyst in the depletion of ozone.

Natural Ozone Cycle (in the stratosphere):

$O_3(g)$+ UV (radiation) $\rightarrow O\bullet + O_2(g)$, followed by $O\bullet + O_2(g) \rightarrow O_3(g)$

Notice the $O\bullet$ and $O_2(g)$ recombine to form $O_3(g)$ once again. This creates a cycle where ozone is regenerated after it absorbs ultraviolet radiation.

Lets compare this to what happens when in the presence of the chlorine atom.

Ozone Depletion Mechanism:

Step 1) $Cl + O_3(g) \rightarrow ClO + O_2(g)$

Step 2) $O + ClO \rightarrow Cl + O_2(g)$

Overall: $O + O_3 \rightarrow 2O_2$

Notice that in this case ozone cannot be regenerated leading to ozone depletion.

What is ClO?

Here you can see that Cl is a catalyst, because it is present at the start of the reaction and at the end of the reaction in the same form. ClO is a reaction intermediate, because it is formed in the first reaction and then consumed by the second reaction. If we look at the overall reaction we can see that two oxygen, O_2 molecules are formed. Oxygen is NOT capable of absorbing the dangerous UV radiation, and the natural cycle has been destroyed.

Estimates predict that one chlorine radical can catalyze the destruction of about one million ozone molecules per second!

In 1996 the manufacture of freons was banned, with a 10 year grace period for developing countries. However, the effects of CFC's will linger for decades.

Why are catalysts important to us?

Catalysts provide us with a safe and efficient way to speed up a reaction.

First it is important to realize that without a catalyst our only method for speeding up a chemical reaction is to raise the temperature. We know this will speed up a chemical reaction. However, sometimes it is not practical to heat a reaction up to a high temperature, this method often causes the products to undergo further reactions that decrease the yield of the reaction. Carrying out a reaction at high temperatures can also be very costly, and even too dangerous. Using a substance that can facilitate the reaction can often be a better method.

Society

Our society depends upon the production of chemicals used for food, shelter and clothing. Industrial chemists work hard to increase the speed of production for chemicals which are in high demand. For developing countries this is a very important factor in the growth of their economy. Approximately 90% of all commercially produced chemical products use catalysts at some point in the process of their manufacture.

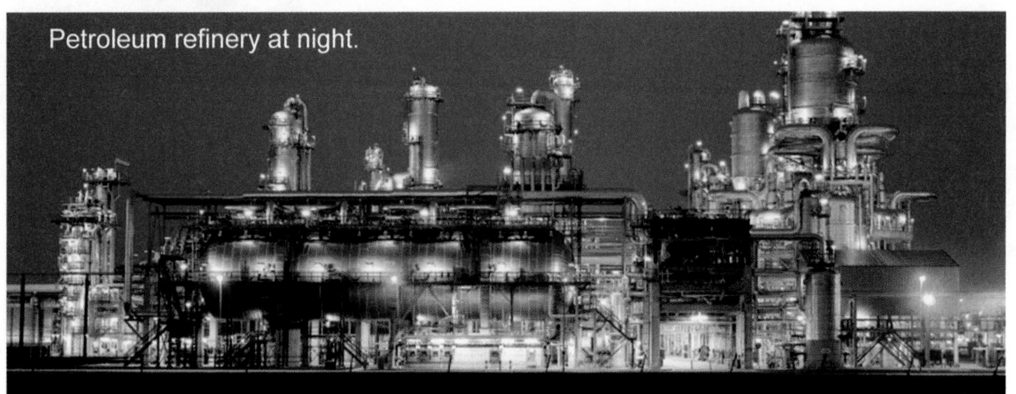

Petroleum refinery at night.

Environment

Environmental chemists also design ways to speed up reactions that decrease the amount of a high level pollutant. Catalytic converters are a good example of this. A catalytic converter speeds up chemical reactions that decrease the amount of harmful gases emitted from the exhaust produced by our automobiles. With a large population of people driving to work the poisonous gases released by our automobiles would accumulate in the air. Los Angeles was the first city to mandate the use of catalytic converters on cars.

Living systems

Catalysts are very important in nature too! Remember that our bodies use enzymes that speed up important metabolic reactions. When our cells burn sugar for energy and heat we are using enzymes that allow us to obtain the potential energy contained in the bonds of the sugar molecules. Burning sugar in the presence of oxygen requires temperatures much higher than body temperatures. Enzymes allow these reactions to take place at the low temperatures found in living animals.

Cave chemistry

Have you ever visited a cave? If so then you have seen equilibrium in action. Stalagmite and stalactites are the long icicle looking formations inside a cave. A stalagmite refers to a "drip" that rises from the floor of a cave. The drip is formed by deposition of calcium carbonate, $CaCO_3(s)$. Stalactites form from drips on the ceiling of the cave. Sometimes the stalagmites and stalactites grow together and form a large column. There are many caves and caverns sprinkled across the United States. The chemistry causing formation of stalactites and stalagmites is a good example of the reversibility of chemical reactions. You will see that Le Chatelier's principle helps us to explain why stalagmites and stalactites form.

There is something beautiful and mysterious about a cave. Why do these interesting looking stalagmites and stalactites form?

Carlsbad Caverns in New Mexico.

Rain absorbs carbon dioxide gas as it moves through the atmosphere, but especially when it moves through the soil. Soil contains some carbon dioxide which is released from decaying animals and plants.

The rain water reacts with the carbon dioxide to form carbonic acid.

$$H_2O(l) + CO_2(g) \rightleftharpoons H_2CO_3\,(aq)$$

Carbonic acid is capable of dissolving some minerals and certain types of rock such as limestone. Limestone is formed from calcium carbonate, $CaCO_3(s)$. Although carbonic acid is a weak acid, it is much more effective at dissolving rock than pure rain water. As the carbonic acid reacts with the limestone rock, it dissolves the calcium ion, Ca^{2+}along with other minerals.

$$CaCO_3(s) + H_2CO_3(aq) \rightleftharpoons Ca^{2+}(aq) + 2\,HCO_3^-(aq)$$

Water seeping through rock.

As this mineral laden water seeps into the rock it follows small cracks and fissures, which allow it to go deeper into the rock. As the water moves along the openings more rock is dissolved and the water is able to get even deeper into the earth. This cycle continues and creates larger and larger openings for the water to pass through. The dissolving of rock slowly opens up into large fissures and small caverns present inside the earth.

There are many different types of rock that are present in caves. Each of these types: limestone , granite, sandstone, and even sea caves, have unique properties. The more colorful stalactites and stalagmites sometimes have dissolved copper ion, Cu^{2+}, which gives them a beautiful shade of blue color.

The formation of stalactites and stalagmites is driven by the process of equilibrium! As the water flowing through the rock reaches any cavity filled with air, the water releases some of the dissolved carbon dioxide it is carrying into the cave air. It does this to try to equilibrate the partial pressure of carbon dioxide between the cave air and the water. Usually, the cave air has a lower partial pressure of carbon dioxide than the water does, because of this some of the dissolved carbon dioxide comes out of solution in an effort to reach equilibrium, and increase the the amount of $CO_2(g)$ in the air. This equilibrium process balances the levels of carbon dioxide gas.

In the reaction below the equilibrium position shifts to the right to products.

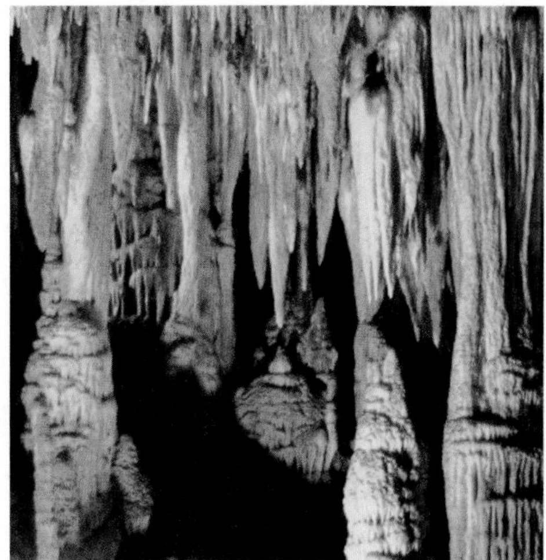

Luray Caverns in Virginia

$$Ca^{2+}(aq) + 2HCO_3^-(aq) \rightleftharpoons CaCO_3(s) + CO_2(g) + H_2O(l)$$

As the carbon dioxide gas is released into the cave air, some calcium carbonate solid, $CaCO_3(s)$ precipitates out along with it. Recall that precipitate is the formation of a solid from aqueous salt solutions. In this example the aqueous solutions are shown on the reactants side above ($Ca^{2+}(aq)$ + $2HCO_3^-(aq)$). As the precipitate forms drips, gravity slowly brings it down to the lowest point. The formation is very similar to that of an icicle on the edge of a roof, the more water that drips the larger the ice sickle that forms.

This section of Carlsbad Caverns is called "the Doll House".

Here is another formation of stalactites.

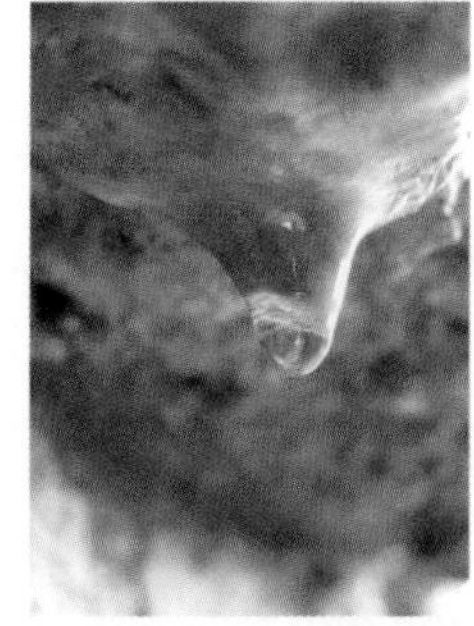

A drop of water slowly drips from a young stalactite.

LeChatelier's Principle explains the precipitation of calcium carbonate, $CaCO_3(s)$ to form stalagmites and stalactites! Because the partial pressure of carbon dioxide is lower in the cave air, than in the water entering the cave, the carbon dioxide molecules leave the water and enter the air. Why do they do this? To try to reach equilibrium! Balance is very important in nature. If you carefully consider the equation above you will see that carbon dioxide appears on the product side of the equilibrium reaction. If the reaction shifts to the right to release some $CO_2(g)$, the precipitate of calcium carbonate, $CaCO_3(s)$ is also formed as well as water liquid. So in essence the release of $CO_2(g)$ forces the equilibrium position to the side that causes deposition of calcium carbonate too.

Chapter 12 Review

Vocabulary

Match each word to the sentence where it best fits.

Section 12.1

reaction rate	activation energy
transition state	activated complex
reaction profile	

1. The ______________ is the smallest amount of energy necessary for molecules to react.
2. The speed or ____________ of a chemical reaction is measured by the change in concentration of a reactant or product, per unit time.
3. A graph that shows the progress of a chemical reaction with respect to energy is called a __________________.
4. Bonds are broken and reformed at this high energy point called the _______________.

Section 12.2

closed system	equilibrium position
equilibrium	LeChatelier's Principle
Law of Mass Action	equilibrium expression
equilibrium constant	

5. The __________________ tells us which direction of a reversible reaction is favored.
6. During _________________ the rate of the forward reaction is equal to the rate of the reverse reaction.
7. In the laboratory we sometimes observe a _________________, where the chemicals are not open to the environment.
8. ___________________ states that when a change is experienced by a chemical system in equilibrium the system will adjust to partially offset the change.
9. A general description of any equilibrium reaction can be explained by the _____________, which shows the relationship of chemicals and their molar coefficients.
10. We use the ________________ to determine unknown equilibrium concentrations.
11. The value of the ___________________ only changes when the temperature changes.
12. This special ratio called the ________________describes the relationship of the products relative to the reactants in a chemical equilibrium system.

Section 12.3

reaction mechanism	intermediate
bimolecular	unimolecular
elementary steps	rate determining step

13. A reaction _____________ is a chemical species formed during the elementary steps, and then consumed, so it is not present in the overall equation.
14. An elementary step that shows the collision of two molecules is called a _____________ collision.
15. A series of ___________________ which represent the overall progress of the chemical reaction.
16. A proposed pathway or ______________ must be supported by experimental evidence.
17. The slow step in a reaction mechanism is called the _________________.
18. An elementary step that involves only one reactant is called a ________________ collision.

Section 12.4

enzyme	catalyst

19. A substance that speeds up a chemical reaction is called a _____________.

20. An _____________ is a biological catalyst the speeds up chemical reactions in our bodies.

Conceptual Questions

Section 12.1

21. List the four factors affecting reaction rate.

22. Give an example of how decreasing the temperature affects the rate of a chemical reaction.

23. Explain how a catalyst like rhodium, Rh, works in a catalytic converter.

24. When strips of magnesium metal are added to a beaker containing 1.0M HCl and another beaker containing 6.0M HCl which one will have the strongest reaction? Explain.

25. How do we measure the "speed" of a chemical reaction?

26. During the course of a chemical reaction the rate changes. Explain how it changes. In other words does it speed up or slow down? and why does it do this?

27. When we measure the "average" rate during a chemical reaction, what formula do we use?

28. Refer to the hypothetical reaction below:

    $A + B \rightarrow 2C$

    a. How does the rate of disappearance of substance A relate to the rate of disappearance of B? Explain.
    b. How does the rate of appearance of substance C relate to the disappearance of substance A? Explain.

29. List the factors that are important during a molecular collision.

30. If molecules collide will they react to form product? Explain.

31. After the activated complex forms during a chemical reaction, what happens next?

32. Look at the graph below and label the following:
    a. reactants, products
    b. activated complex, Ac
    c. the activation energy, Ea
    d) the overall energy change, ΔH in the reaction.

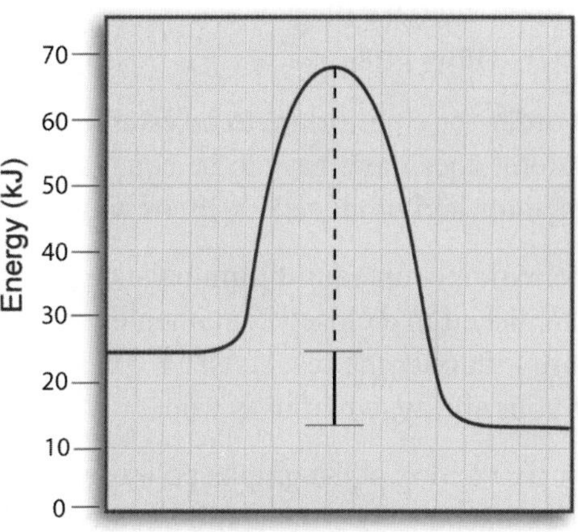

33. For a chemical reaction where $A + B \rightarrow C+D$, the overall energy change is endothermic. Make a sketch of the reaction profile that represents the endothermic reaction, given that the ΔH is 50 kJ.

34. Explain why endothermic reactions are less likely to occur compared to exothermic reactions. To answer this use the information related to a reaction profile.

35. Explain what the graph below is telling us.

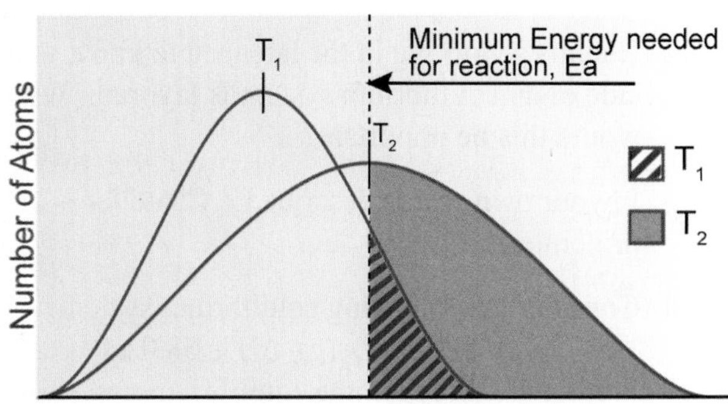

Chapter 12 Review.

Section 12.2

36. Give an everyday example of equilibrium in your daily life. Explain how this example works.

37. Explain how water inside a closed container reaches an equilibrium balance between molecules in the liquid and vapor phase.

38. Explain what we mean when we say "equilibrium is a dynamic process."

39. In order for equilibrium to be established in a system, does there have to be equal amounts of reactants and products? Why or why not?

40. Remarkably an equilibrium balance can be established in a closed container, even when we start with only the reactants! Explain to the best of your ability, why this is true.

41. When we look at the graphs below, **explain** which side (reactants or products) is favored in this equilibrium system
$2SO_2(g) + O_2(g) \rightleftharpoons 2SO_3(g)$.

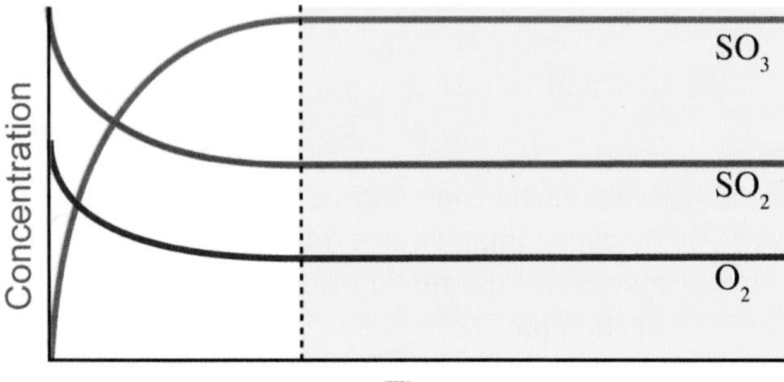

42. Chemists working in the lab need to know which side of an equilibrium system is favored. Why would this be important?

43. In your own words describe Le Chatelier's principle.

44. Consider the following equilibrium system:
$N_2O_4(g) \rightleftharpoons 2NO_2(g)$ $\Delta H = 58.0$ kJ
Predict the direction the equilibrium will shift under the following conditions:
   a. some $N_2O_4(g)$ is removed
   b. some $NO_2(g)$ is added
   c. The temperature is increased.

45. Given the reaction below, predict in which direction the equilibrium will shift under with the following changes.
$PCl_5(g) \rightleftharpoons PCl_3(g) + Cl_2(g)$ $\Delta H = 87.9$ kJ
   a. The pressure of the system is increased.
   b. Some $Cl_2(g)$ is removed.
   c. Some $PCl_5(g)$ is added.
   d. The temperature of the system is decreased.

46. How will each of the following changes effect the equilibrium system shown below?
$2SO_2(g) + O_2(g) \rightleftharpoons 2\ SO_3(g)$ $\Delta H = -\ 320$ kJ
   a. the temperature is increased
   b. some $O_2(g)$ is removed
   c. the pressure of the reaction mixture is decreased
   d. some $SO_3(g)$ is removed.

47. Write the expression for the equilibrium constants for each of the following reactions.
   a. $N_2(g) + O_2(g) \rightleftharpoons 2NO(g)$
   b. $PCl_5(g) \rightleftharpoons PCl_3(g) + Cl_2(g)$
   c. $2SO_2(g) + O_2(g) \rightleftharpoons 2\ SO_3(g)$
   d. $2Cl_2(g) + 2H_2O(g) \rightleftharpoons 4HCl(g) + O_2(g)$
   e. $3NO(g) \rightleftharpoons N_2O(g) + NO_2(g)$

Section 12.3

48. A reaction mechanism is a "series" of elementary steps that make up an overall chemical reaction. Does it make sense that chemical reactions actually happen in a series of smaller steps? Explain your thinking.

49. Given the two elementary steps below:
step 1: $N_2O(g) \rightarrow N_2(g) + O_2(g)$
step 2: $N_2O(g) + O \rightarrow N_2(g) + O_2(g)$
   a. Write the balanced equation for the overall chemical reaction.
   b. Identify any reaction intermediates.

50. Given the elementary steps below:
step 1: $H_2(g) + ICl(g) \rightarrow HI(g) + HCl(g)$
step 2: $HI(g) + ICl(g) \rightarrow I_2(g) + HCl(g)$

a. Write the balanced equation for the overall chemical reaction.
b. Identify any reaction intermediates.

51. Explain why it is unlikely that the following reaction occurs all in one step!
$2SO_2(g) + O_2(g) \rightleftharpoons 2\ SO_3(g)$

52. Give an example of one biological pathway in your body. Explain why you consider it a "pathway."

Section 12.4

53. How does a catalyst speed up a chemical reaction?

54. Does our body use catalysts? If so what are they called?

55. Make a generic sketch of a reaction profile, and label your axis. Clearly show how a catalyst would effect your graph.

56. Why will jello not set-up or "gel" when fresh pinapple is added?

57. Give an example of an enzyme in your body that is involved in the digestion process.

58. Explain one way that catalysts are useful in our society.

59. How does the formation of stalagmites and stalactites in caves relate to our study of equilibrium?

Quantitative Problems

Section 12.1

60. The reaction of nitrogen and oxygen to produce nitrous oxide is shown below:
$N_2(g) + O_2(g) \rightleftharpoons 2NO(g)$
The rate of disappearance of $N_2(g)$ is measured to be 4.3×10^{-5} M/s. What is the rate of formation of NO(g)?

61. The decomposition of NO is shown below.
$3NO(g) \rightleftharpoons N_2O(g) + NO_2(g)$
What is the rate of disappearance of NO if the rate of appearance of $N_2O(g)$ was measured to be 3.6×10^{-4} M/s.

62. Use the data in the table below to answer the following questions about the reaction where $A \rightarrow C$

Time (min)	Moles A	Moles C
0	1.00	0.00
5	0.85	0.15
10	0.74	0.26
15	0.62	0.38
20	0.54	0.46

a. What do you notice about the amounts of reactant relative to the product being formed?
b. Calculate the average rate for the appearance of C during the time period of t = 5min to t = 10 min.
c. Calculate the rate of disappearance of A during the first five minutes of the reaction
d. Calculate the rate of disappearance of A during the last five minutes, where t = 15 min and t = 20 min.
e. What do you notice about the rate at the start of the reaction relative to the rate at the end of the reaction.

Section 12.2

63. For the following reactions look at the value of the equilibrium constant to decide whether the reaction contains mainly reactants or mainly products.
a. $2NO(g) + O_2(g) \rightleftharpoons 2NO_2(g)$, $K = 5.0 \times 10^{12}$
b. $2SO_3(g) \rightleftharpoons 2SO_2(g) + O_2(g)$ $K= 4.0 \times 10^{-10}$
Explain your reasoning.

64. Carbon monoxide and hydrogen gas are catalyzed to commercially produce methanol.
$CO(g) + 2H_2(g) \rightleftharpoons CH_3OH(g)$
Calculate the value of K at a constant temperature of 227°C. Given that there are 0.041 moles of CH_3OH, 0.17 mole CO, and 0.30 moles H_2 present in a 2.0 L reaction vessel.

CHAPTER 13

Acids and Bases

What is an acid? What is a base?

What kinds of chemical reactions occur with acids and bases?

Why are acids and bases important?

You might be surprised to know that your stomach contains a chemical strong enough to burn skin and dissolve metal. That chemical is *hydrochloric acid* (HCl) and it is a necessary part of your body's chemical system for digesting food. The stomach has special adaptations that prevent the strong acid from escaping and damaging other parts of your body. The acid in your stomach is converted to water by a reaction with bicarbonate (HCO_3^-) ions produced in your pancreas. The bicarbonate belongs to a group of chemicals called *bases*. You can think of a base as the chemical opposite of an acid.

The chemical properties we call "acid" and "base" come directly from water itself. Acids and bases are important because they play a huge role in the chemistry of aqueous solutions, including those in your body and also in the environment. In this Chapter you will see how water is both an acid and a base! Some of the acids you probably eat or drink include acetic acid (found in vinegar), citric acid (found in orange juice), and malic acid (found in apples). Acids create the tart and sour flavors in foods. Bases around your household environment include soap, ammonia in cleaning solutions and magnesium hydroxide found in some antacids. Bases create the slippery feel of soaps and the bitter taste in grapefruit, spinach, and other foods.

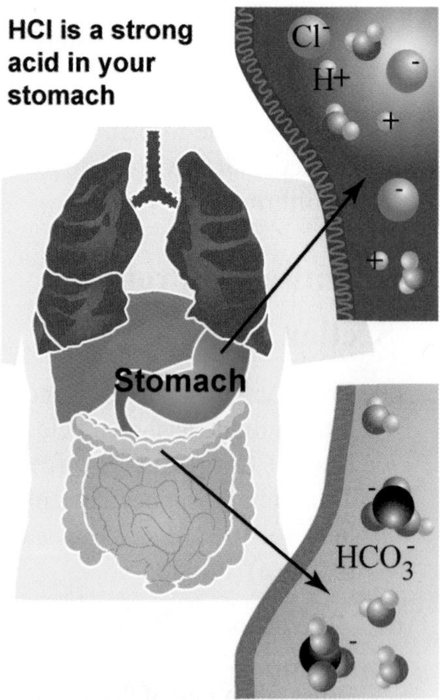

Bicarbonate (HCO_3^-) ions are produced to neutralize any acid leaving the stomach

Chemistry of antacids

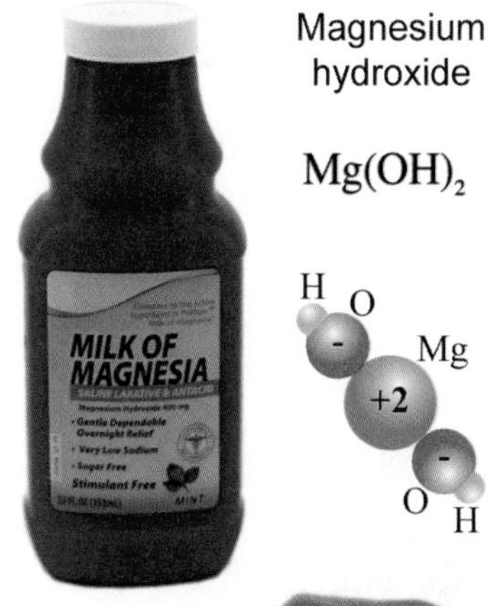

Next time you go to a drug store, look around in the "antacids" section. You will find some interesting chemistry there. Try to find Milk of Magnesia. This medicine is a thick, milky suspension of magnesium hydroxide $Mg(OH)_2$ in water, usually with some sweetener to make it taste better. The medicine is a suspension because magnesium hydroxide is insoluble in water.

People use Milk of Magnesia to relieve acid stomach and as a laxative. The recommended dose is one teaspoon, or 5 mL. If you look at the label on the back, you see that one teaspoon (5 mL) contains 400 milligrams, or 0.4 grams of $Mg(OH)_2$. This information is really telling us the concentration! The formula mass for $Mg(OH)_2$ is 58.3 grams. So that means one teaspoon contains 0.00686 moles of magnesium hydroxide molecules.

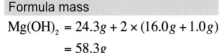

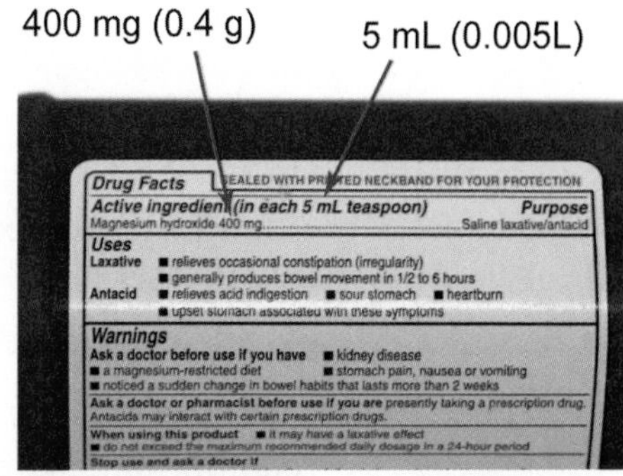

Formula mass

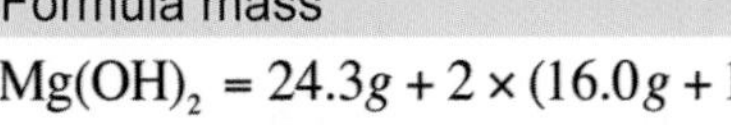

$$Mg(OH)_2 = 24.3g + 2 \times (16.0g + 1.0g) = 58.3g$$

Moles of $Mg(OH)_2$ in 0.400 g

$$\frac{0.400\ g}{58.3\ g/mole} = 0.00686\ moles$$

- Why is the recommended dose 5 mL?
- How does magnesium hydroxide relieve acid stomach?
- How much acid will one teaspoon of this medicine neutralize?

Add
twice as many moles of acid as there are of Mg(OH)2

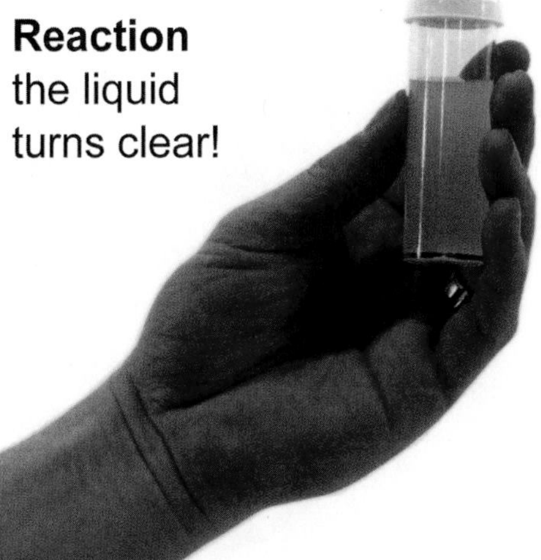

When you add the same number of moles of stomach acid (HCl), nothing visible happens. But when you add exactly twice as many moles, the solution turns from milky to clear! This happens because of a chemical reaction between the hydrochloric acid molecules and the magnesium hydroxide molecules.

13.1 The Chemical Nature of Acids and Bases

The dissociation of water

Acids and bases are important to chemistry because almost every reaction that occurs in an aqueous solution is affected. This includes your body, the ocean, and much of the biosphere of Earth. The reason is partly because water partially *dissociates*. On average, 1 out of every 550 million water molecules is dissociated into a hydrogen (H^+) ion and a hydroxide (OH^-) ion. *The hydrogen ion is chemically powerful, and quite unique.*

$$H_2O \xrightleftharpoons{\text{dissociation}} H^+ + OH^-$$

Neutral means $[H^+] = [OH^-]$

In **neutral** water, the concentration of H^+ ions and OH^- ions is *equal*. This is evident from the balanced equation. In the context of acidity, the word "neutral" means $[H^+] = [OH^-]$.

The dissociation reaction is written with a double arrow to show that the reaction goes both ways. Some H^+ and OH^- ions are always re-combining to make water molecules. In equilibrium, there is a balance between molecules dissociating into H^+ and OH^-, and ions recombining to make whole molecules again.

Properties of acids

An **acid** is a compound that dissolves in water to make a solution that contains *more* H^+ ions than OH^- ions. Some properties of acids are listed below.

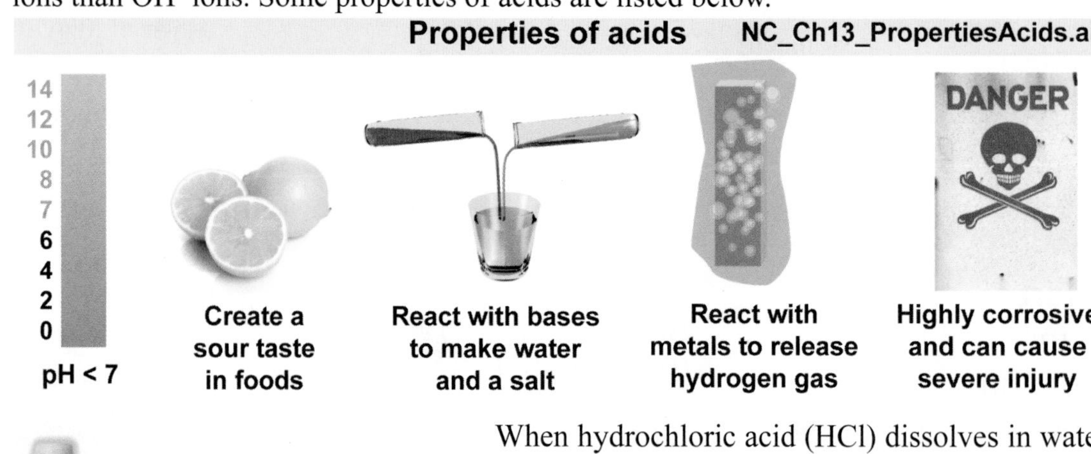

NC_Ch13_PropertiesAcids.ai
14
12
10
8

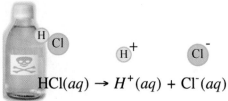

DANGER
Create a sour taste in foods
React with bases to make water and a salt
React with metals to release hydrogen gas
Highly corrosive and can cause severe injury

$$HCl(aq) \rightarrow H^+(aq) + Cl^-(aq)$$

When hydrochloric acid (HCl) dissolves in water it ionizes completely into hydrogen (H^+) and chlorine (Cl^-) ions. The fact that H^+ ions are created is what makes HCl an acid. Some other common acids are acetic acid (in vinegar) and citric acid (in citrus fruits).

Chemistry terms

neutral - in the context of acids and bases, neutral means the pH = 7.0 which also means the concentrations of H^+ and OH^- ions are equal.

6
4

acid - a chemical that dissolves in water to create more H^+ ions than there are in neutral water.

Bases

A base is the chemical opposite of an acids

A **base** is a compound that dissolves in water to make a solution with *more* OH^- ions than H^+ ions. In many ways a base is the chemical opposite of an acid. Acids make more H^+ ions and bases make more OH^- ions. Some of the properties of bases are listed below.

Properties of bases

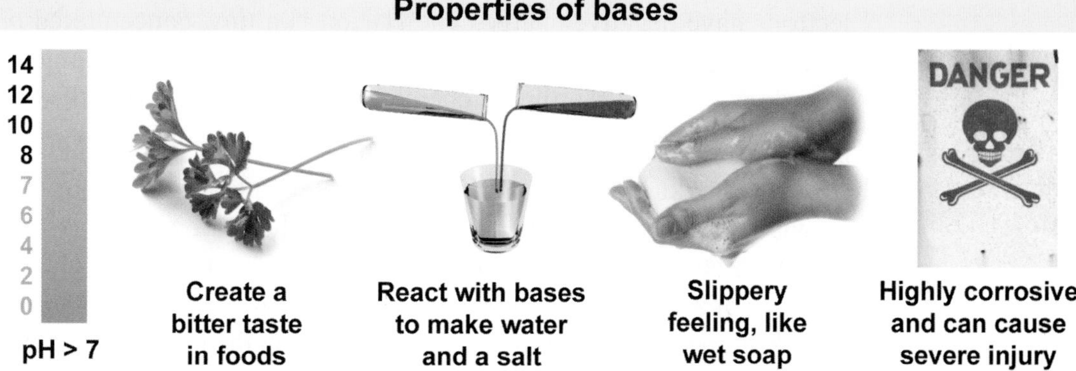

Acid or base but not both

Because H^+ and OH^- can combine to make neutral water, a solution that is not neutral must be either acidic or basic. A solution cannot be both acidic and basic at the same time. The pH scale reflects this either-or behavior. Acids have pH < 7, neutral solutions have pH = 7 and bases have pH > 7.

NaOH is a common base

A common base in both the lab and the hardware store is sodium hydroxide (NaOH). This compound is also known by two common names: *lye* and *caustic soda*. When dissolved in water, sodium hydroxide dissociates into sodium ions (Na+) and hydroxide ions (OH-). A 1.0 molar solution of NaOH has a pH of 14, and is a *strong base*. Concentrated NaOH is a dangerous and powerfully reactive substance which can burn skin and especially eyes.

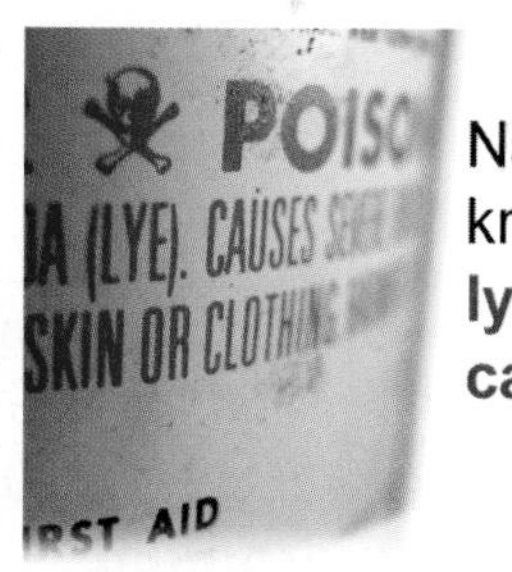

NaOH is also known as **lye** or **caustic soda**

$$NaOH(aq) \rightarrow Na^+(aq) + OH^-(aq)$$

Alkaline substances make basic solutions

An *alkaline* substance is a salt of a group I or group II metal that dissolves in water to make a base. A good example is potassium hydroxide (KOH). This ionic compound is used in alkaline batteries. When dissolved in a paste inside a battery, potassium hydroxide dissociates to make potassium ions (K^+) and hydroxide ions (OH^-).

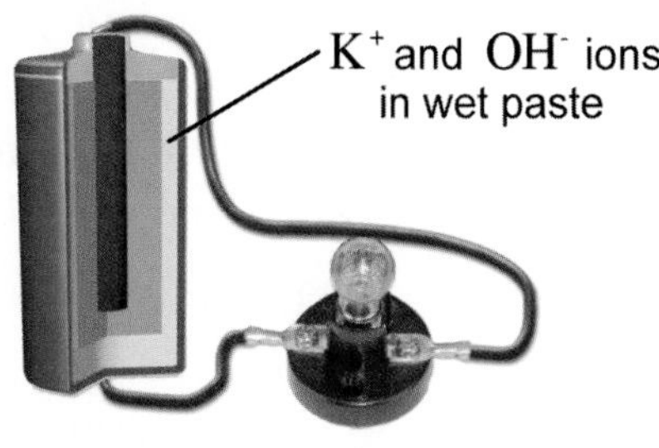

Chemistry terms

base - a chemical that dissolves in water to create less H^+ ions than there are in neutral water, (or equivalently, more OH^- ions).

The importance of the H^+ ion

The H^+ ion is unique because it has no electrons

The H^+ ion is powerful because *it has no electrons*! No other element participates in chemistry as a bare nucleus, a single proton, stripped of its electrons. We learned that chemistry is due to interactions between electrons in atoms. While true, it is ultimately the electrical *energy* linking electrons and protons that drives chemical behavior. Electrons have negative charge. The H^+ ion is a tiny, concentrated *positive* charge with no negative electrons to shield its electrical force. The whole subject of acids and bases has to do with the extraordinary chemical power of the H^+ ion, the "naked proton".

The Arrhenius definition of acid and base

The properties of acids, such as reacting with metals, were known a thousand years before anyone knew *why*. "Acids" were solutions that tasted sour, and dissolved metals. Likewise, bases were identified as solutions that had a bitter taste, were slippery, and neutralized acids. Around 1880, Svante Arrhenius, a Swedish chemist realized that the properties of acids were due to an excess of H^+ ions in aqueous solutions. He also noticed that the properties of bases were associated with an excess of OH^- ions. Arrhenius proposed the first definition of acids and bases.

Arrhenius theory

Acids are chemicals that produce H^+ ions in aqueous solutions

Bases are chemicals that produce OH^- ions in aqueous solutions

H^+ and the hydronium ion (H_3O^+)

The H^+ ion is so chemically attractive that it instantly pairs up with a water molecule to make H_3O^+. This is called the **hydronium ion**. Hydronium is sometimes called a *hydrated proton*, since H^+ is a proton and it becomes "hydrated" by bonding to water molecules. When we talk about the H^+ ion in water, we are really talking about H_3O^+ since H^+ really does not exist by itself for very long.

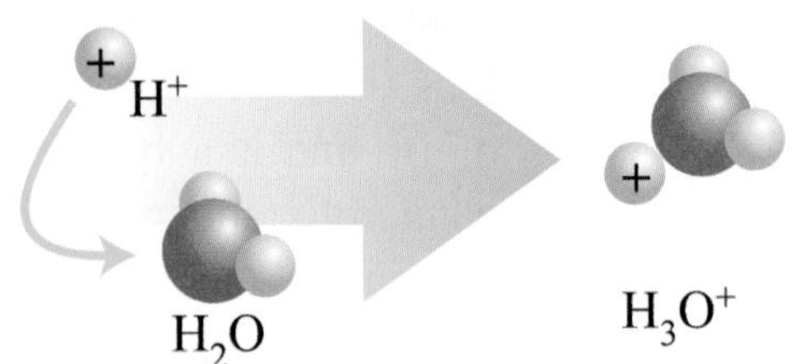

Solitary H^+ ions immediately bond to polar water molecules to form H_3O^+

H^+ is shorthand for H_3O^+

In this chapter (and keeping with history) we will refer the hydronium ion as just H^+ and not H_3O^+. This is really just because H^+ is easier to write and more convenient. However, keep in mind that the hydrated form (H_3O^+) is closer to reality.

Chemistry terms

hydronium ion - the H_3O^+ ion forms when H^+ bonds to a complete water molecule. Hydronium ions are what give acids their unique properties. Any reference to "H^+" in aqueous solution usually means H_3O^+.

Bronsted-Lowry acids and bases

How is NH_3 a base?

Ammonia (NH_3) is a weak base, even though the ammonia molecule does not contain the OH^- ion. When ammonia dissolves in water, the ammonia molecule strips a proton from a water molecule, leaving an OH- ion. This makes NH_3 a base because it dissolves to create OH^- ions, but does it by taking protons (H^+) from water molecules. Ammonia is a *weak* base because only a fraction of the ammonia molecules form NH_4^+ and OH^-.

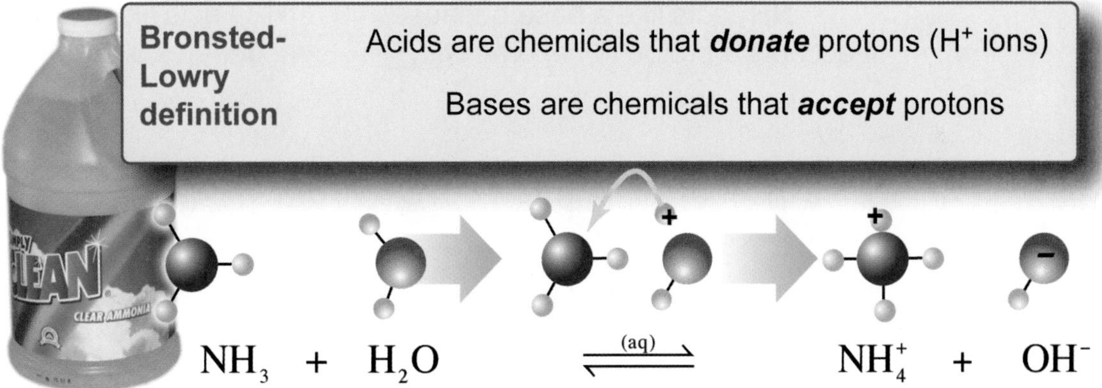

Bronsted-Lowry definition
Acids are chemicals that ***donate*** protons (H⁺ ions)
Bases are chemicals that ***accept*** protons
CLEAN
CLEAR AMMONIA
NH3 + H2O (aq) ⇌ NH4+ + OH-

The Bronsted-Lowry theory

Ammonia acts as a *proton acceptor* when dissolved in water. This give us another, more powerful way to define a base. *A base is a compound that accepts protons*. Ammonia is a base because it accepts a proton from a water molecule, leaving an OH- ion. This idea is known as the **Bronsted-Lowry definition** of acids and bases.

A better way to define acids and bases

According to the Bronsted-Lowry definition, an acid is a compound that donates a proton. A base is a compound that accepts a proton. In many ways the Bronsted-Lowry perspective is a better way to think about acids and bases because it explicitly recognizes the relationship between acids (proton donors) and bases (proton acceptors).

All acids are Bronsted-Lowry acids

All acids and bases can be thought of in the Bronsted-Lowry sense. Nitric acid (HNO_3) dissociates in water to make H^+ ions and NO_3^- ions. The nitric acid molecule donates the proton (H^+), which is why it is an acid.

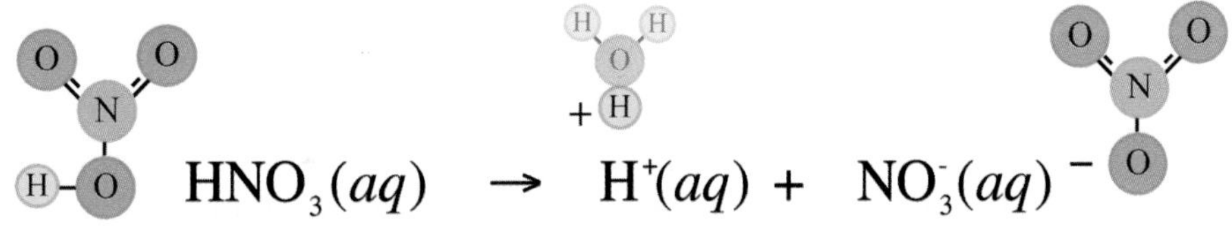

Chemistry terms

Bronsted-Lowry definition - a different way to look at what defines acids and bases. Acids are compounds that donate protons (H^+). Bases are compounds that accept protons.

Acids and bases are always paired

Acids and bases act in pairs

The Bronsted-Lowry definition brings up an important idea: for a molecule or ion to act like an acid by donating a proton, another molecule or ion must act like a base by accepting the proton! The opposite is also true. You cannot have something acting like an acid without something else acting like a base and vice versa.

Water can act like an acid

Going back to ammonia again, the NH_3 molecule acts like a base by accepting a proton to become NH_4^+. Water acts like an acid by donating a proton to become OH^-.

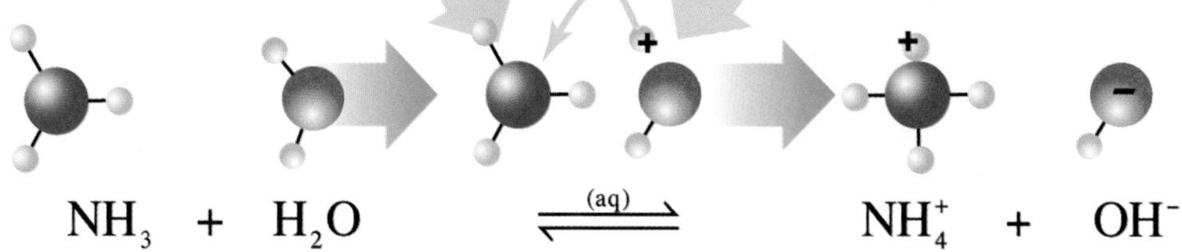

NH3 acts like a base because it accepts a proton to become NH4+
H2O acts like an acid because it donates a proton to become OH-

$$NH_3 + H_2O \underset{}{\overset{(aq)}{\rightleftharpoons}} NH_4^+ + OH^-$$

Water can also act like a base

Now consider a strong acid. When dissolved in water, hydrochloric acid (HCl) is a proton donor, dissociating to make H^+ ions. Water acts basic by accepting a proton and becoming H_3O^+.

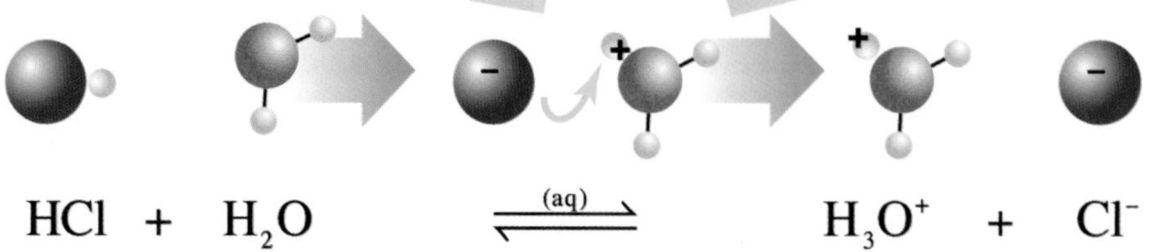

HCl acts like an acid because it donates a proton to become Cl-
H2O acts like a base because it accepts a proton to become H3O+

$$HCl + H_2O \overset{(aq)}{\rightleftharpoons} H_3O^+ + Cl^-$$

Water is amphoteric

One of the most important properties of water is its ability to act as both an acid and a base. In the presence of an acid, water acts as a base. In the presence of a base, water acts as an acid. A substance that can be either acid or base is **amphoteric**. Water is amphoteric because H_2O can both donate protons ($H_2O \rightarrow H^+ + OH^-$) or accept protons ($H_2O + H^+ \rightarrow H_3O^+$). Other substances are amphoteric too.

Chemistry terms

amphoteric - a substance is amphoteric if it can act as either an acid or a base under different circumstances. Water is amphoteric.

Identifying acids and bases

The chemical formula for an acid starts with an "H"

The chemical formula for an acid is always written with an H first (if the acid contains hydrogen). For example, acetic acid is written $HC_2H_3O_2$ instead of $C_2H_4O_2$. Both formulas give the correct molecule, but the first is quickly recognized as an acid because of the leading "H". If the acid dissolves to make 2 H^+ ions, its chemical formula begins with H_2. Ascorbic acid (vitamin C) is written $H_2C_6H_6O_6$ to indicate that it is both an acid, and that it dissolves to make 2 H^+ ions. Table 13.1 lists some common acids

TABLE 13.1. Common acids

Name	Formula	Strength
Hydrochloric	HCl	strong
Nitric	HNO_3	strong
Sulfuric	H_2SO_4	strong
Phosphoric	H_3PO_4	weak

Name	Formula	Strength
Ascorbic	$H_2C_6H_6O_6$	weak
Citric	$H_3C_6H_5O_7$	weak
Carbonic	H_2CO_3	weak
Acetic	$HC_2H_3O_2$	weak

The chemical formula of a base

The chemical formula for a strong base is written with the OH last. A good example is sodium hydroxide, written NaOH. The OH at the end of the name reminds chemists that a chemical is a base. Most weak bases such as ammonia (NH_3) do not contain OH so you have to learn other ways to recognize them as bases.

TABLE 13.2. Common bases

Name	Formula	Strength
Sodium hydroxide	$NaOH$	strong
Potassium hydroxide	KOH	strong
Calcium hydroxide	$Ca(OH)_2$	strong
Ammonia	NH_3	weak

Name	Formula	Strength
Carbonate ion	HCO_3^-	weak
Hypochlorite ion	ClO^-	weak
Caffeine	$C_8H_{10}N_4O_2$	weak
Ephedrine	$C_{10}H_{15}ON$	weak

Strong acids and bases

A **strong** acid or base dissociates completely in water so that each molecule contributes one H^+ ion (acids) or one OH^- ion (bases). Nitric acid is a good example of a strong acid. Every nitric acid molecule dissociates into an H^+ ion and a nitrate (NO_3^-) ion. Sodium hydroxide (NaOH) is a strong base because every mole of NaOH that dissolves creates one mole of OH- ions one mole of Na+ ions.

Weak acids and bases

A **weak** acid or base only dissociates partially or produces relatively few H^+ or OH^- ions in solution. For example a 1M solution of acetic acid creates less than 1% as many H^+ ions as the same concentration of hydrochloric acid.

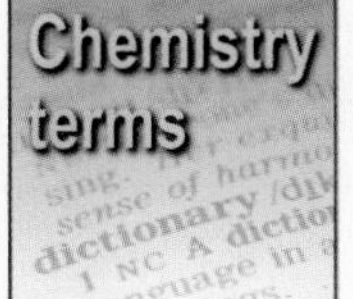

strong acid/base - dissociates completely (or almost completely) usually yielding 1 mole of H^+ or OH^- ions for every mole of acid or base dissolved.

weak acid/base - only partially dissociates in solution, typically only a few percent (or less) of molecules dissociate to yield H^+ or OH^- ions.

13.2 The pH Scale

Why pH is important

Ask anyone who keeps an aquarium about pH. They will tell you pH is *important*. If the pH of their aquarium water is wrong, the fish will die. Ask a serious gardener or a farmer and they will also tell you pH is important. Each kind of plant has a preferred range of soil pH. That's why a good gardener would not plant blueberries next to vegetables. Blueberry bushes need soil with a pH between 4 and 5. Vegetables (like broccoli) need a pH between 6 and 7.

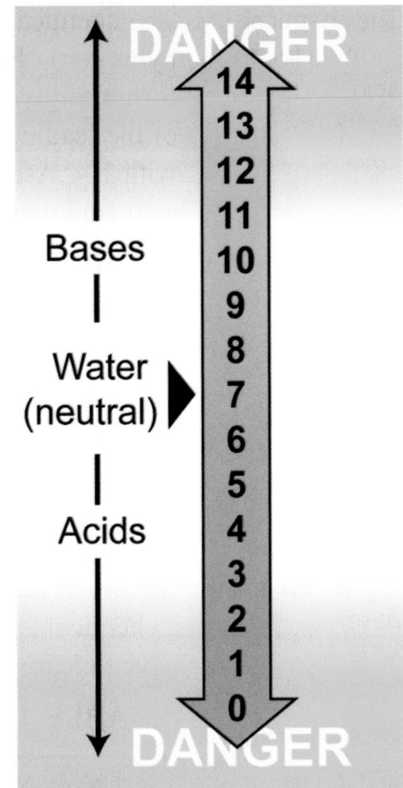

What pH measures

What exactly is pH? The simple answer is that pH tells you if a solution is acid, base, or neutral. Familiar acids and bases have a pH between 0 to 14. Pure water has a pH of 7 and is neutral. The farther away from 7 you get the stronger the acid or base.

Acids have pH less than 7

Acids have a pH less than 7. Natural rain water has a pH of between 5.5 and 6, which is very slightly acidic. A weak acid like tomato juice has a pH of 4. A 0.1M hydrochloric acid solution (strong) has a pH of 1. Anything with a pH of less than about 2 is dangerous and should be handled with care.

Bases have pH greater than 7

A base has a pH greater than 7. Weak bases such as milk of magnesia ($Mg(OH)_2$) and baking soda ($NaHCO_3$) have pH between 8 and 10. A concentrated solution of a strong base such as a 1M sodium hydroxide solution has a pH close to 14. Anything with a pH greater than about 12 is dangerous and should be handled carefully.

Some common acids and bases

Table 13.3 lists the pH of some common substances. Many foods are acidic. Many household cleaning products are bases.

TABLE 13.3. The pH of some common acids and bases

Household chemical	Acid or base	pH
lemon juice	acid	2
vinegar	acid	3
soda water	acid	4
baking soda	base	8.5
bar soap	base	10
ammonia	base	11

Concentration, pH, and scientific notation

The real meaning of pH

The pH tells us much more than just whether a solution is an acid or a base. The pH also tells how *strong* the acid or base is. In this sense, "strong" means concentrated and "weak" means dilute. In fact, *pH is a special way of describing the molarity of the H^+ ion* in a solution. If you know the pH you also know the concentration of H^+ ions, and vice versa. If you know the molarity of H^+ ions you can calculate the pH.

The pH scale is based on scientific notation

Earlier we noted that about 1 in 556 million H_2O molecules in ordinary water dissociates into H+ and OH- ions. There are 55.6 moles of H_2O in one liter. If you work out the math, the concentration of H+ in neutral water is 0.0000001M, or 1×10^{-7}M. The pH of neutral water is 7 because the concentration of H^+ is 10^{-7}M.

Numbers greater than 1 in scientific notation

In scientific notation a number is written as a value multiplied by a power of ten. The power of ten is given by the **exponent**. For example, we can write 500 as 5×100.

The number 100 is a power of ten because $100 = 10 \times 10 = 10^2$. The number 500 can be written $500 = 5 \times 10^2$. The value is 5 and the exponent is 2. Exponents are positive when numbers are greater than one.

$$500 = 5.00 \times 100 = 5.00 \times 10^2$$

Numbers less than 1 in scientific notation

This gets tricky when numbers are less than one. Consider the number 0.05. This is the same as 5/100. The number 100 is a power of ten, but now it is the denominator instead of the numerator. That means we divide by 100 instead of multiply.

The number $0.05 = 5 \times 1/100 = 5 \times 1/10^2 = 5 \times 10^{-2}$. A negative exponent means you divide by the power of ten rather than multiply. *All values less than 1 have negative exponents.*

$$0.05 = 5 \times \frac{1}{100} = 5 \times 10^{-2}$$

A negative exponent means the number is less than 1

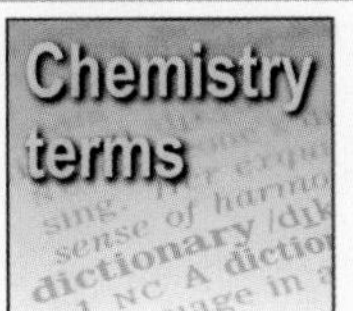

exponent - a number written in scientific notation consists of a value multiplied by a power of ten, such as $500 = 5 \times 10^2$. The exponent is the power of ten. In the example the exponent is 2.
Exponents are negative for numbers that are less than 1. For example, 0.05 is the same as 5×10^{-2}.

Logarithms and pH

The true meaning of pH

On the last page we said the concentration of H+ was 10^{-7} M in neutral water. Neutral water has a pH of 7. Do you see the connection? *The pH value is -1 times the H^+ molarity expressed as a power of ten.* This is the true meaning of pH. Lets explore how it works.

pH = 0 means the H^+ concentration is 1 M

A strong acid like HCl dissociates completely in water to H^+ and Cl^-. That means a 1M solution of HCl also has a 1M concentration of H^+ ions. In scientific notation, $1 = 10^0$. That means the pH of a 1M solution of HCl is zero (pH = 0). A solution with a pH of 0 is a very strong acid. $HCl(aq) \rightarrow H^+(aq) + Cl^-(aq)$

Calculating pH with logarithms

When the number is a perfect power of ten, the pH is the exponent multiplied by -1. Okay, how about numbers that are not perfect powers of ten? Suppose the H+ molarity is 0.05M. The rule still applies, but the exponent is no longer an integer. Instead we ask the question: is there a number N such that $10^N = 0.05$? The answer is yes, and the number is -1.301. If you raise 10 to the power -1.301 you get 0.05.

$$10^{(-1.301)} = 0.05$$

The definition of pH

The number -1.301 is the **logarithm** of 0.05. The logarithm of a number is the power of ten that gives you the same number back. *The pH of the solution is 1.301* because pH is the logarithm multiplied by -1. We can now give a precise definition for pH.

pH is -1 × logarithm of H^+ molarity

Definition of pH

$$pH = -\log[H^+]$$

pH is -1 times the logarithm of the H^+ molarity

Logarithms are easiest to see when numbers are perfect powers of ten. The logarithm of 100 is 2. We write this as 2 = log(100)

$$\log(100) = 2 \quad \text{because} \quad 10^2 = 100$$

Why ordinary pH is between 0 and 14

Almost all the acids and bases you come across in ordinary life have an H^+ molarity between 1 and 10^{-14}. The logarithm of 1 is zero, so when the molarity of H^+ is 1, the pH is zero (strong acid). The logarithm of 10^{-14} is 14. If a solution has an H^+ concentration of 10^{-14}M, then pH is 14 (strong base). However, *pH can be negative* for stronger acids and greater than 14 for exceptionally strong bases.

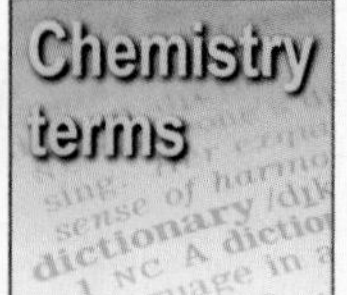

logarithm - the logarithm of a number (A) is another number (B) such that $10^B = A$. For example, the logarithm of 1,000 = 3 since $10^3 = 1{,}000$. Logarithms are positive for numbers greater than one and negative for numbers less than one. For example, the logarithm of 0.01 is -2 since $10^{-2} = 0.01$. Logarithms are often abbreviated "log" so the expression 2 = log(100) means "2 equals the logarithm of 100".

Calculating pH

Why use logarithms for pH?

We use logarithms for pH because the range from 0 to 14 is a lot easier to think about than the range from 1 (or 10^0) to 0.00000000000001 (or 10^{-14}). This is the range of H^+ molarity in ordinary acids and bases. However, the pH scale does not stop at 0 or 14. A concentrated 6M solution of hydrochloric acid has a pH of -0.78.

Calculating pH

To calculate pH, you

1. determine the molarity of H^+ ions
2. $pH = -1 \times \log(\text{ molarity of } H^+)$

If you know the pH, you can also determine the concentration of H+ ions. Use the fact that pH is the power of ten of the H+ molarity. If the pH is 3, then the molarity of H+ ions is 10^{-3} or 0.001M.

A solution of acetic acid ($HC_2H_3O_2$) has an H^+ concentration of 5×10^{-5} M. What is the pH of the solution?

Asked: *What is the pH?*

Given: *Molarity of* H^+ *is* 5×10^{-5}

Relationships: $pH = -1 \times \log(H^+ \text{ molarity})$

Solve: $pH = -\log(5 \times 10^{-5}) = 4.3$, *a relatively weak acid*

You may have noticed a very important rule: changing the pH by 1 changes the concentration by a factor of 10. That means a change of 2 pH units is really a change of 100 in concentration! The converse is also true. If you dilute a solution by 10 to 1, you reduce the H^+ concentration by 10 times and *the pH increases by 1.*

Each pH change of 1 changes the H^+ ***concentration by 10***

Dilution by 10:1 raises the pH by 1 unit

A solution of nitric acid (HNO_3) has a pH of 3. What will the pH be if you add 10 mL of the solution to 90 mL of pure water?

Asked: *What is the pH of the new solution?*

Given: *The old solution had a pH of 3. The new solution has 10 ml of the old solution out of 100 mL of total solution.*

Relationships: *Dilution by 10:1 raises the pH by 1*

Solve: *The new solution has a pH of 4*

Calculating pH for bases

H^+ and OH^- are linked by the dissociation of water

The pH scale works for bases as well as acids. This is because of the one-to-one relationship between the H^+ ion and the OH^- ion, or between proton donors (acids) and proton acceptors (bases). In any aqueous solution, H^+ and OH^- can always recombine to produce neutral water (H_2O). Any change that increases the concentration of OH^- decreases the concentration of H^+ because it forces some H^+ to recombine and form water. Conversely, any change that increases H^+, decreases OH^- for the same reason. Because of the equilibrium.

For this reason the concentrations of H^+ and OH^- satisfy the three rules below.

$$[\text{concentration of } H^+] \times [\text{concentration of } OH^-] = 10^{-14}$$

$$\log[H^+] + \log[OH^-] = -14$$

$$\text{pH} = 14 + \log[OH^-]$$

Mathematically, each of the three equations says the exact same thing! But, the third one is easiest to use because logarithms are numbers you can add and subtract easily.

Calculating pH from OH^- concentration

If you know the concentration of OH^- you can use the rules to quickly find the pH. The diagram below shows how to do it.

Suppose the concentration of OH^- is 0.005M, what is the pH?

Step 1: Calculate the log of $[OH^-]$ → $\log[0.005] = -2.3$

Step 2: Apply the rule $\text{pH} = 14 + \log[OH^-]$ → $\text{pH} = 14 - 2.3 = 11.7$

Find the pH of a 0.012 M sodium hydroxide (NaOH) solution.

Asked: *What is the pH of the solution?*

Given: *NaOH ia a strong base that dissociates 100% in aqueous solution.*

Relationships: $pH = 14 + \log[OH^-]$

Solve: $[OH^-] = 0.012M$

$pH = 14 + \log[0.012] = 14.0 - 1.92 = 12.1$

The solution has a pH of 12.1 and is a strong base.

pH indicators

Measuring pH

You usually can't determine pH by just looking at a solution, or measuring its density or temperature. However, there are three common *indirect* techniques chemists use.

- Do a chemical reaction with a solution of known pH
- Use a chemical that changes color at different pH values
- Measure the electrical properties of the solution

Indicators are chemicals that turn colors at different pH

Certain chemicals turn different colors at different pH. These chemicals are called pH **indicators**. The juice of boiled red cabbage contains chemicals called *anthocyanins*. These chemicals change color at different pH. Red cabbage juice is normally deep purple but becomes deep pink or red in a strong acid and green or yellow in a strong base.

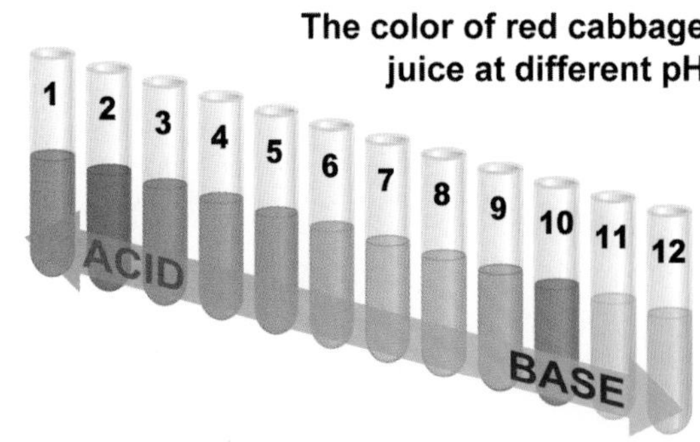

Litmus paper is an indicator

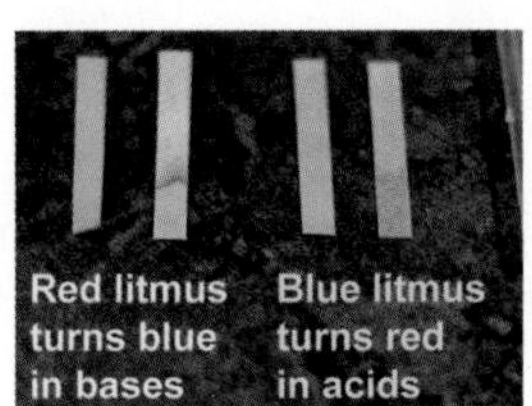

Litmus paper is another pH indicator that changes color. Litmus paper is made from a dye extracted from certain lichens, plants that grow on rocky places. Blue litmus paper turns red in acid. Blue litmus paper stays blue in a basic solution. Red litmus paper turns blue in a base, but stays red in an acid.

Methyl orange and phenolphthalein indicators

In the lab you may use methyl orange, and phenolphthalein, two common pH indicators. The chart below shows the color changes of each one for different ranges of pH. The methyl orange is best for measuring the pH of acids because its color changes between pH 3 and pH 5. Phenolphthalein is best for measuring the pH of bases.

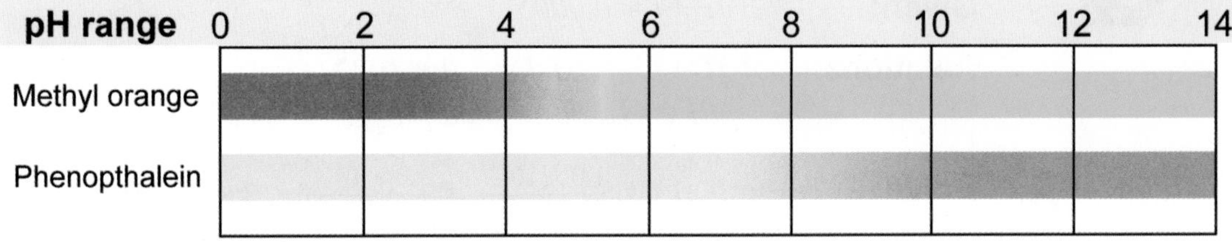

pH meters

There are instruments called *pH meters* which are used to measure pH. These instruments work because solutions that are acids or bases conduct electricity. The ease at which electricity flows through the solution (its *conductivity*) depends on the pH. A pH meter senses the pH by sending a tiny electrical signal through the solution.

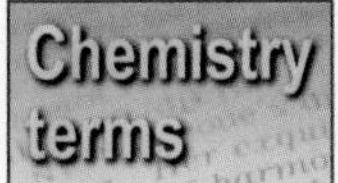

indicator - a chemical that turns different colors at different values of pH. Indicators may be used to determine pH directly, or in reactions with solutions of known pH.

13.3 Acid - Base Equilbria

The relation between H^+ and OH^-

When the concentration of OH^- ions increases, the concentration of H^+ ions decreases and vice versa. This is because excess OH^- ions combine with the H^+ ions normally present in water to make neutral water molecules.

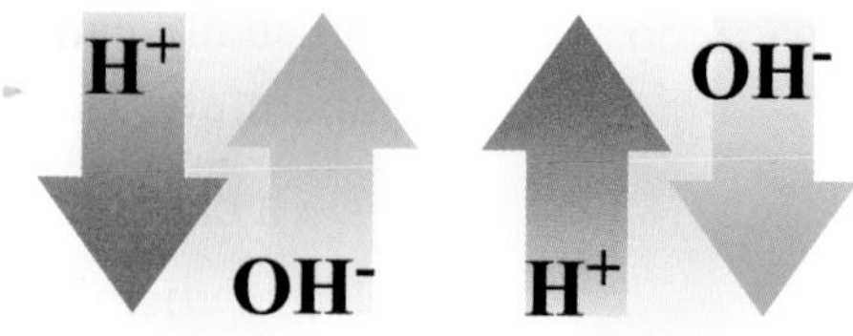

The ion product constant

Because recombination and dissociation are always going on, the concentration of H^+ and OH^- ions in dilute aqueous solutions obeys an equilibrium rule. For the dissociation reaction:

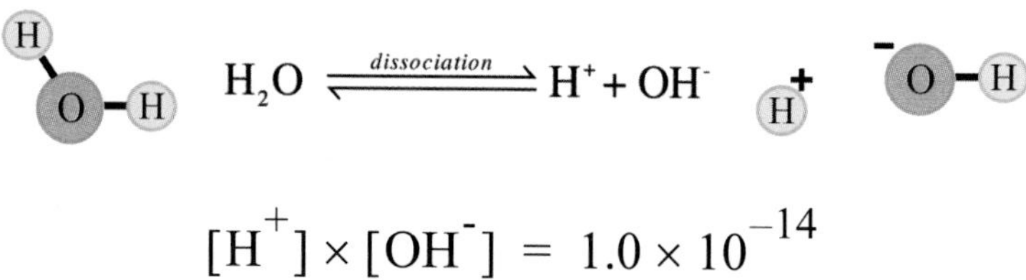

$$[H^+] \times [OH^-] = 1.0 \times 10^{-14}$$

The **ion product constant** for water, K_w, has the value 1.0×10^{-14} at 25°C. When either the H^+ or OH^- concentration is changed (by adding acids or bases) the equilibrium rule is still obeyed. This is why you can calculate the pH of a base from the concentration of OH^-. The pH depends only on the H^+ concentration however, $[H^+]$ and $[OH^-]$ are related by the ion product constant, K_w.

A certain solution has a concentration of H^+ ions of 0.0005M. What is the concentration of OH^- ions? Is the solution acidic or basic?

Asked: *What is $[OH^-]$ and what is the pH of the new solution?*

Given: *$[H+] = 0.0005$*

Relationships: *$[H+] \times [OH-] = 1.0 \times 10^{-14}$*
pH = -1 × logarithm [H+]

Solve: $[OH^-] = \frac{1.0 \times 10^{-14}}{[H^+]} = \frac{1.0 \times 10^{-14}}{0.0005} = 2 \times 10^{-11}M$

pH = -1 × logarithm [0.0005] = 3.3

Answer: *The concentration of OH- ions is 2×10^{-11} M.*
The solution has a pH of 3.3, and is a moderately strong acid.

Chemistry terms

ion product constant - In dilute aqueous solutions, $[H^+] \times [OH^-] = 1.0 \times 10^{-14}$ and the value $K_w = 1.0 \times 10^{-14}$ at 25°C is known as the ion product constant for water.

Weak acids

ADVANCED AP

Comparing strong and weak acids

In a weak acid or base, only some of the molecules actively increase or decrease the concentration of H^+ ions. Acetic acid ($HC_2O_3H_2$), found in vinegar is a good example. If you mix a 0.1M solution of acetic acid you get a pH of 2.87. That immediately tells us not all the acetic acid molecules dissociated to H^+ ions. By comparison, the same molarity of hydrochloric acid has a pH of 1.0, much more acidic.

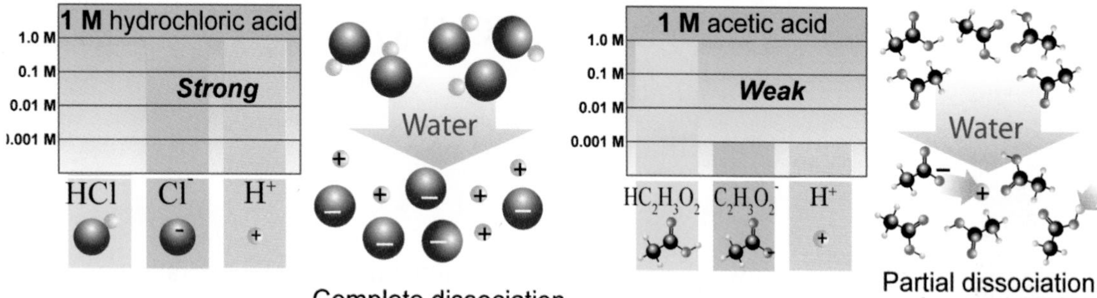

Equilibrium relationship and K_a

The concentration of H^+ in a weak acid (or base) obeys an equilibrium relationship. For acids and bases the equilibrium constant is called the acid/base dissociation constant, K_a. For example, acetic acid has a $K_a = 1.8 \times 10^{-5}$. Like all chemical equilibrium situations, acetic acid molecules are *constantly* dissociating and coming back together again.

Calculating the pH of a weak acid

Let's calculate the pH of a 0.1M acetic acid solution. This is a good example for how to solve similar problems. The calculation has four major steps.

1. Write down the balanced net ionic equation
2. Use the net ionic equation to set up the equilibrium relationship
3. Use the RICE table to solve the relationship for the H^+ ion concentration.
4. Calculate the pH from the H^+ concentration ($pH = -\log[H^+]$)

Start with the balanced net ionic equation

The balanced net ionic equation shows that we need to consider three different species: H^+ ions, acetate ions ($C_2O_3H_2^-$) and whole acetic acid molecules. The concentrations of the three species obey the equilibrium relationship (below). Remember, brackets indicate molarity, so read $[H^+]$ as "the molarity of H^+."

Balanced net ionic equation	Equilibrium relationship
$HC_2H_3O_2 \overset{(aq)}{\rightleftharpoons} H^+ + C_2H_3O_2^-$ acetic acid, hydrogen ion, acetate ion	$K = \dfrac{\text{molarity of products}}{\text{molarity of reactants}}$ $K_a = \dfrac{[H^+][C_2H_3O_2^-]}{[HC_2H_3O_2]}$

Using the RICE table

ADVANCED AP

Setting up variables

Acetic acid is a simple dissociation because one acid molecule dissociates into only two ions of equal concentration. If we let x be the concentration of H^+ ($x = [H^+]$) then the RICE table looks like this. The concentration of $HC_2H_3O_2$ decreases by x and the concentration of H+ and $C_2H_3O_2^-$ increase by x.

RICE table

Reaction	$HC_2H_3O_2$	$\rightleftharpoons$	H^+	+	$C_2H_3O_2^-$
Initial	0.1 M		0.0 M		0.0 M
Change	-x		+x		+x
Equilibrium	0.1 - x		x		x

Writing down the equilibrium relationship yields a quadratic equation.

$$K_a = \frac{[H^+][C_2H_3O_2^-]}{[HC_2H_3O_2]} \Rightarrow K_a = \frac{x^2}{0.1 - x} \Rightarrow x^2 + K_a x - 0.1K_a = 0$$

Quadratic equation

Solving the equation by approximation

This is a weak acid by definition, and therefore the H^+ concentration (x) will always be much less than the initial molarity of acetic acid (0.1M). Mathematically, that means the term $K_a x$ is always *much* less than $0.1K_a$, typically a thousand times less or even smaller. That means $K_a x$ can be neglected without causing much error in the solution of the equation. This makes the equation easier to solve and we get a value of $x = [H^+] = 0.00133$ M.

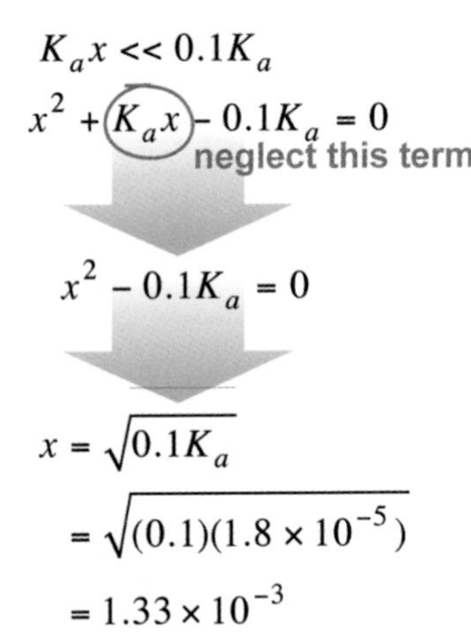

Weak acid/base equilibria

For weak acid or base equilibria of the form $HA \rightleftharpoons H^+ + A^-$ the H^+ concentration is given by the formula below where K_a is the dissociation constant and M is the initial molarity of the solution.

Dissociation of weak acids or bases

If HA is an acid with a net ionic equation of the form

$$HA \rightleftharpoons H^+ + A^-$$

and the ratio M/K_a is greater than 100, then

$$[H^+] = \sqrt{MK_a}$$

The initial molarity [HA] → M; dissociation constant → K_a

Solving equilibrium problems

When the approximation is OK to use

Note that the formula $[H^+] = \sqrt{MK_a}$ is an *approximation*. That means it only gives the right answer if certain conditions are met. The conditions are that the ratio $M \div K_a$ must be greater than 100. For the acetic acid problem, $M \div K_a = 5{,}555$, which is much greater than 100! We can therefore be confident the formula gives us an accurate value for [H+]. If the value of $M \div K_a$ were less than 100, the approximate formula would NOT give us an accurate answer and we would need to solve the quadratic equation.

Carbonic acid is polyprotic

An acid is *polyprotic* if each molecule dissociates to provide multiple H^+ ions. Carbonic acid is a good example. Carbonic acid (H_2CO_3) is an Arrhenius acid because it creates H^+ ions in solution. Carbonic acid dissolves in two steps, creating more than one H^+ ion per acid molecule.

$$H_2CO_3 \rightleftharpoons H^+(aq) + HCO_3^-(aq)$$

$$HCO_3^- \rightleftharpoons H^+(aq) + CO_3^{2-}(aq)$$

Step 1 - provides 1 H^+ ion

H_2CO_3 → HCO_3^- + H^+

Step 2 - provides another H^+ ion

HCO_3^- → CO_3^{2-} + H^+

Weak polyprotic acids

Weak polyprotic acids typically dissociate in several steps with each step weaker than the last. In a 1M solution of carbonic acid, 0.07% of the acid molecules dissociate to make H^+ and bicarbonate (HCO_3^-) ions. Only about 0.005% of the bicarbonate ions dissociate further into a second H^+ ion and a carbonate (CO_3^-) ion. For most purposes it is OK to assume that the pH of a carbonic acid solution is determined solely by the first dissociation, so in fact, carbonic acid acts like a *monoprotic* acid.

Calculate the pH of a 1.0M solution of carbonic acid (H_2CO_3). Assume that only the first dissociation affects the pH ($H_2CO_3 \rightleftharpoons H^+ + HCO_3^-$). The value of K_a for this dissociation is $K_a = 4.3 \times 10^{-7}$.

Asked: *What is the pH?*

Given: *$M = 1.0M$, $K_a = 4.3 \times 10^{-7}$, and it is a simple dissociation.*

Relationships: *$[H^+] = \sqrt{K_aM}$ if $M/K_a > 100$, $pH = -1 \times \text{logarithm } [H^+]$*

Solve: *$M/K_a = (1.0)/(4.3 \times 10^{-7}) = 2{,}300{,}000$ so formula is OK*

$[H+] = \sqrt{(4.3 \times 10^{-7})(1.0)} = 6.6 \times 10^{-4}M$

$pH = -\log(6.6 \times 10^{-4}) = 3.2$, the pH is 3.2

Answer: *Since the pH is less than 7, the solution is an acid.*

Weak bases

Weak bases, such as ammonia

A **weak base** is one that produces relatively few OH- ions per molecule of base. Many important chemicals are weak bases. Ammonia is produced in living organisms as a waste product. You probably smelled ammonia if some one forgot to clean out the cat litter box, or around a chicken coop. Ammonia is also a cleaning product.

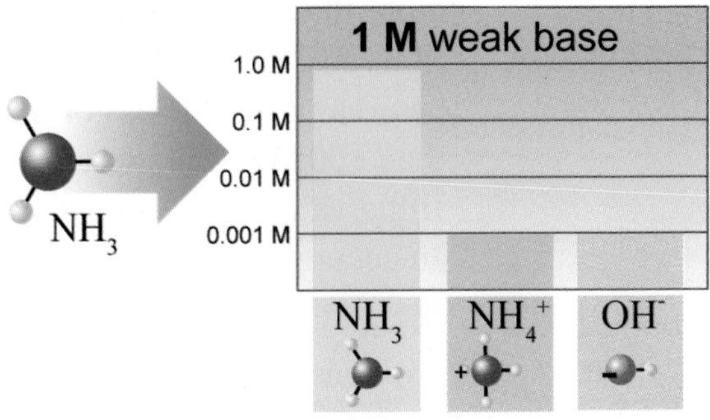

Why ammonia is a base

The basic property of an ammonia solution comes from the ammonia stripping a hydrogen from water to form an ammonium ion (NH_4^+) and a hydroxide ion (OH^-).

$$NH_3 + H_2O \rightleftharpoons NH_4^+ + OH^- \qquad K_b = 1.8 \times 10^{-5}$$

Using the base equilibrium constant

There is a base equilibrium constant K_b for bases just as there is a K_a for acids. For example, to determine the pH of a 1.0 M ammonia solution, we need the concentration of OH- which we get from an equilibrium calculation. The RICE table looks like this.

RICE table

Reaction	NH_3	+	H_2O	$\rightleftharpoons$	NH_4^+	+	OH^-
Initial	1.0 M				0.0 M		0.0 M
Change	-x				+x		+x
Equilibrium	1.0 - x				x		x

Equilibrium expression: $$K_b = \frac{[NH_4^+][OH^-]}{[NH_3][H_2O]} = \frac{[x][x]}{[1.0 - x][1]} = \frac{x^2}{1.0 - x}$$

Approximate solution for [OH-]: $$x = [OH^-] = \sqrt{1.0K_b} = 1.33 \times 10^{-3}$$

Calculate pH: $$pH = 14 - \log[OH^-] = 14 - 2.8 = 11.2$$

Calculating the pH of a base

Since the molarity of water is 1 in an equilibrium relationship, the equation has the same form as the weak acid equilibrium. The approximate solution gives the concentration of OH^- as 1.33×10^{-3}M. Next we know that the pH is 14 - log[OH-] so we get a pH of 11.2.

Chemistry terms

weak base - a chemical that only partially dissociates in solution, typically only a few percent (or less) to produce OH^- ions.

13.4 Acid - Base Reactions

Types of acid base reactions

Long before people knew the chemistry behind why, acids and bases were classified by their reactions with each other and with metals. For example, magnesium hydroxide ($Mg(OH)_2$) is a base. When 1 mole of magnesium hydroxide is added to 2 moles of hydrochloric acid, the resulting solution has a pH of 7 - and is neutral. This is an example of a *neutralization* reaction. Acids and bases neutralize each other though chemical reactions.

Neutralizaton reaction

$Mg(OH)_2$ + HCl
base acid

$MgCl_2$ + H_2O
a salt water

Corrosion and etching

Acids react with metals to produce hydrogen gas. The reaction is an example of corrosion, because the metal is chemically eaten away from its solid form and put into solution. For example, copper reacts with nitric acid (HNO_3). This reaction eats away copper metal and is used to produce fine jewelry and printed circuit boards. A printed circuit board starts out covered in a uniform thin layer of copper. Areas that are not to be etched are covered with an acid resistant material (*resist*). The board is then dipped in nitric acid. The copper is etched away wherever the acid touches it. Areas that were covered with resist are untouched and become the electrical paths for the circuit.

Step 1
Thin copper layer

Step 2
Print resist on areas that are to be copper

Step 3
Etch away areas NOT painted with resist, leaving copper circuit trace

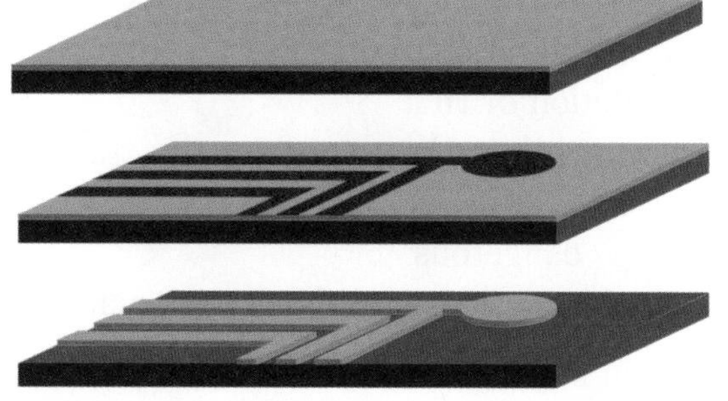

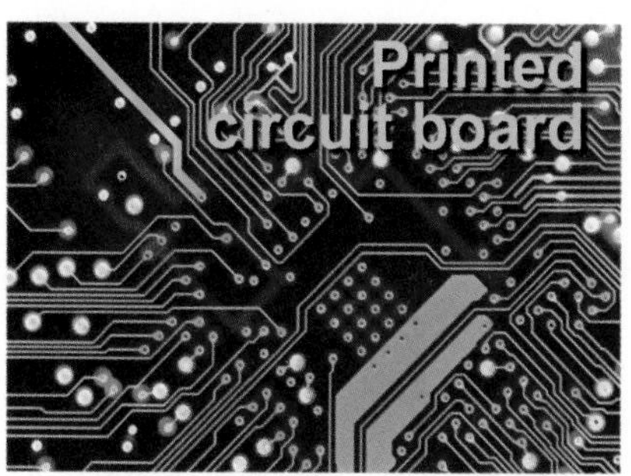

Nitric acid creates H^+ ions $HNO_3(aq) \rightarrow H^+(aq) + NO_3^-(aq)$

Copper metal reacts with H^+ to produce dissolved Cu^{2+} and hydrogen gas $Cu(s) + 2H^+(aq) \rightarrow Cu^{2+}(aq) + H_2(g)$

Electrolysis

Many people believe that hydrogen is the best transportation fuel for the future. Hydrogen burns clean and the product of the reaction is water. In one favorite scenario solar energy is used to split seawater into hydrogen and oxygen. The hydrogen is then liquefied, bottled and used for fuel. The chemical process is called electrolysis and it requires an electric current to flow through water. Acids and bases make excellent electrolytes and one established hydrogen production technique uses potassium hydroxide (KOH) as the electrolyte.

Sunlight

$2H_2O \rightarrow 2H_2 + O_2$

Hydrogen fuel

$2H_2 + O_2 \rightarrow 2H_2O$

Neutralization reactions

Neutralization

In general, adding acid lowers pH and adding base raises pH. Complete **neutralization** occurs when the H^+ ions from the acid combine with the OH^- ions from the base to produce neutral water.

Adding acid lowers pH - Adding base raises pH

Example neutralization of NaOH

Sodium hydroxide is a strong base. When combined with acetic acid ($HC_2O_3H_2$), a neutralization reaction occurs. At a pH of 7.0, we say the solution has been *neutralized.*

$$NaOH(aq) + HC_2O_3H_2(aq) \rightarrow H_2O + Na^+(aq) + C_2O_3H_2^-(aq)$$

Baking soda can partially neutralize strong acids

All strong acids should be neutralized before being disposed of. A simple way to neutralize a strong acid is with baking soda ($NaHCO_3$). When baking soda is added to hydrochloric acid, the H^+ ions from the acid combine with bicarbonate ions to make carbonic acid H_2CO_3. Carbonic acid is a weak acid. A 1M solution of HCl has a pH of zero (strong) while a 1M solution of carbonic acid has a pH of 3.2 (weak), much less dangerous. Over time carbonic acid decomposes into CO_2 and water.

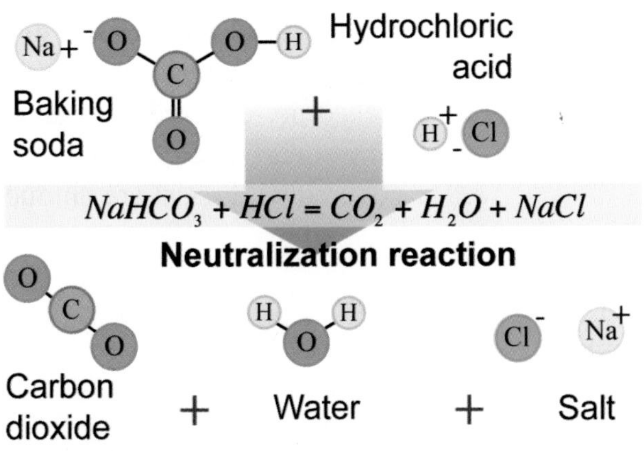

Neutralization in your body

This process also goes on in your body. As food and digestive fluids leave the stomach, the pancreas and liver produce bicarbonate (a base) to neutralize the stomach acid (HCl). Antacids such as magnesium hydroxide (milk of magnesia) have the same effect.

Adjusting soil pH

Neutralization reactions are important in gardening and farming. For example about 1/4 of the yards in the US have soil that is too acidic for grass (pH less than 5.5). Many people add lime, a base, to their yard every spring to raise the soil pH. A common form of lime is ground-up calcium carbonate ($CaCO_3$) made from limestone. For example, sulfuric acid (H_2SO_4) in soil reacts with the calcium carbonate to form water and calcium sulfate ($CaSO_4$) also known as gypsum. Sulfuric acid is created in the atmosphere from pollutants in the air. Many of the walls of buildings and homes are made with "plaster board" which is a sheet of gypsum (plaster) covered with paper on both sides.

Chemistry terms

neutralization - a reaction in which the pH of an acid is raised by combining with a base, or the pH of a base is lowered by combining with an acid. Complete neutralization results in a pH of 7, the same as neutral water.

Titration

Titrations are precise neutralization reactions

A **titration** is a special kind of neutralization reaction used to accurately determine the pH of a solution. Titrations are done with known concentrations of acids or bases. Indicators are used to detect when the pH changes through the neutral region. Titrations can be *much* more accurate than pH papers or indicators used alone.

The equivalence point

For example, suppose you have an acid and you slowly add base to it one drop at a time. Each drop of base raises the pH a proportionally small amount. However, close to the **equivalence point** even a single drop of base can change the pH dramatically. The equivalence point occurs when the moles of H^+ from the acid are exactly balanced by equivalent moles of base. Near the equivalence point, the pH is very close to neutral and that's why pH changes so quickly.

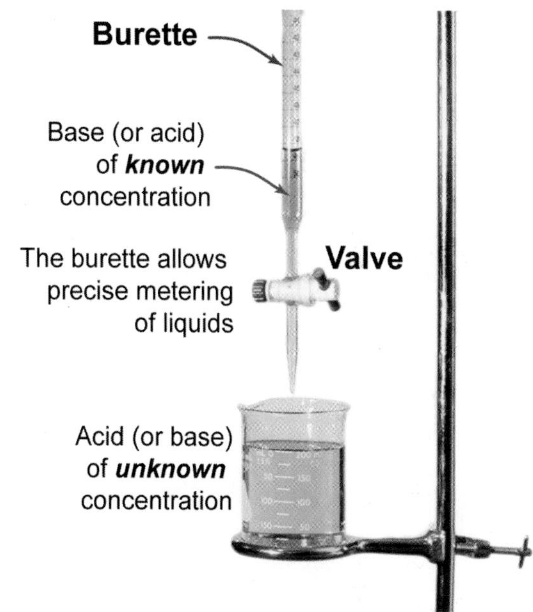

Why titration is accurate

Notice, methyl orange and phenolphthalein both go through their entire color change on the steep part of the curve. That makes it easy to tell that the equivalence occurs when 5 mL of base has been added.

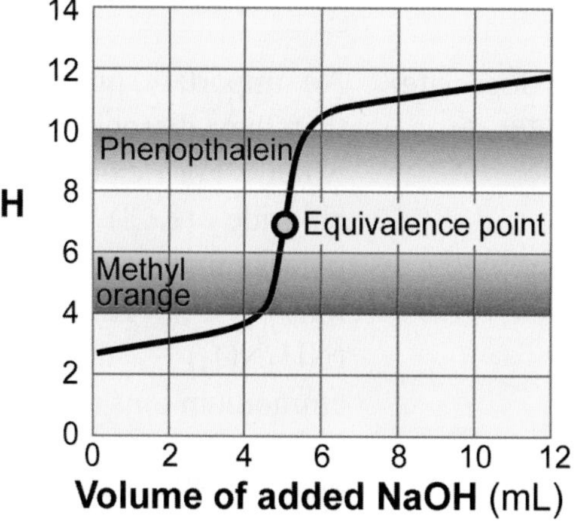

The results of a titration

The volume of the added base gives a precise measure of the number of moles of base added. In the example, 5 mL × 1.0M = 0.005 moles. If sodium hydroxide (NaOH) is the base, each mole of NaOH neutralizes one mole of H^+. That means there were 0.005 moles of H+ in the original solution. The original solution had a volume of 50 mL, therefore its original H+ concentration was 0.1M. This means the original pH must have been 1.0.

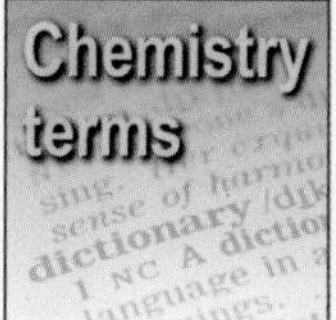

titration - a laboratory process to determine the precise volume of acid or base of known concentration which exactly neutralizes a solution of unknown pH.

equivalence point - in a titration, the point at which the moles of H^+ from the acid and OH^- from the base are exactly equal.

Salts

Acid-base reactions form salts

What types of chemicals are formed when you react an acid and a base? Consider the neutralization reaction between hydrochloric acid (HCl) and sodium hydroxide (NaOH). The product of the reaction is salt: NaCl.

$$NaOH + HCl \rightarrow NaCl + H_2O$$

Salt is an ionic compound, highly soluble in water. Since the whole reaction takes place in solution, the products are dissolved sodium ions (Na^+) and chloride ions (Cl^-).

Base + Acid = a salt + water

$NaOH + HCl = NaCl + H_2O$

Sodium hydroxide | hydrochloric acid | table salt | water

The chemical meaning of "salt"

Chemically, the term "**salt**" means *any* ionic compound in which the positive ion comes from an acid and the negative ion comes from a base. Ordinary table salt (NaCl) is just one example of a whole category of ionic chemicals which form in acid-base reactions.

TABLE 13.4. Common salts

Name	Formula	Name	Formula
Sodium chloride (table salt)	$NaCl$	Sodium sulfate	Na_2SO_4
Sodium acetate	$NaC_2H_3O_2$	Ammonium chloride	NH_4Cl
Potassium Chloride	KCl	Ammonium nitrate	NH_4NO_3
Sodium bicarbonate	$NaHCO_3$	Potassium nitrate	KNO_3

Salts dissociate in water

An important property of salts is that they dissociate completely in water. For example, potassium chloride (KCl) dissociates into potassium ions (K^+) and chloride ions (Cl^-). Ammonium nitrate (NH_4NO_3) dissociates into ammonium ions (NH_4^+) and nitrate ions (NO_3^-). Because salts dissociate into ions, aqueous salt solutions are *electrolytes* - they conduct electricity.

$$KCl(aq) \rightarrow K^+(aq) + Cl^-(aq)$$

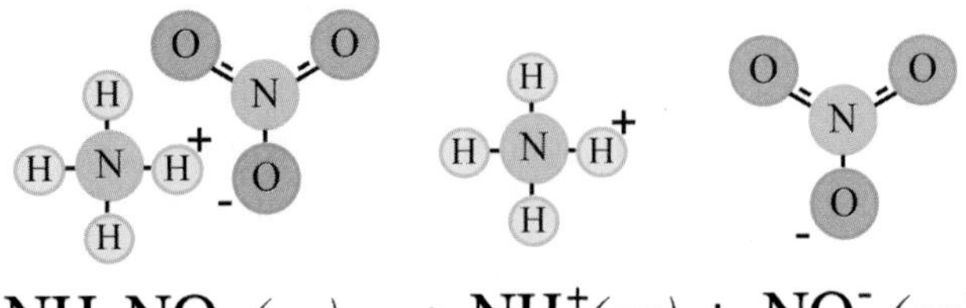

$$NH_4NO_3(aq) \rightarrow NH_4^+(aq) + NO_3^-(aq)$$

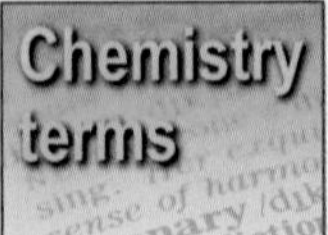

salt - an ionic compound in which the positive ion comes from an acid and the negative ion comes from a base.

Salts of weak acids

Some salts have acid or base properties in solution

Salts form from acids and bases, and it should come as no surprise that some salts can affect the pH of solutions. This is because some salts react with water and change the equilibrium of H^+ and OH^- ions. This effect is strongest when the negative ion of the salt forms a weak acid.

The acetate ion comes from a weak acid (acetic acid)

For example, sodium acetate ($NaC_2O_3H_2$) is a salt that dissociates into a sodium ion (Na^+) and an acetate ion ($C_2O_3H_2^-$). The acetate ion also forms acetic acid ($HC_2H_3O_2$). Since acetic acid is a weak acid, its equilibrium has only a few dissociated acetate ions and mostly whole acetic acid molecules. When sodium acetate is dissolved in water, the acetate ions seek equilibrium by combining with H^+ ions from water to form whole acetic acid.

Why a sodium acetate solution is basic

In the formation of acetic acid OH^- ions are created and H^+ ions are removed. As a result the pH goes up. A solution of sodium acetate in water will be basic, even though sodium acetate is a *salt*!

Some salts act like acids or bases when in solution

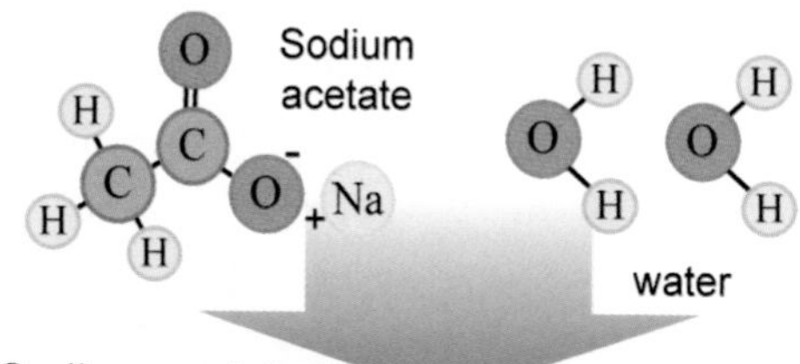

Sodium acetate is the ion from a weak acid (acetic acid)

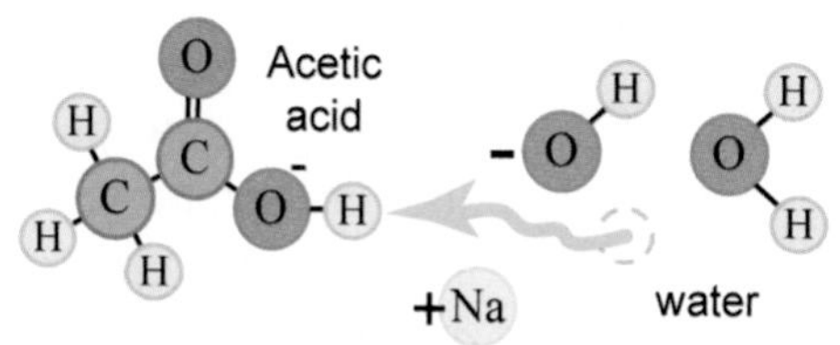

Some of the acetate ions capture H^+ from water and become acetic acid molecules. Some OH^- ions are released from water so the solution is a base!

A salt whose negative ion is from a weak acid raises the pH of a solution (makes it more basic)

Why salts of strong acids do not affect pH

Sodium chloride is also a salt but it has no effect on pH. In solution, NaCl creates positive sodium ions (Na^+) and negative chloride (Cl^-) ions. The acid formed by the chloride ion is hydrochloric acid (HCl), which is a *strong* acid. NaCl has no effect on pH because any Cl^- ions that combine with H^+ ions immediately dissociate again because HCl is a strong acid. When the ions in the salt come from *strong* acids and bases, there is no effect on pH when the salt is dissolved.

A salt whose negative ion is from strong acid has no effect on pH.

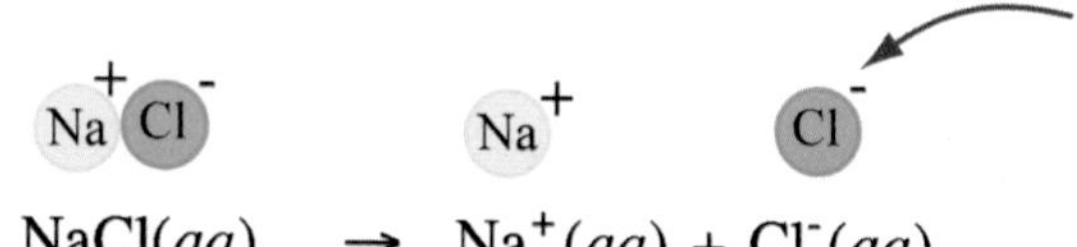

$$NaCl(aq) \rightarrow Na^+(aq) + Cl^-(aq)$$

The Cl^- ion comes from a strong acid (HCl) so it does not combine with H^+ from water - *therefore has no effect on pH*

Buffers

Common ions

Solutions of weak acids and their associated salts have a very important application. Consider a solution containing both acetic acid and sodium acetate (a salt). These two compounds share a **common ion**: acetate ($C_2O_3H_2^-$). A common ion is one that is produced by two different chemicals in the same solution, such as an acid and a salt.

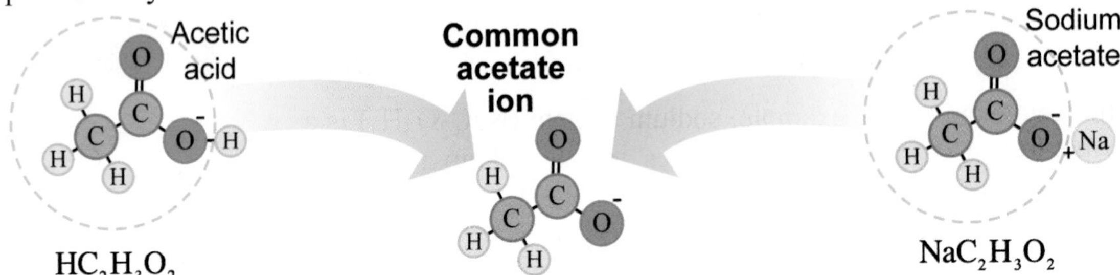
Acetic acid
Common acetate ion
Sodium acetate
$HC_2H_3O_2$
$NaC_2H_3O_2$

Conjugate acids and bases

On the previous page you learned that sodium acetate is a weak base because solutions of sodium acetate have a pH greater than 7. In fact, sodium acetate is a *conjugate base* to acetic acid. Remember, according to the Bronsted-Lowry definition, an acid is a proton donor. Acetic acid contributes protons (H^+ ions) by dissociation. Every proton donor has a proton acceptor which is called its conjugate base.

$$HC_2H_3O_2 \rightarrow H^+ + C_2H_3O_2^-$$

Acetic acid donates a proton (Bronsted-Lowry acid)

$$C_2H_3O_2^- + H^+ \rightarrow HC_2H_3O_2$$

Acetate accepts a proton (Bronsted-Lowry base)

Buffer solutions

A mixture of a weak acid and its conjugate base is an example of a **buffer**. A buffer is a solution that resists small changes in pH by chemical action. Suppose a small amount of acid is added. Acid increases H^+ and would normally lower the pH. However, acetic acid is a *weak* acid so the ratio of H^+ to whole $HC_2O_3H_2$ molecules is small and constant. Because the solution contains an overabundance of acetate ions from the salt, excess H^+ combines with those ions to form acetic acid! *The pH stays about the same*. The excess of common ions allows any additional H^+ to be absorbed.

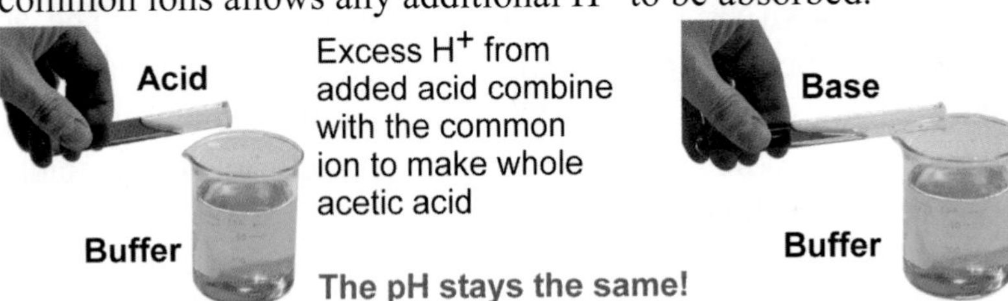
Acid
Buffer
Excess H^+ from added acid combine with the common ion to make whole acetic acid
The pH stays the same!
Base
Buffer
Excess OH^- from added base combine with H^+ from the weak acid to make water. More acid dissociate to keep equilibrium.
The pH stays the same!

When a base is added the pH also stays constant because of neutralization by the weak acid. Excess OH^- ions from the base combine with H^+ from the acid to make water.

Chemistry terms

common ion - an ion which is produced by two or more different chemicals in the same solution.

buffer - a solution that resists small changes in pH by chemical action.

Buffer capacity

A buffer can only absorb a limited pH change

A buffer resists changes in pH because a dissolved ion from weak acid or base supplies a reservoir of ions which can absorb both excess H+ or OH-. The process works very well for small changes in acidity. For example, the graph shows a buffer when increasing amounts of 0.1M HCl are added. Notice that the color, and hence the pH does not change until about 45 mL are added. from 40 to 50 mL of added acid, the color change is dramatic showing that the pH changed substantially.

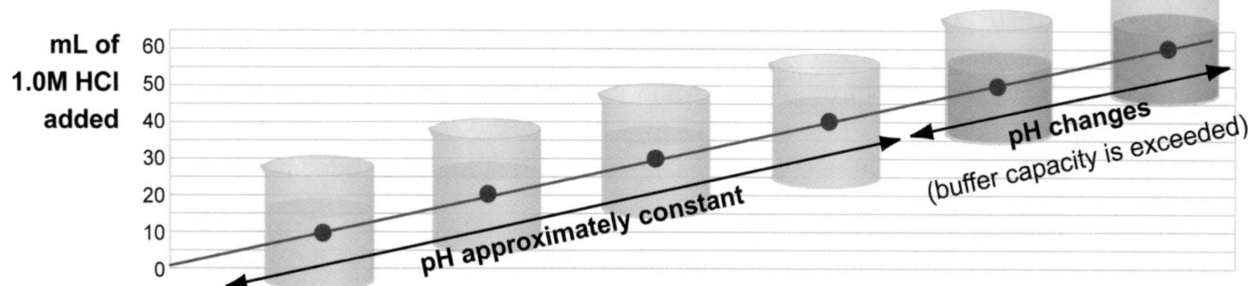

Buffer capacity

Eventually, all the common ions from the salt are combined and there is no longer a reservoir to absorb excess H+ or OH- ions. Once this point is reached, adding additional acid changes the pH just as it would with a non-buffered solution. The **buffer capacity** is describes how much acid or base a buffer can absorb without changing the solution pH.

Seawater is a buffer

Since living things are sensitive to pH changes, buffers are vital to our earth. For example, the ocean maintains a pH of about 8.3 and seawater acts as giant buffer, resisting pH changes from rain (typically acidic) and weathering of substances like limestone (typically basic). The major components of the ocean's buffer are calcium carbonate from shells of marine animals, and carbonic acid from dissolved atmospheric CO_2.

Sea water is a buffer at a pH of 8.3

Blood plasma is a buffer

Your blood is also a buffer, maintaining your body's internal pH very stable and slightly basic with a pH between 7.35 and 7.45. The weak acid is carbonic acid which forms in water from CO_2 generated by respiration. The conjugate base is bicarbonate which is produced in your pancreas. Acidosis can result when excess carbonic acid builds up beyond blood's the buffer capacity. Breathing difficulties can cause acidosis because CO_2 builds up in blood if it cannot be exhaled adequately.

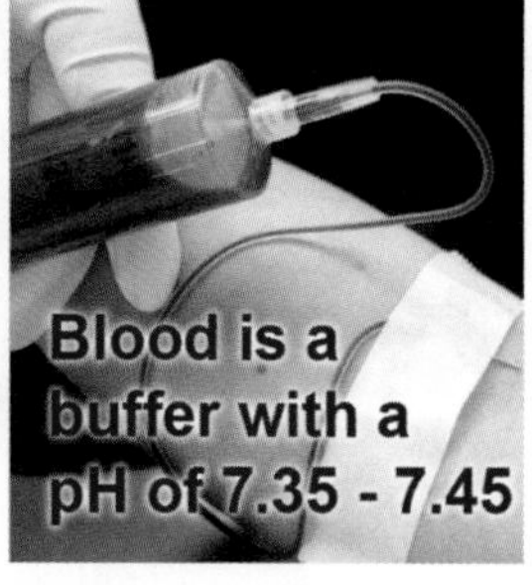

Blood is a buffer with a pH of 7.35 - 7.45

Chemistry terms

buffer capacity - describes the amount of excess acid or base a buffer can neutralize without changing the pH of the solution.

Blueberries prefer highly acidic soil with a pH below 5.5

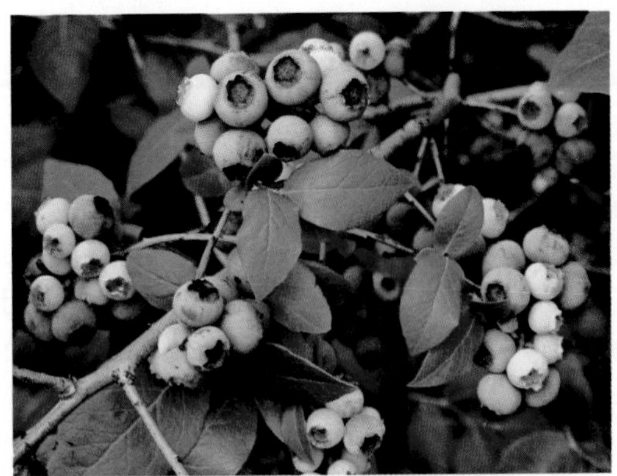

Plants draw trace minerals such as magnesium and phosphorus from dissolved salts in moist soil. The pH of the soil affects the form and concentration of these nutrients for plants. Most plants prefer a slightly acidic soil with a pH between 6.5 and 7.0. Azaleas, blueberries, and conifers grow best in more acid soils with a pH of 4.5 to 5.5. Vegetables, grasses and most other shrubs do best in less acidic soils with a pH range of 6.5 to 7.0.

In highly acid soils (pH below 4.5) too much aluminum, manganese and other elements leach out of solid soil particles. This happens because acids react with metals to produce soluble ions. Metal ion concentrations above a certain threshold are toxic to plants. Also, at strongly acidic pH values, calcium, phosphorus and magnesium are less available to plant roots. Soil pH can also be too high for average plants. At pH values of 6.5 and above, (more basic), iron and manganese become less available.

Every good gardener or farmer pays attention to soil pH. Soil pH depends greatly on average rainfall. Soils in wet or temperate areas with more than 20 inches of rain per year tend to be acidic. Soils in dry climates with less than 20 inches of rain per year tend to be alkaline (basic). Desert soils tend to be alkaline and gardeners in Phoenix, Arizona often have to grow plants in soils with a pH between 7 and 8. A standard remedy for alkaline soil is to add *sulfur*. Over time, sulfur reacts slowly with rainwater to produce a small amount of sulfuric acid to neutralize the alkalinity.

The pH of water directly affects aquatic life. Most freshwater lakes, streams, and ponds have a natural pH in the range of 6 to 8. Most freshwater fish can tolerate pH between 5 and 9 although some negative effects appear below pH of 6. Trout (like the california golden) are among the most pH tolerant fish and can live in water with a pH from 4 to 9.5.

Frogs and other amphibians are even more sensitive to pH than fish. Most frogs prefer pH close to neutral and don't survive below pH of 5.0. Frogs eggs develop and hatch in water with no protection from environmental factors. Research shows that even pH below 6 has a negative effect on frog hatching rates.

Some plants change dramatically with pH. The *hydrangea* flower is a probably the best known example. Hydrangeas have different colored blossoms depending upon whether the soil they are growing in is acidic or basic. The blossoms are blue when in acidic soils and pink in basic soils. Hydrangea growers adjust the soil pH to create the color flower they want.

Basic soil makes hydrangea flowers pink

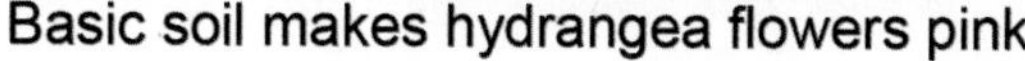

Acid soil makes hydrangia flowers blue

Hydrangea color is affected by the element aluminum! Acidic conditions convert the aluminum in the soil to a form that the shrub can absorb (when it drinks), this turns the flower blue. In basic soil, the aluminum in the soil is insoluble and therefore not able to be absorbed by the hydrangea. Without the aluminum the flower color is pink.

To make a soil more basic (increase pH) gardeners add lime. Lime is calcium carbonate, $CaCO_3$. Calcium carbonate neutralizes some of the acid in the soil by first dissociating into calcium ions and carbonate ions. Carbonic acid is a weak acid so most of the carbonate ions bond with H+ ions in the soil to form whole carbonic acid molecules. This reduces the H+ concentration and makes the soil more basic. The result is pink flowers!

Calcium carbonate (lime) dissociates into calcium and carbonate ions

Step 1 $CaCO_3 \rightarrow Ca^{2+} + CO_3^{2-}$

Carbonate ions combine with H^+ to make bicarbonate ions

Step 2 $H^+ + CO_3^{2-} \rightleftharpoons HCO_3^-$

Bicarbonate ions combine with H^+ to make carbonic acid

Step 3 $H^+ + HCO_3^- \rightleftharpoons H_2CO_3$

To make aluminum available to the plant and get blue flowers, the pH of the soil needs to be between 5.2 and 5.5. Hydrangea growers often add aluminum sulfate, $Al_2(SO_4)_3$ to make their soil more acidic. Adding aluminum sulfate both increases the acidity of the soil and provides a source of aluminum. When dissolved in water $Al_2(SO_4)_3$ ionizes according to the reaction below.

$$Al_2(SO_4)_3(s) + H_2O(l) \rightarrow 2Al^{3+}(aq) + 3SO_4^{2-}$$

The Al^{3+} ion has a very small size, with a radius of 53 pm. The fact that the aluminum ion is small and highly charged means it acts almost like H^+ in water! gives it the ability to act as an acid in aqueous solutions. Al^{3+} is able to draw the electron density of a water molecule away from the OH bond, favoring the release of the H^+ ion. The release of the H+ ion causes the soil become more acidic.

Chapter 13 Review

Vocabulary

Match each word to the sentence where it best fits.

Section 13.1

acid	base
hydronium ion	Bronsted-Lowry
neutral	amphoteric
strong acid	weak base
weak acid	strong base

1. HCl is an example of a strong __________ because it releases hydrogen ions.
2. A solution that is _____________ contains more hydroxide ions than hydrogen ions.
3. A solution that has a pH of seven is considered to be ______________.
4. Water can accept an H+ ion or donate an H+ ion and this makes it _____________.
5. A _____________ only ionizes partially in water and yields fewer hydrogen ions than a _____________ which ionizes 100% in water.
6. Ammonia, NH_3 is an example of a ____________.
7. The __________________ definition of an acid and a base explains which substance is a hydrogen ion acceptor and which is a hydrogen ion donor.
8. When water bonds to a hydrogen ion in an aqueous solution the_ ____________ is formed in solution
9. Sodium hydroxide, NaOH is an example of a _________________.

Sections 13.2 and 13.3

exponent	logarithm
indicatorl	ion product constant

10. The __________ of 10,000 is 10^4.
11. Phenolpthalein is and example of an __________, which is a substance that turns a different color as the pH changes.
12. In the number 2.3 x 10-5, the negative five represents an ____________. It is negative because the number represents a value less than one.
13. The equilibrium constant for water at 25°C and 1 atm is called the ________________ which represents a value called Kw.

Section 13.4

neutralization	equivalence point
titration	salt
common ion	buffer
buffer capacity	

14. A _____________ is an important laboratory method used to determine the concentration of a solution of an acid or base.
15. During a ______________ reaction, the acid and the base react until the H+ ions = the OH- ions.
16. At the _________________ we can make the important assumption that moles base = moles of acid.
17. A solution that is able to resist small changes in pH is called a _____________.
18. Potassium chloride, KCl is a _________ that is formed from the positive ion of a base and the negative ion of an acid.
19. Buffers work based on the ______________ principle, which recognizes that an ion is produced by two or more chemicals in the same solution.
20. The ______________ tells us how much acid or base a solution can neutralize before changing pH.

Conceptual Questions

Section 13.1

21. Compare and contrast an acid and a base. Clearly explain their chemical differences and give an example of each.

22. Give two examples (each) of acidic and basic substances that you use regularly in your daily lives.

23. Explain what a "neutral" solution means in terms of the chemical ions present.

24. Write a chemical equation that represents :
a) How a strong base ionizes in water.
b) How a strong acid ionizes in water.

25. What type of solution is formed when Group I and II metals dissolve in water ? Explain and support your answer using a chemical equation.

26. The hydrogen ion can be thought of as a "naked proton." Explain why this is important in terms of chemical reactions.

27. Explain the main difference between the early Arrhenius theory of acids and base and the Bronsted-Lowry theory of acids and bases.

28. Explain how ammonia, NH_3, acts as a Bronsted-Lowry base when dissolved in water. Show the chemical equation that represents this process.

29. What ion is formed when an acid is dissolved in water? Show the Lewis structure for this ion.

30. In each of the following chemical reactions identify the substance acting as the acid and the substance acting as the base. Label (A)acid or (B)base.
a) $HF + H_2O \rightleftharpoons F^- + H_3O^+$
b) $HNO_2 + H_2O \rightleftharpoons NO_2^- + H_3O^+$
c) $NH_3 + H_2O \rightleftharpoons NH_4^+ + OH^-$
d)$HC_2H_3O_2 + NH_3 \rightleftharpoons C_2H_3O_2^- + NH_4^+$

31. For number 30 (b) above clearly explain answer.

32. Look at 30 (a) and (b) above and explain water's role in each equation.

33. Identify the following substances as strong acids or strong bases
a) KOH
b) HNO_3
c) $Ba(OH)_2$
d) H_2SO_4
e) HCl

34. Write a chemical equation that shows how nitric acid, HNO_3, ionizes in anaqueous solution.

Section 13.2

35. Briefly explain what pH measures in an aqueous solution.

36. Make a sketch of the pH scale and label the numbers zero to 14. Label where on the scale these substances would be located.
a) Lemon juice
b) baking soda
c) dish soap
d) soda water
e) blood
f) milk

37. The pH of a solution is 6.8. From this information can you conclude that the solution is acidic? Explain.

38. Write the equation that relates pH to pOH.

39. If you have a "strong" acid does that mean the pH of the solution of this acid will be low? Explain.

40. Based on the pH scale sustances that are acidic have a pH less than 7. What does this tell you about the amount of [H+] relative to [OH-] in the solution? Explain.

41. Explain how an aqueous solution can have a pH of 7. What would have to be true?

42. If the pH changes by 2 units by what factor does the $[H^+]$ change?

43. Name two different types of indicators.

44. How are indicators used to help determine the pH of a solution? Explain.

Chapter 13 Review.

45. Sometimes the pH of a an acidic solution can be zero or negative. Explain why.

Section 13.3

46. a)Write the chemical equation for the autoionization of water.
b) The equilibrium constant, Kw, for this reaction is 1.0×10^{-14}. Explain what the magnitude of the Kw tells us about the reaction in the forward direction.

47. a) Write the chemical reaction for the strong acid, hydrochloric acid HCl, ionizing in water.
b) Write the chemical reaction for the weak acid, acetic acid $HC_2H_3O_2$, ionizing in water.
c) Compare the weak and strong acid's behavior in water.

48. Give an example of a weak base and a weak acid.

49. Calculate the number of moles of NaOH that are in 8.50 mL of a 0.450 M solution.

50. Predict the products for the following weak acid and base ionizations shown below
a) $HF + H_2O \rightleftharpoons$
b) $CH_3NH_2 \rightleftharpoons$
c) $HNO_2 + H_2O \rightleftharpoons$

51. If you have a 1.0 M soltution of HCl and a 1.0M solution of $HC_2H_3O_2$, List some things that will be different in each of these solutions. If you did not know which was which, how could you determine which one was HCl?

Section 13.4

52. Give an example of the following acid - base reactions, include reactants and products.
a) neutralization
b) metal and acid

53. What are the products of a neutralization reaction between HCl and NaOH?

54. Do you think neutralization reactions always have a pH of 7? Why or why not?

55. In the laboratory baking soda is often kept handy to neutralize the any strong acid spills. How does baking soda neutralize strong acids?

56. a) Show the chemical reaction for the addition of the salt, potassium nitrate KNO_3, to water.
b) Will this aqueous solution conduct electricity ? Explain.

57. List two common salts formed during the neutralization of a strong acid and strong base.

58. Explain how adding 4.0 g of the salt sodium acetate, $NaC_2H_3O_2$, to an acidic solution would affect the pH?

59. Predict how the overall pH would be affected, if some sodium chloride, NaCl solid where added to a neutral solution with a pH of 7.

60. a)What is unique about a buffer?
b) Can acetic acid act as a buffer? Explain.

Quantitative Problems

Section 13.2

61. If a solution has an $[H^+]$ of 3.5×10^{-4}, what is the pH of the solution?

62. A solution is measured to have a pH of 5.60, calculate the hydrogen ion concentration of the solution?

63. Calculate the pH of the following solutions :
a) A glass of lemonade with a 4.30×10^{-5} M H^+ concentration.
b) Salad dressing with an $[H+] = 2.51 \times 10^{-6}$
c) A glass of tomato juice with an $[H+] = 6.40 \times 10^{-5}$
d) A glass of milk with an $[H+] = 7.94 \times 10^{-9}$

64. Calculate the pH for each of the hydrogen ion concentrations. State whether the pH is acidic, basic or neutral.
a) $[H+] = 2.45 \times 10^{-4}$
b) $[H+] = 2.11 \times 10^{-11}$
c) $[H+] = 4.96 \times 10^{-7}$
d) $[H+] = 1.00 \times 10^{-7}$
e) $[H+] = 3.51 \times 10^{-14}$

65. Calculate the pH for each of the hydroxide ion concentrations. State whether the pH is acidic, basic or neutral.
a) $[OH-] = 4.03 \times 10^{-2}$

b) [OH-] = 9.55 x 10^{-9}
c) [OH-] = 5.20 x 10^{-5}
d) [OH-] = 7.26 x 10^{-12}

Section 13.3

66. Calculate the [OH-] concentration for each of the following [H+] concentrations. State whether the pH is acidic, basic, neutral.
a) [H+] = 3.15 x 10^{-4}
b) [H+] = 2.11 x 10^{-10}
c) [H+] = 5.96 x 10^{-7}

67. Calculate the [H+] concentration for each of the following [OH-] concentrations. State whether the pH is acidic, basic, neutral.
a) [OH-] = 3.23 x 10^{-2}
b) [OH-] =8.65 x 10^{-8}
c) [OH-] = 5.41 x 10^{-6}

68. Calculate the hydrogen ion concentration and the pH of each of the solutions of strong acids below.
a) 2.4 x 10^{-5} M HCl
b) 0.0045 M H_2SO_4
c) 5.51 x 10^{-4} HNO_3
d) 6.41 x 10^{-4} HCl

69. Determine the pH of a 0.50 M solution of acetic acid, $HC_2H_3O_2$. Use a RICE table to do your calculation. Ka for acetic acid = 1.8 x 10^{-5}

70. Calculate the pH of a 0.10 M solution of lactic acid. Lactic acid $HC_3H_5O_3$ is a weak monoprotic acid and it is the acid that sometimes makes your muscles sore after you exercise. Ka = 1.4 x 10^{-4}

71. Calculate the pH of a 0.25 M solution of ammonia, NH_3 a weak base. Use a RICE table to do your calculation. Kb for ammonia = 1.8 x 10^{-5}

72. Calculate the pH of a 0.085 M solution of ethylamine, $C_2H_5NH_2$. The Kb for ethylamine is 6.4 x 10^{-4}.

73. Calculate the percent ionization for the acetic acid in number 60) above. Show your work.

74. Codeine, $C_{18}H_{21}NO_3$, is a weak organic base. Calculate the K_b of a codeine solution with an initial molarity of 4.80 x 10^{-3} M, given that this codeine solution has a pH of 9.90.

Section 13.4

75. Calculate the molarity of an NaOH solution, if 26.0 mL of the solution are needed to neutralize 16.8 mL of a 0.298 M HCl solution. Remember that "neutralize" means that the moles $[H^+]$ = moles $[OH^-]$.

76. Calculate the volume in milliliters of a 1.26 M NaOH solution required to titrate the solutions shown below:
a) 25.00 mL of 2.35 M HCl solution.
b) 25.00 mL of 4.30 M HNO_3 solution
c) 30.00 mL of 3.35 M HCl solution.

77. Calculate the volume of 0.80 M HCl solution needed to neutralize each of the following basic solutions.
a) 12.00 mL of a 0.400 M NaOH solution
b) 12.00 mL of a 0.200 M $Ba(OH)_2$ solution

78. A 2.653 g sample of a solid monoprotic acid was dissolved in 25.0 mL of distilled water. The solution required 18.30 mL of a 0.160 M NaOH solution to titrate. Calculate the molar mass of the acid.

79. Acetic acid, CH_3COOH is the acid present in apple cider vinegar. A sample of 25.00 mL of vinegar is titrated using a 1.10 M solution of NaOH. This sample required 2.35 mL of NaOH to neutralize it. Determine the concentration of the acetic acid in the vinegar?

80. What volume of 0.152 M HCl is required to neutralize 3.40 g of $Mg(OH)_2$? Write out the balanced chemical reaction that occurs to help with this calculation.

81. How many milliliters of 0.120 M HCl are needed to neutralize completely 31.00 mL of 0.150M $Ba(OH)_2$ solution? Write out the balanced chemical reaction that occurs as part of your answer.

CHAPTER 14
Gases

Why do my tires look flat on a very cold day?

Why do very high altitude mountain climbers need to bring oxygen tanks when summiting peaks like Mt. Everest?

Why won't my vacuum cleaner work on the moon?

We live at the bottom of an ocean of air. Gases surround us and press on us. We breathe in one mixture of gases and exhale a different one. The oxygen in the air helps us to burn our fuel (food), and keep the pH in our blood balanced by exhaling carbon dioxide. In chemistry, gases were some of the first substances about which careful research was done.

First human balloon flight – 1783.

Gases provided humans with the first chance to fly. In 1783 a balloon made by the Montgolfier brothers lifted off from Paris, France. The balloon flew for 22 minutes and landed 500 feet away. Today the record for the longest distance was set in 1991 with a flight of 4,767 miles from Japan to Canada, and the longest duration was set in 1997, lasting 50 hours and 38 minutes.

Other types of ballooning, using hydrogen or helium instead of hot air for lift extended the ranges for extreme height, distance, and duration. In 2002, the first solo flight around the world was completed by Steve Fosset in approximately 15 days, and in 1960 Captain Joe Kittinger parachuted from a helium balloon after it reached an astounding altitude of 102,000 feet! He was so high at the time of his jump that he was surrounded by the blackness of space, instead of a blue sky.

Shrinking air

When a gas is cooled it contracts and when it is heated it expands again. You can see it for yourself if you take a helium balloon outside on a cold day. The balloon gets noticeably smaller, like it deflates, as it cools. Bring it back into a warm house and it returns to its original size and firmness.

An extreme way to clearly see how temperature affects the volume of a gas is to cool a balloon with *liquid* nitrogen. Liquid nitrogen boils at -196°C! This means for nitrogen to be in the liquid phase it must be at a temperature below -196°C. The pictures below show what happens when liquid nitrogen is poured over a latex balloon at normal room temperature.

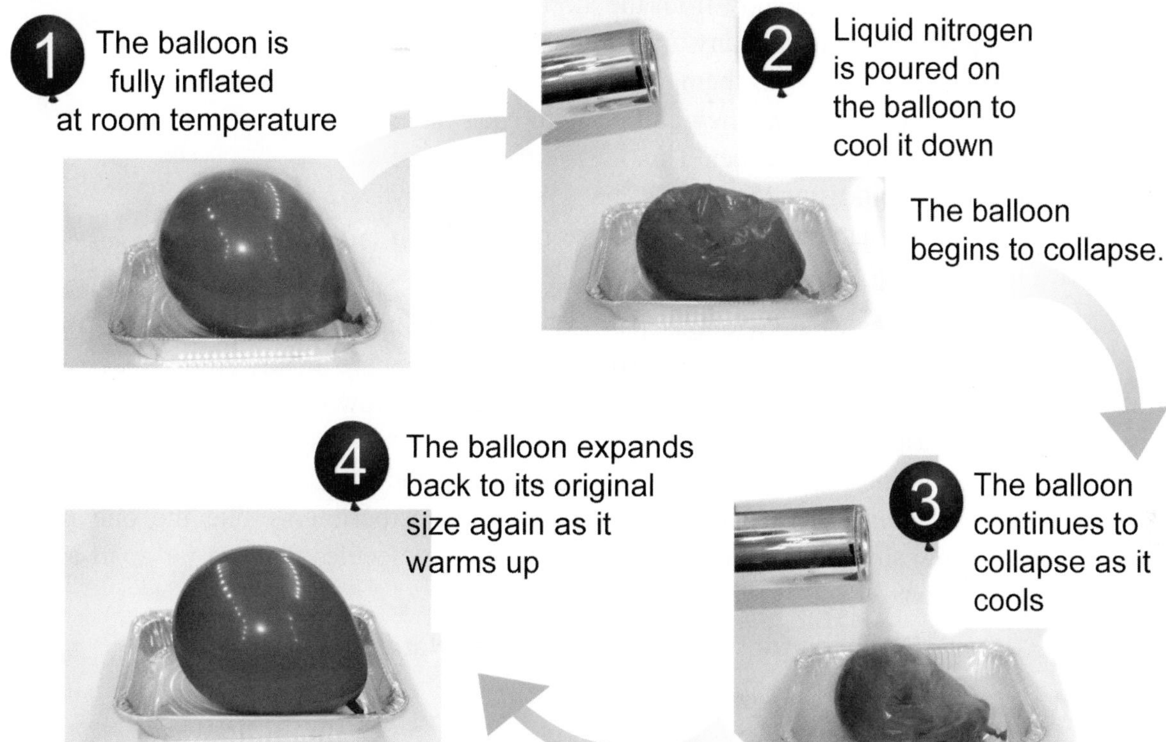

1 The balloon is fully inflated at room temperature
2 Liquid nitrogen is poured on the balloon to cool it down
The balloon begins to collapse.
3 The balloon continues to collapse as it cools
4 The balloon expands back to its original size again as it warms up

Why does the balloon shrink when liquid nitrogen is poured on it? Why does it expand when it warms back up? At first you might think the larger (warmer) balloon has more matter inside. More matter should take up more space, should it not? However, the balloon is sealed so no matter can get in or out. There is the same amount of air in the balloon when it is warm or cold. If you could measure it, the balloon would have the same mass warm or cold.

The real explanation is more subtle. Molecules in a gas are like trillions of tiny bumper cars whizzing about, colliding with each other and with the walls of the balloon. When the balloon is cold, the molecules inside slow down. The molecules in the room are still warm and moving fast. The pressure of molecules from outside air temporarily wins over the chilled molecules inside the balloon and the imbalance of pressure pushes the sides of the balloon inward. As the air inside warms back up, the molecules regain their vigor and push back as strongly as before and the balloon inflates again.

14.1 Pressure and Kinetic Theory

Physical properties of gasses

Gasses are have a unique set of physical properties.

1. Gasses are translucent or transparent.
2. Gasses have very low density when compared to liquids or solids.
3. Gasses are highly compressible compared to liquids and solids.
4. Gasses can expand or contract to fill any container.

Kinetic molecular theory

These properties stem directly from the atomic nature of matter. A gas is made from many tiny atoms or molecules with a lot of space between them. These particles are in continual, frantic motion as they collide with each other and the walls of their container. This idea is the basis of the **kinetic molecular theory**. The kinetic theory provides an explanation for all four of the properties of gases listed above.

Gases consist of atoms or molecules in constant motion.

Experimental demonstration of the kinetic theory.

In a gas the molecules are widely separated from each other. This is the explanation for properties (1) and (2). Gasses are transparent because they are mostly empty space with a large distance between molecules. This also explains the low density. Molecules (mass) are thinly distributed in a gas. The empty space also explains the compressibility and container-filling properties. Compression is reducing the space between molecules and expansion is letting them get farther apart. Simple experiments like the one in the diagram below support the idea that gas is made of molecules that are far apart and in constant motion.

Evidence for the Atomic/Molecular Nature of Matter

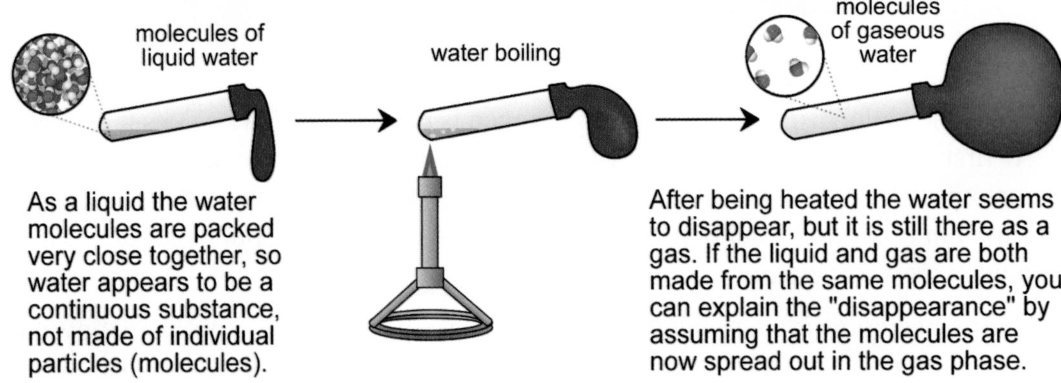

molecules of liquid water
water boiling
molecules of gaseous water
As a liquid the water molecules are packed very close together, so water appears to be a continuous substance, not made of individual particles (molecules).
After being heated the water seems to disappear, but it is still there as a gas. If the liquid and gas are both made from the same molecules, you can explain the "disappearance" by assuming that the molecules are now spread out in the gas phase.

Chemistry terms

kinetic molecular theory - the theory that explains the observed thermal and physical properties of matter in terms of the average behavior of a collection of atoms and molecules.

Brownian motion

Brownian motion

In 1827, botanist Robert Brown made an extraordinary discovery while observing tiny pollen particles suspended in water. The smallest particles seemed to jiggle and move without anything apparently causing the motion. Initially, it was thought that Brown might have stumbled upon the "essence of life." Additional experiments were done with tiny particles of ground glass and other non-living matter. If the particles were small enough, they all showed the same strange jiggling motion. This came to be known as **Brownian motion**. With common skepticism of the existence of atoms and a minimal understanding of heat, the true nature of Brownian motion would remain a mystery for over 70 years.

Brownian motion can be seen by magnifying diluted milk and observing tiny fat globules getting knocked around by the surrounding water molecules.

What Brownian motion tells us

Brownian motion is a result of two fundamental truths about matter.

1. Matter consists of discrete particle (molecules or atoms).
2. Molecules (or atoms) are in constant, vigorous motion due to temperature.

Why Brownian motion occurs

Brownian motion is observed under a microscope with particles smaller than about 10 microns in size (1 micron = 10^{-6} m). The irregular jerky movements are the result of impacts with individual water molecules, which are 1,000 times smaller still. To get a feeling for why brownian motion occurs, consider people throwing pebbles at two balls floating in a quiet pool. One ball is very large and it moves smoothly under the average impact of lots of pebbles. The other ball is small and it moves in a jerky way when struck by individual pebbles. The motion of the big ball is smooth because its mass is much greater than the mass of a single pebble. The smaller ball moves irregularly because its mass is not so much larger than the mass of a single pebble.

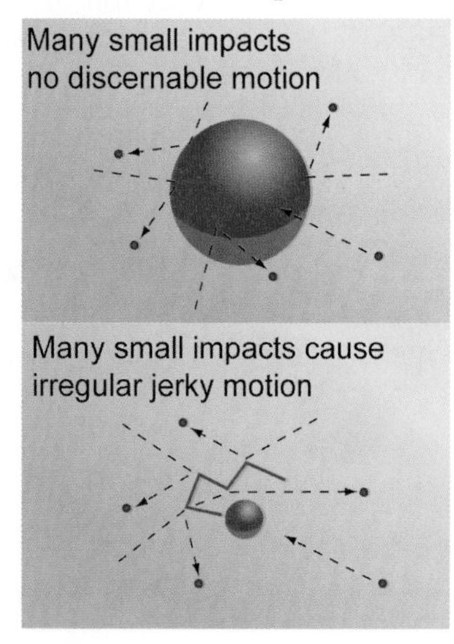

Thermal motion is normally hidden

A ping-pong ball floating in a perfectly still pool does not move at all. There is no indication that water molecules are in constant, agitated motion. Brownian motion is like a peek into the microscopic world of atoms to see details that are normally hidden by the law of averages and the enormous number of incredibly small atoms.

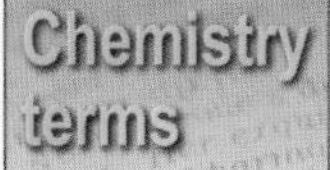

Brownian motion - the random motion of tiny macroscopic particles suspended in a fluid due to individual collisions with the molecules of the fluid.

Pressure

Force in a gas

Think about what happens when you push down on an inflated balloon. The downward force you apply creates forces that act in other directions as well as down. For example, sideways forces push the sides of the balloon out. This is very different from what happens when you push down on a solid ball. A solid ball transmits the force directly down. Because a gas can easily change shape, forces applied to a gas do not act in simple ways as they do on solids.

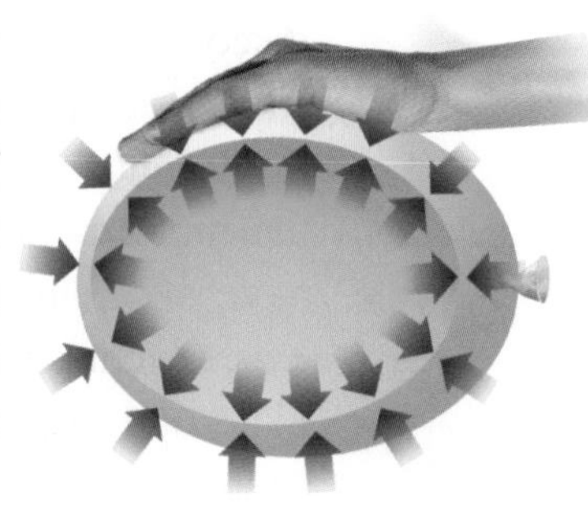

Units of pressure

Gasses have *pressure*. **Pressure** is the force per unit area that the gas exerts on any surface in contact with the gas. The units of pressure are units of force divided by units of area. The SI unit of force is the *newton*. A pressure of 1 N/m^2 means a force of one newton acts on each square meter. The N/m^2 is also named the *pascal* (Pa). One **pascal** is equal to a pressure of one newton of force per square meter of area (1 Pa = 1 N/m^2). The English unit of pressure is pounds per square inch (psi). One psi is equal to one pound of force per square inch of area (lb/in^2). One psi is a much larger unit of pressure than one pascal (1 psi = 6,895 Pa).

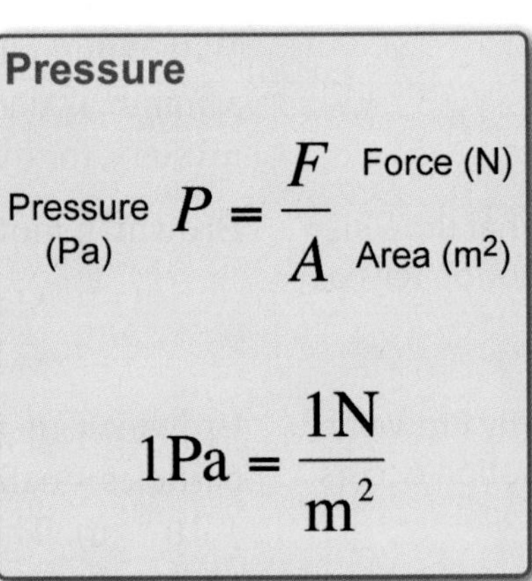

Pressure
Pressure (Pa)
$P = \frac{F}{A}$
Force (N)
Area (m^2)
$1\text{Pa} = \frac{1\text{N}}{\text{m}^2}$

Two jars contain air at a pressure of 35 psi. One jar is 3 inches in diameter and the other is 1 inch in diameter. The larger jar blows its lid off but the smaller one does not. Calculate the force on each lid due to pressure. Hint: $A = \pi r^2$.

Asked: *force on the lid from pressure*

Given: *P = 35 pounds per square inch.*

Relationships: $P = F \div A$ so $F = P \times A$

Solve: *For 3 inch jar,*
$F = (7.07\ in^2)(35\ lbs/in^2) = 247$ pounds
for the 1 inch jar,
$F = (0.79\ in^2)(35\ lbs/in^2) = 27.5$ pounds

Area = 7.07 in^2 Area = 0.79 in^2
3" 1"

Answer: *The lid of the 3 inch jar feels a force of 247 pounds while the lid of the 1 inch jar only feels a force of 27.5 pounds.*

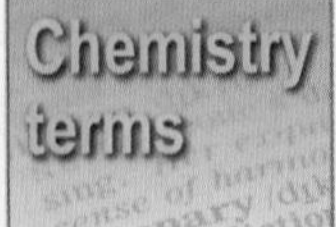

pressure - force per unit area. A gas or liquid exerts pressure on any surface in contact with the gas or liquid.

pascal (Pa) - the SI unit of pressure. 1 Pa = 1 N/m^2.

Atmospheric pressure

Air forms a thin layer around the Earth.

On top of you is a column of air that is about 100 km (60 miles) high being pulled down by gravity. If you could get in a "rocket car" and drive straight up, as you go higher and higher in the atmosphere there is less air on top of you. The pressure decreases and the air gets "thinner," because the molecules are not packed together as closely. The pressure we feel from the air around us is called *atmospheric pressure.*

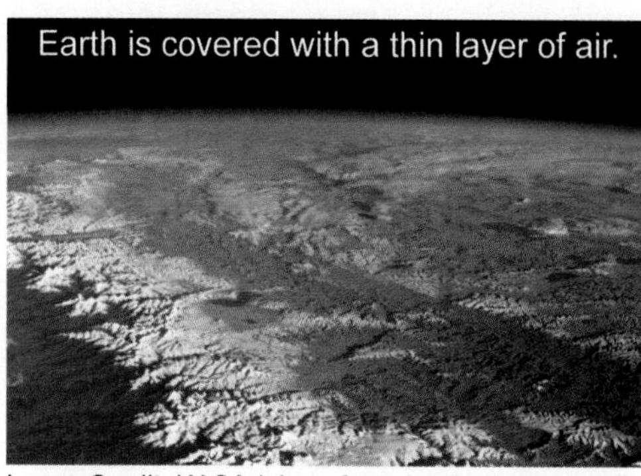

Earth is covered with a thin layer of air.

Image Credit: NASA (photo from Apollo 7 spacecraft)

Standard pressure from the air at sea level is $14.69\frac{\text{lbs}}{\text{in}^2}$

We don't normally feel the pressure from the air around us, because we are used to the constant feeling of it. However, the pressure is significant. Using the units we have used so far, normal atmospheric pressure is $14.69\frac{\text{lbs}}{\text{in}^2}$. Assuming the palm of your hand is approximately 6" x 3", then your palm has a surface area of 18 in^2. If each square inch feels a force of 14.69 lbs from the air, the total force on your hand is about 260 lbs. That is the same pressure you would feel if a 260 lb person stood on one of your hands!

The pressure we feel from the surrounding air on just the palm of our hand is equivalent to the pressure you would feel if a 260 lb person were standing on it.

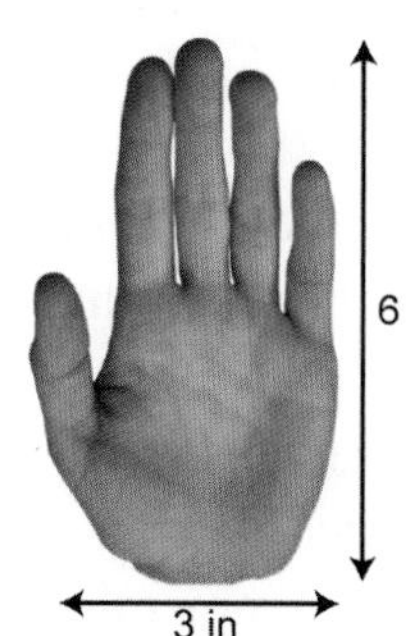

Standard Air Pressure = $14.69\frac{\text{lbs}}{\text{in}^2}$

$A = 6\text{ in} \times 3\text{ in} = 18\text{ in}^2$

$P = \frac{F}{A} \longrightarrow 14.69\frac{\text{lbs}}{\text{in}^2} = \frac{F}{18\text{ in}^2}$

$F = \frac{14.69\text{ lbs} \times 18\text{ in}^2}{\text{in}^2} = \boxed{260\text{ lbs}}$

Pressure is measured in using many different units.

$$\text{Standard Pressure} = 14.69\frac{\text{lbs}}{\text{in}^2} = 14.69\text{ psi} = 1.000\text{ atm} = 760.0\text{ mmHg} = 101305\text{ Pa}$$

Convert $60,000\frac{\text{lbs}}{\text{in}^2}$ of pressure to atmospheres (atm).

Relationships: *Standard Pressure* = $14.69\frac{lbs}{in^2} = 1.000\ atm$

Solve:

$$60000\frac{lbs}{in^2} \times \frac{1\ atm}{14.69\frac{lbs}{in^2}} \rightarrow 60000\frac{\cancel{lbs}}{\cancel{in^2}} \times \frac{1\ atm \cdot \cancel{in^2}}{14.69\ \cancel{lbs}} = 4084\ atm$$

Measuring pressure

Atmospheric pressure drops with altitude

Standard Pressure is defined to be 101,325 Pa. This is the average pressure you would measure from the air if you were standing at sea level. The reason an altitude affects air pressure. The higher you go the less air there is above you, so the weight of the air column decreases. This causes the air pressure around you to decrease as well. The air pressure at the top of Mt. Everest, the world's highest mountain, is only about one third the pressure we typically experience at sea level.

The air pressure at the top of Mt. Everest pictured below is only one third the air pressure at sea level.

The barometer

One of the first instruments used to measure atmospheric pressure was the **barometer**, invented by Italian scientist Evangelista Torricelli in 1643. To understand his invention consider what happens when you place your finger over the top of a straw and then lift the straw out of a glass of water. The liquid stays in the straw, because the pressure exerted by the water in the straw is less than the air pressure outside of the straw. If you had a straw that was over 32 feet tall, then the water would exert a greater pressure than the surrounding air and some water would come out. Torricelli used mercury, which is 14 times more dense than water, causing the column of liquid needed to balance air pressure to be much shorter. It turns out that standard air pressure can support a column of mercury 760 mm tall, and that is where the pressure unit "mmHg" comes from. As air pressure changed, so did the height of the mercury column.

A Barometer

empty space

As air pressure changes, the height of the mercury column it can support also changes.

gravity pulls mercury down the tube

air pressure pushes mercury up the tube

Variations in atmospheric pressure affects the weather.

Most weather maps today show areas of high and low atmospheric pressure. Large areas of slightly varying pressure caused by temperature variations and associated changes in air density, have a significant affect on weather patterns.

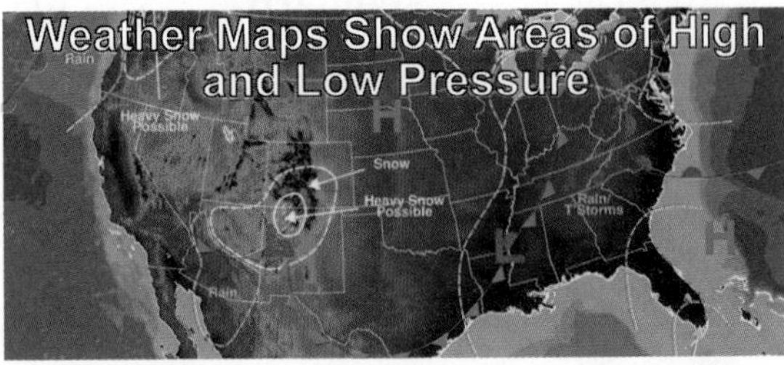

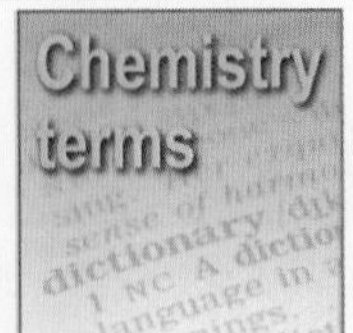

barometer - an instrument that measures atmospheric pressure.

The kinetic theory of pressure

The microscopic view of pressure

On the microscopic level, pressure comes from collisions between atoms or molecules. Think about air in a balloon. The air exerts pressure against the inside of the balloon. On a microscopic level, molecules are moving around and they bounce off the walls of the balloon. It takes force to make a molecule reverse its direction and bounce the other way. The bouncing force is applied to the molecule by the inside surface of the balloon. According to Newton's third law, an equal and opposite reaction force is exerted by the molecule on the balloon. The reaction force is what creates the pressure acting on the inside surface of the balloon. Trillions of molecules per second are constantly bouncing against every square millimeter of the surfaces of the balloon. Pressure comes from the collisions of those many, many atoms, inside *and outside* the balloon.

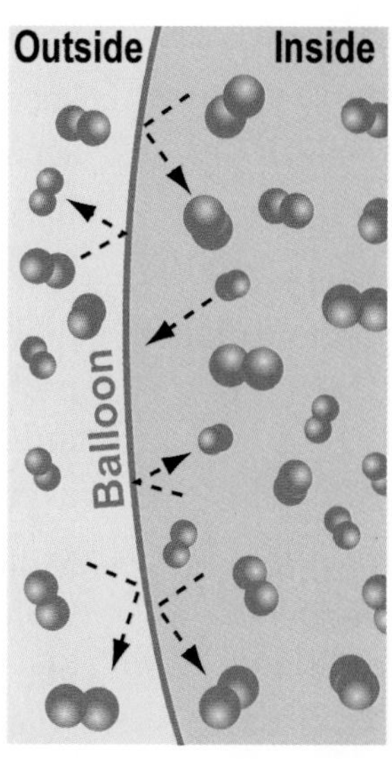

Why gases fill any size container

Bernoulli imagined the molecules of a gas to be like tiny rubber balls continually bouncing off each other and the walls of any container. Each ball was treated as if it followed Newton's laws of motion, moving in a straight line until it collided with another gas molecule or the wall of its container. This would explain why a gas will fill any size container. If you put the gas into a larger container, then the molecules just keep flying straight until they hit the sides of the new container.

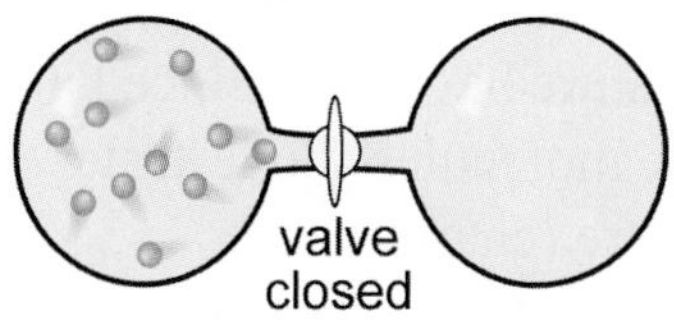

A gas will expand to fill any size (or shape) container

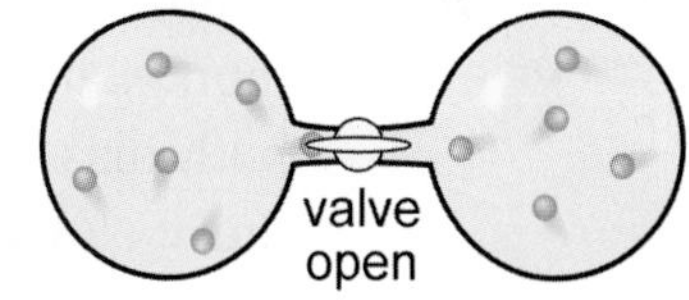

How molecular impacts affect pressure

The average force from molecular impacts depends on two things, and both affect the pressure.

1. Faster (hotter) molecules mean more force per impact, and higher pressure.
2. More molecules per cubic centimeter (higher density) mean more impacts and therefore higher pressure.

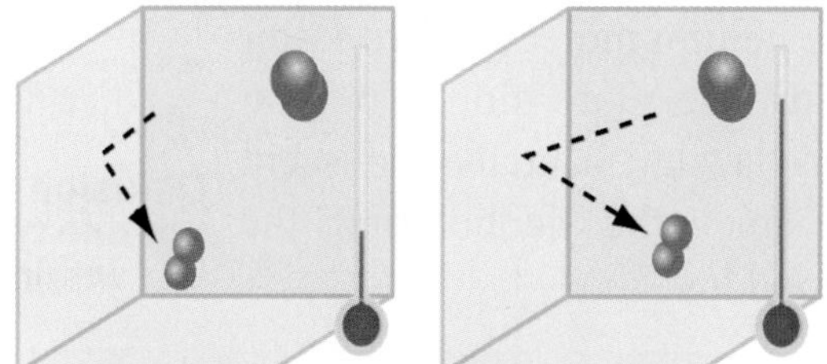

Higher temperature means harder collisions and higher pressure

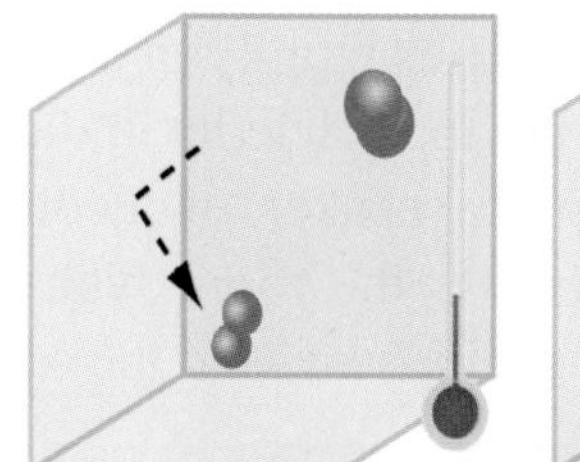
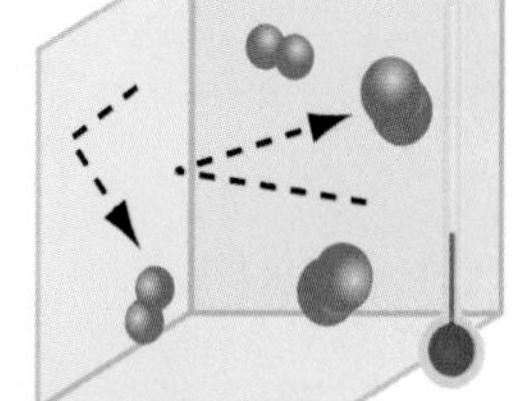

Higher density means more collisions and higher pressure

Diffusion and the speed of molecules in a gas

Molecules move quite fast

The molecules in a gas are moving quite fast! The average speed of a nitrogen molecule in air is 417 meters per second, which is 933 miles per hour! Oxygen molecules move a little slower (390 m/s, 873 mph).

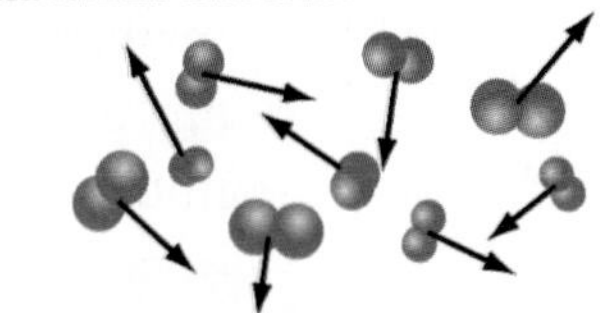

The idea of random motion

A balloon on a table can still be motionless because molecular motion is *random*. Random means there is no preferred direction and molecules are moving in all directions. The average *direction* is zero even though the average speed is not zero.

Non-random motion

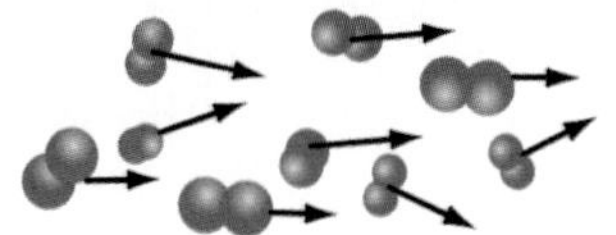

The effect of frequent collisions

Molecules move in the same direction until they collide with the walls of a container or with other molecules. To get a sense of it, consider that air at normal pressure contains 3×10^{19} atoms per cubic centimeter. On average a nitrogen molecule in air has 140,000 collisions with other molecules *in one centimeter*. Each collision changes the direction of both molecules so the frequent collisions completely scramble, or randomize the directions of molecular motion.

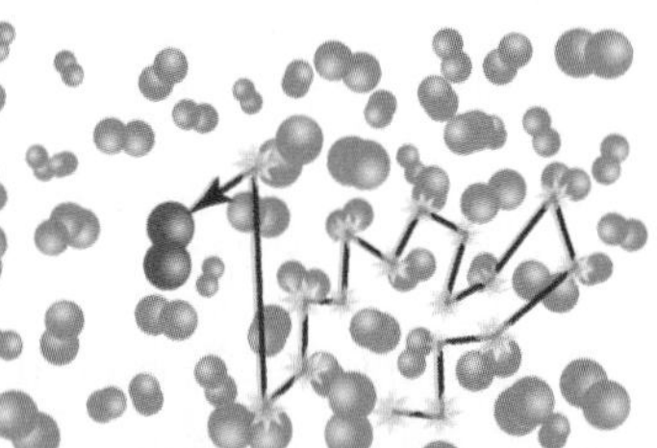

A typical air molecule has 140,000 collisions in a centimeter

Thermal energy depends only on temperature

The kinetic theory tells us that the energy of molecules in a gas depends only on temperature, and nothing else! Molecules at higher temperature have more energy of motion and molecules with lower temperature have less. Since kinetic energy depends on both mass and speed, this means lighter molecules move faster and heavier molecules move slower - even at the same temperature (chart below)

Molecule / atom	He helium	N_2 nitrogen	O_2 oxygen	CO_2 carbon dioxide	$C_8H_8O_3$ vannilin	$C_{15}H_{26}O$ sandalwood
Formula mass (g/mol)	4	28	32	44	152	222
Average speed (m/s)	**1,104**	**417**	**390**	**333**	**179**	**148**

Diffusion

The difference in speeds has a large effect on *diffusion*. **Diffusion** is the rate at which a gas (or other substance) spreads out and mixes with its surroundings. Helium diffuses rapidly in air because helium molecules are faster moving. Heavier molecules diffuse more slowly. If you put a beaker of alcohol, vanilla, and sandalwood scent across the room, you would smell the alcohol first, the vanilla next, and the sandalwood last.

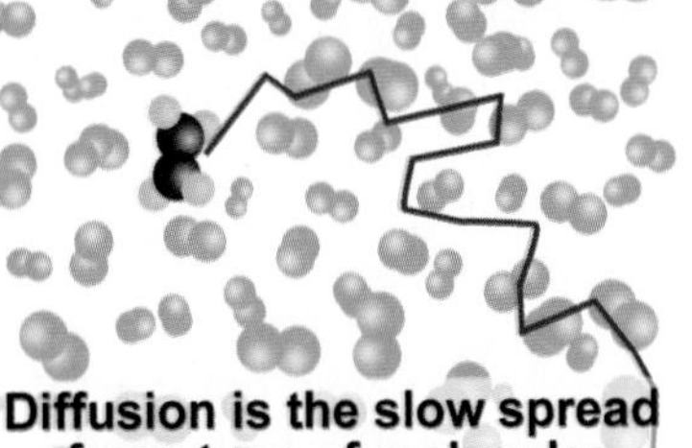

Diffusion is the slow spread of one type of molecules within another type

Chemistry terms

diffusion - the spread of molecules through their surroundings through constant collisions with neighboring molecules.

Boltzmann's constant

ADVANCED AP

Average molecular speed

The energy of molecular motion in a gas is given by Boltzmann's constant, *k*. **Boltzmann's constant** relates the average energy in joules of a single molecule to the temperature of a collection of molecules, and has the value, $k = 1.381 \times 10^{-23}$ J/K. The average molecular speed is derived by assuming all the energy is kinetic energy of motion.

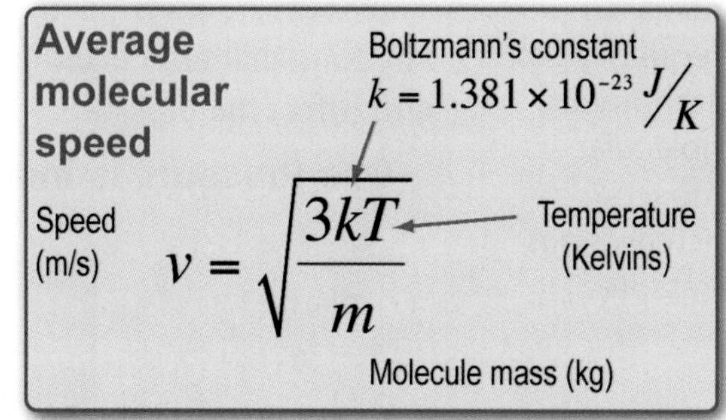

There is a range of molecular speeds

The formula above gives that average speed of a gas molecule but in real gasses there is a wide spread of speeds and energies. Some molecules are going much faster than average and some much slower. The graph shows the distribution of speeds for a sample of 1,000 nitrogen molecules at two different temperatures.

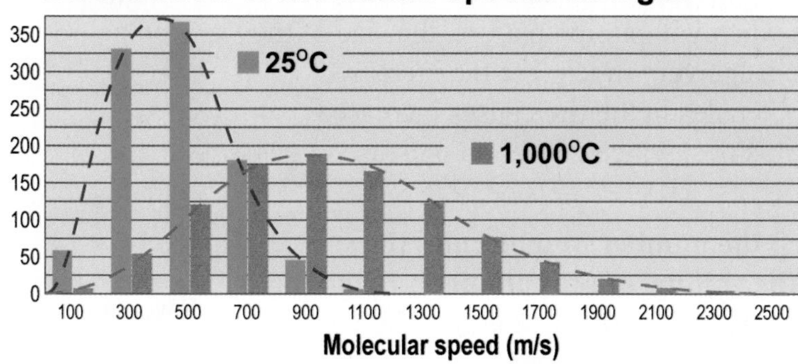

Speed (m/s)	Molecules (25 °C)	Molecules (1,000 °C)
0 - 200	59	7
200 - 400	334	55
400 - 600	370	123
600 - 800	182	176
800 - 1000	48	191
1000 - 1200	7	168
1200 - 1400	1	124
1400 - 1600	0	79
1600 - 1800	0	43
1800 - 2000	0	21
2000 - 2200	0	9
2200 - 2400	0	3
2400 - 2600	0	1
Total molecules	1,000	1,000

Interpreting the graph

Notice two important properties of this graph.

1. As temperature increases, the average speed of molecules increases. At 25°C the largest number of molecules (370) had speeds between 200 and 400 m/s. At 1,000°C, the largest number of molecules (191) had speeds between 800 and 1,000 m/s.
2. As temperature increases, the *spread* of molecular speeds increases. At 25°C, all 1,000 molecules had speeds between 1 and 1,400 m/s. At 1,000°C, the range of speeds went form 0 to 2,600 m/s.

The graph is known as a Maxwellian distribution and it is the basis for the kinetic theory of matter. The average molecular speed is calculated from the Maxwellian distribution.

Chemistry terms

Boltzmann's constant (k) - relates the kinetic energy in molecular motion to temperature. Boltzmann's constant has the value, $k = 1.381 \times 10^{-23}$ J/K.

14.2 The Gas Laws

Increased frequency or strength of molecular impacts increases gas pressure.

The pressure exerted by a gas is due to the average of many molecular impacts, so anything that affects either how frequently the molecules collide or how hard they collide will affect the pressure.

Gas Pressure is Increased by More Frequent and/or Harder Collisions

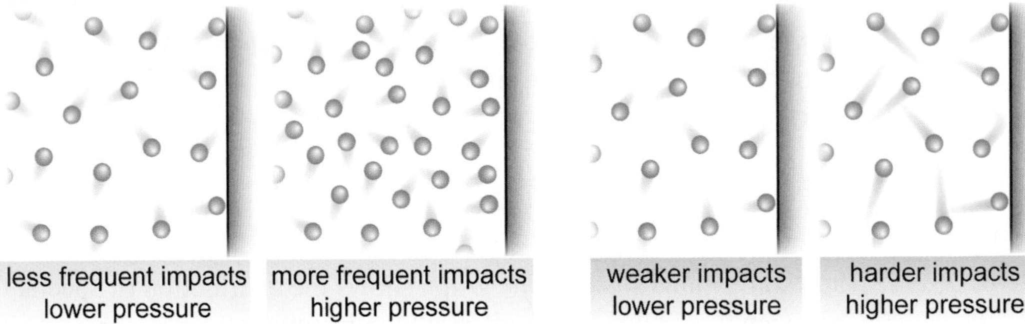

Changes in the amount of gas, the size of the container, and the temperature of the gas all have an impact on the pressure.

You can have more impacts by doing several things:

- **Put more molecules into the same size container.** For example when you pump up a tire, the size of the tire doesn't change very much, but the increased number of molecules in the tire causes increased molecular collisions, which causes an increase in gas pressure.
- **You can keep the number of molecules the same, but reduce the volume of the container.** By giving each molecule less space to move around in, there is an increase in molecular collisions, causing an increase in gas pressure. Every time you exhale, your lungs decrease their volume to increase the pressure inside. Higher pressure inside forces air out.
- **You can heat the gas up.** An increase in temperature causes the kinetic energy of the molecules to increase. This means they move faster. If they are moving faster, then they hit the walls more often *and* they hit harder. Both effects increase pressure. The increase of pressure with heat the reason most aerosol cans have a warning to keep the cans away from heat and fire. If the temperature caused too much of a pressure increase, the can could explode!

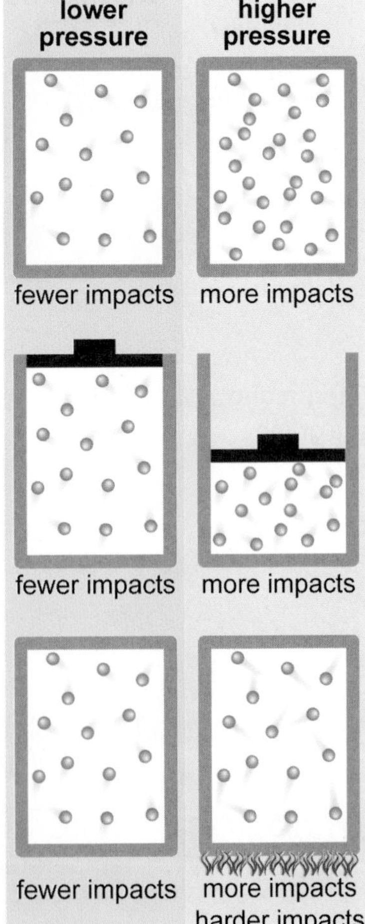

Boyle's law: volume vs. pressure

Robert Boyle explored the relationship of pressure vs. volume at constant temperature and moles of gas.

Robert Boyle explored the relationship between the pressure of a gas and its volume. Using Torricelli's ideas he constructed a device that would allow him to measure the pressure of an enclosed gas by observing the height of a column of mercury supported by that gas. Using a tube shaped like the letter "J" and sealed on one side, he poured mercury into the open end. By measuring the height of the mercury he could determine the pressure, and by taking measurements of the tube and amount of space taken up by the gas he could determine the volume. Because no gas could get in or out, the number of gas molecules was kept constant. Boyle made sure to keep the temperature constant while taking measurements.

J-tube Apparatus to Measure P and V

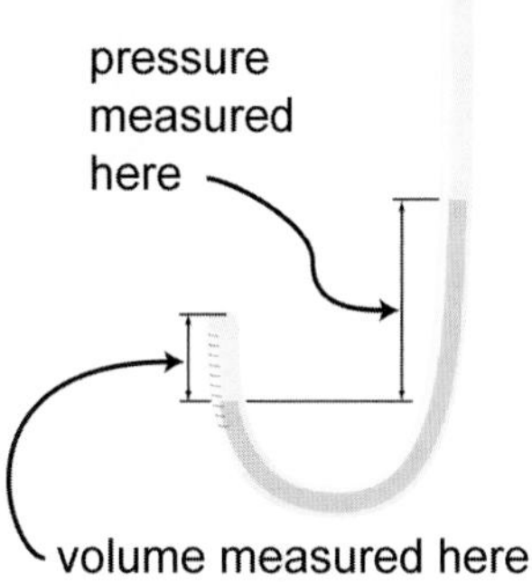

Below is some typical data that could be collected in such an experiment as well as a graph of the data.

TABLE 15.1. Pressure vs. Volume (at constant temp and constant moles)

Pressure (atm)	Volume (mL)	P × V (atm mL)
0.25	200	50
0.50	100	50
1.00	50	50
2.00	25	50
5.00	10	50
10.00	5	50

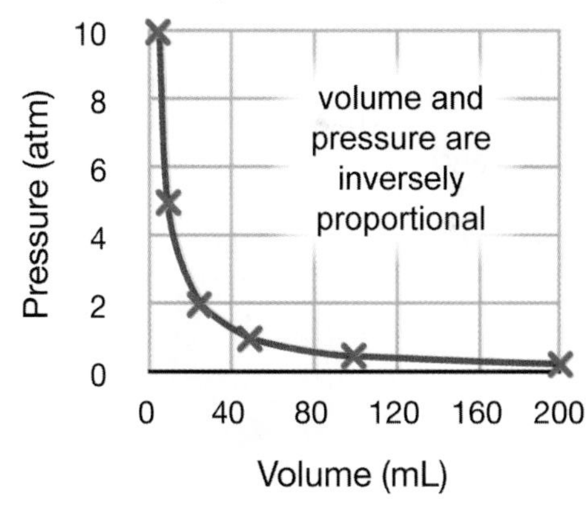

Pressure is inversely proportional to volume.

In 1643 Boyle discovered that doubling the pressure caused the volume to be cut in half, and doubling the volume caused the pressure to be cut in half. This means that pressure is inversely proportional to volume, and can be expressed as PV = some constant. The "constant" could change from one experiment to another if the temperature or amount of initial gas used was different, but as long as the temperature and moles of gas remained the same, the PV=constant relationship would be true.

Boyle's Law

P = pressure
V = volume
k = a constant*
(* value depends on temp. and moles of gas)

$$PV = k \quad \text{or} \quad P_1V_1 = P_2V_2$$

Only true for constant temperature and constant number of gas molecules (moles).

Applications of Boyle's law

Weather balloons expand to huge sizes as they reach high altitudes where the air pressure is low.

Every day, twice a day, at about 1000 stations worldwide, weather balloons are released into the air carrying special instruments to measure wind speed, humidity, and several other properties of the atmosphere. These balloons are filled with hydrogen or helium, and soar high into the sky, reaching an altitude of about 100,000 feet (30 km) before bursting and releasing their parachuting instruments to the ground. As these balloons climb higher and higher the pressure gets lower and lower. Boyle's Law predicts that the balloon should get bigger and bigger. That is why it eventually bursts. The pictures here show a high altitude research balloon being launch. Notice that the balloon looks mostly empty on the ground, but the gas inside has expanded enormously once it reaches high altitude.

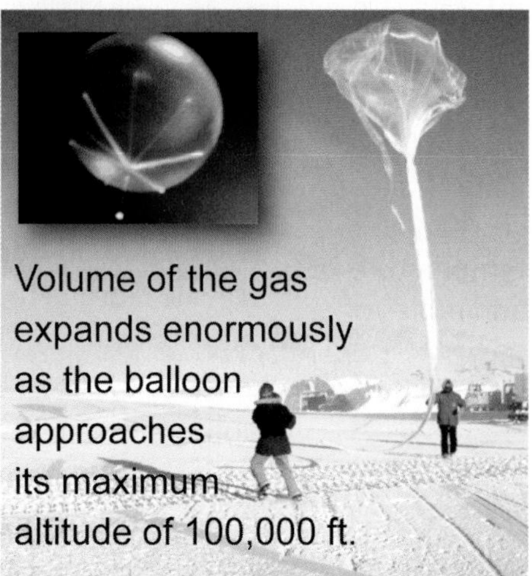
Volume of the gas expands enormously as the balloon approaches its maximum altitude of 100,000 ft.

Changes in volume change the space between molecules and affects the frequency of molecular impacts.

From a molecular standpoint, the inverse relationship between pressure and volume makes sense. If the volume of the gas expands, then there will be more space between the molecules, so there will be fewer impacts per second on the walls of its container, lowering the pressure.

Larger Volume = Fewer Impacts = Lower Pressure

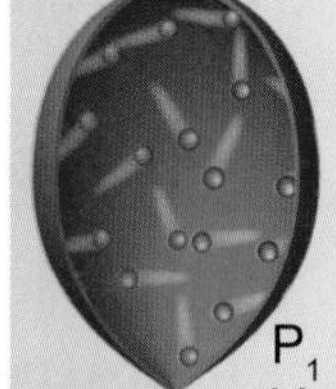

$$P_1V_1 = P_2V_2$$

$$1.00 \text{ atm} \times 3.0 \text{ L} = 0.50 \text{ atm} \times 6.0 \text{ L}$$

$$3.0 \text{ atm} \cdot \text{L} = 3.0 \text{ atm} \cdot \text{L}$$

P_1 = 1.00 atm
V_1 = 3.0 L

P_2 = 0.50 atm
V_2 = 6.0 L

If a weather balloon is released on the ground having a volume of 3.0 m^3 and a pressure of 1.00 atm, how large will it get when it reaches an altitude of 100,000 feet, where the pressure is 0.0100 atm?

Asked: *What is the new volume of the balloon when it reaches 100,000 ft?*

Given: $P_1 = 1.00 \text{ atm}; \; V_1 = 3.0 \text{ m}^3; \; P_2 = 0.0100 \text{ atm}$

Relationships: $P_1V_1 = P_2V_2$

Solve:

$$1.00 \text{ atm} \times 3.0 \text{ m}^3 = 0.0100 \text{ atm} \times P_2$$

$$P_2 = \frac{1.00 \cancel{\text{atm}} \times 3.0 \text{ m}^3}{0.00100 \cancel{\text{atm}}} = 300 \text{ m}^3$$

There is no such thing as suction

Gases can only push.

It seems strange to say that there is no such thing as suction. Of course something happens when you suck air into your lungs. It's just not a pulling force that causes the air to go into your lungs. In almost every case suction can be explained by an imbalance of gas pressures. Gas exerts a pressure by having its molecules PUSH on something, so there is no way for a gas to pull on anything.

Pressures tend to equalize, so we forget we are surrounded by gases.

Remember, we live at the bottom of an ocean of air and the pressure from the air is quite strong, typically $14.69 \frac{\text{lbs}}{\text{in}^2}$. The reason we don't see things getting pushed around by this air pressure is that it is nearly the same everywhere. Take a balloon for example. Once you blow it up and seal it with a knot, it appears as if nothing is happening. However, at the molecular level there is a huge amount of activity with molecules colliding into the balloon both outside and inside the balloon. It doesn't get any bigger or smaller because the pressure inside and outside are equal.

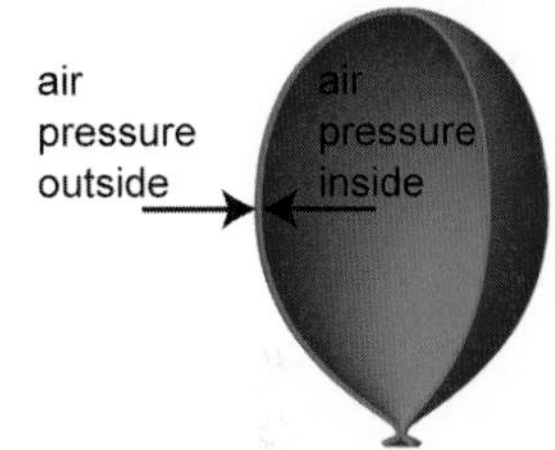

Pressure inside balloon is equal to the pressure outside.

Breathing is possible because the air outside pushes its way into your lungs.

When you inhale, its feels like you are pulling air into your lungs. However, what is really happening is that you are lowering the pressure inside your lungs, so the air outside which is at higher pressure pushes its way in. This is just another application of Boyle's Law. When you breath in, you use muscles to expand the volume of your lungs. If your lungs get bigger then there is more space for the air molecules inside, spreading them out and reducing their impacts. The air outside your lungs is then pushing and colliding with greater frequency, and those molecules push their way into the lower pressure region you created inside of your lungs.

Three Stages of Inhalation

Stage 1: Before inhaling, the pressure inside and outside of your lungs is the same.

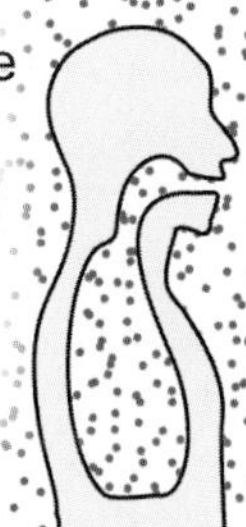

Stage 2: Lung volume is expanded, so molecules spread out, and pressure decreases inside lung.

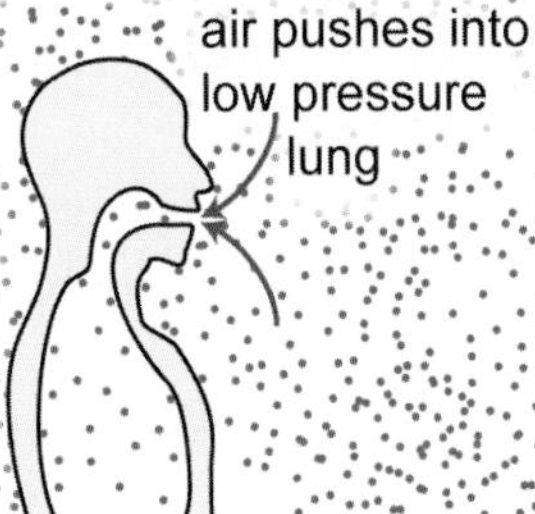

Stage 3: Inhalation is finished when pressure inside and outside of the lung is equal.

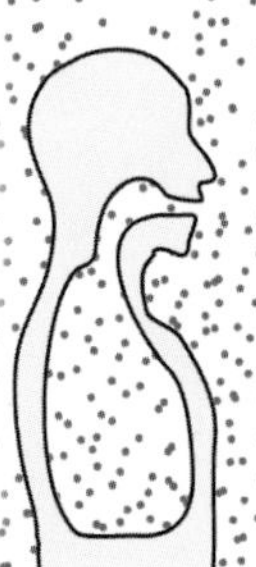

Straws and vacuum cleaners

The same thing happens with a straw, a vacuum cleaner, or anything else that seems to pull something in by suction. When using a straw, you typically expand the volume inside of your mouth, lowering the pressure and allowing the liquid in the straw to be pushed into your mouth by the larger outside air pressure. Vacuum cleaners work by creating a region of low pressure near the mouth of the vacuum. The higher air pressure outside of the vacuum pushes air (and dirt carried by the air) into the machine.

Volume, temperature, and the Kelvin temperature scale

Amontons discovered that heating a gas caused it to expand.

In 1702 Guillaume Amontons demonstrated that a change in temperature caused a change in gas volume. He realized that heating a gas caused it to expand, meaning that an increase in temperature caused an increase in volume. However, at the time, there was no well defined temperature scale, so he was unable to determine the exact relationship between temperature and volume of a gas.

The exact relationship between volume and temperature.

In 1787 Jaques Charles, with the benefit of the Celsius temperature scale, was able to determine a mathematically precise relationship. In his experiments, Charles used a sealed container that could change size easily. By using a container that could change size, the pressure was kept constant, so his experiments assumed a constant number of molecules (or moles) and a constant pressure. Only the temperature and volume were allowed to change. Below is an example of data for such an experiment.

TABLE 15.2. Volume vs. Temperature (at constant pressure and moles)

Volume (mL)	Temp. (°C)	V/(T+273) (mL/°C)
91.6	0	0.335
106.7	45	0.335
125.2	100	0.335
167.1	225	0.335
323.0	200	0.335

**Note: Add 273 to the Celsius temp. for a directly proportional relationship between temp. and vol.*

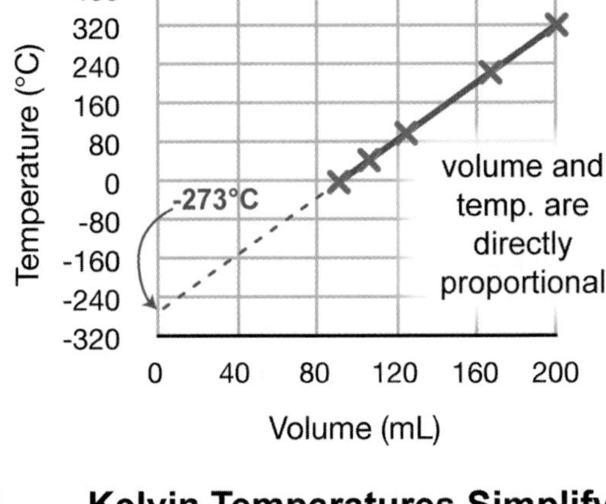

Lord Kelvin discovers absolute zero.

One thing that becomes apparent when you look at the graph is that as temperature decreases, so does volume. Lord Kelvin realized that if you extended the line to the point where the volume of the gas would be zero, the smallest possible volume, that you would also reach the lowest possible temperature. On the graph above the lowest temperature is -273 °C. This lowest temperature came to be known as absolute zero, and a new temperature scale named after Kelvin was invented. It was identical to the Celsius scale except that it started at 0 instead of -273. To convert from Celsius to Kelvin you just add 273 to the Celsius temperature.

Kelvin Temperatures Simplify the V vs. T Relationship

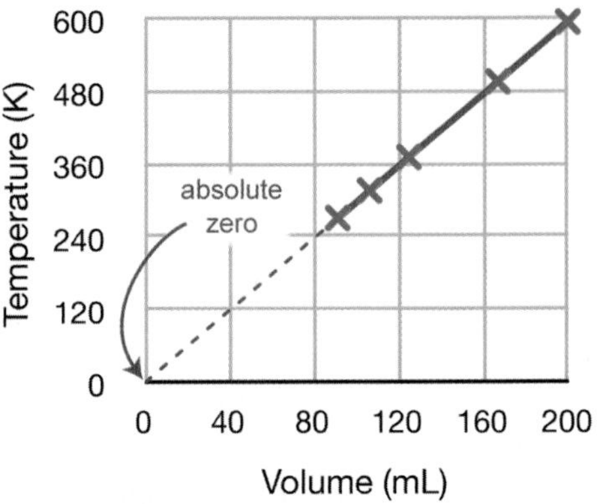

Doubling the Kelvin temperature will double the volume of a gas.

Charles's law: volume vs. temperature

Volume is directly proportional to Kelvin temperature.

If the Kelvin temperature scale is used, the volume vs. temperature relationship is simplified so that doubling the Kelvin temperature of gas causes a doubling of its volume. This means that volume is directly proportional to temperature, and can be expressed as V/T = some constant. The "constant" could change from one experiment to another if the pressure or amount of initial gas used was different, but as long as the pressure and moles of gas remain the same (and the temperature is measured in Kelvins), the V/T=constant relationship will be true.

Charles' Law

T = temperature
V = volume
k = a constant*
(* value depends on pressure and moles)

$$\frac{V}{T} = k \quad \text{or} \quad \frac{V_1}{T_1} = \frac{V_2}{T_2}$$

Only true for constant pressure and constant number of gas molecules (moles).

Changes in temperature

From a molecular standpoint, the direct relationship between pressure and volume makes sense. If the temperature of the gas is increased, then there will be more frequent and harder impacts on the walls of the container. This will cause the gas to expand until the molecules are spread far enough apart so that the pressure inside and outside of the container are now equal again.

Inc. Temp. = Harder & More Impacts = Expanded Vol. to Equalize Pressure

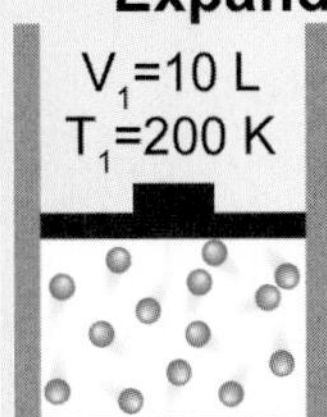

$$\frac{V_1}{T_1} = \frac{V_2}{T_2}$$

$$\frac{10\text{ L}}{200\text{ K}} = \frac{20\text{ L}}{400\text{ K}}$$

$$0.05\frac{\text{L}}{\text{K}} = 0.05\frac{\text{L}}{\text{K}}$$

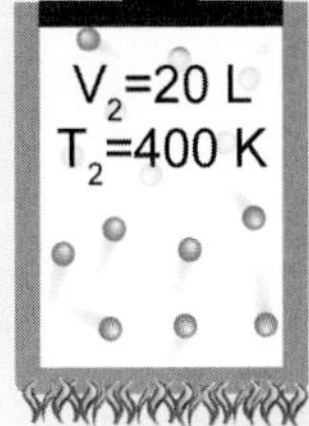

If you inflate a balloon to a size of 8.0 L inside where the temperature is 23 °C, what will be the new size of the balloon when you go outside where it is 3 °C?

Asked: *What is the new volume of the balloon when the temperature drops to 3 °C?*

Given: $V_1 = 8.0\text{ L}$; $T_1 = 23\ °\text{C}$; $T_2 = 3\ °\text{C}$

Relationships: $\frac{V_1}{T_1} = \frac{V_2}{T_2}$ *and Kelvin Temp = Celsius Temp + 273*

Solve: *Convert temps to Kelvin:* $T_1 = 23\ °C + 273 = 296\ K$

$T_2 = 3\ °C + 273 = 276\ K$

$$\frac{8.0\ L}{296\ K} = \frac{V_2}{276\ K} \longrightarrow V_2 = \frac{8.0\ L \times 276\ \cancel{K}}{296\ \cancel{K}} = 7.5\ L$$

Combined gas law

The combined gas law

So far we have looked at Boyle's law which can tell us how pressure varies with volume, but only if you hold the temperature and number of molecules (or moles of molecules) constant. We have also studied Charles's law, which can tell us how volume varies with temperature as long as the pressure and moles of gas are kept constant. However, in most real-life situations all of those things may be changing. The easiest one to keep constant is the moles of gas, because all you need is a sealed container to make sure the moles of gas remain constant. We can combine Boyle's and Charles's laws to give a new formula that will allow for pressure, volume, and temperature to vary. See how this is derived below.

Combined Gas Law

P = pressure
T = temperature
V = volume
k = a constant*
(*depends on moles)

$$\frac{PV}{T} = k \quad \text{or} \quad \frac{P_1V_1}{T_1} = \frac{P_2V_2}{T_2}$$

Only true for constant number of gas molecules (moles).

Boyle's Law (constant T and moles): $P_1V_1 = P_2V_2$

Charles' Law (constant P and moles): $\frac{V_1}{T_1} = \frac{V_2}{T_2}$

Combined Gas Law (constant moles): $\frac{P_1V_1}{T_1} = \frac{P_2V_2}{T_2}$

Imagine you were to hitch a ride on a high altitude research balloon that reaches an altitude of 100,000 feet. At sea level where the pressure is 1.00 atm, and the temperature is 20 °C, you'll need 18 m^3 of helium. What will be the new volume of the gas when you reach altitude where the pressure is 0.0100 atm and the temperature is -50 °C?

Asked: *What is the new volume of the balloon when it reaches 100,000 feet?*

Given: $P_1 = 1.00$ atm; $V_1 = 18.0\ m^3$; $T_1 = 20$ °C $= 293$ K; $P_2 = 0.0100$ atm; $T_2 = -50$ °C $= 223$ K

Relationships: $\frac{P_1V_1}{T_1} = \frac{P_2V_2}{T_2}$ and Kelvin Temp = Celsius Temp + 273

Solve:

$$\frac{1.00\ atm \times 18.0\ m^3}{293\ K} = \frac{0.0100\ atm \times V_2}{223\ K}$$

$$V_2 = \frac{1.00\ \cancel{atm} \times 18.0\ m^3 \times 223\ \cancel{K}}{293\ \cancel{K} \times 0.0100\ \cancel{atm}} = 1370\ m^3$$

Avogadro's law: volume vs. moles

Equal volumes of gas contain equal numbers of molecules

In 1811, Amedeo Avogadro studied how gases chemically combine with each other. For example, we know that hydrogen combines with oxygen in a two to one mole ratio by looking at the chemical equation: $2\,H_2 + O_2 \rightarrow 2\,H_2O$. Avogadro found that volumes of gases would combine in the same kinds of ratios. In other words 2 L of hydrogen would combine with 1 L of oxygen to form 2 L of water vapor. This would only work if each liter of gas contained the same number of molecules regardless of the type of gas.

Two equal volumes of hydrogen react with one volume of oxygen to produce two volumes of water vapor. Gas volumes act like moles becuase the same size container has the same number of molecules (at the same temp. and pressure).

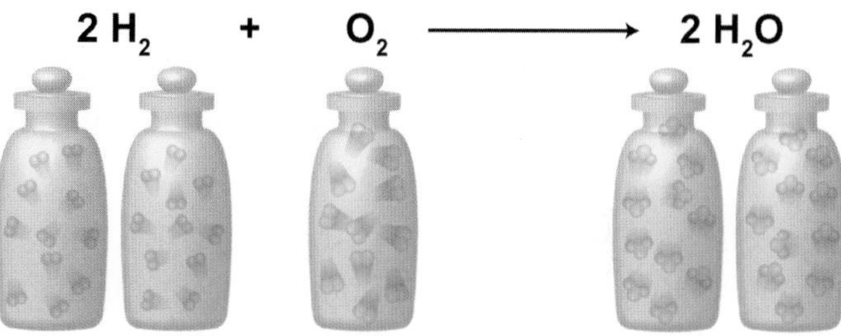

Below is some data that could be collected to see the relationship between volume of a gas and the number of molecules (n) in that volume.

TABLE 15.3. Volume vs. Moles (at constant temperature and pressure)

Volume (L)	Moles (mol)	V/n (L/mol)
200	10	20
300	15	20
400	20	20
500	25	20
600	30	20
800	40	20

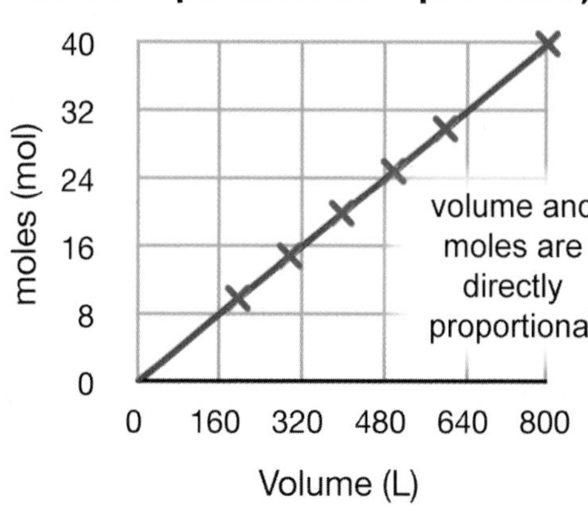

Volume is directly proportional to moles of gas.

The above data shows that doubling the number of moles, doubles the volume. This means that volume is directly proportional to moles and can be expressed as V/n = some constant. The "constant" could change from one experiment to another if the temperature or pressure of the initial gas used was different, but as long as the temperature and pressure remained the same, the V/n=constant relationship would be true.

Avogadro's Law

V = volume

n = moles

k = a constant*

(* value depends on pressure and temp.)

$$\frac{V}{n} = k \quad \text{or} \quad \frac{V_1}{n_1} = \frac{V_2}{n_2}$$

Only true for constant pressure and constant temperature.

Molar volume

Doubling the number of gas molecules doubles the amount of space they require.

At a molecular level Avogadro's Law states that doubling the number of gas molecules (moles) results in doubling the amount of space they take up .

What requires a bit more explanation is why the same volume of gas should contain the same number of gas molecules regardless of the type of gas (as long as the temperature and pressure of both gases are the same).

A volume of gas that is double in size will have double the number of molecules, assuming the temperature and pressure are the same.

This is only true for gases, not liquids or solids.

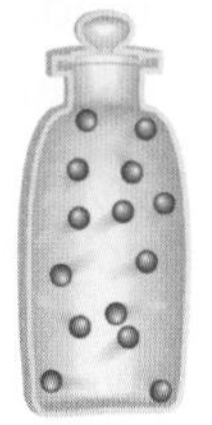

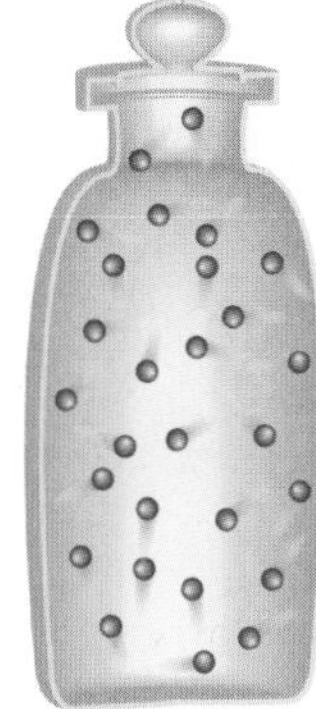

Gas molecules are spread so far apart that the size of a molecule is not important in determining gas volume.

Typically, gases are 1000 times less dense than liquids because there is a lot of empty space between molecules in a gas and almost no empty space between molecules in liquids (or solids). This means that the size of an individual molecule has little to do with the amount of space the gas will take up. Instead, it is the size of the container that defines the volume of a gas.

Liquid volumes are due to the size of their molecules, while gas volumes are due to the size of the container.

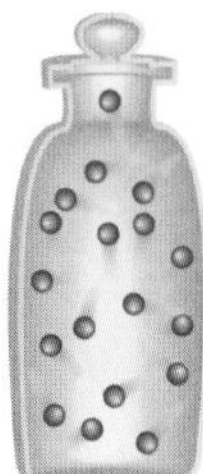

In a gas, the space between molecules is large enough that molecule size doesn't matter.

One mole of any gas occupies a volume of 22.4 L at standard temperature and pressure.

The temperature of a gas is directly related to the kinetic energy of its molecules. This means that two gases that are at the same temperature must have molecules with the same kinetic energy. If one gas has heavier molecules, then its molecules will move more slowly than the molecules of a gas with lighter molecules. The two different molecules will exert the same force when impacting the wall of the container, so both gases will have the same pressure for the same number of molecules and occupy the same volume. At standard temperature (273 K) and standard pressure (1.00 atm) one mole of any gas will occupy 22.4 L. This is considered the molar volume.

Each molecule here has the same kinetic energy.

In gases with the same temperature, lighter molecules move faster, so the force from the impact is the same as the slower molecules that have more mass.

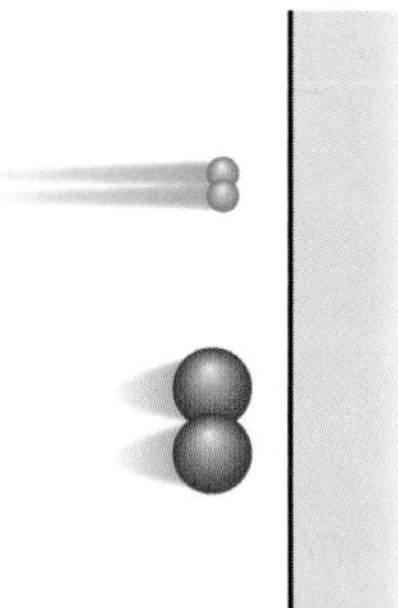

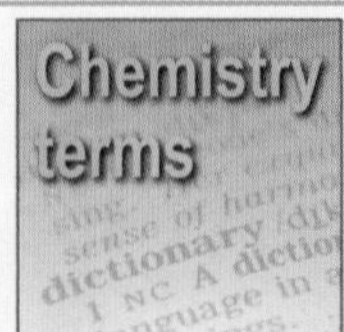

molar volume - the amount of space occupied by a mole of gas at standard temperature and pressure.

Ideal gas law and the universal gas constant

Combining all of the individual gas laws will give you the Ideal Gas Law.

All of the gas laws we have studied so far only work under certain conditions. Boyle's Law only works if the temperature and moles of the gas remain constant. Charles' Law only works if the pressure and volume of the gas remain constant. Avogadro's Law only works if the pressure and temperature remain constant. The problem with these gas laws is that it is more realistic for pressure, temperature, volume, and sometimes moles of gas to change all at the same time. We made some progress by combining Boyle's and Charles' Law. If we also combine Avogadro's Law, then we will have an equation that works for all gases under all conditions: the Ideal Gas Law!

Ideal Gas Law

- P = pressure
- V = volume
- n = moles
- T = temperature
- R = universal gas constant

$$\frac{PV}{nT} = R \quad \text{or} \quad \frac{P_1V_1}{n_1T_1} = \frac{P_2V_2}{n_2T_2}$$

True for all gases under all conditions!

Frequently written as: $PV = nRT$

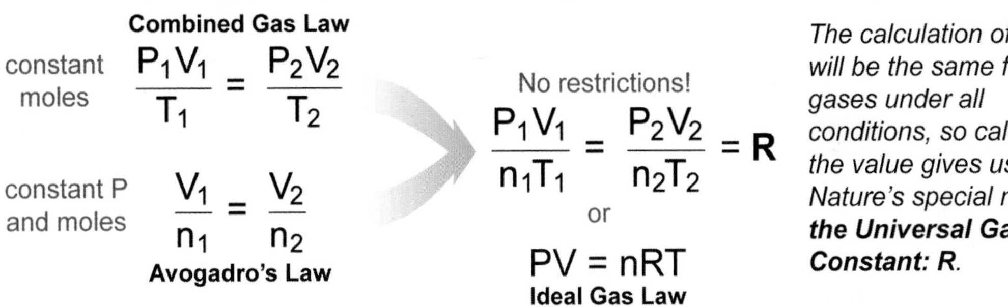

The calculation of PV/nT will be the same for all gases under all conditions, so calculating the value gives us one of Nature's special numbers, ***the Universal Gas Constant: R****.*

The universal gas constant is true for all gases under all conditions.

As with the other gas laws, a constant is calculated. However, unlike the other gas laws in which the constant depended on some starting condition of the gas, this constant is the same value for any gas under any set of initial conditions. It is therefore known as the Universal Gas Constant, and is given a special letter "R" to represent it. This is a fundamental constant of nature like pi (π). The only thing affecting the value of "R" is the units used to calculate it. Consider the following examples:

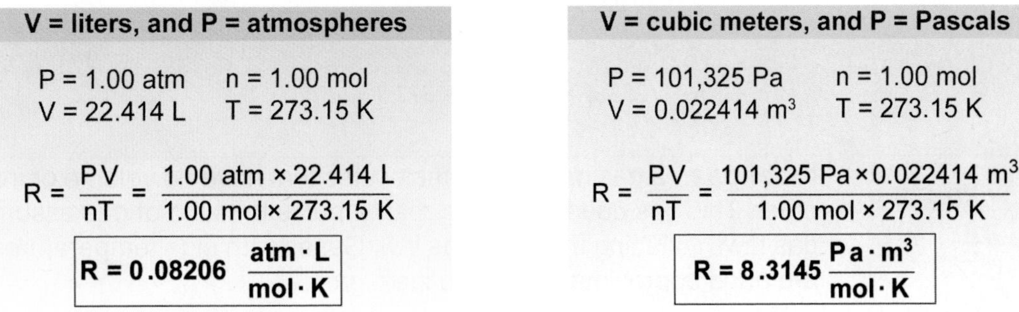

Calculating The Universal Gas Constant Using Various Units
V = liters, and P = atmospheres
P = 1.00 atm n = 1.00 mol
V = 22.414 L T = 273.15 K
$R = \frac{PV}{nT} = \frac{1.00 \text{ atm} \times 22.414 \text{ L}}{1.00 \text{ mol} \times 273.15 \text{ K}}$
$R = 0.08206 \frac{\text{atm} \cdot \text{L}}{\text{mol} \cdot \text{K}}$
V = cubic meters, and P = Pascals
P = 101,325 Pa n = 1.00 mol
V = 0.022414 m³ T = 273.15 K
$R = \frac{PV}{nT} = \frac{101{,}325 \text{ Pa} \times 0.022414 \text{ m}^3}{1.00 \text{ mol} \times 273.15 \text{ K}}$
$R = 8.3145 \frac{\text{Pa} \cdot \text{m}^3}{\text{mol} \cdot \text{K}}$

Using the ideal gas law under uniform conditions

Only gases with low temperatures or high pressures may closely obey the Ideal Gas Law.

There are limitations to how the Ideal Gas Law can be applied. It is only 100% accurate if you are using an **ideal gas**, one in which the molecules take up no space and have no interactions with each other. This is most important when the molecules are close together, which happens for gases under high pressure and/or low temperatures. If the gas is cooled down to very low temperatures the van der Waals attractions will also play a greater role. For gases of sufficiently high temperature and/or low pressure the Ideal Gas Law can be quite accurrate.

Difference Between a Real and Ideal Gas

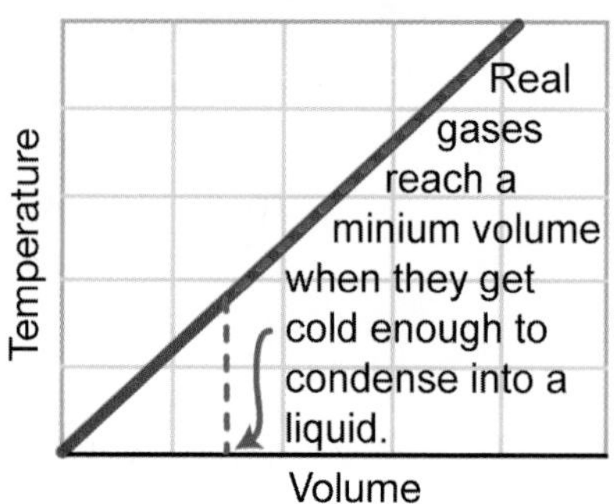
Real gases reach a minium volume when they get cold enough to condense into a liquid.
Temperature
Volume

When calculating that state of an unchanging gas use PV=nRT.

If you are given a problem where the conditions are not changing (i.e. no initial and final pressure, temperature, etc.), then use the PV = nRT version of the Ideal Gas Law. Solving problems using this equation involves the following steps:

Steps for Using PV=nRT		
Step 1	**Step 2**	**Step 3**
Select one version of R.	Convert known values to the same units as R.	Solve for unknown value using PV=nRT.

What would be the pressure inside of a 50.0 L tank containing 1,252 grams of helium at 20 °C?

Asked: *What is the tank pressure under these unchanging conditions?*

Given: $V = 50.0\ L;\ T = 20\ °C;\ mass = 1{,}252\ g;\ He = 4.003\frac{g}{mol}$

Relationships: $PV = nRT \qquad R = 0.08206\frac{atm \cdot L}{mol \cdot K}$

Solve:

Convert to units of R:

$V = 50.0\ L$

$T = 20.0\ °C + 273 = 293\ K$

$$n = 1252\ \cancel{g} \times \frac{1\ mol}{4.003\ \cancel{g}} = 312.8\ mol$$

$$P \times 50.0\ L = 312.8\ mol \times 0.08206\frac{atm \cdot L}{mol \cdot K} \times 293\ K$$

$$P = \frac{312.8\ \cancel{mol} \times 0.08206\ atm \cdot \cancel{L} \times 293\ \cancel{K}}{50.0\ \cancel{L} \times \cancel{mol} \cdot \cancel{K}}$$

$$\mathbf{P = 150\ atm}$$

ideal gas - a gas made from molecules that have no volume or interactions with each other. This gas doesn't exist in reality, but is the kind of gas assumed by the various gas laws including the ideal gas law. Gases with high temperatures and low pressures are good approximations of an ideal gas.

Using the ideal gas law under changing conditions

If any property of the gas is changing, then use the $\frac{P_1V_1}{n_1T_1} = \frac{P_2V_2}{n_2T_2}$ form of the Ideal Gas Law.

Because the Ideal Gas Law contains all the other gas laws within itself, you only need this one formula to solve any problems related to gases. You just need to choose the correct version of the formula.

If you are doing a "changing conditions" problem, then you can use any pressure, and volume units you like as long at the initial and final conditions use the same units. However, you still need to convert your temperatures to the Kelvin temperature scale.

Any properties not mentioned in the problem can be assumed to be held constant, causing the Ideal Gas Equation to be simplified into several of the more basic gas laws. So, solving problems in which one or more properties of the gas are changing involves the following steps:

Steps for Using $\frac{P_1V_1}{n_1T_1} = \frac{P_2V_2}{n_2T_2}$

Step 1	Step 2
Simplify the equation by cancelling out properties that stay constant. For example, if temperature stays constant $T_1=T_2$, so they cancel each other out.	Substitute remaining known quantities and solve for the unknown using the simplified equation.

What would be the new volume of a bubble in bread dough once it goes from a room temperature 20 °C volume of 0.050 cm^3 to a 191 °C oven?

Asked: *What is the volume of the bubble under changing temperature conditions?*

Given: $V_1 = 0.050\ cm^3$; $T_1 = 20\ °C = 293\ K$; $T_2 = 191\ °C = 464\ K$

Relationships: $\frac{P_1V_1}{n_1T_1} = \frac{P_2V_2}{n_2T_2}$

Solve: *Pressure and moles stay constant, so they cancel out.*

$$\frac{\cancel{P_1}V_1}{\cancel{n_1}T_1} = \frac{\cancel{P_2}V_2}{\cancel{n_2}T_2} \rightarrow \frac{V_1}{T_1} = \frac{V_2}{T_2}$$

$$\frac{0.050\ cm^3}{293\ K} = \frac{V_2}{464\ K}$$

$$V_2 = \frac{0.050\ cm^3 \times 464\ \cancel{K}}{293\ \cancel{K}}$$

$$\mathbf{V_2 = 0.079\ cm^3}$$

14.3 Stoichiometry and Gases

Solving stoichiometry problems

In chapter 11 you learned how to do stoichiometry with grams of a substance or liters of solutions. Here we will extend your ability to do stoichiometry when one or more gases are part of the equation. The pattern you previously used followed one of these patterns:

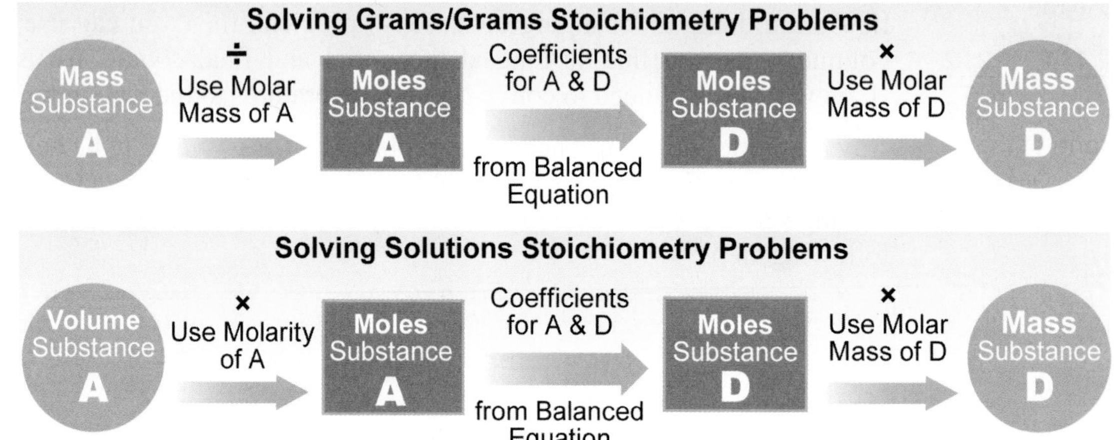

The same pattern can be used for stoichiometry problems that involve gases.

We can generalize these patterns to solve any kind of stoichiometry problem in the following way, including problems that involve gases:

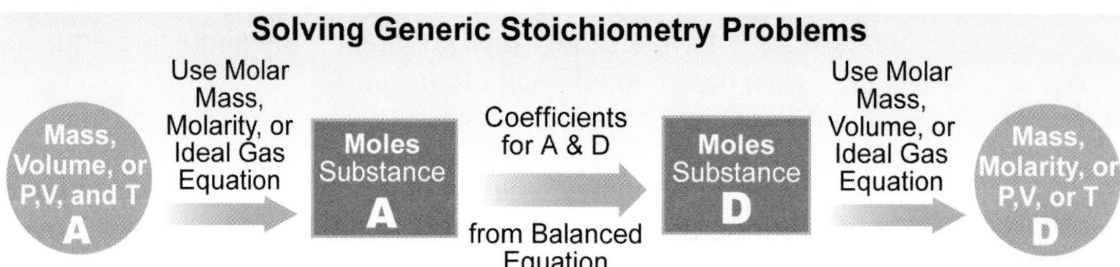

Steps for Stoichiometry Problems

Step 1	Step 2	Step 3
Use what you have about the first substance to convert it to moles. If you have: • grams, use the molar mass. • L of solution, use the molarity. • pressure, volume, and temperature of a gas, use the Ideal Gas Law.	Use the balanced equation to determine a mole ratio between the substance you have information about and the substance you are trying to calculate. Then multiply the moles of the first substance by the mole ratio to give the moles of the second substance.	Now that you have the moles of the second substance, do what ever conversion is necessary to calculate the final answer. To convert moles to: • grams, use the molar mass. • to L of solution, use the molarity. • volume, pressure, or temperature of gas, use the Ideal Gas Equation.

Gas/Gram stoichiometry calculations

In gas/gram calculations use the molar mass and Ideal Gas Law to solve the problem.

In a reaction that involves both a gas and a solid, the gas is typically measured in volume at a certain temperature and pressure, while the solid is measured in grams. As with all stoichiometry reactions the key is to get everything into moles in order to do a ratio between two substances in the balanced equation. In this kind of reaction you will either be given information about the gas and asked to find the grams of the other substance, or you will be given the grams of a substance and asked to find some property of the gas, either temperature, pressure, or volume. Here is one example.

If you burn a 125 g candle made of paraffin wax the peak temperature of the flame is about 1,400 °C. Assuming the carbon dioxide produced is initially at that temperature, what volume of CO_2 is produced at peak temperature and a pressure of 790 mmHg?

$$2\,C_{20}H_{42} + 61\,O_2 \rightarrow 40\,CO_2 + 42\,H_2O$$

Asked: *What is the volume of CO_2 produced?*

Given: *paraffin: mass = 125 g*
carbon dioxide: P = 790 mmHg; T = 1,400 °C

Relationships: *molar mass of paraffin* $= (12.011 \times 20) + (1.0079 \times 42) = 282.6\ \frac{g}{mol}$
mole ratio: $2 \text{ moles } C_{20}H_{42} = 40 \text{ moles } CO_2$
$PV = nRT$

Solve:

Calculate moles of paraffin:

$$125\ \cancel{g\ C_{20}H_{42}} \times \frac{1\ mol\ C_{20}H_{42}}{282.6\ \cancel{g\ C_{20}H_{42}}} = 0.442\ mol\ C_{20}H_{42}$$

Use mole ratio to determine moles of CO_2:

$$0.442\ \cancel{mol\ C_{20}H_{42}} \times \frac{40\ mol\ CO_2}{2\ \cancel{mol\ C_{20}H_{42}}} = 8.85\ mol\ CO_2$$

Convert gas units to the ones used by R:

$$P = 790\ \cancel{mmHg} \times \frac{1.00\ atm}{760\ \cancel{mmHg}} = 1.04\ atm$$
$$n = 8.85\ mol$$
$$T = 1{,}400\ °C + 273 = 1{,}673\ K$$

Solve Ideal Gas Equation: $PV = nRT$

$$1.04\ atm \times V = 8.85\ mol \times 0.08206\frac{atm \cdot L}{mol \cdot K} \times 1{,}673\ K$$

$$V = \frac{8.85\ \cancel{mol} \times 0.08206\ \cancel{atm} \cdot L \times 1{,}673\ \cancel{K}}{1.04\ \cancel{atm} \cdot \cancel{mol} \cdot \cancel{K}} = \mathbf{1{,}170\ L}$$

Gas/Solution stoichiometry calculations

In gas/solution calculations use the molarity and Ideal Gas Law to solve the problem.

In a reaction that involves a gas and a solution, the gas is typically measured in volume at a certain temperature and pressure, while the solution is measured in volume and molarity. As with all stoichiometry reactions the key is to get everything into moles in order to do a ratio between two substances in the balanced equation. In this kind of reaction you will either be given information about the gas and asked to find the volume or concentration of a solution, or you will be given the volume and concentration of a solution and asked to find some property of the gas, either temperature, pressure, or volume. Here is one example.

A common reaction is when an acid reacts with a metal producing hydrogen gas. Assume you are doing a reaction in which 0.050 L of 1.25 M hydrochloric acid reacts with excess magnesium. If the gas produced is captured in a 1.00 L container that starts out empty, what will be the pressure at the end of the reaction when the gas has cooled to 22 °C (room temp.)?

$$Mg + 2\,HCl \rightarrow H_2 + MgCl_2$$

Asked: *What is the pressure of hydrogen produced?*

Given: *acid:* $V = 0.050\ L$; *molarity* $= 1.25\ M = 1.25\ \frac{mol}{L}$

hydrogen gas: $V = 1.00\ L$; $T = 22\ °C$

Relationships: *mole ratio:* 2 moles HCl = 1 mole H_2

$PV = nRT$

Solve:

Calculate moles of HCl:

$$0.050\ \cancel{L\ HCl} \times \frac{1.25\ mol\ HCl}{1\ \cancel{L\ HCl}} = 0.0625\ mol\ HCl$$

Use mole ratio to determine moles of CO_2*:*

$$0.0625\ \cancel{mol\ HCl} \times \frac{1\ mol\ H_2}{2\ \cancel{mol\ HCl}} = 0.0313\ mol\ H_2$$

Convert gas units to the ones used by R:

$V = 1.00\ L$

$n = 0.0313\ mol$

$T = 22\ °C + 273 = 295\ K$

Solve Ideal Gas Equation:

$$P \times 1.00\ L = 0.0313\ mol \times 0.08206 \frac{atm \cdot L}{mol \cdot K} \times 295\ K$$

$$P = \frac{0.0313\ \cancel{mol} \times 0.08206\ atm \cdot \cancel{L} \times 295\ \cancel{K}}{1.00\ \cancel{L} \cdot \cancel{mol} \cdot \cancel{K}} = \mathbf{0.756\ atm}$$

Gas/Gas stoichiometry calculations

Use the Ideal Gas Law to solve gas-only problems.

In a reaction that involves two gases, you will be given the temperature, pressure, and volume of one gas, and asked to calculate the temperature, pressure, or volume of the second gas. As with all stoichiometry reactions the key is to get everything into moles in order to do a ratio between two substances in the balanced equation.

What volume of butane gas is needed at room temperature (23 °C) and typical pressure (0.984 atm) to produce 85.0 L of carbon dioxide at 825 °C and a pressure of 1.04 atm?

$$2\ C_4H_{10} + 13\ O_2 \rightarrow 8\ CO_2 + 10\ H_2O$$

Asked: *What is the volume of butane gas needed to produce a specific volume of carbon dioxide gas?*

Given: *butane: P = 0.984 atm; T = 23 °C*
carbon dioxide: P = 1.04 atm; V = 85.0 L; T = 825 °C

Relationships: *mole ratio: 2 moles C_2H_{10} = 8 mole CO_2*
$PV = nRT$

Solve:

Convert CO_2 units to the ones used by R:

$P = 1.04\ atm$
$V = 85\ L$
$T = 825\ °C + 273 = 1098\ K$

Solve Ideal Gas Equation for moles of CO_2: PV = nRT

$$1.04\ atm \times 85\ L = n \times 0.08206 \frac{atm \cdot L}{mol \cdot K} \times 1098\ K$$

$$n = \frac{1.04\ \cancel{atm} \times 85\ \cancel{L} \cdot mol \cdot \cancel{K}}{0.08206\ \cancel{atm} \cdot \cancel{L} \times 1098\ \cancel{K}} = \mathbf{0.981\ mol}$$

Use mole ratio to determine moles of butane:

$$0.981\ \cancel{mol\ CO_2} \times \frac{2\ mol\ C_4H_{10}}{8\ \cancel{mol\ CO_2}} = 0.245\ mol\ C_4H_{10}$$

Convert butane units to the ones used by R:

$P = 0.984\ atm$
$T = 23\ °C + 273 = 296\ K$
$n = 0.245\ mol$

Solve Ideal Gas Equation: PV = nRT

$$0.984\ atm \times V = 0.245\ mol \times 0.08206 \frac{atm \cdot L}{mol \cdot K} \times 296\ K$$

$$V = \frac{0.245\ \cancel{mol} \times 0.08206\ \cancel{atm} \cdot L \times 296\ \cancel{K}}{0.984\ \cancel{atm} \cdot \cancel{mol} \cdot \cancel{K}} = 6.05\ L$$

High Altitude Training for Athletes

If you are a cycling enthusiast you have likely heard of Lance Armstrong, cancer survivor and seven-time winner of the Tour De France. You may have also heard that the Kenyans dominate in the sport of distance running. The first person to ever reach the top of Mt. Everest was Sir Edmund Hillary. What do all of these people have in common? And what does this have to do with chemistry?

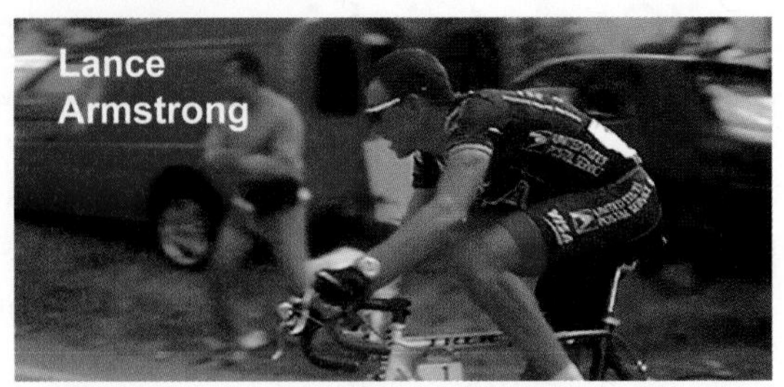

All of these athletes have performed well at high altitudes and this has to do with their body's ability to absorb oxygen, O_2. The body's ability to absorb oxygen is directly related to chemistry! To understand the basics of how the body absorbs oxygen at different altitudes, we must understand the concept of atmospheric pressure and how this influences the partial pressure of oxygen gas. Hemoglobin molecules present in our red blood cells also play a very important role in the body's ability to absorb oxygen.

At sea level where atmospheric pressure is relatively high, more oxygen molecules are able to enter our lungs. At high altitudes the atmospheric pressure is lower, and this results in fewer oxygen molecules entering our lungs per breath. Fewer oxygen molecules enter our lungs, because the partial pressure of oxygen is lower at high altitudes. The partial pressure is the pressure of an individual gas. It is called "partial" because it is one part of a whole group of gases that sum together. Using the atmosphere as an example, we sum the individual partial pressures of nitrogen, oxygen, water and argon, which are the primary gases in the atmosphere. This gives us the total pressure, which we refer to as the atmospheric pressure.

We live at the bottom of an ocean of air.
Credit: NOAA

The surface of the "air ocean" is where our atmosphere meets outer space, so the greatest air pressure is near the ground.

The atmospheric pressure helps push the air molecules into our lungs, and because the pressure exerted by the atmosphere is lower up high, fewer gas molecules enter our lungs. At high altitude the partial pressure of each gas is decreased and when they are added together the result is a lower atmospheric pressure.

Air is less dense (lower pressure) at high altitude.

Effects of Altitude on Oxygen Intake

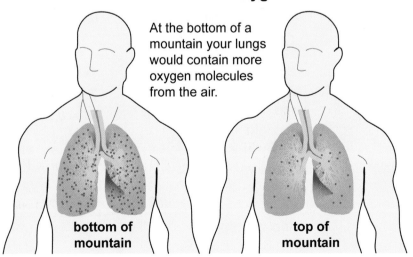

At the bottom of a mountain your lungs would contain more oxygen molecules from the air.

bottom of mountain

top of mountain

Often competitive athletes such as runners, cyclists, and hikers train at high altitudes, way up in the mountains. For example, at 12,000 feet (3,658 meters) the atmospheric pressure is only 0.64 atm, compared to sea level which is 1 atm. Remember the atmospheric pressure is lower up high, because there are fewer air molecules from above pushing down. At an altitude of 12,000 feet approximately 40% fewer oxygen molecules enter your lungs! You may have heard the saying "the air is thinner" up on high mountains, this essentially means the air pressure is not as great.

When athletes first go to train at high altitudes they must allow time (generally 1-3 days) for their bodies to adjust to the lower atmospheric pressure. Why do their bodies require time to acclimate? Because their body needs to function with decreased amounts of oxygen. In order to make sure there is enough oxygen getting to it's tissues the body makes some changes. First, breathing rate increases, this increases the number of oxygen molecules entering the lungs. The body also makes more hemoglobin containing red blood cells, which in turn pick up (bind to) more of the available oxygen molecules and deliver them to the tissues.

When training at high altitude your body makes more oxygen carrying red blood cells (RBCs).

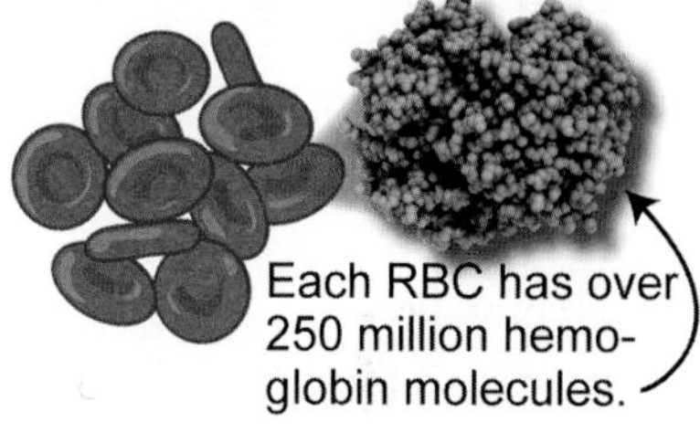

Each RBC has over 250 million hemoglobin molecules.

Athletes perform better when more oxygen is being delivered to their tissues. They experience less fatigue and their bodies are capable of performing better for longer periods of time. In any endurance sport there is a high dependency on oxygen, so there is a definite advantage for athletes that can improve the amount of oxygen being delivered to the working muscles. To take advantage of this many athletes train at higher altitudes. When they return to a lower altitude to complete, they now have an increased amount of circulating oxygen molecules.

Mt. Everest Base Camp

Over time we have noticed the benefits of high altitude training. When the Kenyan runners first began to dominate the international middle and long distance races, it prompted significant interest by society. Much research was done and scientists concluded that the primary factor, which contributed to their repeated success, was the high altitude conditions under which they trained. Interestingly, Lance Armstrong was found to have unusually high levels of oxygen in his blood, when he was tested during his training. Lastly, when Sir Edmund Hillary conquered Mt. Everest back in 1953, he was also documented as having higher than normal circulating levels of oxygen in his blood! All of these amazing people appear to owe a large part of their success to the oxygen molecule! In addition to high altitude training it is believed that there is likely a genetic component involved that may contribute to some people having slightly higher levels of circulating oxygen.

Chapter 14 Review.

Vocabulary

Match each word to the sentence where it best fits.

Section 14.1

kinetic molecular theory	Brownian motion
Bernoulli principle	barometer

1. If you view tiny particles in a liquid or gas using a microscope you can see ________.
2. Using a ________ you can measure atmospheric pressure.
3. Whenever you fly in an airplane you rely on the ________.
4. ________ describes our basic understanding of substances at the molecular level.

Section 14.2

molar volume	ideal gas

5. At standard temperature and pressure you will find 6.02 x 10^{23} molecules in a ________.
6. The gas laws are only 100% accurate for an ________.

Conceptual Questions

Section 14.1

7. Compare a gas, liquid, and solid on a molecular level.
8. What are some common properties of gases?
9. What is kinetic molecular theory.
10. Describe how kinetic molecular theory explains why gases have the properties you mentioned in question 8.
11. How was Brownian motion discovered and what was it caused by?
12. What is the difference between force and pressure?
13. How does a gas exert a pressure if there is lots of empty space between its molecules?
14. Why is it better to lie down on ice that is cracking beneath your feet than to remain in a standing position?
15. What causes atmospheric pressure?
16. Why does atmospheric pressure on top of a mountain less than at the base of that same mountain?

Section 14.2

17. Many aerosol cans have warning labels on the side that say to keep away from heat and fire. Why?
18. One of the first thermometers was made from sealing a gas in a container that would allow the gas to expand or contract. Explain how this could be used as a thermometer.
19. Why does it makes sense that if you squeeze the air in a sealed syringe, that the pressure will go up? Explain on a molecular level?
20. When you drive a car the tires heat up because of friction with the road and internal friction between the molecules of the rubber. What will happen to the gas pressure in the tire as the tire heats up? Explain on a molecular level.
21. When pumping up a bicycle tire with a hand pump it gets harder and harder as you continue pumping. Explain the connection of this phenomenon with Avogadro's Law.
22. Explain why weather balloons, sealed latex balloons filled with helium, eventually pop when they reach extremely high altitudes.
23. Explain why it makes sense to use a temperature scale that starts at zero like the Kelvin temperature scale?
24. What is the relationship between temperature and kinetic energy?
25. Gay-Lussac gave us another law which states that pressure and temperature are directly proportional to each other if volume and moles are constant. Using this law what do you predict will happen to the temperature of a gas with an increase in its pressure? Explain.
26. Plants breathe through the stomata on their leaves, small pores that can be opened or closed as

needed. Plants grown in an environment with a high partial pressure of CO_2 have fewer stomata. Why does this make sense?

27. Describe everything you can about the molecules of two different gases that are at the same temperature.
28. How can the molar volume of ALL gases be the same 22.4 L volume at standard temperature and pressure? Why doesn't the size of the molecule have any affect?
29. What is meant by an ideal gas?
30. Ideal gases have molecules that take up no space. However, real molecules do take up some space. How does fact affect the volume predicted by Boyle's Law and Charles' Law? Will it be larger or smaller than predicted for an ideal gas? Explain.
31. Ideal gases have molecule that don't interact with each other. However, real molecules feel the weak intermolecular attractions felt by all molecules. How does fact affect the volume predicted by Boyle's Law and Charles' Law? Will it be larger or smaller than predicted for an ideal gas? Explain.

Section 14.3

32. If you run an electric current through water it goes through a dissociation reaction, breaking back down into hydrogen and oxygen gas via this reaction: $2\ H_2O_{(l)} \rightarrow 2\ H_{2(g)} + O_{2(g)}$

    If you measure the masses of the hydrogen and oxygen produced you will find a mass ratio of $2\ H_2 : 16\ O_2$. If you measure the volume of the hydrogen and oxygen you will get a volume ratio of $2\ H_2 : 1\ O_2$, the same as the molar ratio in the balanced chemical equation. Why does the volume ratio match the molar ratio, but not the mass ratio?

Quantitative Problems

Section 14.1

33. What is the pressure exerted by an elephant weighing 8,000 lbs if the total area of the bottom of its four feet equals 312 in^2 ?
34. What is the weight of a car that has its tires inflated to 35 $\frac{lbs}{in^2}$ and is touching the ground over an area equal to 285 in^2?
35. Perform the following conversions:
    a. 3.50 psi to Pa
    b. 200435 Pa to atm
    c. 5.67 atm to mmHg
    d. 0.89 atm to inHg (inches of mercury). You will need the following equalities to help: 1 in = 2.54 cm and 10 mm = 1 cm.

Section 14.2

36. Draw a graph of two variables that are directly proportional to each other.
37. Draw a graph of two variables that are inversely proportional to each other.
38. You decide to climb to the tops of some of the tallest mountains. Before you are about to leave on your epic journey, a friend gives you a balloon with a volume of 800.0 cm^3 that was inflated under standard atmospheric pressure.

    You climbed two different mountains with this balloon: Mt. Everest 29,028 ft above sea level which has an average atmospheric pressure of 221 mmHg, and Mt. McKinley 20,320 ft above sea level which has an average atmospheric pressure of 345 mmHg.

    What was the size of the balloon on each mountain top?
39. A 50.0 mL soap bubble is blown at standard pressure. When a thunder storm passes later in the day, the pressure becomes 700.0 mmHg. Will the bubble get bigger or smaller? What is its new volume?
40. A balloon was inflated to a volume of 5.0 liters at a pressure of 0.90 atm. It rises to an altitude where its volume becomes 25.0 liters. Will the pressure around the balloon increase or decrease? What was the new pressure?
41. A SCUBA diver inflates a balloon to 10.0 liters at the surface and takes it on a dive. At a depth of 100.0 feet the pressure is 4.0 'atm. Will the volume of the balloon increase or decrease? What was the new volume of the balloon?

42. The living quarters of the space shuttle has a volume of 20,000 liters (2.00×10^4 L) and is kept at 12.0$\frac{lbs}{in^2}$. If all the air were lost, it would have to be replaced from a compressed air cylinder which has a volume of 50.0 liters. What is the pressure in that tank? (In other words: How much pressure would it take to compress all the air in the shuttle into a 50.0 liter space?)

43. Back to the space shuttle. the living quarters have a volume of 20,000 liters (2.00×10^4 L) at 500.0 mmHg. The shuttle docks with a space-lab which has a volume of 230,000 liters (2.30×10^5 L) and no air in it. What does the air pressure in both become when the hatch is opened between the two?

44. A fountain pen which has a total volume of 2.4 cm^3 is half full with ink at the surface where the pressure is 780.0 mmHg. It is put in a pilot's pocket who flies to an altitude where the pressure is 520.0 mmHg. How much ink leaks out of the pen?

45. A 50.0 mL soap bubble is blown in a 27.0 °C room. It drifts out an open window and lands in a snow bank at -3.0 °C. What is its new volume?

46. A balloon was inflated to a volume of 5.0 liters at a temperature of 7.0 °C. It landed in an oven and was heated to 147 °C. What is its new volume?

47. During the day at 27 °C a cylinder with a sliding top contains 20.0 liters of air. At night it only holds 19 liters. What is the temperature at night? Give the answer in Kelvin and °C?

48. A 113 L sample of Helium at 27 °C is cooled to -78 °C. Calculate the new volume of the Helium.

49. On all aerosol cans you see a warning that tells you to keep the can away from heat because of the danger of explosion. What is the potential volume of the gas contained in a 500.0 mL can at 25 °C if it were heated to 54 °C. In other words if the can could expand to allow the gas to take up a greater volume, what would be the new volume of the gas when heated as previously described?

50. A 0.20 ml CO_2 bubble in a cake batter is at 27 °C. In the oven it gets heated to 177 °C.
    a. What is its new volume?
    b. If the cake had 5,000 bubbles, by how many mL would the cake rise when it was cooked?
    c. What common baking ingredient was used to create the original CO_2 bubble?

51. A 500.0 mL glass filled with air is placed into water up-side-down while at 7.0 °C. The water is heated to 77 °C. How much air bubbles out from under the glass?

52. At one point in history people could measure temperature by looking at the volume of a sample of gas. Suppose a sample in a gas thermometer has a volume of 135 mL at 11.0 °C. Indicate what temperature would correspond to each of the following volumes: 113 mL, 142 mL, 155 mL, and 127 mL.

53. A balloon is filled with helium to a volume of 4.0 liters when the pressure is 1.0 atm and the temperature is 27.0 °C. It escapes and rises until the pressure is 0.25 atm. and the temperature is -23.0 °C. What is the new volume?

54. When a bubble escapes form a sunken ship, it has a volume of 12.0 cm^3 at a pressure of 400.0 atm and a temperature of -3.00 °C. It reaches the surface where the pressure is 1.10 atm and the temperature is 27.0 °C. What is its new volume?

55. A CO_2 bubble in some bread dough had an original volume of 0.30 mL when it formed at 27.0 °C and 750 mmHg of pressure. While baking, its temperature rose to 177.0 °C and a thunderstorm moved in, dropping the pressure to 725 mmHg. What is the new volume of the bubble?

56. If a hot air balloon holds 3000 liters (3.00×10^3 L) of air at 17.0 °C and standard pressure, how much air will escape as the balloon is heated to 67.0 °C and rises to where the pressure is 13.5$\frac{lbs}{in^2}$?

57. "A SCUBA diver's tank holds 200.0 liters of air at 27.0 °C and 150.0 atm. How many 1.50 liter breaths can the diver take where the pressure is 4.00 atm and the temperature is 7.00 °C?

58. You drive to school in a hurricane where the pressure is an abnormally low 720.0 mmHg. However, when you get out of your car you notice that the volume of your tires seems to be a normal 29 L. During the day a high pressure, bright sunshine system moves overhead changing the pressure to 780.0 mmHg. When you come back

outside you notice your tires seem a little flat. What is the new volume of your tires?

59. During the day at 25.0 °C a cylinder with a sliding top contains 20.0 liters of air. At night it only holds 18.0 liters of air. None of the air leaked out. What is the temperature at night? Give the answer in Kelvins and in °C.

60. At what temperature would 2.10 moles of N_2 gas have a pressure of 1.25 atm in a 25.0 L tank?

61. When filling a weather balloon with gas you have to consider that the gas will expand greatly as it rises and the pressure decreases. Let's say you put about 10.0 moles of He gas into a balloon that can inflate to hold 5000.0 L. Currently, the balloon is not full because of the high pressure on the ground. What is the pressure when the balloon rises to a point where the temperature is -10.0 °C and the balloon has completely filled with the gas.

62. What volume is occupied by 5.03 g of O_2 at 28 °C and a pressure of 0.998 atm?

63. Calculate the pressure in a 212 Liter tank containing 23.3 kg of argon gas at 25°C?

64. If you were to take a volleyball scuba diving with you what would be its new volume if it started at the surface with a volume of 2.00 L, under a pressure of 752.0 mmHg and a temperature of 20.0 °C? On your dive you take it to a place where the pressure is 2943 mmHg, and the temperature is 0.245 °C.

65. What is the volume of 1.00 mole of a gas at standard temperature and pressure?

66. A 113L sample of helium at 27 °C is cooled at constant pressure to -78.0 °C. Calculate the new volume of the helium.

67. What volume is occupied by 2.35 mol of He at 25 °C and a pressure of 0.980 atm?

68. An aerosol can contains 400.0 ml of compressed gas at 5.2 atm pressure. When the gas is sprayed into a large plastic bag, the bag inflates to a volume of 2.14 L. What is the pressure of gas inside the plastic bag?

69. At what temperature does 16.3 g of nitrogen gas have a pressure of 1.25 atm in a 25.0 L tank?

70. What mass of CO_2 is needed to fill an 80.0 L tank to a pressure of 150.0 atm at 27.0 °C?

71. At what temperature does 5.00 g of H_2 occupy a volume of 50.0 L at a pressure of 1.01 atm?

72. How many moles of gas would you have if you had a volume of 38.0 L under a pressure of 1432 mmHg at standard temperature?

Section 14.3

73. Use the following chemical equation to answer questions 73 through 76.

    $2\ C_8H_{18} + 25\ O_2 \rightarrow 16\ CO_2 + 18\ H_2O$

74. If 4.00 moles of gasoline are burned, what volume of oxygen is needed if the pressure is 0.953 atm, and the temperature is 35.0 °C?

75. How many grams of water would be produced if 20.0 liters of oxygen were burned at a temperature of -10.0 °C and a pressure of 1.3 atm?

76. If you burned one gallon of gas (approximately 4000 grams), how many liters of carbon dioxide would be produced at a temperature of 21.0 °C and a pressure of 1.00 atm?

77. How many liters of oxygen would be needed to produced 45.0 liters of carbon dioxide if the temperature and pressure for both are 0.00 °C and 5.02 atm?

78. Use the following chemical equation, which shows the acetylene combustion reaction used in acetylene welding torches, to answer questions 78 through:
    $2\ C_2H_2 + 5\ O_2 \rightarrow 4\ CO_2 + 2\ H_2O$

79. If you had 0.564 moles of acetylene, what volume of oxygen would be needed to react with it, if the pressure was 13.9 psi and the temperature was 23 °C?

80. What volume of acetylene did you use if 10.5 grams of water were produced and the pressure was 12.0 atm and the temperature was 18 °C?

81. What is the temperature of the reaction if 3.65 moles of oxygen are used and 500.0 Liters of CO_2 are produced under a pressure of 1.04 atm?

Electrochemistry

What are antioxidants and how do they work?

How do batteries work?

Why are some batteries rechargeable and some are not?

Why do some metals corrode (rust) and others do not?

What does electricity have to do with chemistry? Consider the abundant life on our Earth. Every plant and animal on depends on the energy from the sun. Plants capture the energy of light from the sun and store it as chemical energy in their roots, leaves, fruits and branches as well as in the oxygen that they release. In turn, animals extract the stored energy by eating the plants and breathing the oxygen.

Photosynthesis is not one reaction, but a complex sequence of many chemical reactions. Virtually all the energy in Earthly life originates with photosynthesis. Most of the energy in our technology (oil and natural gas) come from decayed living organisms that lived long ago. So when we burn gasoline we actually use "fossilized photosynthesis" that has been stored as hydrocarbon compounds in the earth for millions of years.

Animals eat fruit and leaves and extract the energy stored in them. This process is called *respiration*. Like photosynthesis, respiration is a complicated sequence of chemical reactions.

Now consider the details of just *how* photosynthesis and respiration work. At the very heart of these reactions are the transfer of electrons between compounds and molecules. Some molecules lose electrons and others gain them. The movement of electrons and the associated chemical change is the subject of *electrochemistry*. Here is the answer to the question we began with. Nature is the ultimate electrochemical engine and in order to understand it we must understand the fundamentals of electrochemistry.

Electricity from a lemon.

Electrochemistry brings electricity and chemistry together. This is the science that relates chemical reactions to the flow of electric charge. It happens in a battery, it happens in your body, and it can be made to happen with a lemon from the grocery store!

To make a lemon battery you need:

- A piece of magnesium (Mg) ribbon
- A piece of copper (Cu) ribbon or thick wire
- A red light emitting diode (LED)
- A lemon
- Two wires with alligator clips
- The Lab-Master unit for measuring the voltage

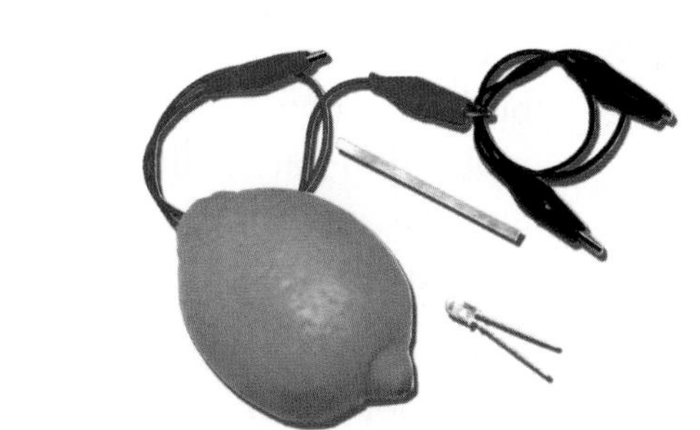

The experiment

The goal of the experiment is to make a lemon light up the LED. However, just poking the LED into the lemon is not going to work. You need to use the electrochemical properties of the magnesium and copper metals.

1. "Knead" the lemon gently to break up the juicy cells inside so the juice can flow easily within the fruit.
2. Next make two small incisions on the lemon about 2 cm apart. Insert the piece of magnesium and the piece of copper in the lemon.
3. Using the alligator clips connect the leads of the LED to the copper and the magnesium.

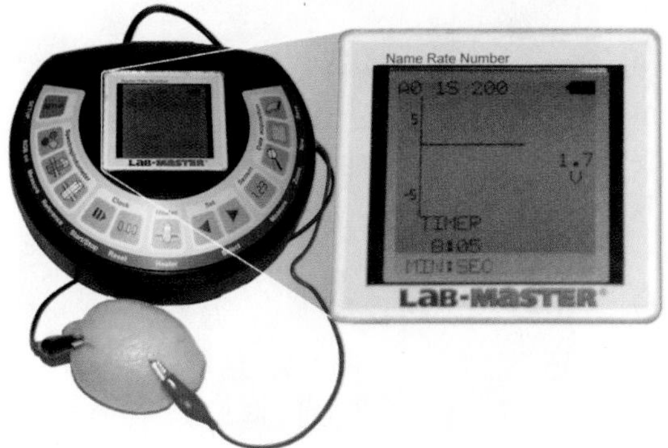

Observations and Questions:

1. Seeing the LED light up implies that electricity is flowing. Where does the electricity come form?
2. Notice the gas bubbles forming at the base of the electrodes. What is that gas and where does it come from?
3. The measured voltage across the electrodes is 1.7 volts.

Everything about this lemon reminds us of a battery. The lemon operates an electrical device (the LED). The lemon causes electricity to flow through wires, and it has a measured voltage like a battery. The chemical processes that generate the electricity the lemon are very similar to the chemistry that goes on inside more familiar batteries.

15.1 Electrochemistry and Electricity

Chemical defense in the body

Our health depends on a complex, self-adjusting dance between many thousands of chemical compounds. In the same way that rust damages the integrity of iron used in a bridge, a living body can be damaged by compounds that are harmful to essential molecules. Part of the chemical processing that every body does is to "deactivate" these harmful molecular compounds. Like most other chemical reactions, the process of "deactivation" is based on electrochemistry.

Rusting is oxidation

Iron rusts via a process called *oxidation*. We will talk a lot about oxidation later in this c hapter. For now let's think about oxidation as the process of losing electrons. In the same way that iron rusts by losing electrons, molecules in our bodies also rust or oxidize. When a molecule in our bodies oxidizes, it loses electrons. An oxidized molecule then tries to grab an electron from some other nearby molecule. These "electron hungry" molecules are called **free radicals**. Some free radicals are generated in the normal operation of our body and are important and useful. Some others are created when we are exposed to toxins, stress and pollutants.

Free radicals

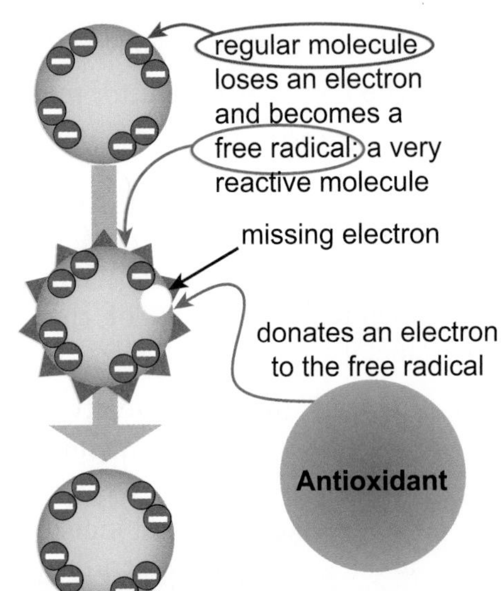
regular molecule loses an electron and becomes a free radical: a very reactive molecule
missing electron
donates an electron to the free radical
Antioxidant

Antioxidants

A healthy body tries to protect itself from free radicals with molecules called *antioxidants*. **Antioxidants** are chemical compounds that fight oxidation. Antioxidants protect the body by donating an electron to the free radicals, and thereby deactivating them. The antioxidant sacrifices itself by becoming oxidized. Since the production and the introduction of free radicals in our bodies is a continuous process the bodies supply of antioxidants need to be continually replenished by eating a healthy diet and exercising.

Chemistry terms

free radicals - oxidized molecules that have lost electrons.

antioxidants - molecules that protect us from the harmful effects of free radicals.

Electric charge and current

The basics of electricity

Being able to generate electricity from a lemon and a couple of pieces of metal is fascinating. Electrochemistry describes the detailed process by which chemical reactions in the lemon generate electricity. Electrochemistry also explains how electricity can cause chemical change. Understanding the basic principles of electricity is the first step in learning electrochemistry.

Charge is measured in coulombs

Consider the hydrogen atom which has a single proton and an electron. The charge of the proton is equal and opposite to the charge of the electron. The unit of charge is the **coulomb (C)**.

$$\textit{Charge of proton} = +\ 1.602 \times 10^{-19}\ C$$

$$\textit{Charge of electron} = -\ 1.602 \times 10^{-19}\ C.$$

Electric charge is conserved

Electric charge, like mass, is a fundamental property of matter and it is conserved. In this context "conserved" means the total electric charge in a closed system remains constant.

In metals such as copper some of the electrons that form the bonds between the copper atoms are free to move. These 'free" electrons move randomly between the fixed positive ions in a solid copper wire. Free electrons are light and fast so they move energy very effectively. This makes copper and other metals very good conductors of heat and electricity. Normally, the average charge of the free electrons does not move because as many electrons are going one way as the other way.

Fixed atom   Mobile electron

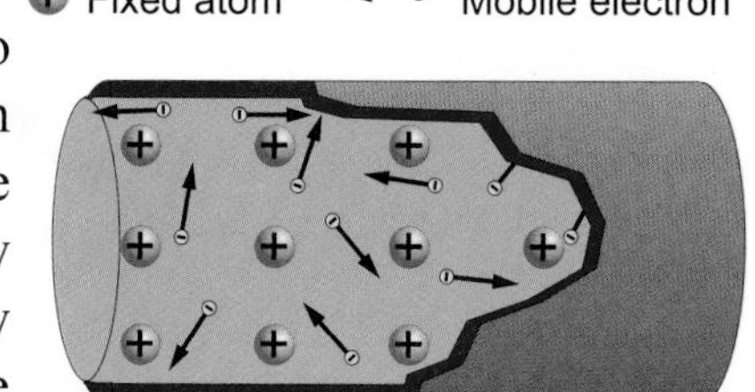

Electric current is the flow of electric charge

When a battery is attached to a wire, the motion of the electrons can change. When the free electrons move in a *collective* way (as a group) their charge also moves. This movement of charge results in the flow of **electric current**. The energy that the electrons need in order to move is provided by the battery. Current in measured in **amperes (A)**, which is a fundamental SI unit like kilogram (kg), meter (m), second (s), Kelvin (K), mole (mol), candela (cd).

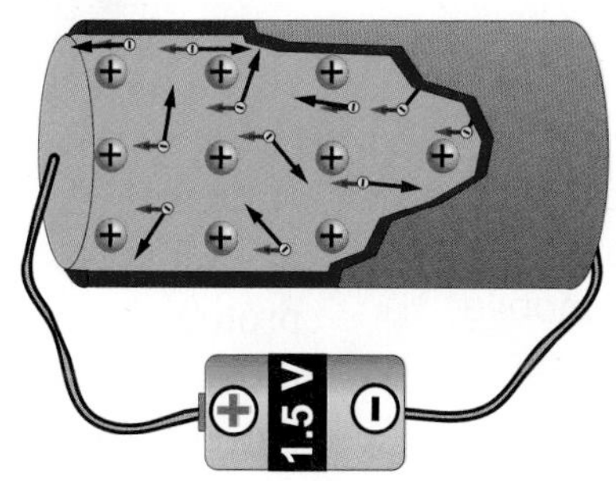

Chemistry terms

coulomb (C) - SI unit of electrical charge.
electric current - is the flow of electric charge.
ampere - SI unit of electrical current.

Electric current in solutions

Current can flow in ionic solutions

In *electrolyte* solutions, mobile ions with positive and negative charges are present. For example, when sodium chloride salt (NaCl) is dissolved in water the Na^+ and Cl^- ions are free to move in the solution.

Movement of either (or both) positive sodium ions (Na+) and/or negative chlorine ions (Cl-) may create an electric current. Any process that creates an average movement of electric charge creates an electric current.

Electro-chemistry generates electric current

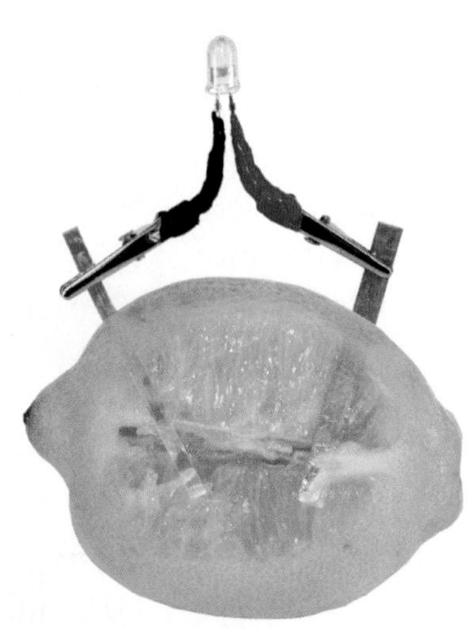

In the lemon battery, electric current is created by chemical reactions at the interfaces between the lemon juice, the copper, and the magnesium metal. The atoms of magnesium interact with the citric acid ($C_6H_8O_7$) in the lemon juice and give up electrons. These electrons flow in the wire, through the light bulb and return into the lemon via the copper metal.

The total number of charges in the entire system (lemon, copper, magnesium, wires) does not change. The number of negative charges is equal to the number of positive charges and the entire system is charge neutral. This property of charges to balance is called **electroneutrality**.

No current can flow if there are no mobile charges

If there are no moveable electric charges, then no electric current can flow. For example, a solution of sugar in water does not conduct electric current. There *are* electrons and protons in sugar however, there are exactly as many electrons as protons, and they cannot move separate from each other. A sugar molecule is *charge neutral* so cannot carry current. Current can only be carried by particles that have a net electric charge and can move independent of each other, such as free electrons or ions.

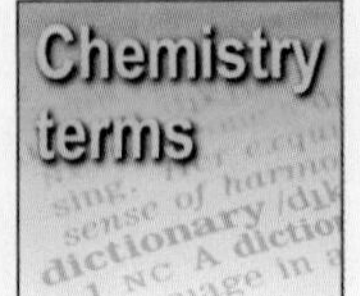

electroneutrality -The property of charges to balance resulting in a charge neutral system.

Voltage

Voltage is the difference in electrical potential

There are always free electrons in a copper wire, but they don't move to make a current unless there is a *force* causing them to move. Think of the wire as a pipe inside which the electrons can flow like water. Water flows in a pipe driven by the difference in pressure.

No height difference, no water flows
Height difference causes water to flow
No voltage difference, no current flows
Voltage difference causes current to flow
+
1.5 V
-

Voltage

Electrons flow in a wire due to a difference in an equivalent *electrical pressure* between two points. This difference in electrical pressure is called **voltage**. The unit for voltage is the **volt (V)**.

Voltage as potential energy

Voltage is a measure of electrical potential energy per unit of charge. One volt is one joule per coulomb.

$$\text{volt (V)} = \frac{\text{joule (J)}}{\text{coulomb (C)}}$$

This means that if one coulomb of charge moves through a difference of one volt it gains or loses one joule of potential energy. The charge loses energy if it goes from a higher voltage to a lower voltage and gains energy if it goes from a lower voltage to a higher voltage.

Resistance

When electrons flow in solid copper they collide with the fixed copper ions. The continuous collisions absorb energy and cause *resistance*. **Resistance** limits how much current flows for any given voltage. If resistance is high, current is low. If resistance is low, current can be high. Electrical resistance is measured in units called **ohms**. A resistance of one ohm means 1 amp of current flows when the voltage is 1 volt. The Greek letter omega (Ω) is used to represent ohms.

Ohm's law

The current, I, depends on the voltage, V, and the resistance, R. **Ohm's law** gives the formula that relates the current, the voltage and the resistance.

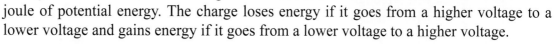

$$I = \frac{V}{R}$$

Current (amps, A) → I; V ← Voltage (volts, V); R ← Resistance (ohms, Ω)

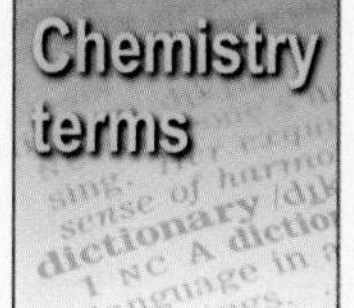
Chemistry terms

voltage -the electrical potential difference.

volt - the unit of voltage.

ohm (Ω) - the unit of electrical resistance.

Ohm's law - the relationship between voltage, current and electrical resistance.

15.2 Oxidation-Reduction (Redox) Reactions

Electro-chemistry is driven by electron transfer

In the early development of chemical knowledge, an important application was the refinement of pure metals such as iron, copper, aluminum and tin. These elements exist naturally as mineral compounds such as aluminum oxide (Al_2O_3). Early chemists talked about "*reducing*" compounds into pure metals long before people understood the chemistry of what was happening. Today we know that "reducing" aluminum oxide to elemental aluminum and oxygen involves transferring electrons from oxygen to aluminum. The atoms that lose electrons are **oxidized** and the atoms that gain electrons are **reduced**. For example, Al^{3+} gains 3 electrons and is *reduced* to aluminum metal. Oxygen gains the electrons and O^{2-} becomes *oxidized* to molecular oxygen.

Oxidation: Loss of electrons. Element charge becomes more positive

Reduction: Gain of electrons. Element charge becomes more negative

redox=short for **red**uction-**ox**idation

Oxidation-Reduction reactions or **redox reactions** are among the most common and most important chemical reactions in everyday life. These reactions always happen in pairs. Oxidation does not happen without reduction and reduction does not happen without oxidation. The process of oxidation-reduction together with the concept of charge conservation are the foundation of electrochemistry.

Even though we can't easily observe the actual transfer of electrons, we can easily see the results. If you dip a zinc (Zn) nail into a solution of copper sulfate ($CuSO_4$) the nail becomes covered with a brownish coating of copper metal.

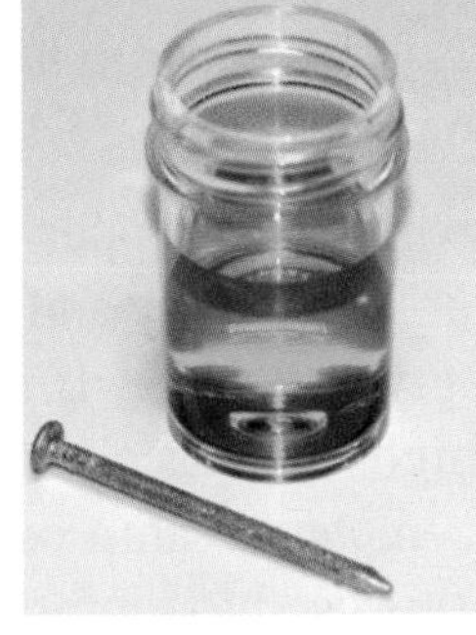

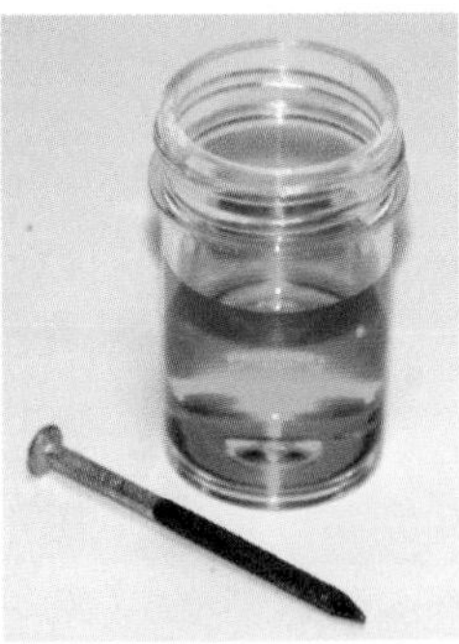

The $CuSO_4$ solution contains Cu^{2+} and SO_4^{2-} ions. The copper ions are *reduced* to pure copper metal by electrons coming from the zinc. The zinc is *oxidized* by losing the electrons to become Zn^{2+} ions in the solution. The chemical equation that describes the complete reaction is

$$Zn(s) + CuSO_4(aq) \rightarrow ZnSO_4(aq) + Cu(s)$$

oxidation - loss of electrons. Element charge increases.

reduction - gain of electrons. Element charge decreases.

redox - abbreviation for oxidation reduction.

Oxidation Numbers

Familiar reactions may be redox reactions

Chemical processes such as the burning natural gas (CH_4) and the rusting of iron (Fe) are very familiar. The chemical equations for these reactions are:

$4Fe(s) + 3O_2(g) \rightarrow 2Fe_2O_3(s)$ Rusting of iron

$CH_4(g) + 2O_2(g) \rightarrow CO_2(g) + 2H_2O(g)$ Burning of methane

How can we determine if these are redox reactions? Which elements are oxidized and which are reduced?

How to determine if a reaction is redox

In order to determine if a reaction is redox we must find out if there is electron loss and gain that takes place among the elements involved in the reaction. If we find that there are some elements that lose electrons and some other elements that gain electrons then we have the answer. The elements that lose electrons are oxidized, the elements that gain electrons are reduced and the overall process is a redox reaction.

Hydrochloric acid (HCl) provides a good example that is easier to analyze. Hydrogen and chlorine combine to form HCl by sharing a pair of electrons. One of these electrons comes from H and the other from Cl and they form a polar covalent bond.

Oxidation number

Since Chlorine is more electronegative than hydrogen, the shared electrons stay closer to Cl than to H. As a result it appears that chlorine has gained an electron and hydrogen has lost an electron. This apparent gain of an electron by the chlorine atom is indicated by an *oxidation number* of -1 which means that Cl is reduced. The **oxidation number** is the effective unit charge an atom in a compound would have if the electron were completely transferred to the more electronegative atom. In this case, chlorine is more electronegative atom, and if it takes the electron from hydrogen, its charge is -1. Therefore its oxidation number in the compound HCl is also -1. In this reaction chlorine is *reduced* since it gains an electron and its oxidation number becomes lower (more negative).

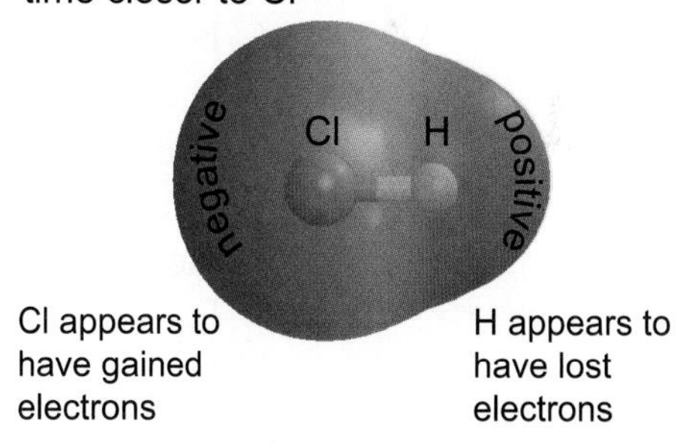
Electrons spend more time closer to Cl
Cl
H
negative
positive
Cl appears to have gained electrons
H appears to have lost electrons

Since HCl is neutral overall, the corresponding oxidation number of hydrogen should be +1. Hydrogen is *oxidized* since it loses an electron and its oxidation number increases (more positive).

Chemistry terms

oxidation number - gives the number of electrons that an element has lost or gained in forming a chemical bond with another element.

Rules for assigning oxidation numbers

Oxidation number and ionic charge

The oxidation number is different from the real charge of the atom. The assignment of electrons to the atoms and the associated oxidation number is just a way to account for the electrons that are associated with the atoms.

ionic charge is denoted 2 -
sign follows the number

The oxidation number is written with a sign followed with the number like -1, +1,-3, 0, etc. This is different than the way we write the ionic charge for which the sign follows the number like 1-, 2-, 3+, etc.

oxidation number is denoted - 2
number follows the sign

Rules for assigning oxidation numbers

By looking at the properties of atoms and the way that they bond with other atoms to form various compounds, scientists have come up with a set of rules that assign an oxidation number to each atom in a compound. These rules are explicit representations of the Lewis formulation and give us the number of electrons - the *oxidation number* - that is assigned to each atom in a compound

TABLE 15.1. Rules for assigning oxidation numbers

1.	The oxidation number of an atom in a pure element is 0. For example, the oxidation number of chlorine in Cl_2 or O in O_2 is 0
2.	The sum of the oxidation numbers of atoms in a neutral molecule is zero. For example the sum of the oxidation numbers of C and O in CO_2 is 0 Oxidation number of carbon (n_C) plus two times the oxidation number of oxygen = 0: $n_C + 2n_O = 0$
3.	The sum of the oxidation numbers of all atoms in an ion is equal to the charge of the ion. For example, the oxidation number of Cu in Cu^{2+} is +2. For SO_4^{2-}, the oxidation number of S (n_S) plus 4 times the oxidation number of O (n_O) = -2: $n_S + 4n_O = -2$
4.	Metals have positive oxidation number according to their group • Group 1A metals (Na, K, Li) have oxidation number +1 • Group 2A metals (Mg, Ca) have oxidation number +2
5.	The oxidation number of non-metals are as follows: • Fluorine (F): -1 • Hydrogen (H): +1 except in hydrides such as LiH and NaH in which the oxidation number is -1. • Oxygen (O): -2 • Group 7A (Cl, Br, I) : -1 • Group 6A (S, Se, Te) : -2 • Group 5A (N, P, As): -3

Application of Oxidation Number Rules

Oxidation rules must be applied in order

The oxidation number rules apply in the order they appear on the list. If there is a conflict, the rule that is higher in the list has priority. For example, consider potassium peroxide (K_2O_2).

From Rule 4 we see that K has oxidation number +1 and Rule 5 tells us that oxygen has oxidation number -2. Obviously there is a conflict since Rule 2 which keeps track of the overall charge balance is violated. To solve this problem we must give priority to the order that the rules appear.

Here are the steps to calculate the oxidation numbers of potassium and oxygen in K_2O_2.

TABLE 15.2.

Step1	Step2	Step3
From Rule 2 we know that the sum of the oxidation numbers of all atoms in K_2O_2 must ne zero: If the oxidation number for potassium is n_K and the oxidation number for oxygen is n_O then $2(n_K) + 2(n_O) = 0$	From Rule 4 we know that the oxidation number of potassium is +1 (n_K= +1). Since Rule 4 has priority over Rule 5, the oxidation number for O (n_O) is still unknown. But now we have an equation that we can use to solve for it. $2(n_K) + 2(n_O) = 0$	Solving the equation $2(n_K) + 2(n_O) = 0$ we obtain n_O = -1 and we are done. In K_2O_2: • the oxidation number of potassium (K) is +1 • the oxidation number of oxygen (O) is -1

The oxidation number of elements in a compound can be found by following these steps.

Find the oxidation number for each element in carbon monoxide (CO)

Asked: *Find the oxidation number of O and C in CO.*

Relationships: *The rules for assigning oxidation numbers*

Solve: *The solution procedure is as follows:*

- *Carbon monoxide (CO) is a neutral compound*
- *Sum of oxidation numbers of C and O must add to zero (rule 2)*
- *The oxidation number of oxygen = -2 (rule 5).*
- *The oxidation number of carbon (n_C) = ?*
- $n_C + (-2) = 0$ *which gives* $n_C = +2$

Answer: *The oxidation number for carbon in carbon monoxide (CO) is +2*
The oxidation number of oxygen in carbon monoxide is -2

Finding oxidation numbers

The same element can have different oxidation numbers

The oxidation number is a way to keep track of the electrons associated with each atom in a compound. *The oxidation number of the same element may be different from one compound to the other.* For example, in carbon dioxide (CO_2) the oxidation number of oxygen is -2. In Potassium peroxide (K_2O_2) the oxidation number of oxygen is -1.

Find the oxidation number for each element in carbon monoxide (CO_2)

Asked: *Find the oxidation number of O and C in CO_2.*

Relationships: *The rules for assigning oxidation numbers*

Solve: *The solution procedure is as follows:*

- *Since carbon dioxide (CO_2) is a neutral compound the sum of oxidation numbers of C and O_2 must add to zero (rule 2)*
- *The oxidation number of O is -2 (rule 5)*
- *The oxidation number of O_2 is 2(-2) = -4*
- *The oxidation number of carbon (n_C) = ?*
- $n_C + 2(-2) = 0$ *which gives* $n_C = +4$

Answer: *The oxidation number for carbon in CO_2 is +4 and for oxygen is -2*

Find the oxidation number for each element in the nitrite ion NO_2^-.

Asked: *Find the oxidation number of an ion with overall charge of -1.*

Given: *The elemental arrangement of the ion.*

Relationships: *The rules for assigning oxidation numbers*

Solve: *Nitrite (NO_2^-) is an ion with a charge of -1 and so the oxidation numbers of the elements must add up to -1.*

- *The sum of the oxidation numbers of N and O_2 must add to -1. (rule 2)*
- *The oxidation number of oxygen is -2 (rule 5)*
- *The oxidation number of nitrogen n_N = ?.*
- $n_N + 2(-2) = -1$ *which gives* $n_N = +3$

Answer: *The oxidation number of oxygen in NO_2^- is -2.*
The oxidation number of nitrogen in NO_2^- is +3.
$+3 + 2(-2) = -1.$

Fractional oxidation numbers

Oxidation numbers may also be fractions

The oxidation number of an atom in a substances is usually a whole number such as +1, -1, +2, -2 etc. However it is also possible for the oxidation number of atoms to be fractions such as +1/2, -1/2, -2/3 etc. This is often the case for the oxidation number of carbon in various compounds with hydrogen (hydrocarbons).

Let's consider methane (C_3H_8) and go through the various steps for calculating the oxidation numbers of C and H in this molecule.

TABLE 15.3.

Step1	Step2	Step3
• From rule 2 we know the sum of the oxidation numbers of all atoms in C_3H_8 must be zero: • If the oxidation number for carbon is n_C and the oxidation number for hydrogen is n_H then the equation that relates these oxidation numbers is $3(n_C) + 8(n_H) = 0$	• From rule 5 we know that the oxidation number of hydrogen is +1 ($n_H = +1$). • The oxidation number for C (n_C) is still unknown but now we have an equation that we can solve for it. $3(n_C) + 8(+1) = 0$	• Solving the equation $3(n_C) + 8(+1) = 0$ we obtain $n_C = -8/3$ • So we see that the oxidation number can be a fraction • Check the answer by verifying that the sum of oxidation numbers is zero. $3(-8/3) + 8 = 0$

Find the oxidation number for each element in the compound C_3H_4.

Asked: *Find the oxidation number of each element in methylacetylyne (C_3H_4)*

Relationships: *The rules for assigning oxidation numbers*

Solve: *C_3H_4 is a neutral compound:*
The sum of the oxidation numbers of H_4 and C_3 must add to zero. (rule 2)

- *The oxidation number of hydrogen (n_H) is +1. $n_H = +1$ (rule 5)*
- *The oxidation number of carbon $n_C = ?$.*
- *$3(n_C) + 4(+1) = 0$ which gives $n_C = -4/3$*

Answer: *The oxidation number of C in C_3H_4 is -4/3.*
The oxidation number of H in C_3H_4 is +1.

Identifying redox reactions

oxidation numbers in redox reactions

Determining the oxidation numbers is the first step to identifying a redox reaction. The next step is to see if the oxidation numbers for any elements have *changed* in the reaction. If we find that any oxidation numbers have changed from the left side (reactants) to the right side (products) *then we know that we have a redox reaction*. If the oxidation numbers of the elements in the reactants and the products do not change the reaction is not redox. The best way to learn this is with an example.

Find the element that is oxidized and the element that is reduced in the reaction of iron with oxygen resulting in rust (iron oxide Fe_2O_3).

$$4Fe(s) + 3O_2(g) \rightarrow 2Fe_2O_3(s).$$

Asked: *Find the element that is oxidized and the element that is reduced*

Given: *The balanced chemical reaction of Fe and O_2 and the rules for assigning oxidation numbers.*

Relationships: *The rules for assigning oxidation numbers to each element in compounds.*

Solve: *We look at both the left and right sides of the reaction:*

- *Let's start with the reactants.*
  Since both Fe and O_2 are free elements the oxidation number of the Fe and O atoms on the left side of the equation is zero.
- *Next let's look at the product (Fe_2O_3) of the reaction.*
  The sum of the oxidation numbers in Fe_2O_3 must add to zero.
  The oxidation number of oxygen is -2.
  Now if we add the oxidation number of oxygen and the unknown oxidation number of iron (n_{Fe}) we have the equation:
  $2(n_{Fe}) + 3(-2) = 0$
  Solving this equation for n_{Fe} we find that the oxidation number of iron in Fe_2O_3 is +3.

Answer: *The oxidation number of Fe has increased; it went from 0 to +3. Iron has lost electrons and so it is oxidized.*
The oxidation number of oxygen has decreased, it went from 0 to -2. Oxygen has gained electrons and it is reduced.

$$3O_2(g) + 4Fe(s) \rightarrow 2Fe_2O_3(s)$$

0   0   +3 -2 ← oxidation numbers

Oxidation

Reduction

Analyzing a redox reaction

Oxidized - oxidation number increases (lose electrons)

Reduced - oxidation number decreases (gain electrons)

Charge balance is maintained

The number of electrons that are received by the reduced element is equal to the number of electrons that have been given up by the oxidized element. The total number of electrons does not change – overall charge neutrality must be maintained.

In the iron - oxygen reaction we saw that the iron is oxidized and the oxygen is reduced. We can also say that "iron is oxidized by oxygen" which is another way of saying that oxygen facilitates the oxidation of iron. We may also say that "oxygen is reduced by iron" oxygen is reduced and iron facilitates this reduction.

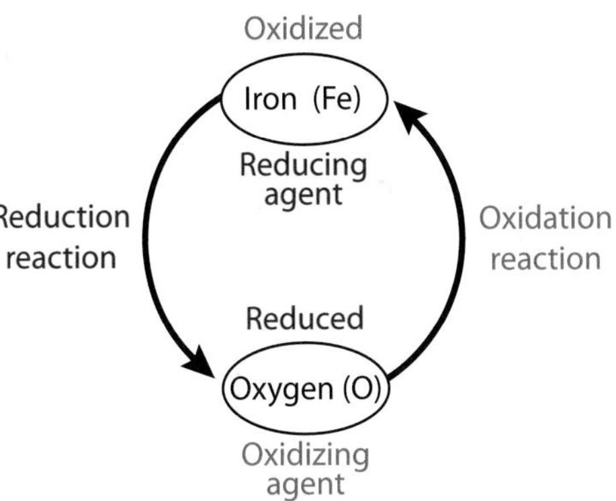

$$4Fe(s) + 3O_2(g) \rightarrow 2Fe_2O_3(s)$$

Oxidizing agent

The element that oxidizes another element is called the **oxidizing agent**. In our example oxygen is the oxidizing agent. The element that reduces another element is called the reducing agent. In our example iron is the **reducing agent**.

Reducing agent

In the reaction $4Fe(s) + 3O_2(g) \rightarrow 2Fe_2O_3(s)$:

Fe is oxidized	O is reduced
Fe is the reducing agent	O is the oxidizing agent

Solved problem

Find the reducing and the oxidizing agents in: $2Mg(s) + O_2(g) \rightarrow 2MgO(s)$.

Relationships: *The rules for assigning oxidation numbers and the chemical reaction*

Solve:

$$O_2(g) + 2Mg(s) \rightarrow 2MgO(s)$$

Oxidation numbers: O in O_2: 0; Mg: 0; in MgO: Mg +2, O -2 ← oxidation numbers

Reduction (O: 0 → -2); Oxidation (Mg: 0 → +2)

The oxidation number of oxygen decreases from 0 to -2 so oxygen is reduced. Therefore, oxygen is the oxidizing agent. The oxidation number of magnesium increases from 0 to +2 so magnesium is oxidized. Therefore, magnesium is the reducing agent.

Chemistry terms

oxidizing agent - the element that oxidizes another element.

reducing agent - the element that reduces another element.

15.3 Balancing Redox Equations

Mass conservation

For redox reactions, like any other type of chemical reaction, mass is conserved. The number and type of atoms in the reactants must equal to the number and type of atoms in the products.

Charge conservation

In addition to mass conservation we must also check to make sure that charge is conserved. We do this by keeping track of the electrons that are exchanged between various atoms and make certain that they balance. The number of electrons associated with the reactants must be equal to the number of electrons associated with the products. Also, the total increase in the oxidation number of some atoms must be equal to the total decrease of the oxidation numbers of some other atoms.

Balance mass and charge

The things that we learned in the previous section about oxidation numbers and redox reactions are now the tools that we use to balance redox reactions.

Balancing redox reactions

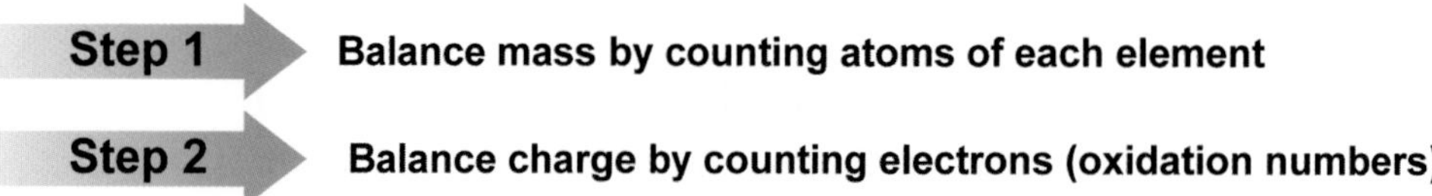

Step 1
Balance mass by counting atoms of each element
Step 2
Balance charge by counting electrons (oxidation numbers)

Methods for balancing redox reactions

Since balancing redox reactions requires both the balance of mass and charge, the procedure is often more complicated than balancing non-redox reactions.

There are two methods for balancing redox reactions.

1. The oxidation number method;
2. The oxidation reaction and reduction reaction method combination. This is also called thehalf-reaction method.

Both of these methods are based on the same fundamental principles and they simply represent structured applications of these procedures.

Balancing by Inspection

In some cases it is possible to balance a redox reaction by the inspection method. These "easy to balance" reactions are usually the ones that do not occur in aqueous solutions. Consider the unbalanced reaction of carbon with iron oxide. This reaction takes place in a furnace where iron oxide (Fe_2O_3) reacts with carbon resulting in pure iron and carbon dioxide as a by product.

$$C + Fe_2O_3 \rightarrow Fe + CO_2$$

This is a redox reaction. Carbon is oxidized and iron is reduced. The oxidation number of oxygen does not change in this reaction. By balancing the mass of the three elements that take part in this reaction we obtain the balanced equation.

$$3C + 2Fe_2O_3 \rightarrow 4Fe + 3CO_2$$

The oxidation number method

Most redox reactions can't be balanced with the inspection method. For such reactions we use the oxidation number method.

Balancing equations using oxidation numbers

The basic rule that drives this method is charge balance.

> Increase in oxidation number for the oxidized atoms
>
> equals
>
> Decrease in oxidation number for the reduced atoms

1) Assign oxidation numbers

The basic steps for balancing a redox reaction with the oxidation number method are:

1. Assignment of oxidation numbers to all atoms
   - Use the rules for assigning the oxidation numbers
   - Write the oxidation number for each element below it

2) Identify oxidation and reduction

2. Identify and label the atoms that are oxidized and the atoms that are reduced.
   - Increase in oxidation number (loss of electrons) means oxidation
   - Decrease in oxidation number (gain of electrons) means reduction

3) Adjust coefficients to equalize oxidation and reduction

3. Adjust the appropriate coefficients in the chemical equation so that the total increase in oxidation number (total loss of electrons) is equal to the total decrease in the oxidation number (total gain of electrons).

4) Check mass balance

4. Check for overall mass balance.
   - Check to see that the number of atoms is the same on both sides of the equation. If not make the necessary changes to the coefficients.

When we apply these steps in order, the resulting balanced equation should satisfy both mass and charge conservation.

Each step has a well defined set of rules. However, applying all of the steps together can be challenging! Balancing redox reactions requires practice in order to become familiar with it.

Example: balancing a complex redox reaction

Using the oxidation number method, balance the equation

$HNO_3(aq) + Cu_2O(s) \rightarrow Cu(NO_3)_2(aq) + NO(g) + H_2O(l)$

Solve: *Follow the steps for balancing redox equations*

Step 1. *Assign the oxidation numbers to each element*

$$\underset{+1\ +5\ -2}{HNO_3} + \underset{+1\ -2}{Cu_2O} \rightarrow \underset{+2\ +5\ -2}{Cu(NO_3)_2} + \underset{+2\ -2}{NO} + \underset{+1\ -2}{H_2O}$$

Green numbers under elements are the oxidation numbers

Step 2. *Identify the atoms that are reduced and oxidized*

$$\underset{+1\ +5\ -2}{HNO_3} + \underset{+1\ -2}{Cu_2O} \rightarrow \underset{+2\ +5\ -2}{Cu(NO_3)_2} + \underset{+2\ -2}{NO} + \underset{+1\ -2}{H_2O}$$

Oxidation (Cu: +1 → +2)
Reduction (N: +5 → +2)

- *The oxidation number of N decreases in NO and so N is reduced.*
- *The oxidation number of Cu increases and so Cu is oxidized.*
- *The oxidation numbers of the other atoms do not change.*

Step 3. *First we balance all atoms whose oxidation numbers have changed*

$$\underset{+1\ +5\ -2}{HNO_3} + \underset{+1\ -2}{Cu_2O} \rightarrow 2\underset{+2\ +5\ -2}{Cu(NO_3)_2} + \underset{+2\ -2}{NO} + \underset{+1\ -2}{H_2O}$$

Oxidation
Reduction

- *N atoms whose oxidation number changes are already balanced*
- *To balance Cu atoms we adjust the coefficients — multiply by 2.*
- *The total increase in oxidation number is 2: Cu looses 2 electrons. The decrease in oxidation number of N is 3: N gains 3 electrons.*
- *Balance the number of electrons lost and gained. Multiply the species that contain Cu with +3 and the species that contain N with +2*

$$2\underset{+1\ +5\ -2}{HNO_3} + 3\underset{+1\ -2}{Cu_2O} \rightarrow 6\underset{+2\ +5\ -2}{Cu(NO_3)_2} + 2\underset{+2\ -2}{NO} + \underset{+1\ -2}{H_2O}$$

-2(3) = -6
+3(2)= + 6

Step 4. *Balance mass.*

- *Add 12 HNO_3 to the left side to balance N.*
- *Add 6 O and 12 H to the right side to balance oxygen and hydrogen.*
- *This is done by adding $6H_2O$ to the right side to balance H and O.*

$$14\,HNO_3 + 3\,Cu_2O \rightarrow 6\,Cu(NO_3)_2 + 2\,NO + 7\,H_2O$$

Half-Reactions

Redox reactions in aqueous solutions

Many interesting and useful redox reactions occur in aqueous solutions. The procedure for balancing the equations for aqueous reactions is based on the fundamental principles described earlier but with an important difference. The difference is the separation of the complete reaction into two *half-reactions*. One **half-reaction** involves the oxidized elements and the other involves the reduced elements. By separating the oxidation from the reduction and writing them separately we make it easier to count electrons and establish charge balance.

Identifying half-reactions

To see where to split the reaction into half reactions, we need to know the oxidation numbers of all the elements involved. Once these are known we can see which are oxidized and which are reduced. Once we know the oxidation and the reduction parts we can separate the complete reaction into half-reactions. When a zinc nail is placed in copper sulfate ($CuSO_4$) solution the overall reaction can be explained with the half-reactions.

Why half-reactions are useful

With half-reactions, it is very direct to see the transfer of electrons. The number of electrons given up by the oxidation half-reaction must be equal to the number of electrons received by the reduction half-reaction.

Find the half-reactions of the reaction $Zn(s) + CuSO_4(aq) \rightarrow ZnSO_4(aq) + Cu(s)$

Given: *The rules for assigning oxidation numbers.*

Relationships: *In solution, $CuSO_4$ and $ZnSO_4$ dissociate as follows:*

- *$CuSO_4$ into Cu^{2+} and SO_4^{2-} ions*
- *$ZnSO_4$ into Zn^{2+} and SO_4^{2-} ions*
- *The reaction may now be writen as*
- $Zn(s) + Cu^{2+} + SO_4^{2-} \rightarrow Zn^{2+} + SO_4^{2-} + Cu$
- *Zn is oxidized and Cu is reduced*
- *SO_4^{2-} is neither oxidized or reduced. It is called a spectator ion.*

Solve: *The half-reactions are:*

- $Zn(s) \rightarrow Zn^{2+} + 2e^-$ *- oxidation*
- $Cu^{2+} + 2e^- \rightarrow Cu$ *- reduction*

half-reactions - the oxidation and the reduction parts of a redox reaction.

The half-reactions method

Steps in the half-reactions method

Redox reaction equations can be balanced using the 2 half reactions. The method is called the *half-reactions* method for balancing redox reaction equations.

Step 1. Write the complete unbalanced reaction showing explicitly all ions

Step 2. Identify which elements are oxidized and which are reduced. Find the spectator ions (oxidation number does not change).

Step 3. Write down the two unbalanced half-reactions.

Step 4. Balance mass with elements other than oxygen and hydrogen. Balance oxygen by adding H_2O then balance hydrogen by adding H^+.

Step 5. Balance the charge for both half-reactions.

Step 6. Make the number of electrons in both reactions equal by adjusting the coefficients.

Step 7. Add the two half-reactions and check that both mass and charge balance.

An example redox reaction equation:

The plating of copper into zinc in a solution of copper sulfate provides a good example of the half-reactions method.

$$Zn(s) + CuSO_4(aq) \rightarrow ZnSO_4\ (aq) + Cu(s).$$

Step 1:

Write the unbalanced equation showing all ions

$$Zn(s) + Cu^{2+} + SO_4^{2-} \rightarrow Zn^{2+} + SO_4^{2-} + Cu$$

Step 2:

Identify oxidation, reduction and spectator ions.

$$\underset{0}{Zn(s)} + \underset{+2}{Cu^{2+}(aq)} + \underset{-2}{SO_4^{2-}(aq)} \rightarrow \underset{+2}{Zn^{2+}(aq)} + \underset{-2}{SO_4^{2-}(aq)} + \underset{0}{Cu(s)}$$

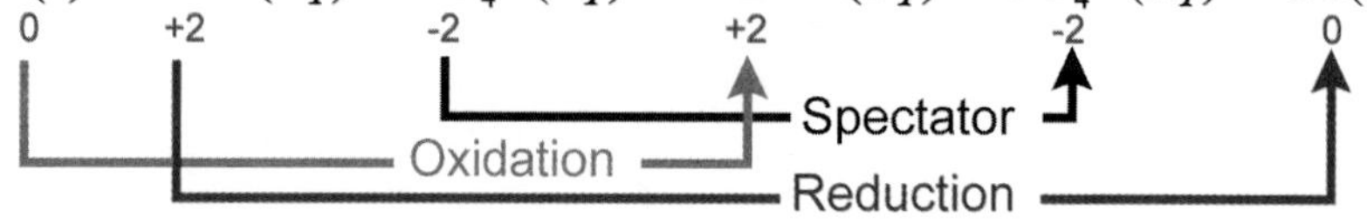

Step 3:

Write down the two unbalanced half-reactions

Oxidation: $Zn\ (s) \rightarrow Zn^{2+}\ (aq)$

Reduction: $Cu^{2+}(aq) \rightarrow Cu\ (s)$

Step 4:

Balance the mass for both half-reactions. It is already balanced.

$$Zn\ (s) \rightarrow Zn^{2+}\ (aq)$$

$$Cu^{2+}(aq) \rightarrow Cu\ (s)$$

Balancing the charge in half-reactions

Why charge balancing is necessary

It is possible, even likely, that mass will balance for each half reactions, but one half-reaction uses more electrons that the other one yields. In a fully balanced electrochemical reaction the electrons given up in the oxidation half-reaction are the same number as the electrons used in the reduction half-reaction.

Step 5:

Balance the charge for both half-reactions. Add electrons to balance changes in oxidation numbers

- $Zn(s) \rightarrow Zn^{2+}(aq) + \mathbf{2e^-}$: Add 2 electrons to the right side.
- $Cu^{2+}(aq) + \mathbf{2e^-} \rightarrow Cu(s)$: Add 2 electrons to the left side.

Step 6:

Make the number of electrons in both reactions equal by adjusting the coefficients.

- $Zn(s) \rightarrow Zn^{2+}(aq) + 2e^-$
- $Cu^{2+}(aq) + \mathbf{2}e^- \rightarrow Cu(s)$

Step 7:

Add the two half-reactions

- $Zn(s) + Cu^{2+}(aq) \rightarrow Zn^{2+}(aq) + Cu(s)$:
  Mass and charge are balanced.
- Return the spectator ions and adjust coefficients to maintain charge neutrality.

$$Zn(s) + Cu^{2+}(aq) + SO_4^{2-}(aq) \rightarrow Zn^{2+}(aq) + SO_4^{2-}(aq) + Cu(s)$$

Step 8:

Simplify equation and check for mass and charge balance.

- Reactants: Zn, Cu^{2+}, SO_4^{2-} , 0 net charge.
- Products: Zn^{2+}, Cu, SO_4^{2-} , 0 net charge.
- Mass and charge are balanced

What the half reactions tell us

At the end of the analysis we could write the final balanced equation as given below. However, this does not tell us directly that 2 electrons were transferred. If this reaction were part of a battery or biochemical process, the fact that 2 electrons were exchanged might create an electrical current!

$$Zn(s) + CuSO_4(aq) \rightarrow ZnSO_4(aq) + Cu(s).$$

A redox reaction with chlorine

Chlorine dioxide (ClO_2) is used for water treatment. The OH^- indicates that the reaction occurs in a basic solution. for which an element that appears in the reactants is both oxidized and reduced.

Using the half-reactions method, balance the equation
$ClO_2 + OH^- \rightarrow ClO_2^- + ClO_3^-$

Solve: *Follow the steps for balancing redox equations.*

Step 1. *Write the complete unbalanced reaction showing explicitly all ions*

- $ClO_2 + OH^- \rightarrow ClO_2^- + ClO_3^-$

Step 2. *Identify which elements are oxidized and which are reduced*

- *ClO_2 is oxidized to ClO_3^-. Cl goes from +4 to +5: Cl is oxidized*
- *ClO_2 is reduced to ClO_2^-. Cl goes from +4 to +3: Cl is reduced*
  Note that Cl is both oxidized and reduced in this reaction.

Step 3. *Write down the two half-reactions*

- *$ClO_2 \rightarrow ClO_3^-$: Oxidation half-reaction*
- *$ClO_2 \rightarrow ClO_2^-$: Reduction half-reaction*

Step 4. *Balance the mass for both half-reactions.*

a. *Cl is balanced in both reactions.*

b. *Balance O and H of the oxidation half-reaction:*
Balance O by adding H_2O and OH^-: $ClO_2 + 2OH^- \rightarrow ClO_3^- + H_2O$

- *O and H of the reduction half-reaction are balanced: $ClO_2 \rightarrow ClO_2^-$*

Step 5. *Balance the charge for both half-reactions. Add electrons to balance changes in oxidation numbers*

- *$ClO_2 + 2OH^- \rightarrow ClO_3^- + H_2O + e^-$: Add electron on the right side*
- *$ClO_2 + e^- \rightarrow ClO_2^-$: Add electron on the left side*

Step 6. *Number of electrons is equal in both reactions.*

Step 7. *Add the two half-reactions*

- *$2ClO_2 + 2OH^- \rightarrow ClO_3^- + ClO_2^- + H_2O$, No spectator ions*

Step 8. *Simplify equation, if needed, and check for mass and charge balance.*

- *$2ClO_2 + 2OH^- \rightarrow ClO_3^- + ClO_2^- + H_2O$*

15.4 Electrochemical Cells

Exchanged electrons and electro-chemical cells

The movement of electrons from the oxidized compounds to reduced compounds is the essence of electro-chemistry. If we physically separate the oxidation and reduction reactions, we can force the electrons to be exchanged by moving through a wire. Electrons moving through a wire are an electrical *current*. A device which creates a separation of oxidation and reduction reactions is an *electrochemical cell*. An **electrochemical cell** is a device in which redox reactions may produce electrical energy or in which electrical energy is used to produce a chemical reaction.

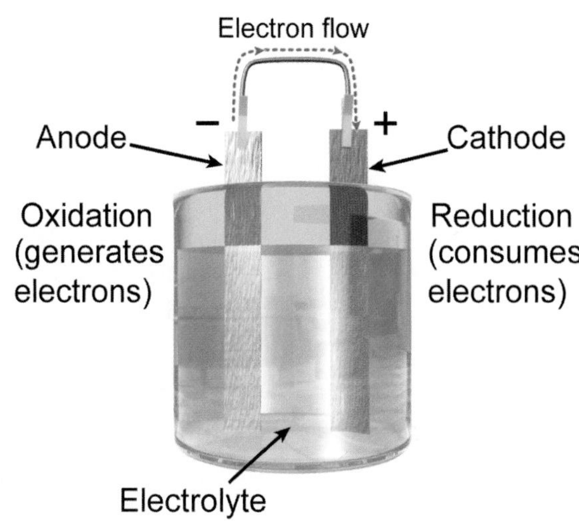

Parts of electro-chemical cells

The various parts of the electrochemical cell are:

1. Two **electrodes**, where the oxidation and reduction reactions occur.
   a. The electrode at which oxidation occurs is called the **anode**
   b. The electrode at which reduction occurs is called the **cathode**.

2. The **electrolyte**, which is a conductive solution inside which the electrodes are immersed.
   a. The electrolyte contains free ions
   b. The electrolyte conducts electricity.

3. The conducting path that connects the electrodes externally.

When we analyze electrochemical cells we must keep in mind the following:

1. The connection between oxidation and reduction
2. The flow of electrons from the anode to the cathode
3. The chemical reactions that occur at the anode and the cathode

electrochemical cell - device in which redox reactions take place.
electrode - the part of a cell where oxidation and reduction reactions occur.
anode - the electrode at which oxidation occurs.
cathode - the electrode at which reduction occurs.
electrolyte - the conductive solution in which the electrodes are immersed.

Types of electrochemical cells

Types of cells

The relation between the chemical reactions and the electrical current divide electrochemical cells into two categories.

Voltaic cells

1. **Voltaic cells**:
   A voltaic cell is an electrochemical cell in which spontaneous chemical reactions at the electrodes generate electrical current. Voltaic cells are named after Italian scientist Allessandro Volta who invented the battery in 1800. Another name for voltaic cell is **galvanic cell** after Luigi Galvani, who is considered the father of electrochemistry. A battery is a voltaic cell.

Electrolytic cells

2. **Electrolytic cells**:
   In an electrolytic cell externally applied electrical current drives non-spontaneous chemical reactions at the interface between the electrolyte and the electrodes. Electroplating and electrolysis are done with electrolytic cells.

Half-cells

Both voltaic and electrolytic cells are constructed with two half-cells. One of these half-cells corresponds to the anode which is related to oxidation half-reaction. The other half-cell corresponds to the cathode which is related to the reduction half-reaction.

Batteries are voltaic cells

Batteries are voltaic cells. The positive and the negative terminals of the battery correspond to the electrodes of the cell. The positive terminal is connected to the cathode half-cell where reduction occurs and the negative terminal is connected to the anode half-cell where oxidation occurs.

When an external circuit such as a light bulb is connected between the electrodes, electrons flow from the anode electrode to the cathode electrode.

The flow of electrical current is completed inside the battery. Here negative ions move from the cathode to the anode and positive ions move in the opposite direction, from the anode to the cathode, in order to maintain electroneutrality.

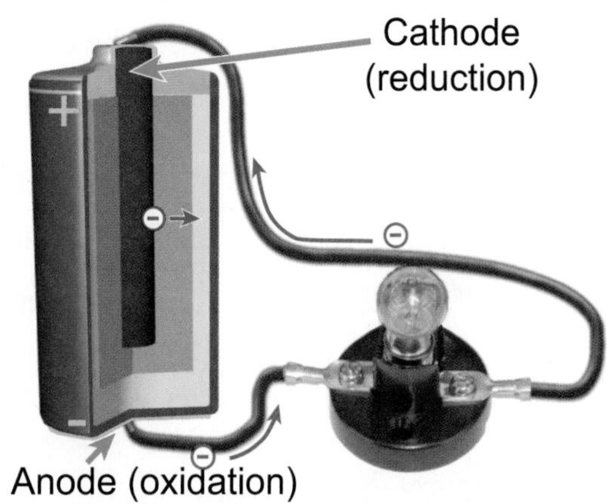

Chemistry terms

voltaic cell - electrochemical cell in which chemical reactions generate electricity.
galvanic cell - another name for voltaic cell.
electrolytic cell - a cell in which chemical reactions are driven by electricity.

Chemistry in a voltaic cell

Constructing a voltaic cell

In order to construct an electrochemical cell we start with the two half-cells. In the oxidation part (anode) electrons are released. Electrons are absorbed in the reduction part (cathode). We connect the two cells so that electrons flow from the anode to the cathode.

Mg-Cu voltaic cell

To illustrate the process, consider a voltaic cell using Mg and Cu as the electrodes. The magnesium electrode is immersed in a $MgSO_4$ solution that contains Mg^{2+} ions. The copper electrode is immersed in a $CuSO_4$ solution that contains Cu^{2+} ions. The half-reactions that describe the oxidation of Mg and the reduction of Cu are:

$Mg(s) \rightarrow Mg^{2+}(aq) + 2e^-$	Oxidation at the anode
$Cu^{2+}(aq) + 2e^- \rightarrow Cu(s)$	Reduction at the cathode

Magnesium loses electrons more easily than Cu. When the two electrodes are connected by a wire (or bulb) the electrons released by Mg at the anode move through wire to the cathode. This flow of electrons gives electrical current, *but only for a very short time*! The current stops because the Mg^{2+} that are generated at the anode have nowhere to go. The electrons and the Mg^{2+} ions are attracted to each other and the reaction stops as soon as it starts.

No path to complete the circuit.
No current flow

No reactions
No electron release

In order to solve this problem we must provide a path for positive ions to move from the anode half-cell to the cathode half-cell. The addition of positive ions to copper sulfate balances the addition of electrons and maintains charge neutrality. You might think that such an effective path can be created by mixing the two solutions. If we mix the solutions, Cu^{2+} ions will come in contact with Mg generating localized redox half-cells which are not going to produce any useful electrons *flowing in the wire*. The electrons in this case are exchanged locally and do not pass through the external circuit. We need a better solution.

Salt Bridge

Salt bridge

An effective source of ions is provided by a *salt bridge*. A **salt bridge** is a connection between the oxidation and the reduction half cells. An ideal salt bridge provides a path for ions to flow between the half cells but does not permit mixing of the half cell solutions. A common salt bridge is a tube that contains an electrolyte such as potassium chloride (KCl) whose ends are plugged with a porous material. The bulk of the material can't leak out but ions such as K^+, Cl^- can pass through. For short experiments a strip of filter paper soaked in KCl makes an effective salt bridge.

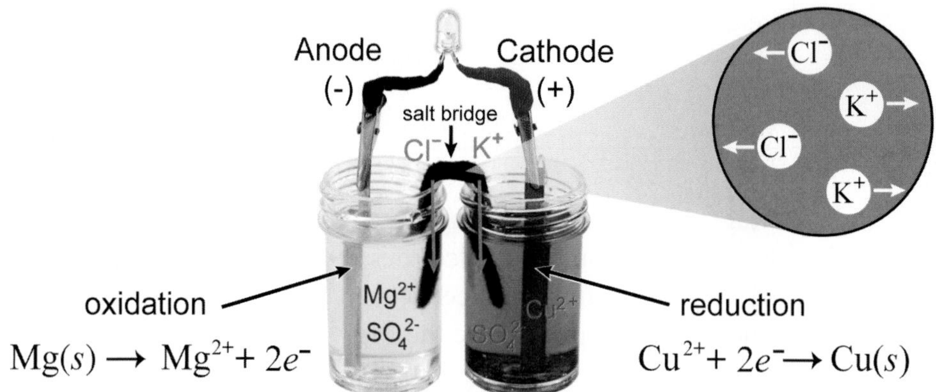

There is a compact way for representing the reaction and the electrochemical cell in general. For the magnesium, copper voltaic cell this representation is:

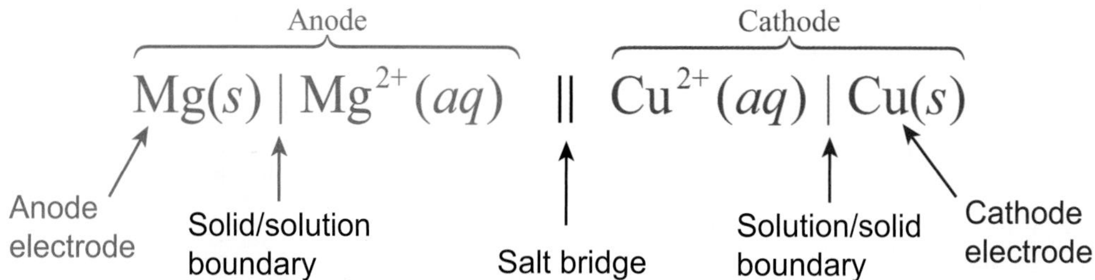

The double lines in the middle indicates the salt bridge. The anode or the oxidation half-cell, is always written on the left of the double lines and the cathode or reduction half-cell, is written to the right of the double lines.

The electrodes of the cell are written on the extreme left (anode) and the extreme right (cathode) of the expression. The two single vertical lines on the left and the right represent the boundaries between the solid and the solution.

Chemistry terms

salt bridge - an electrical connection between the oxidation and the reduction half-cells of an electrochemical cell.

Electromotive force

Electrical potential

The movement of electrical charge from one point to another requires that there is a difference in the potential energy between these points. Electric charge moves from a point of higher potential to a point of lower potential. The energy that is required in order to move a certain amount of charge depends on the amount of charge moved and the potential difference.

Electron movement

In an electrochemical cell the electrons that flow from the anode to the cathode are "pushed" by the force that is generated by the potential difference between the anode and the cathode.

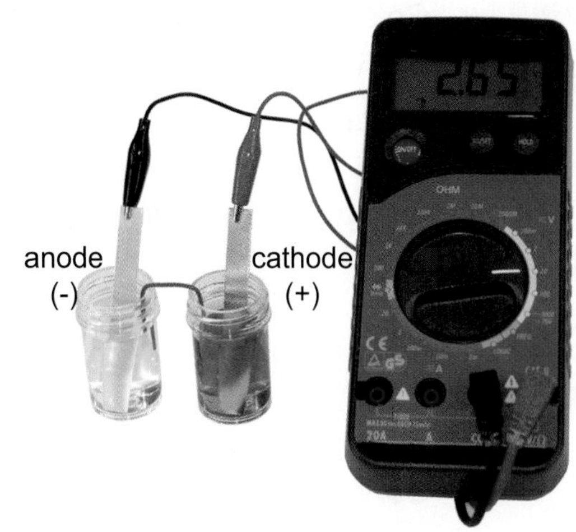

Electromotive force

This difference in potential between the anode and the cathode is called the **electromotive force (emf)** of the cell and is denoted by E_{cell}, also called **cell emf**.

Cell energy

Once we know the emf of an electrochemical cell we can calculate the amount of energy that the cell can provide when a certain amount of reactants are consumed. The energy is given by:

$$Work_{cell} = n\,F\,E_{cell}$$

F is the Faraday constant which is equal to 96,500 C and n is the number of moles of electrons that are produced in the reaction.

Voltaic cells generate energy

If we are given a certain cell reaction we can calculate the electrical energy that the cell can provide when a certain amount of reactants are consumed. All we need to do is determine the number of moles of electrons that are released by the anode as the electrode is consumed. We then multiply the number of moles with Faraday's constant and the voltage of the cell E_{cell}.

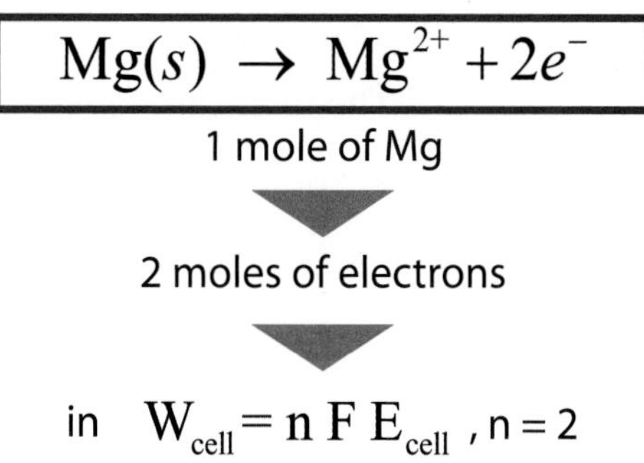

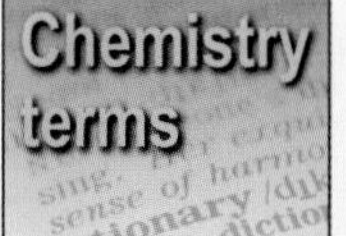

electromotive force (emf) - the difference in the electrical potential between the anode and the cathode of an electrochemical cell.

cell-emf - abbreviation for electromotive force of a cell.

Cell voltage and energy

The cell emf, E_{cell}, gives us a measure of the driving force of the cell reaction. As we know, the cell reaction is the sum of two half-reactions: the oxidation reaction which happens at the anode and the reduction reaction which happens at the cathode.

The Mg/Cu cell has E_{cell} = 2.71 Volts.
The cell reaction is: $Mg(s) + Cu^{2+}(aq) \rightarrow Mg^{2+}(aq) + Cu(s)$
Calculate the total energy that can be obtained from this cell if 2.43 g of Mg are consumed in the reaction.

Given: *The cell reaction, the emf voltage and the amount of reactant consumed*

Relationships: *The fundamental relationships and their contributions are:*

- *The equation that relates the number of mol of electrons and the cell voltage with the energy released*
  $W_{cell} = n\,F\,E_{cell}$
- *The half-reactions are:* $Mg(s) \rightarrow Mg^{2+}(aq) + 2e^-$ *and* $Cu^{2+}(aq) + 2e^- \rightarrow Cu(s)$

Solve: *From the half reactions we see that 1 mole of Mg generates 2 mole of electrons.*

- *One mole of Mg weighs 24.3 g.*
- *Therefore, 2.43 g of Mg corresponds to 0.1 mole of Mg.*
- *0.1 mol of Mg corresponds to 0.2 mole of electrons*
- *and so we have n = 0.2*
- *Substitute into the energy equation:*
  $W_{cell} = 0.2 \times 96{,}500 \times 1.10 = 21{,}230$ *Joules.*

Answer: *2.43 g of Mg release 21,230 Joules*

The total cell voltage (E_{cell}) is a combination of the potential at each half-cell.

We define E_{cell} as

$$E_{cell} = \text{reduction potential} + \text{oxidation potential}$$

$$E_{cell} = E_{reduction} + E_{oxidation}$$

For example if the voltage associated with the reduction half-reaction

$$Mg^{2+}(aq) + 2e^- \rightarrow Mg(s) \text{ is -2.37 Volts,}$$

then the voltage associated with the reverse reaction

$$Mg(s) \rightarrow Mg^{2+}(aq) + 2e^- \text{ is } -(-2.37) = +2.37 \text{ Volts.}$$

Standard reduction potentials

Since $E_{oxidation} = -E_{reduction}$, chemists have decided for convenience to only keep track of the $E_{reduction}$ voltages. Since the rate of chemical reactions depends on conditions such as temperature, pressure and concentration the reduction potentials are tabulated for defined or standard conditions. These standard conditions are:

Temperature:	25°C
Pressure:	1 atmosphere (for gases)
Concentration:	1 M (for solutions)

Standard reduction potential

The cell voltage under these conditions is called the **standard reduction potential** and it is labeled as E^o_{cell}. Since it is not possible to measure the potential of single electrodes the standard reduction potentials are given with respect to a well defined reference half-cell. The reference used is the hydrogen half-cell which has the half reaction: $2H^+(aq) + 2e^- \rightarrow H_2(g)$

The standard potential assigned to this half-reaction is 0 Volts and the potential of every other cell is measured with respect to it. The standard reduction potentials are given on Table 15.4.

Using the standard reduction potentials calculate the cell voltage (E^o_{cell}) of the cell Zn(s) | Zn^{2+}(aq) || Cu^{2+}(aq) | Cu(s)

Asked: *Find E^o_{cell} of the reaction* $Zn(s) + Cu^{2+}(aq) \rightarrow Zn^{2+}(aq) + Cu(s)$

Given: *The cell reaction and the standard reduction potentials.*

Relationships: *The half-reactions and their associated standard potentials are:*

- *The standard potential of the reduction half reaction $Zn^{2+}(aq) + 2e^- \rightarrow Zn(s)$ is -0.76V. Therefore the corresponding oxidation reaction $Zn(s) \rightarrow Zn^{2+}(aq) + 2e^-$ is -(-0.76V) = 0.76V $E^o_{oxidation} = 0.76V$*
- *$Cu^{2+}(aq) + 2e^- \rightarrow Cu(s)$: $E^o_{reduction} = 0.34V$*
- *$E^o_{cell} = E^o_{reduction} + E^o_{oxidation}$*

Solve: *Substitute the values for $E^o_{oxidation} = 0.76V$ and $E^o_{reduction} = 0.34V$*

- *$E^o_{cell} = 0.34\ V + 0.76\ V = 1.10\ V$*

Answer: *The cell voltage is 1.10 Volts*

standard reduction potential - the potential of a cell measured under standard conditions of temperature, pressure and concentration.

Standard reduction potentials

TABLE 15.4. Standard Reduction Potentials

Reduction half reaction	E° (V)	
$Li^+(aq) + e^- \rightarrow Li(s)$	-3.04	Strong reducing agent, strong tendency to be oxidized, strong tendency to give up electrons, weak oxidizing agent. Tendency to occur in the reverse direction
$Na^+(aq) + e^- \rightarrow Na(s)$	-2.71	
$Mg^{2+}(aq) + 2e^- \rightarrow Mg(s)$	-2.38	
$Al^{3+}(aq) + 3e^- \rightarrow Al(s)$	-1.66	
$2H_2O(l) + 2e^- \rightarrow H_2(g) + 2OH^-(aq)$	-0.83	
$Zn^{2+}(aq) + 2e^- \rightarrow Zn(s)$	-0.76	
$Cr^{2+}(aq) + 2e^- \rightarrow Cr(s)$	-0.74	
$Fe^{2+}(aq) + 2e^- \rightarrow Fe(s)$	-0.41	
$Ni^{2+}(aq) + 2e^- \rightarrow Ni(s)$	-0.23	
$Sn^{2+}(aq) + 2e^- \rightarrow Sn(s)$	-0.14	
$Pb^{2+}(aq) + 2e^- \rightarrow Pb(s)$	-0.13	
$Fe^{3+}(aq) + 3e^- \rightarrow Fe(s)$	-0.04	
$2H^+(aq) + 2e^- \rightarrow H_2(g)$	0	Reference half cell. Assigned zero potential
$Cu^{2+}(aq) + 2e^- \rightarrow Cu(s)$	0.34	
$O_2(g) + 2H_2O(l) + 4e^- \rightarrow 4OH^-$	0.40	
$Cu^+(aq) + e^- \rightarrow Cu(s)$	0.52	
$Ag^+(aq) + e^- \rightarrow Ag(s)$	0.80	
$ClO_2(g) + e^- \rightarrow ClO_2^-(aq)$	0.95	
$O_2(g) + 4H^+(aq) + 4e^- \rightarrow 2H_2O(l)$	1.23	
$Cl_2(g) + e^- \rightarrow 2Cl^-(aq)$	1.36	
$PbO_2(s) + 4H^+(aq) + 2e^- \rightarrow Pb^{2+} + 2H_2O(l)$	1.46	
$Au^{3+}(aq) + 3e^- \rightarrow Au(s)$	1.50	
$H_2O_2(aq) + 2H+(aq) + 2e- \rightarrow 2H_2O(l)$	1.78	
$F_2(g) + 2e^- \rightarrow 2F^-(aq)$	2.87	Strong oxidizing agent, strong tendency to be reduced. Strong tendency to attract electrons, weak reducing agent. Tendency to occur in the forward direction

Reaction spontaneity

ADVANCED AP

Spontaneous and non-spontaneous reactions

Using the standard potentials given on Table 15.4, we can calculate the cell emf (E^o_{cell}) of any redox reaction. The sign of E^o_{cell} indicates if the reaction is **spontaneous** or **non-spontaneous**. A spontaneous reaction is one that happens in the indicated direction, reactants to products. A non spontaneous reaction is one that does not proceed in the indicated forward direction.

A positive E^o_{cell} indicates spontaneous - product favored - reaction. A negative E^o_{cell} indicates a non spontaneous - reactant favored - reaction.

$E^o_{cell} > 0$: Spontaneous reaction.

$E^o_{cell} < 0$: Non spontaneous reaction.

Determine if the reaction $Zn(s) + Ni^{2+}(aq) \rightarrow Zn^{2+}(aq) + Ni(s)$ is spontaneous under standard conditions.

Given: *The reaction and the table of standard reduction potentials*

Relationships: *The half-cells and their half-cell potentials are:*

$$Zn(s) \rightarrow Zn^{2+}(aq) + 2e^- \quad E^0_{ox} = +0.76\ V$$

$$Ni^{2+}(aq) + 2e^- \rightarrow Ni(s) \quad E^0_{red} = -0.23\ V$$

Solve: *Now we need to calculate the total cell potential.*

$$Zn(s) \rightarrow Zn^{2+}(aq) + \cancel{2e^-} \quad E^0 = +0.76\ V$$

$$Ni^{2+}(aq) + \cancel{2e^-} \rightarrow Ni(s) \quad E^0 = -0.23\ V$$

By adding these two equations the total cell potential is:

$$E^0_{cell} = E^0_{ox} + E^0_{red} = -0.23\ V + 0.76\ V = +0.53$$

Answer: *Since the total cell voltage is a positive number, the reaction is spontaneous and proceeds as indicated.*

The reactions at the bottom of Table 15.4 ($F_2(g) + 2e^- \rightarrow 2F^-(aq)$) have a tendency to occur in the forward direction. The reactions at the top of Table 15.4 ($Li^+(aq) + e^- \rightarrow Li(s)$) have a tendency to occur in the reverse direction.

The relative position of the half-reactions on the table tells us if the redox reaction is spontaneous or not. Any reduction half-reaction will be spontaneous if it is paired with reverse of the half-reaction above it on Table 15.4.

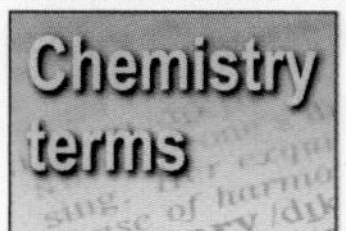

spontaneous - a reaction that occurs in the indicated direction without any energy input.

non spontaneous - a reaction that does not occur in the indicated direction.

Nernst equation

ADVANCED AP

Batteries run out of energy

The battery is a voltaic cell. The voltage of a battery is related to the emf of the battery cell. We are all familiar with the fact that batteries run out. Why does this happen? Well, the energy stored in the battery runs out. Or to say it another way, the chemical energy stored in the battery runs out. Since the chemical energy is released by redox reactions something happens to these redox reactions inside the battery as time goes on.

As the redox reactions proceed in an electrochemical cell, the relationship between reactants and products changes. Reactants are consumed by the reaction and products are generated. This implies that the concentration of every substance in the cell changes as reactions proceed. As the reactants decrease, their concentration decreases and equivalently the concentration of the products increases. We expect that the voltage output of a battery depends on the concentration of reactants and products.

The relationship that describes this dependence is the **Nernst Equation** and is given by:

$$E_{cell} = E^0_{cell} - \frac{0.0592}{n} \log(Q)$$

Standard cell potential (E^0_{cell})

number of moles (n)

$$Q = \frac{\text{product concentration}}{\text{reactant concentration}}$$

The parameter **n** is the number of moles of electrons generated by the reaction and Q is the ratio of product to reactant concentrations.

Under standard conditions, $Q = 1$ and since $\log(1) = 0$, $E_{cell} = E^0_{cell}$.

Using a battery

As a battery is used the amount of concentration of the products is increased and the concentration of reactants is decreased. This makes Q greater than 1 and $E_{cell} < E^0_{cell}$. The battery voltage is decreased: The battery is running out.

Recharging a battery

When a battery is recharged the concentration of reactants increases and the concentration of products decreases. This decreases Q and drives E_{cell} towards E^0_{cell}. The battery is then fully recharged when $Q = 1$ and $E_{cell} = E^0_{cell}$.

Battery Discharging	Battery Charging
[reactants]: decrease [products]: increase	[reactants]: increase [products]: decrease
Q increases log(Q) increases E_{cell} decreases	Q decreases log(Q) decreases E_{cell} increases

Chemistry terms

Nernst equation - the mathematical equation that relates the electrical potential of a cell to the standard reduction potential and the state of the reaction as given by the concentration of reactants and products.

Electrolytic cells

Voltaic vs. electrolytic cells

In the voltaic cell, spontaneous redox reactions are used to generate electrical current. In an **electrolytic cell**, externally applied electricity is used to drive a non spontaneous redox reaction.

The result of the induced chemical reaction in an electrolytic cell is called **electrolysis**.

When a battery is used it acts as a voltaic cell: chemistry generates electrical current. When the battery is recharged we are switching the type of cell from voltaic to electrolytic.

A rechargeable battery is both a voltaic cell and an electrolytic cell.

Many useful substances, including aluminum and chlorine, are commercially produced using electrolysis.

Water decomposition

The decomposition of water into its constituent elements is a non spontaneous reaction. This means that the reaction

$$2H_2O(l) \rightarrow 2H_2(g) + O_2(g)$$

does not proceed without the addition of energy.

The oxidation number of hydrogen in H_2O is +1 and the oxidation number of hydrogen in H_2 is 0. Therefore H_2O is reduced to H_2. Similarly, since the oxidation number of oxygen in H_2O is -2, H_2O is oxidized to O_2.

This is a redox reaction for which the oxidation and reduction half reactions and the associated reduction potentials are:

Reduction:	$2H_2O + 2e^- \rightarrow H_2 + 2OH^-$	$E^o_{red} = -0.83V$
Oxidation:	$2H_2O \rightarrow O_2 + 4H^+ + 4e^-$	$E^o_{ox} = -1.23V$

The cell potential of this redox reaction is E^o_{cell} = -2.06 V. Since this is a negative number, the reaction is non spontaneous. We must provide energy in order to break up water into hydrogen and oxygen.

Hydrogen generation

Hydrogen is currently considered as an energy source for the future. Hydrogen gas does not exist as a free gas on earth. Hydrogen has to be produced from water, fossil fuels or biomass. The production of hydrogen from water is the cleanest but most expensive way of producing hydrogen. The production of hydrogen corresponds to energy storage in hydrogen. The energy is extracted when hydrogen gas is oxidized to form water.

Chemistry terms

electrolytic cell - an electrochemical cell in which chemical reactions result from the application of electrical current.

electrolysis - the result of a chemical reaction in an electrolytic cell.

Catalytic converters: Chemistry for a cleaner environment

Catalytic converter is a device in the exhaust system of automobiles. It is located between the engine and the muffler and it changes harmful pollutants released by the engine into less harmful compounds.

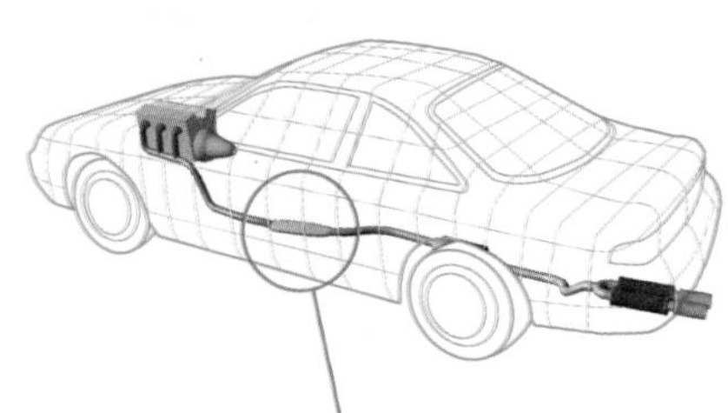

Catalytic Converter

Catalytic converters play a major role in the reduction of air pollution in populated areas. Automobiles produce three major types of air pollutants:

1. nitrogen oxides (NO and NO_2, also labeled NOx),
2. uncombusted hydrocarbons (uncombustable fuel) and
3. considerable amounts of carbon monoxide (CO).

Automobile engineers have searched for ways to reduce these harmful pollutants with careful engine design, but they have found that under normal driving conditions it is impossible to decrease them to an acceptable level. The catalytic converter was designed to reduce the quantity of these pollutants in the exhaust before they are released into the air. Catalytic converters are very effective in removing the harmful gases! Approximately 96% of the uncombusted hydrocarbons and carbon monoxide are converted to CO_2 and H_2O, and approximately 76% of the nitrous oxides (NOx) are reduced to harmless nitrogen gas (N_2). You might wonder why automobile exhaust is such a source of pollution in our major cities if our catalytic converters work so well? This is because there are so many people driving! Our population is soaring and this is increasing the overall amount of pollution.

The warm temperatures and the mountains surrounding Los Angeles trap the harmful gases in the valley, causing them to reach dangerous levels. Los Angeles was the first city in the U.S. to use catalytic converters to reduce smog.

Catalytic converters selectively oxidize and reduce dangerous gases in our exhaust.

As part of the exhaust system of a car, catalytic converters perform two very distinct functions:

1. Reduce nitrogen oxides to nitrogen gas.
   NO and NO2 $\rightarrow$ N2(g) (Reduction)
2. Oxidize carbon monoxide and uncombusted hydrocarbon fragments to CO_2 and H_2O.
   CO and CxHy $\rightarrow$ $CO_2 + H_2O$ (Oxidation)

In effect, catalytic converters work like antioxidants in our cars.

In the catalytic converter nitrogen in NO or NO_2 is reduced because it's oxidation number goes from being +2 or +4 to zero in N_2 gas.

Carbon in CO or methane CH_4, which is a type of hydrocarbon, is oxidized because it goes from having an oxidation of +2 or -4 to a +4 oxidation state in CO_2.

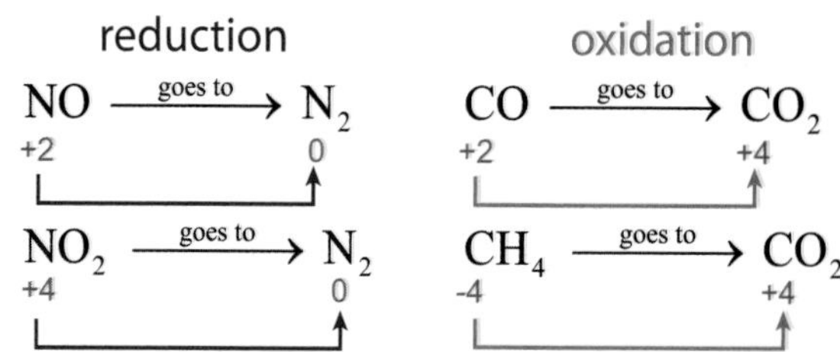

These redox reactions take place that the surfaces of metals such as platinum, Pt, palladium, Pd, and rhodium, Rh. In the periodic table these are called transition metals. The role of these metals is to increase the rate of the redox reactions and so they are called catalysts.

The processes of oxidation and reduction require two different catalysts in the same way that the electrodes of electrochemical cells were made of different materials. These catalysts are heterogeneous, because the catalyst is in the solid phase and the reacting molecules are in the gas phase.

Companies that make catalytic converters mix these metals with metal oxides such as CuO and Cr_2O_3 and other materials to make a "wash coat." The wash coat decreases the amount required of the very expensive transition metal. The wash coat is sprayed onto a brick like ceramic substrate that looks like a sponge or honey comb structure. This acts as a support structure for the catalyst. This ceramic structure is engineered to increase the surface area of the catalyst, so it is very porous.

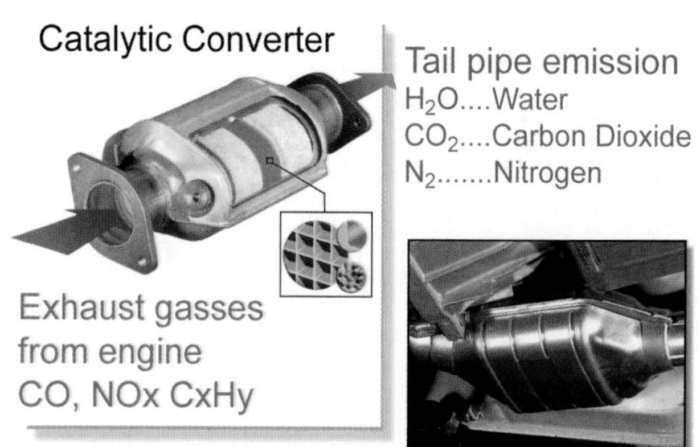

By spraying the noble metals onto this surface more of the exhaust comes into contact with the catalyst. This increased contact of gas molecules is very important in the catalytic process, because the gases are moving through the exhaust very quickly.

Typically these chemical reactions facilitated by the catalyst occur in four steps.

1. Adsorption and activation of the reactant molecules, such as CO, or NO_2. Adsorption is the binding of molecules to a surface.
2. Migration of the adsorbed molecules along the surface.
3. Reaction among the adsorbed substances. The catalyst promotes dissociation (breaking apart) of the molecules such as NO or CO. For NOx the NO bonds weaken and this allows for the nitrogen atoms to bond and form N_2. For CO and CxHy the CO or CH bonds are weakened and oxygen can then bond to form CO_2 and H_2O molecules.
4. Escape or desorption of the products.

Preparation of these catalysts has a tremendous effect on their properties. The development of catalytic converters is always being improved and is a large area of research in the automotive and environmental areas of science. Without catalytic converters our air, particularly in urban areas, would be much more polluted.

Chapter 15 Review

Vocabulary

Match each word to the sentence where it best fits.

Section 15.1

electrochemistry	antioxidants
free radicals	rust

1. Molecules that have lost electrons are called ____________________.
2. The science that combines chemistry and the flow of electricity is called __________________.
3. Iron _________ results from a process called oxidation.
4. ________________ are molecules that protect us from the harmful effects of free radicals.

Section 15.2

electron charge	electroneutrality
coulomb	voltage
proton charge	volt
electrical current	resistance
Ampere	ohm
electrical potential difference	Ohm's law
	electron charge

5. The unit of electrical charge is called __________.
6. The charge of the proton is equal but opposite to the charge of the _____________.
7. The ________________________ is -1.602×10^{-19} coulomb.
8. The movement of electrical charge gives ______________________.
9. The ________________________ is $+1.602\times10^{-19}$ coulomb.
10. Electrical current is measured in _____________.
11. The property of __________________ describes a balance between the positive and the negative charges.
12. The difference in electrical potential is called ____________________.
13. Voltage is measured in ___________.
14. Electrical charges experience a _____________ as they move in a conductor.
15. Electrical resistance is measured in ohms.
16. ________________ describes the relationship between current, voltage and resistance.

Section 15.3

redox	oxidation number
oxidized	oxidizing agent
reduced	reducing agent
charge conservation	

17. An element is _____________ when it loses electrons.
18. An element is _____________ when it gains electrons.
19. Chemical reactions in which elements are oxidized and reduced are called __________ reactions.
20. Redox reactions obey the principle of ____________________________.
21. The charge state of an atom is described by the ______________________.
22. An element that oxidizes another element is called the ____________________.
23. An element that reduces another element is called the ____________________.

Section 15.4

reduction half	balance charge
balance mass	half reactions
	oxidation half

24. Balanced redox equations must _______________ and _________________.
25. Redox reactions can be separated into two ___________________.
26. One of the half reactions is called the ____________________ and the other is called the ___________________.

Section 15.5

electrochemical cell	electromotive force
electrode	cell emf
electrolyte	standard reduction potential
anode	spontaneous
cathode	non spontaneous
voltaic cell	Nernst equation
galvanic	electrolysis
electrolytic cell	hydrogen reference half cell
salt bridge	

27. A device in which redox reactions produce electrical current is called a _________________.
28. A device in which electrical energy produces redox reactions is called a _________________.
29. Voltaic and electrolytic cells are called ______________________.
30. The part of a voltaic cell on which oxidation occurs is called the ________________.
31. The part of a voltaic cell on which reduction occurs is called the ________________.
32. The cell electrodes are immersed in a solution called _____________.
33. Redox reactions occur at the interface between the electrolyte and the ________________.
34. A battery is a _________________.
35. A voltaic cell is also called a ____________ cell.
36. A _____________ is the connection between the oxidation and the reduction half cells.
37. __________ means cell electromotive force.
38. The electrical potential difference between the anode and the cathode is called ______________________.
39. The cell voltage under standard conditions of temperature, pressure and concentration is called ______________________________.
40. The standard reduction potentials are measured with respect to the _________________________.
41. When the cell voltage is greater than zero the reaction is _________________.
42. When the cell voltage is less than zero the reaction is ______________________.
43. The ____________________ gives the cell potential under non standard conditions.
44. The result of the induced chemical reaction in an electrolytic cell is called _______________.

Conceptual Questions

Section 15.1

45. What are free radicals?
46. Why are free radicals harmful?
47. How do antioxidants protect us from free radicals?
48. Describe what is the science of electrochemistry.
49. Give four examples of devices or natural process that are based on electrochemistry.

Section 15.2

50. What is electrical current?

Chapter 15 Review.

51. What is electroneutrality and how does it affect electrical current flow?

52. What is voltage and how is it related to electrical energy?

53. If the voltage increases what happens to the current flow? Does it increase, decrease or stay the same and why?

54. If the electrical resistance increases what happens to the current if the voltage stays the same?

55. If the electrical resistance increases by a factor of 2 and the current decreases by a factor of 2 what happens to the voltage? Does it increase decrease or stay the same? Explain your answer.

Section 15.3

56. When we say that an element is oxidized what do we mean?

57. What happens when an element is reduced?

58. What is the oxidation number of a free element.

59. How is the oxidation number of an ion related to the charge of the ion?

60. What is a redox reaction?

61. Describe what happens to charge conservation during redox reactions.

62. What is the sum of the oxidation numbers of all individual elements in a neutral molecule?

Section 15.4

63. Describe the basic steps that we have to follow when balancing redox reaction equations.

64. What are half reactions and how are they used when balancing redox reactions.

Section 15.5

65. What are the three major components of a voltaic cell?

66. Describe the role of the electrodes in an electrochemical cell.

67. Describe the differences between a voltaic and an electrolytic cell.

68. Describe the role of the salt bridge in an electrochemical cell.

69. What is electrolysis and how is it used to extract hydrogen and oxygen from water?

70. What is cell voltage and how is it related to electromotive force?

71. Describe how a voltaic cell generates energy.

72. What are standard reduction potentials and how are they measured?

73. Explain how the reaction spontaneity is related to cell voltage.

74. Describe how changes in the concentration of reactants and products affects the voltage of a voltaic cell.

Quantitative Problems

Section 15.2

75. A voltage of 10 V is applied across a resistor that has a resistance of 5 Ω. What is the amount of current that flows through the resistor?

76. If we want to move a charge of 1 coulomb from a potential of 1 V to a potential of 2 V, how much energy must we provide?

Section 15.3

77. Find the oxidation number for each element in the following compounds:

    a. CO_2

    b. HCl

    c. $NaCl$

    d. HNO_3

78. What is the oxidation number of each atom in the following ions?

    a. ClO_2^-

    b. Ca^{2+}

c. ClO_4^-
d. HNO_3

79. What is the oxidation number of carbon in each of the following compounds?
a. CO
b. CH_4
c. C_3H_4
d. CO_3^{2-}

80. What is the oxidation number of nitrogen in the following compounds?
a. N_2O
b. N_2
c. NO_2^-

81. For each of the following reactions identify the elements that are oxidized and the elements that are reduced.
d. $4Fe(s) + 3O_2(g) \rightarrow 2Fe_2O_3(s)$
e. $Fe + CuSO_4 \rightarrow FeSO_4 + Cu$
f. $2H_2O(l) \rightarrow 2H_2(g) + O_2(g)$
g. $2Na(s) + 2H_2O(l) \rightarrow 2NaOH(aq) + H_2(g)$

82. For each reaction of the previous problem identify the oxidizing and the reducing agent.

Section 15.4

83. Determine which of the following reactions are redox and which are not.
a. $NaOH(l) + HNO_3(l) \rightarrow NaNO_3(l) + 2H_2O(l)$
b. $4Li(s) + O_2(g) \rightarrow 2Li_2O(s)$
c. $2NO(g) + 5H_2(g) \rightarrow 2NH_3(g) + 2H_2O(g)$
d. $2Al(s) + 3H2SO_4(l) \rightarrow Al2(SO4)_3(l) + 3H_2O(g)$

84. Balance the following reactions?
a. $CH_4(g) + O_2(g) \rightarrow CO_2(g) + H_2O$
b. $Cu_2S(s) \rightarrow Cu(l) + SO_2(g)$
c. $NaOH + HCl \rightarrow NaCl + H_2O$
d. $NaOH + Cl_2 \rightarrow NaCl + NaClO + H_2O$

85. Balance the following redox reactions using the oxidation and reduction half reactions.
a. $Al(s) + Fe^{2+}(aq) \rightarrow Al^{3+}(aq) + Fe(s)$
b. $Zn(s) + Sn^{2+}(aq) \rightarrow Zn^{2+}(aq) + Sn(s)$

86. Balance the following redox reaction equations using the oxidation and reduction half reactions. All reactions occur in acidic solutions.
a. $Cl_2(g) + S_2O_3^{2-}(aq) \rightarrow Cl^-(aq) + SO_4^{2-}(g)$
b. $Cl_2(g) \rightarrow HClO(aq) + Cl^-(aq)$
c. $Fe^{2+}(aq) + Cr_2O_7^{2-}(aq) \rightarrow Fe^{3+}(aq) + Cr^{3+}(aq)$

87. Balance the following redox reaction equations using the oxidation and reduction half reactions. All reactions occur in basic solutions.
a. $O_3(aq) + Br^-(aq) \rightarrow O_2(g) + BrO_3^-(aq)$
b. $MnO_4^-(aq) + S^{2-}(aq) \rightarrow S(s) + MnO_2(s)$

Section 15.5

88. Make a sketch of the electrochemical cell with the following reaction equation:
$Mg(s) + Ni^{2+} \rightarrow Mg^{2+}(aq) + Ni(s)$
Label the anode, the cathode. Use a KCl salt bridge and label the direction of movement for the K^+ and Cl^- ions. Label the direction of the electron flow thought an external path.

89. Write the cathode and the anode half reactions of the following cells.
a. $Ni(s) | Ni^{2+} || Ag^{2+}(aq) | Ag(s)$
b. $Cr(s) | Cr^{2+} || Au^{2+}(aq) | Au(s)$
c. $Fe(s) | Fe^{2+} || Ag^{2+}(aq) | Ag(s)$

90. Calculate the cell voltage for all the cells of the previous problem under standard conditions.

91. An electrochemical cell has the following reaction occurring at the anode.
$Mg(s) \rightarrow Mg^{2+}(aq) + 2e^-$
Which of the following reduction reactions occurring at the cathode will produce the highest cell voltage under standard conditions?
a. $Au^{3+}(aq) + 3e^- \rightarrow Au(s)$
b. $Fe^{3+}(aq) + 3e^- \rightarrow Fe(s)$
c. $Ni^{2+}(aq) + 2e^- \rightarrow Ni(s)$
d. $Cu^{2+}(aq) + 2e^- \rightarrow Cu(s)$

CHAPTER 16
Solids and Liquids

How are solids and liquids different from gases?
How are solids different than liquids? How are they similar?
What makes some solids different from each other?
What makes some liquids different from each other?

At 1,500°C steel flows like thin syrup, and can be poured from the furnace into molds of any shape. Liquid steel has no *strength* at all, just like liquid water. However, at room temperature, steel is hard, tough, and extremely strong. A single strand of wire the thickness of a thread can easily support your whole weight.

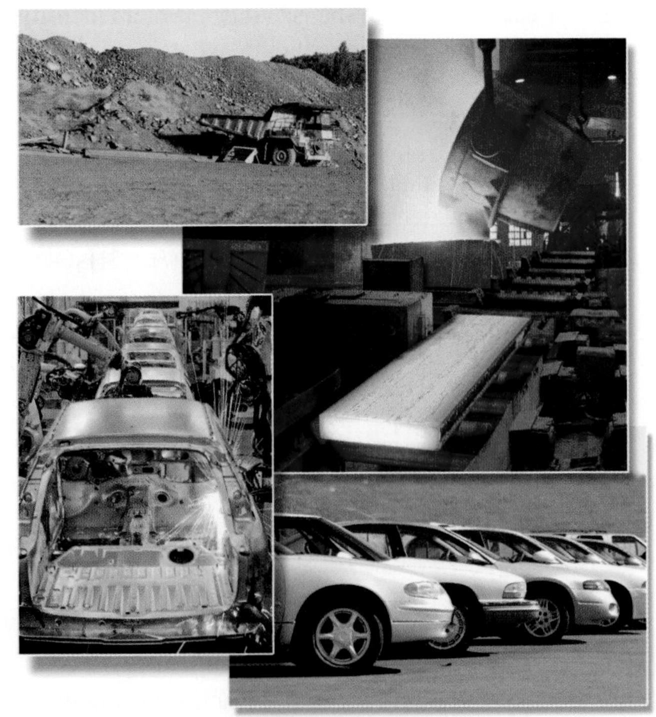

You might be surprised to know that steel is not one substance, but many. All steel contains mostly iron, and a little carbon but there are many variations. Car frames are made of #1010 carbon steel which is 99% iron with 0.1% carbon and a few other elements. Silverware is made of #304 stainless steel, also known as 18/8 because it contains 18% chromium and 8% nickel. In the last year you probably came in contact with more than 100 types of steel. Each one is a mixture (or alloy) with its own specific chemistry matched to the uses for which the steel is intended.

Melt metal in your hands

Melting steel in your hand would be foolish and painful. However, melting chocolate or butter in your hand is probably something you have already done and felt no ill effects from at all! The difference is the melting point, of course. Most metals melt well over 1,000°C however, there are some that melt at much lower temperatures. Pure gallium metal has a very low melting point, an incredible 29.8°C (85.6°F). *This is lower than the temperature of your body.*

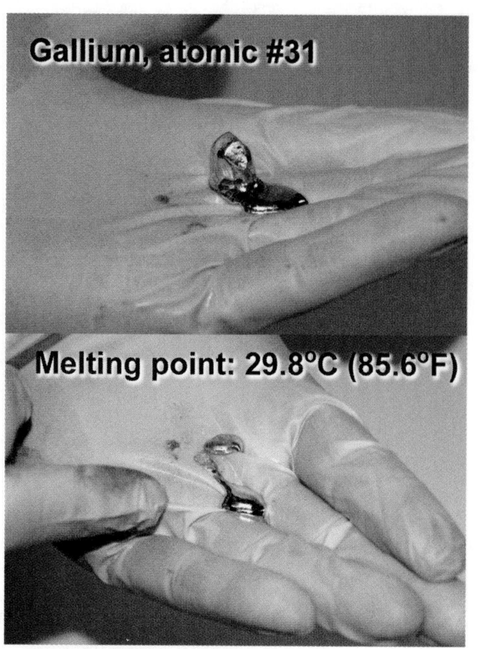

Gallium is a heavy metal, and like mercury, should not be handled with bare hands. Here is what we used.

1. Safety glasses and gloves
2. One gram of pure gallium
3. A petri dish for catching any excess gallium.

First, put the solid gallium in the petri dish. Next pour some hot water over it. Watch what happens as the metal warms to the water temperature. Poke it with your gloved finger. How does the solid gallium deform? How does it deform when it melts?

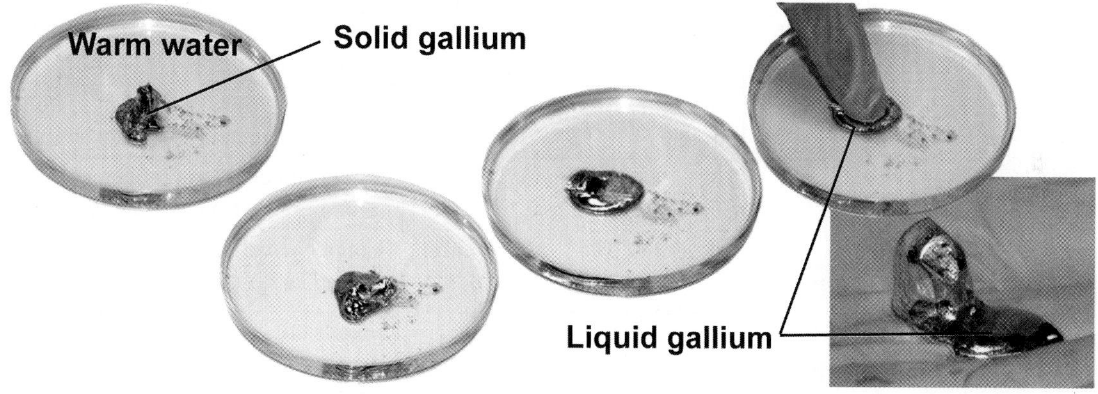

How does the gallium change under the force of gravity? Now hold the gallium in your hand over the tray. How does the gallium deform if you poke it? How does it deform under the force of gravity? Finally, add a few ice cubes to the petri dish and the gallium freezes (solidifies).

(1) The solid gallium should hold its shape under the forces of gravity and your finger.

(2) The liquid gallium should deform under external forces, including gravity and your finger.

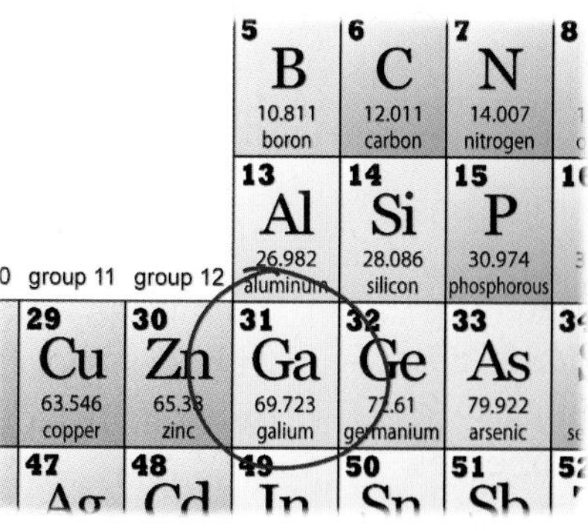

16.1 The Properties of Solids

Properties of some typical solids

How would you describe the differences between rubber, glass, and copper? How about the differences between aluminum and copper, or between soft and hard plastic? You might think of aspects such as color, hardness, flexibility, clarity, and density. These are some of the many physical properties that distinguish one solid from another.

These are all solids.
How are they different?
How are they similar?
Hard plastic
Soft plastic
Rubber
Glass
Copper
Aluminum

Examples of physical properties

Property	Description	Examples
Density	The amount of matter per unit volume	Steel has a density of 7.8 grams per cm^3 while polyethylene plastic has a density of 2 g/cm^3
Hardness	A harder material will scratch a softer material	Glass is harder than copper, but copper is harder than plastic
Elasticity	An elastic material will stretch and rebound without breaking.	Rubber is often very elastic and glass is hardly elastic at all.
Transparency	The ability for light to pass through.	Glass and plastic can be transparent while rubber and metals are typically opaque
Strength	The amount of force a material can withstand before breaking	Steel is much stronger than aluminum, and aluminum is stronger than rubber or plastic.
Electrical conductivity	The ability for electrical current to flow through a material.	Copper and aluminum are excellent conductors while glass, rubber and plastic are insulators.
Thermal conductivity	The ability for heat to flow through a material.	Metals are good heat conductors, rubber and plastics are moderate heat conductors.

The microscopic causes of physical properties

Macroscopic physical properties ultimately come from the microscopic world of molecules and chemistry. That connection is the subject of this chapter. How do the properties of molecules explain the observable physical properties of liquids and solids? The converse statement is even more interesting. *How can we affect the molecular structure and chemistry of materials to achieve the physical properties we desire?*

Material properties vs. object properties

The properties of a *material* are different from the properties of an object. For example, rubber is not that strong but a rubber rope as thick as your arm could easily support your weight. A steel wire the thickness of a heavy thread can achieve the same feat with far less matter. Steel as a material is intrinsically stronger than rubber. When we talk about *material properties* we need to separate from our thinking any considerations of shape, size or design of objects.

Density

Materials have a wide range of density

Solid materials have a wide range of densities. One of the densest metals is platinum with a density of 21.4 g/cm^3. Platinum is twice as dense as lead and almost three times as dense as steel. A ring made of platinum has three times as much mass as a ring of the exact same size made of steel. Rocks have lower density than metals, between 2.2 and 2.7 g/cm^3. As you might expect, the density of wood is less than rock, ranging from 0.4 to 0.6 g/m^3.

TABLE 16.1. Densities of some common materials

Material	Density (g/cm^3)	Material	Density (g/cm^3)
Platinum	21.4	Graphite	2.1
Lead	11.3	Polyethylene	2.0
Steel	7.8	Brick	1.6
Titanium	4.5	Rubber	1.2
Diamond	3.5	Liquid water	1.0
Aluminum	2.7	Ice	0.92
Glass	2.7	Oak (wood)	0.60
Calcite ($CaCO_3$)	2.7	Pine (wood)	0.44

Factors affecting density

The density of a substance is affected by two factors:

1. The mass of the molecules or atoms of which the substance is mode
2. How closely the atoms or molecules are packed together

The atomic weight of platinum is 195 g/mol and that of lead is 207 g/mol. Platinum has a lower atomic mass but nearly twice the density of lead. That tells us platinum atoms must be packed together more closely than lead atoms. In a more extreme case the molecular weight of polyethylene is more than 200,000 g/mol yet its density is much lower than platinum. The conclusion is that density is strongly dependent on how tightly atoms or molecules can be packed together.

The packing fraction

The atoms in most metals are packed as tightly as they can get. The *packing fraction* is the volume taken up by atoms compared to the total volume of a substance. The atoms in most metals are arranged in a hexagonal pattern that nests each layer into the adjacent layer for the highest possible packing fraction. This is the primary reason for the high density of metals compared to other substances. Polyethylene is a long chain molecule. It is not possible to pack large molecules efficiently therefore molecular substances tend to have low density.

packing fraction = 0.68%

packing fraction = 0.74%

Hardness

Hardness

The **hardness** of a material describes how easy it is to move molecules out of their place by scratching the surface with a relatively small force. Diamond is the hardest known substance. Diamond is hard because it is a network covalent substance. A diamond is like a single huge molecule because every carbon atom is bonded to four other carbon atoms. It is very difficult to deform a diamond at all. In fact, the only thing that can cut a diamond is *another diamond.* Diamond polishers use special disks, like fine sand paper, made of diamonds in order to polish bigger diamonds.

The Mohs scale of hardness

The **Mohs hardness scale** is a scale used for describing hardness. The Mohs scale goes from 1 to 10, with 10 being the hardest. The scale is based on the relative hardness of other materials. Your fingernail is a 2.5. If something is harder than your fingernail, then it can scratch your nail, and has a higher Mohs hardness. If a substance is softer than your fingernail, then your fingernail can scratch it! That substance would have a lower Mohs hardness. Since diamond is a 10, it can put a scratch in anything!

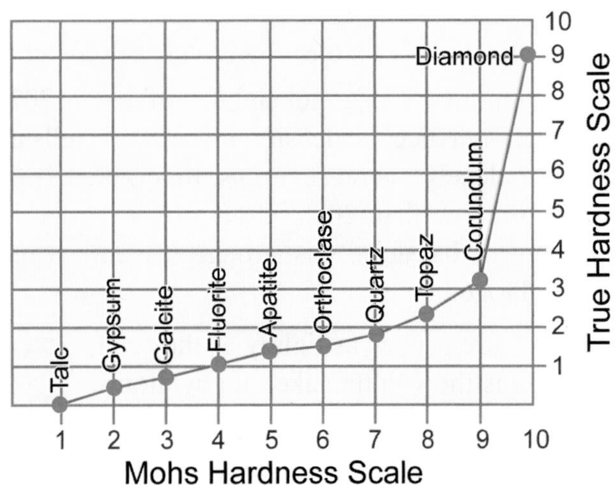

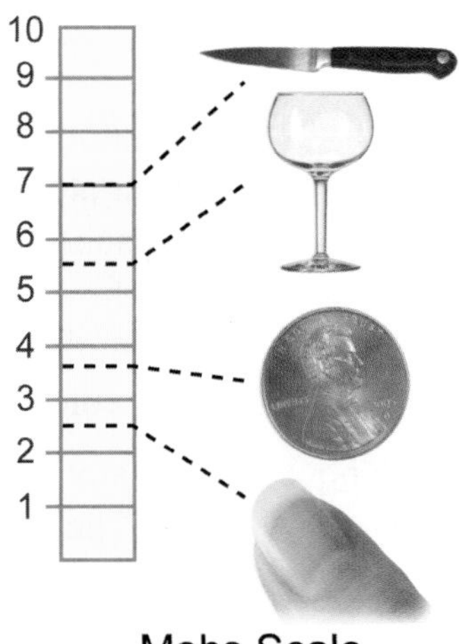

Chemistry terms

hardness - The ease of deforming a material by a small amount.

Mohs hardness scale - A hardness scale describing how easy it is to scratch one material with another.

Strength

The meaning of "strength"

The strength of a solid material really is about failure. The only way to know a material is strong enough is to push it until its limits, where it isn't strong enough. Strength has three primary aspects which are described by the following questions.

1. How much will a material deform under a given force?
2. Does a material stretch before breaking or does it snap suddenly?
3. How much force can a material ultimately withstand before it breaks?

All of these may be strongly affected by chemistry.

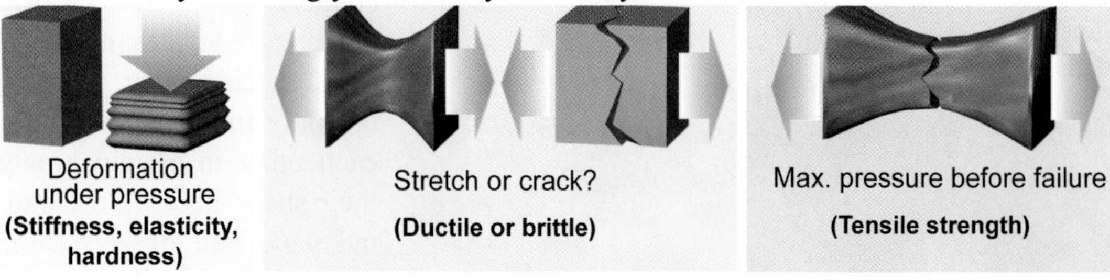

Deformation under pressure

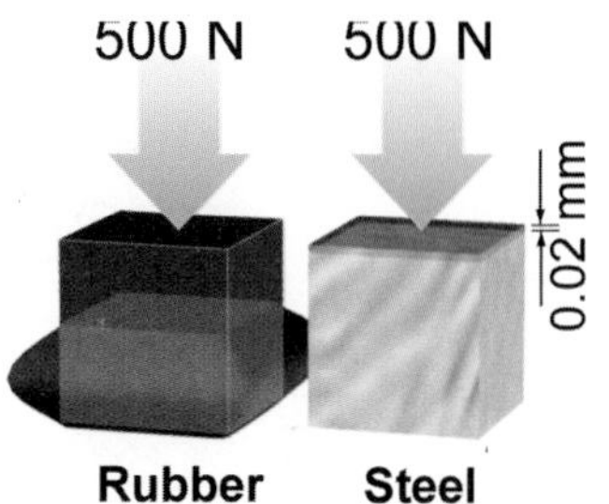

Elastic materials such as rubber and soft plastics can deform a great deal under a relatively small pressure. These are materials you think of as "soft." Metals deform much less for the same amount of pressure. For example, 500 N of force (about 100 lbs) on a 1 cm cube of rubber will compress the cube to 0.5 cm. The same pressure applied to a 1 cm cube of steel will deform the steel only 0.02 millimeters, less than the thickness of your hair!

Ductility and brittleness

A **ductile** material will bend or deform substantially before breaking. Think about stretching a roll of soft clay. The clay deforms a lot before it finally breaks. Metals are typically ductile. For example, gold is so ductile it can be hammered out into very thin sheets without breaking. A **brittle** material will break before it deforms much. Ceramic and glass tend to be brittle. Ionic solids, including most mineral crystals and rocks tend to be brittle materials.

Tensile strength

The tensile strength of a material is the maximum pressure it can sustain in tension before breaking. Tension is a straight pulling force, like a weight hanging from a wire. Most materials will stretch a certain amount, then snap.

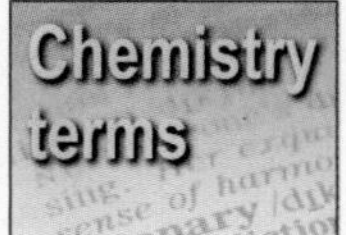

tensile strength - A measure of the energy needed to break a material in two.

brittle - Describes materials that require little energy to fracture.

ductile - Describes materials that require much energy to fracture.

16.2 The Microstructure of Solids

Molecules in a gas

In a gas, the molecules are free to go where they please. They fly around at very high speeds, colliding with each other and everything in their way. However, except for the brief instants of their collisions, molecules do not interact with each other.

Condensed matter includes liquids and solids

Solids and liquids are collectively called *condensed matter*. **Condensed matter** is made of molecules (or atoms) that are essentially touching each other. In a solid, the molecules are not only touching, but they are stuck tightly to each other in a rigid arrangement. Solids hold their shape because the molecules cannot move independently of each other. In a liquid, the molecules are still touching each other, but the connections are looser and more slippery than in solids. Liquid gallium flows because the atoms can slide around each other. It does not expand to fill space the way a gas would because the atoms are not energetic enough to completely break away from each other.

Competition between thermal agitation and attractive forces

Liquids and solids are "condensed" because the attractive intermolecular forces acting between molecules (or atoms) are stronger then the disruptive forces of thermal motion. In a solid, thermal agitation is much too low to overcome attractive forces. In a liquid, thermal agitation is almost enough, but not quite. As a liquid is heated to its boiling point however, thermal agitation wins and molecules escape to become gas.

TABLE 16.2. Properties of Solids, Liquids and Gases

	Solids	Liquids	Gases
Fixed Volume	Y	Y	N
Fixed Shape	Y	N	N

Chemistry terms

condensed matter - matter in which atoms or molecules are closely packed together and strongly interacting. Both liquids and solids are condensed matter.

Crystalline and amorphous solids

Two types of solids

Solids are solid because the molecules are bound in one of two possible *microstructures*. **Microstructure** means the detailed arrangement of molecules (or atoms) in a material. In a crystalline solid, the molecules have an ordered pattern, like bricks in a wall. In an amorphous solid, there is not a repeating pattern and molecules are randomly squeezed together without any long-range order. **Crystalline** solids include minerals and gemstones you think of as "crystals" but they also include metals and even some plastics. **Amorphous** solids include glasses and most plastics. Biological solids such as wood and bone have both crystalline and amorphous components and are mixtures of the two.

Crystals have ordered structures

A **crystal** is defined by its crystal structure. Crystal structures describe how the molecules or atoms bond to each other. They tell us how far apart neighboring atoms are, and at what angles the bonds are. You can see examples of natural crystals all around you. Look closely at an aluminum traffic pole, and you will notice "grains" of the aluminum metal. Each grain is a single crystal. Under a magnifying glass the salt in your salt shaker is in little cubic crystals!

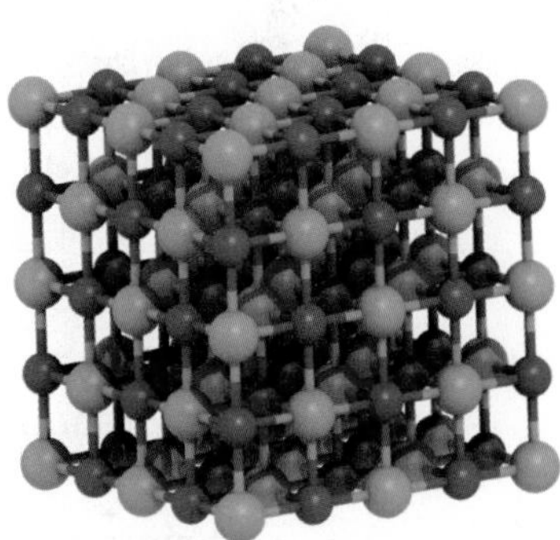

Crystal structure and shapes

The microscopic arrangement of molecules in a crystalline solid will often show up in the shape of an object. Snowflakes are a lovely example. Snowflakes have six-way symmetry because water molecules align in six-sided rings.

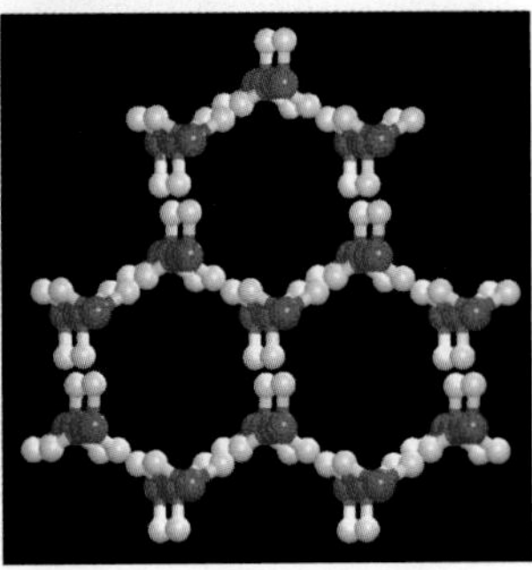

Chemistry terms

microstructure - the spatial arrangement of atoms and molecules in matter.

crystalline - a microstructure with a repeating, ordered pattern.

amorphous - a microstructure that does NOT have any ordered pattern.

crystal - a piece of crystalline matter in which the microstructure is uniform and continuous over the entire piece.

Glasses

Glasses have disordered structures

Glasses do not have an ordered structure. There is no "brick-like" order to the molecules inside a glass - they just bond to each other wherever they can. The bond lengths are all different, and the angles are all different too.

Given enough time, glasses will sag under their own weight. This slow deformation is common to all amorphous substances. They are *deforming*, since the lack of order means they can't hold their shape forever. Also, the higher the temperature, the faster the deformation.

Amorphous

On the molecular level, an amorphous solid is like an "immobile" liquid. There is a fixed shape and volume like a solid. But the molecules are loosely packed and disordered, like a liquid. The hotter an amorphous substance gets, the more it behaves like a liquid. Amorphous things do not 'melt' in the same way crystals do. Crystals melt at only one temperature. Amorphous materials just get softer as it gets hotter.

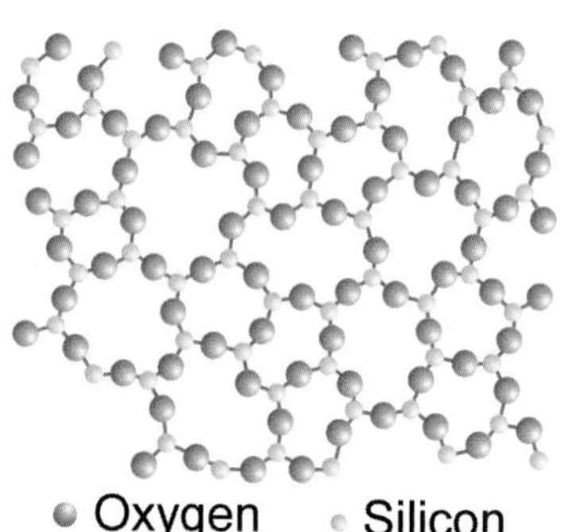

Metallic glasses

Most glasses we use everyday are made of sand, or silica. They are called *oxide glasses* because they are made mostly of silicon dioxide (silica). There are also some very special glasses called **metallic glasses**. By cooling a metal down very, very fast, you can freeze the atoms into a glass. These metallic glasses have amazing properties. They transmit energy very well. That's why the fanciest golf clubs have heads made out of metallic glass.

Chemistry terms

glass - a solid structure with no long-range order for bond lengths and angles.
amorphous - a structure that behaves like both a solid and a liquid.
metallic glass - a special form of glass made by cooling metal extremely quickly.

Crystal structures

Crystal Structures

Crystals can look quite different from each other. Some are square (table salt), some are rectangular, some are hexagonal (quartz, snowflakes), and some have other geometric shapes. The similarity is a well defined **crystal structure**. A crystal structure describes exactly the 3 dimensional repeating pattern of atoms and molecules.

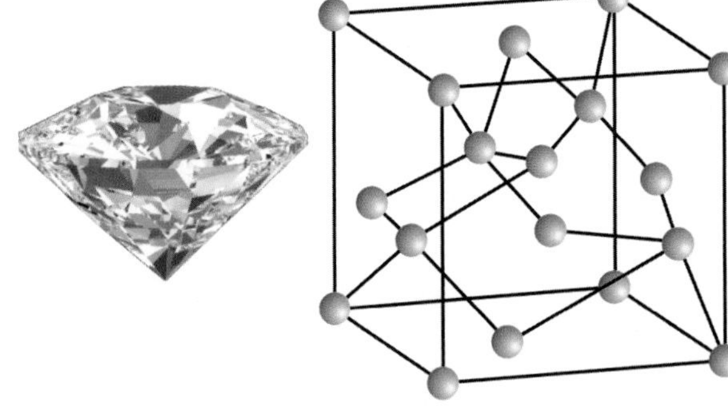

Bravais Lattices

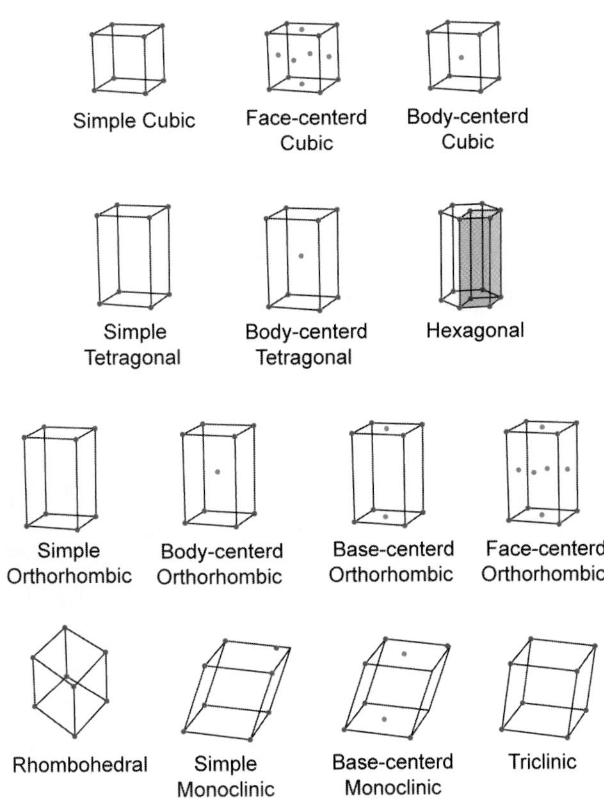

There are 14 geometrically unique crystal structures known as the 14 **Bravais lattices**. The Bravais lattices are used to classify crystal structures into groups with common properties.

If it is undisturbed while cooling crystalline materials tend to form along the same geometry as their structure. A good example is calcite a mineral crystal with a rhombohedral lattice. . .

Chemistry terms

crystal structure - system which describes the distances and angles between atoms in a crystal.

bravais lattices - fourteen groups of crystal structures used to help group the many different types of crystals.

Crystal defects

Perfect crystals are never found in nature. There are defects even in the most flawless diamonds. Because crystals are three dimensional structures, defects can have up to three dimensions to them. Some common defect types are 0-D (point), 1-D (line) and 2-D (surface). Each type of defect has a different effect on the properties of the crystal.

Point Defects (0-D)

Point defects occur when one atom is somehow different than the rest of the crystal. A *vacancy*, occurs when an atom is missing from the crystal structure. An *interstitial* occurs when there is an extra atom wedged in between others. A third point defect is a *substitutional defect*, where there is a different type of atom or molecule present than in the rest of the crystal. Substitutional defects are the reason why rubies and sapphires have their colors. Both rubies and sapphires are made of nearly pure Al_2O_3. Their color comes from chromium substitutional defects. Very small changes in the chemistry of Al_2O_3 are what make beautiful gems.

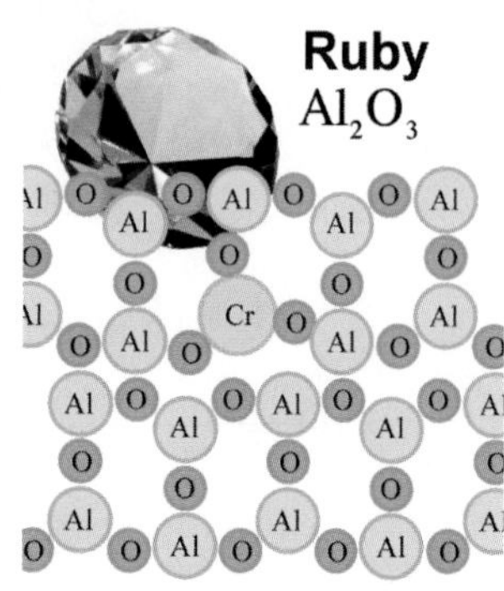

A chromium substitution defect gives ruby its red color

Line defects (1-D)

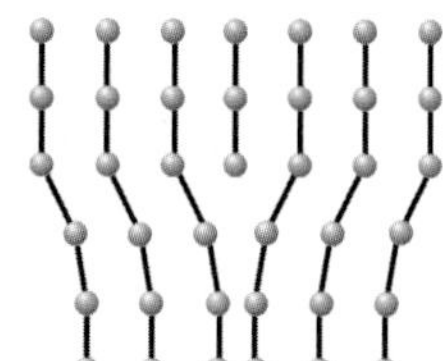

Line defects occur when a line of atoms is somehow different than the rest of the crystal. These most often take the form of **dislocations**, or an extra "half plane" of atoms in a crystal. Dislocations form the main mechanism of deformation in crystals.

Grains and grain boundaries

You don't normally think of metal as being a crystal. That is because most crystalline materials are polycrystalline. Polycrystalline materials contain millions of tiny crystals, called *grains*, connected to each other. Almost all metals are fine grained, meaning the grain size can be seen only with a microscope. The image on the right shows the grains in a sample of metal. The surfaces between grains are called **grain boundaries**, and are the most common 2-D defect in a crystal. If a crystalline solid has only one large grain, it is a *single crystal*, or *monocrystalline*. The mineral "crystals" you see are actually large monocrystals.

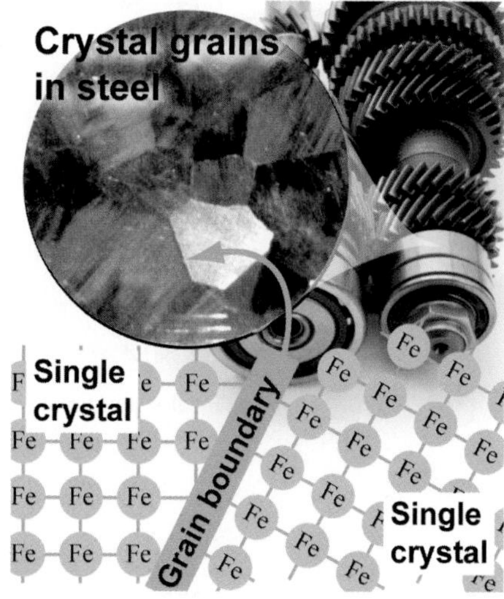

Chemistry terms

point defects - single atom defects in a crystal. These are 0-D defects.

dislocation - a missing half-plane of atoms in a crystal. This is a 1-D defect.

grain boundary - the interface between two grains. This is a 2-D defect.

16.3 Metals and Alloys

Metallic bonding

Metals can be very strong, are good conductors of heat and electricity, and are easily formed by melting and bending. All of these characteristics come from the way metallic bonds behave. Because their ionization energies are low, the valence electrons in metals are not strongly bound to their parent atoms. In solid metal the valence electrons become detached but do not reattach to another atom. A good analogy is that a metal is a fixed lattice of positive ions in a sea of free electrons.

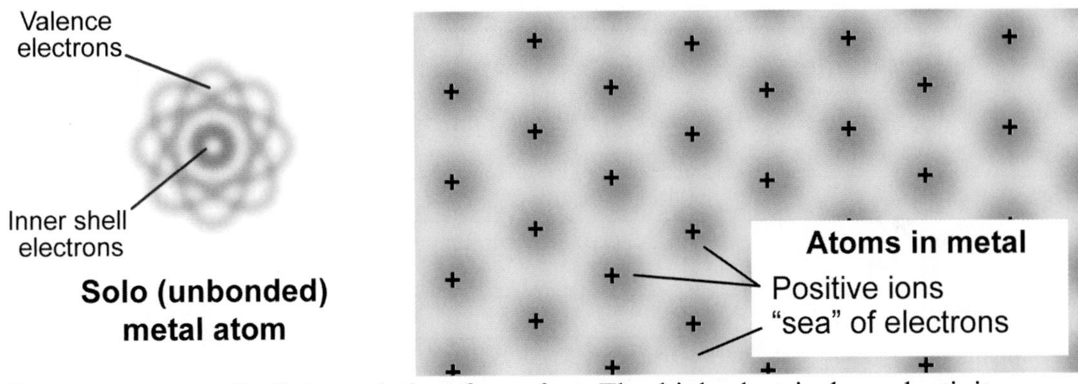

Thermal and electrical conductivity of metals

Electrons are small, light and therefore, *fast*. The high electrical conductivity comes directly from the fact that metals have many charged electrons that are free to move around. The thermal conductivity of metals is high for the same reason. Moving electrons can carry energy quickly and when they get heated, the heat spreads rapidly through the metal. The *thermal conductivity* describes how many watts of *heat* would flow through a 1 cm cube of a material if there were a 1°C temperature difference between its opposite sides. The *electrical conductivity* tells us how many *amps* of electrical current would flow through a 1 cm cube if 1 volt was maintained across the sides.

Thermal conductivity
1 w/cm°C means
1 watt of heat flows through 1 cm for 1°C temperature difference

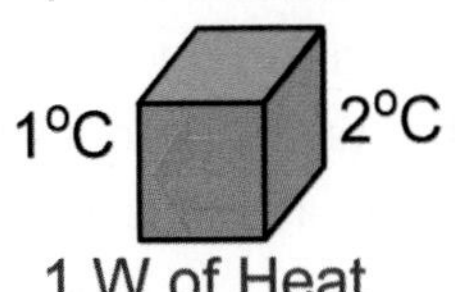

Electrical conductivity
1 A/cmV means
1 amp of current flows through 1 cm for 1 volt potential difference

Material	Thermal conductivity (W/cm-°C)	Electrical conductivity (A/V-cm)
copper	4	600,000
aluminum	2.2	380,000
polyethylene plastic	0.004	0.1
glass	0.01	10^{-10}

Comparing metals and non-metals

Note how different the numbers are between metals and non-metals such as polyethylene, and glass. The electrical conductivity of copper is 6 million times higher than it is for polyethylene plastic. The thermal conductivity of copper is 1,000 times higher than it is for polyethylene.

Alloys – solid solutions of elements

Alloys - Solid solutions of elements

Alloys are a special class of solid mixtures of two or more elements where the element in the largest proportion is a metal. The atoms in an alloy are not bound up in molecules, but are bound to each other with metallic bonds. The table below lists some common alloys and their uses. For example, bronze was perhaps the first important alloy discovered by humans. Bronze is an alloy of copper and tin. Other common examples of alloys include brass (copper and zinc) and steel (iron and carbon).

Alloy name	Composition	Uses
Bronze	Copper and tin	Coins, tools, bells, statues
Brass	Copper and zinc	Musical instruments, plumbing
Steel	Iron and carbon	Buildings, bridges
Stainless steel	Iron, carbon, chromium, vanadium	Wherever corrosion resistance is needed
6061 aluminum	96-98% Al, 1% Mg, 05-1% Si	Bicycle frames, aircraft, boats
6-4 titanium	6% Al, 4% V, 90% Ti	Bio-implants
14k gold	58% Au, 42% of Ni, Cu, Ag	Jewelry,
Sterling silver	92% Ag, 8% Cu	Jewelry, silverware

6061 Aluminum

Steel

Brass

Bronze

Alloys improve physical properties

Early metalworkers discovered that by melting copper and tin together, they could create bronze - a metal that was stronger than both. This is perhaps the most important property of alloys. Alloys often have better physical properties than any of their constituent pure elements. Most pure metals, even iron, are naturally very soft. As a result almost everything made of metal today is made of an alloy. Some of the most important discoveries in the rise of modern civilization were the alloys of bronze and iron. The Bronze Age and the Iron Age were periods of civilization named after them.

Chemistry terms

alloy - solid material made up of two or more elements (usually metals) that is evenly mixed on the microscopic level.

Alloys - chemistry changes properties

Chemistry changes properties

Adding two or more elements together in the form of a metal alloy creates a material with different physical properties than any of the original elements. Simple alloys made of two elements are called **binary alloys**. Sometimes it only takes a small amount of the **alloying element** in a binary alloy to greatly affect the properties of the alloy. For example, adding less than 0.5% carbon by weight to iron can create incredibly strong steel.

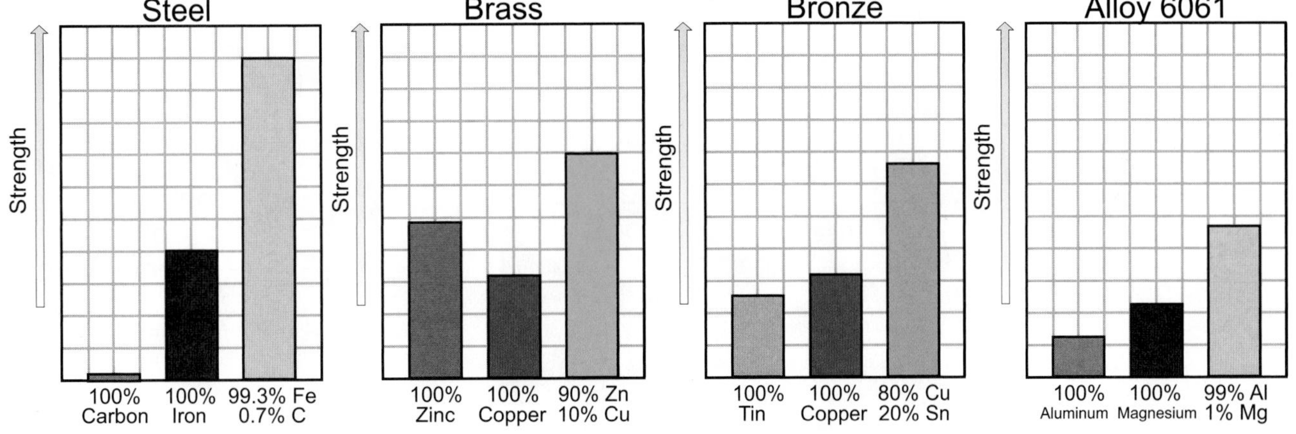

Complex alloys

Three or more alloying elements can be combined to produce complex alloys with even more interesting properties. For example, steel is made of iron and carbon. Adding more elements to the mix further changes the properties. This is explored in the Chemistry Connection in this chapter.

The latest military aircraft are made of complex alloys of titanium, vanadium and zirconium. Mixing them in the correct proportions creates an alloy that is incredibly strong and tough. If the chemistry of the alloy is off by even a tiny bit, the properties of the aircraft alloy will be greatly reduced.

Chemistry terms

binary alloy - An alloy made up of only two different elements.

alloying element - One of the elements that make up a certain alloy.

Binary phase diagrams

Binary phase diagrams

The properties of alloys depend on their composition in unusual ways. For example, fitting two *different* atoms into a crystal lattice changes the crystal structure and that also changes how an alloy melts or freezes. A **binary phase diagram** shows how the composition of an alloy changes its melting point. You may have heard of "sterling silver." Sterling silver is an alloy in which 12% of the atoms are copper (Cu) and 88% are silver (Ag). This works out to 7.5% by mass of copper.

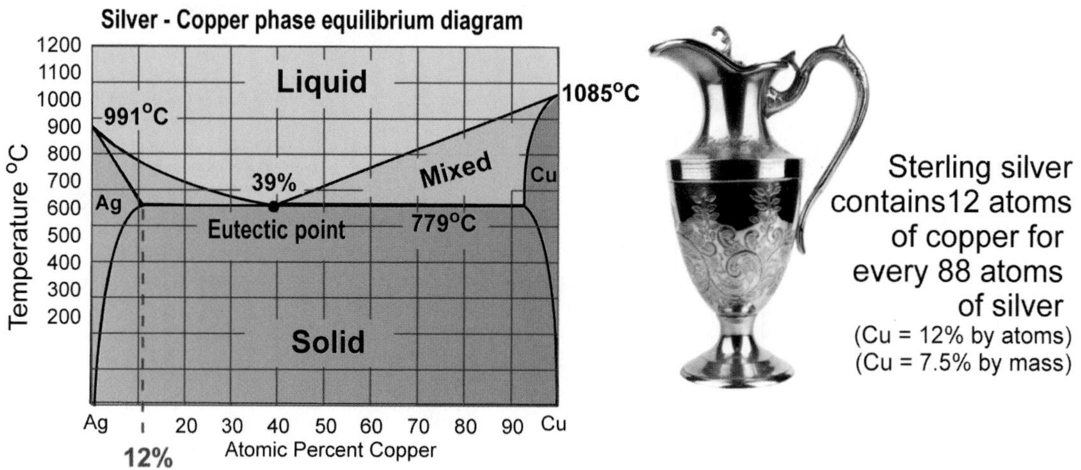

Interpreting the diagram

A silver-copper alloy is a liquid solution of copper and silver in the blue-shaded region of the diagram. It is a solid solution below 779°C. Above 779°C the mixture may be a mixed state, or a liquid, depending on its composition. The mixed state is neither fully solid or fully liquid.

Alloys may have a lower melting point

Pure silver melts at 991°C and pure copper melts at 1085°C. Sterling silver melts at 779°C! The phase diagram shows that a silver-copper alloy becomes liquid at a much lower temperature than the typical freezing point of either pure metal.

The eutectic point

A silver-copper alloy has a single melting point at 779°C when the composition is 39% copper (by atoms). This is called the *eutectic point*. The **eutectic point** is the composition with the lowest melting point. *Solders* are metals that are designed to melt at low temperatures for assembling electronic devices or sealing water pipes. The composition of solder is chosen so it corresponds to the eutectic point.

binary phase diagram - a graph showing the phase of an alloy at different temperatures and compositions.

eutectic point - the temperature and composition where an alloy's melting point is the lowest.

16.4 Physical Properties of Liquids

Liquids are more weakly bound than solids

Liquids, unlike solids, do not have a network of permanent intermolecular bonds. Instead, intermolecular bonds are continuously made and broken. The making of bonds allows liquids to hold together, and the breaking of bonds allows liquids to **flow**. It is a characteristic of liquids that they flow under a force (like gravity) rather than just fall

Energy of liquid vs. solid

Because they can flow, liquids lack a rigid molecular structure, like a crystal. On the molecular level, liquids are like amorphous solids, except that intermolecular bonds are not strong enough to overcome thermal agitation. The energy of a liquid is higher than that of a solid. The higher energy allows the bonds formed in a liquid to break quickly. The speed at which these bonds are made & broken, along with the strength of these bonds, lead to the unique physical properties of different liquids.

Freezing rate and solid microstructure

It is important to note that the same liquid can freeze (solidify) into a crystal or a glass. For example, sand and quartz are *crystalline* forms of SiO_2. Window glass is an *amorphous* form of SiO_2. You get liquid SiO_2 if you melt quartz or window glass and the molten liquid doesn't remember whether it came from a crystal or a glass! Depending on how it is cooled, the same liquid SiO_2 can become a glass (if it is cooled fast) or a beautiful quartz crystal (if it is cooled VERY slowly). Virtually all materials display differences in their solid microstructure depending on how fast they solidified. This is the basis for heat treatment of steel, glass, aluminum and other materials.

flow - The ability of a liquid to move and change shape under a force, like gravity (weight).

Cohesion

Cohesion in liquids

Liquids tend to stay together, even if forces like gravity are acting on them. We call this property *cohesion*. **Cohesion** is an attractive force between molecules in a liquid. Cohesion is quite striking in the weightless environment of the orbiting space station. In microgravity, cohesion causes liquid drops to come together as tight as possible in perfect *spheres*. A sphere is the shape that results in the least surface area for the most volume. A liquid drop assumes this shape to maximize the number of liquid molecules that can bond together.

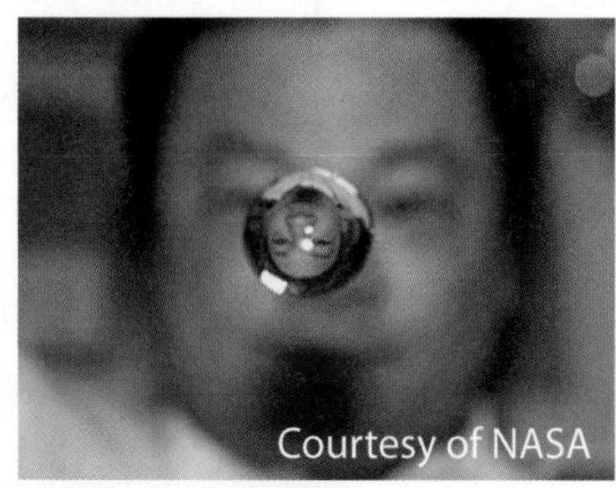
Courtesy of NASA

Cohesion in nature

Cohesion explains why rain falls in raindrops, and how clouds form. Water vapor in the atmosphere tends to stick together because cohesion keeps liquids together. These droplets get bigger and bigger until they become visible as clouds. When it rains, tiny water droplets in clouds come together to make bigger and bigger droplets until they are so heavy that they fall as rain.Cohesion keeps these raindrops together as they fall to the ground.

Cohesion in mercury

Mercury is a liquid metal at room temperature

Some liquids are more cohesive than others. For example, water is a fairly cohesive liquid. Liquid mercury, a liquid metal, is very cohesive. The drops in liquid mercury are held together so tightly by cohesion that they form almost perfect spheres.

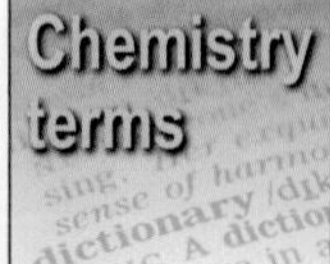

cohesion - the property of a liquid that causes it to hold together.

Adhesion

Adhesion

You may have also noticed that liquids tend to stick to things. We call this property **adhesion**, and it describes how liquids stick to surfaces. When something is 'wet,' we mean that liquid is *adhering* to it. Different surfaces can also *be wet* by the same liquid differently - we call this property *wettability*. Liquids that adhere to things very well are called *adhesives*, and are commonly used as glues or put onto tape to make it sticky. Geckos use the property of adhesion to stick to surfaces, like the walls and ceilings of houses.

Meniscus

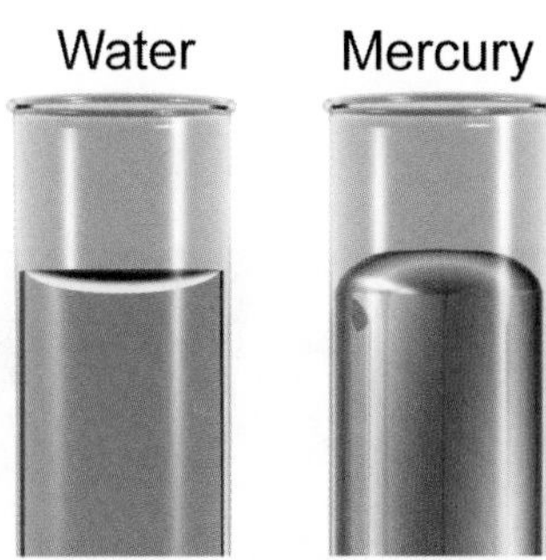

A good way to see if a liquid is more cohesive or adhesive is to put it into a test tube and observe the **meniscus**, or the top of the column of liquid. A cohesive liquid will try to stay together instead of sticking to the walls of the cylinder. A more adhesive liquid will stick to the walls of the cylinder and form a "U" shape. Water is a more adhesive liquid, while mercury is more cohesive. Adhesion also explains why droplets of water stick to glass, such as windows. Liquids like water are very adhesive, and adhesion is strong enough to hold large drops of water from falling.

Adhesion and capillary action

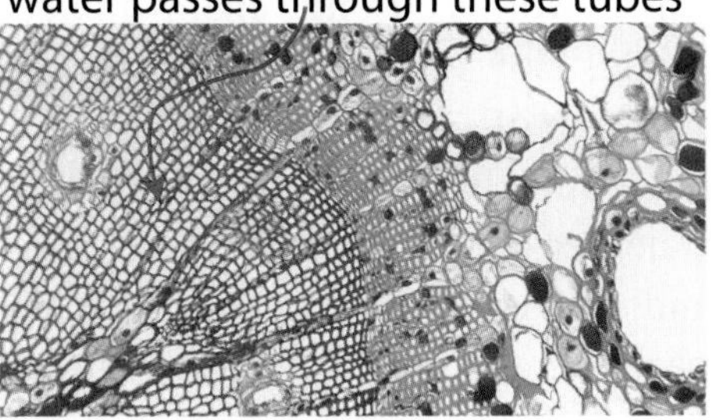

Plants use adhesion to draw water up from roots to leaves. The water adheres to the tubes inside the plant, and **capillary action** causes the water to be drawn up the empty tube to the top of the plant. Even very high trees move water this way. Adhesion is the force that causes capillary action to happen. Without adhesion, plants could not use water.

Chemistry terms

adhesion - the property of a liquid that causes it to stick to surfaces.
meniscus - the curved top of a column of liquid.
capillary action - an effect where liquid is pulled up a thin tube by adhesion.

Viscosity

Viscosity is resistance to flow

While the making and breaking of bonds allow a liquid to flow, the frequency and strength of the bonds controls how *easily* it flows. The "thick" or "runny" aspect of a liquid is called *viscosity*. **Viscosity** is a measure of *resistance to flow*. A liquid with low viscosity is very runny and will flow very easily. Water and alcohol are good examples of liquids with low viscosity. A liquid with a high viscosity flows slowly. Syrup and corn oil are liquids with high viscosity.

Viscosity in nature

We can find liquids with very different viscosities in nature. Snails and slugs use mucus, a *very* high-viscosity substance, to help them stick to surfaces and move. Magma, molten rock has different viscosity depending on the silica (SiO_2) content. High silica magma is sticky and viscous. It is sticky enough to allow pressure to build up and potentially cause a violent volcanic eruption. Low silica magma is runny and has a lower viscosity. This type of magma forms the lava in the spectacular fire fountains in Hawaii.

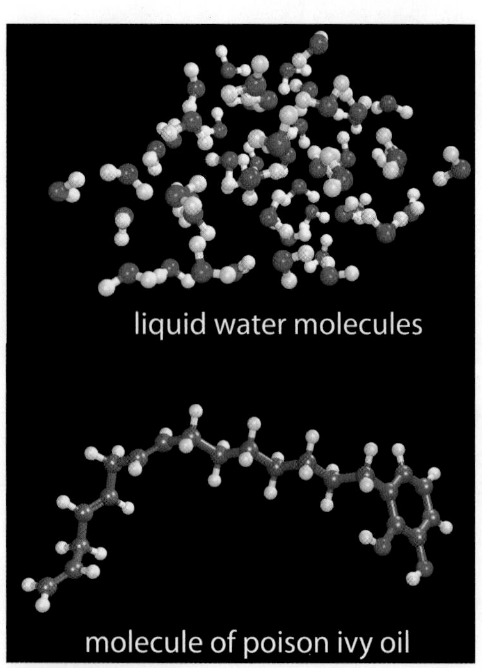

Viscosity is determined mostly by two variables – the strength of the intermolecular bonds, and the shape of the molecules. If there are relatively strong intermolecular bonds, a liquid will be more viscous. Molten metals have a high viscosity because the bonds are strong. Liquids with polar or hydrogen bonding, like water, will be more viscous than nonpolar liquids, like liquid nitrogen.

Viscosity is also affected strongly by molecular size and shape. Water has a lower viscosity than oil because water molecules are small can move around each other easily. The molecules in oils are very long, like worms. Long chain molecules are be easily entangled and this slows down the flow. A good analogy is to think about pouring a cup of cooked spaghetti noodles and a cup of sand. The sand flows quickly because the particles are small and slide over each other easily. The noodles get tangled and pour very slowly. Liquids made of long chain molecules tend to flow like entangled spaghetti, and therefore have high viscosity.

Chemistry terms

viscosity -the resistance of a liquid to flow under an applied force.

viscous - having a high viscosity.

non-viscous - having a low viscosity.

Surface tension

The strength of a free surface

When you pour water on the table, you may have noticed that it forms a *surface*. It takes energy to actually *break* that surface. Obviously not too much energy, since you are able to poke your finger in without any trouble. However, there is a measurable energy required to break the surface of a liquid, and the physical property that measures this energy is called the **surface tension** of the liquid. You may be able to break the surface of water easily, but insects like the *water strider* are so light and spread their weight out so far that they can actually rest *on top* of the surface of the water! They are not floating on the water, but resting upon it!

What surface tension does

Surface tension is what causes liquid to actually stay together. It makes the entire surface of a liquid act like a big elastic band, pulling back when the liquid is stretched out. On Earth, gravity is much stronger than surface tension so water spilled on a flat table spreads out into a thin sheet. Without surface tension, gravity would squish the liquid into a perfectly thin sheet of water. In space, where there is *no* gravity, spilled water pulls itself into the tightest possible shape - a *sphere, and sits in a ball on top of the table!*. Surface tension is also what causes raindrops to fall in *drops*, rather than in sheets or any other shape.

Soaps

Soaps work partly by lowering surface tension. The active molecules in soap are medium length chains with one end that is polar and one that is non-polar. The polar end tends to adsorb to the water surface and this breaks up the surface tension. Traditional soap consists of sodium or potassium salts of fatty acids. Soap is made by reacting common oils or fats with a strong alkaline solution, usually sodium hydroxide or lye.

surface tension - a property of liquids to resist having their surface broken. The energy needed to remove a certain amount of area on a liquid surface, measured in J/m^2 .

How to measure surface tension

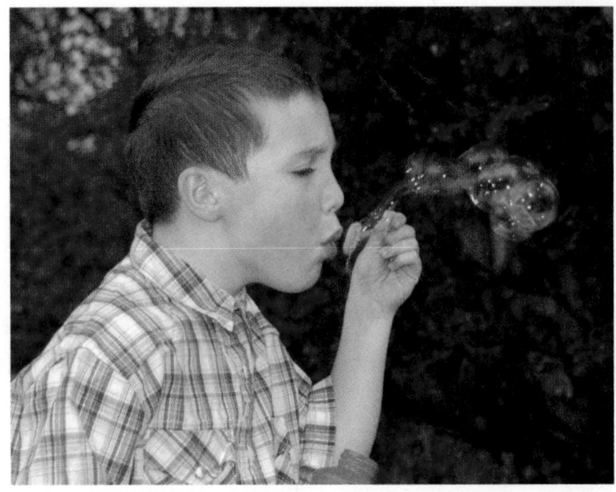

Surface tension is measured in energy per unit area [J/m^2]. It represents the energy needed to break a certain amount of area of a liquid's surface. A very easy way to measure surface tension is to blow a bubble of a liquid. If you carefully measure the pressure needed to pop the bubble, and the size of the bubble when it pops, then you can calculate the surface area of the bubble, and then the surface tension.

Surface tension experiments

You can observe surface tension's amazing effects with a few simple materials. Pour water into a beaker to just near the top. Now keep adding water with a dropper. Notice how the water level is *above* the top of the beaker! Now carefully place a paper clip on the surface of the water with a pair of tweezers. Metal is much denser than water, but surface tension can support the weight of the paper clip. How many paper clips can you add before they sink?

How we use surface tension

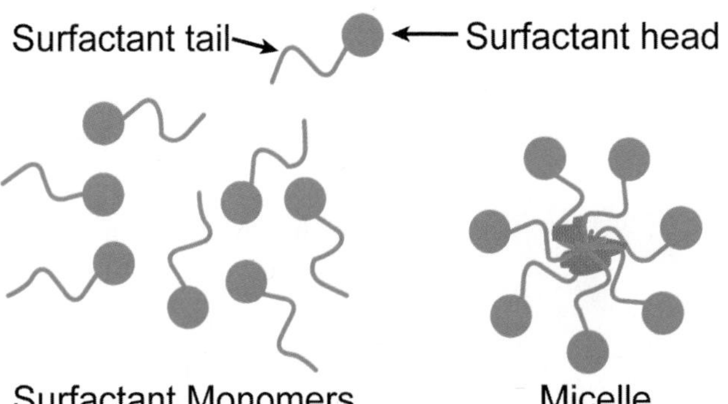

We can cause liquids to break up more easily by reducing their surface tension. Materials that do this are called **surfactants**, and they are most commonly found in soaps and detergents. They cause the sticky liquids to break up into smaller and smaller pieces, which are more easily washed away.

Chemistry terms

surfactant - a substance that reduces the surface tension of a liquid.

Steel: The most useful alloy

Steel is perhaps the most important alloy we use today. Steel is the building block for thousands of different things in our everyday lives, from cars to buildings to household objects. The discovery and development of steel marked one of mankind's greatest achievements.

All steels must contain both iron and carbon. In its simplest form, *carbon steel*, carbon and iron are the only ingredients in the alloy. The more carbon, the harder the steel can become. Over time and with much ingenuity, metallurgists have developed thousands of different steel alloys with very different physical and chemical properties. The two most common methods to change the properties of steel are (1) changing the size, type, or distribution of crystal defects, and (2) changing the chemistry by adding or adjusting component elements in the alloy.

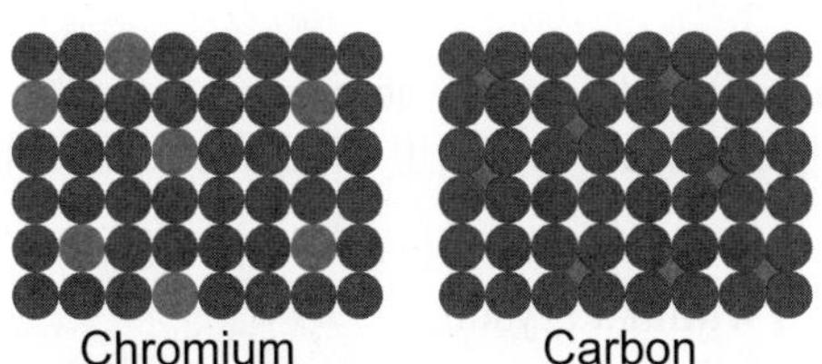

Carbon atoms are much smaller than iron atoms, so they exist as *interstitials* in an iron crystal. The chromium atoms in stainless steel exist as *substitutional* defects, because the Cr and Fe atoms are close to the same size. Adding chromium makes the steel very *corrosion resistant*, meaning that it is very hard to make it rust. This is why it is called *stainless*, because it is difficult to get anything to react and stain the metal.

The various defects present in steel can *greatly* change its properties. Metallurgists have added and controlled the types of defects in steel to change its properties. By changing grain size and the frequency of crystal defects, a skilled metallurgist can control hardness, strength, corrosion resistance, hardenability, and many other properties. For example, working the steel by forging, hot working, cold working, drawing, quenching, annealing and tempering it will change the shapes and sizes of the grains. Dislocations move well through grains, but not through grain boundaries. Working the steel and quenching it will make smaller grains, and therefore more grain boundaries. The more grain boundaries, the harder it is for dislocations to move, and the harder the steel. On the other hand, annealing (slow cooling) the steel will increase the grain size, and make the steel softer.

Steels can also be joined together in fascinating ways. *Pattern Welded Steel* is made of two or more types of steel, folded together in different ways. Objects made from pattern-welded steel often have properties of both the original steels, and often have artistic patterns due to the folding, like in this pattern-welded knife.

Changing the chemistry of steel greatly affects its properties.
Just changing the amount of carbon in steel completely changes the nature of the steel. Very little carbon produces *mild steel*, or just iron with carbon. Adding more carbon produces *pearlite*, a microstructure of steel that is both strong and hard. Adding even more carbon can produce *carbides*, or particles of iron carbide (Fe_3C). These are very hard, but can make the steel brittle.

Chapter 16 Review.

Vocabulary

Match each word to the sentence where it best fits.

Section 16.1

solids	gases
liquids	

4. ________ have a fixed volume, but no fixed shape.
5. ________ have no fixed volume and no fixed shape.
6. ________ have both a fixed volume and a fixed shape.

Section 16.2

crystal	slip systems
glass	close-packed directions
amorphous	hardness
crystal structure	Mohs hardness scale
Bravais lattices	strength
metallic glass	brittle
point defects	ductile
dislocations	alloy
grain boundaries	alloying elements
inclusions	binary alloy

7. The property of a solid that describes the energy needed to break it is called ________.
8. __________ are 1-D defects in a crystal that consist of a missing half-plane of atoms.
9. A/an ________ is a solid solution of two or more elements.
10. A material that breaks easily is described as being ________.
11. _________ materials behave like both solids and liquids, and have no long-range order to their molecular structure.
12. Interstitials, vacancies and substitutionals are examples of __________ in a crystal.
13. A/an ________ is an ordered arrangement of atoms in a solid.
14. A/an ________ is an example of an everyday amorphous material.
15. _________ describes the ability of a material to resist deformation.
16. A material that stretches a lot before breaking is said to be _________.
17. A/an ___________ is an alloy made up on only two elements.
18. Crystals often deform by activating _________, where planes of atoms will slide over each other.
19. A/an __________ is a 3-D defect in a crystal, like a crystal within a crystal.
20. The __________ compares materials by their scratch resistance, with diamond at the top.
21. ____________ are the elements that go into an alloy.
22. Slip systems are often activated on ____________ in a crystal.
23. A/an _________ describes a specific arrangement of atoms.
24. The fourteen _________ describe the fourteen general possible ways to arrange one type of atom.
25. ___________ are 2-D defects in a crystal.
26. ___________ are a special type of material, formed by cooling metal at a fantastic rate.

Section 16.3

phase diagram	binary phase diagram
triple point	eutectic point
critical point	

27. The _________ is the temperature and pressure where a material can be a solid, a liquid and a gas at the same time.

28. A/an __________ shows the phase and crystal structure of an alloy depending on chemistry and temperature.

29. A/an __________ shows the phase of a material depending on the temperature and the pressure.

30. The _________ on a binary phase diagram shows the alloy composition with the minimum melting point.

31. The __________ on a phase diagram shows the point where the line between liquid and gas begins to blur.

Section 16.4

flow	viscosity
cohesion	viscous
adhesion	non-viscous
meniscus	surface tension
capillary action	surfactant

32. The resistance of a material to flow is called __________.

33. A _________ material has a high viscosity, and a _________ material has a low viscosity.

34. _________ describes when a group of molecules in a liquid hold together. It describes why water in space forms a sphere.

35. The movement of a liquid under an external force such as gravity is called_________.

36. You can observe the balance between cohesion and adhesion in a liquid by pouring it into a test tube and looking at the __________.

37. Plants draw water up their stems by __________.

38. Water can stick to a glass surface because of __________.

39. A water strider and some lizards can walk on water because of _________.

40. _________, such as soaps and detergents, are effective at encapsulating dirt and oil.

Conceptual Questions

Section 16.1

41. Why is bonding between molecules stronger in solids? Why don't the molecules escape?

42. Why does adding energy in the form of heat melt a solid? What does it do to the bonds?

43. What would happen to the molecules in a water balloon if you popped it in outer space, where there is no gravity?

44. What would happen to the molecules in a balloon filled with air if you popped it in outer space? How far would the molecules go?

Section 16.2

45. Why are NaCl (salt) crystals cubic? Why are snowflakes hexagonal? (Hint: The answers are the same)

46. What would happen to a glass if you left it alone at room temperature for one year? How about at 1000F for one year? Why?

47. What would happen to a glass if you left it alone at room temperature for millions of years? Why?

48. Why do we say that glasses have no 'long-range order?' Do glasses have a crystal structure? Why / why not?

49. How do you make a metallic glass? Why do we use it on energy-impact surfaces, like golf club heads?

50. Try and find an example of each of the fourteen Bravais lattices in nature. Look at different kinds of crystals, rocks and gemstones.

51. Name the three types of point defects in a crystal. Describe each one.

52. Describe the four types of dimensional defects in a crystal.

53. What is a slip system? Why do crystals deform using slip systems? What does it mean if a crystal breaks without using a slip system at all?

Chapter 16 Review.

54. Why do slip systems happen first on close-packed directions?
55. What does it mean for a material to be hard? Does it have to be strong to be hard?
56. How would you use the Mohs hardness scale to determine the relative hardness of a new material?
57. What does it mean for a material to be strong? Does it have to be hard to be strong?
58. Name some materials that you know are brittle. Name some materials that you know are ductile. How would you figure out if a new materials is brittle or ductile?
59. Why do we use alloys? Why don't we just use pure metals?
60. Describe two alloys where changing the chemistry changes the properties.
61. Research a complex alloy that is interesting to you. Describe what each alloying element does to change the properties of the alloy.

Section 16.3

62. Describe how to use a phase diagram to tell what phase would exist in a material at a given temperature and pressure.
63. What does the triple point mean? What does the supercritical point mean?
64. How do you tell if an alloy is a solid or a liquid, depending on its temperature and chemistry?
65. What is the eutectic point for an alloy? Why is it useful? Name three places that we use eutectic alloys in our lives.

Section 16.4

66. Why do liquids flow? Why don't solids flow?
67. Why does rain fall as drops instead of any other shape?
68. What property allows geckos to crawl on ceilings? How do we use this property in our everyday lives?
69. Design an experiment to measure the ratio of cohesion to adhesion for a number of liquids.
70. How do we measure viscosity? Design an experiment to measure viscosity.
71. Describe how the molecular shape of a molecule can lead to its viscosity. Why is water non-viscous? Why is oil so viscous?
72. How can a something heavier than water 'float' on top?
73. How would you reduce the surface tension of a liquid? What would you add to it?
74. Why can a water strider walk on water?
75. Why can some lizards run on water? What would happen if they stood still?
76. What would happen if you ran on water? Why?

Chemistry Connection

77. What is steel in its most basic form?
78. Why is steel such an important alloy for us?
79. Name five alloying elements (besides carbon) that can be added to steel, and what they do.
80. Name five places that steel is used in our everyday lives. What alloying elements would you add to steel in each situation to improve its properties?

Quantitative Problems

Section 16.2

81. Research the names of ten common alloys on the internet or in the library. Make a graph of their strength vs. their hardness. Do you see a pattern?
82. Research aluminum alloy 6061. Make a graph of the amount of magnesium vs. the strength. Do you see a pattern?

Section 16.3

For these next problems, examine the phase diagram of water:

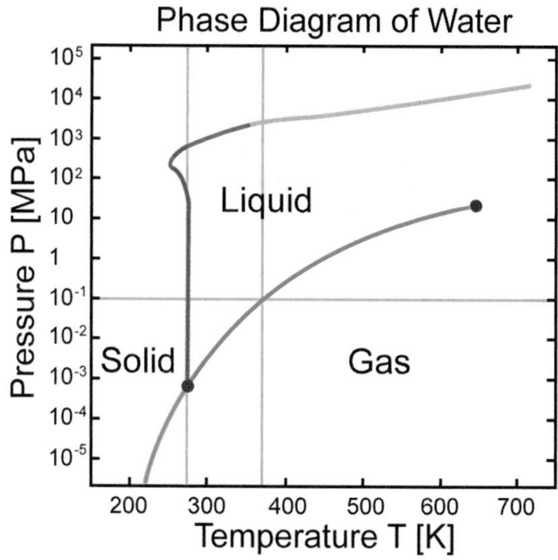

83. What phase is water at STP (25C, 0.1MPa)?
84. What is the triple point of water? Give the temperature and pressure
85. What is the melting point of water on Earth (0.1MPa)? What is the melting point of water on Mars (0.01MPa)? What is the melting point of water in outer space (~0MPa)? Does it melt?
86. What is the critical point of water? What happens at this temperature and pressure?

For these next problems, examine the phase diagram of CO_2:

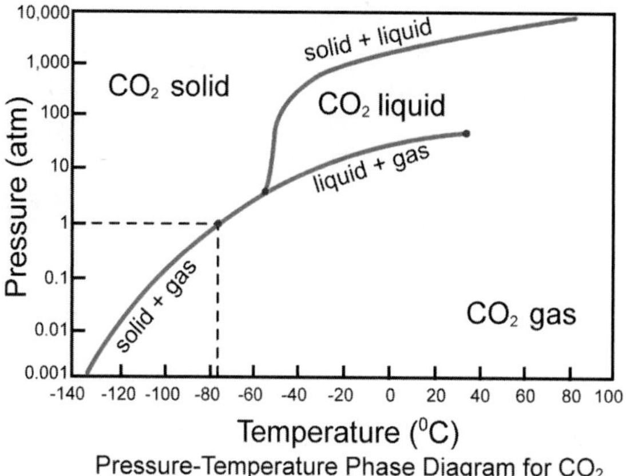

87. What is the freezing point of CO2 on Earth (1atm)? What is its melting point? Does it freeze? Does it melt?
88. When can CO2 exist as a liquid?
89. Frozen CO2 is called 'dry ice.' Why do we call it 'dry' ice? Show why on the phase diagram.

For these next questions, examine the binary phase diagram for lead and tin shown below:

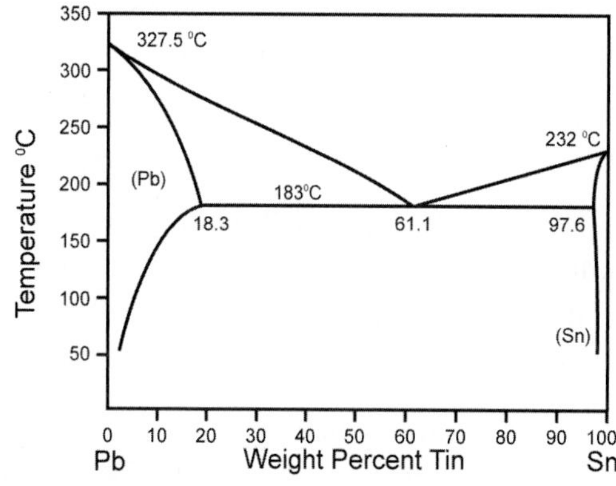

90. What is the melting point of lead? Of tin? Of the eutectic point?
91. Where do we use lead-tin eutectic? Why is it so useful to us?

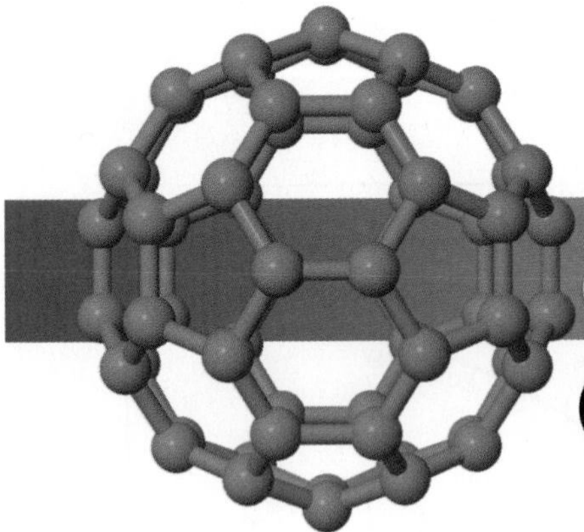

CHAPTER 17
Organic Chemistry

What is so special about carbon?

Why is carbon so abundant in nature?
Are organic molecules "alive?"

Does "organic" chemistry study foods grown without harmful pesticides?

What exactly is a plastic made of?

If you look around your classroom or home you might be surprised by the fact that almost everything you see is made up of carbon containing compounds! Several million carbon containing compounds are known to exist, and the number continues to grow. It is remarkable that carbon can make so many compounds. Organic chemistry is the study of carbon containing compounds and their properties. These carbon containing compounds vary widely in their size and complexity. The term "organic" in chemistry does *not* mean foods without harmful pesticides. You will find that you are familiar with many organic chemicals. One important group of organic compounds are fossil fuels such as gasoline, kerosene, natural gas, and coal. We use organic compounds to heat our homes and provide us with energy in our daily lives. Fossil fuels provide us with energy when we break the covalent bonds holding the molecules together.

Organic substances have changed peoples lives in very significant ways. For instance, can you imagine our society without nylon, velcro, and plastic? We have become dependent on many synthetic organic substances.

We consume organic chemicals all the time! Artificial sweeteners, flavorings, and foodstuffs are produced by organic chemists.

Lastly, and perhaps most importantly, our understanding of the human body and living systems centers around the study of organic chemistry.

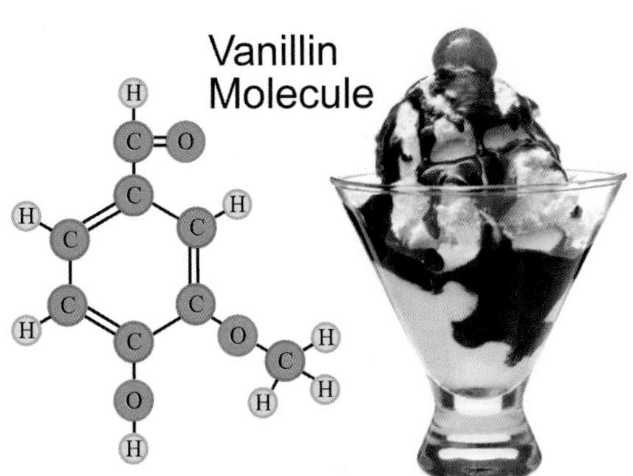

Vanillin is used as a flavoring for vanilla ice cream.

Mixing molecules

Properties of organic compounds.

What chemicals are "organic?" To begin our study of organic chemistry we will start by observing some of the important properties of these chemicals. Here we will examine the solubility of three substances: candle wax, vegetable oil, and petroleum jelly. We will place each of them in contact with a different solvent. To each watch glass we will add drops of alcohol, mineral oil, and water.

"Like dissolves like"

After making some observations we will mix the substances with a toothpick.

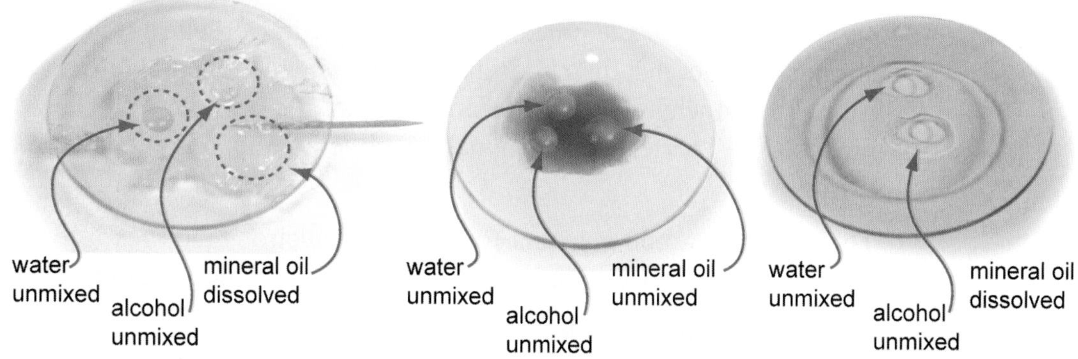

In the first picture we can see that mineral oil dissolved in the petroleum jelly but the alcohol and water did not. In the second one candle wax does not appear to dissolve in anything. In the third case vegetable oil dissolved in mineral oil but not in alcohol or water.

What do our observations tell us about organic compounds such as wax, vegetable oil, and petroleum jelly?

Molecules of oil and water do not mix

Organic compounds must not be attracted to water or alcohol molecules. In order for substances to mix the attractive forces must be similar between molecules. In this chapter we will learn more about the molecular structure of these molecules. Why these substances do not mix will be understood in more detail. The environmental impact of an oil spill is a good reason to understand these principles!

17.1 Carbon Molecules

Carbon has the ability to make four bonds

Carbon has a unique ability to form many strong bonds to itself, and create carbon networks, no other element on the periodic table shares this ability. Carbon can also form strong bonds to other nonmetals such as hydrogen, and nitrogen. In this section, you will learn about some simple organic molecules. Understanding their molecular structure will help you to understand their properties and why they are important in our daily lives.

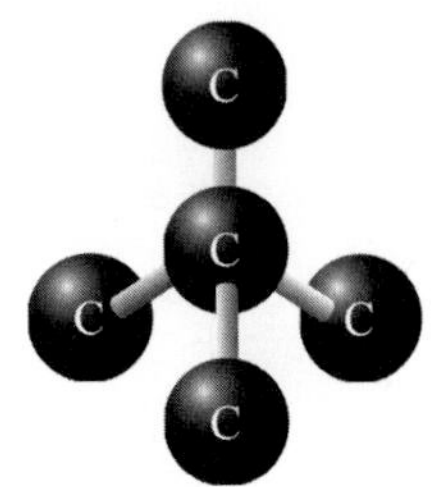

To begin this section it will be helpful to review some of the concepts covered in the bonding chapter. Carbon given its placement on the periodic table has the ability to make four bonds. When carbon is bonded to four atoms it will always take the geometric shape of a tetrahedron. This tetrahedral structure is very strong! The fact that diamonds are arranged in a tetrahedral lattice is one reason why they are the hardest substance we know of. The non polar, or "perfect" covalent nature of the C-C bond is another factor in the strength of diamonds.

The fact that carbon contains four valence electrons requires it to make four bonds. As you know it can bond to fewer than four atoms if it makes a multiple bond. For instance:

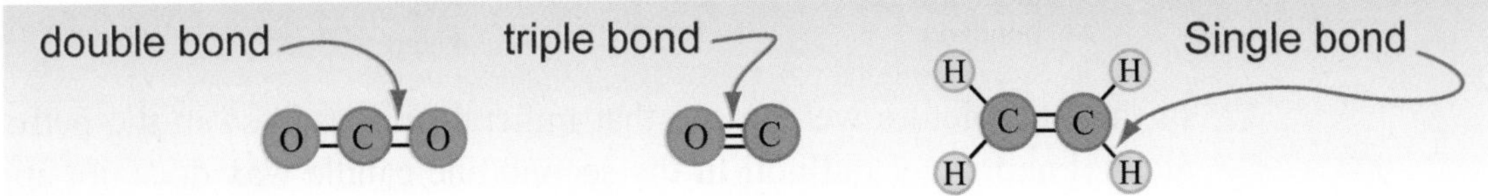

Hydrocarbons contain carbon and hydrogen

The simplest organic compounds are called **hydrocarbons**, and they contain only the elements carbon and hydrogen. There are two types of hydrocarbons, saturated and unsaturated. In a saturated hydrocarbon each carbon atom is making four bonds the maximum number possible. **Saturated** hydrocarbons contain only single carbon to carbon bonds. In an **unsaturated** hydrocarbon the carbon to carbon bonds contain double and triple bonds. Due to the multiple bond it is possible for carbon to bond to one more atom, so these hydrocarbons are considered "unsaturated."

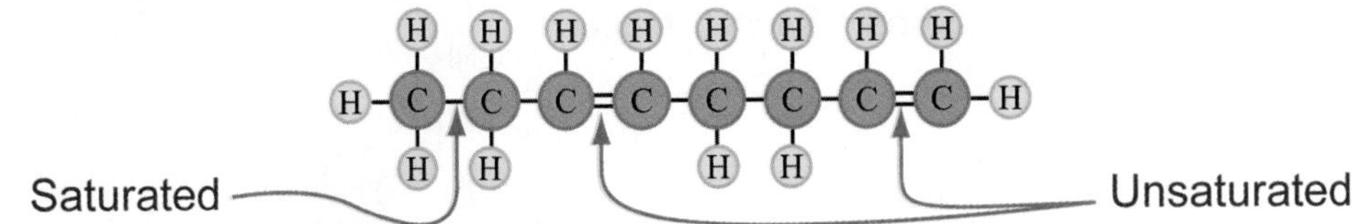

Chemistry terms

hydrocarbon - an organic compound containing only hydrogen and carbon.

saturated hydrocarbon - a hydrocarbon containing only single carbon to carbon bonds.

unsaturated hydrocarbon - a hydrocarbon containing multiple bonds between the carbon atoms.

Alkanes

Saturated hydrocarbons are called **alkanes**. Alkanes can form straight chains or branched chains. The simplest alkane is methane, because it contains only one carbon.

Pattern for alkanes

Ethane, C_2H_6 has two carbons attached by a single bond. Propane, C_3H_8 has three carbons attached by a single bonds. Butane, C_4H_{10} has four carbons attached by a single bonds. As the length of the carbon chain grows we can see a pattern in the formula for alkanes. General Formula = $C_nH_{(2n+2)}$ where n represents the number of carbon atoms

TABLE 17.1. Formulas of the First Ten Straight Chain Alkanes.

Name	Condensed Formula	Structural Formula
Methane	CH_4	CH_4
Ethane	C_2H_6	CH_3CH_3
Propane	C_3H_8	$CH_3CH_2CH_3$
Butane	C_4H_{10}	$CH_3CH_2CH_2CH_3$
Pentane	C_5H_{12}	$CH_3CH_2CH_2CH_2CH_3$
Hexane	C_6H_{14}	$CH_3CH_2CH_2CH_2CH_2CH_3$
Heptane	C_7H_{16}	$CH_3CH_2CH_2CH_2CH_2CH_2CH_3$
Octane	C_8H_{18}	$CH_3CH_2CH_2CH_2CH_2CH_2CH_2CH_3$
Nonane	C_9H_{20}	$CH_3CH_2CH_2CH_2CH_2CH_2CH_2CH_2CH_3$
Decane	$C_{10}H_{22}$	$CH_3CH_2CH_2CH_2CH_2CH_2CH_2CH_2CH_2CH_3$

Observe the structure of butane

The best way to visualize an alkane is to build it with molecular models. Build a molecule of butane.

When you observe the structure of butane, you will see that a "straight" chain is really NOT straight or linear

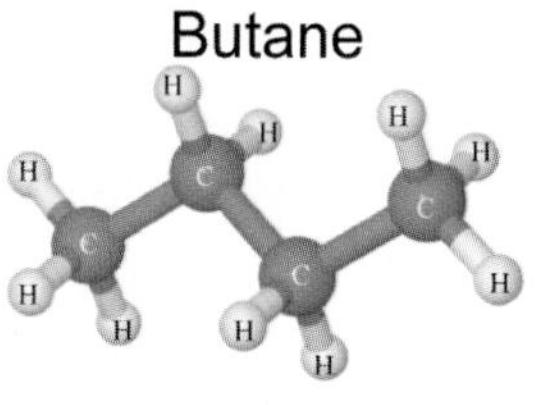

Why is this? Basically because carbon makes a three dimensional tetrahedral structure. Remember this is because it wants the single covalent bonds to be spaced out as far apart as possible to minimize the electron repulsions between the bonds. Look closely at your molecule of butane. Can you move the atoms around? Can the hydrogens change positions? Did you notice that the atoms can rotate around the single bonds? It is important to see that the hydrogens can occupy different positions as the carbon atoms rotate. The carbon atoms can also occupy different positions that make the ends of the chain closer together or further apart. You will see that as the alkane gets longer and the carbon chain grows the molecule will have different properties.

Naming alkanes

Naming hydrocarbons with single bonds

The naming process is very systematic and can easily be applied following simple rules. The prefixes for the first ten alkanes are listed below.

To name an alkane we simply count the number of carbon atoms in the longest chain and add the suffix –ane. For example $CH_3CH_2CH_2CH_2CH_3$ has five carbons in the chain and is called pentane.

TABLE 17.2

Number	Greek Root
1	Meth
2	Eth
3	Prop
4	But
5	Pent
6	Hex
7	Hept
8	Oct
9	Non
10	Dec

Pentane

5 carbons — alkane

Organic chemists call this n-pentane, where the n indicates a straight chain, and single bonds between the carbons.

Count longest continuous chain of carbons

For a branched chain alkane, we look for the longest continuous chain of carbons and that is the name of the **parent compound**. The parent compound tells us the base alkane name. As you count the carbons do not lift your pencil off the paper this helps you identify the *longest continuous chain*. For example:

Longest chain

$$CH_3—CH_2—CH_2—CH(CH_2CH_3)—CH_2—CH_2—CH_3$$

Shortest chain

This may also be written with an **R** representing the CH_2CH_3 branch as shown below.

$$CH_3—CH_2—CH_2—CH(R)—CH_2—CH_2—CH_3$$

Here the longest carbon chain contains seven carbons. We call the parent compound heptane. The two carbon chain that is not included is called an ethyl group. Eth- standing for 2 carbons and -yl indicates it is a “substituent,” or an add-on. For the exact name of the compound we count the number of carbons over to the ethyl group. Here it is on the fourth carbon, so we name it 4-ethyl-heptane. The ethyl group can also be referred to as an **R-group**. The R-group notation is used as a shorthand way of writing a hydrocarbon side chain attached to the parent compound.

Chemistry terms

parent compound - the longest continuous chain of carbons in an organic compound. The parent compound tells us the base alkane name.

R-group - in an organic formula the R- group can stand for an H, hydrogen atom, or a hydrocarbon side chain, of any length. It can also refer to any group of atoms.

Structural isomers

Branching of the carbon chain forms new structures

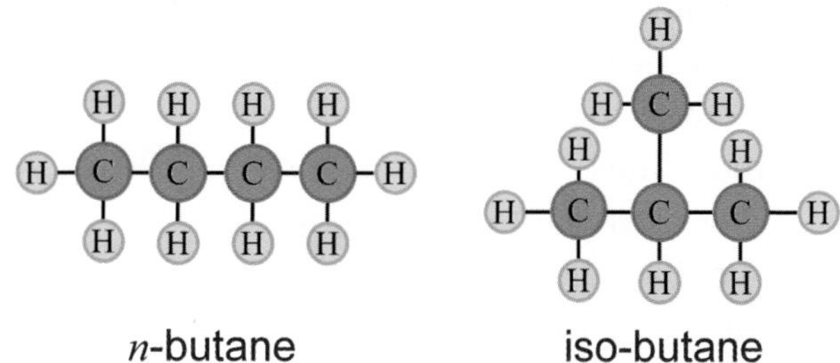

When the carbon chain contains four or more carbon atoms it may form a "branched" chain alkane. For example lets take the butane molecule and remove one atom of carbon by pulling apart the covalent bond. If we place it in the middle of the remaining carbon chain with 3 atoms, we have changed the bonding arrangement. If we try to place it anywhere else, but in the middle, we end up with straight chain butane again.

When a molecule has the same number and type of atoms, but a different bonding pattern it is a **structural isomer**. If you compare the two butane molecules above you will see that they have the same structural formula of C_4H_{10}.

Pentane has five carbon atoms

Lets draw the structural isomers of pentane, C_5H_{12}.

First begin by making a chain of five carbon atoms. -C-C-C-C-C-

Now we add the hydrogen atoms so that each carbon atom makes 4 bonds total.

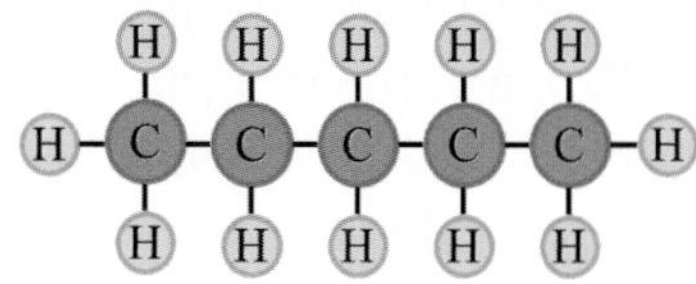

This is called n-pentane or normal pentane, which indicates the straight chain form. Now we can remove one carbon and form a branch in one position. We can also remove two carbons and form a symmetrical branched structure.

Structural isomers for pentane

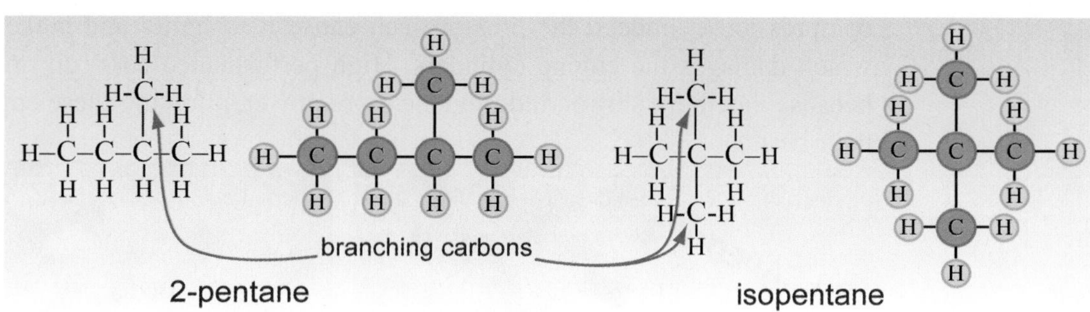

Chemistry terms

structural isomer - molecule with the same number and type of atoms as another molecule but a different bonding pattern.

alkane - a hydrocarbon containing only single bonds. A saturated hydrocarbon.

Petroleum chemistry

Octane or Gasoline forms structural isomers!

Petroleum undergoes reactions, at high temperatures with specific catalysts, which create structural isomers. These reactions cause some linear octane molecules to become branched. Branched chain hydrocarbons burn better in your car engine, because they burn more slowly and smoothly. Much research has been focused on how to get gasoline to burn effectively, and one of the discoveries was that increased branching allowed for better combustion by the engine. The more branched the fuel is the more complete the combustion of the fuel, because it burns slower.

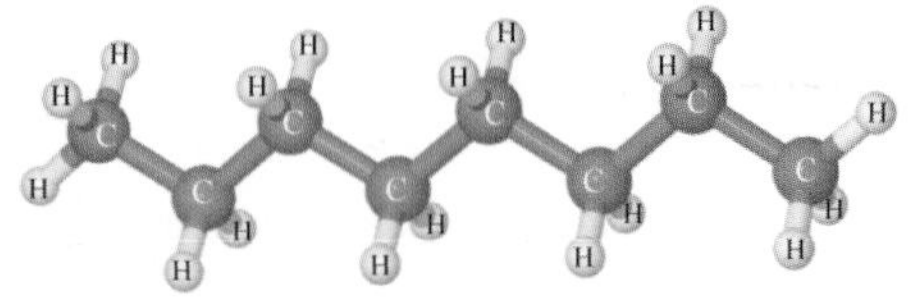

Octane

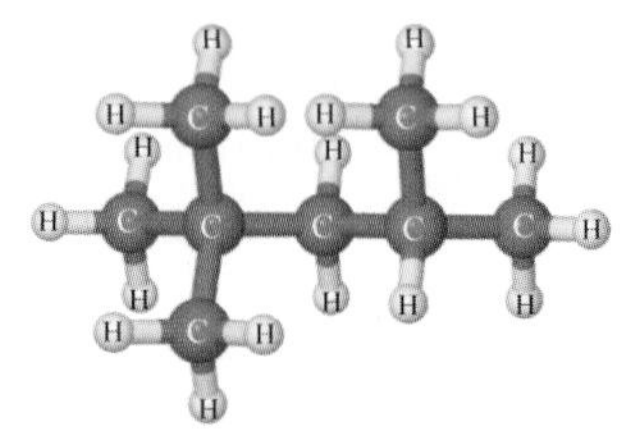

Isooctane

Branched chain hydrocarbons have higher octane ratings. At the fuel pump the octane ratings are based on a ratio of branched isooctane to linear heptane. Linear heptane has an octane rating of zero, and isooctane has an octane rating of 100%. Gasoline is a mixture of hydrocarbons. When we look at the gas pump and see the number 87 this tells us the ratio of 87% isooctane to 13% heptane. Isooctane is a branched alkane formed during the refining of petroleum. *Isomerization reactions are used to produce large amounts of branched chain alkanes.* This is an important application of structural isomers.

Performance cars require high octane fuels

Branched chain hydrocarbons such as isooctane have higher octane ratings. This means these molecules can withstand the high compression of the cylinders in a high performance car. High performance cars have engines that are designed to operate under a higher compression. The high compression produces more energy when the fuel is burned, this gives the performance car more power. If normal gasoline is used in high compression cylinders, the pressure can cause it to ignite and make small explosions, which damages the engine cylinders. High performance cars run on high octane fuels because they burn slower and produce more power, allowing these engines to perform at their best.

Properties of hydrocarbons

Hydrocarbons are nonpolar

Thinking back to chapter eight when you studied intermolecular forces you might remember that methane, CH_4 was a nonpolar molecule. Do you remember why? This was because carbon has four of the same C - H bonds, all at equal angles, which share the electrons relatively equally between the nuclei. Symmetrical molecules such as methane have no net dipole or charge separation therefore they are considered non polar. Nonpolar molecules are held together with London dispersion forces. London dispersion forces can be a relatively "weak" attractive force between hydrocarbon molecules that have low molecular weights such as methane, because of this many of these exist as gases at room temperature.

The longer the hydrocarbon chain the stronger the attractive forces between molecules

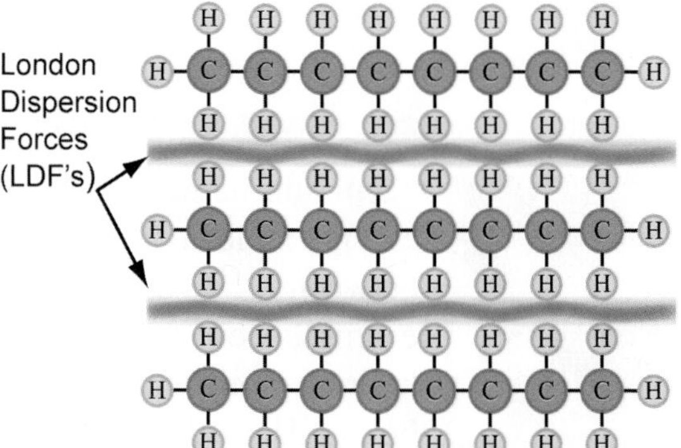

Name	Molecule	Boiling Point(°C)
Methane	CH_4	-161
Ethane	C_2H_6	-88.5
Pentane	C_5H_{12}	36
Hexane	C_6H_{14}	68.7
Octane	C_8H_{18}	125.6

However, as the hydrocarbon chain grows in length the London dispersion forces *increase significantly.* As the length of the chain increases it allows for a larger surface area, which allows the molecules to be in greater contact. The increased attraction causes many longer chain hydrocarbons to be liquids at room temperature.

To clean up an oil spill surfactants and bacteria are used

These nonpolar properties have been given important consideration so that we can better help our environment. A thorough understanding of these properties has allowed scientists to design effective ways to clean up oil spills. Initially, some of the oil is skimmed off the surface of the ocean water. Chemicals called surfactants are then added to the oil to help it mix with the ocean water. Surfactants are detergent like chemicals which break apart the nonpolar attractions, causing the oil to be dispersed in smaller portions. Bacteria are then added to help engulf or eat the smaller, more manageable amounts of oil.

Image courtesy of NOAA.

The pink color comes from oil eating bacteria.

Unsaturated hydrocarbons

Multiple bonds require more energy to break

Unsaturated hydrocarbons contain multiple bonds between the carbon atoms. You may remember that double bonds take more energy to break than single bonds and triple bonds require more energy to break than double bonds. This is because the distance between the nuclei of the atoms is shorter. Just remember the more bonds between the atoms the more energy required to pull them apart. Ethene is used to make the plastic polyethylene. Polyethylene is simply a long chain formed by joining ethene molecules. Ethene is an example of an **alkene**. Alkenes are hydrocarbons that contain one or more double bonds in their structure.

Acetylene gas, used in welding torches contains ethyne, C_2H_2. Acetylene is the "common name" or slang term used for the ethyne molecule. Ethyne has a triple bond between the two carbon atoms. Ethyne is an example of an **alkyne**, which is a hydrocarbon containing one or more triple bonds. Here the prefix eth- still means two carbons and the -yne ending tells us there is a triple bond. When ethyne is burned in air the triple bond is broken releasing a large amount of energy. Acetylene torches reach temperatures up to 3300 °C (6000 °F)! This is enough to melt steel!

Ethyne contains triple bonds

Naming alkenes and alkynes

All alkenes end by using - ene, the parent names are the same ones we used for the alkanes (refer to table 17.1). For example, the name propane means a three carbon chain alkane, when there is a double bond we replace "-ane" with "-ene". If there is a triple bond we use "-yne." We will always use the name of the longest parent chain, so it is important to identify that correctly if the molecule is long.

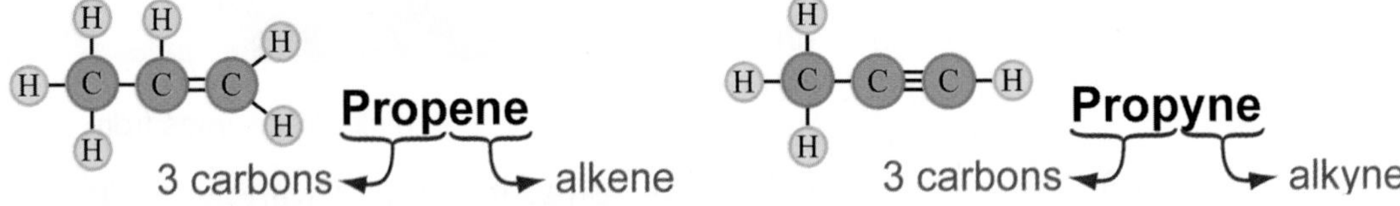

Chemistry terms

alkene - a hydrocarbon containig one or more double bonds.

alkyne - a hydrocarbon containing one or more triple bonds.

Stereoisomers

Spacial relationship of atoms can also result in isomers

In addition to the structural isomers you learned about earlier, there are two more types of isomers. In stereoisomers, atoms are bonded in the same order, but the three dimensional positions of the atoms in space is different.

The first type are called **geometric isomers**. Geometric isomers form when a double bond is present between two carbon atoms. The double bond prevents rotation and the carbons are in "fixed" positions relative to each other.

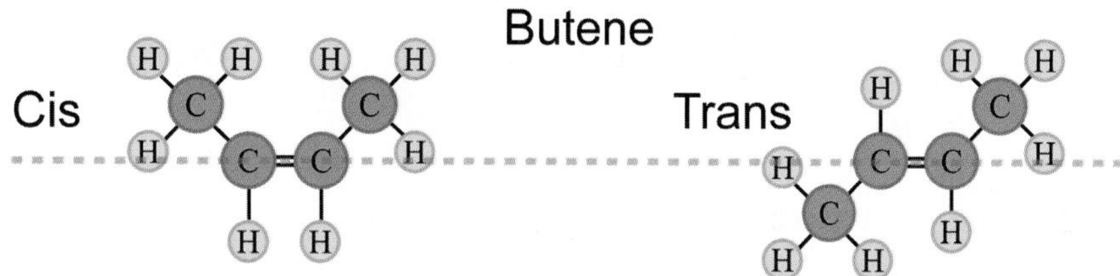

The cis configuration results when $-CH_3$ groups of the molecule are both on the same side. The trans configuration results when the $-CH_3$ groups of the molecule are across from each other. Here is helpful to make these molecules with molecular models! This allows you to observe the differences first hand.

The second type of isomer is called an **optical isomer**. Optical isomers result when the central carbon atom has four different atoms or groups attached to it. Optical isomers rotate light differently with respect to each other. When plane polarized light, or light from only one direction, shines on an optical isomer the molecule either rotates the light clockwise or counter clockwise.

Make two mirror image tetrahedrons with molecular models

Here we will consider 4 different atoms(H, F, Cl, Br) attached to a central carbon atom. These isomers look the same at first glance. But place one model on top of the other and you will see that no matter how you turn them they cannot be super imposed! This is like your left and right hand. If you hold them facing each other they are mirror images. If you place one on top of the other your thumbs do not line up. Compare your tetrahedrons and see.

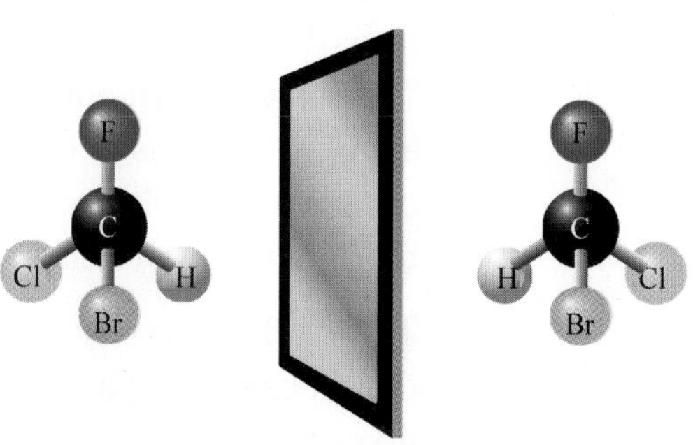

geometric isomers - are molecules with the same sequence of atoms, but different placement of groups around a double bond.

optical isomers - are formed when the carbon atom has four different groups attached to it. These isomers form "mirror images" of one another.

Aromatic hydrocarbons:

Vanilla and cinnamon have pleasant aromas

This unique class of organic compounds were named for their distinct aromas. Originally these compounds were discovered in spices and flavorings. Some of these compounds include: vanillin, cinnamon, cloves, and wintergreen.

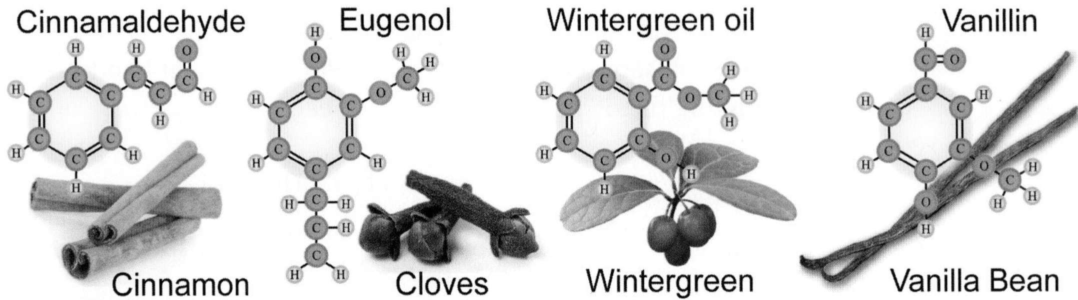

Benzene is common to aromatic hydrocarbons

The **aromatic hydrocarbons** contain at least one cyclic ring structure, and in some cases more than one ring is present in their structure. When scientists examined these substances to see what gave them their pleasant odor they discovered that they all had one feature that was the same! Each contained a six membered ring structure known as **benzene**. Benzene, C_6H_6, is a six membered carbon ring with alternating double bonds between the carbon atoms.

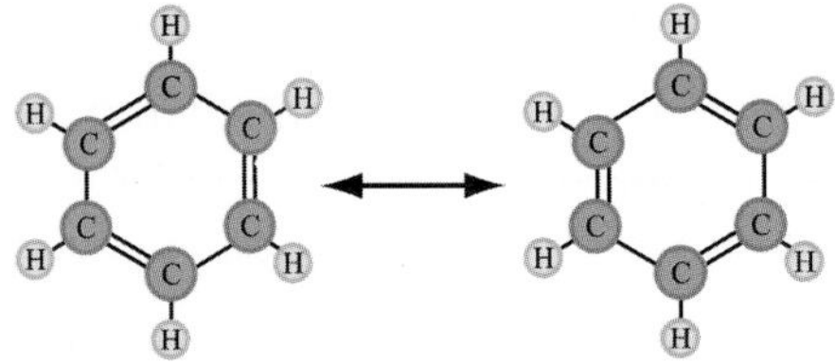

Because electrons are always moving benzene's double and single bonds are never actually static, or fixed in place. The alternating double bond structures represent two "resonance" structures that help us to visualize the bonding pattern. Benzene never actually exists as either of the resonance structures, but is a "blend" of the two structures, because the electrons are evenly distributed throughout the structure. A better representation of benzene is a six membered ring with a circle inside it.

The circle represents the electrons being shared equally between the six carbons. This equal distribution and sharing of the electrons gives benzene a special stability. The electrons in benzene's double bonds are referred to as "delocalized", because they have the ability to move about the entire molecular structure.

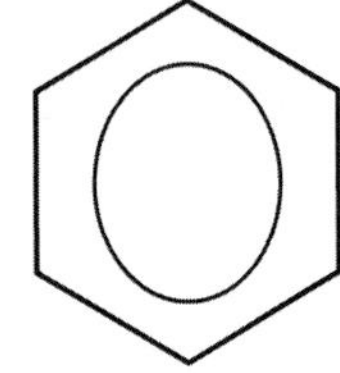

Chemistry terms

aromatic hydrocarbon - a hydrocarbon containing one or more cyclic ring structures.

benzene - an aromatic hydrocarbon C_6H_6, with alternating single and double bonds between the carbon atoms in the ring.

17.2 Functional Groups

Functional groups give a molecule special properties

The term "functional group" refers to a group of elements that give an organic compound very specific properties. In fact, most organic chemistry is dictated by the type of functional groups attached to the molecule.

Many common products that we use contain functional groups that give each product specific characteristics. These characteristics make each product useful in a different way.

Let's use an analogy to help us understand how the functional groups might influence the behavior of the hydrocarbon chain. Humans can perform different activities depending upon what we have attached to our feet! If we have on skis we can slide down a snow covered mountain, if we have on roller blades we can roll along a paved street. Without skis or roller blades humans have the same capabilities to walk or run. Hydrocarbon chains, are for the most part, similar to each other in their chemical behavior, but when functional groups are added, the organic compounds have special abilities. Functional groups give organic molecules the ability to do different things, just like skis and roller blades give us different capabilities. The different types of functional groups that we will cover in this chapter are listed below in the table.

Remember R refers to any hydrocarbon chain. Shown here it represents CH_3 each time. R = CH_3

TABLE 17.3 Common Functional Groups

Name	Functional Group	Example
Alcohol	R–OH	CH_3CH_2OH
Ether	R–O–R	CH_3–O–CH_3
Aldehyde	$\mathrm{R{-}\overset{\overset{\large O}{\|}}{C}{-}H}$	$\mathrm{CH_3\overset{\overset{\large O}{\|}}{C}{-}H}$
Ketone	$\mathrm{R{-}\overset{\overset{\large O}{\|}}{C}{-}R}$	$\mathrm{CH_3\overset{\overset{\large O}{\|}}{C}{-}CH_3}$
Carboxylic Acid	$\mathrm{R{-}\overset{\overset{\large O}{\|}}{C}{-}OH}$	$\mathrm{CH_3\overset{\overset{\large O}{\|}}{C}{-}OH}$
Ester	$\mathrm{R{-}\overset{\overset{\large O}{\|}}{C}{-}O{-}R}$	$\mathrm{CH_3\overset{\overset{\large O}{\|}}{C}{-}O{-}CH_3}$

Alcohols and ethers

Alcohols contain a hydroxyl group

Hydrocarbons that contain an -OH group are classified as **alcohols**. Methanol and ethanol are likely the most familiar to you. Rubbing alcohol has methanol in it and wine contains ethanol. Alcohol has chemical properties that allow it to be used as a disinfectant on the surface of the skin. It breaks apart lipids and precipitates proteins, because of this it kills organisms (bacteria, fungi and most viruses) on our skin. It evaporates easily leaving a cool feeling. The -OH group is often referred to as a "hydroxyl group." Ethanol is used in medical wipes and antibacterial hand sanitizers. It is used extensively as a fuel source or fuel additive to gasoline. Most american gasolines have 10% ethanol content.

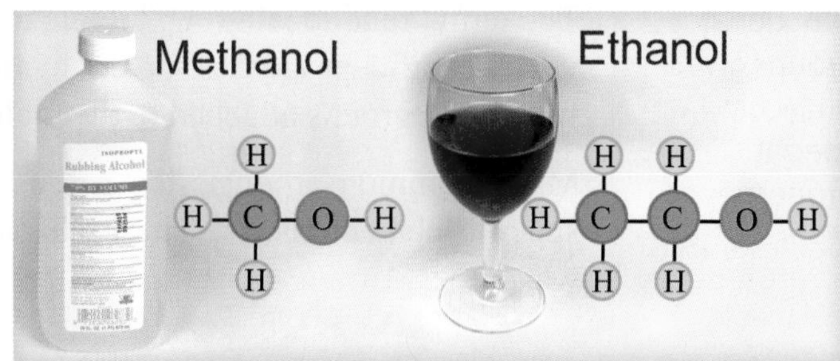

Alcohols can form hydrogen bonds

The - OH or hydroxyl group is capable of hydrogen bonding and this makes short chain alcohols able to easily dissolve in water. The strong hydrogen bonds cause these alcohols to be liquids at room temperature, without the -OH functional group the one and two chain hydrocarbons would be gases. However, alcohols like methanol and ethanol do vaporize easily because they absorb heat from their surroundings.

Ethers readily vaporize

Ether molecules have hydrocarbon chains on either side of an oxygen molecule. Ethers are volatile, meaning they easily form vapors. Recall that a substance can go from being a liquid to a vapor easily if the attractive forces between the molecules are weak. One type of ether, diethyl ether, has been used as an anesthetic in medical surgery for over 150 years. It was first used in surgery by an American doctor in the mid 1800's. When inhaled, diethyl ether caused patients to become unconscious for a period of time, allowing surgery to be performed. Dimethyl ether is used as a "propellant" in aerosol cans. The gas suspends the particles inside the can and allows a mist-like spray.

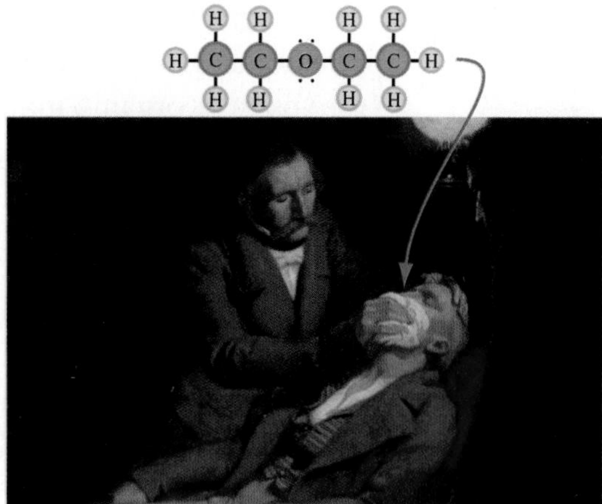
Early use of diethyl ether as an anesthetic.

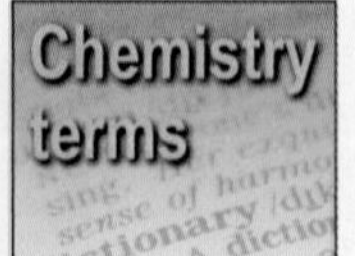

alcohol - a hydrocarbon with a hydroxyl group, -OH, attached.

ether - a molecule where oxygen is bonded to two carbon groups. The two carbon groups are often hydrocarbon chains.

Aldehydes and ketones

Nature provides us with fragrant smells

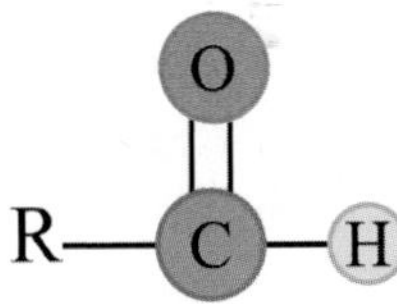

A wide variety of familiar smells and flavors come from aldehydes and ketones. These organic compounds are similar in structure. In both the **aldehyde** and the **ketone**, functional groups contain a central carbon atom double bonded to an oxygen. In an aldehyde the carbon always has one hydrogen bonded to it, and on the other side the carbon is bonded to a hydrocarbon. In a ketone, the carbon is bonded on both sides to a hydrocarbon. The carbon double bonded to the oxygen (C=O) is referred to as a **carbonyl group**. This carbonyl group shared by both the aldehydes and the ketones is quite polar which increases the reactivity of these molecules.

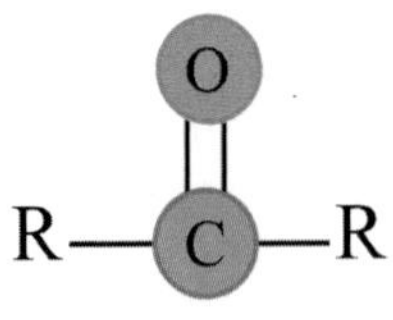

Flavoring molecules found in nature are also produced artificially

Many different aldehydes and ketones have been isolated from plants and animals in nature. They are so popular, that larger quantities of the molecules are often produced industrially. For example vanillin is found in vanilla beans and used as a flavoring in ice cream and cookies. You have heard of "artificial" vanilla flavoring and "pure" vanilla flavoring from extract. The pure extract costs more to purchase, but most agree that it tastes better than the artificial vanilla flavoring. .

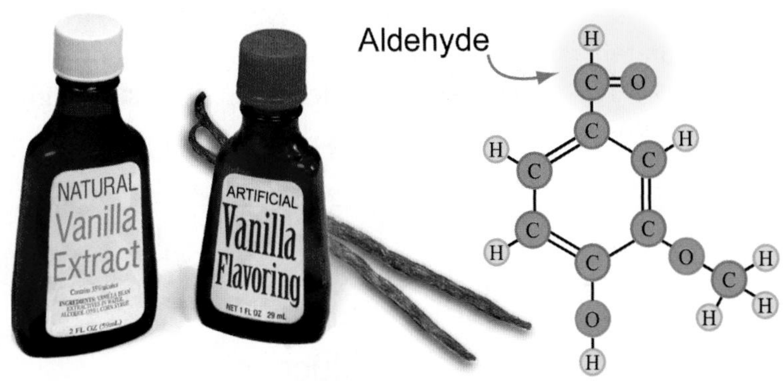

Cinnamon spice is also an aldehyde. It comes from the bark of cinnamon trees! The smell and flavor of cinnamon comes from the molecule cinnamaldehyde, which as it's name indicates, is an aldehyde. Formaldehyde, is used to preserve tissues and as a fixative for viewing tissues under the microscope for medical diagnosis. It also has a distinct smell.

aldehyde - a functional group in which the central carbon makes a double bond with oxygen and then is bonded to hydrogen on one side and to another hydrocarbon.

ketone - a functional group in which the central carbon makes a double bond with oxygen and then is bonded to two hydrocarbons one on each side.

carbonyl group - formed when a carbon atom is double bonded to an oxygen atom (C=O). Common organic reference.

Aldehydes and ketones

Acetone and testosterone are ketones

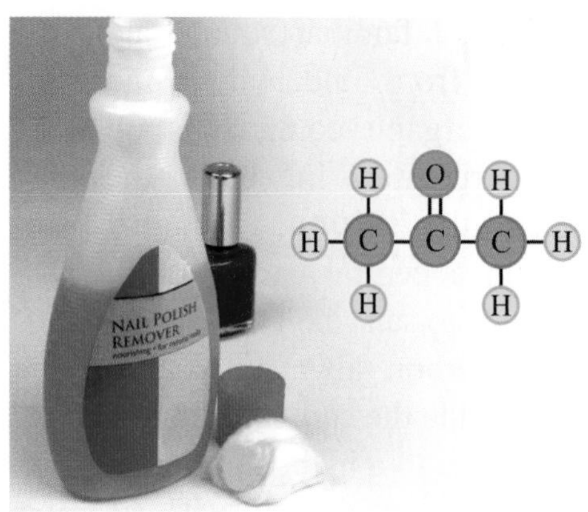

There are also many ketones that you are familiar with. A biological example of a ketone would be testosterone the male sex hormone. Another common ketone is acetone, which is also the simplest molecular example of a ketone molecule. Acetone is found in finger nail polish remover and paint thinner. The smell you notice when you open the fingernail polish remover is actually acetone. Interestingly, our bodies make small amounts of acetone that is used as a building block for other molecules. Chemically, acetone is widely used as a solvent because it mixes well with water and dissolves organic substances. Acetone is widely used in the lab and in industry as a cleaning agent. Acetone is synthesized in large amounts for use in the production of polymers.

Carbohydrate molecules are aldehydes and ketones

Perhaps one of the most important examples of aldehydes and ketones in nature are carbohydrate molecules. The sugar, glucose, $C_6H_{12}O_6$, is a 6 carbon aldehyde sugar,

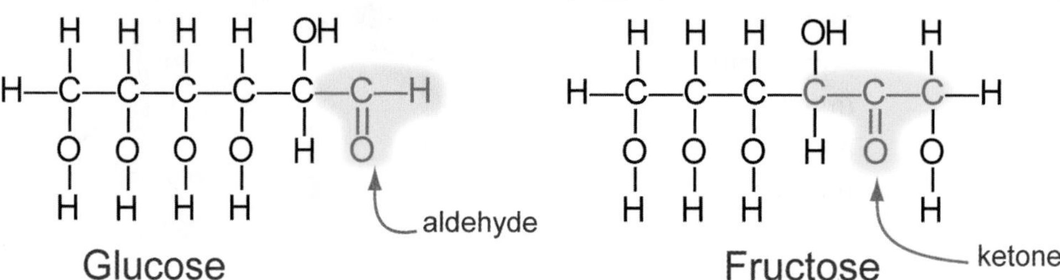

that contains several hydroxyl groups. In contrast, fructose which is the sugar commonly found in fruit, is a ketone. Although it is much easier to identify the aldehyde and ketone functional groups when looking at the linear structures, these sugars are generally found in the ring form shown below.

Carbohydrates are the primary source of energy for most animals. These aldehyde and ketone sugars make disaccharides. A disaccharide is a 2-unit sugar molecule, hence the name di-meaning two and saccharide-meaning sugar. Sucrose, or table sugar is a disaccharide formed by connecting a glucose molecule to a fructose molecule. Starch molecules from pasta and potatoes are also formed by connecting many glucose molecules in a chain.

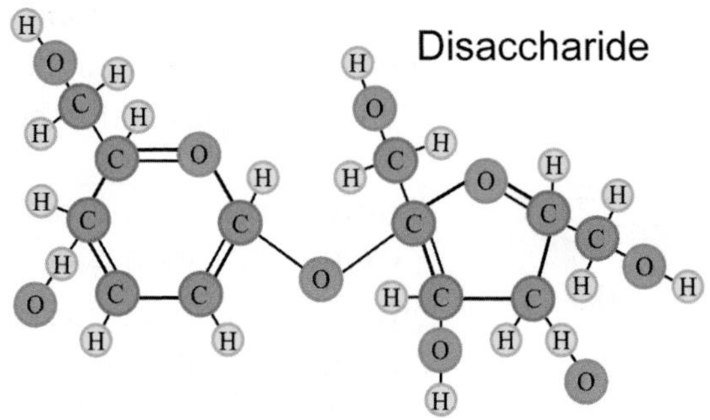

Carboxylic acids

Carboxylic acids are widespread in nature. A **carboxylic acid** contains a functional group composed of four elements, COOH, called a carboxyl group.

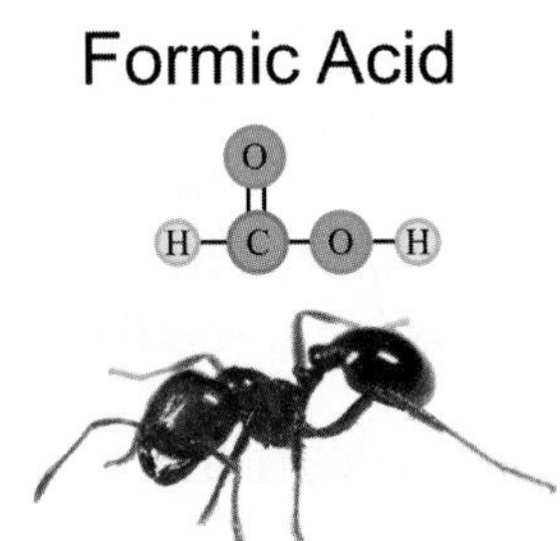

Acetic acid is an example of a carboxylic acid. You can see its structure on the right. Acetic acid is present in sour foods like vinegar and citrus fruits. Red ants contain another type of carboxylic acid, named formic acid. This is what makes their bite sting. Spoiled foods also contain carboxylic acids.

Carboxylic acids are weak acids

As the name indicates carboxylic acids have acidic behavior. You may remember that acids are proton, H^+, donors. The structure of the carboxyl group gives it the ability to lose the hydrogen attached to the oxygen molecule, and this is responsible for the acidic properties of these molecules. The electronegative oxygens draw the electrons in the bonds more toward themselves, this leaves the hydrogen bonded to the oxygen with a slight positive charge. The positive charge is attracted to a nearby water molecule, which causes it to be pulled off, or ionized. Remember this only happens approximately five percent of the time, but this partial ionization still produces significant amounts of H^+ ion in solution.

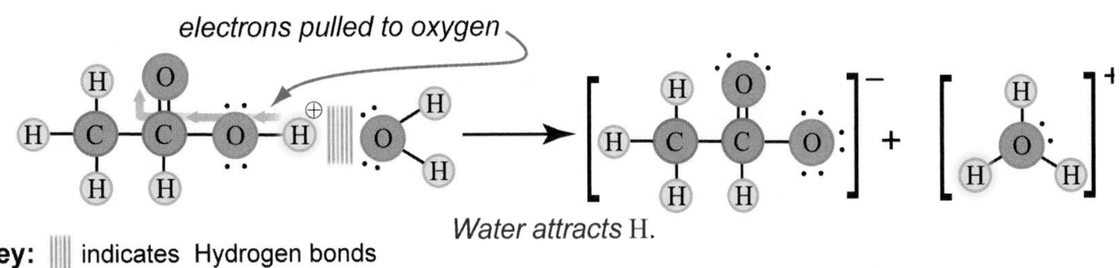

Carboxyl groups form hydrogen bonds

Another important property of molecules containing carboxyl groups is that they can form strong hydrogen bonds and this increases their melting and boiling points.

Carboxylic acids are one of the key functional groups of all amino acids. In the next chapter you will see that this group is involved in the bonding of amino acids to make proteins. Carboxylic acid functional groups are also present on long chain fatty acid molecules.

Chemistry terms

carboxylic acid - an organic molecule containing a carboxyl group, COOH.

Amines

Found on amino acids

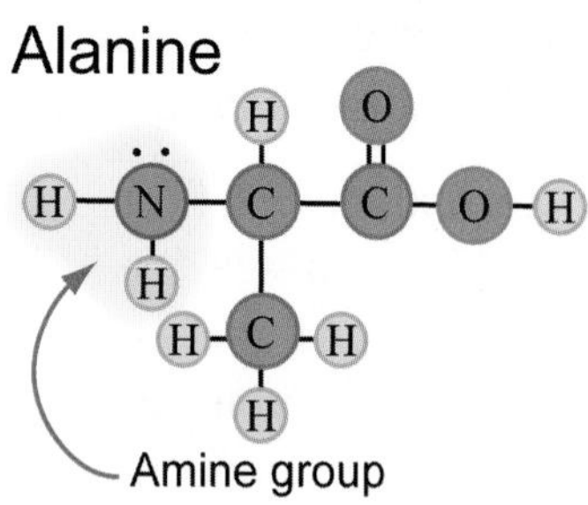

Amines are organic compounds that contain an amino group, NH_2. The amino group is an important functional group on amino acids. You may recall that amino acids make up all of the proteins in our bodies! The amino group forms a bond with the neighboring carboxyl group which links the amino acids together in a chain. As you will see in the next chapter these proteins regulate most of the chemical reactions in our bodies.

As a class of molecules the amines have several important applications. The first being protein synthesis. However, amines are also used to make dyes, and to treat allergies. Chlorpheniramine, shown here, is the active ingredient in many allergy medications. This medication is also effective in reducing the side effects of other drugs, which may cause allergic reactions.

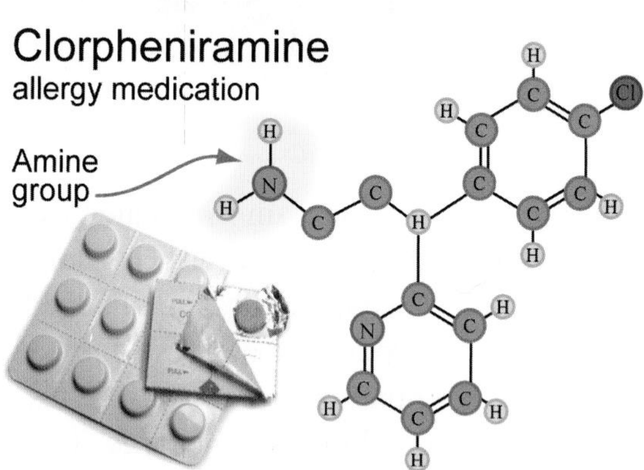

Amines are weak bases and also form hydrogen bonds

Amines have basic properties, because the lone pair on the top of the nitrogen atom allows them to accept a proton or hydrogen ion (H^+). The high electronegativity of the nitrogen atom also contributes to the molecules high polarity, and makes it capable of forming strong hydrogen bonds. The amine functional group causes amino acids to have basic behavior in aqueous solutions, because of this pH plays an important role in the functional group behavior.

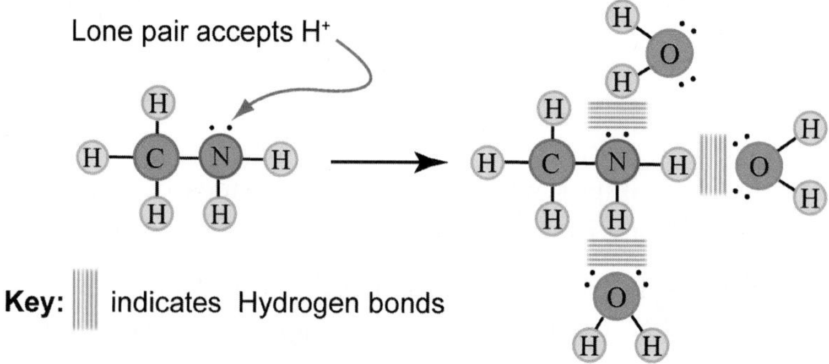

Chemistry terms

Amines - an organic molecule containing an amino group, NH_2.

Esters

Esters smell good!

Esters are very common in nature. They are often responsible for the lovely fragrances of flowers and ripened fruit. On the right you can see the molecule responsible for the smell of lavender. The **ester** functional group resembles a carboxyl group, except that it has a carbon group attached to the oxygen instead of a hydrogen atom, H.

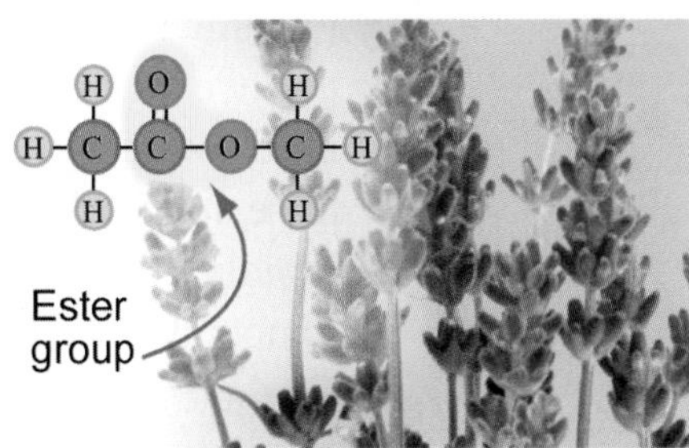

Esters are added to foods and drinks.

Flavor comes from a combination of taste and odor that is unique. Fragrances and tastes are often due to a complex blend of several esters, instead of just one. Manufacturers often add specific esters to make their foods and drinks taste better. Esters are also added to shampoos to give the desired fragrance. Using the appropriate mixture you can make things smell like strawberries with no actual strawberries!

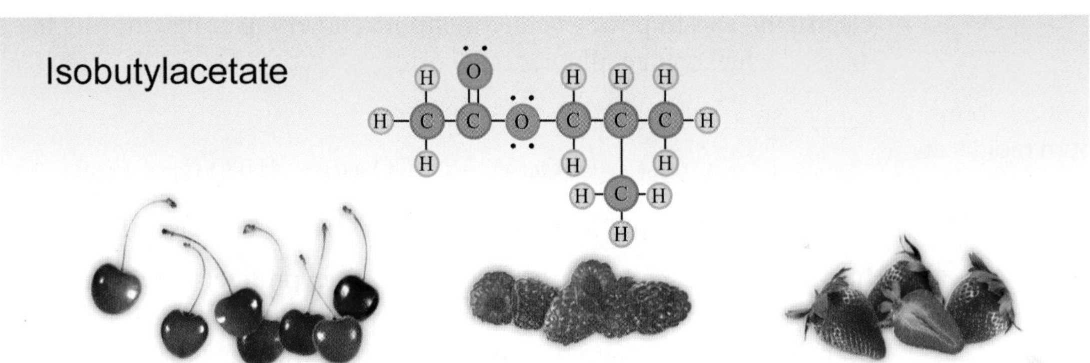

Fats and oils are esters too. Long chain fatty acids are attached by ester bonds to the glycerol molecule. Ester bonds can be formed by joining carboxylic acids and alcohols.

Esters cannot hydrogen bond among themselves because they have no hydrogen atom that is directly attached to an electronegative atom of nitrogen or oxygen. Esters can however, accept a hydrogen bond on the oxygen's lone pairs. Esters cannot make a hydrogen bond to donate. Due to the lack of hydrogen bonding, esters vaporize more readily than carboxylic acids.

Polyesters are polymers with the ester group attached.

Polyesters are commonly used in clothing and in foods. Polyesters are polymers that contain the ester functional group on the hydrocarbon chain. One of the most common synthetic fibers is polyethylene terephthalate, or PET as it is abbreviated. Polyester clothing made from these fibers is very popular because it does not easily wrinkle.

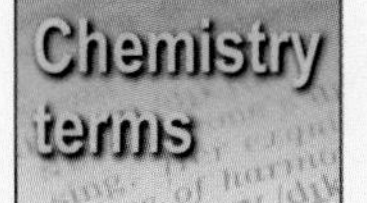

ester- an organic molecule containing a -COOR group, is an ester. Esters are responsible for many smells and tastes in fruits and flowers.

17.3 Organic Reactions

Organic compounds undergo some important chemical reactions that we will discuss in this section. Organic reactions have been used to create heat and light since early human history. Humans have long taken advantage of the fact that hydrocarbons burn rapidly in the presence of oxygen at sufficiently high temperatures. Industrial production of solid shortenings, such as Crisco$^{(TM)}$ and of textiles and fabrics are dependent upon organic reactions. Plastics are widely used in our society, and the formation of plastics takes place through polymerization reactions, which are an important type of organic reaction.

Lets begin with combustion reactions since you are already familiar with them. Earlier we studied combustion reactions in several areas including: stoichiometry, chemical reactions, and thermochemistry. Combustion reactions have been used to generate electricity, and to power vehicles and machinery. Because of this the alkanes are widely used as a fuel source all over the world!

Hydrocarbons burn rapidly at high temperatures

<u>Combustion reactions:</u>

1. $C_3H_8(g) + 5O_2(g) \rightarrow 3CO_2(g) + 4H_2O\ (l) + \text{Heat}$
2. $2C_4H_{10}(g) + 13O_2(g) \rightarrow 8CO_2(g) + 10H_2O\ (l) + \text{Heat}$

In all "complete" combustion reactions the products are always carbon dioxide and water. Here we are assuming complete combustion. In real life when hydrocarbons are burned there are uncombusted hydrocarbon fragments remaining after the combustion process. These fragments contribute to our air pollution.

Lets look at what happens to the alkane molecule structurally as we burn, or combust it. Some of us use natural gas to heat our homes. Natural gas contains several alkanes, but primarily methane, CH_4. High temperatures, or heat is necessary to break the stable carbon to hydrogen bonds, and initiate the reaction. However, when the more stable bonds of CO_2 and H_2O form as products, energy is given off, or released to the surroundings. The amount of heat released is *more* than the amount necessary to initiate

C_3H_8 + O_2 in air —heat→ CO_2 + H_2O + heat

the reaction, so overall the reaction provides us with energy.

Reactions of alkanes

Alkanes have low reactivity

Overall as a group the alkanes are considered to be relatively unreactive! This low reactivity comes in part from the strength of the C - C and the C - H bonds. These bonds are very stable. At room temperature alkanes do not react easily with substances like acids and bases that often cause chemical reactions. However, when light energy is added or when the temperature is raised their reactions are very useful to us.

A new atom is substituted for hydrogen.

Alkanes undergo substitution reactions. In **substitution reactions** one or more hydrogen atoms on the alkane are removed and replaced by different atoms. The name substitution reaction comes from the fact that a different atom is "substituted" for one or more of the hydrogen atoms. For example, using a halogen such as Cl_2 or Br_2, methane will undergo substitition.

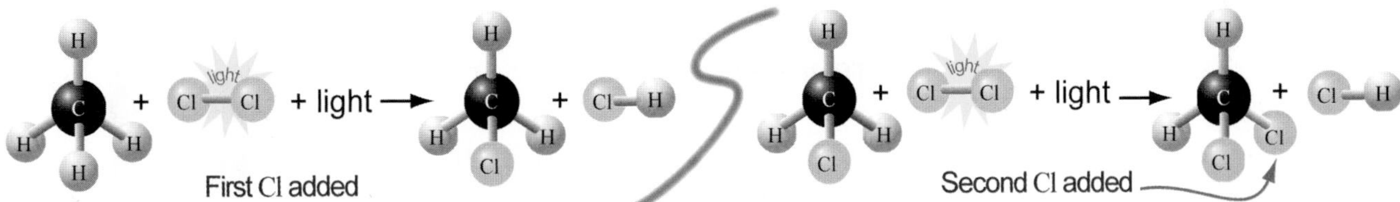

The light used in the equations above indicates that energy in the form of light, $E = h\nu$ is required to break the covalent Cl - Cl bond, and form free chlorine radicals (Cl•) with an unpaired electron. $Cl_2(g) + h\nu \rightarrow Cl\bullet + Cl\bullet$ The chlorine radical is very reactive and is able to break the C - H bond on the hydrocarbon molecule. The overall result is that a chlorine atom is substituted for a hydrogen atom. This process can be repeated until several or all of the hydrogen atoms are replaced on the hydrocarbon. Interestingly, alkanes become more reactive when modified by substitution reactions. This is because the bond between carbon and the new element is not as strong as the original carbon to hydrogen bond. The difference in electronegativity causes the new bond to be more polar and therefore more reactive.

Removal of H_2 forms double bond(s)

Alkanes also undergo **dehydrogenation reactions**. Dehydrogenation reactions remove hydrogen, H_2 from the alkane causing formation of double bonds Dehydrogenation causes areas of unsaturation along a hydrocarbon chain.

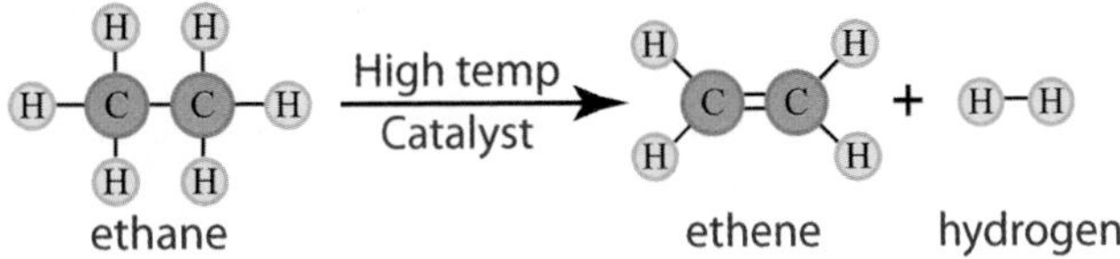

substitution reaction - hydrogen atoms on an alkane are removed and replaced by different atoms. The name substitution reaction comes from the fact that a different atom is "substituted" for one or more of the hydrogen atoms.

dehydrogenation reaction - remove hydrogen, H_2 from the alkane causing formation of double bonds. This process forms unsaturated alkenes.

Reactions of alkenes and alkynes

Double and triple bonds are more reactive!

The presence of double and triple bonds in their molecular structures make alkenes and alkynes much more reactive than alkanes. If compounds are more "reactive" it means they are more likely to react with other chemicals. The presence of these multiple bonds allows them to undergo addition reactions, and polymerizations. Both of these reaction types are widely used for industrial purposes, to make plastics and synthetic fibers. In an **addition reaction** atoms are added to the two carbon atoms forming the multiple bond. **Hydrogenation** is one of the most common types of addition reactions. In hydrogenation reactions hydrogen atoms are added to the carbon atoms forming double bonds. The process of hydrogenation is used to make solid shortenings, such as Crisco(TM). The removal of the double bonds allows the hydrocarbon molecules to pack tighter allowing the intermolecular forces have a stronger associations and this causes the liquid oil to become a solid.

Hydrogenation reaction

```
H       H H                                      H H H H
 \      | |          catalyst (Ni)               | | | |
  C=C-C-C-H + H2(g)  ------------->          H-C-C-C-C-H
 /    | | |             at 500°C                 | | | |
H     H H H                                      H H H H
```

Predict the product of the hydrogenation of 2-pentene, C_5H_{10}

```
  H H H H H
  | | | | |
H-C-C=C-C-C-H
  |     | |
  H     H H
```

Given: *Structure of 2-pentene*

Relationships: *In hydrogenation H atoms are added across the double bond.*

```
  H H H H H
  | | | | |
H-C-C-C-C-C-H
  | | | | |
  H H H H H
```

Discussion: *The hydrocarbon 2-pentene is now saturated and contains only single bonds.*

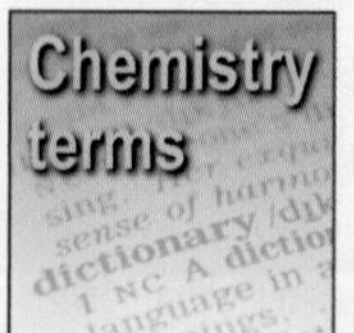

addition reaction - a chemical reaction that adds atoms to the carbon atoms forming multiple bond(s) in a molecule.

hydrogenation reaction - a type of addition reaction where hydrogen gas, H_2, is added to a double or triple bond in a hydrocarbon, causing the structure to become saturated.

Transfats and "partial" hydrogenation

Now that you understand hydrogenation, we will review how transfats are formed. Transfats were created by accident through the **partial hydrogenation** of unsaturated fats. The term "partial" means that some double bonds still remain in the structure of the fat, and that the fat is not completely hydrogenated. Partial hydrogenation causes the *remaining double bonds* to have a "trans" orientation on either side of the double bond.

Cis vs. Trans configuration

In their natural form fats have cis double bonds. Below you can see elaidic acid and oleic acid. These fats are geometric isomers, which you may remember from earlier in the chapter, means they have identical structures EXCEPT for the double bond configuration. Elaidic acid is the main ingredient of partially hydrogenated vegetable oils, and it has the trans configuration. Oleic acid is found in olive oil, and has the cis configuration shown below. Remember the partial hydrogenation process does not fully

```
Oleic acid      H H H H H H H H H H H H H H H H O
Cis             | | | | | | | | | | | | | | | | ||
              H-C-C-C-C-C-C-C-C=C-C-C-C-C-C-C-C-C
                | | | | | | | |     | | | | | | |  \
                H H H H H H H H     H H H H H H H   OH

Elaidic acid    H H H H H H H H   H H H H H H H H O
Trans           | | | | | | | |   | | | | | | | | ||
              H-C-C-C-C-C-C-C-C=C-C-C-C-C-C-C-C-C
                | | | | | | | | |   | | | | | | |  \
                H H H H H H H H H   H H H H H H H   OH
```

hydrogenate all of the double bonds. This gives the fat a semi-solid texture. Unsaturated fats such as corn oil, olive oil, and peanut oil are organic hydrocarbons. Due to the presence of double bonds in their structure these fats are liquids at room temperature.

Unsaturated fats are liquids at room temperature

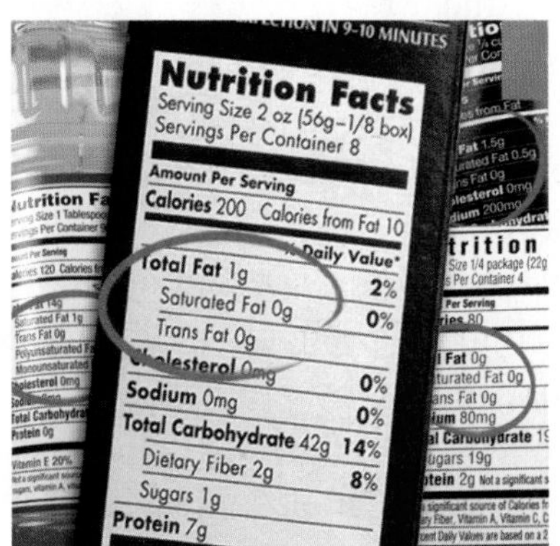

The process of partial hydrogenation causes food stuffs that contain oils, like potato chips, to last longer on the shelf in a grocery store. Manufacturers use this process to cut down on spoilage in foods that contain unsaturated fats. Now that research has shown transfats to be harmful to people's health, scientists have had to come up with ways to process oils that does not produce the harmful transfats. In fact, most manufacturers have advertised no transfats on the label of their products, so that consumers will purchase them .

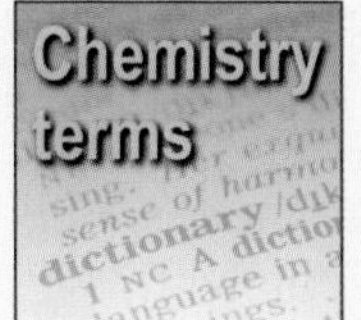

partial hydrogenation - a chemical reaction where only some of the unsaturated fat is hydrogenated. In this case some double bonds remain in the structure, causing the formation of transfats in place of the naturally occurring cis configuration.

Petroleum refining

Organic matter formed petroleum

Today's petroleum was formed from the remains of prehistoric plankton and algae! This organic material settled to the bottom of an ancient lake or sea in large amounts, and over time under conditions of extreme pressure and temperature formed crude oil.

Crude oil, or **petroleum** is a blend of hydrocarbons. Most of the mixture is composed of straight chain and branched chain alkanes. There are also small amounts of aromatic hydrocarbons, such as benzene, and some organic compounds that contain sulfur, oxygen and nitrogen. Petroleum in its crude form is not very useful! It needs to be refined, and separated into parts called fractions. Fractions have different sizes, weights, and boiling temperatures depending on the length of the carbon chains.

Petroleum must be separated into fractions

The refining of petroleum begins by distillation. The distillation process heats the crude oil to temperatures above 1000°F(600°C). The mixture boils and forms vapor which then enters the fractionating column. The fraction condenses at the point in the column equal to it's boiling point. Fractions are removed and separated from the crude oil mixture as they reach their boiling point. Each fraction is a mixture of hydrocarbons.

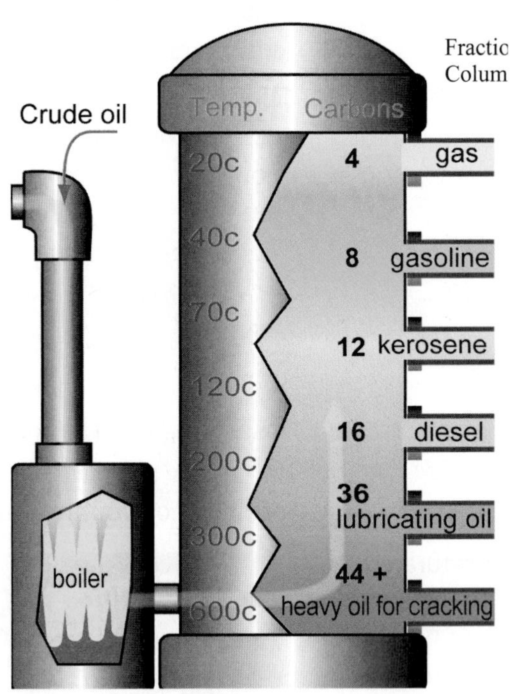

A process called **cracking** is used to break larger hydrocarbons down into smaller hydrocarbons. For example, when the demand for gasoline is high, fractions with longer hydrocarbon chains and higher molar mass, are *broken down or cracked* to form gasoline and kerosene. The cracking process requires very high temperatures and a catalyst. Ethylene and other alkenes are produced during the cracking process, these are used in the production of plastics. Low molar mass alkanes and hydrogen gas are also produced as by-products of the cracking process. These are useful as raw materials in the chemical industry. Crude petroleum provides us with many different molecules which serve as fuels or as building blocks for new materials.

Chemistry terms

petroleum - a blend of hydrocarbons, formed from prehistoric organic matter, containing mainly straight and branched chain alkanes.

cracking - a process used to break long chain hydrocarbons down into smaller hydrocarbon fragments. This process requires high temperatures and a catalyst.

Polymerization

Long chain molecules

Many important biological molecules are polymers. **Polymers** are long chain molecules. A polymer literally means many parts. The word comes from two Greek words, "poly" which means many, and "meros" meaning parts. Here we will review some information covered in the early bonding chapter 8. We make a chain by connecting the units called **monomers** together. Monomers are small molecules that make up the polymer chain. The small molecules are joined together by covalent bonds. The chemical reaction that joins the monomers is called **polymerization**.

Where are polymers?

For example starch is a polymer of sugar molecules, called a polysaccharide, which is contained in foods such as pasta, potatoes, and wheat. Many polysaccharides are also complex carbohydrates. Polymers formed in nature make up the majority of polymers we are exposed to in our daily lives. Cotton, silk and rubber are all examples of natural polymers. Synthetic polymers began to be produced during World War II after our supply of raw materials became limited. Our country needed to produce materials that could be use in place of the scarce, but naturally available polymers. Polyethylene and polyester fabrics were among the first synthetic polymers.

Plastics are also polymers. Can you imagine what we would do with out the different forms of plastic? Plastics have revolutionized our modern society! Each of these products shown uses some form of polymer. Think about the plastics you use each day and imagine how things would change if we did not have plastic. Making a large molecule by connecting lots of small monomers together is very practical. The process of connecting molecules to make a long chain occurs in two different types of chemical reactions. We will learn about these in the next few pages.

polymer - "poly" many and "meros" parts. A long chain molecule formed by connecting small repeating units with covalent bonds.

monomer - small molecule used to make the chain of a polymer.

polymerization - chemical reaction that joins monomers to produce a polymer.

Addition polymers

Polyethylene is the most widely used plastic.

One of the simplest and most widely used polymers in the world is polyethylene. Polyethylene is used to make plastic shopping bags, shampoo bottles, children's toys, and many other consumer products. It is a strong, but flexible plastic.

Polyethylene forms through a simple polymerization reaction called **addition polymerization**. Addition polymerization attaches monomers using their double bonds. Initially, the double bond is broken by a **free radical**. A free radical is a molecule with an unpaired electron, that is very reactive. The double bond then opens up and the two electrons are used to make new carbon to carbon single bonds. In this way The unsaturated monomers are added to the right side of the chain where the unpaired electron is. The stepwise addition of these monomers creates a long saturated polymer. The polymerization reaction shown below uses ethylene monomers.

Polymerization of ethylene

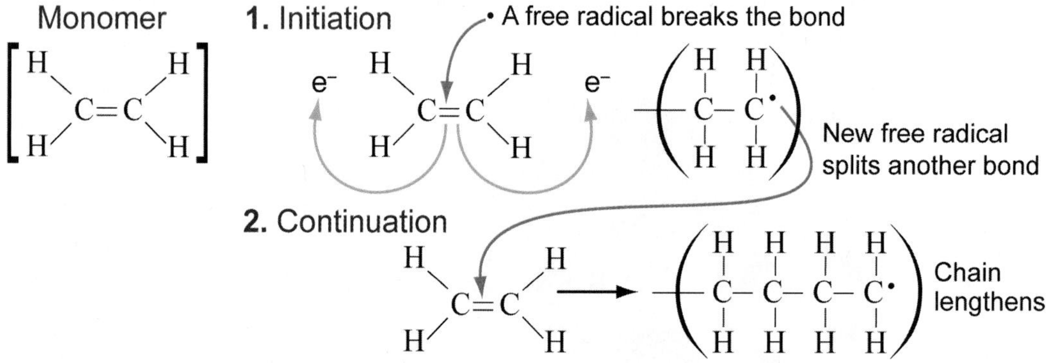

Monomer

$[H_2C{=}CH_2]$

1. Initiation

• A free radical breaks the bond

e^- $H_2C{=}CH_2$ e^- $\rightarrow$ $-(CH_2-CH_2^{\bullet})$

New free radical splits another bond

2. Continuation

$H_2C{=}CH_2 \longrightarrow -(CH_2-CH_2-CH_2-CH_2^{\bullet})$

Chain lengthens

The softness of the polyethylene can be controlled by the length of the chain. If the chain is short, about 100 units long the plastic will be soft. If the chain is long, 1000 units the plastic is more rigid. The branching of the polymer chain also effects it's properties. HDPE or high density poly ethylene is formed from polymer chains without many branches. The density increases because the polymer can pack closer to the other nearby chains. In LDPE, or low density polyethylene the polymer chains are more branched and the plastic is thinner, more like a film. Sandwich bags are made from LDPE.

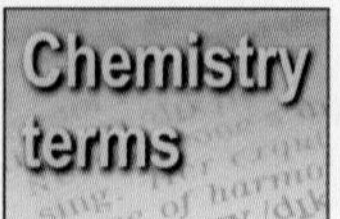

addition polymerization - type of polymer formed by adding monomers together using their double bonds.

free radical - molecule or atom with an unpaired electron.

Condensation polymers

Monomers connected by loss of small molecule

Another general type of reaction used by nature and industry to synthesize polymers is called **condensation polymerization**. Here the polymer is formed by eliminating a smaller molecule such as water. This type of polymerization process does not involve double bonds and can occur between any two monomers. The monomers are joined in a step wise fashion from head to tail. Nylon is an important industrial polymer, and a good example of condensation polymerization. Nylon is used for clothing, carpets, and ropes. Over 1 million tons of nylon are produced in the U.S. each year!

$$H_2N-(CH_2)_6-N(H)-H \;\; H-O-C(=O)-(CH_2)_4-C(=O)-O-H \longrightarrow H_2N-(CH_2)_6-N(H)-C(=O)-(CH_2)_4-C(=O)-O-H \; + H_2O$$

In this reaction a water molecule is removed, and this is what joins the polymer together. The bond formed is called an **amide linkage**. If you think back to section 17.2 to the amine functional group was- NH_2 and an amide is formed when an amine is linked to a carbon double bonded to an oxygen, or a carbonyl (C=O) group. The chain continues to grow adding on to the carboxylic acid and amino ends of the chain. A water molecule is lost each time a monomer is joined. Take a careful look at the above reaction.

Kevlar is another polymer formed through condensation polymerization. Kevlar is a very strong material, about five times the strength of steel. It is best known for its use in bullet proof vests. Kevlar is unique in because it is light weight but also very strong. Prior to the discovery of Kevlar, materials that were strong were also very heavy which limited movement and made them uncomfortable. Kevlar's strength is partially due to the crystallinity of the structure. Its molecules are very organized, with a high degree of internal symmetry. The individual polymer strands of Kevlar are held together with hydrogen bonds. Stephanie Kwolek, a researcher for Dupont, discovered Kevlar in the late 1960's. It is likely the most important polymer discovery to date.

Chemistry terms

condensation polymerization - type of polymer formed by linking monomers through the loss of a small molecule such as water.

amide linkage - type of bond formed between an amine group ($-NH_2$) and a carbonyl (C=O) group.

Diapers, milkshakes, and polymers

Polymers called hydrogels have many different real life applications. These polymers are very good at absorbing water and are thus termed "hydrogels." Did you know that the same type of polymers used in baby diapers were also used in our fast food milk shakes? In diapers these polymers are used for their absorbent properties. They can hold up to 300 times their own weight in water. In milk shakes a small amount of these polymers can act as a gelling agent. When the polymer is added to the milk shake it forms a gel, sort of like jello, but not as thick. Have you ever noticed that your milk shake does not melt, even at warm temperatures?

There are several different types of hydrogel polymers. Chemically hydrogels are cross-linked polymers that contain strong hydrophillic groups. *Hydrophilic means water loving*! The most common synthetic hydrogel is sodium polyacrylate. A common natural hydrogel is sodium alginate. How do these hydrogels work?

Hydrogels work by forming hydrogen bonds with the water molecules in the aqueous solutions. The water molecules are drawn into the hydrogel by osmosis. Osmosis is the flow of water from a low to a high concentration. Water always moves into the polymer because there is no water inside of it initially. There are negative portions along the polymer chain that attract the positive ends of the water molecule. As water is added to the hydrogel it is slowly pulled in and hydrogen bonds are formed. This process continues until the polymer is saturated or "full" holding as many water molecules as it is capable of holding.

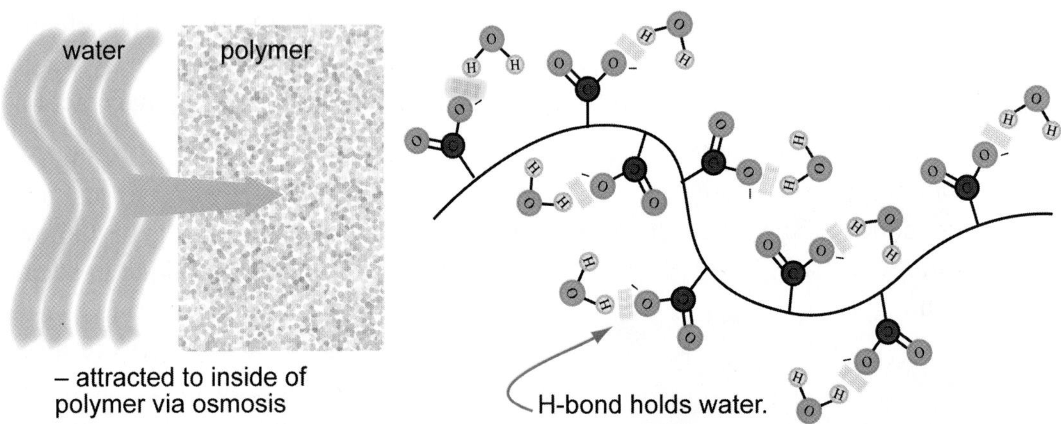

Once the water molecules are hydrogen bonded to the oxide anion, they are "held" by the polymer, or "absorbed." These hydrogel polymers can hold between 300 to 500 times their own weight in water. The absorbent range depends upon the amount of salt in the solution. When salt is present, the positive ion, can selectively bind to the oxide anion leaving less room for hydrogen bonds to form. What does this mean? Basically that the absorbent properties of hydrogels are greater in pure water or in solutions with dilute salt concentrations. Large amounts of salt can cause the absorbing to decrease by roughly 80%.

Today sodium polyacrylate is most commonly used in baby diapers. If you open up a baby diaper, and look under the layer of material, you will see the small crystalline beads of this super absorbent polymer. As water is absorbed the little beads swell and become heavier. Urine does contain some salt, but the amount is so dilute that the liquid is still absorbed effectively.

Many uses have been developed for these super absorbent polymers. One very important application is their use in farming. In very dry regions of the world this polymer is worked into the soil so that when it rains it absorbs water. The polymer absorbs much more water than the soil would normally hold on its own. The polymer retains the water until the soil is once again dry. It then slowly releases the water back into the soil allowing crops to continue to grow. A major benefit of this super absorbent polymer technology is that it also allows these communities to conserve on precious drinking water, because they now use less of it to water their crops and gardens.

The small animals that children love to watch "grow" in water are also made from sodium polyacrylate! These "grow" toys have become very popular. As they grow each day they are absorbing water, until they reach a point of saturation and at this point they are full size. Remember the chemical properties of the polymer, hydrogen bonding and osmosis, are what allow these toys to "grow." Without dye, sodium polyacrylate is white and this polymer is frequently used to make a wintery scene in a broadway play, where real snow is not available!

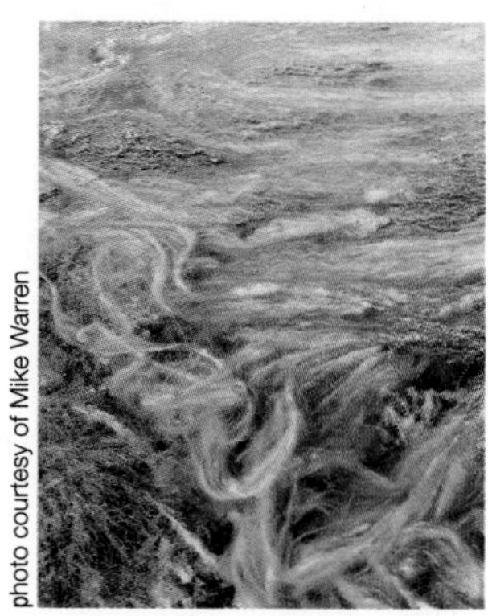
photo courtesy of Mike Warren

Today in your milkshakes, fast food restaurants commonly use polysaccharide polymers such as carrageenan to gel their milk shakes. Carrageenan is a linear polymer of about 25,000 galactose units. Galactose is a natural sugar made by our bodies and it is also found in dairy products and vegetables. Carrageenan comes from a red seaweed native to the coast of Ireland. When the seaweed is boiled for 20 to 30 minutes and removed there is a gelatinous substance left behind. Carrageenan has the ability to form complexes with calcium and milk proteins. These complexes help to thicken and "stabilize" products such as ice cream and milk shakes, protecting them, so that they do not thaw easily. Vegetarians use carrageenan in many products that commonly use gelatin, so they can avoid animal by products.

Chapter 17 Review

Vocabulary

Match each word to the sentence where it best fits.

Section 17.1

hydrocarbon	parent compound
R-group	saturated hydrocarbon
alkene	unsaturated-hydrocarbon
alkane	alkyne
structural Isomer	benzene
optical isomers	geometric isomers
aromatic hydrocarbon	

1. The longest continuous chain of carbons is called the _____________ .
2. An organic compound containing only hydrogen and carbon is called a _________.
3. Sometimes an organic formula is written with an ____________ which a short hand notation that represents a hydrocarbon side chain.
4. A hydrocarbon compound which contains multiple bonds is said to be ______________.
5. Some molecules have the same chemical formula but a different bonding pattern, these molecules are referred to as ______________.
6. A hydrocarbon that contains only single bonds is called an ____________.
7. An aromatic ring structure, with the formula of C6H6, is called ____________.
8. Ethene is an example of an __________.
9. Cis and Trans isomers are also called __________________ because their atoms are arranged differently with respect to their double bonds.
10. ____________ hydrocarbons contain one or more cyclic rings in their structures.
11. Isomers that rotate light differently with respect to one another are called ____________.
12. An organic molecule such as butyne belongs to the class of molecules called __________.

Section 17.2

aldehyde	ether
alcohol	carboxylic acid
ketone	amines
carbonyl group	esters

13. An _______ is an organic molecule where oxygen is bonded in the middle, with single bonds, with hydrocarbon chains on either side.
14. Methanol is an example of an ___________.
15. Cinnamon and vanilla are types of ______________(s), giving us familiar smells and flavors.
16. Fructose is an example of a ___________, which is a functional group containing a carbon to oxygen double bond with hydrocarbon groups on both sides.
17. An organic acid containing COOH is called a _______________.
18. The ___________ functional group is an important part of amino acids, and also has the ability to act as weak base in aqueous solutions.
19. ___________ are responsible for the fragrances of many flowers and ripened fruits.

Section 17.3

dehydrogenation	substitution
hydrogenation	addition
partial hydrogenation	petroleum
cracking	polymer
monomer	polymerization
amide linkage	condensation polymers

20. Hydrogenation is a type of ___________ reaction, that adds hydrogen atoms to the double bonded areas of an unsaturated hydrocarbon molecule.

21. During a _____________ reaction hydrogen atoms are removed from an alkane and replaced with different atoms, often a halogen.

22. The process of ______________________ was used to make food stuffs containing oils last longer on the shelf at the grocery store. It was later discovered that this process also created trans fats.

23. Removal of hydrogen through the process of ________________ creates double bonds in the structure of a saturated alkane.

24. A long chain molecule formed by connecting many small repeating units with covalent bonds is called a _____________.

25. The process used to break long chain hydrocarbons into smaller fragments is called _________.

26. The small repeating unit of a polymer is called a ____________.

27. ______________ is a blend of hydrocarbons formed from prehistoric organic material.

28. A chemical reaction that joins monomers together is called __________________.

29. Some polymers form linkages by the loss of a water molecule between the monomer units these are called _________________.

30. A bond formed between an amine group and a carbonyl group is called an _____________.

31. Polyethylene is formed through the common process of ________________ polymerization, where double bonds are opened up, and a growing chain is created.

Conceptual Questions

Section 17.1

32. a) Write the formula for the straight chain alkane with seven carbons.
b) Name the compound.
c) Show one branched form of this compound.

33. Using the formula $C_nH_{(2n+2)}$, write the compound containing 12 carbon atoms.

34. a) Write the formula for the straight chain alkane with 4 carbons.
b) Name the compound.

35. Draw the formula for two structural isomers of pentane.

36. What is the name of the parent compound in the structure below?

$$CH_3{-}CH_2{-}CH_2{-}CH_2{-}\underset{\underset{\displaystyle CH_3}{|}}{\underset{\displaystyle CH_2}{\underset{|}{CH}}}{-}CH_3$$

37. The alkane butane is referred to as a straight chain alkane. Is this an accurate description of what the molecule looks like? Explain.

38. How many hydrogen atoms are bonded to a carbon atom in a saturated alkane?

39. Write the structures for two unsaturated alkanes.

40. Which alkane is referred to as gasoline? Write it's chemical formula.

41. When we read the numbers on the gas pump and we see the numbers 87, 89, and 93,
a)what do these numbers tell us about the gasoline? Explain.
b)Which number would be the best choice in a race car? Why?

Chapter 17 Review.

42. Which is the smallest alkane that has a structural isomer? Explain and show structures to support your answer.

43. Compare the boiling points of ethane and hexane. Clearly explain why one is so much higher than the other.

44. What types of intermolecular forces are present between hydrocarbon molecules?

45. Water and oil do not mix. Explain why this is so using your knowledge of intermolecular forces .

46. What is the structural feature that indicates a hydrocarbon is an alkene?

47. Write the chemical structure that represents each of the following compounds.
a) 2-butene
b) 1-heptyne
c) 3-propene
d) 2-methyl-butane

48. Identify which of the following geometric isomers has the trans configuration .

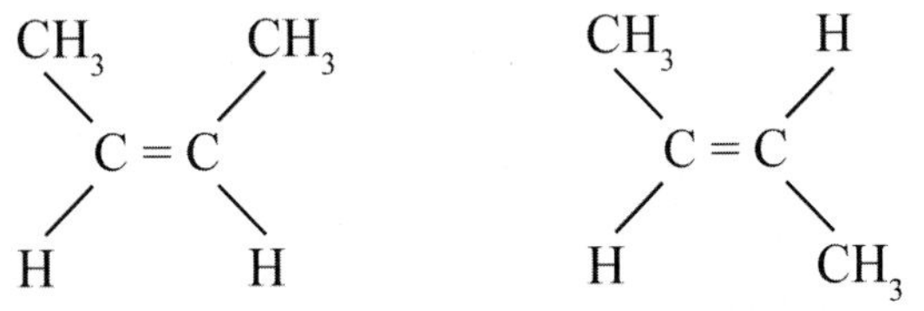

```
CH3      CH3          CH3      H
   \    /                \    /
    C = C                 C = C
   /    \                /    \
  H      H              H      CH3
```

49. Write the cis and trans isomers for 3-hexene.

50. What is an optical isomer? Sketch an example using a tetrahedral geometry.

51. Explain how an optical isomer is different from a structural isomer.

52. How did aromatic organic compounds get their name?

53. List the names of two of aromatic compounds that are commonly used in the kitchen.

54. Why is the benzene ring, C_6H_6, written using a six sided ring structure with a circle in the middle of it?

Section 17.2

55. Without looking back in the chapter try to write the structural formula and the name for each of the functional groups listed below.
a) alcohol
b) aldehyde
c) ketone
d) amine
e) ester
f) ether
g)carboxylic acid

56. Which functional group does ethanol belong to?

57. Which functional group does acetic acid belong to?

58. Look at the compounds shown below and try to identify the functional group to which each molecule belongs.
a) $CH_3CH_2CH_2CH_2OH$
b) CH_3CH_2-O-CH_2CH_3
c) CH_3CH_2COOH

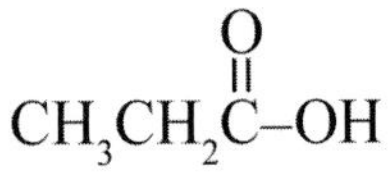

```
        O
        ||
CH3CH2C–OH
```

d)CH_3NH_2

59. a)What feature distinguishes the difference between and aldehyde and a ketone?
b) What do their structures have in common?

60. List one example of a commercial use for an aldehyde and a ketone.

61. Sketch the structure for acetylsalicylic acid, the active ingredient in aspirin, and circle the portion of the structure that it is an ester. You may look the structure up on-line.

62. Acetone is the active ingredient in finger nail polish remover. Draw it's structural formula and label which functional group it belongs to .

63. Using a sketch, show why carboxylic acids, like acetic acid in vinegar, behave as weak acids in water.

64. List two functional groups that would NOT be very soluble in water. Explain your choice.

65. Amines have basic properties. Using a sketch show how methylamine can act as a base in aqueous solution.

Section 17.3

66. Identify the type of alkane reaction shown below.
a)C_3H_8 + $O_2(g)$ $\rightarrow$ CO_2 (g) + $H_2O(l)$ + Heat
b) What are some uses for this type of reaction?

67. Name the type of reactions that the equations shown below represent.
a) CH_3-CH_2-$CH_3(g)$ + $F_2(g)$ $\rightarrow$ CH_3-CH_2-CH_2F + HF

b) CH_3-CH_2-CH_3 $\rightarrow$ CH_3-CH=CH_2 + H_2

c) CH_3-CH=CH-CH=CH_2 + $H_2(g)$ $\rightarrow$ CH_3-CH_2-CH_2-CH_2-CH_3

68. Predict the products and write the balanced chemical equation for the following reactions.
a) C_3H_8 (g) + $O_2(g)$ $\rightarrow$
b) $C_6H_{14}(g)$ + $O_2(g)$ $\rightarrow$

69. Explain how corn oil can be converted into a solid form that could be used for shortening?

70. Write the chemical equation that represents each of the following reactions shown below.
a) Ethene is burned in air.
b) Propene is hydrogenated with the help of a platinum, Pt catalyst.

71. In a couple of sentances describe what happens in a substitution reaction.

72. Write the chemcial reaction that represents the dehydrogenation of ethane.

73. Name the process in which transfats are formed. Why was this process originally used?

74. Look on three food labels at home and record the type of food (and brand) and the amount of total fat, saturated fat and transfat.

75. Are fats ok to eat as long as they are not transfats? Explain briefly.

76. The process called "cracking" is widely used in petroleum refining. What is its overall purpose?

77. What types of substances originally formed petroleum?

78. Using plastic snap beads as an example explain how polymers are formed.

79. List at least five of the plastic polymers you use on a daily basis.

80. Write out the steps for addition polymerization.

81. How are addition and condensation polymerization different? Explain the specifics.

82. List two different materials formed from condensation polymers.

83. Select 2 different types of polymers and go on-line and look up two large chemical manufacturers that make them. What are the building blocks that they use?

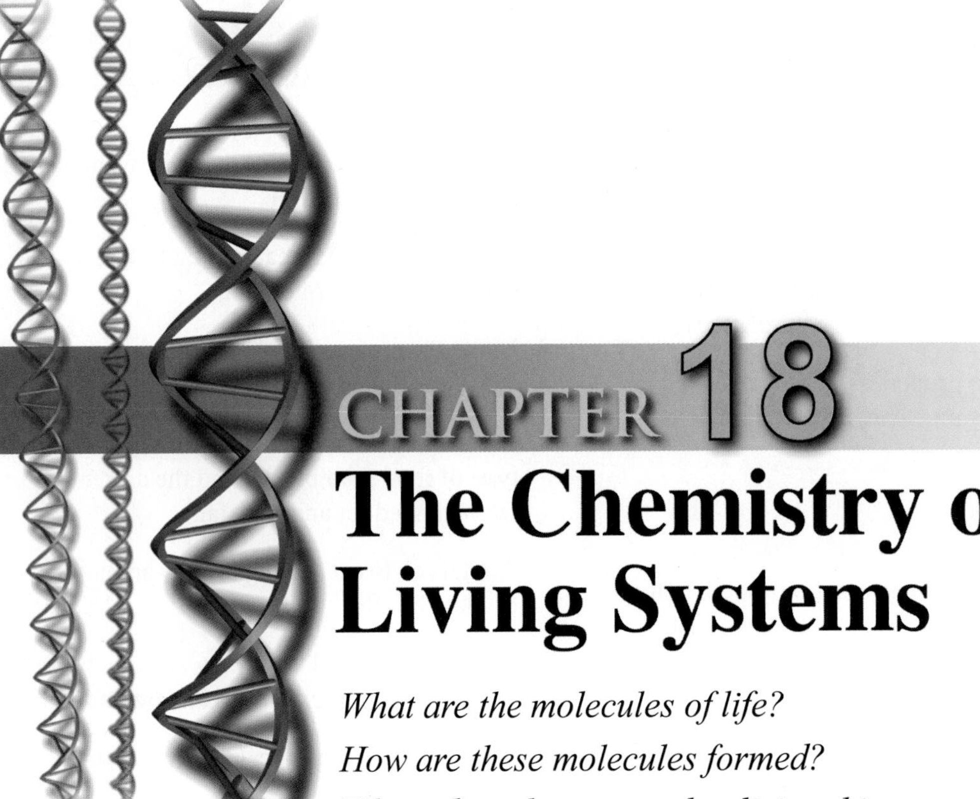

CHAPTER 18
The Chemistry of Living Systems

What are the molecules of life?
How are these molecules formed?
Where does the energy that living things need come from?
Do all living things have the same molecules?

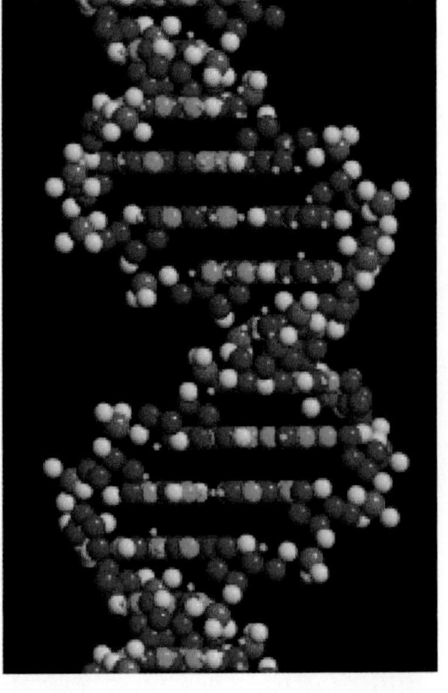

Are you familiar with the molecules that make up living things? You may have learned about them in biology class. Often our first encounter with biological molecules is overwhelming, do not let that discourage you! When you study the molecules of life from a chemistry perspective you will see that what looks like a very large molecule is really made of small repeating units. Does this remind you of another type of molecule? Yes, they are polymers. Most important biological molecules are polymers. The molecules that make up living things are very large. One of the truly amazing things about living systems is that they are able to make these extraordinary molecules. For instance how do organisms get the energy they need to make large molecules? What do we eat? Proteins, fats, carbohydrates, and sugars. Our daily diet determines how well our body can build these important molecules. The expression "you are what you eat" is actually true to a large extent. Molecules also carry important information that allow organisms to reproduce. The molecules that are passed on to our offspring are read by other molecules. Our growth and development is dependent upon molecular information. Enzymes, which are actually proteins, help to regulate how many and what type of molecule is made. The molecular systems in living organisms work together in a very specialized way. We are alive because of molecules! In this chapter you will see the importance of many of the concepts you have learned through out the year.

Chlorophyll: Protein in leaves

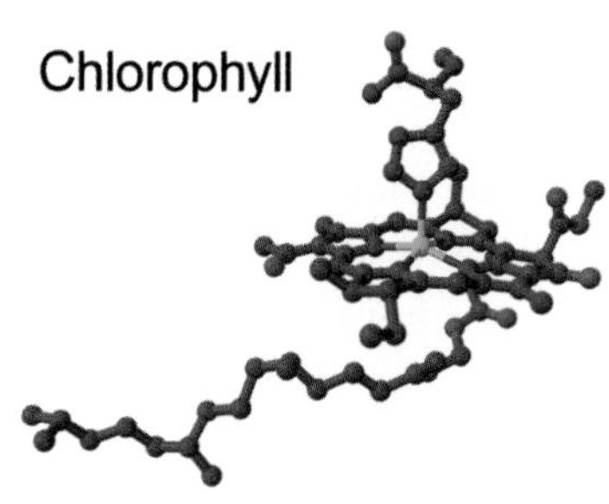

Do you know how plants absorb sunlight? By using a special molecule called chlorophyll. In this experiment we will learn about this very important molecule. Chlorophyll allows plants to use the energy from sunlight to carry out chemical reactions that make sugars and oxygen. The molecular structure of chlorophyll is similar to another molecule that is also very important to us - hemoglobin. Chlorophyll is made up of a four membered ring structure called a porphyrin ring, which contains alternating double bonds. These alternating bonds allow this molecule to strongly absorb energy in the visible region of the spectrum. Magnesium metal, Mg, is held on the inside of the molecule.

Using paper chromatography we are going to separate chlorophyll pigments from leaves. First we will need 3 large leaves, you can use spinach leaves for this too. The leaves are cut, or chopped into very small pieces using scissors or a knife. Pour a small amount of rubbing alcohol (isopropyl) over the pieces of leaves. Just enough to cover the surface. Smaller pieces of leaves will allow the chlorophyll pigment to be absorbed in the alcohol. Cover the jar loosely with plastic wrap and let it sit for 30 minutes.

After a 30 minutes swirl the glass beaker and check to see if the alcohol has absorbed color. The darker the color the better your results will be. Place a long strip of filter paper into the jar so that it just touches the surface of the alcohol tape the other end to a pencil or glass stirring rod and place it horizontally over the glass beaker. The alcohol will be absorbed by the filter paper and it will bring the pigment molecules with it. Watch the filter paper and you will see color start to rise up and form a "band". The filter paper will allow all of the pigments dissolved in the alcohol to migrate up the paper and they will be separated according to their molecular weight. The smaller molecules will travel the furthest.What do we see? The olive green colored band represents the chlorophyll b. The brighter colored green band is chlorophyll a. The two other pigments we see are carotene which is light orange and xanthophyll which is yellow. There are two types of chlorophyll in plants, chlorophyll a and chlorophyll b. Each of these absorb light in different regions of the visible spectrum. Both of the chlorophyll molecules absorb light the best in the blue(400-500nm) and red(610-700nm) regions of the visible spectrum. Each chlorophyll captures light in slightly different regions of the red and blue spectrum, this accounts for their different colored bands. Leaves appear green because that is the color that is reflected to our eyes! Life exists because of the chlorophyll molecule! Without chlorophyll plants could not carry out photosynthesis.

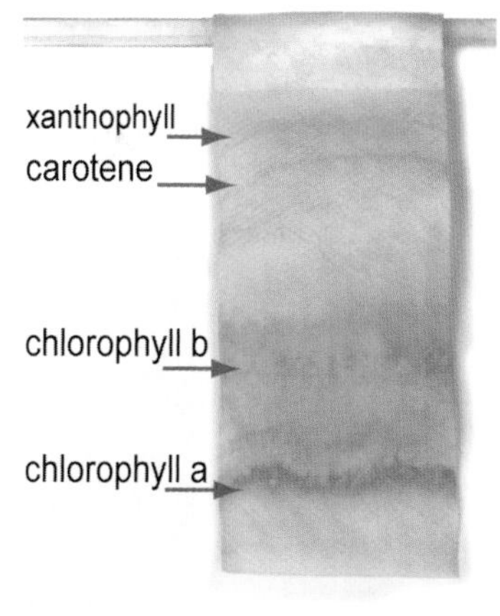

18.1 Fats and Carbohydrates

Fats and carbohydrates store energy

All animals need to eat in order to survive. If this seems obvious then the reason why must be obvious too! Living things get their energy from food. Without energy for our body processes we would not survive. We get a large portion of our daily energy from **fat** and **carbohydrate** molecules that we consume. Our bodies are capable of breaking apart fat and carbohydrate molecules and obtaining energy from them. A fat molecule is made of fatty acids and a glycerol molecule. Fats are used and formed by animals as an efficient means of storing energy. Carbohydrates are made of both short and long chain polymers of sugar molecules. All organisms, including plants, release energy from carbohydrate molecules.

Nature uses fat and carbohydrate molecules for energy and for structural purposes

Lets look at a few examples. If we consume a big meal of spaghetti, we are filling up on starch. Starch is a type of carbohydrate. Athletes often eat large meals of pasta, especially before a big race so that they have full energy stores, which allows them to perform longer. Polar bears on the other hand prefer to eat fatty meats, such as seals. Seals fill up the fat stores on the polar bear's bodies, this gives them energy during periods with no food. Plants use cellulose, which is a also a type of carbohydrate, for structural purposes. The cellulose found in plant cell walls allows trees and some plants to grow tall without falling over!

Carbohydrates and fats have very different molecular structures. In the next few pages we will explore how their molecular structure allows for their different functions in nature.

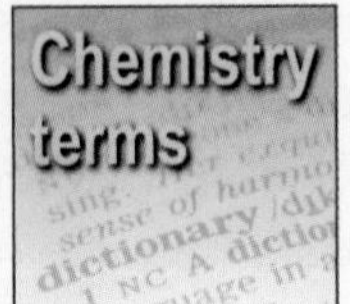

fat - or lipids, are nonpolar molecules made from fatty acids and a glycerol molecule. Fats molecules are an efficient way to store energy for many types of animals. Fats are not soluble in water.

carbohydrate - short and long chain polymers of sugar molecules. All organisms use carbohydrates as a main source of energy.

Carbohydrates

Carbohydrates are the most abundant biomolecule

Carbohydrates are the most abundant type of biomolecule! The word carbohydrate literally means carbon atoms with water molecules. You may remember from the last chapter that carbohydrates are aldehydes or ketones with many hydroxyl groups attached. Simple carbohydrates refer to sugar molecules and "complex" carbohydrates refer to starches or grain products. Plants make starch as a storage form of energy, and animals make glycogen to store some of the energy from carbohydrates.

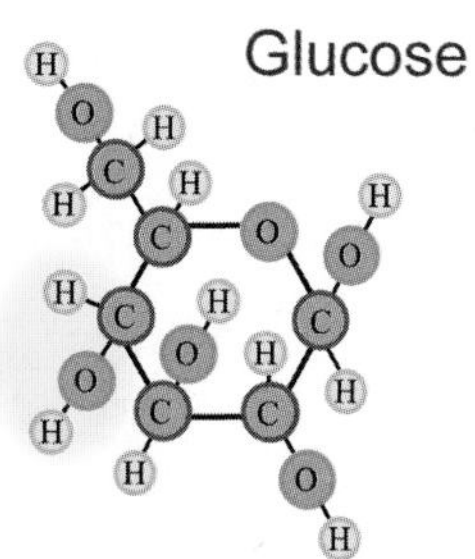

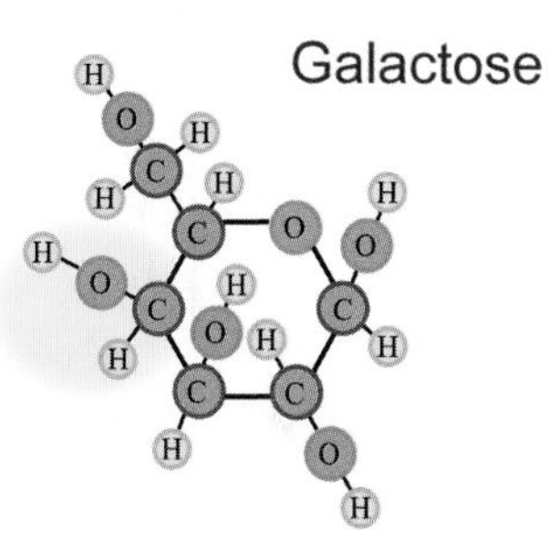

The smallest repeating unit of a carbohydrate is called a **monosaccharide**. Monosaccharides are individual sugar molecules. The chemical properties of these sugar units are similar. Most monosaccharides have a sweet taste and are crystalline, colorless and water soluble. Glucose, and galactose are examples of monosaccharides. Galactose is found in certain vegetables like sugar beets, and also in dairy products. When glucose and galactose are linked together, through condensation polymerization, they form the sugar lactose which is found in milk. Small chains are formed by linking two to ten monosaccharides units together, these small chains are called **oligosaccharides**. Lactose is an example of an oligosaccharide. Both monosaccharides and oligosaccharides are referred to as sugars, which is a much easier word to use!

Long chains of more than ten monosaccharides form **polysaccharides**. Polysaccharides form starches, glycogen, and cellulose. These naturally occurring, long chains of sugars serve many purposes in nature. Starch is the primary ingredient in rice, cereal, potatoes, and corn. Glycogen is the back up supply of carbohydrate in animals. It is stored in the liver and in muscle tissue, where it can be broken down and used as a fuel source. Cellulose is the primary structural material in plants. It is the main component of straw, wood, and cotton. Cellulose can not be digested by humans, because we do not contain the necessary enzymes to break the molecules apart. Some ruminant animals such as cows and termites can break apart cellulose and use it as a food source.

monosaccharide - a single sugar unit. Mono - meaning one and saccharide meaning sugar molecule.

oligosaccharides - two to ten monosaccharides linked together in a chain.

polysaccharide - a long chain, greater than ten, monosaccharides connected together. Poly- means many and saccharide means sugar units. Examples are starch, cellulose and glycogen molecules.

Monosaccharide ring structures

Sugars like glucose form a ring structure

When carbohydrates are made, the monomer units or monosaccharides, are not linear molecules. The sugar molecules form a ring or a cyclic structure, and it is the ring structures that are joined in a long chain polymer. *These ring structures contain carbons with optical centers, or chiral carbons.* Remember from the previous chapter that carbons form stereoisomers when four different groups are attached to it. In most living things the isomers that *rotate light clockwise* are the ones that are utilized.

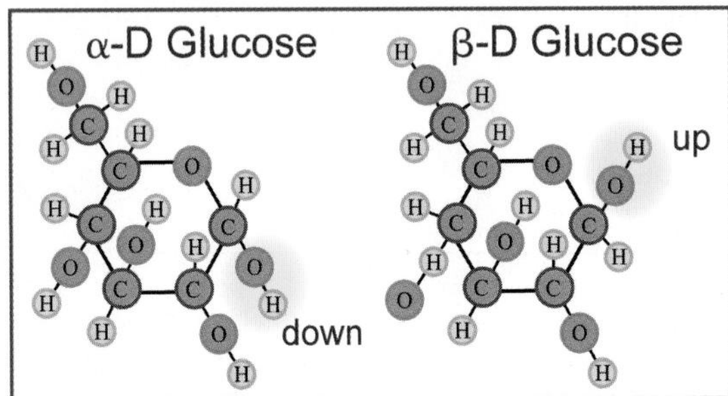

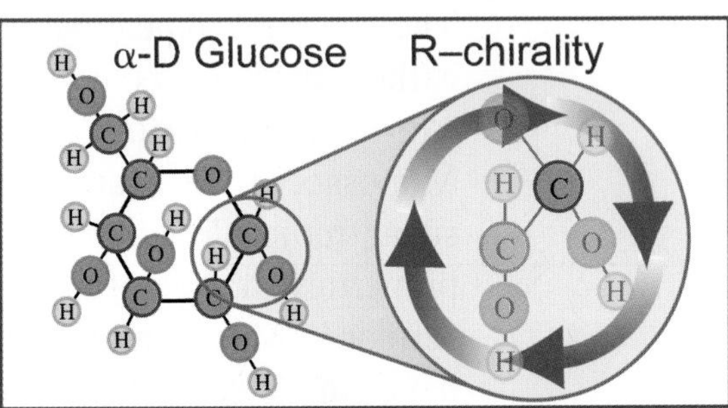

If you look carefully at the structures above you will see that the hydroxyl group on the first carbon is oriented downward in alpha -D-glucose and it is positioned up in beta-D-glucose. *This single difference is an important aspect of carbohydrate chemistry!* D-glucose is an optical isomer that rotates light clockwise, or to the right. Our bodies cannot metabolize L- glucose, the mirror image of D-glucose which rotates light to the left, or counter clockwise direction.

α-D-Glucose is connected in carbohydrates using a condensation reaction

When these cyclic sugar molecules are used to make polysaccharides they are joined in a very specific way. For example in starch, which is formed from repeating glucose units, alpha-D- glucose is connected exactly the same way each time.

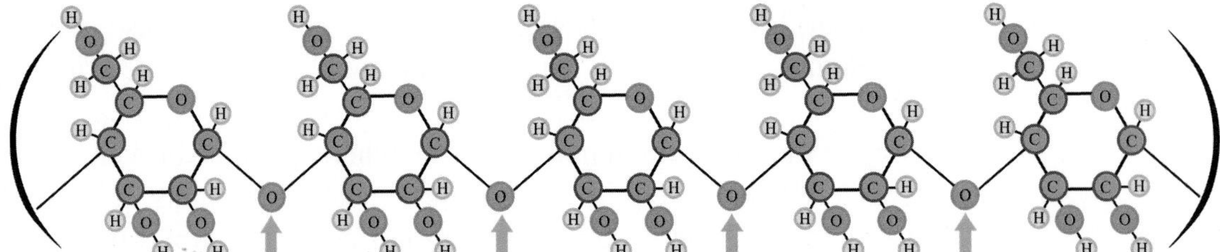

In sucrose, which is a disaccharide, the glucose and fructose rings are connected. Sucrose or sugar occurs naturally in every fruit and vegetable. It is the primary way that plants transform the suns energy into food. Notice here that glucose is the six membered ring and fructose is a five membered ring. This helps you to recognize the structure for sucrose!

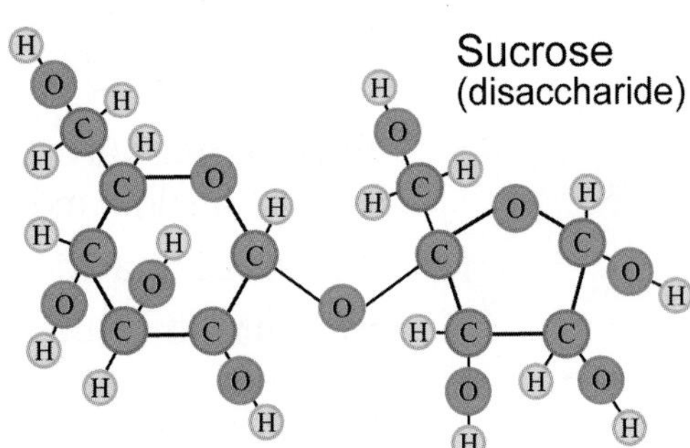

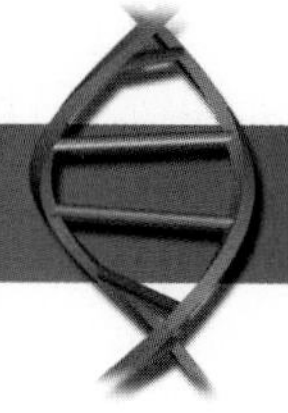

Carbohydrates in our diet

Excess carbohydrates are stored as fat

When we consume carbohydrates in our food our body utilizes what it needs, and stores the excess. Some of the excess is stored as glycogen in our muscles and liver tissue, however they are not capable of storing large amounts excess carbohydrate. Once the stores of glycogen are restocked the excess carbohydrate is *converted to fat* and stored on our bodies. So it is important to understand that overeating on carbohydrates will increase your body fat. You do not need to eat lots of "fatty" foods to become over weight.

Some carbohydrates are digested quickly, while others are slow

When we consume foods we take in two types of carbohydrates. Simple carbohydrates and complex carbohydrates. Simple carbohydrates are digested quicker. They do not contain much fiber and their molecular structure allows our digestion process to get glucose molecules into the blood stream quickly. Sometimes we get a "sugar" rush from consuming these on an empty stomach. Complex carbohydrates such as brown rice, *pasta, whole wheat bread, and vegetables take much longer to digest. They have a high fiber content and they are rich in vitamins and minerals. The slow digestion of these carbohydrates leaves us feeling "full" longer. Complex carbohydrates take longer to break down due to their longer chains and branched molecular structures. Selecting the right types of carbohydrates is an important part of our daily nutrition. You have likely heard the phrase "empty calories." These are calories from which we get little or no nutrition. Foods high in sugar like candy bars and donuts do not fill us up, and leave us feeling hungry again in a short time.

TABLE 18.1 Carbohydrates

	Simple Carbohydrates	Complex Carbohydrates
GOOD	fruit, fruit juice, yogurt, molasses	brown rice, whole wheat pastas, vegetables, whole grain breads and cereals
BAD	processed sugars, white cane sugar, candy, high fructose corn syrup	white enriched flour, white breads, instant rice, sugared and processed cereals

There is much information based on current research that suggests that 55% of our daily diets should come from carbohydrates. It is highly recommended that we eat a variety of carbohydrates to ensure that we get the adequate nutrients. The complex carbohydrates are really good for our bodies overall health. Whole grains are put through fewer chemical and mechanical processes, so more of the vitamins and minerals remain for our bodies to absorb and use.

Fats and oils

Fats come mainly from animals

Fats and oils come from plant and animal tissue. For example butter is made from milk fat and soybean oil comes from soybeans. Plants generally make oils. The fact that fats are solids at room temperature while oils are liquids at room temperature, is one helpful distinguishing feature. Both fats and oils are an efficient way for plants and animals to store energy. By efficient we mean that they provide more energy per gram than carbohydrates. You may already be aware of the fact that some fats are better for you than others.

Why does it make any difference what type of fat you consume?

Unsaturated fats are best for us

The answer to this lies in the *molecular structure of the fat molecule* and how our bodies can break it up. For example fried foods are considered "bad" for us while fish oil is considered quite healthy. The primary difference in the molecular structure of good and bad fats lies in the number and position of the double bonds between the carbon atoms. Saturated fats contain no double bonds and are solid at room temperature. These types of fats are not as good for our bodies, because they are harder for us to split apart for energy. Transfats are not as good for the same reason. Nutritionally we want to consume fats that our bodies can readily use for energy. Unsaturated fats have double bonds in their molecular structure are much easier for our bodies to metabolize.

Fats have useful body functions

Certain fat molecules help to protect us from diseases and keep our bodies functioning properly. For example, our nerve cells and spinal cord have a fatty coating that helps to protect and insulate them. Our cell membranes contain fats, or a lipid bilayer, that helps to maintain a barrier between the inside and outside of our cells. Fats that make up adipose tissue in our bodies also provide a cushion that protects our vital organs.

In this scan of normal fat distribution in a human body the adipose tissue is yellow.

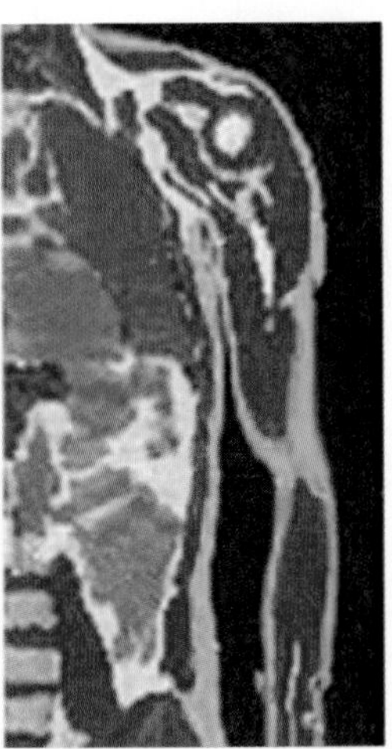

Lastly, fats have been important in our existence throughout history. Oil was used in lanterns long before electricity was discovered and fats were also used to give clothing and fabrics water proof properties. This helped humans to survive during periods of bad weather while living and working outdoors.

Fat molecules

Fats dissolve in nonpolar solvents

We have already established that fats are important molecules in life systems. But what are they, and what do they look like? Lets look closely at the molecular structure of a fat molecule. You can see the ester bonds of attachment for the fatty acid chains.

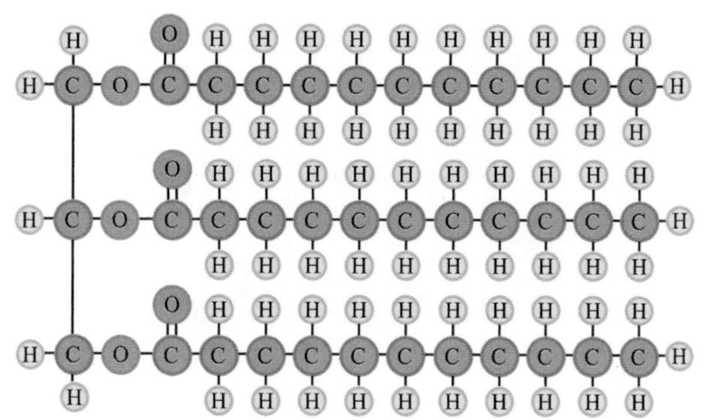

The non polar glycerol molecule is attached to three fatty acids to form a fat molecule. Fatty acids have long nonpolar hydrocarbon chains. Along the length of the hydrocarbon chain there are London dispersion forces which attract other fatty acid chains. Fats tend to aggregate together for this reason.

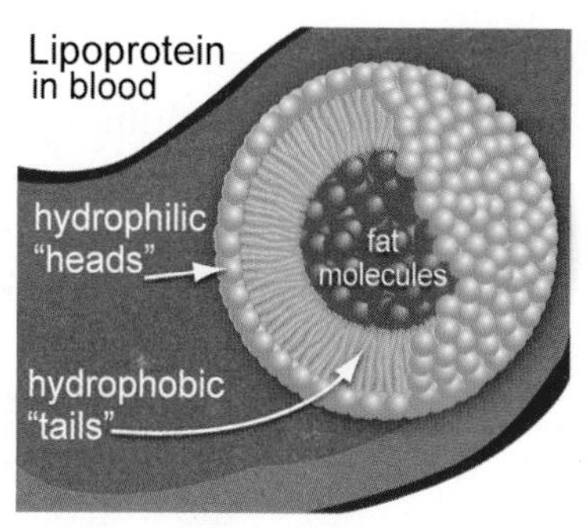

Fats, or lipids as they are often called, are actually classified by scientists based on their solubility in nonpolar solvents. This means they dissolve in organic liquids such as benzene, hexane or other liquid oils (vegetable oil). Remember "like dissolves like."

Remember that our blood and body fluids are mostly water. In order for a molecule to dissolve in water it must have the ability to hydrogen bond or it must carry some form of a charge, either a partial dipole-dipole or an ionic charge. To carry fat molecules in our blood stream we use lipoproteins. Lipoproteins are small spherical assemblies with hydrophilic groups on the exterior which allow them to be soluble in our blood. The fat is carried inside the lipoprotein.

Fats with unsaturated double bonds tend to be liquids at room temperature

Fats can vary depending upon the fatty acid molecules that are attached. The fatty acids can all be the same molecule, or they can be different in length. Fatty acid molecules range from 4 to 24 carbons in length. In humans fatty acid chains of 16 to 18 are the most common. The fatty acid chains can also contain saturated bonds or unsaturated double bonds. The reason fats are solid at room temperature is because they contain more saturated bonds. Oils are liquids at room temperature because they contain more unsaturated double bonds. You might remember that oils are healthier for us to consume, because of the unsaturated double bonds.

Omega-3 fatty acid

Triglyceride

Triglycerides store energy for humans

A fats and oils are made of **triglyceride** molecules. Fats are transported by our blood stream in the form of triglyceride molecules. A triglyceride molecule is made of three fatty acid molecules attached by ester bonds to a glycerol molecule.

The image on the right shows tristearin, which is the main type of fat in contained in beef. Fatty acids are long chain, C_{12} to C_{24}, hydrocarbon molecules with a carboxyl group at the end.

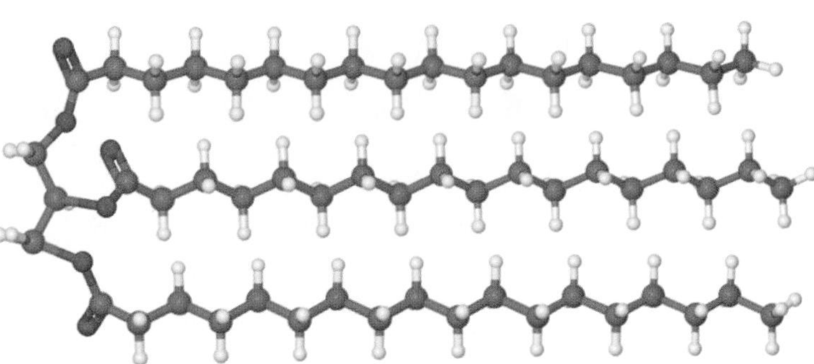

The special texture of chocolate

Most naturally occuring triglycerides have 16,18 or 20 carbon chain fatty acids. Natural fats contain a mixture of these fatty acids, not just one type. Because of this they have a melting range not a melting point. A good example of a popular fat, cocoa butter, comes from the cacao tree. Chocolate contains only three triglycerides and this allows it to melt without feeling greasy. The smooth chocolate texture is due to a blend of fats melting slowly in your mouth.

We obtain a good deal of energy from fat molecules. Long chain fatty acids yield large amounts of ATP when they are oxidized. For every two carbons on the fatty acid chain of the triglyceride we can make more than 15 ATP molecules. Longer fatty acid chains yield more energy when they are metabolized. This is because there are more C-C bonds to break, which in turn release more energy.

If we consume more fat than our body needs for energy some of the excess is stored on our bodies in adipose (or fat) tissue. We obtain most of our fat in our diet, but the body can make triglycerides in the liver and fatty tissues when necessary. When we eat foods that contain triglyceride molecules, our body digests them in our intestines and breaks them apart into smaller molecules some of which are glycerol and fatty acid molecules once again. If we consume a high fat diet our blood will have a larger amount of triglycerides floating around in it. The number of fat molecules circulating in the blood stream has been shown to contribute significantly to blocked blood vessels and arteries. For this reason, our blood should be tested during routine physical exams to determine the triglyceride levels.

triglyceride - is a fat or lipid molecule used for energy storage. It is made of a glycerol molecule and three long chain fatty acids.

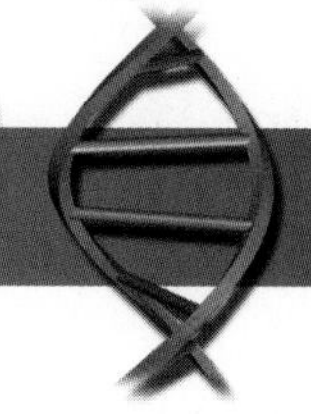

Phospholipids

Phospholipids are important in our cells

One very important type of lipid molecule in our bodies is called a **phospholipid**. These fat molecules are important because they make up our cell walls, which protects the internal environment of our tissues. A phospholipid has a long hydrocarbon chain with a phosphate group for a head. You may remember this structure from chapter 8. These molecules have a unique ability to orient themselves in water, because of the difference in polarity on different ends of the molecule. The phosphate portion of the molecule is polar and attracted to water. This portion is called hydrophilic, or water loving. The hydrocarbon tail is nonpolar and repels, or pushes away, the water molecules. The hydrocarbon tail is hydrophobic, or water hating. As a result these molecules are able to organize themselves in a spherical fashion in water forming **micelles**. The sphere forms because the nonpolar hydrocarbon tails want to be together, near each other, and away from water!

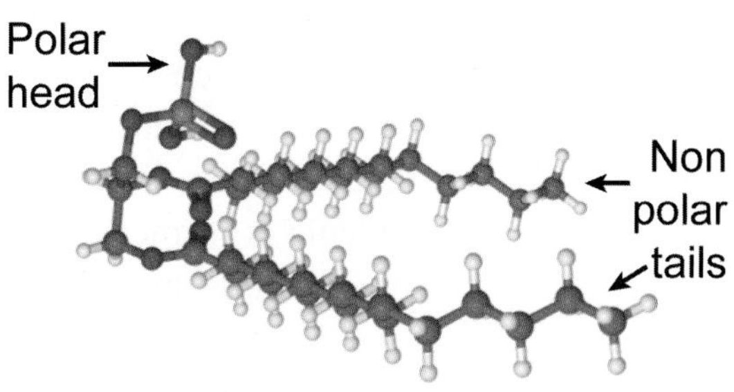

Micelles

Phospholipid molecules have a polar and a nonpolar region

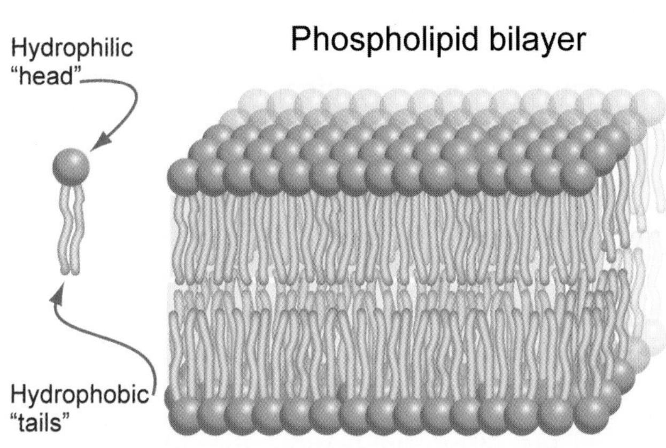

In an aqueous environment such as the one in our bodies. The hydrophobic tails come together and form a layer that excludes water, while the phosphate heads are pushed outwards to be near the water. This causes a "layer" to form. In our cells these molecules make up our lipid bilayer.

The fact that phospholipids have long nonpolar regions and small polar heads causes them to aggregate together, or stick near one another. The layer that is formed *acts as a barrier in our cells to keep needed nutrients inside and to keep unwanted substances from mixing with the cells interior.* There is some fluidity to our cell membranes because these phospholipid molecules can move or migrate within their own layer, however they cannot change from one layer to another layer very easily.

phospholipid - a lipid that has a phosphate group attached, giving it a polar head and a nonpolar tail. The phosphate group is the polar region of the molecule and the hydrocarbon chain is the nonpolar region.

micelle - a collection of phospholipid molecules that form a small sphere which allows the nonpolar tails to be away from the aqueous solvent in the center of the sphere.

18.2 Photosynthesis and Respiration

Plants use sunlight to make sugars and starches

In the previous section we established that the energy living things need comes from food. The sugar molecule, glucose, is the most important energy yielding molecule in living systems. Here we will look at how the glucose molecule is formed in a process called **photosynthesis**. Plants harness sunlight and use it's energy to convert water and carbon dioxide molecules into sugars and starches during the process of photosynthesis. Photosynthesis is the ultimate first step in this natural pathway that uses carbon atoms from carbon dioxide as building blocks to eventually make complex organic molecules. Plants use the energy they harness to grow and transport nutrients. Photosynthesis also makes food energy available to animals that eat plants. Animals that consume plants take the energy harnessed in the molecules and use it to carry out life processes. The overall equation for photosynthesis is:

Plant absorbing sunlight.

$$6CO_2(g) + 6H_2O(l) \xrightarrow{\text{Light}} C_6H_{12}O_6(s) + 6O_2(g)$$

This equation may look familiar if you remember it from biology or from the chemical reactions chapter 10, when you were learning about endothermic reactions.

Carbon dioxide and water are the building blocks

Plants, algae and some bacteria are capable of using sunlight to make glucose. Plants use water and carbon dioxide gas as the building blocks for sugar molecules. These sugar molecules are turned into starch for more efficient storage. Lucky for us oxygen is also produced! Plants get the carbon dioxide that they need from the air or from water depending on where they grow. Plants need a certain amount of sun and water to flourish. Many of us who have raised plants have seen this first hand.

Chlorophyll

Energy is required to make sugars from stable molecules like CO_2 and H_2O. So how does the plant use the energy of sunlight to make larger six carbon molecules? The answer to this lies in a pigment molecule called chlorophyll. **Chlorophyll** is a large molecule capable of absorbing sunlight in the blue, violet and red wavelengths of the visible spectrum. Chlorophyll transfers this absorbed light energy to electrons within its molecular structure, and these electrons become higher in energy. These high energy electrons make the process of photosynthesis work by indirectly providing the energy to break chemical bonds and reform new chemical bonds.

photosynthesis - the process where plants and algae, capture sunlight and use it's energy to convert water and carbon dioxide into glucose (sugar) and oxygen.

chlorophyll - pigment molecule that absorbs sunlight and uses it to excite electrons in it's molecular structure, thereby storing energy to aid in photosynthesis.

Overview of photosynthesis (part 1)

How does a plant put together carbon dioxide and water molecules to form glucose and oxygen? The first part of this process involves water molecules being split apart using the energy absorbed by sunlight. These are often referred to as the "light dependent" reactions of photosynthesis, because they depend upon sunlight. Chlorophyll molecules absorb light which allows the water molecules to be split apart. Each water molecule is split apart into 2 electrons, $2H^+$ ions, and 1 oxygen atom. As electrons continue to be removed from water, oxygen is released into the air. The reaction looks like this:

$$2H_2O(l) \rightarrow 4H^+(aq) + O_2(g) + 4 \text{ electrons.}$$

NADPH

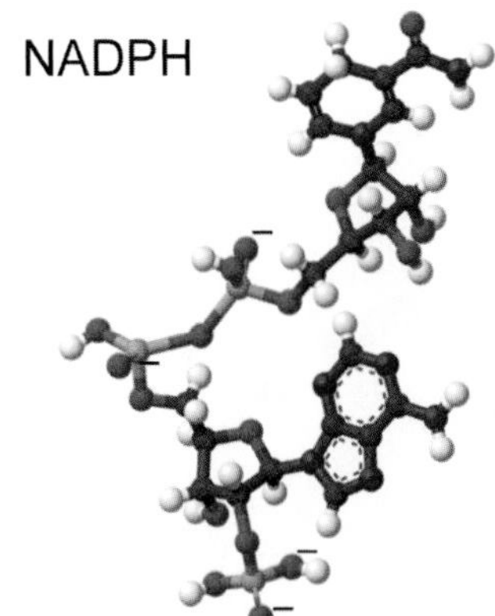

These four electrons are passed onto special "carrier molecules" that are capable of *transferring the energy* obtained while loosing very little of it. It takes very special molecules to do this. Nicotinamide Adenine Dinucleotide Phosphate ($NADP^+$), is such a molecule. It is able to pass these high energy electrons through a series of oxidation and reduction reactions. One $NADP^+$ molecule is capable of holding two high energy electrons and a hydrogen ion to form **NADPH**. The formation of NADPH is one way sunlight can be stored in a chemical form. The energy from sunlight is stored in the chemical bonds of a "carrier molecule" like NADPH. NADPH can then be oxidized to once again form of $NADP^+$.

ATP molecule

Another very important molecule that stores energy is formed during these light dependent reactions. This molecule is adenosine triphosphate or **ATP**. ATP also "holds" and carries energy in it's bonds. Specifically in the bonds between each phosphate group, look carefully at the picture on the right. The phosphate groups are broken off by the addition of water molecules, and the energy released is used to power chemical reactions that require energy.

ATP

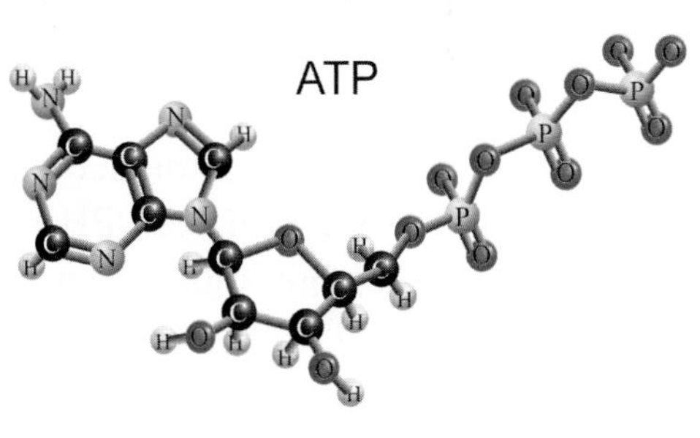

ATP & NADPH carry energy to chemical reactions

The important piece to understand about ATP and NADPH is that they carry energy from spontaneous reactions to non-spontaneous reactions. These molecules that are formed provide the energy for the next step of photosynthesis.

NADPH- molecule that carries two high energy electrons and stores sunlight as chemical energy.

ATP- adenosine triphosphate. A molecule that carries chemical energy from spontaneous chemical reactions to non spontaneous reactions.

Overview of photosynthesis (part 2)

CO_2 is added to carbon molecules to make glucose

In the second phase of photosynthesis the **Calvin cycle**, sunlight is not required and these reactions are sometimes referred to as the "dark reactions" in contrast to the light reactions.

During the Calvin cycle, carbon dioxide is added to other carbon molecules and the six carbon molecule of glucose is formed

However, the steps of this process require energy and are non spontaneous. They could not occur with out the NADPH and the ATP molecules formed in the first part! The energy harnessed in these molecules helps to put together the six carbon sugar of glucose.

NADPH and ATP provide the energy for these reactions

The way nature works in simple yet interconnected ways is amazing.

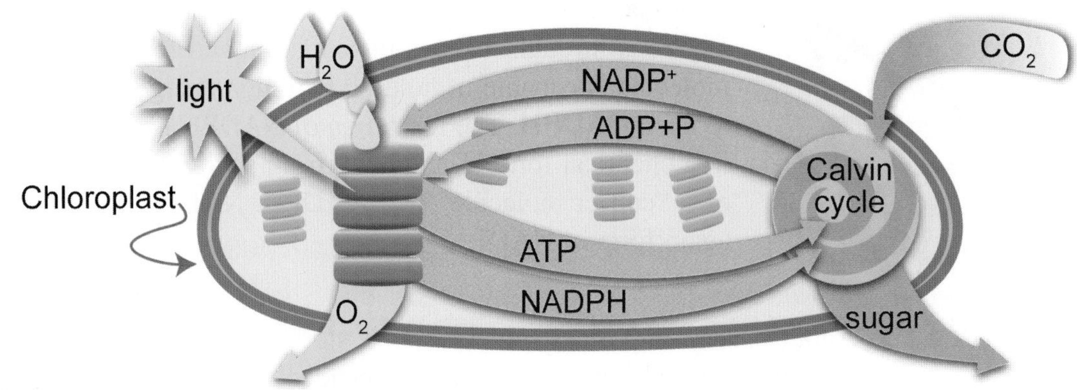

Here you can see that the light dependent reactions take in water and light and release ATP and NADPH molecules that feed into the Calvin cycle. The Calvin cycle takes in carbon dioxide and the high energy molecules to make the sugar glucose. The Calvin cycle recycles the high energy molecules back to the light dependent reactions. Together the light and dark reactions use six carbon dioxide molecules to make a single six carbon molecule of glucose.

Plants work continuously removing CO_2 from our atmosphere and making energy rich sugar molecules

Plants use glucose for their own energy needs and to make more complex molecules like cellulose and starch. These more complex molecules are used for structure and growth.

Calvin cycle - makes energy rich sugars from CO_2 by using the high energy molecules ATP and NADPH formed in the light dependent reactions.

Cellular respiration

Cellular respiration provides energy for life processes

We now need to understand how we use food molecules to give us energy. How is the six carbon sugar, glucose, metabolized and utilized by our bodies? We know that chemicals store energy in their bonds, but how do living systems obtain this energy and use it for life processes? The glucose molecule gives us energy through the process of **cellular respiration**. During cellular respiration, glucose and other food molecules are broken apart in the presence of oxygen and energy is released.

Athletes get the energy they need from cellular respiration.

The overall chemical equation for cellular respiration is :

$$6O_2(g) + C_6H_{12}O_6 \rightarrow 6CO_2(g) + 6H_2O(l) + \text{Energy}$$

Oxygen is a powerful electron acceptor in this process. Without oxygen cellular respiration cannot take place. This equation is greatly over-simplified, because it makes the process appear to happen all at once. Similar to other chemical reactions it actually occurs in a series of steps.

Slow release of energy from glucose allows it to be captured

Cellular respiration involves three steps. The three step process helps to control the slow release of energy from glucose

If the release of energy from the bonds of the glucose molecule is not slow, too much of the energy will be lost as heat and therefore unavailable to use for other life processes. There must be a way to capture some of the energy released when the chemical bonds in glucose are broken. Each of the three steps in cellular respiration captures energy. The energy captured by forming ATP molecules provides the energy for transporting nutrients across cell membranes, powering muscle contractions during movement, and maintaining body temperature. Many of the bodies chemical reactions require energy and the ATP molecule provides this energy.

cellular respiration - breaks down glucose and other food molecules in the presence of oxygen and releases energy that is used to carry out life processes.

Respiration: Glycolysis

The first step in cellular respiration is **glycolysis** (gly-KOHL-ih-sis). Glycolysis is carried out in the cytoplasm of plant and animal cells. During glycolysis one molecule of glucose is broken in half yielding two 3 carbon molecules.

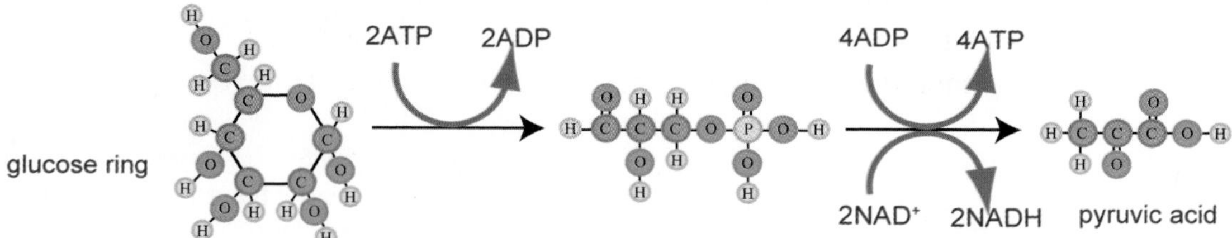

Some energy is required to split the glucose molecule.

You'll notice in the first step to split the glucose molecule apart 2ATP's are required, but then in the second step 4 ATP's are produced. *A little energy needs to be added at the start of the reaction, but more energy is obtained in the overall process*. This is similar to charging the battery on your cell phone. You have to plug it in and use energy to charge your phone when the battery is low, but once charged you get many more hours of power back. The amount of energy you get back from the charged battery represents a larger return. Glycolysis also gives a larger return!

In the second step you will see that 2 **NADH,** nicotinamide adenine dinucleotide molecules are formed. Like NADPH in photosynthesis NADH is a high-energy electron carrier molecule. NADH is made of two nucleotides, adenine and nicotinamide, which are connected through their phosphate groups. In it's oxidized form, after it loses the 2 electrons it carries, NADH becomes NAD^+. Each NAD^+ molecule is able to accept two high-energy electrons, becoming reduced to NADH once again. NADH acts as a strong reducing agent! In cell respiration, NADH sends the electrons to a pathway, where they are used to make ATP molecules.

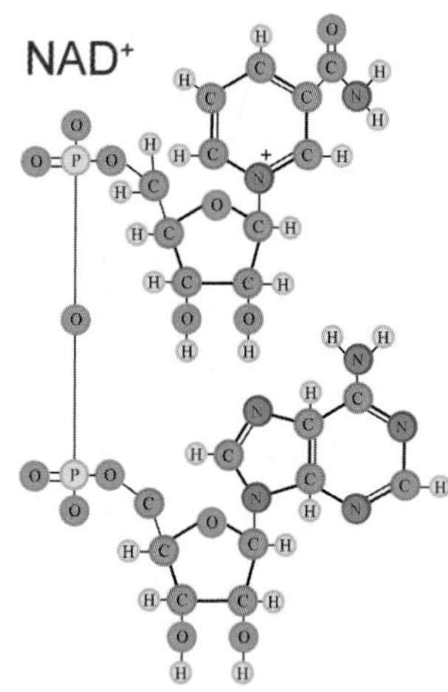

Overall energy yield of glycolysis is small

The overall energy yield of glycolysis is small, however it can produce energy quickly, and without oxygen. When oxygen is present, a second step called the Krebs cycle is able to proceed. The 3 carbon molecules of purveyed acid are passed on to the Krebs cycle, where they release more energy.

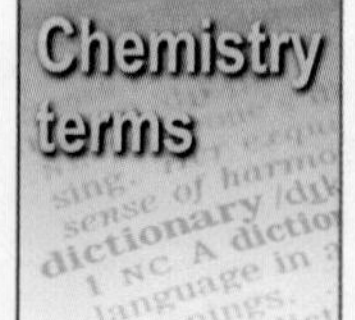

glycolysis - (gly-KOHL-ih-sis) first step in cellular respiration. During glycolysis one molecule of glucose is broken in half yielding two 3 carbon molecules (of pyruvic acid).

NADH - nicotinamide adenine dinucleotide; is a molecule capable of carrying high-energy electrons and transferring them to another pathway.

Respiration: The Krebs cycle and electron transport

The Krebs cycle removes energy while breaking chemical bonds

Glycolysis only uses a small portion of the energy available in the glucose molecule, about 10%. The majority of the energy still remains locked inside the pyruvic acid molecules. These 3 carbon molecules enter the **Krebs cycle**. The Krebs cycle is a sequence of chemical reactions that remove energy while breaking pyruvic acid down to carbon dioxide.

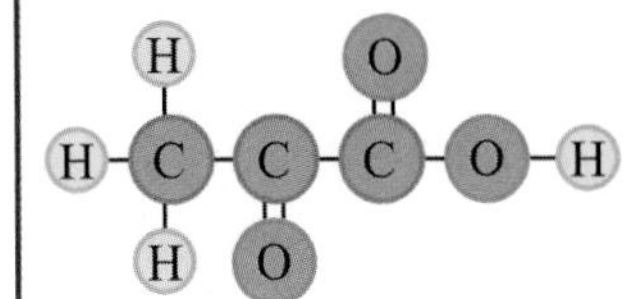

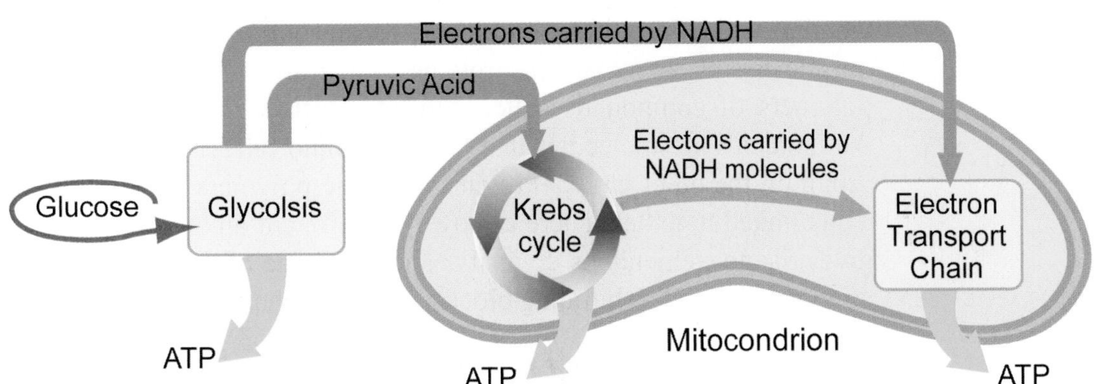

During the steps of the Krebs cycle the three carbons in pyruvic acid are lost as single carbons in the form of carbon dioxide. The breakdown of pyruvic acid yields several more ATP molecules for the cell to use. The rest of the energy is yielded in the form of high energy electrons. NADH and other carrier molecules bring these high-energy electrons to the **electron transport chain**. The electron transport chain uses these high energy electrons from the Krebs cycle to make more ATP molecules. To accomplish this the electron transport chain uses special proteins called cytochromes to transfer the electrons, which we will discuss in the next section about proteins. In general, each pair of electrons that moves along the chain has enough energy to form 3 ATP molecules.

Oxygen is the final electron acceptor

At the end of the electron transport chain, hydrogen ions, H^+ and electrons combine with oxygen to form water. *Oxygen serves as the final electron acceptor.* Oxygen is necessary to remove the low-energy electrons, and hydrogen ions which are considered waste molecules. $4H^+ + 4e^- + O_2 \rightarrow 2H_2O(l)$

Overall the process of cellular respiration uses roughly 38% of the energy in the glucose molecule. The other 62% is lost as heat.

Krebs cycle - a series of energy extracting chemical reactions that break pyruvic acid down into $CO_2(g)$, during cellular respiration.

electron transport chain - uses high-energy electrons from the Krebs cycle to make ATP molecules.

18.3 Proteins

Proteins: the work horse polymers in our body

Proteins are another very important molecule that perform numerous functions inside our bodies and all living organisms. Perhaps you have heard about how important it is to have enough protein in your diet? Proteins are involved in *every* cell process within our bodies! For example, our bodies break apart the protein we digest and use it to build and repair our muscle tissue. What are proteins? They are polymers of compounds called **amino acids**. Amino acids are molecules that have an amino (NH_2) group on one end and a carboxyl group on the other end. Proteins are large polymers containing more than 100 amino acids linked together. There are over 20 different amino acids found in nature, and this gives rise to tremendous diversity. The large size of the protein chain and the variety of amino acids used as building blocks allows for many different combinations 20^{100}!

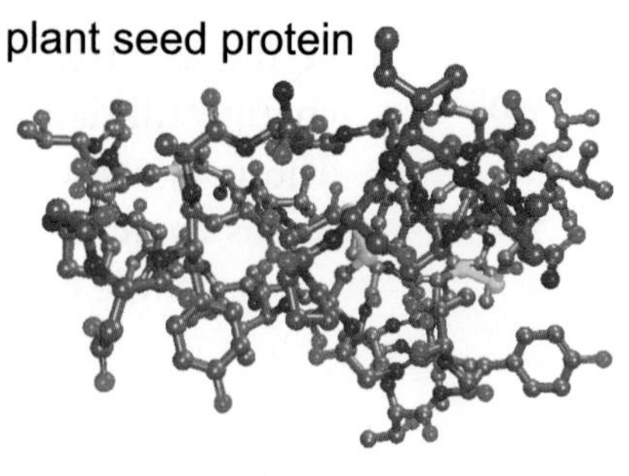

Proteins are required for almost all chemical reactions that take place in our bodies!

Proteins help make chemical reactions faster

One of the primary functions of proteins is to facilitate chemical reactions. You may recall from chapter 12 or from biology class that many proteins are catalysts, and that catalysts speed up the rate of chemical reactions. Chemical reactions would be *so slow at body temperatures* that we could not survive without these protein catalysts.

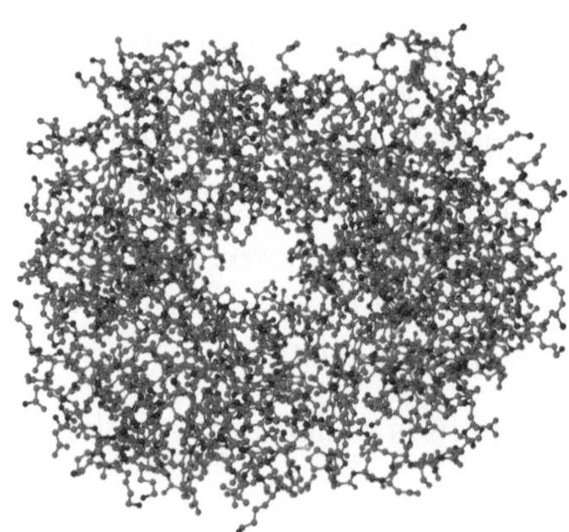

Proteins also make up our hair, muscles, and cartilage. In addition, they transport and store oxygen and nutrients to our tissues. For example, hemoglobin (Hb) is a bundle of four proteins, responsible for transporting oxygen from our lungs to our tissues. This protein can bind to an oxygen molecule and hold onto it until it gets to a place in the body where it is needed. In muscle tissue for instance, hemoglobin changes its shape and releases the oxygen molecule, so that the muscles have the oxygen that they need.

Chemistry terms

protein - polymers of 100 or more amino acids.

amino acid - chiral molecules that have an amino group on one end and a carboxyl group on the other. There are 20 different naturally occurring amino acids.

Amino acids

To understand how proteins are formed it is important to have a basic understanding of what amino acid molecules are. There are 20 different naturally occurring amino acids commonly found in most proteins (see table on the next page). Each amino acid has an amino group (NH_2) and a carboxyl group (COOH) covalently bonded to the central carbon atom.

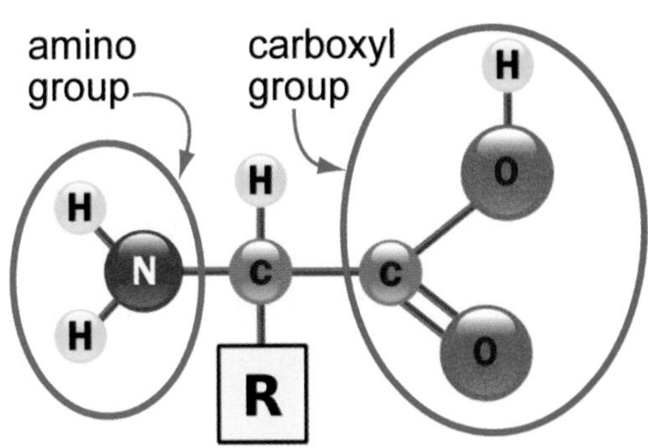

The R group determines the chemical properties of the amino acid

Because the *central carbon atom has four different things attached to it, it is chiral and forms an optical isomer.* Each amino acid also contains a hydrogen atom and a side chain(R- hydrocarbon chain). The side chain significantly affects the chemical properties of the amino acid. These R-groups are unique for each amino acid.

Chemically some R groups are nonpolar, some are polar. Amino acids such as alanine or valine, contain R groups that are nonpolar, and in an aqueous environment they would be found on the interior of a protein molecule. Inside the protein these hydrophobic R groups are "shielded" or protected from an aqueous environment. Other amino acids such as serine or aspartic acid contain polar R groups and these are found on the exterior of a protein molecule. These amino acids would form hydrogen bonds with water molecules. In solution the R group determines the functionality of a protein.

Alanine

COOH
H_2N–C–H
CH_3
nonpolar

Serine

COOH
H_2N–C–H
CH_2OH
polar

Side chains can be either polar or nonpolar.

Some amino acids have acidic and basic properties

COOH
H_2N–C–H
CH_2
CH_2
COOH
acidic

COOH
H_2N–C–H
CH_2
CH_2
CH_2
NH
C=NH_2
NH_3
basic

Some amino acids contain acidic and basic R groups such as glutamic acid and arginine. In the body these side chains are exposed to a neutral pH of approximately 7.35. In the laboratory pH changes are used to separate proteins, because these amino acids can be made to come apart and "denature" under conditions of high or low pH. Biochemists take advantage of this when want to isolate proteins.

Amino acid table

Twenty Amino Acids

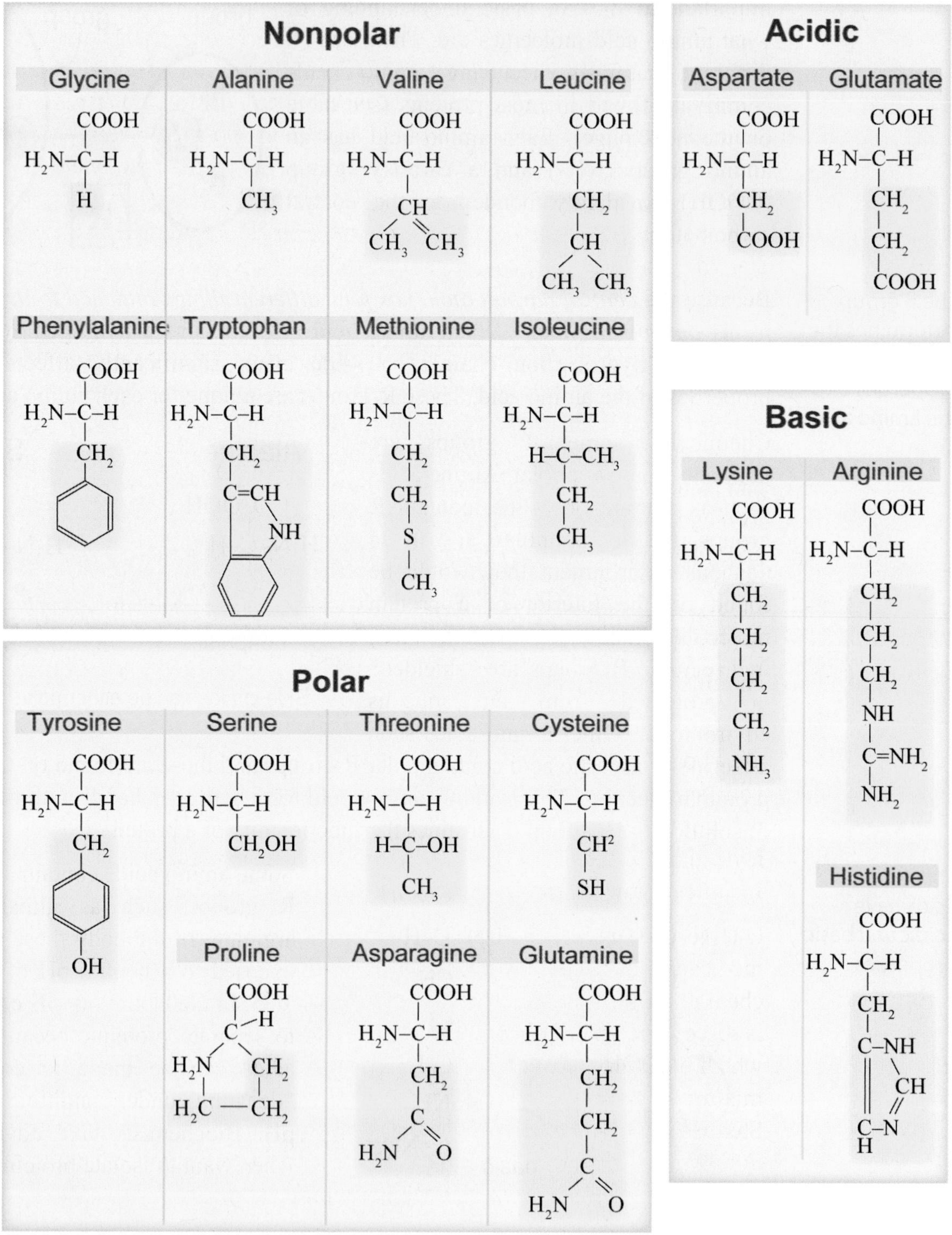

Nonpolar

Glycine	Alanine	Valine	Leucine
COOH H_2N–C–H H	COOH H_2N–C–H CH_3	COOH H_2N–C–H CH CH_3 CH_3	COOH H_2N–C–H CH_2 CH CH_3 CH_3

Phenylalanine	Tryptophan	Methionine	Isoleucine
COOH H_2N–C–H CH_2	COOH H_2N–C–H CH_2 C=CH NH	COOH H_2N–C–H CH_2 CH_2 S CH_3	COOH H_2N–C–H H–C–CH_3 CH_2 CH_3

Acidic

Aspartate	Glutamate
COOH H_2N–C–H CH_2 COOH	COOH H_2N–C–H CH_2 CH_2 COOH

Polar

Tyrosine	Serine	Threonine	Cysteine
COOH H_2N–C–H CH_2 OH	COOH H_2N–C–H CH_2OH	COOH H_2N–C–H H–C–OH CH_3	COOH H_2N–C–H CH^2 SH

Proline	Asparagine	Glutamine
COOH C–H H_2N CH_2 H_2C — CH_2	COOH H_2N–C–H CH_2 C H_2N O	COOH H_2N–C–H CH_2 CH_2 C H_2N O

Basic

Lysine	Arginine
COOH H_2N–C–H CH_2 CH_2 CH_2 CH_2 NH_3	COOH H_2N–C–H CH_2 CH_2 CH_2 NH C=NH_2 NH_2

Histidine
COOH H_2N–C–H CH_2 C–NH CH C–N H

Chirality in nature

Most amino acids have the L or left-handed (S) configuration

Interestingly, it seems as though nature almost exclusively uses one form of amino acid stereoisomer. Nature is said to exhibit homochirality, which means it utilizes only one type of isomer. There are many theories as to why one particular form is favored. One theory is based upon the fact that the L configuration amino acids are favored when dissolved in water. This would make them readily available in aqueous environments, which is where life forms are thought to have evolved. Another intriguing theory is that the L amino acids were brought to earth by meteorites. Meteorites, such as the murchinson meteorite that landed in Australia in 1969, contain primarily L amino acids. Scientists think this is because they were formed by circularly polarized light from the stars! If one chiral form was available back when life was first developing perhaps nature adapted itself to utilize that one form?

The chirality of the amino acids commonly found in nature are almost all left handed

Other things in nature exhibit what we call handedness. Most sea shells for example are formed in the right handed direction. Snails and plants also show tendency toward one specific chirality. This has been of great interest to biologists. The snail shells either twist to the right, or to the left. Which type of chirality is expressed is determined by genetic factors.

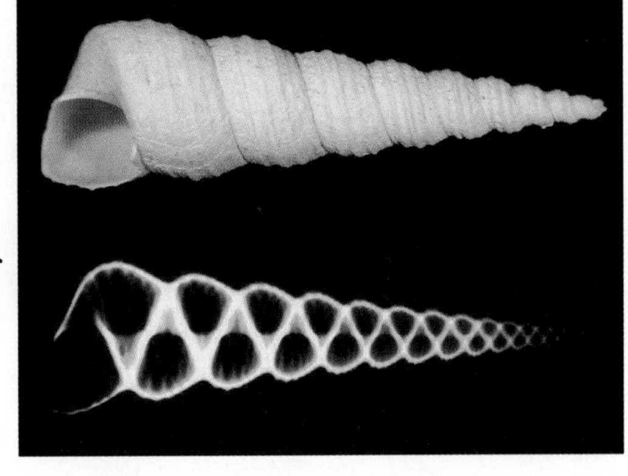

Our body chemistry functions are designed with specificity toward one stereoisomer. As mentioned earlier sugar molecules utilized by the body for energy are all of the D or right handed conformation. *Chemical reactions are carried out in the body by recognizing very specific molecular structures.* This is what allows our bodies to function properly.

One very important application of chirality can be seen in medicine. Many widely used medications are derived from nature. For example Taxol, which is a drug used for chemotherapy, was discovered from the bark of the Pacific yew tree. When a useful drug is discovered in nature, scientists try to isolate the different active molecules to identify which one works. When we try to synthesize these molecules artificially we find that a mixture of both the isomers are made. But nature specifically makes only one isomer! Steroisomers have been shown to have very different effects in the body! *Molecules of different chirality have completely different biological activity.* Because of this over 50% of medications must be specifically one chiral form.

Functions of proteins

One example is muscle tissue

Proteins are used in numerous ways by our bodies. Perhaps you have been told to "eat your meat" so that your muscles will grow. Well it is true that muscles are mainly made up of protein. Athletes and body builders who have to increase their muscle mass often consume protein shakes. The protein helps them to build up their muscle fibers. For a while creatine was a popular body building supplement taken by athletes. It helped to build muscle mass faster and without the damaging side effects of steroids. Creatine is a protein naturally found in meats and taking a supplement provided muscles with more energy. However, it was later found that taking creatine causes liver and kidney damage, because our bodies cannot break it down it effectively.

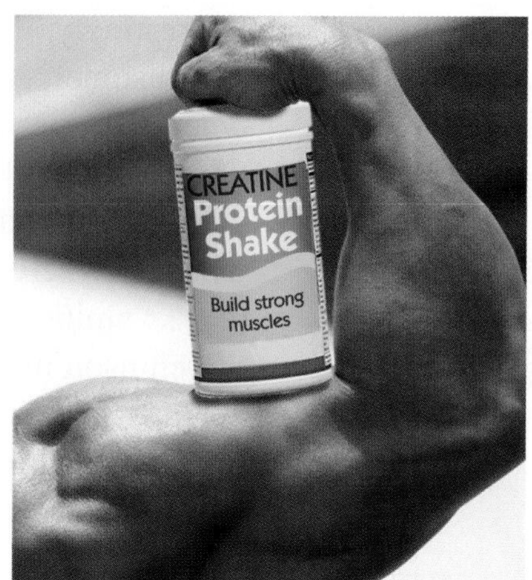

Functions of Proteins	
Function	**Example**
Movement	Muscle is primarily made up of protein. Protein molecules allow muscle cells to contract.
Structure	Tendons, skin, bones, fingernails and claws are made up of structural proteins. Also fibers such as wool, or hair are proteins.
Catalysis	Enzymes which are proteins, catalyze almost all chemical reactions in living systems.
Transport	Oxygen is carried by proteins to our tissues by hemoglobin.
Storage	Proteins store or hold onto minerals needed by the body.
Protection	Blood clotting proteins such as fibrinogen protect us from bleeding too much. Antibodies such as interferon are proteins that protect cells from viral infection.
Energy Transfer	Living cells contain cytochromes which are proteins. Cytochromes transfer electrons in a series of oxidation-reduction reactions during cellular respiration (metabolizing food molecules).

Protein composition - Primary structure

The information for building proteins comes from our hereditary information. Our genes contain the instructions for which amino acids to link together when forming specific proteins.

Chemically amino acids are linked together by a **peptide bond**. A peptide bond joins the carboxyl group and the amino group of adjacent proteins.

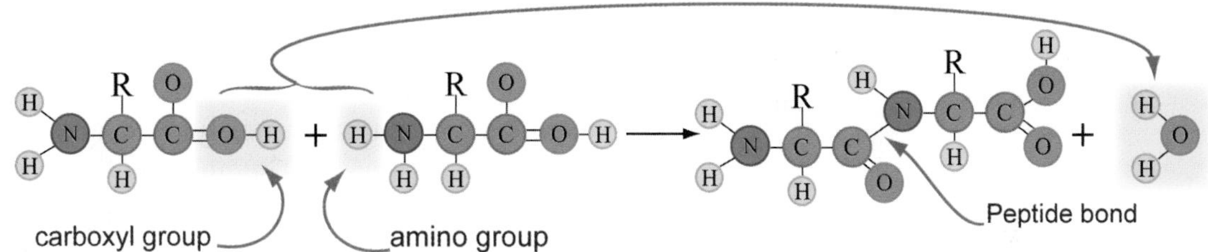

Amino acids are linked together with peptide bonds

The formation of a peptide bond involves the loss of a water molecule. As you saw in chapter 17 this is called condensation polymerization. A bond is formed between two monomers, in this case amino acids, and a polymer is formed. Each amino acid in the sequence is added on the right end of the polymer as it continues to grow. All of the amino acids have a carboxyl group and an amino group, which allows them to be put together in any combination.

Amino acid sequence determines protein function

The amino acid sequence is what makes one protein have different chemical properties than another. Remember that a protein sequence must be at least 100 amino acids long! While these proteins are long, a sequence for a protein is very specific. A change in just one amino acid in the protein chain can result in very different chemical properties. The sequence of amino acids in a protein chain is called its **primary structure**.

Our bodies break apart and reuse the amino acid building blocks.

You might wonder where we get all the amino acids necessary to build a protein. Our bodies can make ten of the necessary amino acids, but the other ten we need to get from our diet. When we eat protein containing foods these proteins must be broken apart and put back together to make new proteins for our bodies to use.

Chemistry terms

peptide bond - a bond formed between the amino group and the carboxyl group of two amino acids. The peptide bond is formed through condensation polymerization or the loss of water molecules.

primary structure - sequence of amino acids in a protein chain.

Secondary structure of proteins

Proteins form a 3-dimensional shape

The way a protein is put together is very important. The sequence of amino acids influences the way a protein functions. The next piece to this puzzle is to understand how the chain of amino acids orient themselves three dimensionally. There are two spacial arrangements that are commonly seen in nature, the **alpha helix** and the **pleated sheet**. The alpha helix resembles a spiral staircase or a slinky. It slowly winds around in a coil that is primarily held in place by hydrogen bonds. The alpha helix is one of the most important and common secondary structures. The pleated sheets resemble a folded paper fan or an accordion.

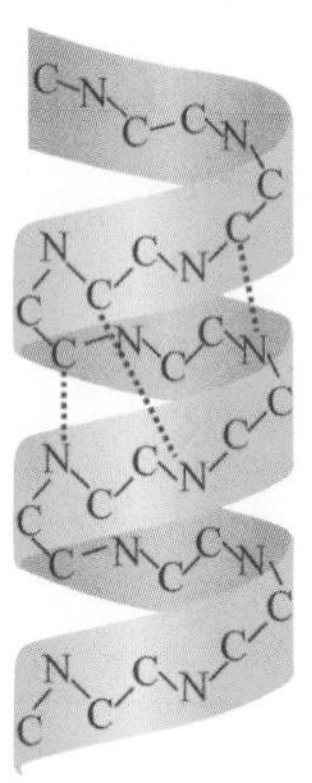

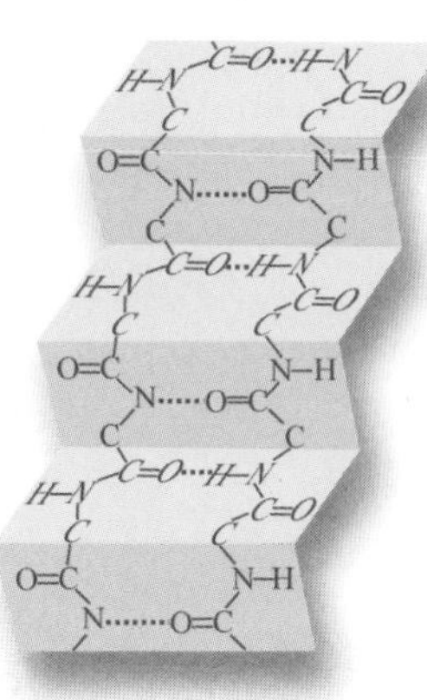

Hydrogen bonds are important in forming and holding the protein's shape

The regular coils of the alpha helix give it some elasticity and "stretchiness." These properties influence the protein function. For example, many fibers such as wool, hair and tendons have the alpha helix shape. The alpha helix shape is held together by hydrogen bonds formed between a carbonyl group (C=O) and a nearby amino group (N-H). The formation of a helix is perfectly designed so that the hydrogen bond associations are perfectly aligned. The twist of the helix causes the side chains or R-groups to be on the outside of the helix, where they have more room and can interact with their environment. Some amino acids, such as tyrosine and tryptophan, have large and bulky side chains that do not fit well into an alpha helix. These amino acids prefer to form pleated sheets.

The pleated sheet is made up of two or more protein chains located side by side. This spacial arrangement causes the protein to be *flexible and strong*, but resistant to stretching. Pleated sheet proteins are found in muscle tissue and in natural fibers like silk. Both muscle fibers and silk have a very high tensile strength, in fact stronger than steel! Similar to the alpha helix, hydrogen bonds hold the pleated sheet together, but these hydrogen bonds are formed *between adjacent proteins*, not between groups along the same chain.

alpha helix - a common form of the secondary structure of a protein that forms a spiral or twist similar to a slinky or spiral staircase. A single protein chain is held together by hydrogen bonding attractions.

pleated sheet - a secondary structure of protein that forms a folded or pleated pattern similar to an old fashioned fan. Formed by two or more side by side protein chains held together by hydrogen bonding.

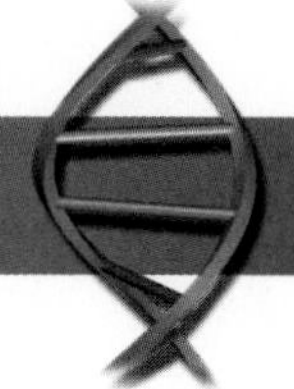

Tertiary structure of proteins

Tertiary structure

A protein's shape involes a great deal of chemistry. You have learned about hydrogen bonds, electrostatic attractions between charged or polar molecules, and you have learned about hydrophobic and hydrophilic interactions. Here we will apply all of this knowledge to our study of tertiary protein structure. Tertiary structure defines the functionality of the protein. Proteins fold or bend their secondary alpha helix structures into what is called a **tertiary structure** or a "globular" protein. The term globular is a good one because the protein looks like a glob of amino acids stuck together. A globular protein has a unique tertiary structure that makes it capable of performing a very specific function. To imagine this structure think of a phone cord wrapped around itself with several twists and turns until it is all knotted together. This is what a globular protein looks like.

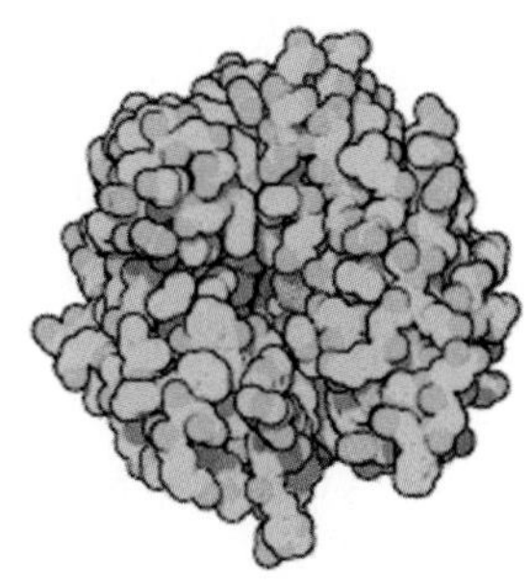

Intermolecular forces stabilize the tertiary structure

The interactions of the side chain, R-groups, greatly affect the shape of the tertiary structure. Most proteins in living systems are in aqueous water environments. The nonpolar side chains tend to be tucked inside the tertiary structure and away from the polar water molecules, while the polar side chains are on the outside so that they can interact with the aqueous environment. These intermolecular forces are what stabilize the tertiary structure and keep it in the proper shape so that it can perform its role in living organisms.

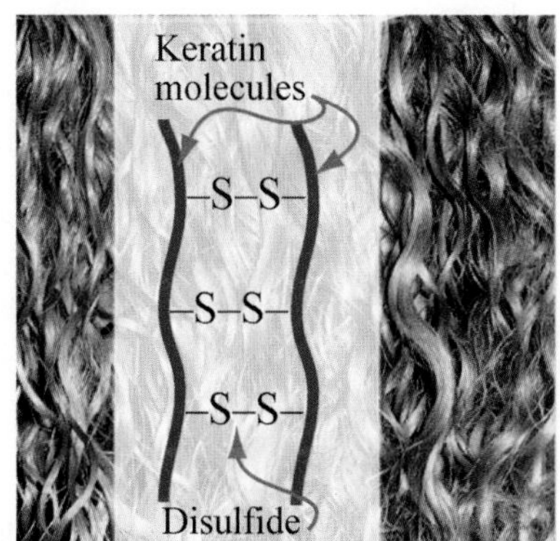

The amino acid cysteine plays an important role in stabilizing tertiary structure as well. Two nearby cysteine groups can form a covalent disulfide (S-S) bond, which is very strong and can hold two parts of a protein chain together forming a bend. The keratin in our hair contains these disulfide bonds. When we get our hair "permed" at the hair dresser they are breaking some of these disulfide bonds in our hair shafts and reforming new ones that hold a curl. As our hair grows new protein is formed that does not have the same bonds holding it's shape.

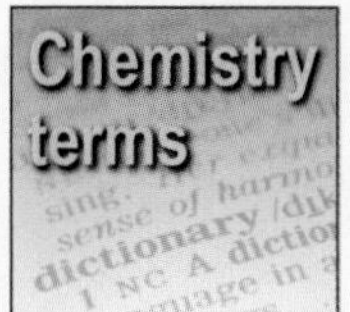

tertiary structure - protein structure that forms from the bending or folding of the secondary alpha helix. Tertiary structures form specific globular proteins. The structures are held together by intermolecular attractions, hydrophobicity of nonpolar regions, and ion-dipole interactions.

Enzymes

Enzymes are proteins that orchestrate chemical reactions in living systems

Now that you understand a little more about proteins, we can study an important class of proteins called enzymes. Almost all enzymes are proteins! Enzymes serve a wide variety of roles inside living systems. As mentioned earlier in chapter 12 enzymes dramatically speed up chemical reactions. Living systems have an incredibly complex system of integrated chemical reactions. Enzymes orchestrate these complex systems so that living organisms can survive. Enzymes are able to turn molecules "on" and "off" at appropriate times as they are needed by living organisms. Enzymes were mentioned earlier in chapter 12 when we were studying the rates of chemical reactions. Enzymes lower the energy needed for a chemical reaction to take place, because of this they speed up chemical reactions.

Bio-luminescence

Some enzymes such as luciferase are responsible for "bioluminescence." If you have ever seen a firefly light up in the dark then you have seen this enzyme in action.

The shape of the enzyme determines what molecules bind to it

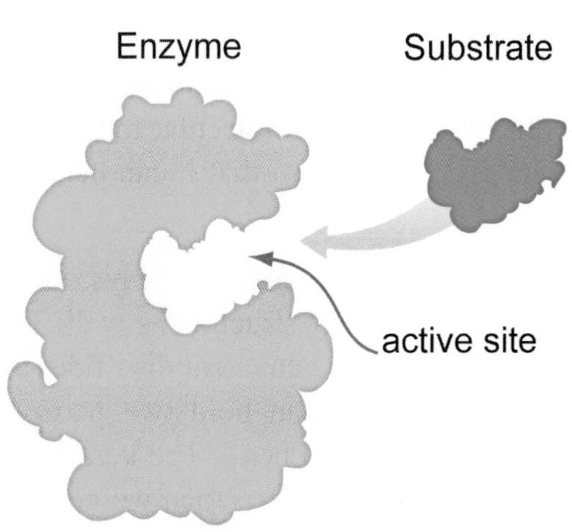

Enzymes are large proteins with molecular weights ranging from approximately 10,000 amu to 1 million amu. Enzymes bind other molecules called **substrate** molecules. The substrate is the "reactant" in the chemical reaction taking place. The substrate binds in a small region of the enzyme called the **active site**. The active site makes up only a small section, about 3 to 4 amino acids, of the enzyme molecule. When these substrate molecules are bound by the enzyme, the enzyme and the substrate can change shape, molding to each other. These large enzyme molecules have very high specificity for the substrates that attach. Generally an enzyme will only allow one type of substrate molecule to fit in its active site. *The shape of the active site is determined by the tertiary structure of the protein.*

Chemistry terms

substrate - a reactant molecule that is able to be bound by a particular enzyme.

active Site - a small region of an enzyme molecule that is responsible for "binding" or holding on to the substrate molecules.

Cytochrome proteins

Cytochrome proteins transfer electrons in the electron transport chain

The **cytochrome proteins** involved in cellular respiration are one of the most important proteins in nature. A cytochrome molecule is any type of iron containing protein, with heme groups. The electron transport chain contains several types of cytochromes. The cytochrome protein molecules are responsible for the transfer the electrons in the electron transport chain. You may remember from earlier in the chapter, the electron transport chain makes ATP using high energy electrons originally from food molecules. The electron transport chain slowly releases the energy, with the help of cytochrome molecules, the electrons are passed slowly along, and the energy is released in manageable amounts. Cellular respiration relies on the slow release of energy achieved by these cytochrome proteins, which are located in the inner mitochondrial membranes inside our cells.

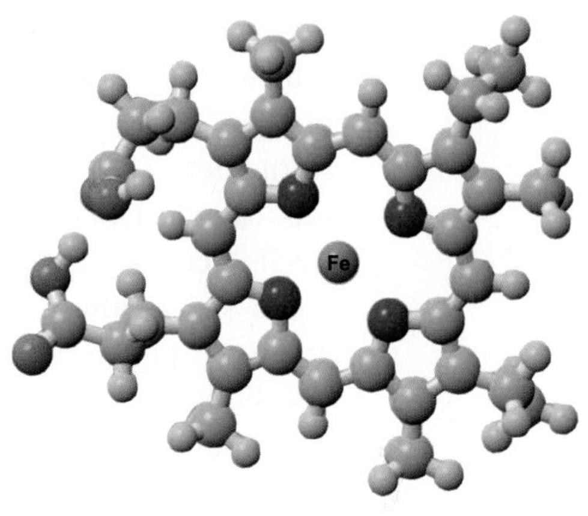

All organisms that carry out respiration use cytochrome proteins

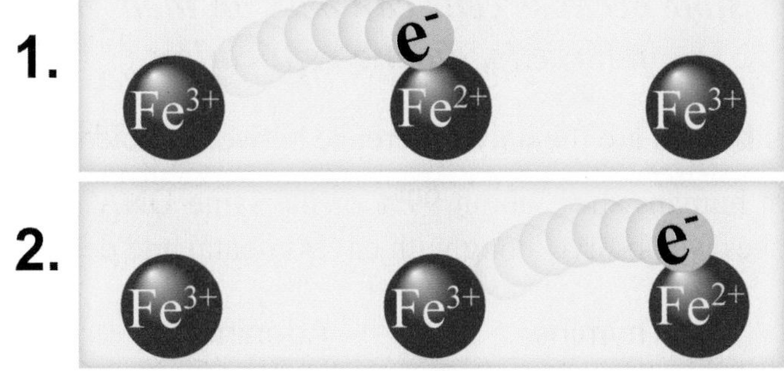

The cytochrome system functions to catalyze oxidation of iron(II) to iron (III) and the reduction of iron(III) to iron(II). The molecular structure of a cytochrome is made up of a large organic ring structure called a porphyrin with an iron atom held at its center. The cytochrome molecule picks one electron up and gets reduced from Fe^{3+} to Fe^{2+} , when the electron is transferred to the next carrier protein, Fe^{2+} is oxidized back to Fe^{3+}. These electron transfers are used to pump H^+ ion into the cell against the concentration gradient. When the H+ ions diffuse back out of the cell they go through a protein channel, which couples the energy released when the H^+ flows out, to the formation of ATP. Each pair of electrons produces roughly three ATP molecules.

cytochrome protein- any iron containing protein with heme groups, these proteins are import as catalysts in oxidation-reduction reactions in the electron transport chain.

18.4 DNA and Molecular Reproduction

DNA nucleotides contain 3 parts

What is a DNA molecule? How can it possibly carry enough information to code for a living organism? These questions will help us focus on this unique molecule. **DNA** is a polymer that stores and transmits genetic information. The DNA polymer is made up of units called nucleotides. A **nucleotide** has three molecular parts to it. The first part is a 5 carbon sugar called deoxyribose, the second part is a phosphate group, derived from phosphoric acid, H_3PO_4 and lastly a nitrogen containing organic base. DNA molecules are HUGE with their molar masses ranging from 5 to 20 million amu's, and possibly more!

Nucleotide Structure

Phosphate group

Nitrogenous base (guanine)

Sugar

Even with only four different building blocks, there are an incredibly large number of possible sequences of DNA nucleotides. DNA molecules carry all of the hereditary information necessary for an organism to grow and survive. Almost every cell in the human body has the same DNA!

Life is only possible because cells can pass on their genetic material to new cells, during replication.

Nitrogen containing bases are different between each of the nucleotides

The **nitrogenous bases** are the only difference between nucleotides.

Did you know that humans share about 99% of the same DNA? That means only 1% causes the differences we observe along with enviromental and developmental effects.

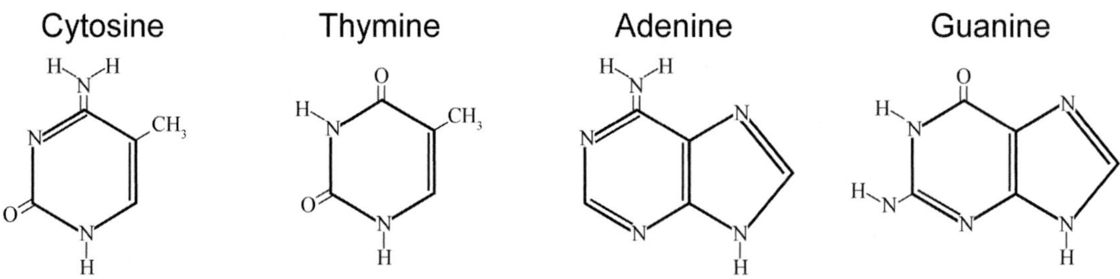

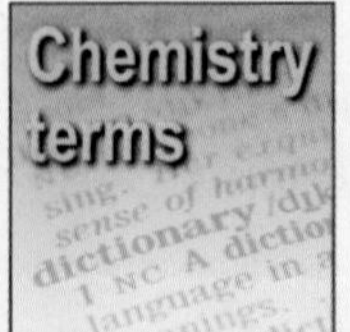

DNA - long polymer chain of nucleotides that stores and transmits genetic information.

nucleotides - a molecular unit composed of a 5 carbon sugar, a phosphate group, and a nitrogenous base.

nitrogenous base - nitrogen containing bases, one component of nucleotides.

DNA double helix

Nucleotides are linked by a condensation reaction.

Individual strands of DNA polymers are formed by joining nucleotides, together in a condensation reaction. The nucleotides are linked together when a phosphate group on one nucleotide bonds with the hydroxyl, -OH group, on a deoxyribose sugar molecule of the other nucleotide. A water is lost in this condensation reaction when the two monomers are joined. As you can see *each DNA polymer has a central backbone formed from the sugar and the phosphate group on each nucleotide.* This backbone holds the polymer together. Notice that the nitrogenous bases stick out from the backbone.

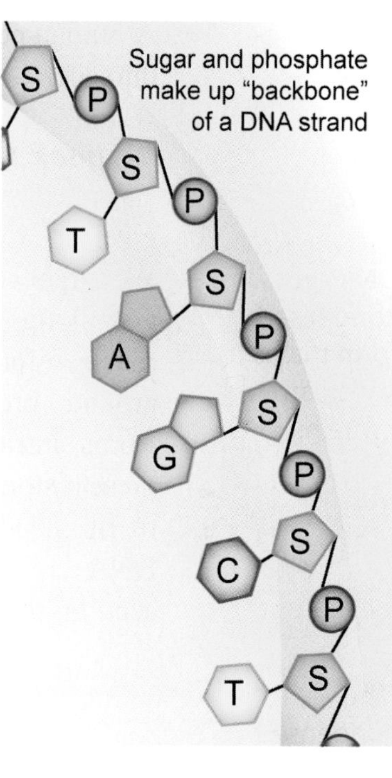

In the double helix, which consists of two individual strands of DNA. The nitrogenous bases play a key role in DNA's structure and function. Each nitrogenous base has a partner or a "complementary base", which is determined by the number of hydrogen bonds formed. Adenine(A) always bonds to thymine(T) forming two hydrogen bonds, and guanine(G) bonds to cytosine(C) forming three hydrogen bonds between them.

Nitrogenous bases (G,C,A,T) play a key role in the double helix structure.

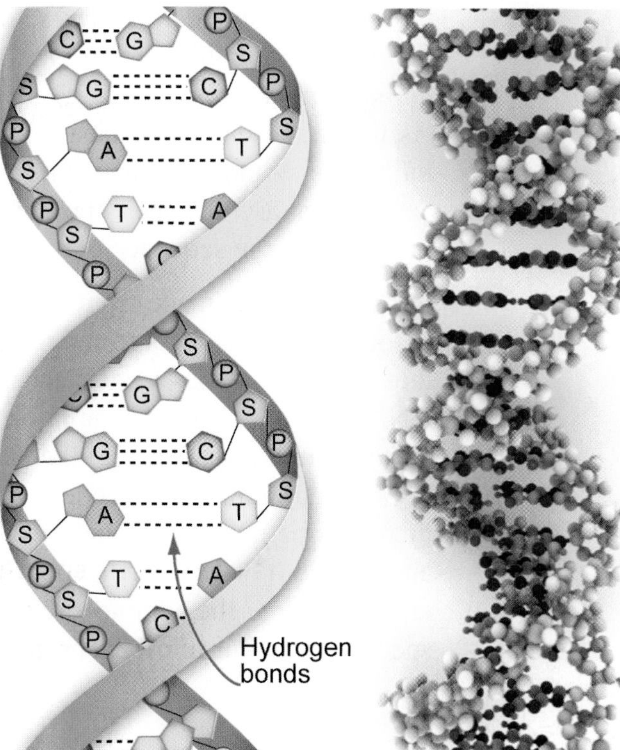

It is interesting to see how the nitrogenous base structures on one strand match up with the opposite strand allowing them to form uniform hydrogen bonds of two or three depending upon the base pairs. The formation of the double helix, and the spacing in the turns of the helix, are just right, allowing for only these specific pairings of the bases. Pairings like A to G do not fit properly!

Hydrogen bonds and london dispersion forces hold together the DNA double helix.

DNA replication

A human cell contains approximately 1 meter of DNA. How could a strand that long fit into such a tiny cell?

To pack DNA effectively inside a cell it needs to be folded many times and coiled tightly

DNA exists in a condensed form that is coiled around proteins

We might expect the DNA molecules to be very long because they are carrying so much information. Human DNA coils around proteins called histones. DNA forms intricate weaving patterns around these histone proteins, this allows the DNA to fit inside the cell. Only some of our DNA codes for our genes, thre are long sections that only have structural purposes.

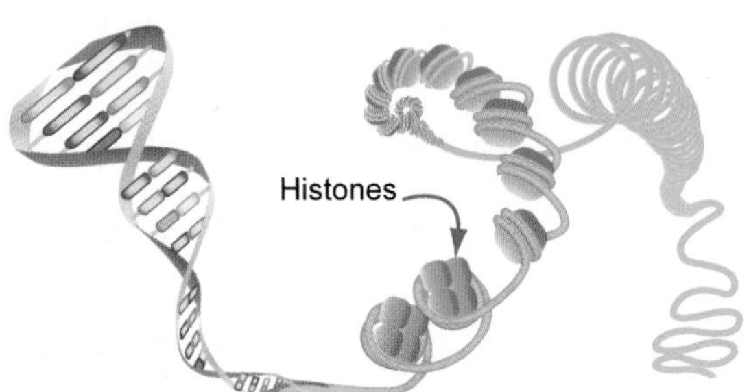

DNA coils around Histones.

The condensed DNA "opens up" to replicate

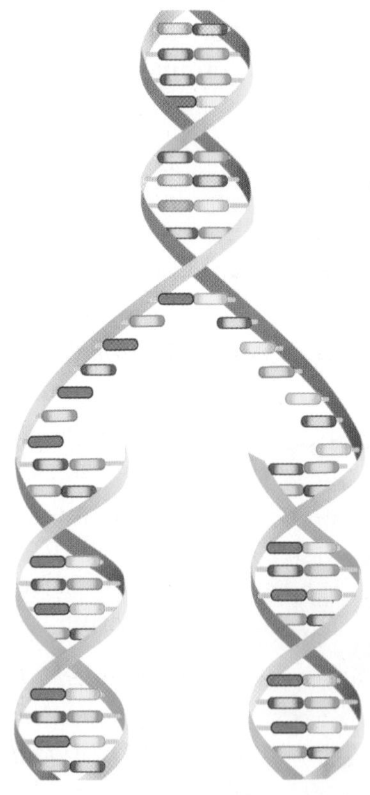

The fact that this long polymer can replicate itself with relatively few mistakes is truly amazing. During cell division the double helix unwinds so that new complementary strands can be formed from each of the original strands. *Enzymes cause the hydrogen bonds between the base pairs to come apart, this allows for the "unzipping" of the DNA double helix.* As the strand unzips new base pairs are added to the individual strands. Because the pairing of the bases is complementary between the same partners we can predict what the sequence of nucleotides will be.

Each new cell is made from and an existing cell. The ability of these strands to replicate allows for each of the two cells to have an exact copy of the original DNA molecule. The genetic information is stored like a "code" in the sequence of the base pairs, A, T, G, C. For each amino acid the code is three letters long, but there can be more than one three letter code for each amino acid. For example CGA, CGG, CGC and CGT all code for the amino acid alanine. Scientists speculate that this allows for some flexibility and error when the sequences are formed.

DNA and protein synthesis

DNA also stores the information necessary to make proteins

DNA also carries the code to make proteins that an organism needs to live. As discussed in the previous section, proteins are necessary for all life processes. All chemical reactions that occur in the body require proteins. A segment of DNA called a **gene** is the code for a specific protein. All organisms, including humans, consume proteins in the foods that they eat, but these proteins are not the same ones required by the organism to survive. Individual organisms require specific instructions for the proteins they need.

The DNA "code" stores the information for a sequence of amino acids that is passed on to the cell "machinery." Recall that the sequence of amino acids represents the primary structure of the protein.

RNA carries the information from DNA to the cell machinery where proteins are made

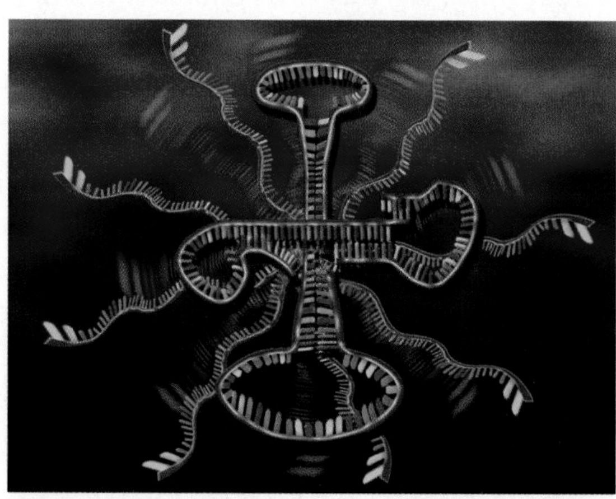

How is the information stored in DNA used by the cell to make protein molecules? Depending upon the type of organism the DNA is stored in the cytoplasm or in the nucleus of the cell. Another nucleic acid called **RNA (ribonucleic acid)** is responsible for transmitting the information to the cell machinery where proteins are made. RNA is similar to DNA except for a few distinguishing features. It contains a different sugar called ribose that contains a hydroxyl -OH group on the second carbon in the 5 carbon ring structure. It also uses one different nitrogenous base called uracil (U). Uracil is used in place of thymine(T) when RNA is formed. RNA molecules are also much smaller than DNA polymers. RNA molar masses are typically between 20,000 to 40,000 grams while DNA's molar mass is in the billions.

RNA is made following the pattern of the gene for a specific protein. RNA carries this information to a ribosome. Ribosomes are little factories where the proteins are made. Once the instructions for the pattern reaches the ribosome it makes the protein. Specific amino acids are then brought to the ribosomes following the instructions from the RNA. Scientists have discovered that RNA has a wide variety of functions, which has lead to the possibility that perhaps it evolved before DNA. For example, RNA also plays an important role as a catalyst of chemical reactions, similar to enzymes.

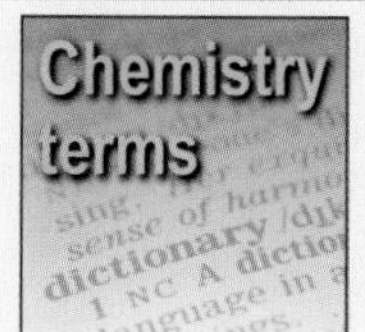

gene - segment of DNA that contains the code for a specific protein.

RNA - a polymer composed of nucleotides that contain a 5 carbon sugar called ribose, a phosphate group, and a nitrogenous base. This molecule transmits the information stored in DNA to cell "machinery."

Farming and green chemistry

Over the years, commercial farming has relied on chemical pesticides to grow healthier crops and to increase crop yields. Now with the rise of "green chemistry" there has been a dramatic shift in how pesticides are being made and used. By definition "green chemistry" is the design of chemical products and processes that reduce or eliminate the use of hazardous substances. Since the late 1990's consumer awareness has caused people to begin to buy "organic." Many of us know that organic fruits and vegetables are good for us. In fact, in our society buying organic has become very important. Consumers have been so influenced by the importance of purchasing organic commodities that grocery stores offer complete shopping areas dedicated to just these products.

But what is so important about buying organic vegetables anyway? Of course it has to do with chemicals! Over the past 50 to 60 years chemicals used by farmers as pesticides have become of great concern. We have learned that these chemicals used to kill insects and pests sometimes cause cancer and other illnesses in our bodies, and they also have adverse effects in our environment.

Science has shown that some widely used pesticides can persist in the soil and in the environment for long periods of time. The fact that these chemicals do not biodegrade, (break-down into less harmful by-products in the soil) allows them to accumulate every year until they reach dangerous levels. For example, Dichloro-Diphenyl-Trichloroethane, DDT is a synthetic pesticide that was once widely used by U.S. farmers during the 1950's and 1960's. This product is now banned (since 1972) in the U.S. It was found that DDT remained active in the soil for up to 30 years after it was used. DDT's effects on wildlife and the environment were of significant concern to biologists and environmental scientists.

Pesticides that remain in the environment for long time periods have large energy barriers, this is why it takes a long time for the compounds to degrade and become harmless. Recall that the energy barrier is the amount of energy necessary for the chemical to react, which means breaking and reforming new bonds.

We are replacing chemicals like DDT with pesticides that have lower energy barriers and this means they persist in the environment for shorter periods of time.

An example of one of these chemicals is glyphosate. Glyphosate is an analog of glycine, which means it is similar to the amino acid glycine. Glyphosate interferes with plants ability to make amino acids.

Glyphosate is replacing a herbicide known as Simazine, which disrupts plant photosynthesis. Simazine has been associated with several human health risks, and it persists in the environment for relatively long time periods. The environmental protection agency, EPA has placed a limit of 4 ppb in drinking water for Simazine. In contrast, the drinking water limit of glyphosate is 700 ppb, this would indicate that glyphosate is less toxic than caffeine. This kind of progress replacing harmful chemicals like Simazine with less harmful alternatives like glyphosate, is the focus of "green chemistry."

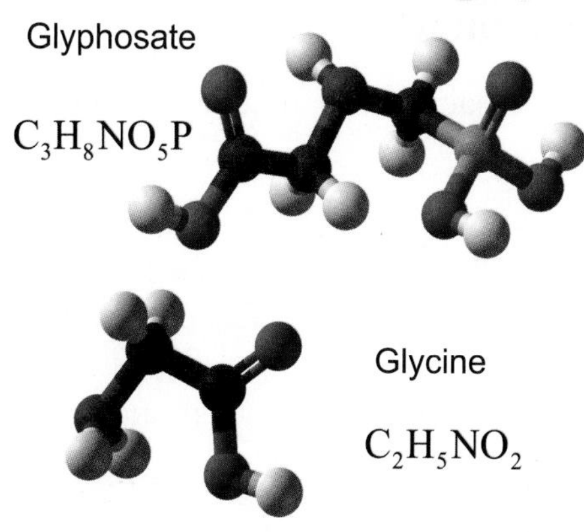

You will be happy to know that currently in the U.S. pesticide use is declining. This is due to a dramatic shift in how pesticides are being made and used. First, better pesticides are being made. These pesticides have have lower toxicity, are more selective to specific insects, and are effective in lower amounts, allowing less to be applied to crops. Farmers are also becoming more educated about the life cycles of pests and how they affects the growth of their crops. This allows farmers to use pesticides in more responsible ways. Lastly, organic farming deliberately avoids the use of pesticides!

What does it mean to be organic? Organic foods must be certified by the U.S. food and drug administration, USDA. This requires that crops are grown without synthetic pesticides or chemical fertilizers. Regulations also specify that produce must not be irradiated (treated with radiation to kill bacteria), and that it is not grown from bioengineered crops. Farmers have to meet these standards for three years before they can be certified organic. However, organic farmers do use natural forms of pesticides. Some chemicals such as sulfur, copper sulfate, pyrethrins and compounds from a soil bacterium have been approved for organic use. Sulfur and copper sulfate are used all over the world as fungicides, especially on grapes, berries and melons. Pyrethrins are a pair of natural organic compounds, that act as a neurotoxin to insects. Pyrethrins come from the seed cases of the chrysanthemum flower, Chrysanthemum cinerariaefolium. Pyrethrins break down easily when exposed to sunlight and oxygen, becoming inactivated.

So although organic foods are grown by natural methods farmers still need to rely occasionally on some naturally occuring chemicals to help them achieve good crop yields.

Overall, green chemistry is successfully helping farmers improve their methods for treating insects and weeds. These achievements directly affect our food supply, our bodies and our surrounding environment in a very positive way!

Chapter 18 Review

Vocabulary

Match each word to the sentence where it best fits.

Section 18.1

fat	carbohydrates
oligosaccharides	monosaccharides
polysaccharides	triglyceride
phospholipid	micelle

1. All organisms use ________________ as a primary source of energy.
2. A ___________ is a single sugar unit.
3. Lipid or ________ molecules are nonpolar and therefore they do not dissolve in water.
4. Starch, and cellulose are examples of ________________.
5. A fat molecule made of glycerol and three long chain fatty acids is called a _________.
6. A chain of two to ten monosaccharide units linked together in a chain is called an ___________.
7. A lipid that has a phosphate group attached to it is called a __________________.
8. A small sphere made up of phospholipids is called a _______________.

Section 18.2

Chlorophyll	Calvin cycle
ATP	cellular respiration
photosynthesis	Glycolysis
NADPH	NADH
Krebs cycle	Electron transport chain

9. The process of _______________ uses sunlight to convert carbon dioxide and water into glucose and oxygen.
10. ________ is an electron carrier molecule used by plants to store the energy from sunlight.
11. Adenosine triphosphate or ________ is a molecule that carries chemical energy to nonspontaneous chemical reactions.
12. Plants contain a pigment molecule called _____________ that absorbs energy from the sun to be used in photosynthesis.
13. In the presence of oxygen, _____________________ breaks down glucose and other food molecules releasing energy for life processes.
14. The ______________ is also sometimes referred to as the "dark" reactions, because sunlight is not required for this part of photosynthesis.
15. During ______________ which is the first step in cellular respiration glucose is broken into two three carbon molecules.
16. In cellular respiration, a high energy electron carrier molecule called _________ is used to transfer electrons.
17. The _________________ is a series of chemical reactions that break pyruvic acid down into carbon dioxide during cellular respiration.
18. High energy electrons from the Krebs cycle are transferred to the _________________ where they are used to make ATP molecules.

Section 18.3

protein	primary structure
amino acids	pleated sheet
peptide bond	tertiary structure
alpha helix	cytochrome protein
substrate	active site

19. ____________ are polymers of amino acid molecules, which are used in every cell process in our bodies.

20. There are twenty different naturally occuring ________________.

21. The ____________________ of a protein is the sequence of amino acids.

22. A bond formed between the amino group of one amino acid and the carboxyl group of another amino acid is called a __________________.

23. The secondary structure of a protein can have two different structural forms called an ____________ and a _________________.

24. An enzyme has an _______________ that can change shape once the specific substrate molecule is bound to it.

25. A reactant molecule that can bind to an enzyme is called a ________________.

26. A ____________ is an iron containing heme protein, that is important in the electon transport chain.

27. Proteins fold or bend their secondary structrues into a _________________ that is the result of molecular interactions.

Section 18.4

nucleotide	DNA
gene	RNA
nitrogenous bases	

28. Different pairs of __________________ match up because of the number of hydrogen bonds they can make to each other.

29. The __________molecule forms an alpha helix that is a long polymer chain of nucleotides used to transmit genetic information.

30. A________________unit is made up of a five carbon sugar, a phosphate group, and a nitrogenous base.

31. A segment of DNA that contains the code for a specific protein is called a ________.

32. The ____________ molecule uses uracil(U) instead of thymine(T).

Conceptual Questions

Section 18.1

33. Give three examples of how living systems use fats and carbohydrate molecules.

34. Distinguish between a monosaccharide, and an oligosaccharide. Give one example of each.

35. Sketch the structure of glucose and galactose. Explain how these molecules are different from one another.

36. Do our bodies use alpha-D-glucose or beta-D-glucose? Explain why.

37. When forming a polysaccharide which carbon atoms in the glucose molecule are used to bond the molecules together as the chain forms? Explain.

38. Explain how glucose and fructose are different from each other.

39. When glucose and fructose are linked together what do they form?

40. List four examples of carbohydrates that are considered "good" for you.

41. Why is it better to consume complex carbohydrates rather than simple carbohydrates?

42. What is cellulose?

43. Where is cellulose most commonly found?

44. When you eat more carbohydrates than your body needs, what happens to these extra carbohydrates? Explain.

45. a)How much of our daily diet should come from carbohydrates?
b) How much of our daily diet should come from fats?

46. Contrast the basic differences between fats and oils.

Chapter 18 Review.

47. Explain the diffference in the molecular structure of a saturated fat and an unsaturated fat.

48. Why are unsaturated fats better for us to consume? Explain, and be sure think back to the organic chapter for some help with the specifics.

49. Fats serve several different useful purposes in our bodies. List three examples and briefly explain the purpose of each.

50. Do animals store most of their energy in the form of fats or carbohydrates? Briefly explain why.

51. a) Name the types of molecules make up a typical fat molecule.
 b) How are these molecules joined together? What is the bond called?

52. What physical property is used to classify fat molecules from other biological molecules like proteins or carbohydrates?

53. a) What are lipoproteins?
b) Explain why lipoproteins are needed to carry fat molecules in our blood ?

54. Based on the molecular structure, why are some fats solid at room temperature while others are liquid?

55. Give 2 examples of foods you could eat that contain unsaturated fatty acids.

56. What type of fat comes from beef?

57. Longer fatty acid chains yield more energy than shorter ones. What is it about the molecular structure of longer fatty acids that makes this true?

58. a) What type of fat molecule are miscelles made up of?
b) Draw a simple sketch that describes what you think a miscelle would look like.

59. a) Which end of the phopholipid molecule is attracted to water? Explain why.
b)Describe the purpose of the phospholipid bilayer in our cells.

Section 18.2

60. a) Write the chemical reaction for photosynthesis.
b) What living systems carry out photosynthesis?
c) Show the lewis structure for the gas that is produced.

61. a)What molecule is responsible for the process of photosynthesis?
b) In what regions of the visible spectrum does this molecule absorb light?

62. a) In the first part of photosynthesis how is the energy in sunlight captured?
 b) What carrier molecule holds onto the high energy electrons ?
 c) Where does this carrier molecule "hold" or store the energy?

63. Why are these carrier molecules so important in the overall process of photosynthesis?

64. In a general way, describe what happens in the Calvin cycle.

65. For what purposes do plants use glucose?

66. Where is the energy stored in the ATP molecule ?

67. Once the carrier molecules tranfer their energy in the Calvin cycle, what happens to them?

68. Write the oveall equation for cellular respiration.

69. What is oxygen's role in this chemical reaction? Explain.

70. List the three steps involved in cellular respiration.

71. Why is it important that the energy from the glucose molecule is released in 3 steps?

72. What happens to the glucose molecule during glycolysis?

73. How many ATP molecules are produced as a result of glycolysis?

74. Is the first step in glycolysis spontaneous? Explain.

75. In the second step of glycolysis two NADH molecules are formed. Is NADH a strong reducing or oxidizing agent? Explain.

76. Do the processes of glycolysis and the Krebs cycle require oxygen in order to occur?

77. What is the net result of the Krebs cycle? Explain.

78. Write the reduction half reaction where oxygen acts as the final electron acceptor.
What are the "waste" molecules that oxygen gets rid of?

79. Does cellular respiration use most of the energy in the glucose molecule? Explain.

Section 18.3

80. Draw an amino acid molecule and label the three parts that all amino acids have in common.

81. What is the primary function of proteins in our bodies?

82. How do animals get the protein they require?

83. How many different amino acids are there?

84. Give an example of an acidic amino acid. Label the portion of the molecule that makes it acidic.

85. Give an example of a basic amino acid. Label the portion of the molecule that makes it basic.

86. List four examples of where proteins are found in our bodies.

87. What type of amino acid stereoisomer is commonly found in nature? Why do scientists think nature prefers to use use that form?

88. What is chirality? Give two examples in nature where chirality is seen.

89. List the six different functions of proteins in our bodies.

90. Select two amino acid molecules and show how they come together to form a peptide bond. Label the peptide bond.

91. What is a primary structure of a protein?

92. Create a ten amino acid protein sequence.

93. Name and contrast the differences between the two secondary structures found in proteins.

94. How are enzymes important to living systems?

95. Explain briefly what an active site is.

96. What happens when to an enzyme when the substrate attaches?

97. What is the enzyme in saliva responsible for breaking down carbohydrates?

98. In what part of cellular respiration are the cytochrome proteins important ?

99. What metal is bound to the inside of the heme groups in a cytochrome protein? How is this metal important in the transfer of electrons from one cytochrome to another?

Section 18.4

100. What are the three molecular components of a nucleotide?

101. What part of the nucleotide structure forms the structural "backbone" of a DNA strand?

102. What types of attractive forces hold together a DNA molecule?

103. Which nitrogeous base pairs with guanine, G ? Why do these two bases "match-up?"

104. Explain how large amounts of DNA are able to fit inside a small cell?

105. During cell replication how does the DNA double helix come apart? Explain.

106. Make a small table and list the differences between DNA and RNA.

CHAPTER 19

The Chemistry of the Earth

How do chemical reactions in the atmosphere affect life?

What is the role of water on earth?

How are elements and molecules recycled on earth?

How is chemistry applied to environmental issues?

Our home is the planet earth. All living things are made up of matter and matter is made from elements. How do the elements used by us become available in our environment? If we take a walk in nature and pay close attention to all that is around us, we become aware of the great diversity of living things. All of these life forms exist because of the intricate balance of chemicals that are present here on earth. In this chapter we will study how these chemical balances are maintained.

Chemicals cycle between the atmosphere, ocean, land and living things. Factors that affect one of these things will eventually affect all of them, because they are all connected. For this reason green chemistry and the study of "sustainability" have come to the fore front of scientific research today. The earths resources need to be consumed at a rate at which they can be replaced. We observe the effects of chemical imbalance when conditions like global warming occur. The difficult part is that the effects of chemical imbalances are not readily apparent so our awareness of them comes when significant changes have take place. The lag time between noticeable changes and the initial imbalance makes it hard to correct for problems. As a global society we are now learning that we need to plan ahead and study the chemical balances of our planet. In order to do this we need to understand the earths composition and it's cycles.

Investigating the water cycle

In this chapter you will learn about the atmosphere, the land, and the ocean. The water molecule plays an important role in each of these systems! Water is recycled and redistributed through each of these earth systems. Without this recycling process our living systems could not be supported.

Have you ever wondered why rain does not have mud in it? Think about it. If rain falls to the earth and flows through the ground, could it possibly have mud, or soil in it? Here you will get to see first hand how the water molecule is able to be transferred easily from one phase to another and what happens to the mud.

1)Fill one large test tubes, about 2.3 of the way up with some muddy water. Place the test tube in the Lab-Master probe system heater, and attach the heater device. Set the temperature initially for 140°C. As the water heats observe what happens to the water. Record your observations for the first minute.

What naturally heats the water molecules here on earth?

Where do you think most of the water molecules vaporize?

If you guessed over the ocean you would be right!

2) As the water warms, place the condenser over the top of the test tube. Place a small beaker under the condenser to collect your condensate.

How will the condenser unit affect your system?

Record your observations.

What happens to the mud as water changes from one physical state to another?

3)Turn off your heater once you have evaporated most of the water.

How does the water look? Describe the appearance of the condensate.

If we consider water here on earth, what causes condensation to occur naturally?

Here on earth water condenses in the clouds.

3) Look carefully at your system. Can you tell what happens to the water molecules after they condense? Support your answer with observations.

4) Where did the water in the test tube go? Explain.

Now let's consider how these changes occur in nature and think about the recycling of water. We know energy is required to loosen and separate the hydrogen bonds holding the water molecules together. The addition of energy in the form of heat is used to cause these changes, in nature the sun provides this energy. When we remove heat, the molecules come closer together and the hydrogen bonds reform. The water cycle is a physical process. The molecule itself remains intact throughout the cycle. The fact that it can undergo physical changes and remain together as a molecule is what allows for water cycle to work.

19.1 Chemistry of the Atmosphere

Our atmosphere determines the environment in which we live. Have you ever wondered what is in the air? Or what our atmosphere is made of? Most of us tend to take our atmosphere for granted. We live and breathe it everyday, and for the most part as far as we know it's composition stays the same.

We live in the troposphere

The atmosphere around the earth has four layers of gases that have been given specific names depending on their distance from the earth. The **troposphere** is is the layer closest to the earths surface, and it extends for about 10 kilometers. The weather that we experience all happens in the troposphere. The troposphere is more turbulent than the other layers which causes gases in this layer to mix more rapidly. When you fly on large jet airplanes, you are flying at the top of the troposphere.

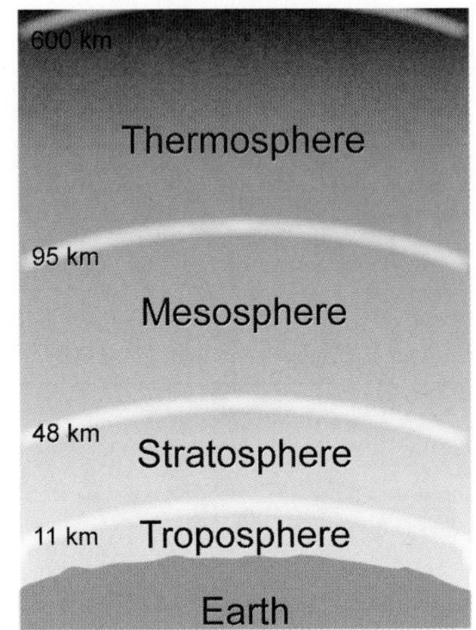

The next layer of gases is called the **stratosphere**. The region of gases called the stratosphere begins at approximately 10 km above the earth and goes to about 50 km. Very important chemical reactions take place in the stratosphere. The most significant being the formation of the ozone (O_3) molecule. Approximately ninety percent of the ozone that protects the earth from high energy photons is formed in the stratosphere.

Meteors burn up in the mesosphere

The mesosphere is the "middle" layer of gases around the earth. In this layer the temperatures are very cold. The higher in altitude the colder the temperature becomes, due to decreased heating from the sun. Waves created by gravitational pull in the mesosphere cause gases to circulate around the globe. Meteors burn up in the mesosphere when they collide with the gas particles contained in this layer. The energy created by these collisions produces a lot of heat causing most meteor fragments to vaporize before reaching the earth. The mesosphere contains high concentrations of metal atoms such as iron, Fe, because of these meteor collisions.

The thermosphere, is the outermost layer, beginning at 85 km above the earth. In this layer ultraviolet light from the sun causes ionization of the gases contained there. For example oxygen loses an electron when absorbing UV radiation.

$O_2(g) + \text{photon}(hv) \rightarrow O_2(g)^+ + e-$ These ionized gases are attracted to the poles of the earth and are responsible for the auroras seen at the polar regions.

Chemistry terms

troposphere - the layer of gases closest to the earth, extending up to approximately 10 km above the earths surface.

stratosphere - the second layer of gases after the troposphere. This region of gases extends from 10 to 50 km above the earth.

Composition of the atmosphere

Heavier gas molecules stay closer to the earth

The atmosphere contains primarily nitrogen(78%) and oxygen(21%) molecules. The remainder of the atmosphere is made up of carbon dioxide and noble gases. Most of the mass of the atmosphere is contained in the troposphere, because the majority of the heavier molecules are found close to the earth's surface. The earths gravitational field causes the lighter molecules and atoms to rise higher in the atmosphere. The sun plays an important role in the chemistry of our atmosphere. The molecules contained in the atmosphere are constantly being hit by radiation from the sun. These energetic collisions have significant chemical effects. We will learn about these in this section.

TABLE 19.1
Composition of dry air

Element	% by Volume
N_2	78.08
O_2	20.95
Ar	0.93
CO_2	0.037
Ne	0.0018
He	0.00052
CH_4	0.00020

Why are these gases so abundant in our atmosphere? Lets look at the four most abundant gases. We think most of the nitrogen gas in our atmosphere was brought out from inside the earth by volcanoes. Nitrogen gas is heavy with a molar mass of 28.02 g/mole causes it tends to stay near the earth's surface and not float up to the higher regions of the atmosphere. Photosynthesis is responsible for the oxygen gas present in our air today. The production of oxygen by plants has greatly influenced the chemistry of the earth. Oxygen is a more reactive molecule than nitrogen, because of this it is responsible for many important chemical reactions in the atmosphere. Argon is also a heavy gas, with a molar mass of about 40 g/mole. Most of the argon in the air was produced by the radioactive decay of other elements in the earths crust such as potassium, (K). Lastly, the carbon dioxide gas present was created during respiration of plants and animals.

Water vapor in the air keeps the earth warm

Water is also present in the atmosphere as moisture. It enters the air through evaporation from the oceans and from the transpiration of plants. Water is an important contributor to the earths relatively moderate temperature, because it absorbs infrared radiation and holds heat near the earth.

The weather patterns we experience in the troposphere are the result of how the sun heats the earth. Different places on the earth are heated at different angles and this unequal heating is what causes the wind and ocean currents to circulate around the earth. Winds are formed when warm air masses rise and cool air sinks, this process sets up movement of air currents around the globe. Weather patterns help to cycle chemicals between the atmosphere, living organisms, and the ocean. This continuous movement of air distributes nutrients to living systems.

These cycles can be disrupted by pollution or environmental changes, which are cause for much concern. The majority of the pollutants found on the earth come from the widespread use of combustion reactions.

Chemical reactions in the air

Nitrogen and oxygen gas molecules make up about 99% of the atmosphere! These molecules are involved in many of the chemical reactions that take place in our atmosphere, because they are present in such high concentrations. To help you better understand some of the chemical reactions that take place in our atmosphere we will review some of the properties of nitrogen and oxygen molecules.

Nitrogen is nonreactive because of it's triple bond

You may remember that nitrogen, N_2, contains a triple bond and that oxygen, O_2 has a double bond holding its atoms together. The triple bond in nitrogen has a high bond energy and requires 946 kJ/mole of energy to break. The strength of the triple bond makes nitrogen a relatively inert, nonreactive gas close to the earth. Under high temperatures such as in a combustion engine nitrogen can be broken apart. The double bond in oxygen requires significantly less energy to come apart because it has a bond energy of 495 kJ/mole. *In the atmosphere, oxygen is a more reactive molecule than nitrogen because it requires less energy to split apart it's atoms.* Remember, in order to react atoms have to come apart so they can interact with new atoms.

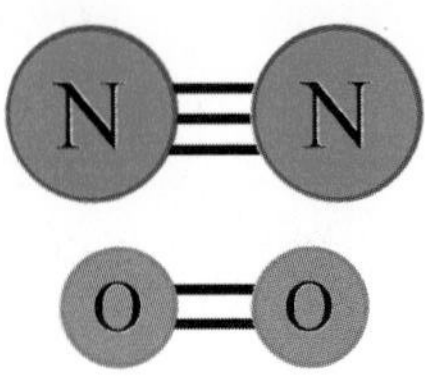

Cars produce nitrous oxide

In the troposphere, there are several chemical reactions that are important to understand, primarily because they are close to us and we experience their effects directly. For instance, most of us have seen smog. Smog is formed when pollutant gases are trapped in a highly populated city or urban area. The smog effect is created in these areas, because the exhaust emissions contain nitrous oxides. Cars or vehicle engines provide the energy necessary for some of the nitrogen in the air to react with the oxygen in the air producing nitrous oxide, NO(g). $N_2(g) + O_2(g) \rightleftharpoons 2NO(g)$ $\Delta H = 181$ kJ. In the air, nitrous oxide, NO is quickly converted to $NO_2(g)$ through oxidation.

$2NO(g) + O_2(g) \rightleftharpoons 2NO_2(g)$ $\Delta H = -113$ kJ

Los Angeles was the first city to experience the full effects of smog. The temperature in the city is hot and the mountains trap the gases down in the valley causing them to become concentrated. When photons from the sun, with wavelengths of 393 nm or less, strike the NO_2 molecules and they dissociate into NO(g) and atomic oxygen O. This atomic oxygen can react with oxygen gas to make ozone, which is another primary pollutant in smog. $O(g) + O_2(g) \rightarrow O_3(g)$ Ground level ozone is very toxic and is dangerous for us to breathe. Ozone in the upper atmosphere protects us, but ozone in the troposphere is a serious health hazard.

Chemical reactions in the troposphere

Formation of sulfur dioxide comes primarily from fossil fuels

Similar to smog other gases in our atmosphere have significant effects even though they are only present in small amounts. Sulfur dioxide, SO_2 is one of the most harmful pollutant gases in our air. How is it formed? The combustion of coal and oil by industry releases significant amounts of sulfur in the form of sulfur dioxide gas. The combustion process causes the sulfur present in these fuels to react with oxygen in the air to produce SO_2. $S(s) + O_2(g) \rightarrow SO_2\ (g)$ Small amounts of sulfur are released naturally into the environment, primarily through the decomposition of organic materials. However, most of the sulfur is released from the use of fossil fuels.

Approximately 80% of the sulfur dioxide in the air comes from the burning of coal used to provide us with electricity. The sulfur dioxide in the air readily reacts with oxygen in the air to form sulfur trioxide, which in turn reacts with water vapor to form sulfuric acid.

$$SO_3(g) + H_2O(l) \rightarrow H_2SO_4\ (aq)$$

Acid rain is formed from sulfur and nitrogen oxides

Sulfuric acid is a strong acid! This has significant environmental and health impacts. Sulfuric acid is one of the acids responsible for acid rain, which has damaged many ecosystems in the northeast, and great lakes regions of the U.S. For example in New York, there are no fish in hundreds of lakes and ponds because of the low pH caused by acid rain. Certain lakes and streams contain bicarbonate, HCO_3^- , buffer which helps to counteract the effects of acid rain. In areas where the amounts of acid rain are extremely high the pH can get as low as 4.0 pH units. It is important to note that even naturally occurring rain is acidic, having a pH of approximately 5.5-5.8. This happens because natural rain water absorbs carbon dioxide in the air reacting to form the weak acid carbonic acid, H_2CO_3. Carbonic acid is responsible for lowering the pH of our rain water.

Natural rain water is acidic due to the absorption of CO_2

$$CO_2(g) + H_2O(l) \rightarrow H_2CO_3(aq)$$

Nitrous oxides also react with water vapor to form the strong acid called nitric acid, HNO_3.

$$3NO_2\ (g) + H_2O(l) \rightarrow 2HNO_3(aq) + NO(g)$$

On an environmental level, acid rain can damage forests, and crops and it also makes the pH of lakes and streams too low. You may remember that strong acids react with carbonates, like limestone, $CaCO_3$, and this is why it causes damage to stone and buildings. These strong acids also react with many different metals causing corrosion. Overall these effects cause a serious economic impact as well as environmental.

Chemical reactions in the upper atmosphere

The outer layers of the atmosphere absorb dangerous UV radiation

The outer layers of the atmosphere act like a shield protecting us from damaging forms of radiant energy. These outer layers of the atmosphere, beyond the thermosphere, are constantly being struck by high energy particles from the sun. The gas molecules here are exposed to continuous high levels of radiation. The sun produces photons with a wide range of wavelengths. The short wavelength photons in the ultraviolet region of the spectrum (below 400nm) are very damaging to living organisms. These high energy photons can damage cellular DNA causing mutations and possibly cancer. However, the outer layers of our atmosphere are able absorb most of these high energy photons and they do not travel down close to the earth and cause harm.

Chemical changes are caused by high energy photons

How are these photons absorbed? The gas molecules in the upper atmosphere absorb these high energy photons. Chemical changes are caused by high energy photons colliding with the molecules in these layers. One of these important chemical changes is called **photodissociation.** Photodissociation is the breaking of a chemical bond in a molecule caused by the absorption of a photon. Because of nitrogen's high bond energy the oxygen molecule, O_2, is the principle absorber of high energy photons in the upper atmosphere. When a bond is broken by the process of photodissociation, no ions are formed. The bonding electrons are divided equally between the two atoms and two neutral particles are created. The amount of energy required to split the double bond between the oxygen atoms is equal to the bond energy of 495 kJ/mole.

Let's calculate the wavelength of a photon emitted by the sun that contains enough energy to dissociate the oxygen molecule.

Asked: *Calculate the wavelength of a photon that can split an O_2 molecule.*

Given: *The bond dissociation energy for O_2 is 495 kJ/mole.*

Avogadro's number = 6.022×10^{23}, Planck's constant = $6.63 \times 10^{-34} J\,s$

Relationships: $E = hc/\lambda$.

Solve:

$$\frac{495\text{ kJ}}{\cancel{\text{mole}}} \times \frac{\cancel{\text{mole}}}{6.022\times10^{23}\text{molecules}} = 8.22\times10^{-22}\frac{kJ}{\text{molecule}} = 8.22\times10^{-19}\frac{J}{\text{molecule}}$$

$$\lambda = \frac{hc}{E} = \frac{(6.63\times10^{-34}\cancel{J}\cancel{s})\left(3.0\times10^{8}\left(\frac{\text{m}}{\cancel{s}}\right)\right)}{8.22\times10^{-19}\cancel{J}} = 242\times10^{-9}\cancel{m}\left(\frac{10^{9}nm}{1\cancel{m}}\right) = 242\text{ nm}$$

Discussion: *A photon with a wavelength of 242 nm is energetic enough to break the double bond in the oxygen molecule.*

Chemistry terms

photodissociation - the breaking of a chemical bond in a molecule caused by the absorption of a photon.

Chemical reactions in the upper atmosphere

Molecules absorb short λ of UV light causing them to lose an e^-

Another type of reaction that is very important in absorbing high energy radiation is called **photoionization**. When a molecule absorbs a photon of sufficient energy it can eject an electron and ionize. Photoionization in the upper atmosphere absorbs photons in the high energy region of the ultraviolet spectrum. These short wavelengths would be very harmful if they reached the earth. It is important to understand that when a molecule absorbs a photon with lots of energy a chemical change is created. Here the change is a loss of an electron. This chemical change in the molecule "uses up" the energy in the photon, so that it no longer has this damaging energy to pass on. Some examples are:

$$O_2(g) + \text{photon}(h\nu) \rightarrow O_2^+(g) + e^-$$

$$N_2(g) + \text{photon}(h\nu) \rightarrow N_2^+(g) + e^-$$

In both of these cases photons with a wavelength between 80 and 100 nm are absorbed, and a positively charged ion is formed.

Stratospheric ozone, O_3, acts as a protective shield against longer λ UV light

The other important chemical reaction that occurs just above the stratosphere is the formation of ozone, $O_3(g)$. Ozone forms when atomic oxygen, formed from photodissociation, collides with molecular oxygen: $O(g) + O_2(g) \rightarrow O_3(g)$. Ozone acts as a shield protecting us from the longer ultraviolet wavelengths. The ozone molecule absorbs high energy radiation that is between 240nm and 310nm. Ozone has a natural cycle of formation and decomposition. Once it collides with a ultraviolet photon it splits apart into atomic oxygen, O and molecular oxygen, $O_2(g)$. These two molecules are then available to react again and form ozone. Basically the ozone molecule splits when it absorbs sufficiently energetic light and then comes back together by undergoing molecular collisions.

Ozone

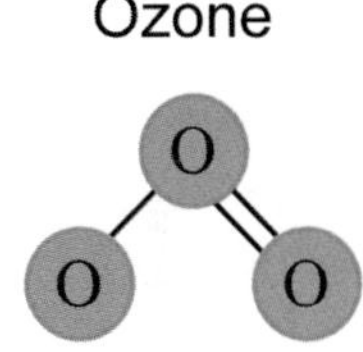

Discussions about the "ozone hole" have been in the news since the late 1980's. The production and use of chlorofluorocarbons in aerosols and air conditioners were causing destruction of the ozone layer. The highest concentrations of ozone are in the stratosphere, here it protects us. Remember near the earth it is toxic and considered a pollutant! But up in the stratosphere we do not breathe it, and it has a useful purpose.

Chemistry terms

photoionization - when a molecule absorbs a photon and ejects an electron causing it to become an ion.

The climate and global warming

Some greenhouse effect is a good thing

In order for a planet to be inhabited by living organisms it must have a climate that is reasonably moderate so that life forms can adapt. A planet that has extreme temperature fluctuations would not support life. In our atmosphere, carbon dioxide and water are the two most important molecules in maintaining a balanced temperature near the surface of the earth. The earth takes in energy from the sun and it radiates that energy back out through the atmosphere. The greenhouse effect actually is necessary to maintain life on earth. The gases close to the earth allow visible light to enter, but they absorb the longer wavelengths (infrared radiation) of light that are reflected by the earths surface. Now when we discuss the greenhouse effect we are focusing on the fact that the earths gases are trapping too much heat and not allowing it to escape back into the atmosphere. We need the gases to trap some heat, but with the increased use of fossil fuels these gases are trapping too much heat.

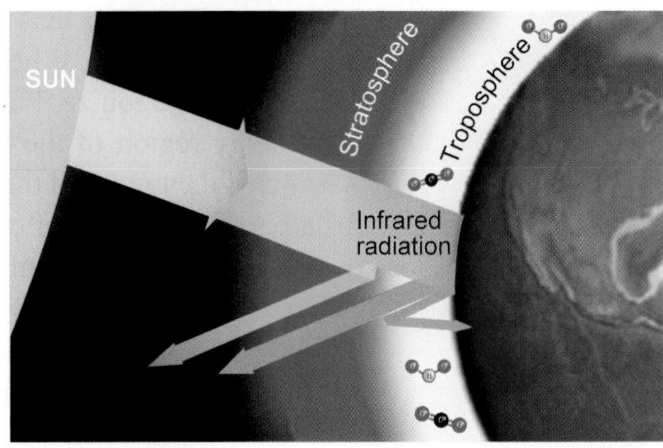

Molecules trap infrared radiation in their bonds

How do water and carbon dioxide molecules trap heat? They absorb longer wavelength photons in the infrared region of the spectrum in their bonds. When these molecules absorb the energy their bonds they are able to stretch, vibrate and rotate. These motions allow these molecules to hold onto the energy for a short time before re-radiating it back to the earths surface. When the energy is released, or re-radiated it can go in any direction, about half of it returns to the surface of the earth.

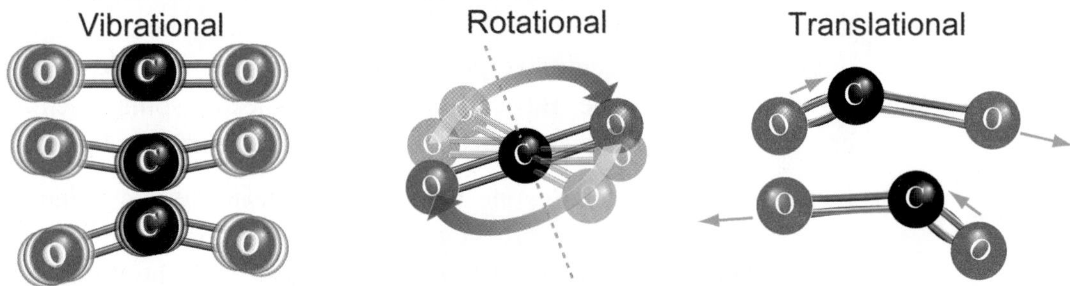

Water molecules in our troposphere keep the earth warm at night, when there is no solar energy. In dry climates, like the desert, it can gets very cold at night because there are few water molecules absorbing warmth from the earth.

Chemistry terms

global warming - the warming of the surface of the earth by gas molecules in the atmosphere which trap heat and reflect it back to the surface of the earth.

Trace amounts of chemicals in our atmosphere

Gases are measure in ppm

When measuring trace amounts of chemical substances in our air we typically measure them using parts per million (ppm). You many recall we used this same unit to measure amounts of chemicals in aqueous solutions in chapter 2. The difference here is that we are dealing with gases and not water based solutions.

TABLE 19.1. Concentrations of Gases (Natural and man-made)

Molecule	Sources	Typical Concentrations
Carbon dioxide, CO_2	fossil fuel combustion, decomposition of organic materials, ocean	375 ppm in troposphere
Methane, CH_4	decomposition of organic matter	1-2 ppm in troposphere
Carbon monoxide, CO	industrial pollution, fossil fuels, and decomposing organics	0.05 ppm 1- 56 ppm in urban areas
Nitric oxide, NO	combustion, lightning, electricity	0.02 ppm in nonpolluted air; 0.25 ppm in urban areas
Ozone, O_3	photodissociation, electricity	0.55 ppm in urban areas 0.01 ppm in nonpolluted air
Sulfur dioxide, SO_2	fossil fuel combustion, industry, volcanos, fires	0.01 ppm in nonpolluted air 0.15-2 ppm in urban areas

Atmospheric chemists measure the partial pressure of pollutant gases

When we wish to measure the relative amounts of gases we measure the pressure of each individual gas, this is directly related to the moles of the gas given the ideal gas equation, PV = nRT. Atmospheric chemists measure the partial pressure of pollutant gases in our atmosphere and decide whether the amount is dangerous. By using the partial pressure measurements of a gas relative to the total pressure of the atmosphere, scientists can determine the mole fraction of a particular compound in our air. One part per million for a pollutant would be one mole of the compound per one million moles of total gas.

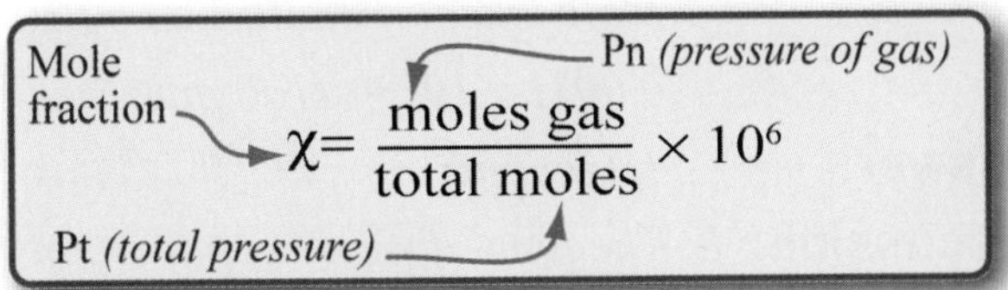

Mole fraction
$\chi = \frac{\text{moles gas}}{\text{total moles}} \times 10^6$
Pn (pressure of gas)
Pt (total pressure)

Amounts of chemicals in our atmosphere

It is helpful to be able to quantitatively measure these small amounts of gases so that we can determine whether or not they are harmful to our health. Lets calculate the concentration in parts per million for carbon monoxide in a small city, that has a total air pressure (P_t) of 702 torr. The partial pressure of CO for that day was measured to be 3.1 x 10^{-3} torr. To begin we first need to calculate the mole fraction, X. The mole fraction is simply the number of moles of the gas as compared to the total moles in the air sampled. This tells us how much of the gas is present relative to the total gases in the air. Mole fraction means the "fraction" or amount of moles present in the total sample.

How to calculate ppm of CO_2

$X_{CO} = \frac{P_{CO}}{P_{Total}} = \frac{3.1 \times 10^{-3}}{702} = 4.42 \times 10^{-6}$, next we multiply the mole fraction by 10^6 to calculate the parts per million. $4.42 \times 10^{-6} \times 10^6 =$ 4.42 ppm of CO is the concentration of carbon monoxide in this city.

In the troposphere carbon monoxide is present in small concentrations, as our calculation just showed us. These concentrations vary in different areas due to population differences and industrial activity, refer to table 19.1 on the previous page. In some places the levels are quite low, but the gases can still travel in the air currents and effect more rural regions.

Lets try calculating the concentration of water vapor in the air, in ppm.

Calculate the concentration of water vapor in ppm for a sample of air that contains a pressure of 0.83 torr of water vapor, and a total air pressure of 729 torr.

Asked: *Find the ppm concentration of water vapor*

Given: *Total pressure of air = 729 torr*
Partial Pressure of H_2O vapor = 0.83 torr

Relationships: $X_{H2O} = \frac{P_{H2O}}{P_{Total}}$

Solve: *Find Mole fraction of water.* $X_{H2O} = \frac{0.83 \text{ torr}}{729 \text{ torr}} = 0.0011$

Mole fraction multiplied by 10^6 is the ppm.

$0.0011 \times 10^6 = 1100$ ppm

Answer: *1100 ppm*

Discussion: *The ppm of water vapor in this sample of air is 1100.*

19.2 Chemistry of the Oceans

97% of earths water is in the oceans

The majority of the earths water is contained in it's oceans. Out of all the water on earth only about 0.5% is NOT in the oceans! Roughly about 97% of the water on earth is in the oceans and another 2.3% is contained in ice caps and glaciers. The remaining 0.7% of our water is in groundwater, lakes, rivers and salt marsh areas. Water is the most abundant liquid on earth, and as mentioned earlier it is crucial to living systems.

You may recall that water's unique ability to dissolve other substances makes it known as the "universal solvent." Water's polarity allows it to form strong hydrogen bonds which accounts for water's solvent properties and the majority of it's chemistry. Water is present in almost all chemical reactions that happen in nature. Let's review how water plays a key role in chemical reactions.

Water plays a key role in chemical reactions

First, as a solvent it can separate ionic compounds or salts, by surrounding them with water and hydrating them. This hydration process allows ions to be mobile and therefore transported to different areas, in our bodies and in the oceans. Many minerals are key players in biological reactions and hydration makes them available for organisms to use.

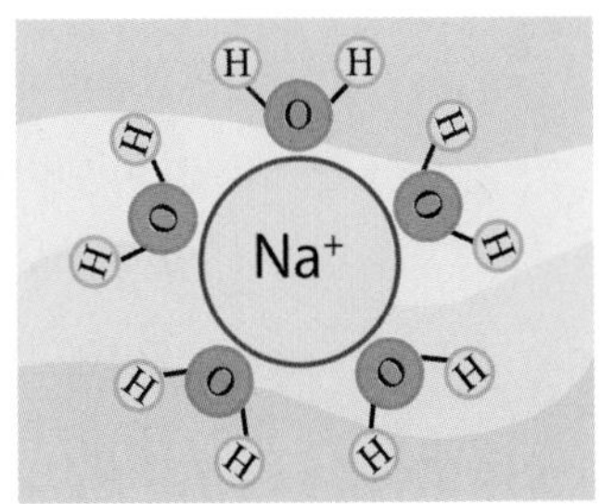

Secondly, water is important in the acid-base chemistry of the ocean, because it is amphoteric. Water can accept a hydrogen ion from an acidic compound or it can donate a hydrogen ion to a basic compound. In the ocean, water reacts with carbon dioxide forming carbonic acid. $H_2O(l) + CO_2(g) <=> H_2CO_3(aq)$. Once formed the carbonic acid ionizes and is present the majority of the time as bicarbonate ion, HCO_3^- and hydrogen ion, H^+. In the ocean about 88% of inorganic carbon is in the form of the bicarbonate ion.

Lastly, water is also able to accept and donate electrons in oxidation-reduction reactions. The ocean is an electrolyte solution containing dissolved minerals and salts. These dissolved ions give the ocean water conductivity and allow for electron transfer reactions.

Chemical reactions in the ocean are affected by temperature

Chemical reactions in the oceans are significantly effected by the temperature of the water. The ocean water absorbs energy from the sun and carries this solar energy to different places in the ocean. Where this energy goes is dictated by the ocean currents. Gases are more easily dissolved in the cold waters of the ocean. Just like your soda when it is cold! In warm conditions the increased motion of the molecules allows more of the gases to escape into the air. Density is a key factor here as well. The warm waters of the ocean have more kinetic energy and the molecules are more spread out. In cold water the decreased kinetic energy allows the molecules to be closer together. In the ocean when ice freezes the salt ions are left behind, this increases the concentration of salt in the water, which also increases the density of the water. More dense water sinks.

Chemicals in the ocean

Have you ever gone swimming in the ocean and experienced it's saltiness? Ocean water is known for it's **salinity**, or saltiness. The salinity of the ocean is approximately 3.5%. This means that out of one hundred percent about 3.5% of the water is composed of salt ions. Of the salt ions chloride, Cl^- is the most abundant followed by the sodium ion, Na^+.

TABLE 19.2.

Ion	Amount (g/kg)	Concentration (M)
Chloride, Cl^-	19.40	0.56
Sodium, Na^+	10.75	0.48
Sulfate, SO_4^{2-}	2.70	0.030
Magnesium, Mg^{2+}	1.29	0.054
Calcium, Ca^{2+}	0.41	0.011
Potassium, K^+	0.40	0.010
Bicarbonate, HCO_3^-	0.145	2.75×10^{-3}
Bromide, Br^-	0.068	8.5×10^{-4}
Boric acid, H_3BO_3	0.026	4.1×10^{-4}
Strontium, Sr^{2+}	0.0080	9.2×10^{-5}
Fluoride, F^-	0.0014	6.9×10^{-5}

Ocean water has a relatively constant composition

The concentration of ocean water remains relatively constant around the world. The concentration is maintained partly because all of the oceans are connected and because there is constant mixing of the water. These effects circulate the salts and chemicals in the water giving an even distribution.

How do these minerals and compounds find their way into the ocean water? Many come from erosion and run off from the soil, some come from the ocean sediments, or are absorbed from the nearby air. When earth was first formed volcanoes brought many minerals, such as iron to the oceans. Volcanoes brought minerals up from deep in the earth and introduced them to ocean waters. Dissolved minerals are critical to living organisms who depend upon the dissolved minerals in ocean water to provide them with the necessary nutrients. It is widely believed that early life developed in the ocean waters. Geothermal vents deep in the ocean floor provide sulfur for bacteria and other life forms that use it for energy.

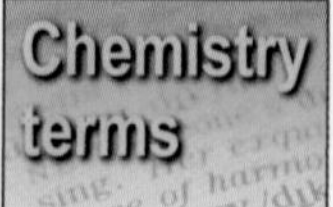

salinity - is a measure of how salty water is. The concentration of salt in ocean water is a measure of it's salinity.

The water cycle

Water is constantly being recycled

Where does all the water that living things require come from? Well most of it comes from the ocean, but it cycles around the earth moving from the ocean to the air, and from the air to the land. The fact that water readily changes from a liquid to a vapor allows it to be carried by the atmosphere from the oceans to the land where it condenses and returns to the land as precipitation in the form of snow or rain. The patterns of the movement of water are described by the **water cycle** diagram shown below.

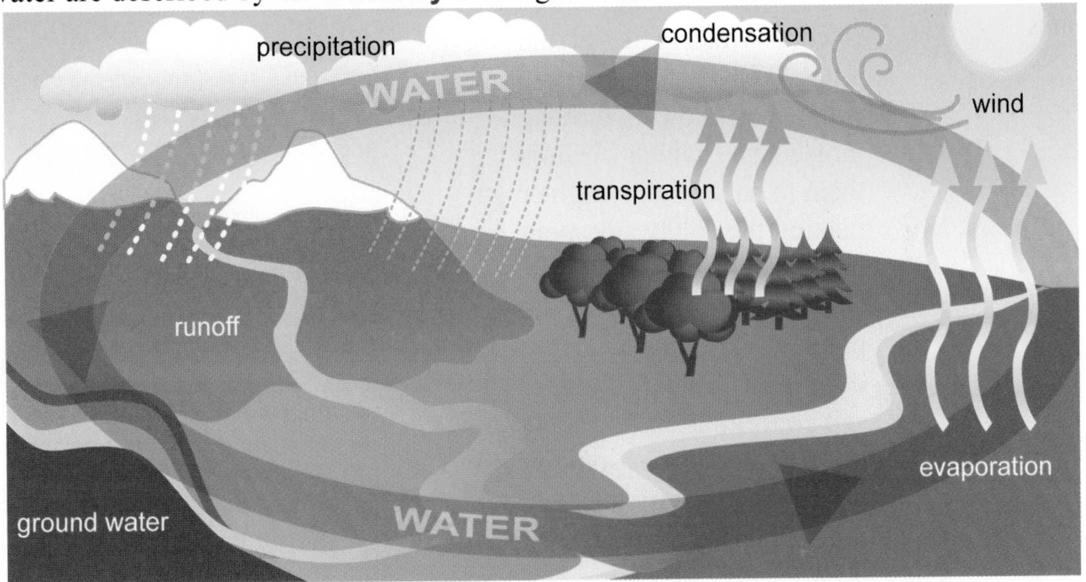

Evaporation over the ocean forms clouds

From looking at the diagram above you can see that water evaporates from the ocean and from the leaves of plants in a process known as **transpiration**. Over the course of a day the suns radiation is absorbed by the atmosphere this warms the water vapor in the air. The warm air rises to a point in the atmosphere where it begins to cool, as it cools it condenses into small droplets which forms clouds. The clouds bring the "weather" which depending on the location and time of year brings either snow, sleet or rain. Evaporation exceeds precipitation over the ocean and this is what allows water vapor to be carried by the wind to the inland areas. On land precipitation exceeds evaporation which sets up ground water systems that eventually flow back to the oceans.

With the exception of photosynthesis the water molecules that cycle through this process are not chemically changed. The water cycle is mainly a physical process rather than a chemical one.

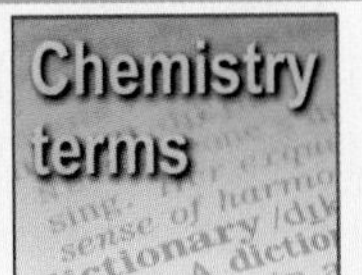

transpiration - is the release of water from the leaves of plants as they carry out photosynthesis.

water cycle - the process showing how water molecules move between the atmosphere, the land, and the ocean.

The ocean currents and climate

Here on earth the ocean currents have a tremendous affect our climate. The global currents are driven by density differences. The density of ocean water is largely affected by two factors, temperature and the concentration of salt.

Cold ocean water is more dense and it sinks

To start to think about currents lets begin at the poles where the temperature is cold. Water sinks at the poles. Why? Because cold water has slower moving molecules, which causes them to pack slightly tighter making the cold water more dense. This causes it to sink. The cold ocean water at the poles is also more dense because it contains a higher salt concentration. This happens because as some of the cold water forms ice crystals some of the salt does not freeze with it. The salt that is left behind increases the concentration of the remaining water. *The fact that cold salty ocean water sinks has a huge impact on our global climate!* As the cold water sinks more water flows in to replace it, when this new water is pulled in it is cooled, which in turn causes it to slowly sink. It is this overall motion that sets up a current. This current moves very slowly, but it circulates large amounts of ocean water every year. The circulation brings minerals and nutrient rich water near the surface of the ocean where algae and plankton depend on them to survive.

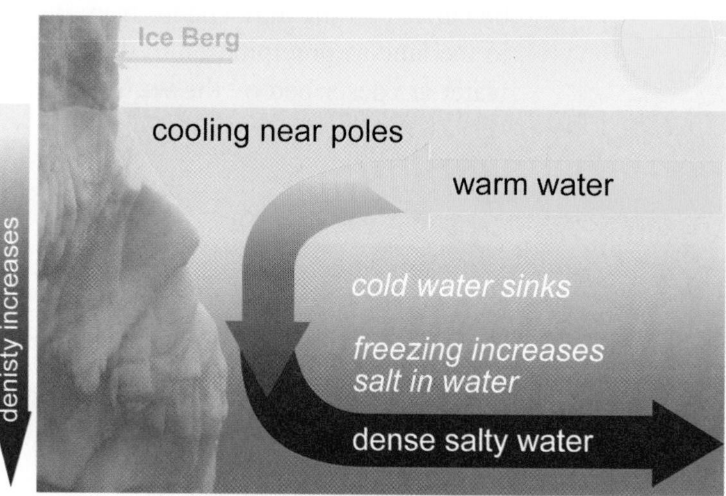

As cold water sinks more warm water is pulled in above it to take it's place.

Warm water rises to the surface and some evaporates creating precipitation

Ocean currents carry cold water down and away from the North Atlantic, as these waters reach the arctic they are cooled once again. Once the water flows back up to the equator it begins to be heated by the warm climate. The warmer the water becomes the less dense it is and the water rises to the surface where it can absorb more heat directly from the sun. The oceans surface water plays a key role in our weather, because it evaporates and provides water molecules for precipitation here on land.

If the glacial ice melts due to the warming of the earth by greenhouse gases this introduces fresh water to the salty ocean water. As large quantities of fresh water are added to the ocean the overall density of the surface water decreases. This means that the water will not sink as much. Overall changes such as this would lead to differences in the motion of the deep ocean currents. If we no longer have the deep ocean currents returning the warmer water north the North Atlantic regions would become much colder.

The carbon cycle

Carbon is the 4th most abundant element in the universe

Carbon is used as the building block for all living organisms. It is present in our earth, ocean and atmosphere. In fact, carbon is so widely used that it makes sense it is recycled. The **carbon cycle** below shows us how carbon moves between the ocean, land and atmosphere. Carbon *does* change its chemical form throughout the different aspects of the cycle. For example when it goes from CO_2 to glucose, $C_6H_{12}O_6$, and bicarbonate, HCO_3^-, each of these forms contain carbon as it cycles through its natural processes.

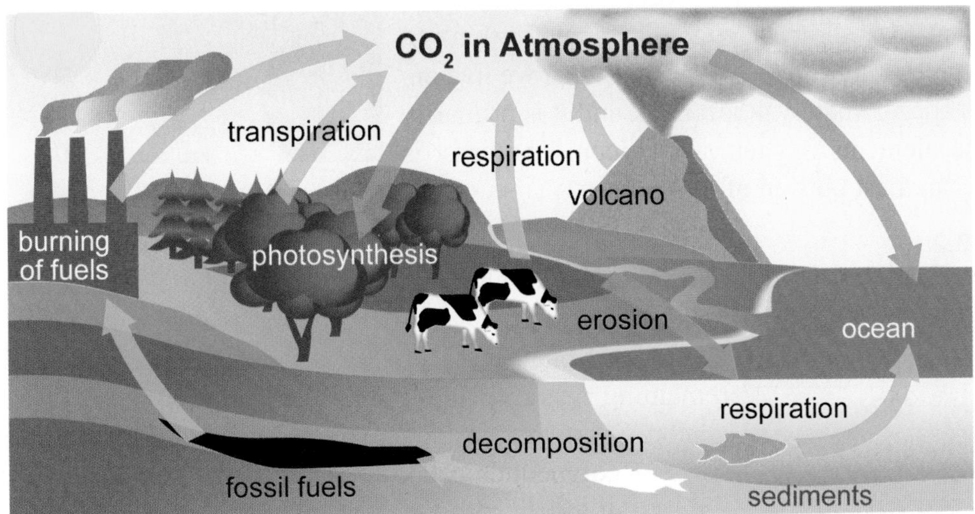

The largest amounts of carbon are present in the form of carbon dioxide contained in the atmosphere and in the ocean. The four major areas that hold carbon are sediments and fossil fuels, plants, the ocean and on land fresh water and soil. *The surface waters of the ocean contain the largest amount of carbon that is actively cycling around the earth.* The cycling of carbon can take millions of years when we include all of the geological aspects or it can take days to thousands of years if we consider the biological and physical cycles.

Movement of carbon between the storage areas occurs because of biological activity and geological changes, both of these processes bring about chemical changes along with some physical changes. The amount of carbon dioxide released by the earth has roughly balanced the amount the earth absorbs. Lately this has not been the case due to the increase of carbon dioxide from fossil fuels.

Bicarbonate, HCO_3^- is essentially dissolved CO_2 in the ocean

The ocean chemistry revolves around formation of bicarbonate from carbon dioxide in the familiar reaction of carbon dioxide dissolving in water.

$CO_2(g) + H_2O(l) \rightleftharpoons H_2CO_3(aq)$, which ionizes in water:

$H_2CO_3(aq) \rightleftharpoons H^+(aq) + HCO_3^-(aq)$.

The formation of the bicarbonate ion is favored and this allows the ocean to hold large amounts of carbon dioxide.

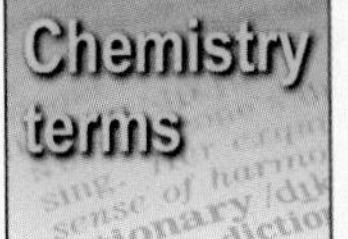

carbon cycle - the process showing how carbon moves between the atmosphere, the land, and the ocean. Carbon changes it's chemical form as it moves through the cycle. Some of the common chemical forms of carbon are CO_2, $C_6H_{12}O_6$, and H_2CO_3.

19.3 Chemistry of the Land

Scientists believe the earth formed from a solar nebula

What types of matter and atoms is the earth made of? The actual composition of the earth was formed from relatively few elements. Scientists think that the earth was formed 4 to 5 billion years ago by a **solar nebula**. A solar nebula is a region of dust and gases that accompanied the formation of the sun. This flat circular area of dust orbiting the sun eventually formed the planets. When we refer to the "land" we tend to think of the crust, which is the surface of the earth that we live on. The crust is approximately 40km thick and it is primarily made of the elements oxygen and silicon. The crust is actually the smallest portion of the earth.

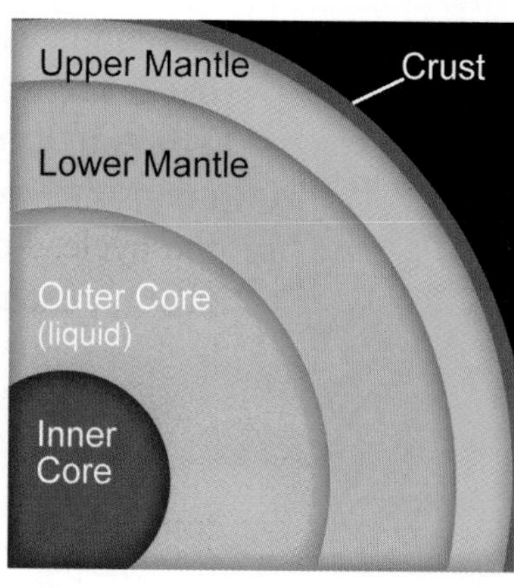

Si and O form the silicates which make up most of our rocks

TABLE 19.2
Elements Earth's Crust

Element	% by Mass
O_2	46.6
Si	27.7
Al	8.1
Fe	5.0
Ca	3.6
Na	2.8
K	2.6
Mg	2.1

Understanding the composition of the earth requires an understanding of chemistry. There are eight elements that make up most (98.5%) of the earths matter, with small amounts of other elements making up about 1.5% of the remaining mass. Of these eight elements oxygen, O and silicon, Si are the most abundant. They combine to form a class of compounds called the **silicates**. *Silicates make up most of the different types of rocks and minerals on the earth.* These silicates contain different amounts of the other eight elements and these variations give different properties to the types of rocks they form.

If the earth is made up of mainly the same elements why are the layers different? As the earth cooled and condensed it separated into layers as shown above. *The elements that were more dense sunk to the center of the earth forming the core.* Iron, Fe is much more dense than the silicates formed from oxygen and silicon,

Conditions of high temperature and pressure affect the arrangement of atoms.

this caused it to be concentrated in the center of the earth or the **core**. The conditions of high temperature and pressure below the surface of the earth have significant effects on the arrangement of the atoms. For example, under conditions of high temperature and pressure carbon forms diamond which has a tetrahedral lattice and is very strong. On the earths surface carbon forms parallel planes of soft flaky graphite. Graphite has a trigonal planar arrangement of atoms with in a layer and is held loosely between layers.

solar nebula - disk shaped region of dust and gases around a forming star. The sun's solar nebula is where the earth formed.

core - is center of the earth, where temperature and pressure are very high.

Composition of the earth

Magma is the source of most of our rocks

You may wonder where all of the different types of rocks and soils that make up the earth come from! There is a great variety here at the earth's surface. Around the world we observe such variation in the land and rock formations that it may seem almost impossible that just a few elements give rise to such diversity. But they do! Different types of rock are formed from a substance called **magma**. Magma is molten rock, made up of silicates(Si and O). Magmas contain suspended crystals and dissolved gases too. Most magma is formed just below the upper mantle region. As magma cools by crystallization it forms **igneous rock**. Most of the earths upper crust is made of igneous rock. Igneous rock is widely studied for clues about the earths inner compostition.

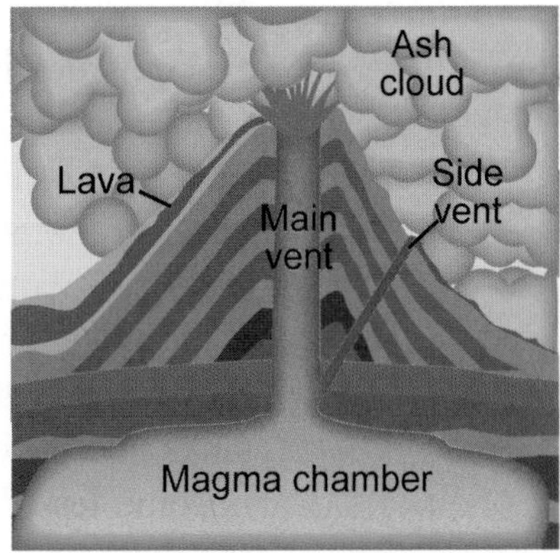

Temperature plays an important role in rock formation

Feldspar

Courtesy of TreasureMountainMining.com photo

Quartz

Temperature differences experienced as the rock crystallizes are important in the final composition of the rock. Basically all igneous rock is formed primarily from silicates. The variation seen in these rocks comes from the other six elements present in significant amounts. The term "rock" applies to **minerals** of several different types. What is a mineral? Minerals are naturally occurring solids, not made up of organic mater, that have an organized internal structure and a definite chemical make-up. In other words their atoms are arranged in a definite pattern. Most minerals are made of silicate materials. Quartz, feldspar and mica are examples of silicates. Carbonates are an example of a common nonsilicate mineral, as well as oxides. Some minerals are prized as gems, because of their beautiful appearance, hardness and color. For example rubies are made of aluminum oxide, Al_2O_3, with the impurity of chromic oxide, Cr_2O_3, a mineral known as corundum. Some elements are found in their pure form on earth such as gold, silver, copper, lead, and carbon in the form of diamond. .

magma - molten rock found below the surface of the upper mantle.

igneous rock - formed from crystallized magma, covers more than 90%of the earths crust.

minerals - a natural non-organic solid with an organized internal structure, where the atoms have a definite pattern.

Volcanoes

Openings in the earth are created where rigid crust layers meet

Most of the earths crust is made up of igneous rock, and igneous rock is made from magma. Well how does magma get to the surface of the earth? The answer is through a Volcano! A **volcano** is an opening in the earths crust that allows magma, gases and other materials to escape from inside the earth. The earth has about twelve rigid plates that move independently of each other. These plates are in contact with each other and as they move and shift some rock is pulled underneath. This rock sinks into the earth and as it does it heats up, which causes it to become less dense. Because it is less dense, it slowly rises to the earths surface again. As the molten magma rises to the surface and is released it changes into **lava**.

Melting of rock is affected by pressure, temperature, and composition

Deep inside the earth the temperatures are very hot, reaching 1000°C at a depth of 100 km. However, because of the high pressure experienced at that depth most of the rock is not melted! The melting of rock depends not only on the temperature, and pressure, but also the composition of the rock. As mentioned earlier rock is mainly silicon and oxygen, but these elements vary in their relative percentages. For example basalt is one of three common types of lava. Basalt is made up of 45 to 50% SiO_2. Basalt is considered to be a low silica type of lava. Low silicate lavas flow the furthest, sometimes going for miles before they solidify. The chemical composition of the lava plays a key role here. In molten lavas and magmas silicates tend to form SiO_4^- anions, and these link together to form polymers. The higher the silicate content the greater amount of polymerization. This polymer formation makes the lavas "thicker" or more viscous, which does not allow them to flow as far.

Water significantly decreases the melting temperature of rock

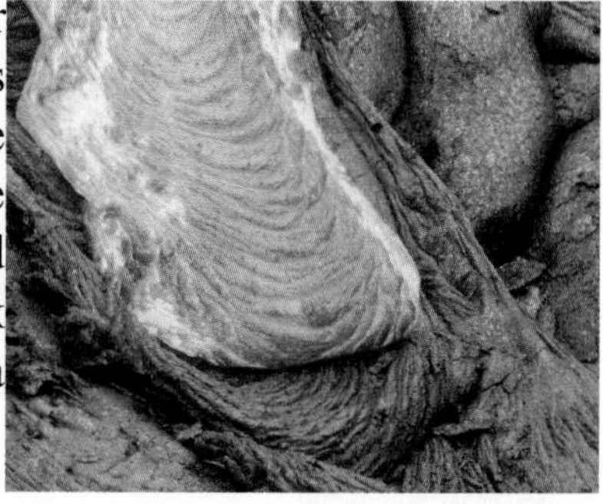

Most of our volcanoes appear at the edge of continents or on islands in the ocean. This is because water decreases the melting temperature of rock! Water has the opposite effect of pressure. Water finds it's way deep into the earth, particularly near or in the ocean where the rigid plates meet. Even small amounts of water cause the rock to melt more quickly, forming magma. As this magma rises it is released from volcanoes as lava.

Chemistry terms

volcano - opening in the earth were magma, gases and other materials can escape from inside the earth.

lava - molten rock or magma released from the vent(s) of a volcano.

The rock cycle

Rocks are recycled within the earth

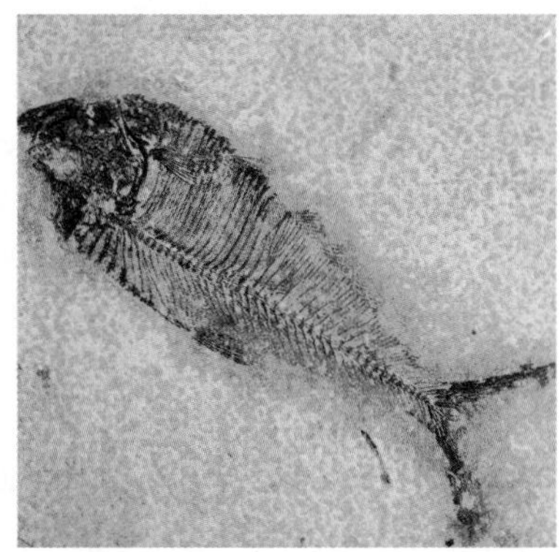

The study of rock has greatly influenced our knowledge of the earth. Rock tells a story about the earth that reaches far beyond our (man-made) history record. Rocks cycle through a process that is ongoing and in this way rocks can be considered to be "recycled" in the true sense of the word! The rock cycle shown opposite highlights the three main types of rock, igneous, sedimentary, and metamorphic. Each of these types of rock come originally from magma. **Geologists** are scientists who study the matter that makes up the earth. The study of the earth has many applications that include identifying areas where our natural resources can be found, which may include oil, coal or precious minerals. Learning about the earth also gives us valuable information for understanding earthquakes and other natural dangers. In general, the information geologists gather can be used to protect and improve our environment.

Motion of rock layers occurs underneath the earths crust

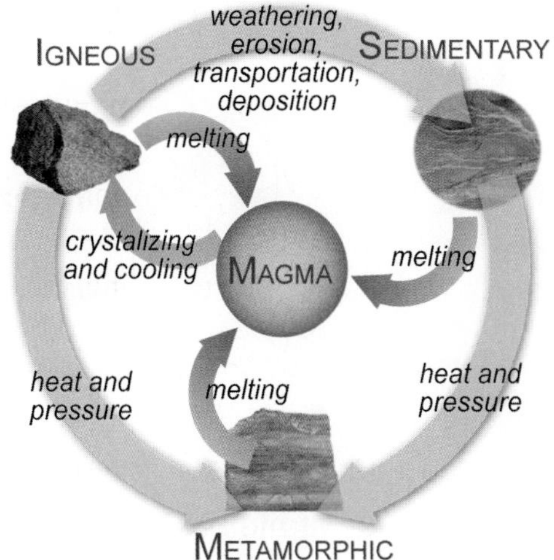

By studying different types of rock geologists have discovered how they are all interconnected. Over time the earth changes a rock from one type to another. How does the earth do this? Motion under the earths crust is responsible for the recycling of rock materials. The mantle layer slowly moves in response to density changes within the layers, and the rigid areas above called "plates" shift and cause some rock matter to move underneath them, where it experiences more pressure. Magma forms igneous rock and this begins the cycle. Igneous rocks are eroded over time by the environment, these small pieces of igneous rock are deposited in layers and these are eventually compressed into rock, forming **sedimentary rock**. Fossils are primarily found in the layers of sedimentary rocks. The other type of rock, **metamorphic**, is created when heat and pressure are applied to igneous or sedimentary rocks. The conditions of heat and pressure cause the rocks to "change form" into a new type of rock. All three types of rock can be melted and form a new type of magma, starting the cycle over once again.

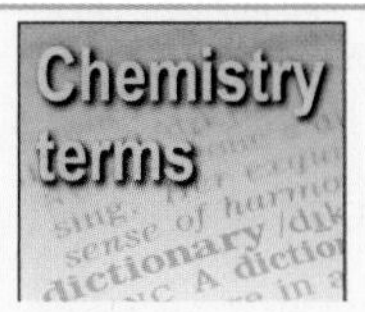

geologists - scientists that study the matter making up the earth.

sedimentary rock -is formed from the compression of small pieces of igneous rock.

metamorphic rock - is created from igneous and sedimentary rock under conditions of heat and pressure.

Nitrogen cycle

Nitrogen in the air, N_2, cannot be used by most living organisms

Nitrogen is a very important element. Essentially all living organisms use amino acids to make proteins, and amino acids contain nitrogen. In addition to amino acids, nitrogen is used to make nitrogenous bases for DNA and RNA molecules, and it is also used in the molecular structure of the porphyrin ring. You may recall that the porphyrin ring is important in the cytochrome systems and in the hemoglobin molecule. Even though nitrogen is abundant here on earth in the form of nitrogen gas, living organisms cannot use this form of nitrogen. Why can't we use the $N_2(g)$ in the air? The strong triple bond holding nitrogen atoms together requires too much energy, because of this we cannot use them to make other molecules. We need nitrogen in a chemical form that our bodies can break apart and use to make our biological molecules. Animals simply inhale and exhale nitrogen gas.

Nitrogen cycle

How do we obtain the types of nitrogen our bodies can use? We eat plants and meat that contain nitrogen, and we reuse it to make our own proteins. Living organisms can use nitrogen in the form of ammonia, NH_3, nitrate, NO_3^- and nitrite, NO_2^-.

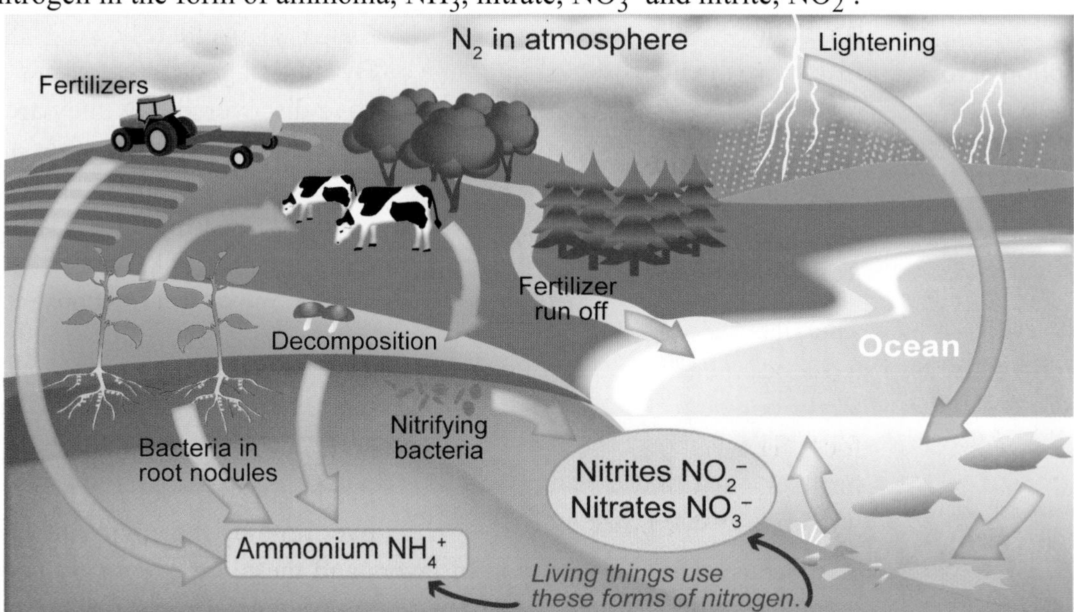

As you can see from the diagram above there are three ways for nitrogen to become available to living organisms. Atmospheric **nitrogen fixation**, bacterial nitrogen fixation, and through the production of synthetic fertilizers to grow plants. We will study the chemistry behind nitrogen fixation on the next page.

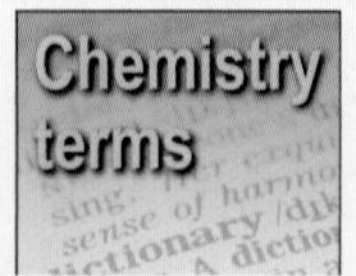

nitrogen cycle - the process that shows how nitrogen moves from the atmosphere to the ocean, land and living systems.

nitrogen fixation - the ability to use nitrogen gas from the atmosphere and convert it into a nitrogen compound such as ammonia or nitrate.

Nitrogen fixation

Nitrogenase allows bacteria to use atmospheric N_2

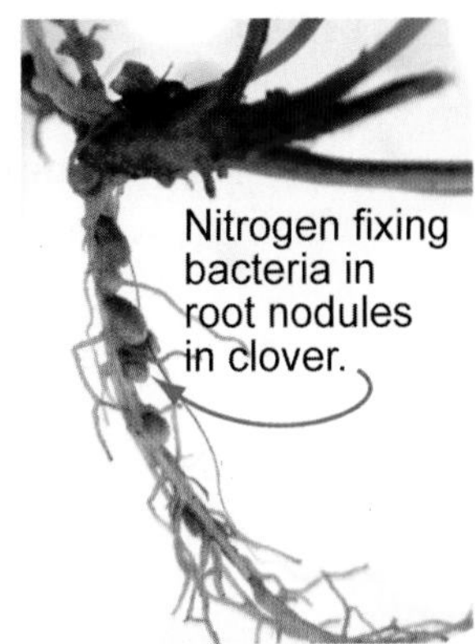
Nitrogen fixing bacteria in root nodules in clover.

The ability to take nitrogen gas and make compounds requires either special enzymes, lots of energy, or high pressure and a catalyst. Each of these processes is unique. In nature only certain types of soil bacteria can convert gaseous N_2 into a usable form of nitrogen compounds. These bacteria live in the soil and on the roots of plants. The reason they can break this strong double bond in the nitrogen molecule is because they have a special enzyme called nitrogenase. The nitrogenase enzyme catalyzes the reaction shown below.

$$N_2 + 8H^+ + 8e^- + 16\ ATP \rightarrow 2NH_3 + H_2 + 16ADP + 16\ P_i$$

As you can see this reaction is carried out in acidic environment and it requires energy from ATP molecules. The nitrogenase enzymes can be damaged by the presence of oxygen, and for this reason the bacteria that carry out this process live in the soil and create environments where the amount of oxygen gas is minimal. Prior to the creation of synthetic fertilizers farmers would fertilize with manure, because it added the nitrogen that the animals consumed by eating grass, back into the soil.

Nitrogen fixation also occurs when lightning reacts with the air. The energy necessary to split the triple bond is provided by the energy in the lightning, and this yields nitrogen containing compounds like NO_3^- and NO_2^-. The combustion of fossil fuels also can provide the energy necessary to convert atmospheric nitrogen into compounds such as NO and NO_2, which you learned about earlier with respect to acid rain.

The Haber-Bosch process is the largest nonbiological means of nitrogen fixation

The Haber-Bosch process is one of the most important commercial chemical processes used today, because we need more fertilizer than can be produced by nature alone. As the earths population has grown it has become necessary for us to produce larger amounts of food. The Haber-Bosch process produces approximately 100 million tons of nitrogen fertilizer each year, and this provides food for more than one third of our population. This process was discovered and refined by two German chemists Fritz Haber and Carl Bosch during the early 1900's. The production of ammonia from nitrogen and hydrogen gas requires high pressure conditions and the use of a catalyst and was *very challenging* to overcome.

Haber and Bosch earned Nobel Prizes for their work

$$N_2(g) + 3H_2(g) \rightleftharpoons 2NH_3(g)\ \Delta H = -92\ kJ/mole$$

Look carefully at the equation above. Can you see why conditions of high pressure would favor the formation of product? Correct, with increased pressure on the system the reaction shifts to the products side where there are fewer moles of a gas, which decreases the overall pressure of the system. This is in agreement with Le Chatelier's principle. Chemical reactions that require high pressure can be very dangerous to carry out on a large industrial scale, but by using an iron oxide catalyst it was accomplished.

Phosphorous cycle

Similar to nitrogen, phosphorous (P) is another very important element in living systems. In living organisms phosphorous is found in the form of the familiar polyatomic ion compound called phosphate, PO_4^{3-}, where it is bound to oxygen. You may recall that phosphorous is part of many biological molecules, DNA and RNA use phosphate molecules to form the backbone of their molecular structures. Important energy molecules like ATP and ADP also incorporate phosphorous in the form of phosphate. The phospholipid bilayer of a cell membranes also incorporates the phosphate ion.

Phosphates PO_4^{3-} are important in living systems

Where do living organisms obtain this essential element?

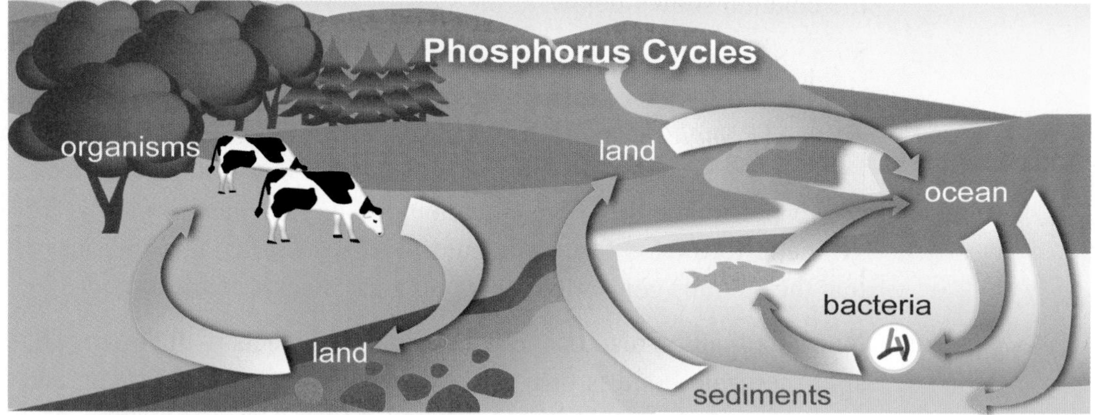

Phosphorous cycles between the land and the ocean

As you look carefully at the **phosphorous cycle** diagram above you will notice a few interesting features. First phosphorous does not actually enter the atmosphere like the other cycles we studied did. The phosphorous cycle takes place between the land, living organisms and the ocean. Phosphorous is found on land mainly in the form of rocks and minerals in the soil. A large amount of phosphorous, in the form of inorganic phosphate salts, is also found on the ocean floor as sediment. Plants are able to absorb phosphate that is dissolved in soil and water. When phosphate is taken up by the plants it is bound in organic compounds. Phosphate bound by plant or animal tissue is considered organic. Organic phosphate makes it's way through the food chain. Soil and rocks release phosphate, plants absorb phosphate from the soil and water, animals eat the plants and obtain phosphate. In the ocean, sediments can slowly bring phosphates to the land and to organisms who live in the sea. Living organisms eventually return the phosphorous back to the earth and sea at the end of their life cycle. Run-off can eventually return some phosphate back to the ocean. We can cause a disruption of the phosphorous cycle with the over use of phosphates in our fertilizers and detergents these phosphates make their way into the environment, creating levels that are too high.

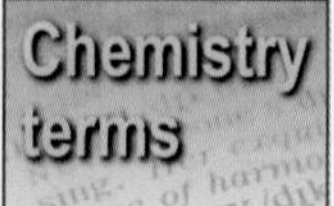

phosphorous cycle - the process that shows how phosphorous moves between the land, the ocean, and living systems.

Soil chemistry

Soil is a mixture of minerals, organic matter, water and air

The chemical reactions that take place in soil play an important role in our environment. Originally, soil chemistry was primarily important to gardeners and farmers, but now it has become a very important developing field of science. Our environment is suffering due to the increased global population and the significant increase in industrial production. Today knowledge of soil science is essential for the clean up of contaminated areas near industrial sites. What exactly is soil? Soil is a mixture of minerals (inorganic solids), organic matter (living microorganisms and decaying plant and animal matter), water, and air. Plants need soil to grow, and plants are required to support most living organisms here on earth. Soils are formed initially from bare rock, which can come from volcanic lava or be exposed from melting glaciers. Soil develops over time from lichen and algae growing on these rocks. As lichen and algae grow they cause some of the rock to break down and erode. Over time these organisms also add to the formation of soil. Gradually the soil becomes deep enough for small plants to grow and as vegetation dies it continues to build the soil layers. The soil that develops over time is influenced by several factors including climate, the types of minerals in the soil, organisms living in the soil, and the geography of the land itself.

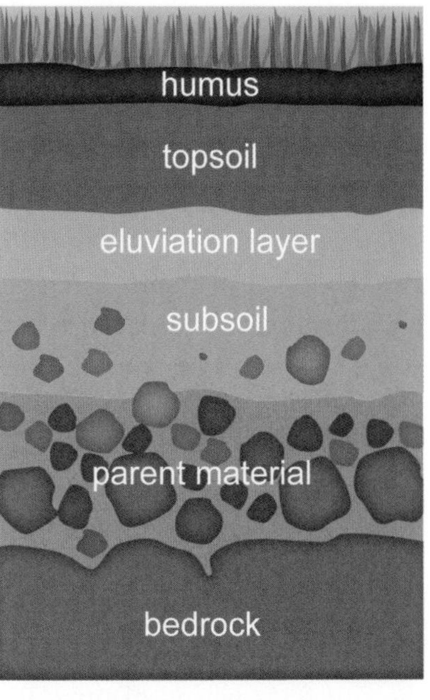

pH plays a key role in soil chemistry

The chemical properties in soil are largely influenced by the pH of the soil. This is because the nutrient availability in soil is regulated by pH. Nutrients that are required by plants include cations such as Mg^{2+}, Ca^{2+}, Zn^{2+}, Fe^{2+}, Co^{2+}, Mn^{2+}, Cu^{2+} and Al^{3+}. Elements such as nitrogen (N) and phophorous (P) are also important and plant growth. The pH range of 6 to 7.5 makes these elements and divalent cations available to plants in the soil. When the pH gets too high nitrogen and phosphorus can form solid precipitates in the soil that make them unable to be absorbed by plants. Overly acidic conditions can make too many cations available to plants, which can lead to the leaching of nutrients from the soil too rapidly and to toxicity. Both of these conditions negatively impact plant growth. There are several ways soil can become acidic, these include the decay of organic matter, large amounts of rain fall, added fertilizers, as well as the type of minerals present in the soil originally.

Carbon sequestering

The earths systems all play a role in the recycling of carbon, as you learned by studying the carbon cycle. One area of cutting edge research today is how to decrease the amount of carbon dioxide being released into our atmosphere. Global efforts to slow down global warming trends are focusing on many different technologies to capture carbon dioxide. Carbon sequestering means to hold carbon underground for long periods of time. The capturing of carbon in the form of carbon dioxide is one form of carbon sequestration.

If you think back to the carbon cycle there are many natural ways that nature removes carbon dioxide gas from the air. The largest amount of carbon dioxide is absorbed by our oceans in the form of the bicarbonate ion, HCO_3^-. Research is now being done to determine how much more carbon dioxide the ocean can hold. It is believed that in some warm climates the ocean has reached it's full capacity to absorb carbon dioxide. Atmospheric chemists are measuring the partial pressure of carbon dioxide just above and below the surface of the ocean to see which is higher. The process of photosynthesis also absorbs carbon dioxide, because of this we have become more aware of how important natural resources such as large forests actually are. Lastly, decaying organic matter remains in the soil for a period of time, this also keeps carbon in the earth.

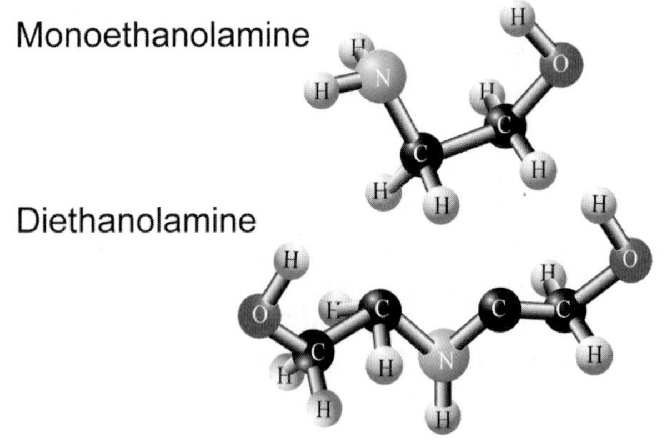

Beyond these natural means of absorbing carbon, scientists have been studying ways to decrease the release of carbon dioxide into our atmosphere. One of the biggest challenges of sustainability is how to decrease industrial emissions of carbon dioxide. One method currently being used is called carbon capture and sequestering (CCS). This method requires chemicals and significant energy to pump carbon dioxide deep into the earth. In the first step of this process, the chemicals added are amine based solvents, such as MEA (mono ethanol amine) or DEA (diethanolamine). These are added to the emission gases (flue gases) of natural gas, coal fired power plants, or other industrial emissions containing carbon dioxide or hydrogen sulfide. The gases that would normally be eliminated into the air through the stack are bubbled through an aqueous amine solution.

$$CO_2 + H_2O \rightleftharpoons HCO_3^- + H^+ + H_2NCH_2CH_2OH \rightarrow HCO_3^- + H_3N\text{-}CH_2CH_2OH^+$$

In this process cold aqueous solutions of amines are used to treat gases with carbon dioxide, CO_2 and hydrogen sulfide, H_2S. You may recall that these gases are both acidic gases. Chemically the amine solvent acts as a weak base, because it contains an amino, NH_2 group. Remember it is the lone pair on the nitrogen that attracts the H^+. The weak base effectively neutralizes the weak acid contained in the dissolved gas, this forms ionic compounds that are much more soluble in the cold solution. The use of amine solvents helps to concentrate the CO_2 so that it can be collected. The first large-scale CO_2 sequestration project was in 1996 called Sleipner, and is located in the North Sea where Norway's StatoilHydro strips carbon dioxide from natural gas with amine solvents and disposes of this carbon dioxide in a deep saline aquifer.

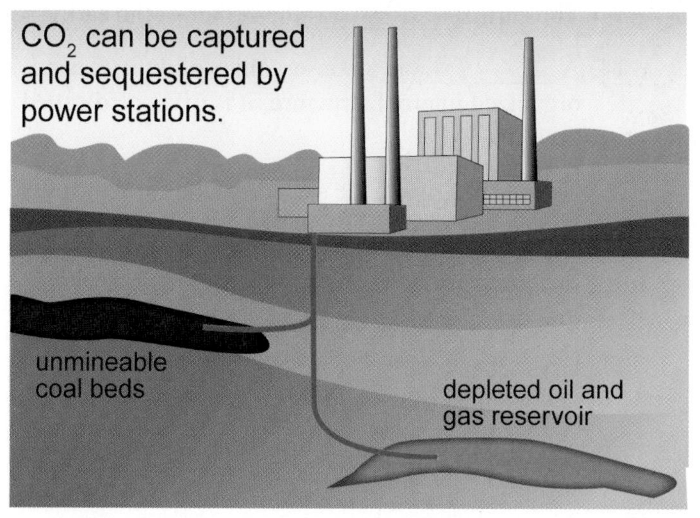
CO_2 can be captured and sequestered by power stations.
unmineable coal beds
depleted oil and gas reservoir

The overall effect of this process is to keep the carbon dioxide dissolved in the solution in the form of bicarbonate ion. This allows the carbon dioxide to be "captured" and therefore it is not released out into our atmosphere. However, after it is captured the amine solution is warmed and the carbon dioxide gas is released from it. The carbon dioxide is then injected deep into the ground. The capture of the carbon dioxide gas can be expensive and was initially challenging because CO_2 only makes up about 15% of the overall combustion gases produced. Geologists are studying ways to safely inject carbon dioxide back into the earth. Currently we use old oil wells or coal seams that cannot be mined. Geologically these areas appear to be ideal sites because carbon dioxide is soluble in oil. The United states has used the injection of CO_2 in depleted oil wells since the early 1970's. Injecting CO_2 into the oil well improved the ability of the oil to flow by increasing the adhesive forces between the surface molecules of the oil and the CO_2 gas molecules. Old oil fields also have a geologic barrier that helps to contain the injected CO_2. An estimate predicts that a new 1000 W coal powered electrical plant will produce about 50 million barrels of carbon dioxide in one year!

Another option is making natural processes occur at a faster speed. Some believe this to be the best and safest way to handle the removal of CO_2 from the atmosphere. One such process is fertilization of the ocean with iron, Fe. Adding iron to the ocean would increase the rate of photosynthesis by plankton and algae. This is a good idea, but it is not fully tested yet. There are environmental concerns about how the added iron would affect the nutrient balance and the overall ecosystem. As you can see, many chemical reactions are important in our pursuit to decrease the CO_2 in our air.

Chapter 19 Review

Vocabulary

Match each word to the sentence where it best fits.

Section 19.1

troposphere	stratosphere
photoionization	photodissociation
Global warming	

1. Most of the ozone that protects the earth from high energy photons is formed in the __________.
2. When a molecule breaks a bond due to the absorption of a photon this is process is called __________.
3. Most of the mass of the atmosphere is contained in the ____________.
4. Gas molecules in the atmosphere trap heat and reflect it back to the earth causing ___________.
5. Sometimes a molecule undergoes ____________which causes it to become an ion by ejecting an electron.

Section 19.2

carbon cycle	water cycle
salinity	transpiration

6. Plants can release water in a process known as _______ when they carry out photosynthesis.
7. The measure of __________ tells us the salt content in water.
8. The process that shows water moving through the atmosphere, ocean, and land is known as the ___________.
9. Carbon moves between the different areas of our earth in a process known as the ___________.
10. How were the majority of our fossil fuels, including coal and oil formed?

Section 19.3

solar nebula	core
igneous rock	magma
mineral	volcano
lava	sedimentary rock
geologists	metamorphic rock
nitrogen fixation	nitrogen cycle
phosphorous cycle	

11. Molten rock found below the surface of the upper mantle is called ___________.
12. The earth formed in the sun's ___________, which is a region of dust and gases.
13. The earth's ________ is the center of the earth.
14. A __________ is a non-organic solid that has an organized internal structure, it is often also called "rock."
15. Magma can rise to the surface of the earth, through an opening in the earth called a _________.
16. Scientists who study the matter making up the earth are called ___________.
17. Crystallized magma forms _____________ which covers most of the earth's crust.
18. As igneous rock erodes, small pieces collect and form ______________ when compressed.
19. Magma is called __________ after it is released from the vent of a volcano.
20. Under conditions of high heat and pressure ____________ can from igneous rock or sedimentary rock.
21. Plants can use atmospheric nitrogen, N_2 in a process called _______________ .
22. Nitrogen moves between the atmosphere, land and ocean in a process known as the __________.
23. The ____________, is unique in that it cycles between the land and the ocean and never enters the atmosphere.

Conceptual Questions:

Section 19.1

24. List the different layers of our atmosphere.

25. Explain why the troposphere has a larger total mass than the stratosphere, when the stratosphere is so much bigger.

26. Explain why the photodissociation of O_2 is more important in our atmosphere than the photodissociation of nitrogen gas?

27. Draw the Lewis structure of ozone, O_3 and for nitrogen dioxide, NO_2.

28. Where do these pollutant gases come from? List naturally occurring and man made sources.
a) NO
b) CO
c) CH_4
d) SO_2
e) O_3

29. What is the pH of naturally occurring rain water? Explain why it is lower than the neutral pH of 7.

30. What happens in the thermosphere that causes the auras seen in the polar regions?

31. Explain what important role the mesosphere plays in protecting the earth.

32. List the four most abundant gases in the atmosphere .

33. What gases are important in keeping the earth warm?

34. Explain how each of the following gases came to be present in our atmosphere.
a) CO_2
b) Ar
c) O_2

35. a)What are the primary gases responsible for "smog?"
b)What causes smog?

36. Explain why natural rain water is acidic. Show a chemical reaction to help explain.

37. a)List the two gases responsible for the formation of acid-rain.
b)Show the chemical reactions for the formation of the two strong acids that form acid rain.
c)What are some of the detrimental effects of acid rain on our environment?

38. What are two important chemical reactions initiated by the sun that occur in the upper atmosphere.

39. a)What molecules are responsible for absorbing the short wavelength UV light.
b) What molecules are responsible for absorbing the longer wavelength UV light.

40. a) What kind of electromagnetic radiation in the atmosphere does carbon dioxide trap?
b)Explain how carbon dioxide molecules area able to trap heat.

41. Make a sketch of the earth and show with arrows how infrared radiation passes through our atmosphere, is reflected and re-radiated back to earth.

42. What unit do we use to measure trace amounts of gases in our atmosphere?

Section 19.2

43. Explain how temperature can affect the density of the ocean waters.

44. Write the chemical reaction that shows how the ocean is able to absorb large amounts of CO_2.

45. List the four most abundant ions in ocean water.

46. Even though the ocean is vast it maintains a relatively constant composition. What factors help it to do this.

47. a)Make a sketch of the water cycle.
b)Label the key features of the cycle.
c) Where does most of the evaporation take place?
d) Do water molecules change form chemically as they move through the cycle? Explain.

48. What are the two factors that affect the density of ocean water?

Chapter 19 Review.

49. Explain how the ocean currents are influenced by the cold air near the poles of the earth.

50. a)How does the melting of glaciers affect the density of the water at the poles?
b) How might this affect the deep ocean currents?

51. a)Make a sketch of the Carbon cycle.
b) Label the processes that bring carbon into the atmosphere with a red arrow.
c) Label the processes that return carbon to the earth with a blue arrow.
d) Do these processes appear to balance one another? Explain.

52. Where is the largest amount of carbon available to cycle around the earth contained?

53. a)Explain under what conditions photosynthesis and respiration would be able to balance each other.
b) What human factors may influence this relationship?

54. Explain what will happen to the ocean's pH as it continues to absorb CO_2.

Section 19.3

55. a) Make a sketch and showing the different layers of the earth.
b)What conditions caused these layers to form?

56. What two elements compose most of the earth's crust?

57. Explain why there is a high concentration of iron in the earths core.

58. What two factors affect the arrangement of atoms deep within the earth?

59. a)What type of rock is the most abundant on the earth's crust?
b)What is the common source for all types of rock?

60. a)What is another common term for minerals?
b) List three common types of minerals.

61. a)Some minerals for "gem stones," list the chemical formulas for two different gem stones.
b) What types of conditions form diamonds?

62. Explain how the silicate content affects the "flow" of the lava.

63. Why are most of our volcanoes near the ocean on the edge of continents?

64. What are some of the ways that geologists are helpful to us?

65. What are the two ways that metamorphic rock can be formed?

66. In what rock "layer" are fossils commonly found?

67. What is the primary source of nitrogen here on earth?

68. Make a sketch of the nitrogen cycle and label the methods by which nitrogen can enter the land.

69. What are the forms of nitrogen that animals and humans can metabolize?

70. What are the three ways that nitrogen can become available to living organisms?

71. Why is N_2 not a usable form of nitrogen for animals?

72. a)What do bacteria require to form ammonia, NH_3 from nitrogen?
b) Write the equation for the bacterial fixation of nitrogen.
c) What molecule is damaging to the bacteria who carry out this process?

73. Write the equation that represents the Haber - Bosch process.

74. What important biological molecules use the element phosphorous?

75. Where does most of our inorganic phosphorous come from?

76. Phosphates are commonly used in fertilizers and detergents. How can this upset the balance of phosphorous in our environment ?

77. What is the primary source of phosphate in the ocean?

78. Compare the nitrogen and phosphorous cycles.
a) List one major difference
b) List two similarities
c) What role do plants play in each of the cycles?

79. What is soil made up of?

80. What pH range is the best for most plants to grow in?

81. Contrast the chemical effects on nutrient availability when the soil is too acidic or too basic.

Quantitative Problems:

Section 19.1

82. Does a photon with a wavelength of 170 nm, emitted by the sun, have enough energy to dissociate an oxygen molecule? The bond dissociation energy of an O_2 molecule is 495 kJ/mole. Show a calculation to support your answer.

83. Formaldehyde, CH_2O is formed in car exhaust when fuels containing alcohol are combusted. Formaldehyde photodissociates and contributes to smog formation.
A photon with a wavelength of 336 nm is energetic enough to cause the C-H bond to rupture and dissociate.
a) In what part of the electromagnetic spectrum is this light found? (visible, UV, IR, etc.) Explain.
b) Calculate the bond energy in kJ/mole of the bond that can be broken by a photon of 336 nm.

84. In polluted areas where there is a larger amount of NO_2 in smog the formation of NO through photodissociation is an important factor.
$NO_2(g) + hv \rightarrow NO(g) + O(g)$
Light with a wavelength of 420nm can cause this dissociation process.
a) Where in the electromagnetic spectrum is light with a wavelength of 420 nm found?
b) Calculate the bond energy in kJ/mole of the bond that can be broken by a photon of 420 nm.

85. In one sample of air from an urban area, the concentration of carbon monoxide, CO was 4.5 ppm. What is the partial pressure of the CO? The atmospheric pressure(total air pressure) is 702 torr.

86. Magnesium, Mg is reclaimed from seawater by precipitation with calcium oxide, CaO. Given by the equation show below.
$Mg^{2+}(aq) + CaO(s) + H_2O(l) \rightarrow Mg(OH)_2(s) + Ca^{2+}(aq)$
Calculate the amount of CaO in grams, needed to precipitate 1.0 kg of $Mg(OH)_2$.

87. In one sample of air from New York City, the concentration of carbon monoxide, CO was 3.7 ppm. What is the partial pressure of the CO? The atmospheric pressure (total air pressure) is 718 torr.

88. The average concentration of carbon monoxide in air in Chicago, Illinois, was 4.1 ppm. Calculate the number of CO molecules in 2.0 liters of this air. The atmospheric pressure was 738 torr and the temperature was 25°C

89. The average concentration of nitrous oxide in the air in Los Angeles California was 0.30 ppm. Calculate the number of NO molecules in 1.0 liter of this air. The atmospheric pressure was 756 torr and the temperature was 26°C

90. Calculate the concentration of water vapor in ppm for a sample that has a water vapor pressure of 0.76 torr, and a total air pressure of 709 torr.

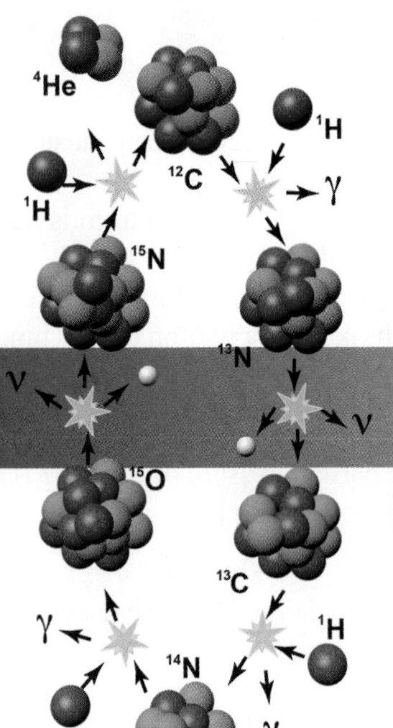

CHAPTER 20

Nuclear Chemistry and Radioactivity

How do we write nuclear reactions?

How are nuclear reactions different from chemical reactions?

What is radioactivity and how is it used?

What are the biological effects of radioactivity?

Where does nuclear energy come from?

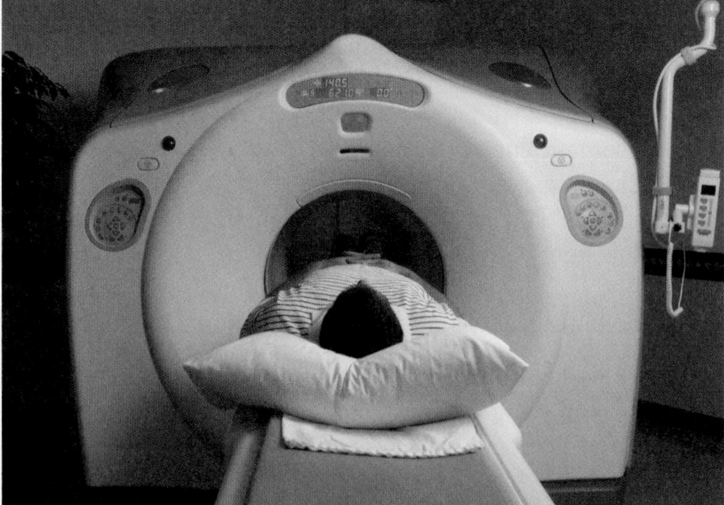

Nuclear chemistry and nuclear processes are central to many products and devices that are used in our daily lives. Nuclear chemistry is at the core of technologies ranging from the smoke detector in your home to medical procedures like x-rays. The generation of energy from nuclear reactors is also based on the science of nuclear chemistry.

Archaelogists are able to date ancient objects by using nuclear chemistry techniques.

The development of new drugs depends on nuclear procedures that are used to investigate and evaluate chemical reaction mechanisms.

Radioactivity is the energy that is associated with nuclear proecsses. The discovery of radioactivity and the eventual development of the science of nuclear chemistry is one of the great accomplishments of humanity. It gave us tools to diagnose and treat conditions such as cancer and heart disease. However, it also led to the discovery of nuclear weapons.

Nuclear chemistry is a technology that presents us with the ultimate trade-offs between benefit and risk. In this chapter we will learn the fundamentals of nuclear chemistry science and so will be able to accurately evaluate its impact on our lives.

Radioactivity: The invisible energy

Radioactivity is a natural phenomenon that occurs all around us, and has since the Solar System condensed out of interstellar gasses 4.3 billion years ago. Uranium, with 92 protons and 146 neutrons is a fossilized remnant of the ancient exploded stars that contributed matter to our own Earth. Uranium is a radioactive element and today it is mined and used in nuclear reactors.

Rocks that contain uranium have some very special properties. These properties lead to the discovery of the nucleus and the dawn of the atomic age. The crucial experiment in which radioactivity was discovered was first done in 1896 by French scientists Antoine-Henri Becquerel and Marie Curie. You can do it too!

For the experiment we use three rocks and a piece of black and white photographic paper. One of the rocks is a mineral called pitchblende. The other two can be any type of rock you can pick up from the field. We perform our experiment in a dark room so that the photographic paper is not affected by light. We place the three rocks on the paper and let them sit in a dark room, or in a light sealed box for 24 hours.

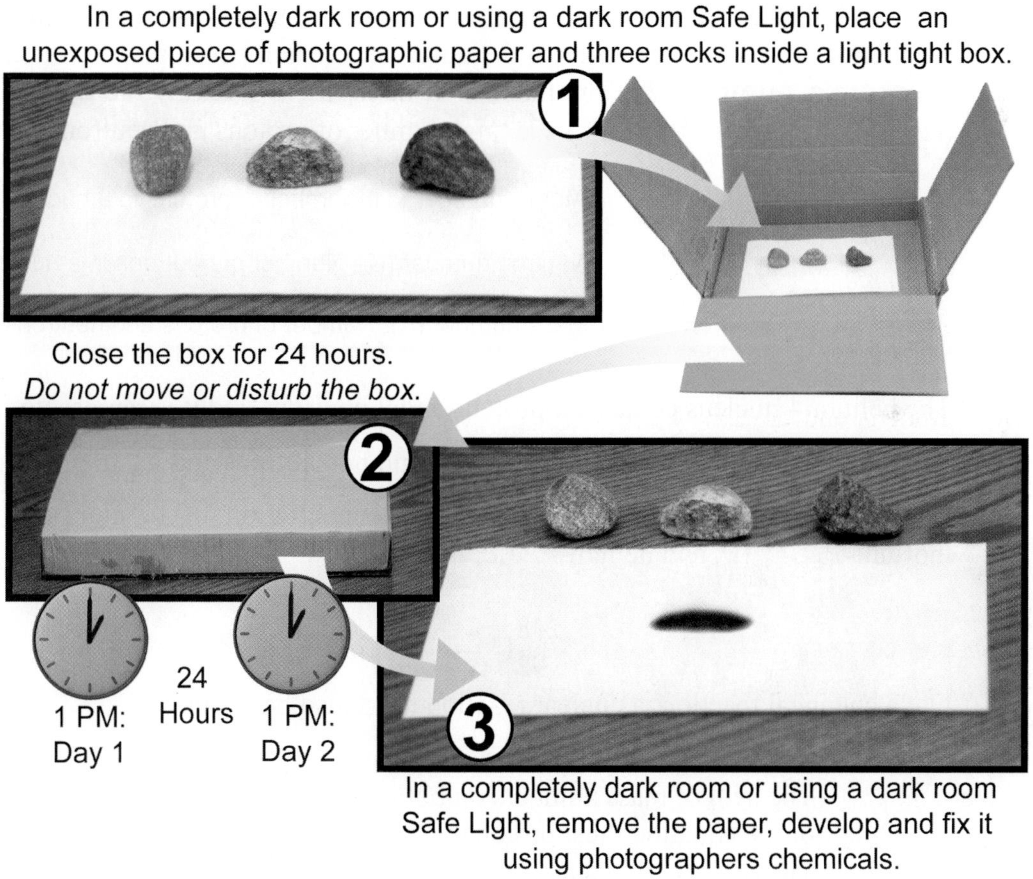

In a completely dark room or using a dark room Safe Light, place an unexposed piece of photographic paper and three rocks inside a light tight box.

1

Close the box for 24 hours.
Do not move or disturb the box.

2

1 PM: Day 1

24 Hours

1 PM: Day 2

3

In a completely dark room or using a dark room Safe Light, remove the paper, develop and fix it using photographers chemicals.

After twenty fours we remove the rocks and develop the photographic paper. We notice there is an *image* under the pitchblende rock, as it the rock gave off its own light. Some type of energy must have made this image. What type of energy? Where did it come from?

The invisible energy that exposes photographic paper comes from uranium in the pitchblende rock. The source of the energy is a sequence of nuclear reactions that occur in uranium. In this chapter we will study these reactions and learn how energy is released by nuclear reactions

20.1 Nuclear Equations

nuclear versus chemical

Unlike chemical reactions, which only involve the outer electrons of atoms, nuclear reactions involve and affect the nuclei of atoms. Nuclear reactions can change an element into a different element. This change means that the the number of protons and neutrons are changed during a nuclear reaction. Therefore, in order to represent a nuclear reaction we must explicitly indicate the number of protons and neutrons for each nucleus involved in the reaction.

Chemical reactions create new compounds from elements

Nuclear reactions create new elements from other elements

representing nuclei

The symbols that we use to represent the atomic nucleus contain information about the number of protons and neutrons in the nucleus. The number of protons is represented by the **atomic number (Z)**. The number of neutrons is given by the **neutron number (N)**. The **mass number (A)** is the total number of protons and neutrons in a nucleus.

Mass number → A; *Atomic number* → Z; *Element* → X: $^{A}_{Z}X$

Atomic number (Z): Number of protons in a nucleus

Neutron number (N): Number of neutrons in a nucleus

Mass number (A): Number of protons and neutrons. A = Z + N

The helium-4 nucleus contains 2 protons and 2 neutrons and it is represented by: $^{4}_{2}He$.

The hydrogen nucleus contains one proton and no neutrons and it is represented by : $^{1}_{1}H$.

nuclear symbols

The nuclear equation that represents the disintegration of uranium-238, $^{238}_{92}U$, into thorium-234, $^{234}_{90}Th$, and helium-4, $^{4}_{2}He$, is written as:

$$^{238}_{92}U \rightarrow {}^{234}_{90}Th + {}^{4}_{2}He$$

balanced nuclear equation

like a chemical reaction, a nuclear reaction must also be balanced:

1. There must be mass number balance. The sum of mass numbers on both sides of the equation must be equal.
2. There must also be charge conservation. The sum of atomic numbers on both sides of the equation must be equal.

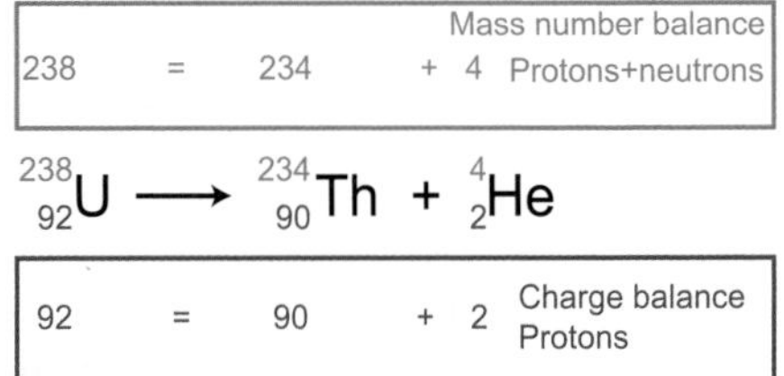

Chemistry terms

atomic number (Z) - the number of protons in a nucleus.
neutron number (N) - the number of neutrons in a nucleus.
mass number (A) - the number of protons and neutrons in a nucleus.

20.2 Nuclear Reactions: Radioactivity

Nuclear versus chemical

Nuclear reactions can change one element into a different element. They can also change an isotope into a different isotope of the same element. Chemical reactions do not change the types of atoms involved. They only rearrange atoms to form different molecular compounds.

Energy of nuclear reactions

Nuclear reactions involve much more energy than chemical reactions. The energy released by a chemical reaction is related to the rearrangement of the electronic structure of atoms which involves electrical forces. For nuclear reactions the energy released is related to the rearrangement of the atomic nucleus which involves the strong nuclear force, the strongest force in the universe.

Reaction types

There are two main types of nuclear reactions.

1. Decay reactions: during which a nucleus breaks up (dissintegrates) spontaneously
2. Bombardment reactions: during which a nucleus is struck by another nucleus or some nuclear particle such as a proton or a neutron.

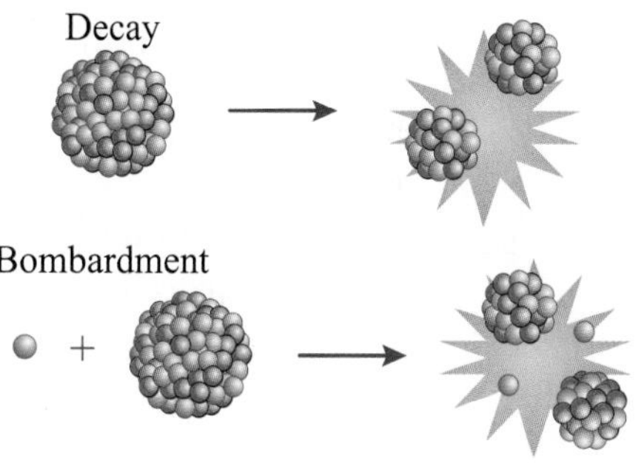

Radioactivity

The four most common types of nuclear decay are: alpha (α) decay, beta (β) decay, gamma (γ) decay, and positron emission (β^+) decay. The elements that decay by α, β, γ, β^+ decay are said to be radioactive. Alpha, beta and positron emission decay release energy that is carried by particles. Gamma decay releases electromagnetic energy. Light, radio waves, microwaves and x-rays are forms of electromagnetic energy. **Radioactivity** is a general term used to describe the property of some elements to break up and release energy associated with matter or waves.

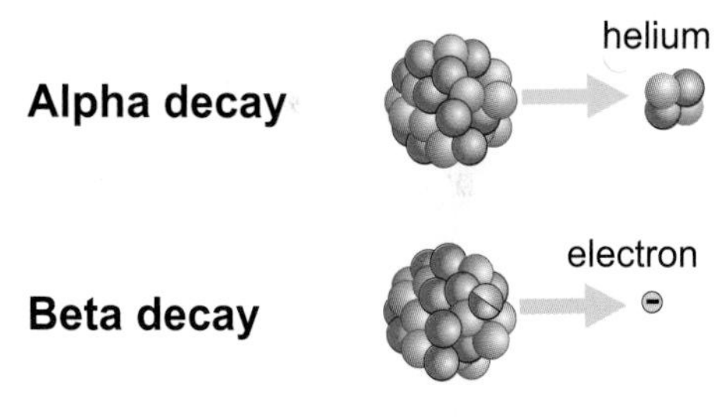

Gamma decay

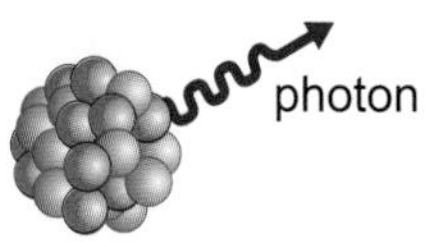

Radiation

The transmission of energy, matter or waves, through space is called **radiation**. Radiation can be dangerous if it has high enough energy to break chemical bonds in molecules. Exposure to radiation over a long period of time can be harmful. Ultraviolet radiation from the sun is an example of radiation that can be harmful to living organisms.

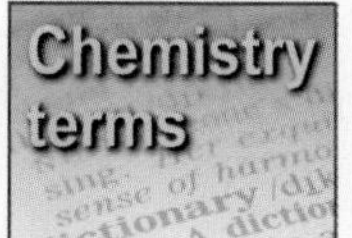

nuclear reactions - involve the nuclei and may change one element into another.
radioactivity - property of some element to break up and release energy.
radiation - the transmission of energy through space.

Intensity of Radiation

ADVANCED AP

Intensity

The **intensity of radiation** measures how much energy flows per unit of area per second.

$$\text{Intensity (W/m}^2) = \frac{\text{Power (W)}}{\text{Area (m}^2)}, \quad I = \frac{P}{A}$$

Intensity

Intensity (W/m²) $I = \frac{P}{A}$ *Power (W)* / *Area (m²)*

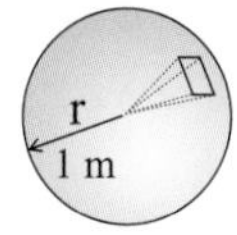

Area, $A = 4\pi r^2 = 12.6\ m^2$

Intensity, $I = \frac{100\ W}{12.6\ m^2} = 7.96\ W/m^2$

On a clear day, the intensity of sunlight on the surface of the Earth is about 1,000 watts per square meter (W/m^2). A good flashlight produces an intensity of about 100 W/m2 in the brightest part of the beam at a distance of a half- meter.

Inverse square law

When radiation comes from a single point, the intensity decreases inversely over the square of the distance. This is called the **inverse square law**. If you get two times farther away from a radiation source, the intensity goes down by a factor of 1/4, which is one divided by the square of two ($1/4 = 1/2^2$). If you get 10 times farther away, the intensity goes down by one divided by 10 squared ($1/100 = 1/10^2$).

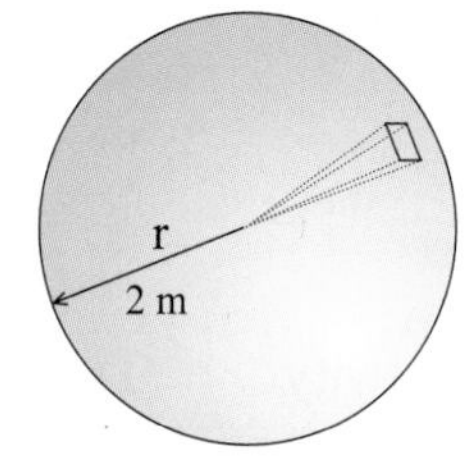

Area, $A = 4\pi r^2 = 50.4\ m^2$

Intensity, $I = \frac{100\ W}{50.4\ m^2} = 1.99\ W/m^2$

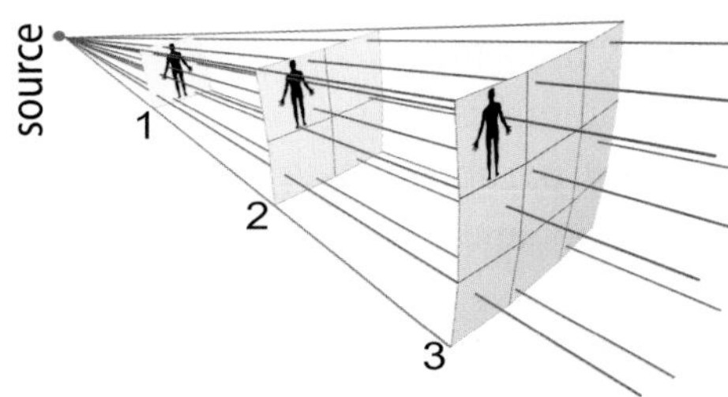

The further we are from a source the smaller the amount of radiation exposure. Increasing the distance by 3 decreases the radiation by 9.

The inverse square law is a property of geometry. Think of a 100-watt light bulb at the center of a sphere of radius r. The intensity at the surface of the sphere is 100 watts divided by the area of the sphere ($4\pi r^2$). At twice the distance, the area of the sphere is four times greater. Since the same amount of power is spread over a larger area, the intensity goes down. Mathematically, the area goes up like the square of the radius, so the intensity goes down like the inverse square of the radius. This is the inverse square law.

Chemistry terms

intensity of radiation - the amount of energy that flows per unit area per second.

inverse square law - the intensity of radiaiton emitted from a point source decreases inversely over the square of the distance.

Decay Reactions: Alpha Decay

Alpha decay

In **alpha decay** the original nucleus ejects two protons and two neutrons which is the nucleus of helium-4, ${}^{4}_{2}\text{He}$, also called alpha (α) particle. During alpha decay the emitted radiation is also called **alpha radiation** which is a fast moving ${}^{4}_{2}\text{He}$ nucleus.

Uranium-238 decays by releasing an α particle, ${}^{4}_{2}\text{He}$, and thorium-234, ${}^{234}_{90}\text{Th}$.

The original uranium atom is called the **parent nuclide** and the thorium atom is called the **daughter nuclide**.

$${}^{238}_{92}\text{U} \rightarrow {}^{234}_{90}\text{Th} + {}^{4}_{2}\text{He}$$

Alpha decay

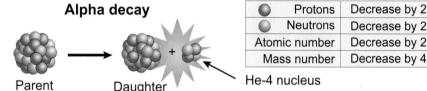

Parent → Daughter + He-4 nucleus

Protons	Decrease by 2
Neutrons	Decrease by 2
Atomic number	Decrease by 2
Mass number	Decrease by 4

alpha particles deposit their energy over small distances

When the α particle hits a molecule it transfers energy to it and damages it. Damaged molecules in the cells of biological systems may result in cell death or the abnormal reproduction of cells. Since the α particles are very large, their ability to penetrate into matter is very limited. α radiation can be stopped by a sheet of paper. If an element that undergoes α decay is ingested it is very dangerous. As the α source is carried throughout the body the emitted α particles are in direct contact with the molecules in the cells of organs and can deposit their energy causing great damage.

Write the complete nuclear equation for the α decay of radium isotope ${}^{226}_{88}\text{Ra}$.

Relationships: *We need to find the type of the daughter nucleus X, the atomic number Z and the mass number A.* ${}^{226}_{88}\text{Ra} \rightarrow {}^{A}_{Z}\text{X} + {}^{4}_{2}\text{He}$

Solve: *First: Balance the atomic number and the mass number.*

- *By balancing the mass number we obtain A=222*
- *By balancing the atomic number we obtain Z=86*

Second: Look at the periodic table to determine the identity of the unkown daughter element.

- *Since the atomic number is 86 the daughter nuclide is radon (Rn).*

Answer: *The complete α decay equation is* ${}^{226}_{88}\text{Ra} \rightarrow {}^{222}_{86}\text{Rn} + {}^{4}_{2}\text{He}$

alpha decay - happens when a nucleus decays by releasing a helium nucleus.
alpha radiation - the radiation associated with alpha decay.
parent nuclide - the original nucleus involved in a nuclear reaction.
daughter nuclide - the nucleus resulting from a nuclear reaction.

Beta Decay

Beta decay

Beta decay happens when an unstable nucleus emmits an electron. We know that the nucleus does not have any electrons in it. So where does the emitted electron during beta decay come from? It turns out that the electron is formed when a neutron in the nucleus changes into a proton and an electron. The radiation released during beta decay is called **beta radiation**.

Beta particle is an electron

Since the charge of the nucleus increases by +1 as the neutron changes to a proton, the formation of the electron, which has a charge of -1, maintains the overall charge of the nucleus. The electron released during beta decay is called beta particle.

Beta decay
Nucleus converts a neutron to a proton

Parent → Daughter + electron

Beta symbol $^{0}_{-1}e$

The beta (β) particle is an electron. For consistency with the genral nuclide symbol and in order to help us with the balance of mass and atomic numbers the β particle is written with the symbol $^{0}_{-1}e$.

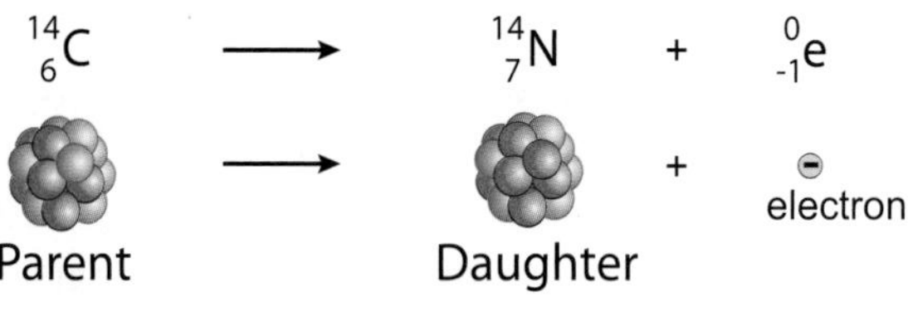

Radium $^{228}_{88}Ra$ could also decay by β-decay according to: $^{226}_{88}Ra \rightarrow ^{228}_{89}Ac + ^{0}_{-1}e$

Write the complete nuclear equation for the β decay of Radium isotope $^{228}_{88}Ra$

Given: *The parent nuclide and the type of reaction*

Relationships: *We need to find the type of the daughter nucleus X, the atomic number Z and the mass number A.* $^{228}_{88}Ra \rightarrow ^{A}_{Z}X + ^{0}_{-1}e$

Solve: *First: Balance the atomic number and the mass number.*

- *By balancing the mass number we obtain A = 228*
- *By balancing the atomic number we obtain 88 = Z-1 or Z = 89*

Second: Look at the periodic table to determine the identity of the unkown daughter element.

- *The atomic number is 89 and the daughter nuclide is actinium, Ac.*

Answer: *The complete β decay equation is* $^{228}_{88}Ra \rightarrow ^{228}_{89}Ac + ^{0}_{-1}e$

Chemistry terms

beta decay - when an unstable nucleus releases an electron.
beta radiation - the radiation resulting from beta decay.

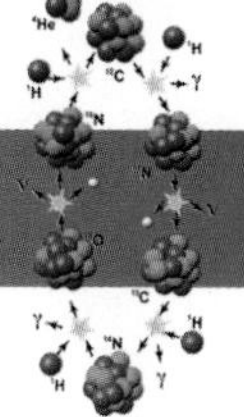

Gamma Decay, Positron Emission

Gamma decay

Radioactive decay may also happen with the emission of electromagnetic radiation. **Gamma decay**, abreviated γ-decay, happens when an excited nucleus goes to a lower energy state by emitting high energy electromagnetic radiation. The wavelength of this radiation is about 10^{-12} m. In some cases the emission of γ radiation, also called γ-rays and denoted by ${}^{0}_{0}\gamma$, follows some other radioactive decay such as β-decay or α-decay.

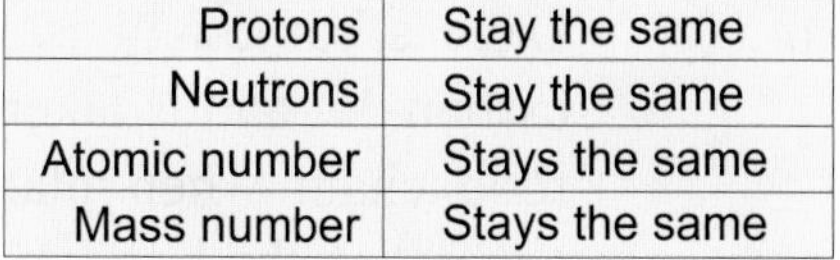
Nucleus emits gamma radiation and lowers its energy.
Gamma decay
Gamma ray

Protons	Stay the same
Neutrons	Stay the same
Atomic number	Stays the same
Mass number	Stays the same

γ-ray energy

The number of protons and neutrons does not change during γ-decay. Gamma radiation has high enough energy (greater than 10^{-13} joules per disintegration) to break apart other atoms making them dangerous to living organisms. The best way to stop γ-rays is by using a thick shielding material made of lead or concrete.

Positron

Positron, denoted with the symbol ${}^{0}_{+1}e$ in nuclear equations, is a nuclear particle that has the same mass as the electron, but it has a positive charge.

Positron emission

Nuclear decay by **positron emission** happens when an unstable nucleus emits a positron. In doing so it converts a proton into a neutron.

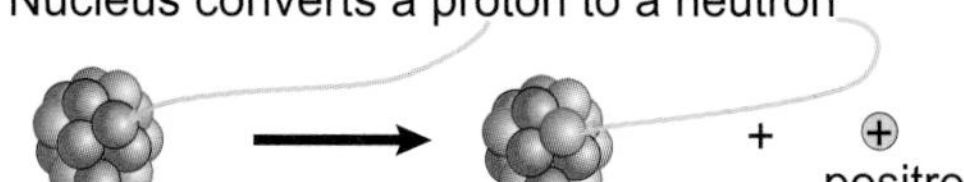
Nucleus converts a proton to a neutron
+
positron
Positron emission

Protons	Decrease by 1
Neutrons	Increase by 1
Atomic number	Decreases by 1
Mass number	Stays the same

Solved problem

Determine the type of decay for the reactions:

1) ${}^{234}_{90}Th \rightarrow {}^{234}_{90}Th + {}^{A1}_{Z1}X1$. 2) ${}^{95}_{43}Tc \rightarrow {}^{95}_{42}Mo + {}^{A2}_{Z2}X2$

Given: *We are given the mass and atomic numbers of the parent and daughter nuclides. Balance the equations to find unknowns X1 and X2.*

Solve: *Balance the atomic numbers to find Z and mass numbers to find A*

- *For reaction 1: 234 = 234 + A1, A1=0. And 90 = 90 + Z1, Z1=0*
- *For reaction 2: 95 = 95 + A2, A2=0. And 43 = 42 + Z2, Z2=+1*

Answer: *For reaction 1:* ${}^{A1}_{Z1}X1 = {}^{0}_{0}X1$ *which denotes γ-decay.* ${}^{A1}_{Z1}X1 = {}^{0}_{0}\gamma$

For reaction 2: ${}^{A2}_{Z2}X2 = {}^{0}_{+1}X2$ *denotes positron decay.* ${}^{A2}_{Z2}X2 = {}^{0}_{+1}e$.

Chemistry terms

gamma decay - when a nucleus decays releasing electromagnetic energy.
positron - a particle that has the same mass as the electron and positive charge.
positron emission - when a nucleus decays by releasing a positron.

20.3 Rate of Radioactive Decay

Carbon-14, ${}^{14}_{6}C$, is an isotope of carbon that is produced when the neutrons generated by the cosmic rays - high energy radiation from the sun and elsewhere in the universe - bombard nitrogen in the upper atmosphere.

$$ {}^{1}_{0}n + {}^{14}_{7}N \rightarrow {}^{14}_{6}C + {}^{1}_{1}H $$

Carbon-14, ${}^{14}_{6}C$

Once a carbon-14 isotope is formed it is oxidized into a ${}^{14}CO_2$ molecule, just like carbon-12, and is taken up by plants during photosynthesis. In turn carbon-14 enters the food chain when plants are consumed by animals. Eventhough ${}^{14}_{6}C$ is continously generated, the fraction of carbon-14 to carbon-12 present in living organisms is constant at about 1 to 10^{12} ${}^{14}_{6}C$ to ${}^{12}_{6}C$ atoms.

Half-life

The reason for this constant ratio is that ${}^{14}_{6}C$ decays back to ${}^{14}_{7}N$ by β decay at a fixed rate. The rate at which an atom decays may be estimated by collecting a large number of them and then observing how many have remained after a certain period of time. The time that it takes for half of the initial atoms to disintegrate is called the **half-life** ($t_{1/2}$).

Half-life of carbon-14

For ${}^{14}_{6}C$ the half life is 5,730 years. For example, if we start with 1000 carbon-14 atoms, after 5730 years we will have 500 ${}^{14}_{6}C$ and 500 ${}^{14}_{7}N$ atoms. After another 5730 years we will have 250 atoms of ${}^{14}_{6}C$, 750 ${}^{14}_{7}N$ atoms and so on.

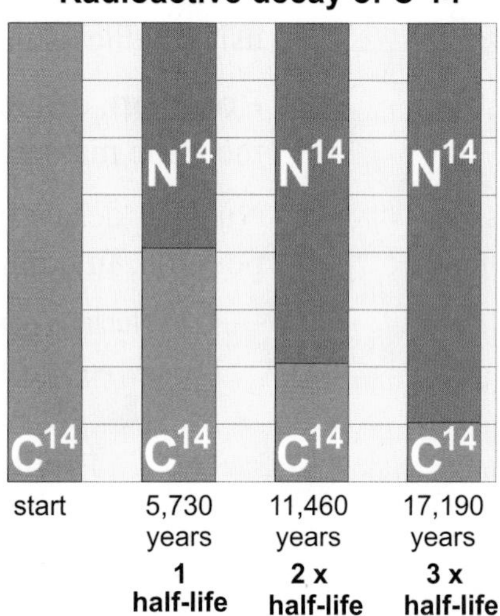

The half-lives of different isotopes varies greatly

Nuclide	Half-life	Decay type
${}^{238}_{92}U$	4.5 x 10⁹ yr	alpha
${}^{131}_{53}I$	8.05 days	beta
${}^{18}_{9}F$	110 min	positron
${}^{220}_{86}Rn$	55.6 sec	alpha

Every radioactive element has a different half-life. Some of them, like uranium-238, ${}^{238}_{92}U$, have a half-life of 4.5 billion years. ${}^{238}_{92}U$ was created in the nuclear reactions of exploding stars and it has been around since the formation of our solar system. Others like like iodine-131, ${}^{131}_{53}I$, have a half-life of 8.05 days. Some others like fluorine-18, ${}^{18}_{9}F$, have a half-life of 110 minutes. The isotopes with short half lifes, like fluorine-18 do not occur naturally and have to be generated in the laboratory.

Chemistry terms

carbon-14 - a radioactive isotope of carbon.

half-life - the time it takes for half of the atoms in a sample to decay.

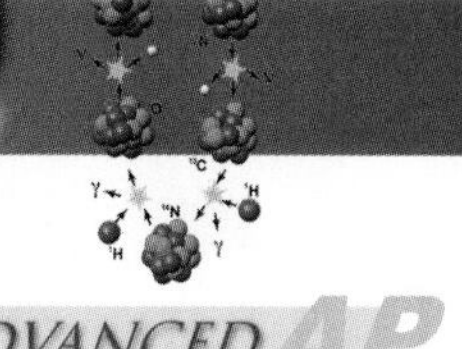

The Mathematics of Radioactive decay

ADVANCED AP

Rate of decay

The number of radioactive nuclei that decay per unit time, called the rate of decay, is directly proportional to the number of nuclei that are present in the first place. The number of decays per unit time is called the **rate of decay** and it is given by:

$$Rate = k\,N$$

How many nuclei decay per unit time (Rate); *Rate constant* (k); Number of nuclei in sample (N)

Rate constant

The **rate constant**, k, is a parameter that describes how fast the radioactive nuclides decay. The rate constant k has the units (1/time) and is related to the half life, $t_{1/2}$. The longer the half-life the fewer radioactive decays will happen over a period of time and the smaller the rate constant.

- Short half-life implies large rate constant.
- Long half-life implies small rate constant.

The equation that relates k to $t_{1/2}$ is $t_{1/2} = \frac{0.693}{k}$

C-14, and Ra-220 have half-lives of 5730 yr and 1 minute respectively. Calculate the rate constants for their decay and give them in units of 1/sec.

Given: *The half life $t_{1/2}$ of each radioactive decay process.*

Asked: *Calculate the rate constant k.*

Relationships: *The equaiton that relates $t_{1/2}$ to k:* $k = 0.693/t_{1/2}$

Solve: *For C-14:* $t_{1/2} = 5730$ *years and*

$$k = \frac{0.693}{5730\cancel{yr}} \cdot \frac{1\cancel{yr}}{365\cancel{days}} \cdot \frac{1\cancel{day}}{24\cancel{h}} \cdot \frac{1\cancel{h}}{60\cancel{min}} \cdot \frac{1\cancel{min}}{60\text{sec}} = \frac{3.84\times10^{-12}}{\text{sec}}$$

For Ra-220: $t_{1/2} = 1$ *min and*

$$k = \frac{0.693}{1\cancel{min}} \cdot \frac{1\cancel{min}}{60\text{sec}} = \frac{1.155\times10^{-2}}{\text{sec}}$$

Discussion: *Note that a small $t_{1/2}$ gives a large k. k gives us an indication of the number of decays over a certain period of time.*

Chemistry terms

rate of decay - the number of decays per unit time.

rate constant - a number that descrubes how fast a radioactive nuclide decays.

The Law of Decay Rate

ADVANCED AP

Activity

The rate of decay of a radioactive sample is also called the **activity** of the sample. The activity of a sample is proportional to the number of nuclei present in the sample at any time. As time passes the activity of radiation decreases since the number of parent nuclei decreases.

When making radioactive decay calculations it is important to be able to answer the following question:

If we start with a radioactive sample that has a half-life, $t_{1/2}$, *that initialy contained* N_0 *nuclei, what is the number, N, of radioactive nuclei present in the sample after time t has passed?*

The answer to this question is given by the equation: $N = N_0e^{-kt}$

$$N = N_0\, e^{-kt} = N_0\, e^{\frac{-0.693\,t}{t_{1/2}}}$$

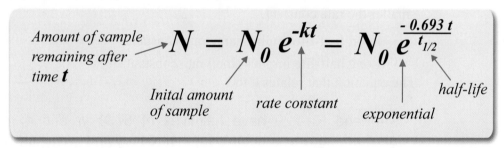

Amount of sample remaining after time ***t*** (N); *Inital amount of sample* (N_0); *rate constant* (k); *exponential* (e); *half-life* ($t_{1/2}$)

The parameter *e* is the number 2.718 that is the base of the natural logarithm. This is one of the important mathematical constants used in many areas of science and engineering.

Plutonium-236 decays by emitting an alpha particle and has a half life of 2.86 years. If we start with 10 mg of Pu-236, how much remains after 4 years?

Given: *The half life* $t_{1/2}$ *, the inital amount,* N_0*, and the elapsed time, t.*

Asked: *Calculate, N, the amount left after 4 years.*

Relationships: *The equation that relates* $t_{1/2}$ *and* N_0 *to N is* $N = N_0e^{-\frac{0.693t}{t_{1/2}}}$

Solve: $N = 10\text{ mg } 2.718^{\frac{-(0.693)4\text{yr}}{2.86\text{ yr}}} = 3.79\text{ mg}.$

Discussion: *After 4 years the initial 10 mg is reduced to 3.79 mg, which is 37.9% of the inital amount of Pu-236.*

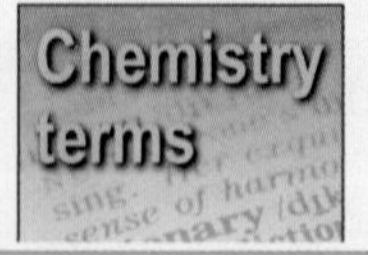

activity - the rate of decay of a radioactive sample.

Carbon Dating

$^{14}_{6}C$: Dating organic matter

Living things contain a large amount of carbon. The isotope carbon-14, $^{14}_{6}C$, is used by archeologists to determine the age of things that were once alive. The ratio of carbon-14 to carbon-12 in the environment is a constant, determined by the balance between production and decay of carbon-14. As long as an organism is alive, it constantly exchanges carbon with the environment. Therefore, the ratio of carbon-14 to carbon-12 in the organism that is alive stays the same as in the environment.

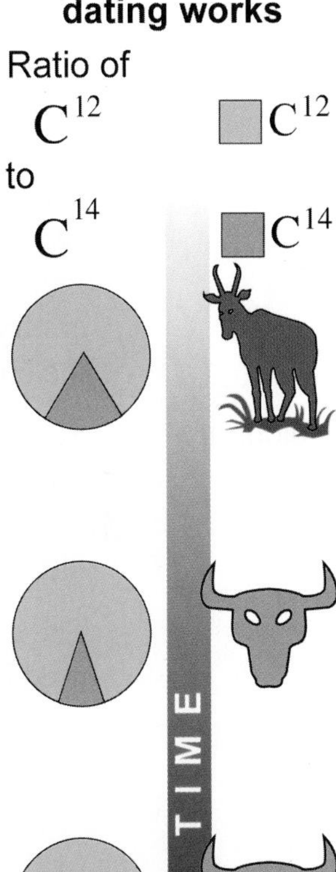

When an organism dies it stops exchanging carbon with the environment. All the carbon-12 in the organism remains because it is a stable isotope. Almost no new carbon-14 is created because most cosmic rays do not reach the ground. As the carbon-14 decays, the ratio of carbon-14 to carbon-12 slowly gets smaller with age. By measuring this ratio, an archeologist can tell how long it has been since the material was alive. Carbon dating works reliably up to about 10 times the half-life, or 57,300 years. After 10 half-lives there is not enough carbon-14 left to measure accurately. Carbon dating only works on material that has once been living, organic materials such as tissue, bone or wood

Dating non-organic matter

It is also possible to date non living matter by using other nuclear decay reactions. For example, the decay of uranium-238, which has a half life of 4.5 billion years has been used to determine the age of the earth.

About 18% of the mass of a live animal is carbon. 1 g of live bone contains about 90 billion carbon-14 atoms, which have a half life of 5,730 years. How many C-14 atoms remain in 1 g of bone 17,190 years after the animal dies?

Given: *The half life and the number of carbon-14 atoms.*

Solve: *17,190 years is 3 half-lives and so the initial amount must be reduced by a factor of* $2\times2\times2=8$.
90 billion/8 = 11.25 billion

Answer: *After 3 half-lives the amount of carbon-14 atoms is reduced by a factor of 8, from 90 billion to 11.25 billion.*

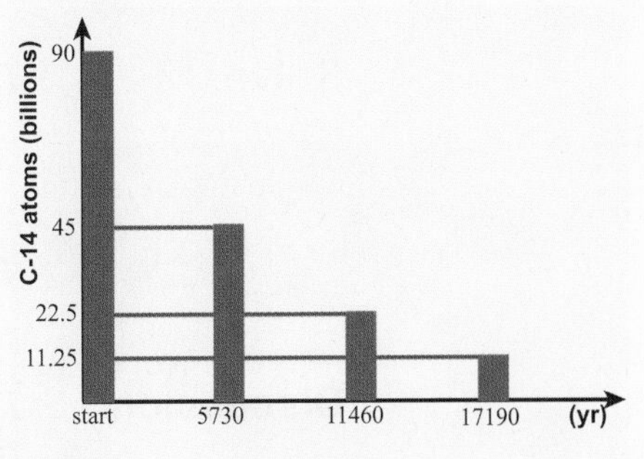

Applications of radioactive dating

ADVANCED AP

Radiometric dating

The technique of dating archeological finds by carbon-14 dating was developed in 1949. Since then, this technique of radiometric dating has been a very important tool for archeology and other sciences.

A radiometric technique is also used by climate scientists that are working to understand the earth's atmosphere over time. Scientists are able, by detecting differences in the concentration of oxygen isotopes (O-18, O-16) that are present in ice cores and ocean sediments, to measure the composition of the atmosphere over time. This gives an indication of climate change over time. This is a very important area of research as we are trying to understand how the release of various gases in the atmosphere could affect the climate of the earth.

Geological dating

Scientists are also able to determine the age of rocks by measuring the ratio of U-238 to Pu-209 that is present in the rock. Since U-238 decays with a half-life of 4.5 billion years, this technique allows geologists to determine the age of rocks that are billions of years old.

The equations: $N = N_0 e^{-\frac{0.693t}{t_{1/2}}}$ and $k = \frac{0.693}{t_{1/2}}$, $N = N_0 e^{-kt}$ form the basis for all calculations used in radiometric dating.

- If we know the half-life, $t_{1/2}$, the initial amount N_0 and the current amount N then we can calculate the time t from the equation: $t = -\ln\frac{N}{N_0}/\text{k}$

An ancient Greek scroll written on an animal skin is discovered by archeologists.
They isolate 10 g of it and measure the C-14 decay rate to be 111 disintegrations/min. Calculate the age of the scroll.
(Living organisms have a C-14 decay rate of 15 disintegrations/min/gr-C)

Given: *The initial decay rate of C-14: N_0=15 dis/min/gr. The present decay rate of C-14 is 11 dis/min/gr. The half-life of C-14 is 5,730 years.*

Solve: *The rate constant k = 0.693/$t_{1/2}$ =0.693/5730 yr =* 1.209×10^{-4}/yr

$$t = \frac{-\ln\frac{N}{N_0}}{k} = \frac{-\ln\frac{111}{150}}{1.209\times10^{-4}} yr = 2{,}491 \text{ years}$$

Discussion: *The animal skin on which the scroll was written is 2,491 years old (in 2008). It was written in about 483 BC.*

20.4 Nuclear Energy

Energy of nuclear reactions

For nuclear reactions the energy released is related to the rearrangement of the atomic nucleus. The energy released by a chemical reaction is related to the rearrangement of the electronic structure of atoms. A chemical explosive such as dynamite results from a chemical reaction and it is millions of times weaker than a nuclear explosion which results from the rearrangement of nuclei.

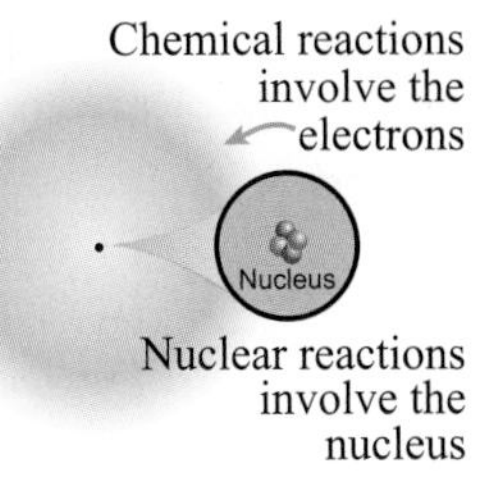

Nuclear energy vs chemical energy

For example, the combustion of 1.0 g of carbon produces about 33 kJ of energy. By comparison, a nuclear reaction of 1.0 g of uranium produces about 8.2×10^7 kJ, more than 2 million times as much.

So where does this large amount of energy come from? We know that energy cannot be created out of nothing. It must be somehow stored in the nucleus. When a nuclear reaction happens the nucleus is rearranged. As we have seen, the number of protons and neutrons in the nucleus changes. A result of this change is the release of energy.

Nuclear size and energy

The energy of the nucleus depends on the number of neutrons and protons that it has. In other words it depends on the atomic number and the mass number. The nucleus with the lowest energy is iron-56 ($^{56}_{26}Fe$). Any nucleus that is larger or smaller than iron contains more energy. **Nuclear energy** is released when the nuclear reaction produces a nucleus that is lower down on the graph of energy.

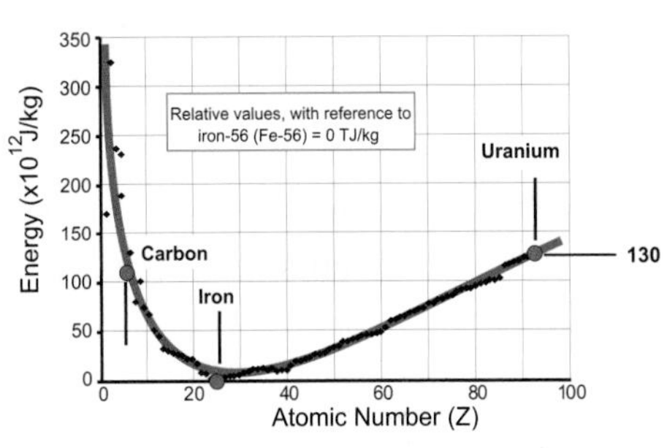

Nuclear force and electromagnetic force

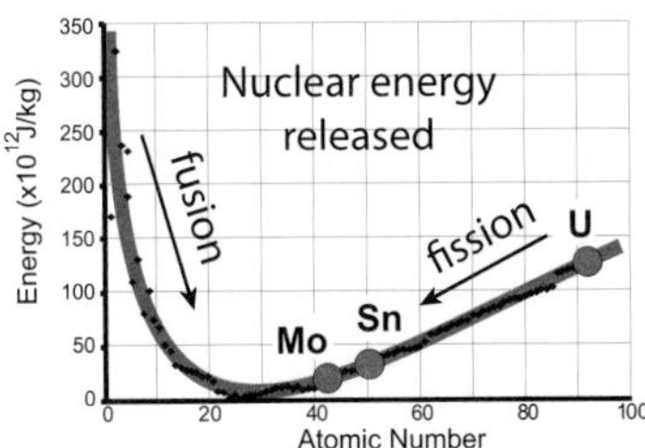

The shape of the energy curve is due to the competition between two forces that act on the protons and neutrons: The electromagnetic force which causes protons to repel and the strong nuclear force which causes protons and neutrons to attract each other. Nuclei that are lighter or heavier than iron have more energy than iron because their nucleons are less tightly bound. You may think of the nuclei as trying to hold themselves together by balancing the electromagnetic force with the nuclear force. When they break up, either by bombardment or decay, they release some of the energy that they were using to hold themselves together.

Chemistry terms

nuclear energy - the energy released during nuclear reactions.

Energy - mass equivalence

Mass and energy are related

The formula $E = mc^2$ is one of the most important formulas in science. It was developed by Albert Einstein in 1908 and it says that energy and mass are related.

Energy (Joules) → $E = mc^2$
Mass (kg)
Speed of light (3×10^8 m/sec)

The equation tells us that for any mass, *m*, there is an associated energy, *E*, or to any energy, *E*, there is an associated mass, *m*.

If we consider the general chemical equation that represents an exothermic reaction

$$Reactants \rightarrow Products + Energy,$$

the amount of energy released must be related to the difference between the mass of the reactants and the mass of the products acoording to Einsteins formula.

$$Energy = (mass_{reactants} - mass_{products})c^2$$

In other words, the change in mass ($\Delta m = mass_{reactants} - mass_{products}$) is related to the change in energy (ΔE) by

$$\Delta E = \Delta m\, c^2$$

As we have seen in Chapter 10, when carbon burns with oxygen it releases energy in the form of heat and light. Each mole of C, or 12 g of C, releases 393.5 kJ of energy.

$$C + O_2 \rightarrow CO_2 + 393.5\ kJ/mole$$

Chemical reactions obey $E=mc^2$

The energy that is released by this chemical reaction is energy that is given off by the carbon and the oxygen. This energy is thus removed from the carbon-oxygen system. The loss of energy (change in energy) means that there should also be a loss of mass (change in mass). The total mass of C and O_2 is less than the mass of CO_2.

According to $\Delta E = \Delta m\, c^2$ the change in mass is

$$\Delta m = \frac{\Delta E}{c^2} = \frac{3.935 \times 10^5 (kg \cdot \cancel{m^2}/\cancel{s^2})}{(3.0 \times 10^8 \cancel{m/s})^2} = 4.37 \times 10^{-12} kg$$

This mass of 4.37×10^{-12} kg or 4.37×10^{-9} g is a very small amount of matter and it can't be detected even by the most accurate balance. However, the energy released by the reaction is very pronounced and it is easily detected.

Mass change in nuclear reactions

The energy that is released during a nuclear reaction is related to a change in mass acoording to Einstein's formula.

We could for example create an oxygen $^{16}_{8}O$ nucleus in two different ways:

1. Combine 8 protons and 8 neutrons.
2. Combine $^{4}_{2}He$ with $^{12}_{6}C$ to obtain $^{16}_{8}O$.

Nuclear reactions change mass

Each one of these ways is a nuclear reaction. Let's calculate the energy that is released in both of these cases. In the first case when we combine 8 neutrons and 8 protons the difference in mass between them and the mass of the $^{16}_{8}O$ nucleus is:

$$8\ ^{1}_{1}H + 8\ ^{1}_{0}n \longrightarrow ^{16}_{8}O + \text{Energy1}$$

$$\Delta m_1 = (\text{mass}_{8\text{neutrons}} + \text{mass}_{8\text{protons}} - \text{mass}\ ^{16}_{8}O)$$
$$= (8\times1.0087 + 8\times1.0073 - 15.999)\text{amu}$$
$$= 0.129\ \text{amu}$$

The molar mass for oxygen, 15.999, includes the mass of the electrons. However, since the mass of the electrons is about 1000 times less than the mass of the protons it is ignored in elementary calculations

Mass is not conserved in nuclear reactions

When we combine combine $^{4}_{2}He$ with $^{12}_{6}C$ the difference in mass is:

$$^{4}_{2}He + ^{12}_{6}C \longrightarrow ^{16}_{8}O + \text{Energy2}$$

$$\Delta m_2 = (\text{mass}\ ^{4}_{2}He + \text{mass}\ ^{12}_{6}C - \text{mass}\ ^{16}_{8}O)$$
$$= (4.0028 + 12.011 - 15.999)\text{amu}$$
$$= 0.015\ \text{amu}$$

$\Delta m_1 = 8.7\ (\Delta m_2)$ and so Energy1 = 8.7× Energy2.

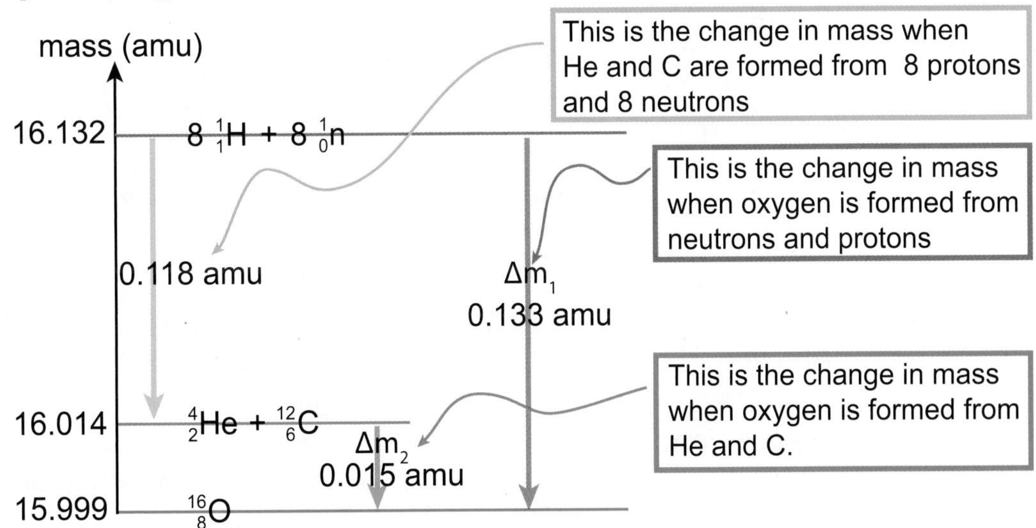
mass (amu)
16.132
$8\ ^{1}_{1}H + 8\ ^{1}_{0}n$
16.014
$^{4}_{2}He + ^{12}_{6}C$
15.999
$^{16}_{8}O$
0.118 amu
Δm_1
0.133 amu
Δm_2
0.015 amu
This is the change in mass when He and C are formed from 8 protons and 8 neutrons
This is the change in mass when oxygen is formed from neutrons and protons
This is the change in mass when oxygen is formed from He and C.

Nuclear energy

Let's look at the nuclear equation that represents the disintegration of radium-226 by alpha particle decay.

$$^{226}_{88}\text{Ra} \rightarrow {}^{222}_{86}\text{Rn} + {}^{4}_{2}\text{He}$$

From the periodic table we see that the masses of the nuclei involved in this reaction are: Ra: 225.9771 amu, Rn: 221.9703 amu, He: 4.0015 amu.

Therefore the change in mass that results from this reaction is

$$\Delta m = (225.9771 - 221.9703 - 4.0015)\ \text{g} = 0.0053\ \text{amu}$$

In grams the change in mass is

$$0.0053\ \cancel{\text{amu}} \times 1.66\times10^{-27}\frac{kg}{\cancel{\text{amu}}} = 8.798\times10^{-30}\text{kg}$$

The energy associated with this change ion mass is

$$E = \Delta m\, c^2 = 8.798\times10^{-30}\ \text{kg}\ (3.0\times10^{8}\text{m/s})^2 = 7.92\times10^{-13}\ \text{Joules}$$

This is the change in energy for each Ra nucleus. To find the energy released for each mole of Ra, we just multiply by Avogadro's number 6.02×10^{23} .

We find that for each mole of Ra the energy released is 4.77×10^{11} Joules.

Given: Calculate the energy released from the nuclear reaction per mole of He-4.

$$^{2}_{1}\text{H} + {}^{3}_{1}\text{He} \longrightarrow {}^{4}_{2}\text{He} + {}^{1}_{0}\text{n}$$

2.0134   3.0149   4.0015   1.0073 amu

Given: *The nuclear reaction and the molar mass of the nuclei involved*

Relationships: $E = \Delta m\, c^2$

Solve: *Calculate the change in mass* Δm

- $\Delta m = (2.0134 + 3.0149 - 4.0015 - 1.0073)\text{amu} = 0.0195\ \text{amu}$
- *Convert amu to grams*
  molar mass in grams is equal to the nulceus mass in amu. Therefore the change in mass for each mole of He-4 is: $\Delta m = 0.0195\ g$
  $0.0195\ g = 1.95\times10^{-5}\ kg$
- $E = \Delta m\, c^2 = 1.95\times10^{-5}\ kg\ (3.00\times10^{8}\ m/s)^2 = 1.76\times10^{12}\ J$

Answer: *The energy released by this reaction is* 1.76×10^{12} *Joules per mole of helium. (4 grams of helium release* 1.76×10^{12} *Joules)*

Discussion: *This is one of the reactions that takes place in the sun. The energy released is in the form of kinetic energy of the nuclei. Since hydrogen is lighter than He-4, it has most of the energy and moves faster.*

Nuclear binding energy

Binding energy

When protons and neutrons are combined to form oxygen the mass changes. The mass of the oxygen nucleus is less than the total mass of the separated protons and neutrons. The energy that is released as a result of this change in mass is called the **binding energy**. The larger the binding energy, the tighter the nucleons are bound together.

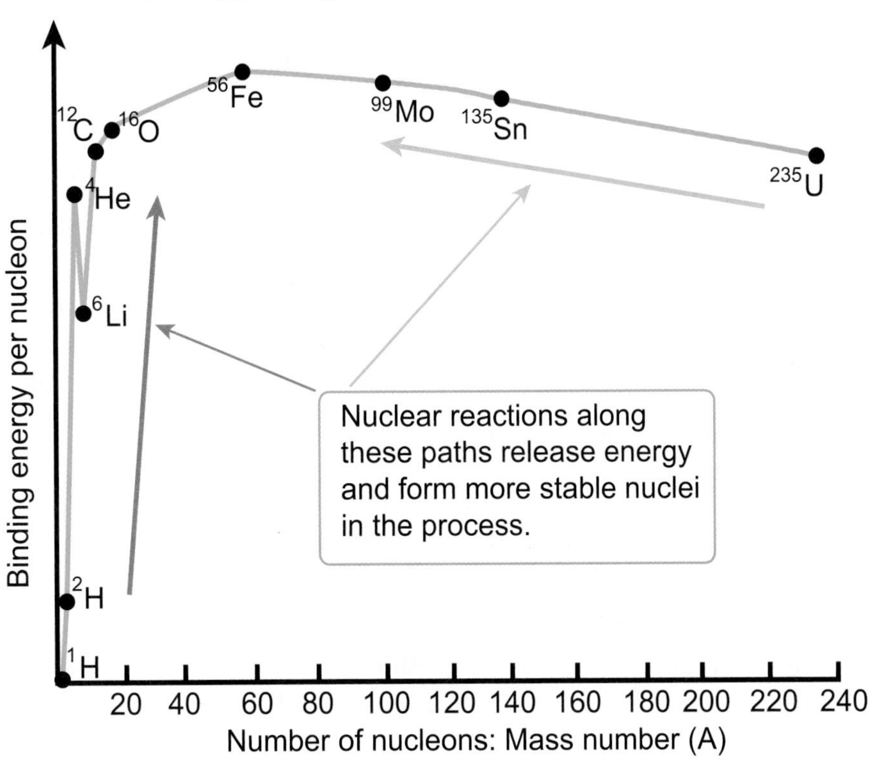

Iron is the most stable element

The nucleons are bound most tightly near iron. The higher the binding energy per nucleon (BE/n), the higher the stability of the nucleus. Iron which has the highest binding energy per nucleon, is the most stable element. This is why iron is such an abundant element.

We can't get any nuclear energy out of iron. This is because the nucleons (protons and neutrons) in iron have the highest binding energy. Any other nucleus is less tightly bound. We can only get energy from nuclei to the left or to the right of iron.

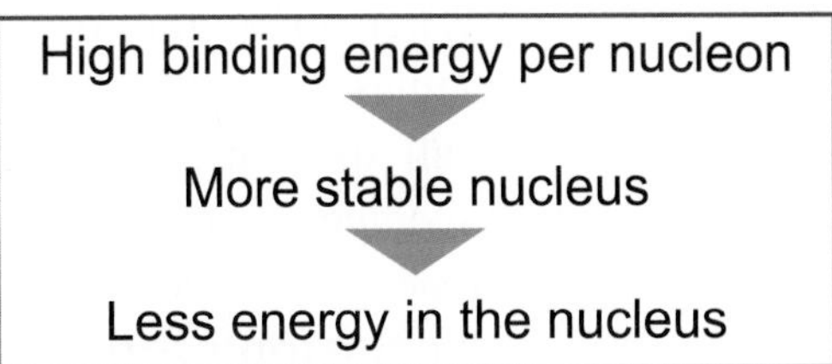

Chemistry terms

binding energy - is the process of chemical change.

Fission reactions

Fission reaction

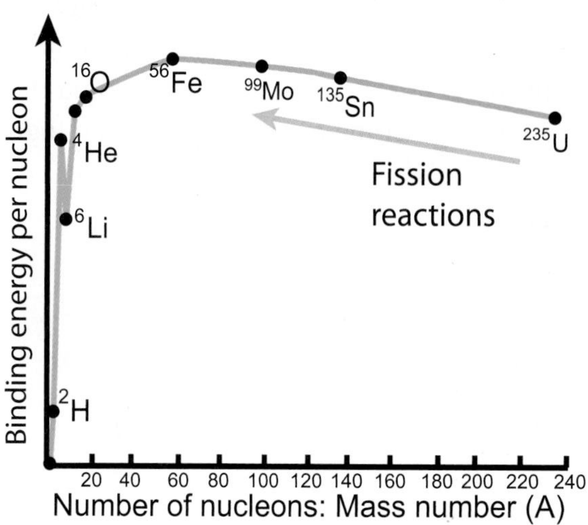

A **fission reaction** splits up a large nucleus into smaller pieces. For elements heavier than iron, breaking the nucleus up into smaller pieces releases nuclear energy. For example, if a uranium nucleus is split up we end up with smaller nuclei.

These smaller nuclei have lower mass numbers and their nucleons are bound tighter than the nucleons in the uranium nucleus. The energy that is released is equal to the difference in the binding energies of the uranium and the products of the reaction.

When ${}^{235}_{92}U$ is bombarded with a neutron, ${}^{1}_{0}n$, it fissions according to:

$$ {}^{1}_{0}n + {}^{235}_{92}U \rightarrow {}^{99}_{42}Mo + {}^{135}_{50}Sn + 2{}^{1}_{0}n + energy $$

This is a bombardment reaction and it happens when a neutron hits the nucleus with enough energy to make the nucleus unstable. The unstable nucleus then breaks up "fissions" into two smaller pieces and releases two neutrons.

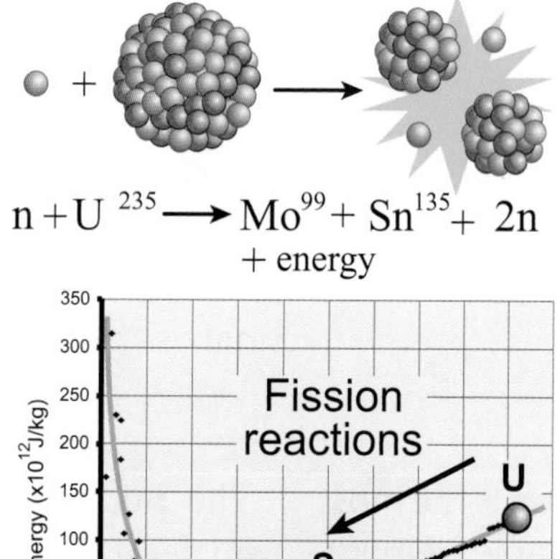

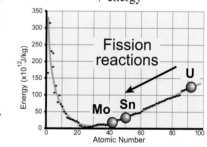

Fission energy

Some of the fission energy released by the reaction appears as gamma rays and some as kinetic energy of the smaller nuclei and the two neutrons.

Fission products

When a kilogram of ${}^{235}_{92}U$ fissions it releases 98 trillion Joules of energy. This amount of energy from a golf-ball-sized piece of uranium is enough to drive an average car 19 million miles! The by-products of this reaction are the ${}^{99}_{42}Mo$ and ${}^{135}_{50}U$ nuclei which are called the **fission products**.

+123 TJ	Energy of uranium (U) nucleus.
-25 TJ	Average energy of nuclei of molybdenum (Mo) and Tin (Sn)
+98 TJ	Energy released by fission of uranium into Mo and Sn.

fission reaction - is the process of chemical change.

fission products - is the process of chemical change.

Energy from fission reactions

Chain reaction

A **chain reaction** occurs when the fission of one nucleus triggers fission of many other nuclei. In a chain reaction, the first fission reaction releases two (or more) neutrons. The two neutrons hit two other nuclei and cause fission reactions that release four neutrons. The four neutrons hit four new nuclei and cause fission reactions that release eight neutrons.

The number of neutrons increases rapidly. The increasing number of neutrons causes more nuclei to have fission reactions and enormous energy is released. This is the basic energy release mechanism that takes place in the cores of nuclear reactors.

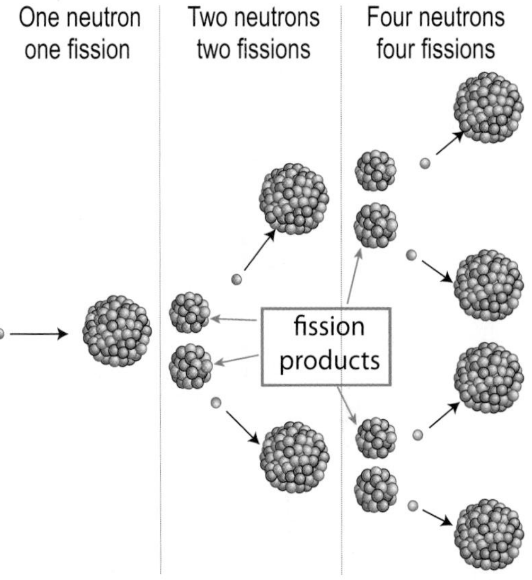

Nuclear power plants

In a **nuclear power plant**, the $^{235}_{92}U$ fission reaction is used to generate electricity. The process of getting electricity from nuclear reactions takes many steps. First, nuclear reactions in uranium produce heat in the reactor core. The heat is carried through high pressure hot water into the steam generator. Heat in the steam generator boils water and makes steam. The steam turns a turbine. In the last step, the turbine is connected to an electric generator that makes electricity. The steam is condensed and pumped back to the steam generator.

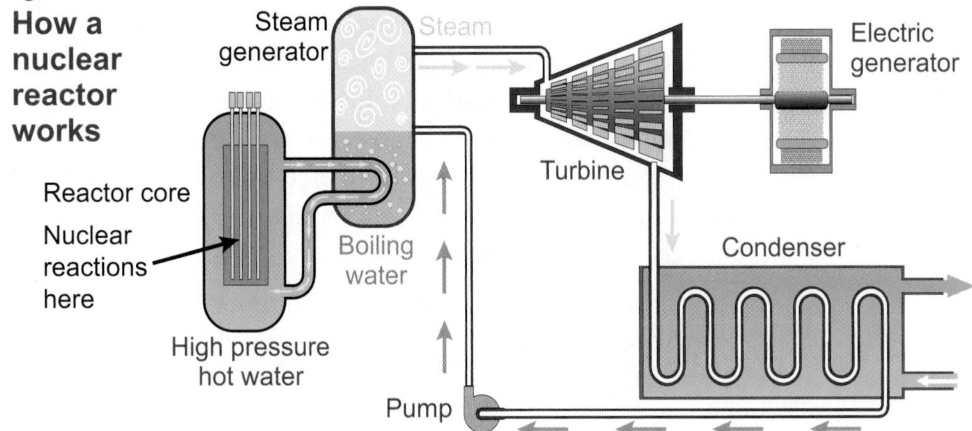

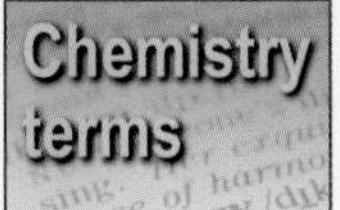

chain reaction - is the process of chemical change.
nuclear power plant - is the process of chemical change.

Fusion reactions

Fusion reactions

Fusion energy from the sun

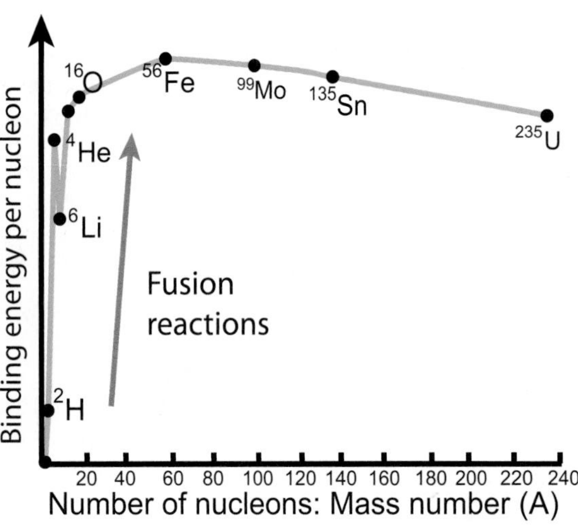

A **fusion reaction** is a nuclear reaction that combines, or "fuses", two smaller nuclei into a larger nucleus. Fusion reactions can release a very large amount of energy. The sun and other stars make their energy from fusion reactions. The temperature at the core of the sun is about 15 million Celsius. Its density is so high that a tablespoon of the sun weighs more than a ton. The primary fusion reaction that happens in the sun combines hydrogen nuclei to make helium, converting two protons and two electrons into two neutrons along the way. All of the energy reaching the Earth from the sun comes from fusion reactions in the sun's core.

Fusion reactions release a lot of energy

For example, two carbon-12 (C-12) nuclei can fuse together to create a magnesium-24 (Mg-24) nucleus. The neucleons in the Mg nucleus are bound tighter than the nucleons in C-12. The Mg-24 nucleus has a lower energy. If the nuclei in a kilogram of carbon are combined to form magnesium, about 56 trillion joules of energy is released.

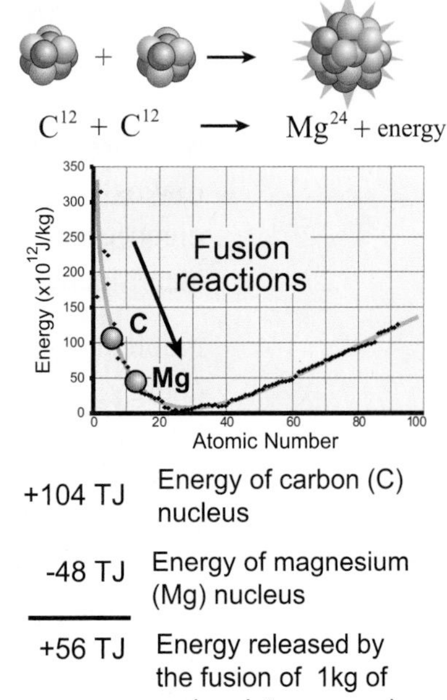

Fusion reactions need very high temperatures

It is difficult to make fusion reactions occur because positively charged nuclei repel each other. The attraction from the strong nuclear force has a very short reach. Two nuclei must get very close for the attractive strong force to overcome the repulsive electric force. If the temperature is high enough, nuclei slam together with enough energy to almost touch, allowing the strong force to take over and cause a fusion reaction. Temperatures greater than 10 million degrees Celsius are required for nuclei to get close enough for fusion reactions.

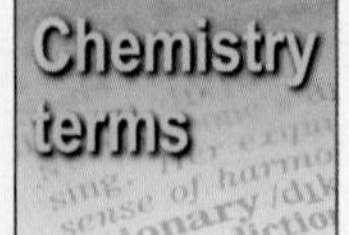

fusion reaction - is the process of chemical change.

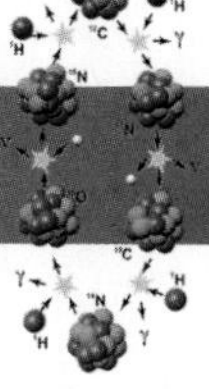

Energy from fusion reactions

Using nuclear energy

Today about 20% of electric power in the USA comes from nuclear fission plants. In some other countries like France, 70% of their electricity comes from nuclear fission power. Nuclear fission is an energy source that does not use any fossil fuels and thus it does not generate carbon dioxide or other air polutants. However, it generates waste which is highly radioactive.

Issues with nuclear energy

Even though the volume of waste generated by nuclear reactos is small, its radiation intensity is very high making it extremely dangerous. In addition, the half-life of this radiatioactive waste is very long, making it very difficult to safely store for a long time.

Fusion: The future of nuclear energy

In the future it might be possible to use the energy from fusion reactions to generate electricity. Fusion reactions have all the benefits of fission and in addition generate a small amount of waste. It is a very challenging thing to do since we have to create the conditions in the sun in order to make fusion reactions possible.

Fusion fuel comes from sea water

The basic fuel for fusion energy are hydrogen isotopes. The nuclear reaction between Deuterium, ${}^{2}_{1}H$, which is an isotope of hydrogen with one neutron and tritium, ${}^{3}_{1}H$, which is an isotope of hydrogen with 2 neutrons is the basic fusion reaction. Deuterium is found naturally in sea water.

$$ {}^{2}_{1}H + {}^{3}_{1}H \rightarrow {}^{4}_{2}He + {}^{1}_{0}n + energy $$

Notice that the only by-products of this reaction are heluim and a neutron. The reaction uses tritium which is a highly radioactive but short lived isotope ($t_{1/2}$=12.3 yr). Tritium does not occur naturaly but it can be produced from lithium which is a very common element on earth.

Nuclear fusion is in many ways an ideal energy source since it uses inexpensive, abundant fuel and it generates little harmful waste. However, it is very difficult to build a fusion reactor capable of generating electricity. One difficulty is heating the deuterium and tritium fuel to more than 50 million °C, about five times hotter than the core of the sun. Fusion fuel, in the form of a hot plasma, must be contained in a magnetic force field and kept isolated from solid materials. The tokamak reactor is the most successful experimental fusion reactor yet constructed so far.

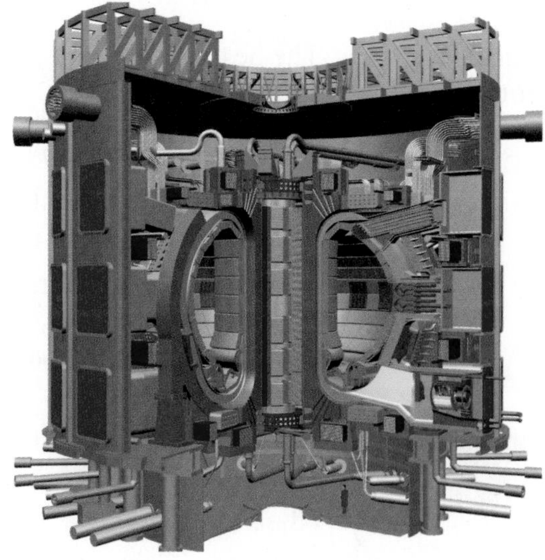

20.5 Biological Effects of Radiation

Radioactivity is a dangerous natural phenomenon

Radioactivity is a natural phenomenon which can be both dangerous and useful. Exposure to a large amount of radiation can damage biological tissues. The damage could range from genetic defects, to increased cancer risk, to death. The use of controlled radiation sources can be a very useful tool for the diagnosis and treatment of many diseases.

Non-ionizing radiation
Energy is absorbed by electrons.

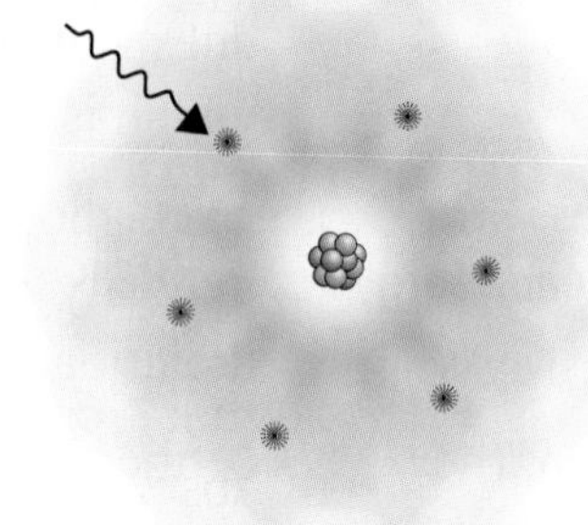

Ionization

Ionizing radiaiton

Non-ionizing radiation

The process by which radiation affects biological tissues is based on the principle of **ionization** of atoms. An atom is ionized when electrons are knocked out of their orbits. The energy that is required to remove electrons from their orbits comes from radiation. When the radiation energy is high enough to knock electrons out of their obrits it is called **ionizing radiation**. When the energy is not high enough it is called **non-ionizing radiation**.

Ionizing radiation
Energy is enough to knock electrons out of the atom.

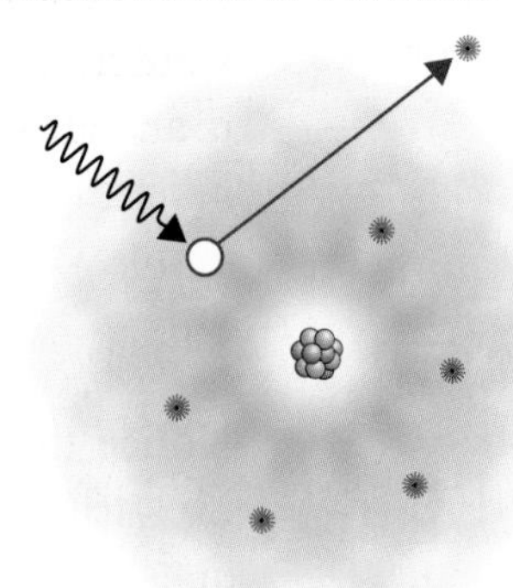

Chemical bonds and radiation

Chemical bonds, which form by sharing or exchanging electrons, are affected when electrons are knocked out of their orbits. When the chemical bonds of biological molecules such as DNA break, the biological function of the molecule is affected. The human body and other biological systems have evolved to deal with low-level damage caused by the naturally occuring low-level radiation.

Radiation protection: —
—Shield
—Distance
—Time

The best ways to protect ourselves when working with radioactive materials are:

1. Use a shield such as lead to block the radiation. The radiation energy is absorbed by the shield and not by our body.
2. Limit the amount of time that we are close to a radiation source by working efficiently and quickly
3. Keep as large of a distance as we can from the radiation source. Increasing the distance by 3 decreases the exposure by 89% (the inverse square law). $\left(1-\frac{1}{3^2}\right)\cdot 100 = 89$

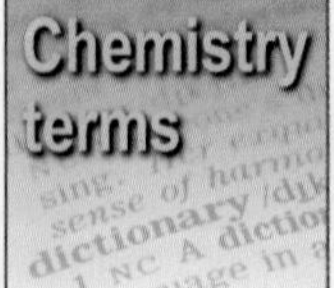

ionization - the removal of outer electrons from their orbits.

ionizing radiation - radiation with high enough energy to remove electrons from their orbits.

non-ionizing radiation - radiation that cannot cause ionization of atoms.

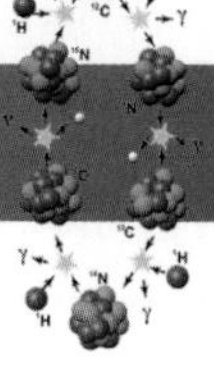

Radiation units - Dose

Radiation dose

A **dose** of radiation refers to the amount of radiation absorbed by the body. The units for measuring radiation dose are the **rad** and the **rem**.

rad

The *rad* corresponds to 0.01 J of energy absorbed by 1 kg of body tissue.

$$1\ rad = 0.01\ \text{J/kg body tissue}$$

rem

The *rem* measures the effectiveness of radiation in causing biological damage. For example, alpha radiation has a much higher biological effectiveness compared to gamma radiation. The reason for this is that alpha radiation is absorbed over a very short distance. This results in a very concentrated energy deposition which in turn gives a large number of ionizations. A large number of ionizations gives a large number of damaged tissue molecules. By comparison, gamma radiation of the same initial energy travels much further before it is absorbed by the tissue. This means that the energy of the radiaiotn is spread over a longer path. The longer path means lower energy density, fewer ionizations and fewer damaged molecules.

Average annual radiation exposure mrem/yr	
Radiation source	**Dose**
Ground	30
Inhaled source	200
Medical	50
Consumer products	10
Body	35
Space	25
Total	**350**

Alpha radiation is 20times more damaging than β or γ radiation

Alpha radiation, compared to β and γ radiation, is 20 times more effective in causing biological damage. The rem is obtained by multiplying the dose in rad by a number that is related to effectivenss of radiation in causing biological damage.

The average annual radiation exposure for a person is about 350 mrem or 0.35 rem. Most of this radiation comes from natural sources. People are exposed to different amounts of radiation depending on their occupation and where they live. For example, a pilot that flies at 10,000 meters for 10 hours is exposed to the same amount of radiation as a person on the ground would be exposed to in 200 days.

People that work around radiation sources carry special monitors that measure their exposure to radiation. These monitors are checked regularly to make certain that the total amount of radiation exposure does not exceed a maximum allowable limit. In general, the limit The limit for all occupations is that the total radiation dose should be below 5 rem/year.

ionization - the removal of outer electrons from their orbits.

rad - radiation with high enough energy to remove electrons from their orbits.

rem - radiation that can't cause ionization of atoms.

Nuclear medicine. Positron Emission Tomography

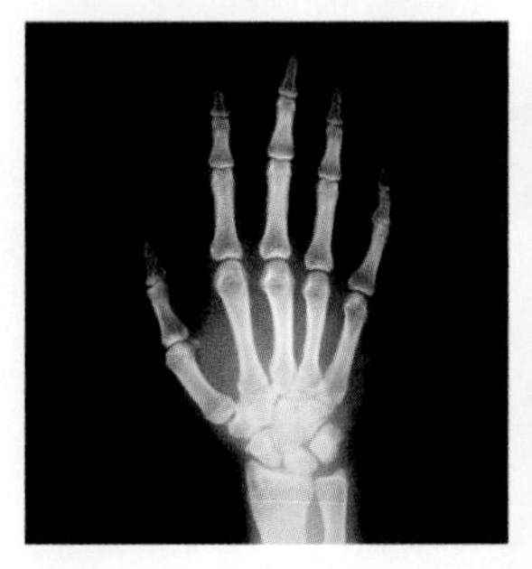

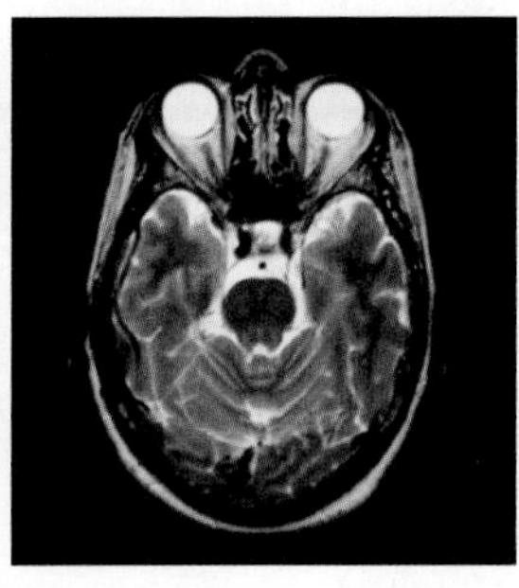

Some of the greatest contributions of nuclear technology is in the area of medical diagnosis and treatment. Nuclear scientists have developed a number of techniques that give medical doctors a way to look inside the human body and diagnose a number of diseases; from cancer tumors to broken bones. Recently nuclear techniques are also used in mapping the functioning of the brain and other organs as we try to understand conditions such as Alzheimer, Parkinson and vascular diseases.

In general, the use of radioactivity in medical diagnosis can be grouped in two categories:

1. *Detect the change in absorption* of radioactivity as it goes through the body;
2. *Map the location* inside the body where radioactivity was released.

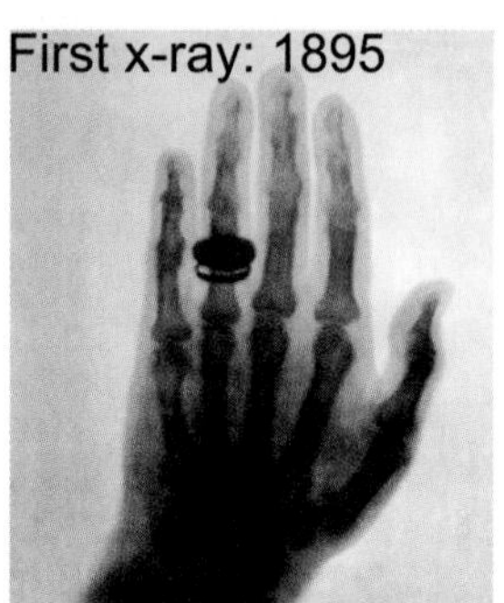

The traditional x-ray works by detecting the change in the absorption of the radioactivity as it passes through various parts of the body. Dense tissues and bone attenuate x-rays more than soft tissues. This is a great technique and it has been used since its discovery in 1895 by professor Wilhelm Roentgen.

Modern techniques such as computed tomography (CT), and positron emission tomography (PET) are at the forefront of medical diagnosis. CT is based on the principle of radiation absorption and has the ability to create three dimensional images which are very useful in detecting structural abnormalities in the body.

PET technology was first demonstrated in 1976 and since then it has become one of the most important tumor diagnostic tools. Early and precise diagnosis of cancerous tumors is the best first step in treating such conditions.

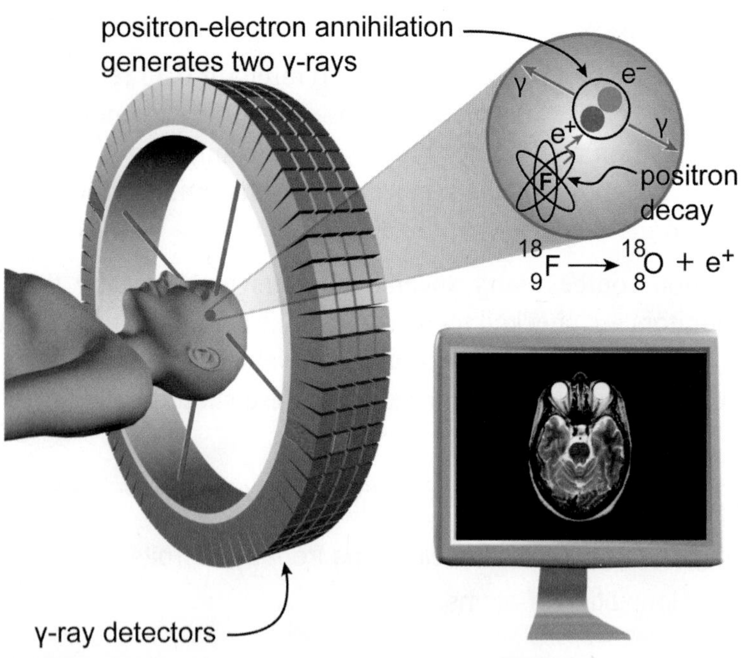

PET is based on the decay of certain nuclei by positron emission. As we have learned, a positron is a nuclear particle that has the same mass as an electron but a positive charge. Positron is the antimatter counterpart of an electron.

When a positron and an electron come in contact they immediately disappear or annihilate. Since energy and mass are related by Einstein's equation, the mass of the electron and the positron is converted into energy. This energy appears as two γ-ray photons traveling in opposite directions.

The PET imaging method is based on the detection of these γ-ray photons and their precise mapping in the body.

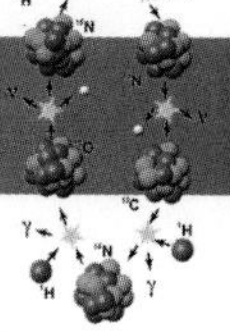

Scientists had to overcome many challenges in order to develop the PET technology.

- First, they had to understand the details of nuclear decay and position emission.
- Then they had to understand the interaction between the positron and the electron.
- Once they knew the science, they had to deal with major mathematical and technological challenges such as:
    - Develop the complicated mathematics that is required in order to be able to create an image from the detection of the γ-rays.
    - Create the nuclear isotopes that have the appropriate half-life.
    - Construct the machine to perform all of these operations.

To achieve these goals many chemists, physicists and engineers worked for many years.

PET uses the radioisotope fluorine-18, ${}^{18}_{9}F$ which has a half life of 110 minutes. Since the half life of 110 minutes is relatively short, ${}^{18}_{9}F$ does not exist naturally and so it has to be produced.

The production is done by hitting a target oxygen-18 nucleus, ${}^{18}_{8}O$, with a proton which is obtained in an accelerator. The complete nuclear reaction is:

$${}^{1}_{1}H + {}^{18}_{8}O \rightarrow {}^{18}_{9}F + {}^{1}_{0}n$$

The relatively short half-life of the ${}^{18}_{9}F$ isotope means that it has to be produced close to the hospital where it is used.

Once the ${}^{18}_{9}F$ isotope is produced, it is then attached to a glucose molecule and injected into the blood stream. The glucose molecule and the attached ${}^{18}_{9}F$ isotope, is taken up by cells and it concentrates in cells with higher activity such as the brain and tumor cells.

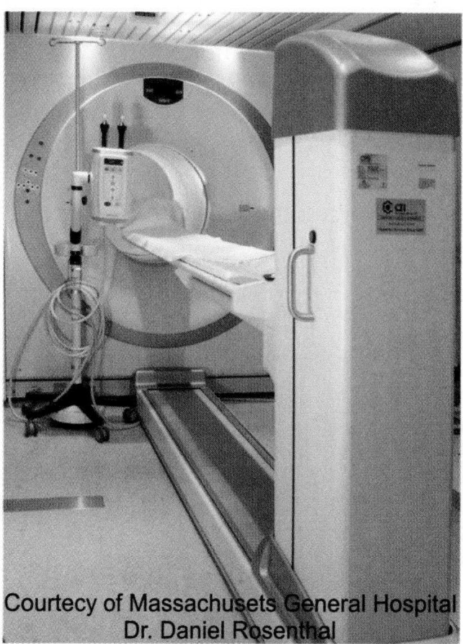
Courtecy of Massachusets General Hospital
Dr. Daniel Rosenthal

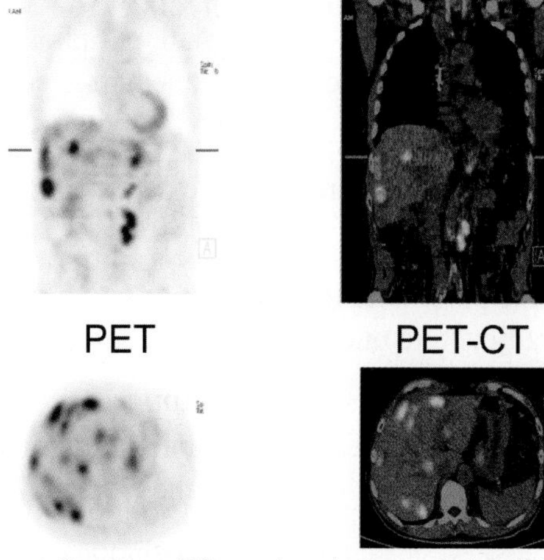

Courtecy of Massachusets General Hospital
Dr. Daniel Rosenthal

Pet images have a very high resolution and so doctors are able to see details in the range of mm.

The combination of PET and CT technologies gives doctors the ability to detect and observe even finer details at every part of the body as shown on the images on the left.

These nuclear technologies help save lives every day.

Chapter 20 Review

Vocabulary

Match each word to the sentence where it best fits.

Section 20.1

atomic number	mass number balance
neutron number	nuclear charge
mass number	

1. The number of protons in a nucleus is given by the ________________.
2. The ________________ gives the number of neutrons in a nucleus.
3. The total nuimber of protons and neutrons in a nucleus is called the ________________.
4. Balanced nuclear equations must conserve __________.
5. There must be______________________ in balanced nuclear equations.

Section 20.2

nuclear reactions	daughter nuclide
radioactivity	beta decay
radiation	beta radiation
intensity	gamma decay
inverse square law	positron
alpha decay	positron emission
parent nuclide	

6. ______________ can change one element into a different element.
7. ______________ descibes the property of some elements to break up and release energy.
8. The transmission of energy through space is called ______________.
9. ____________________ measures how much energy flows per unit area per second.
10. The decrease of radiation intensity with distance is given by the ________________________.
11. During ________________ the nucleus releases a helium nucleus.
12. The original nucleus during a nuclear reaction is called the ______________________.
13. The Th-234 nucleus resulting during the alpha decay of U-238 is called the ________________.
14. When an unstable nucleus emmits an electron we say that it undergoes ________________.
15. __________________is released during beta decay.
16. When a nucleus undergoes ________________ it releases electromagnetic radiation.
17. The nuclear particle that has the same mass as the electron and a positive charge is called________________.
18. When a nucleus decays by the emmission of a positron it undergoes ________________decay.

Section 20.3

carbon-14	carbon dating
half life	rate constant
rate of decay	activity

19. ________________ is an isotope of carbon.
20. The time that it takes for a pure radioactive substance to decrease by one half is called the ________________.
21. The number of radioactive decays per unit time is called the ________________.
22. The parameter that describes how fast a radioactive nuclide decays is called ________________.
23. The rate of decay of a radioactive element is called the ______________ of the element.
24. ________________ is a techique used to determine the age of organic materials.

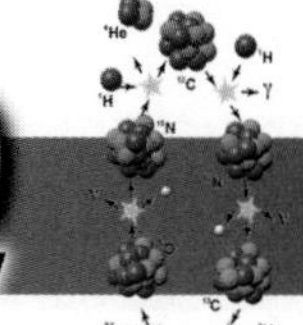

Section 20.4

nuclear energy	fission reaction
mass-energy equivalence	fission products
mass change	chain reaction
binding energy	nuclear power plant
	fusion reaction

25. Nuclear reactions release _______________.
26. Einstein's formula, $E = mc^2$, is an expression of _________________________.
27. The energy that is released during nuclear reactions results from the _______________ between the reactants and the products of the reaction.
28. The energy that is released as a result of the change in mass during a nuclear reaction is called ____________________.
29. A ____________________ splits up a large nucleus into smaller pieces.
30. The byproducts of a fission reaction are called ______________________.
31. A nuclear ____________________ occurs when one nucleus triggers many other fission reactions.
32. The energy generated by a _______________ comes from fission reactions.
33. the energy of the sun comes from ____________________.

Section 20.5

ionization	dose
ionizing radiation	rad
non-ionizing	rem

34. _______________ happens when radiation ejects electrons out of their orbits.
35. Radiation that induces ionization is called ______________________.
36. __________________ radiation does not have enough energy to induce ionization.
37. the amount of raditaion absorbed by our bodty is called the ____________.
38. The _______ is a unit of measuring radiation and corresponds to 0.01 J of energy basorbed by 1 kg of body mass.
39. The ________ is a dose unit that measures the effectiveness of radiaiton to cause biological damage.

Conceptual Questions

Section 20.1

40. Describe the differences between chemical and nuclear reactions.
41. Describe the procedure for balancing nuclear equations.
42. How do we describe atomic nuclei? How do we represent the atomic nucleus?

Section 20.2

43. What are the two major types of nuclear reactions?
44. Give four examples of decay reactions.
45. Which nuclear decay reactions release energy in the form of particles?
46. Which nuclear decay reactions release energy in the form of electromagnetic radiation.
47. Why do nuclear reactions release much more energy than chemical reactions?
48. How does the intensity of radiation decrease with distance from the source?
49. Describe the general nuclear equation that represents alpha decay.
50. Describe how the energy released during alpha decay is distributed among the various particles involved in the reaction.

Chapter 20 Review.

51. What are the differences and similarities between alpha decay and beta decay?
52. Describe what happens to a nucleus during gamma decay.
53. Compare and contrast beta decay and decay via positron emission.

Section 20.3

54. Describe the concept of half life. Explain what happens to a radioactive sample after one half-life.
55. Why is carbon-14 useful for dating archeological finds?
56. What is rate of decay and how is it related to activity?

Section 20.4

57. Give a general explanation for the source of nuclear energy.
58. Explain each symbol in Einstein's equation and describe how it is used in making nuclear energy calculations.
59. What is binding energy and how is it related to the energy that is released in a nuclear reaction?
60. Why do fission reactions release energy?
61. What is the form of the nergy released during fission reactions?
62. Describe the operation of a nuclear power plant.
63. What the differences and similarities between a fission reaction and a fusion reaction?
64. Why is it difficult to obtain fusion reactions?

Section 20.5

65. Why is radiation harmfull to biological systems?
66. What happens when radiation interacts with
67. How do we protect ourselves when working with radioactive materials?
68. How do we msure the amount of radiation absorbed by our body?

Quantitative Problems

Section 20.1

69. Balance the following nuclear equations
    a. $^{238}_{92}U \rightarrow \quad + ^{4}_{2}He$
    b. $\rightarrow ^{217}_{85}At + ^{4}_{2}He$

Section 20.2

70. A radiation source releases 1000 W from a point. What is the intensity of radiation 4 meters from the point and 8 meters from the point?
71. A person stands at a distance of 2 meters from a source of radiation that releases 100 J per second. What is the intensity of radiation experiemced by the person?
72. Write the complete nuclear equation for the alpha decay of:
    a. $^{237}_{93}Np$
    b. $^{230}_{90}Th$
    c. $^{210}_{84}Po$
    d. $^{220}_{86}Rn$
73. Complete the following nuclear equations:
    a. $^{241}_{95}Am \rightarrow ^{237}_{93}Np +$
    b. $^{19}_{11}Na \rightarrow ^{19}_{10}Ne +$
    c. $^{15}_{8}O \rightarrow \quad + ^{0}_{+1}e$
    d. $^{15}_{8}O \rightarrow \quad + ^{0}_{+1}e$

Section 20.3

74. Iodine-131 decays with beta decay and has a half life of 8.02 days.
    a. If we start out with a sample that contains 10 g, how many grams are left after 16.04 days?
    b. how many beta decays happen in 8.02 days?
75. Potassium-40 has a rate constant $k=5.3\times10^{-10}$ y. Calculate the half-life.

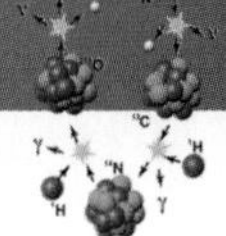

76. If we have a sample of potassium-40 that contains 10×10^{10} atoms, how many of these nuclei will decay in one second?

77. What percentage of C-14 sample, that has a half life of 5730 years, remains after 3000 years? What is the percentage after 20000 years?

78. The half life of tritium is 12.3 years, what is the rate constant?

79. Krypton-85 (Kr-85) decays with a half life of 10.8 years. If we start out with 10 mg of Kr-85 how much would remain after 20 years?

80. A sample that contains Fe-59 is measured with a radiation counter to give about 11 decays per minute. After 20 days the same sample is measured to give about 8 decays per minute. Calculate the half life of Fe-59 in days.

81. An archeologist finds an ancient bone. After cleaning the bone she isolates 5 g of it and measures the C-14 decay rate to be about 22 dissintegrations per minute. What is the age of the bone? (Living organisms have C-14 decay rate of about 15 dissintegrations/minute/g of carbon. You may assume that 80% of the bone is made of carbon)

Section 20.4

82. By using Einstein's equation calculate how much energy would be released if we convert 1 g of mass into energy.

83. How much energy in joules would be released if we convert
    a. the mass of ome electron into energy?
    b. the mass of one proton into energy

84. Every second the sun releases about 4×10^{-10} joules of energy. How much mass is lost each second?

85. Calculate the nuclear binding energy per nucleon for the following nuclei.
    a. ${}^{2}_{1}H$ (atomic mass = 2.0135 amu)
    b. ${}^{7}_{3}Li$ (atomic mass = 7.0160 amu)
    c. ${}^{16}_{8}O$ (atomic mass = 15.9949 amu)
    d. ${}^{58}_{28}Ni$ (atomic mass = 57.9353 amu)
    e. ${}^{129}_{54}Xe$ (atomic mass = 128.9048 amu)
    f. ${}^{232}_{90}Th$ (atomic mass = 128.9048 amu)

    Which of these nuclei is the most stable?

86. Calculate the energy released when one gram of deuterium, ${}^{2}_{1}H$, reacts with one gram of tritium, ${}^{3}_{1}H$, according to the reaction. ${}^{3}_{1}H + {}^{2}_{1}H \rightarrow {}^{4}_{2}He + {}^{1}_{0}n$
The atomic masses of the reactants and products are:
${}^{2}_{1}H$=2.0134 amu, ${}^{3}_{1}H$=3.0149 amu,
${}^{4}_{2}He$=4.0015 amu, ${}^{1}_{0}n$=1.0073 amu

87. Calculate the energy released for each ${}^{4}_{2}He$ nuncleon produced in the reaction:
${}^{6}_{3}Li + {}^{1}_{0}n \rightarrow {}^{4}_{2}He + {}^{3}_{1}H$
The masses of the reactants and products are:
${}^{6}_{3}Li$=6.0151 amu, ${}^{3}_{1}H$=3.0149 amu,
${}^{4}_{2}He$=4.0015 amu, ${}^{1}_{0}n$=1.0073 amu

88. The follwoing reaction takes place inside the core of a nuclear power plant.
${}^{1}_{0}n + {}^{235}_{92}U \rightarrow {}^{144}_{54}Xe + {}^{90}_{38}Sr + {}^{1}_{0}n$
How much energy is released for each gram of uranium burned?
Atomic masses are: U-235(235.0439 amu), Xe-144 (143.9385 amu), Sr-90 (80.9077 amu), neutron (1.0073 amu)

Section 20.5

89. A one kg sample exposed to a sourfe emitting beta radiation absorbs 0.2 J of energy. What is the dose in rads?

90. A 70 kg person is exposed to 10 rad of radiation. How much energy in joules is absorbed by the person's body?

91. A laboratory mouse weighing 35 g is exposed to 10 rad of radiation. How much energy in joules is absorbed by the mouse's body?

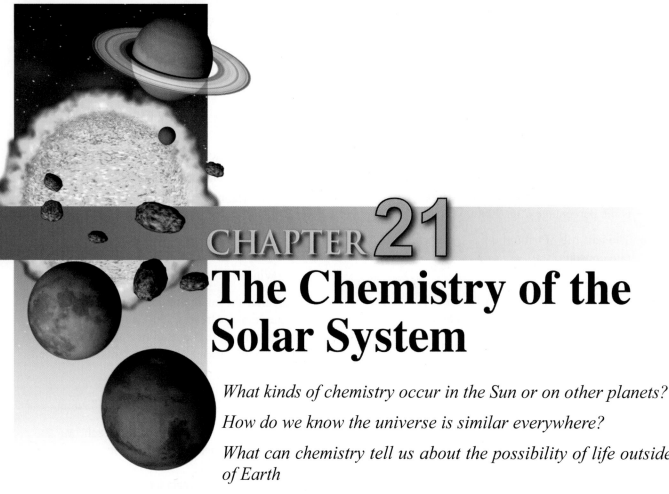

CHAPTER 21

The Chemistry of the Solar System

What kinds of chemistry occur in the Sun or on other planets?

How do we know the universe is similar everywhere?

What can chemistry tell us about the possibility of life outside of Earth

What is the Sun? Is it literally a ball of fire as believed by many ancient peoples? On earth, fire is a chemical reaction: combustion in the presence of oxygen. The Sun is something totally different and flammable compounds like gasoline or methane do not exist there. However, there are *oceans* of liquid natural gas on Titan, Saturn's largest moon.

The planet Venus is named for the goddess of love because on a clear night Venus is second only to the moon as the brightest object in the evening sky. Not long ago scientists imagined Venus might be like Earth. On Earth, we breath an atmosphere of 78% nitrogen and 21% oxygen. However, one deep breath of the atmosphere on Venus would be instantly fatal. The atmosphere there is mostly carbon dioxide but with clouds of hydrochloric acid (HCl), hydrofluoric acid (HF) and sulfuric acid (H_2SO_4)!

Earth is just right for living things. People have long wondered if other planets have life. Mars and Europa (a moon of Jupiter) are good candidates for having extraterrestrial life, but are only just candidates. Space probes have explored only a tiny fraction of Mars looking. The small amount of evidence collected gives no definite answers. The most common type of life on Earth is bacteria. If microscopic life existed on Mars, how would we know? This Chapter will pose many such questions, and many of them have no answers yet. Our human curiosity about the unknown is part of what makes science so intriguing.

Eye in the sky

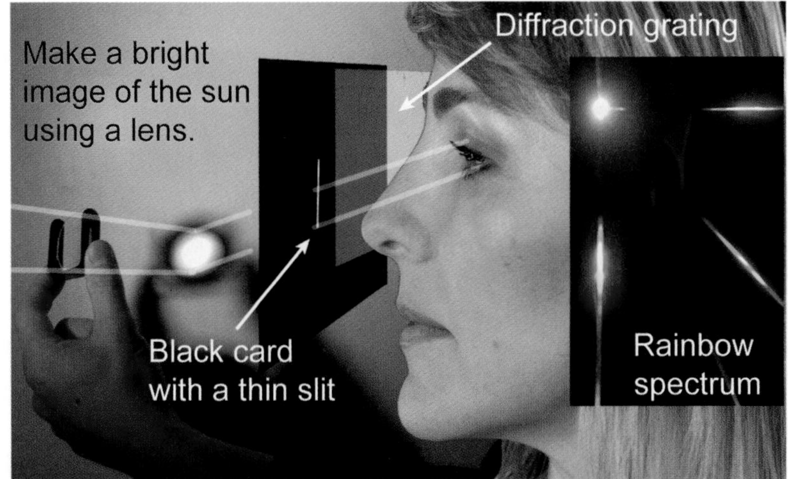

What is the Sun? You should NEVER look directly at the sun, but you can look at the image of the sun! On a sunny day you can observe the sun by making an image with a lens on a white surface. The light appears bright white. Look at sunlight from the image through a diffraction grating. A diffraction grating shows a rainbow spectrum, which tells you that sunlight is made of many colors. However, a diffraction grating or even a classroom spectrometer does not have the ability to show fine detail in a spectrum.

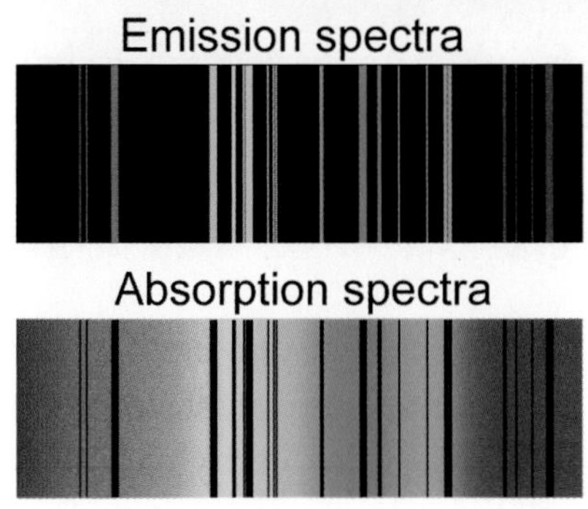

Atoms emit light in spectral lines but they also absorb light in discrete spectral lines. Absorption lines are dark against the rainbow of the visible spectrum.

NASA's high resolution solar spectrometer breaks each part of the solar spectrum into a strip of color. The diagram below shows an actual solar spectrum broken up into 47 bands. The many dark lines are absorption lines from different elements in the surface layers of the Sun.

In fact, on element (helium) was first discovered not on Earth but in the sun! In 1868, Sir Norman Lockyer observed yellow lines in the solar spectrum which had not been seen in any laboratories on Earth. Sir Lockyer correctly deduced that the mystery lines must come from an unknown element he named *helium*, from the Greek helios (ήλιος) (sun). Helium wasn't proven to exist on Earth for 25 more years.

Spectroscopy and telescopes together allow us to identify elements and compounds that exist outside of our Earth-bound laboratories.

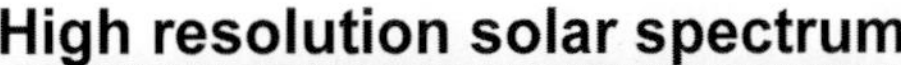

High resolution solar spectrum

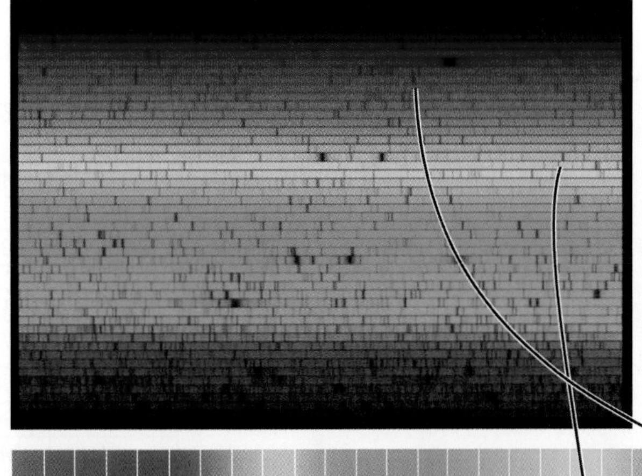

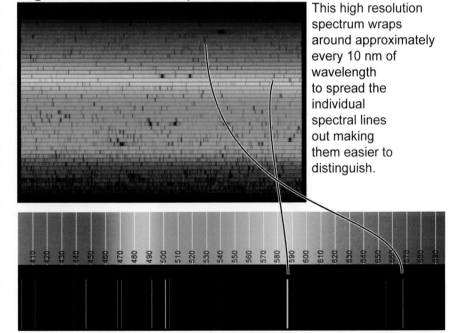

This high resolution spectrum wraps around approximately every 10 nm of wavelength to spread the individual spectral lines out making them easier to distinguish.

21.1 The Sun and the Stars

What is in the Universe?

Scientists know a lot about what we can see with telescopes. However, there are vast mysteries regarding what we cannot see, but know is out there. For example, all matter attracts other matter through gravity. When we look out at the universe we can estimate the amount of matter needed to explain the apparent strength of gravity. However, we can "see" only 15% of the matter! The remaining 85% is "dark matter" which is another way of saying we don't know what it is. There are many theories and little proof so dark matter is a mystery.

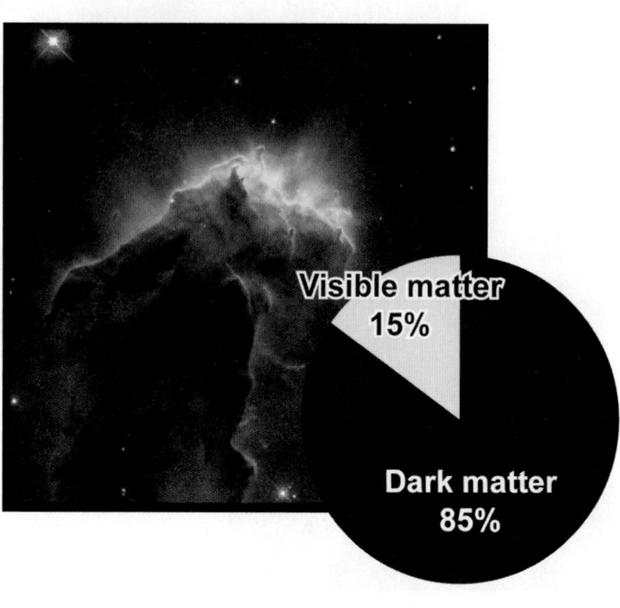

Hydrogen and helium are the most common elements

Of the matter we can see, Hydrogen is 75% of the matter in the known universe, by far the most common element. Helium is 23% of the rest, and the second most common element. The third most common element is oxygen, at about 1% followed by carbon at 0.5% and heavier elements. The chart below gives the relative abundance of the first dozen most common elements.

Elemental composition of the visible universe

Atomic #	Element	Parts per million
1	Hydrogen	750,000
2	Helium	230,000
8	Oxygen	10,000
6	Carbon	5,000
10	Neon	1,300
26	Iron	1,100
7	Nitrogen	1,000
14	Silicon	700
12	Magnesium	600
16	Sulfur	500

Planets concentrate the heavier elements

The distribution of elements in the universe outside Earth is very different from our rocky, water covered planet. This is because planets constitute only a tiny fraction of the mass of the solar system; the Sun has more than a million times the mass of Earth. Hydrogen and helium make up only a tiny fraction of the Earth's total mass.

The sun

Composition of the Sun

The solar system is the empire of the Sun. Our sun contains more than 99.8% of the total mass of the Solar System (Jupiter contains most of the rest). Today, the mixture of elements in the sun is 74% hydrogen and 24% helium by mass. Everything else amounts to less than 2%. The chemical makeup of the Sun changes slowly over time as the Sun converts hydrogen to helium in its core. This conversion has been going on for about 4.5 billion years and will continue for another 4.5 billion years.

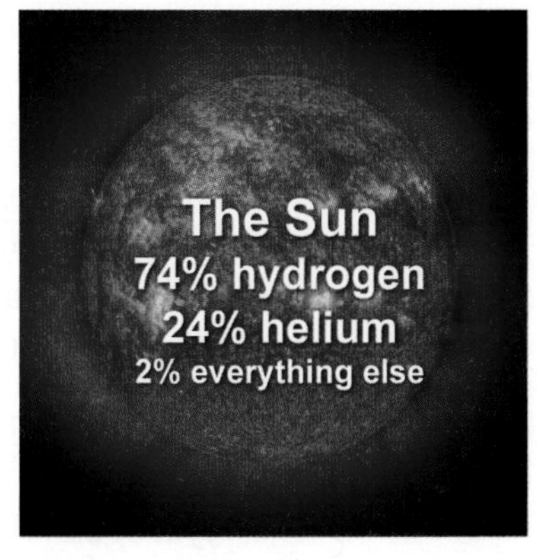

Only pure elements exist in the Sun

On earth, most elements are combined in compounds. That is not true in the Sun. Most elements in the Sun exist in their pure form and *no compounds exist at all*. The reason is the extreme high temperature and in the interior of the sun, extraordinary pressure as well. In the Sun's core, scientists estimate the temperature is 15.6 million Kelvin. The thermal energy is so high that chemical bonds cannot exist.

The Sun's internal structure

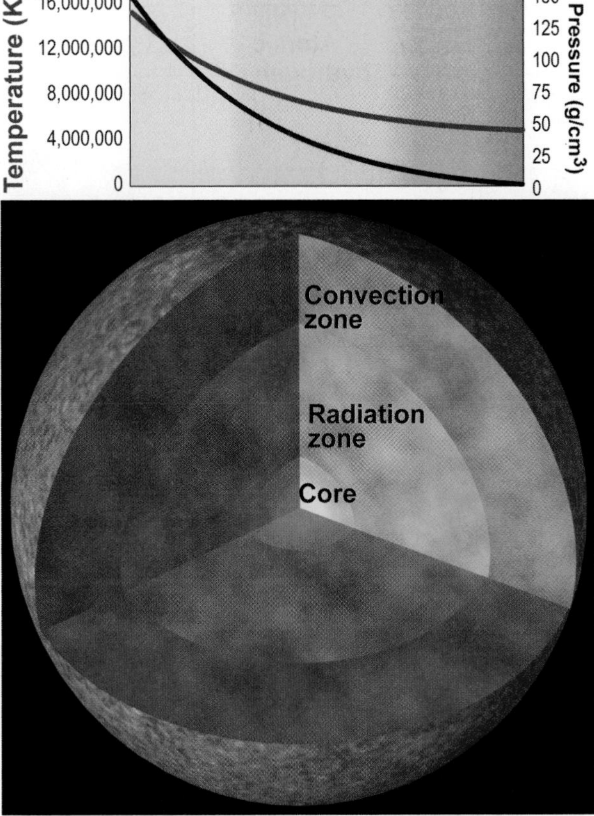

The photosphere that we see, often called the sun's "surface" is a hot, low density gas. The temperature is about 5,800K and the density averages about 10 grams per cubic meter. This is about 1/10 the density of earth's atmosphere at sea level.

Just below the photosphere is the convection zone. Here huge circulating currents of gas move heat to the surface. Next in depth is the radiation zone. In this region of the Sun energy from the core is carried primarily by radiation (light). From the surface to the bottom of the radiation zone, the temperature rises from 5800K to more than 10 million K at the edge of the core. The density also rises from less than the density of air to over 150 g/cm^3. This is 150 times the density of water. Near the core a *single cup* of solar matter would have a mass of 34 kilograms, (75 pounds)!

Plasma - the fourth state of matter

Ionization and plasmas

Inside the sun, the thermal energy per atom exceeds the energy it takes to remove one or more electrons. Atoms come apart into positive ions and free electrons. This state of matter is called a *plasma*. A **plasma** is a gas in which the atoms are split apart, or *ionized* into positive ions and free electrons. **Ionization** is the process of turning a gas into a plasma. Lightning is a plasma ionized by electricity flowing between the clouds and the ground. The Sun is a hot ball of plasma ionized by the energy released by nuclear reactions at the Sun's core. In fact, 99% of the visible matter in the universe is plasma.

Ionization as a function of temperature

At room temperature, a mole of hydrogen gas contains about 3,600 joules of thermal energy. Above a few thousand degrees, molecular hydrogen (H_2) dissociates into atomic hydrogen (H). Above 10,000°C, atomic hydrogen starts to become ionized from gas into plasma. The graph shows that hydrogen is fully ionized above about 20,000°C. Other elements may ionize at lower temperatures since their outer electrons are farther from the nucleus and take less energy to remove. Hydrogen has only one electron that can be ionized, but helium has two and other elements have more. Very hot plasmas may include double, triple or even higher energy ionized atoms.

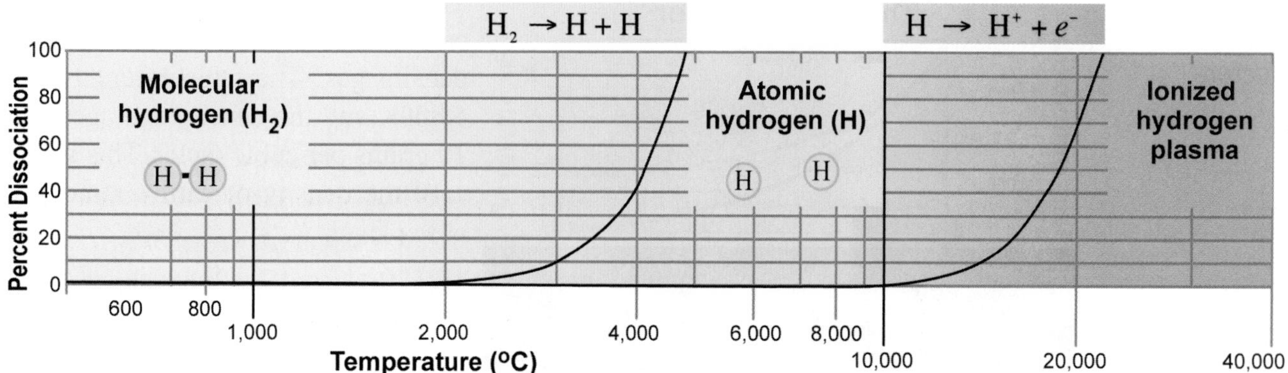

Chemical bonds cannot exist in a hot ionized plasma

To understand the reactions that occur in stars, we need to consider interactions between the particles in hot, ionized plasmas. The chemistry of an ionized plasma is hardly chemistry in the normal sense because *chemical bonds cannot form between energetic, ionized atoms*. Since 99% of the visible universe is plasma, you might think that "astronomical" chemistry would be a sparse subject, but you would be quite mistaken! The 1% of visible matter that is not plasma includes all the planets, ourselves, and the matter between the stars that might one day become planets.

plasma - a hot, energetic phase of matter in which the atoms are broken apart into positive ions and negative electrons that move independent of each other.

ionization - the process of removing one or more electrons from an atom, creating a positive ion and a free electron.

Energy production in the sun

Nuclear reactions power the sun

The core temperature of the sun is approximately 15 million Kelvins. Matter is a dense plasma, too hot for chemical bonds to exist so chemical reactions do not occur. However, at this high temperature protons are hot enough to slam together with sufficient force to touch and undergo nuclear reactions. Nuclear reactions *do* occur in the Sun and are the source of the Sun's prodigious output of energy.

The proton-proton chain

Like chemical reactions, the chance of three nuclear particles hitting each other at the exact same instant to have a reaction is negligible. The reactions in the sun fuse hydrogen to helium in a sequence called the *proton-proton chain*. Each step in the chain involves only two particles. The first step fuses two hydrogen nuclei (protons) into deuterium, releasing a positron and a neutrino as one proton changes into a neutron. The positron immediately annihilates with an electron, and the energy is carried off by gamma rays.

Step 1

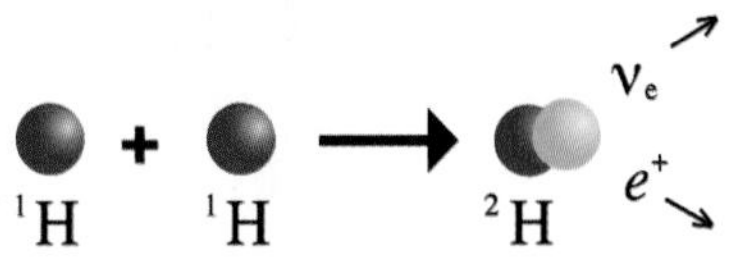

$^1H + ^1H \rightarrow ^2H + \nu_e + e^+$

The deuterium fuses with another hydrogen to produce helium-3 (^3He).

Step 2

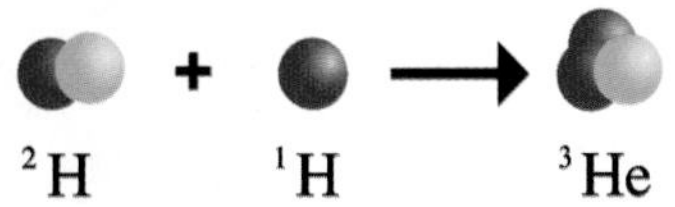

$^2H + ^1H \rightarrow ^3He$

Final steps of the proton-proton chain

Three possible reactions occur to produce helium-4 (^4He). Approximately 86% of the ^3He nuclei fuse with another ^3He nucleus to make ^4He and two protons. About 14% of the ^3He nuclei fuse with ^4He in a second reaction sequence that results in two ^4He nuclei and a single proton. The third reaction (rare, only 0.11%) also fuses ^3He and ^4He nuclei together but with a slightly different outcome (an extremely rare fourth branch is not shown). The energy production of each reaction sequence is shown in the diagram.

Step 3

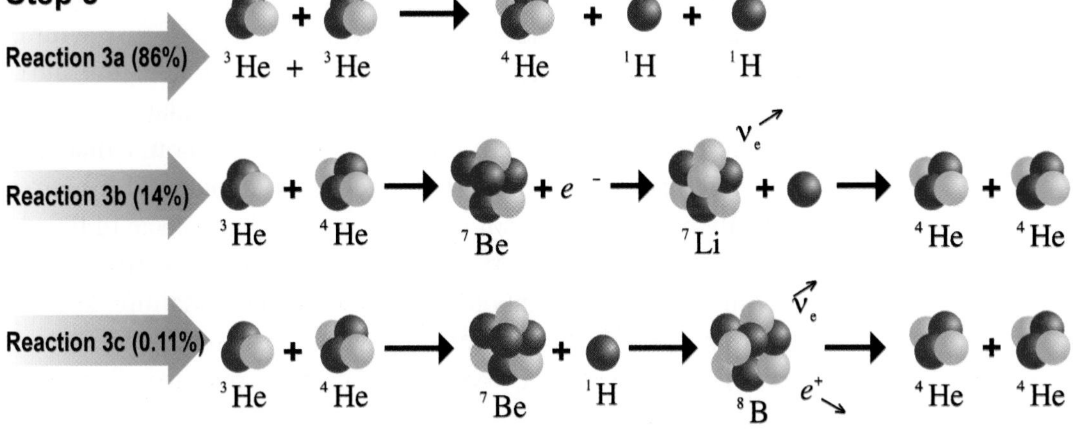

Reaction 3a (86%): $^3He + ^3He \rightarrow ^4He + ^1H + ^1H$

Reaction 3b (14%): $^3He + ^4He \rightarrow ^7Be + e^- \rightarrow ^7Li + \nu_e + ^1H \rightarrow ^4He + ^4He$

Reaction 3c (0.11%): $^3He + ^4He \rightarrow ^7Be + ^1H \rightarrow ^8B \rightarrow \nu_e + e^+ + ^4He + ^4He$

The Big Bang theory

The universe is expanding

Our Milky way galaxy contains some 200 billion stars. Outside the Milky way are more galaxies, billions more. Over the past half century, repeated observations show that all the billions of distant galaxies are moving away from each other (and us). The farther away a distant galaxy is, the faster it is receding. The inescapable conclusion is that *our universe is expanding*. As it expands, the total energy remains constant so the universe is also cooling down.

The beginning of the universe and the Big Bang

If we think backwards, an expanding universe must have been smaller and hotter in the distant past. Taking this idea to its logical beginning, scientists believe that 13.7 billion years ago all of the matter and energy in the universe must have been concentrated in one infinite point. The universe we know exploded from that infinite point in an extraordinary explosion nick-named "The Big Bang." As strange as it seems, there is a very convincing body of evidence from many areas of physics and astronomy that consistently supports the big bang theory.

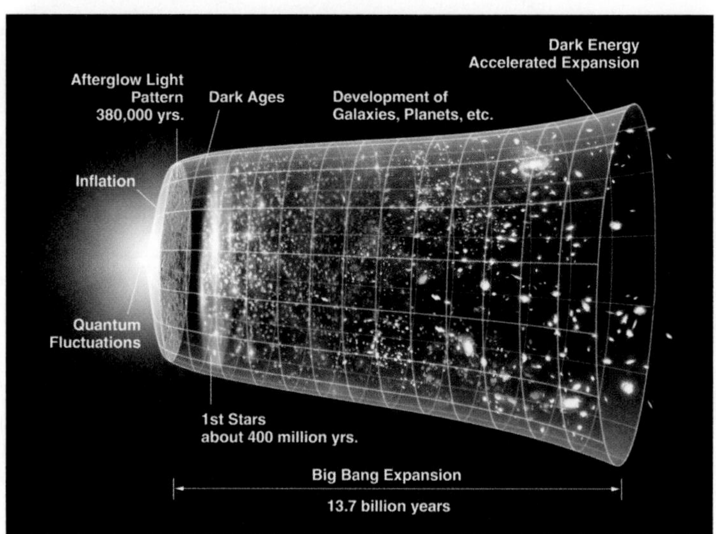

Credit: NASA/WMPA Team

The first three minutes of the universe

Initially, the universe was so hot that protons and neutrons could not stay bound together to form nuclei. After about one minute, the universe cooled enough for nuclei to begin forming. When the universe was 3 minutes old, the temperature had dropped far enough that any free neutrons were bound to protons, and any nuclei already formed had insufficient energy to overcome coulomb repulsion and fuse with each other any more. At the three minute mark, the elemental composition of the universe was essentially "frozen" in place. Detailed quantum mechanics calculations predict that the "freezing out" of the elements from the hot nuclear plasma should theoretically have produced 75% hydrogen and 25% helium. Only very tiny amounts of lithium and beryllium would theoretically be created. This is almost exactly what we see in the elemental composition of visible matter in the observed universe of today. Hydrogen accounts for 75% of all atoms, while helium accounts for almost all of the remaining 25%. That leaves us with a puzzle. If only four different elements were made in the early universe, where did all of the heavier elements such as oxygen and carbon come from?

Nucleosynthesis

Stars are factories for making elements

The early universe contained essentially hydrogen and helium. The Sun, and other stars convert hydrogen into helium in nuclear reactions. To discover the source of elements heavier than helium, we need to look at the life cycle of stars. In the Sun, gravity tends to pull matter inward. When matter falls inward, energy gained from gravity becomes thermal energy and the temperature goes up. In the Sun today, an equilibrium exists between gravity (inward force) and pressure (outward force). The pressure is sustained by the energy released by nuclear reactions in the Sun's core. As long as the Sun's core contains enough hydrogen this balance is maintained.

Helium is fused into carbon and oxygen

Over the next 4.5 billion years the Sun will slowly convert the hydrogen at its core into helium. Energy production will slow down, and the Sun's core will get hotter and denser as gravity compresses it. When the core temperature reaches 100 million K, a new series of nuclear reactions will begin that combine helium into carbon and oxygen.

Synthesis of Carbon Through Nuclear Fusion

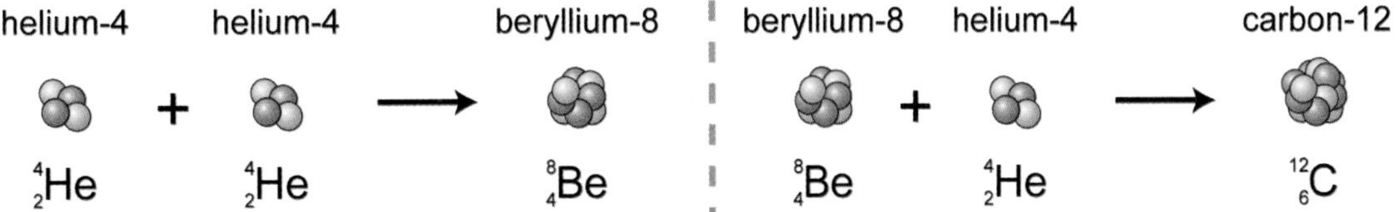

As the helium in the core is gradually converted to oxygen and carbon, the core is again compressed and heated by gravity. At higher temperatures helium is fused with oxygen and carbon to make even heavier elements such as sodium and magnesium.

Supernovae expel element-enriched matter back into the universe

Eventually the core temperature gets hot enough to fuse iron. Iron is the most tightly bound of all nuclei. That means any nuclear reactions involving iron *use* energy rather than release energy. Once the core fuses to iron, the Sun will become unstable and explode in a cataclysmic outburst called a *supernova*. A supernova is such a powerful explosion that for a short time the exploding star can outshine the entire galaxy containing it, releasing as much energy in the initial explosion as our sun will generate during its entire 9 billion year lifetime. To the right is an image of the remnants of a supernova that was spotted by very early astronomers back in the year 1054. It was so bright that it was visible during the day for almost a month.

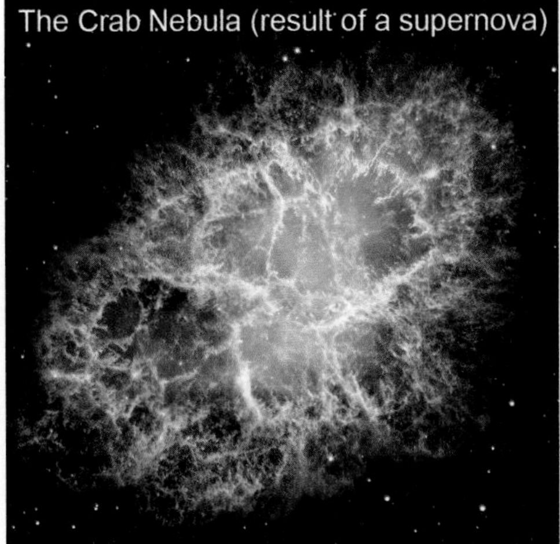

Credit: NASA, ESA and Allison Loll/Jeff Hester (Arizona State University).

Matter in between the stars and planets

The interstellar medium

In between the stars lies a vast, cold, and nearly empty space that astronomers call the *interstellar medium*. We say "nearly empty" because there *is* matter in the interstellar medium and in fact, all the matter in our precious Solar System was once part of the interstellar medium. The interstellar medium consists of an extremely dilute mixture of ions, atoms, molecules, and larger dust grains. Astronomers estimate about 99% gas and 1% dust by mass. Densities range from a few thousand to a few hundred million particles per cubic meter. The average mass density in our own Milky Way Galaxy is about a million particles per cubic meter.

Over time, heavy elements are produced

In the 13 billion years since the universe was formed, countless stars have been born, illuminated the emptiness of space with their light, then exploded when their cores reached the iron burning phase. Each such explosion has enriched the interstellar medium with heavy elements. Today we have evidence that suggest the interstellar medium is roughly 89% hydrogen, 9% helium and 2% heavier elements including oxygen, carbon, and the rest of the periodic table.

Molecular clouds

Stars and galaxies are in constant slow motion and over millions of years the random gravitational tugs from this motion sweep the interstellar matter into vast, cold molecular clouds. Hundreds or thousands of light years across, these clouds typically have temperatures of 10 - 50 Kelvins and contain millions of times as much matter as the Sun. Their composition is mostly molecular hydrogen (H_2) as well as helium and a small percentage of other elements and compounds including water (H_2O) carbon dioxide (CO_2), ammonia (NH_3) and methane (CH_4). Spectroscopic evidence also suggests that more complex organic compounds such as methanol (CH_3OH) and even amino acids exist in molecular clouds.

New stars and planets condense out of molecular clouds

Astronomers believe molecular clouds are the birthplaces of stars (and planets). This is a natural consequence of their low temperatures and high densities, because gravitational force acting to collapse the cloud can exceed the internal pressures that acts "outward" to expand the cloud. The photograph shows a cluster of hot, young stars that has formed from a molecular cloud in the giant nebula NGC 3603. The stellar wind from the new stars has partially blown a bubble of clear space around the cluster, likely triggering the collapse and formation of more stars at the bubble's edge.

Courtesy NASA / ESA - Hubble Space Telescope

21.2 The Planets

The inner and outer planets

The planets naturally fall into two groups. The inner planets are rocky and include Mercury, Venus, Earth, and Mars. They have relatively high densities, slow rotations, solid surfaces, and few moons. The outer planets are made mostly of hydrogen and helium and include Jupiter, Saturn, Uranus, and Neptune. These planets have relatively low densities, rapid rotations, thick atmospheres, and many moons.

TABLE 21.1. Properties of the planets

Property	Mercury	Venus	Earth	Mars	Jupiter	Saturn	Uranus	Neptune
Diameter *(km)*	4,878	12,102	12,756	6,794	142,796	120,660	51,200	49,500
Mass *(Earth = 1)*	0.06	0.82	1.00	0.11	316	94.7	14.5	17.1
Orbit radius *($\times 10^6$ km)*	58	108	150	228	778	1430	2870	4500
Major moons *(#)*	0	0	1	2	39	30	21	8
Gravity, (Earth = 1)	3.7	8.9	1	3.7	23.1	9.0	8.7	11.0
Surface temp. *(°C)*	-170 / +400	430 / 460	-88 / +48	-89 / -31	-108	-139	-197	-201
Atm. press. *(Earth = 1)*	0	92	1	0.01	Ranges from 0 to >10,000 in interior			

Scale of the Solar System

The orbits of the inner planets are separated by about 50 million kilometers. The outer planets are much more spread out. The distance between Mars and Jupiter's orbits is 550 million kilometers, almost ten times the distance between Earth and Mars' orbits. Notice that the size and mass of the planets is quite different for the inner and outer planets as well. The inner planets range from 6% of Earth's mass to Earth itself. The outer planets are all more then 10 times more massive than Earth. Both the size, mass, and distance differences arose from how the solar system formed some 4.5 billion years ago.

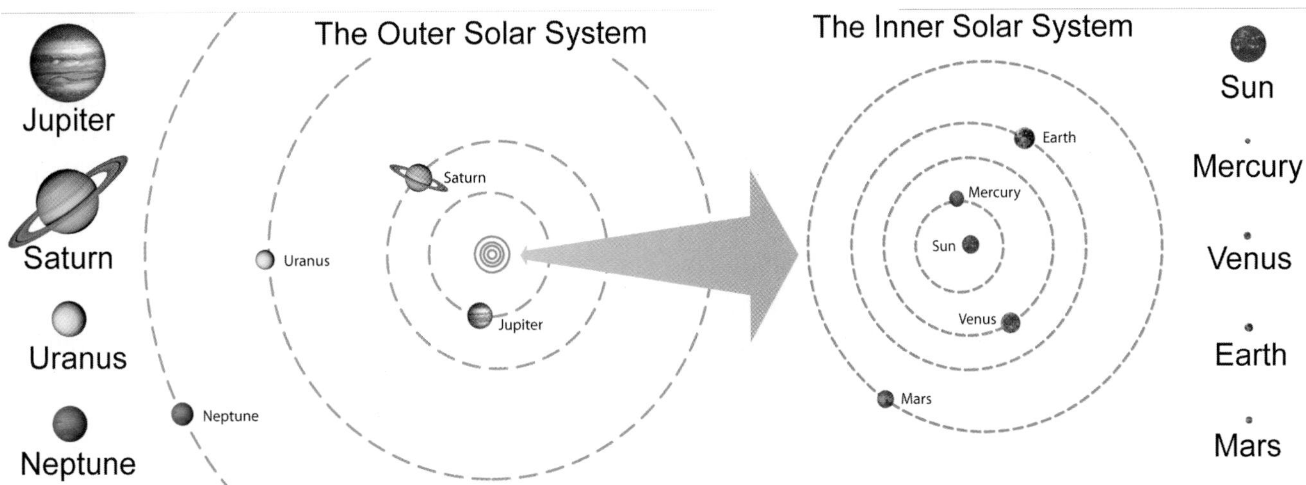

The formation of the solar system

Birth of the Solar System

The solar system began as a vast, cold molecular cloud of interstellar material. Under its own weak gravity the cloud pulled in on itself slowly. Because it was rotating slowly, the cloud spun faster as it slowly collapsed inward, like a figure skater pulling in her arms to make herself spin faster. Once the center of the cloud had collapsed to stellar density the sun ignited. Around the new sun was a slowly rotating disk of left-over matter that coalesced into the planets.

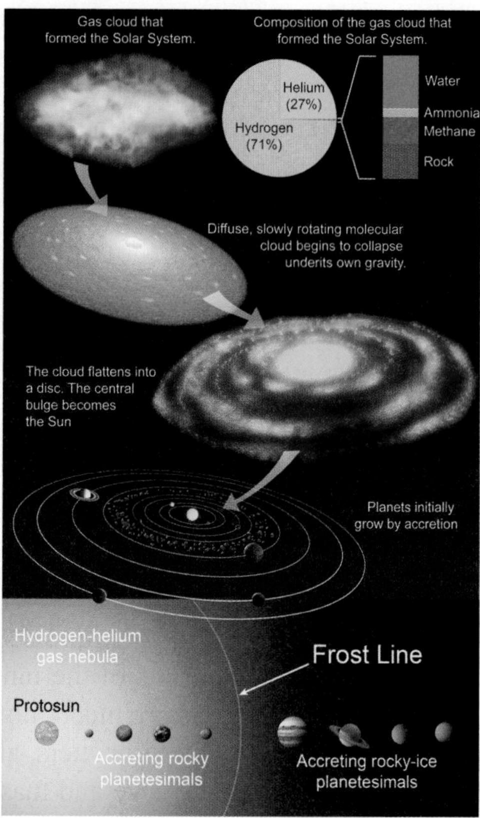

Planets formed initially by accretion

The currently accepted theory is that the planets formed first by *accretion*. Over a few million years, dust particles and molecules formed into clumps between 1 and 10 m in diameter. The clumps collided to form planetesimals of ~5 km in size. These gradually increased through further collisions, growing at the rate of centimeters per year over a few million more years. The differences between the inner and outer planets arose because of differences in melting point, density, and temperature.

Formation of the inner planets

Closer to the sun than about 300 million km the temperature was too high for volatile molecules like water and methane to condense. Planetesimals close to the Sun could only accrete materials with high melting points, such as metals (like iron, nickel, and aluminum) and rocky silicates. These compounds are relatively rare in the universe, and astronomers believe they made up only 0.6% of matter that formed the Solar system. This is one reason the inner planets are small compared to the outer planets. Scientists believe accretion could build up about 0.05 Earth masses, or 5 times larger than the Moon. Over the next few million years, collisions and mergers between these planet-sized bodies allowed terrestrial planets to grow to their present sizes.

Formation of the outer planets

The outer, gas giant planets formed out beyond the "frost line" approximately between the orbits of Mars and Jupiter. This imaginary boundary is where the temperature was cool enough for volatile icy compounds such as water, methane, and ammonia to remain solid. The outer planets grew large because the ices which formed them were much more abundant than the metals and silicates which formed the terrestrial planets. Once they reached a sufficient mass and gravitational strength, the outer planets were able to capture and retain hydrogen and helium, the lightest and most abundant elements. Today, the four gas giants contain almost 99% of the total mass orbiting the Sun.

Energy from the sun

Distance from the Sun strongly affects temperature

The chemistry of each planet is strongly dependent on its temperature. With the exception of Venus, temperature is mostly dependent on distance from the sun. The Sun radiates 3.8×10^{26} watts of power equally in all directions. As you get farther from the sun, that energy is spread out over a larger and larger area. The diagram below shows how the intensity of sunlight changes with distance in the Solar system. The horizontal axis is in *astronomical units* (AU):1AU=150×10^6 km, the distance from Earth to the Sun.

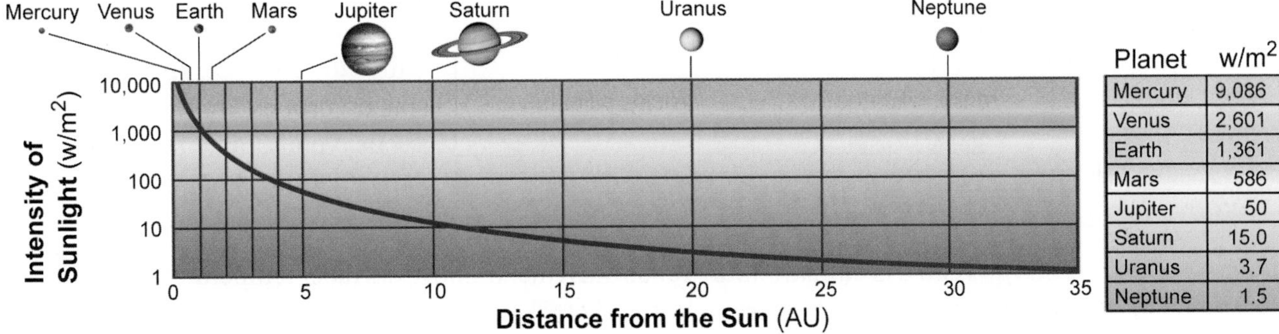

Planet	w/m^2
Mercury	9,086
Venus	2,601
Earth	1,361
Mars	586
Jupiter	50
Saturn	15.0
Uranus	3.7
Neptune	1.5

Intensity is power per unit area

Intensity means how much power falls on each square meter of surface. At the top of Earth's atmosphere, the intensity of sunlight is 1,360 watts per square meter. This is sufficient to maintain the average temperature of Earths surface at a comfortable 15°C. The second graph (below) shows the average surface temperature of each of the planets. For the outer planets that do not have a true surface, the temperature is just below the cloud tops of the upper atmosphere.

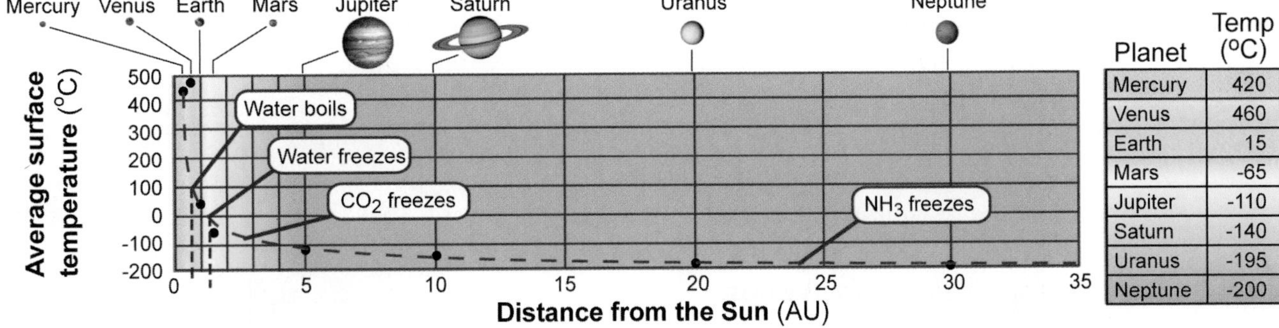

Planet	Temp (°C)
Mercury	420
Venus	460
Earth	15
Mars	-65
Jupiter	-110
Saturn	-140
Uranus	-195
Neptune	-200

Only a small range of distances allow liquid surface water

On the chart are marked the boiling points of several important compounds in planetary atmospheres. At one atmosphere of pressure, water boils at 100C and freezes at 0 C. Carbon dioxide freezes -78C. Methane freezes -182C and nitrogen freezes at -210C. Water, carbon dioxide, methane and nitrogen have similar molecular weights (18, 44, 16, 28). Water has an extraordinarily high melting point, and also a wide range over which it remains liquid. Earth falls within a narrow range of orbits for which water is liquid. Notice that Venus is much too hot for its distance from the Sun. This has to do with the particular chemistry of Venus's atmosphere.

Venus

Why the extreme temperature difference?

	Venus	Earth
Distance from Sun	108 million km	150 million km
Diameter	12,100 km	12,760 km
Composition	rocky, silicates	rocky, silicates
Total CO_2	9.6×10^{-5} %	16×10^{-5} %
Total H_2O	2×10^{-5} %	28×10^{-5} %
Total N_2	2×10^{-6} %	2.4×10^{-6} %

Venus is often thought of as Earth's "sister planet," because Earth and Venus are similar in size, gravity, and composition. They are at comparable distances from the Sun. Like Earth, Venus has a substantial atmosphere however Venus is perpetually covered in dense clouds. Only in the last fifty years have space probes revealed the truth about Venus, and how it's climate and chemistry are very different from Earth.

Venus is extraordinarily hot

Today, Venus is *hot*, in fact Venus has the highest surface temperature in the Solar System, averaging a blistering 460°C (860 °F). Venus is hotter than Mercury even though Venus is nearly twice as far from the Sun and receives only 25% as much solar energy. The reason for Venus extreme temperature is the chemistry of its atmosphere, particularly with respect to carbon dioxide and water.

Venus' atmospheric pressure is 92 times greater than Earth's

Venus atmosphere is 96% carbon dioxide. The atmospheric pressure on Venus' surface is 92 times higher than the pressure at sea level on Earth. On Venus' surface the atmospheric pressure is equivalent to being a kilometer deep under water in Earth. In terms of mass, each cubic meter on Venus contains 65 kilograms of matter. On Earth, a cubic meter of air contains about 1 kilogram of matter.

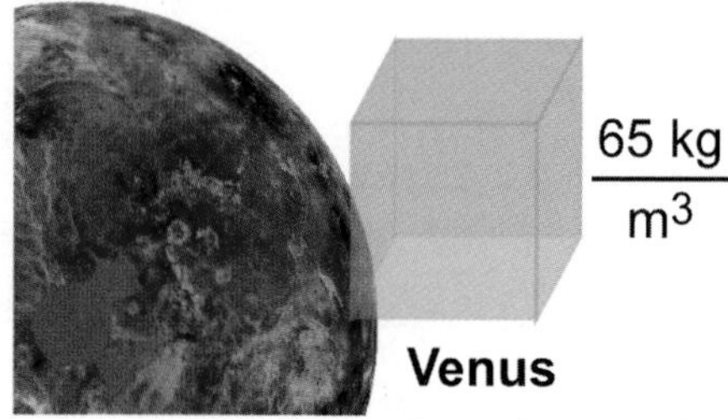

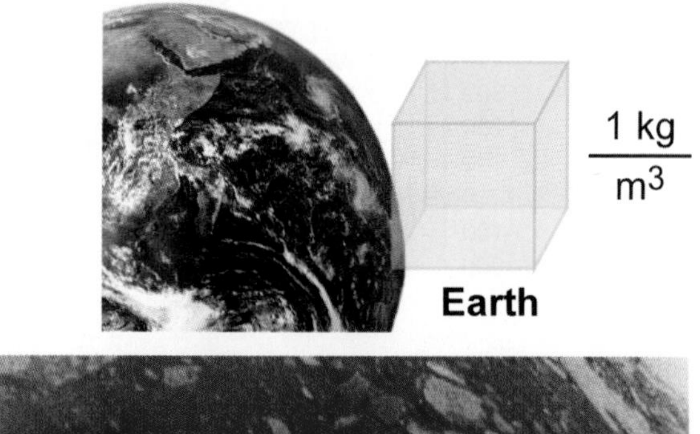

Venus has corrosive, acid weather

Venus upper atmosphere is perpetually surrounded in dense clouds. However, the clouds on Venus are sulfuric acid, not water! The hot acid fog makes a highly corrosive atmosphere. The photograph (right) was taken by the Venera 13 space probe, which lasted only an hour before being destroyed.

Blackbody radiation and planetary energy balance

Absorption and radiation

To understand why Venus is so hot, we need to look at two competing effects that occur on all planets. *Absorption* of energy heats a planet up. *Radiation* of energy cools a planet down. Absorption is something you feel directly on a sunny day. Your skin absorbs energy and heats up.

The blackbody spectrum

Thermal radiation from the sun, loosely called *light,* is how the energy gets from the Sun to your skin. ALL objects with a temperature above absolute zero give off thermal radiation, just like the sun. Most objects are not as hot as the sun, so the radiation is both lower in intensity and at different wavelengths. The amount of power radiated at different wavelengths (colors) is called the *blackbody spectrum.* The curve for the Sun shows that at 5,500°C the radiation has its highest power in the visible range of wavelengths. Room temperature objects (20°C) radiate much less power and mostly in the infrared.

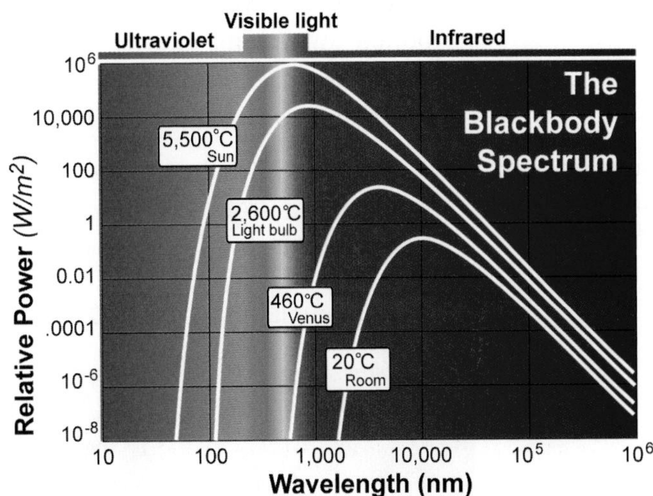

Power increases like T^4

Notice the power scale on the blackbody spectrum. The amount of energy radiated depends on the temperature to the fourth power. If the temperature doubles, the radiated power is multiplied by 16!

Planetary energy balance

In steady-state, a planet must re-radiate all the energy it absorbs from the Sun. Or else, the planet will get warmer. If absorbed energy exceeds radiated energy, the planet's temperature slowly increases, also increasing the energy the planet radiates. The temperature stops increasing when radiated energy equals absorbed energy from sunlight.

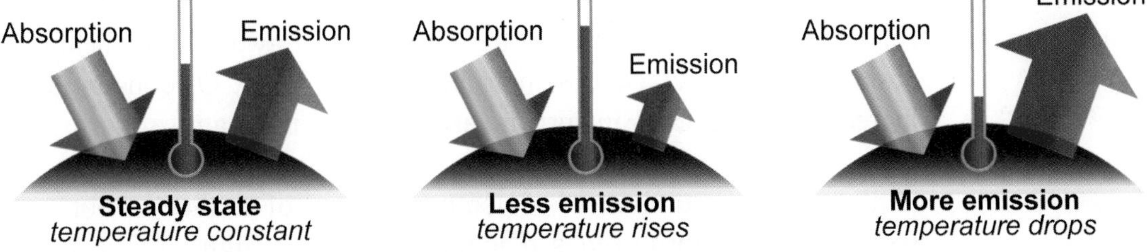

If absorbed energy is *less* than radiated energy, the planet surface cools down. Cooling down lowers the radiated energy until the two are again in balance and the temperature stabilizes. The average temperature of a planet adjusts itself up or down to keep balance between absorption and radiation.

The greenhouse effect

Carbon on Venus and Earth

While Venus and Earth have similar amounts of carbon, on Venus the carbon is gas (CO_2) in the atmosphere. On Earth, carbon dissolves into the oceans where it is entrapped by organic life and deposited on the ocean floor as solid calcium carbonate and other minerals. The Earth's carbon cycle keeps carbon out of the atmosphere. There is evidence that Venus has surface water in its distant past. However, any water on Venus boiled away long ago as a result of a *runaway greenhouse effect*. Over time Venus' atmospheric water was broken apart by ultraviolet light from the sun. The light hydrogen escaped so water could not form again and today Venus is completely dry.

Sunlight and absorption

To explain the greenhouse effect, consider that sunlight contains a spectrum of wavelengths from infrared to visible. Both Venus' and Earth's atmosphere are mostly transparent to sunlight so it passes through and is absorbed by the ground.

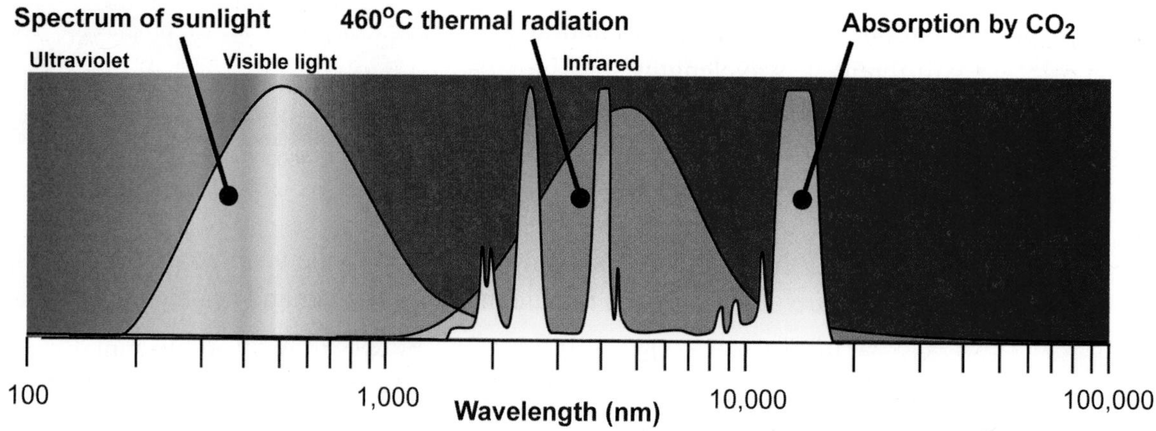

CO_2 absorbs in infrared

The orange shaded areas show wavelengths of light that are strongly absorbed by CO_2 gas. Notice that CO_2 is transparent to sunlight but absorbs strongly in infrared. Like Earth, energy from the Sun is absorbed by Venus. However, because of the CO_2, part of the energy radiated by Venus surface is absorbed by the atmosphere before escaping to space. Because the energy could not escape, Venus slowly became warmer and warmer.

Explanation of Venus high temperature

As Venus got warmer, its surface water boiled away. That meant there was no way to absorb CO_2 from volcanoes and other sources. The CO_2 concentration in the atmosphere increased. More CO_2 in turn meant even stronger atmospheric absorption of radiated heat. More absorbed heat, meant Venus temperature increased even more in a self-reinforcing cycle until most of Venus inventory of surface carbon was in the atmosphere. Eventually equilibrium was reached at the very high surface temperature we observe today. Atmospheric CO_2 concentration is strongly correlated to a planet's average temperature because it affects the overall planetary energy balance. Many scientists are concerned because human activities have increased the concentration of CO_2 in Earths atmosphere 35% since the industrial revolution.

Mars

Mars

Mars is a cold and relatively small planet with a mass only 11 percent the mass of Earth and surface gravity only 38% as strong as Earth. Like Venus, Mars's atmosphere is mostly carbon dioxide (95%) however, unlike Venus, Mar's atmosphere is very thin. The atmospheric pressure on Mar's surface is 100 times lower than the pressure at sea level on Earth. Because of the thin atmosphere and the planet's distance from the sun, Martian temperatures are below 0°C most of the time.

Martian surface
Viking 2 (NASA)

Mars was wetter in the distant past

Grey hematite

Red hematite *(rust)*

Mars is cold and dry today, but there is strong evidence that Mars was much wetter and had a thicker atmosphere in the past. Aerial photos of the Martian surface show erosion and patterns of riverbeds similar to those formed by flowing water on Earth. Even today, there is evidence of water beneath the Martian surface. Much of the evidence is chemical in nature, such as the discovery of a grey variety of the iron oxide mineral called hematite (Fe_2O_3). Hematite comes in two forms: red and gray. The red kind is ordinary rust which forms when iron is in contact with air. Red rust is everywhere on Mars and its what gives the planet its reddish color. Gray hematite has a dark gray metallic luster. Unlike red hematite, gray hematite usually precipitates from iron-rich water over long periods of time. The discovery of gray hematite near the Martian equator supports the hypothesis that Mars once had deep, liquid water on its surface.

Searching for life on Mars

We know living organisms invariably leave a chemical signature. In 1976, the Viking mission landed a robot probe on Mars to search for evidence of life, but the results were ambiguous. In 2013 a new mission will land on the red planet to specifically look for amino acids. Amino acids can form with right-handed or left-handed structures. Non-living processes produce a 50-50 mix of the two. All life on Earth uses only left-handed amino acids. If amino acids are found with a preference for either right, or left handed structures, it will be strong evidence for the presence of life.

Mars Rover - 2004 - NASA

The outer planets

The outer planets are much larger

The inner planets are tiny compared to the outer planets both in size and mass. Jupiter is by far the largest object in the Solar System after the Sun. Jupiter contains more than twice the mass of all the other planets combined.

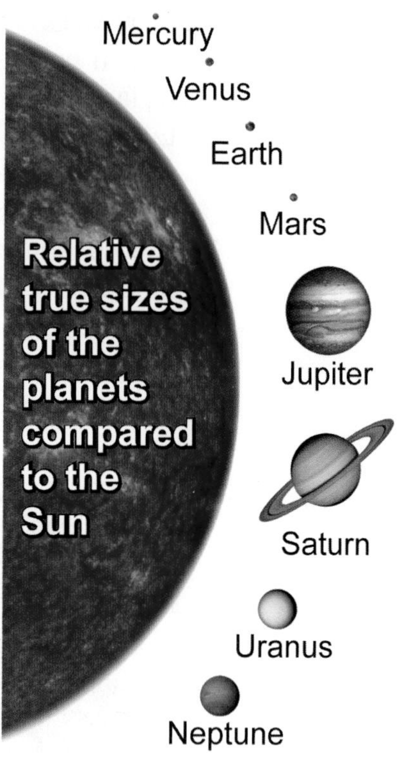

The outer planets compared to Earth

Jupiter's diameter is 11 times the diameter of Earth. The smallest of the outer planets, Neptune is four times as large as Earth. No longer considered a planet, Pluto is half the size of Mercury, and made of ice, not rock. It belongs to neither the inner or outer planets, but instead is the closest member of a group of frozen ice balls called Kuiper Belt Objects.

Comparing composition

In composition, all the outer planets are much different from the rocky, inner planets like Earth. Notice that the outer planets have densities between 0.67 and 2.67 g/cm^3. The inner planets are much denser, ranging from 3.91 to 5.52 g/cm^3. The density difference tells us the outer planets are mostly light gasses, liquids and ices such as frozen ammonia and methane. The inner planets are mostly rock.

TABLE 21.2. Comparing inner and outer planets

Property	Mercury	Venus	Earth	Mars	Jupiter	Saturn	Uranus	Neptune
Diameter *(km)*	4,878	12,102	12,756	6,794	142,796	120,660	51,200	49,500
Mass *(Earth = 1)*	0.06	0.82	1.00	0.11	316	94.7	14.5	17.1
Avg. Density *(g/cm^3)*	5.44	5.25	5.52	3.91	1.31	0.69	1.21	1.67

The outer planets have multiple moons and rings

The outer planets have many moons and are more like miniature solar systems themselves. This is very different from the inner planets. Earth has 1 large moon, Mercury and Venus have none, and Mars has two tiny orbiting asteroids that hardly qualify as moons. By comparison, as of the writing of this book, Jupiter has 63 known moons. Four of them are comparable to Earth's moon in size. Saturn has 61 moons, Uranus has 27 and Neptune has 13. Saturn is best known for its rings, but all the outer planets are now known to have rings, even Jupiter.

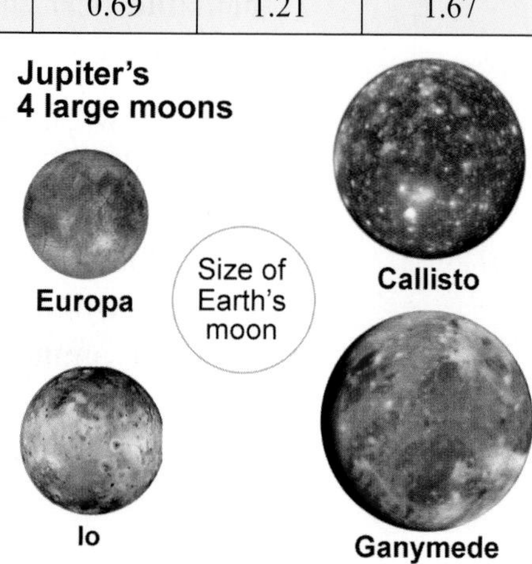

Composition of the outer planets

Jupiter, Saturn, Uranus, and Neptune are gas planets

The word "planet" usually elicits a mental picture of a spherical object (like the Moon) with a hard surface on which you could imaging walking. The outer planets of our solar system don't really fit this description. Instead, the outer planets are giant balls of cold gas and liquid without a definite surface.

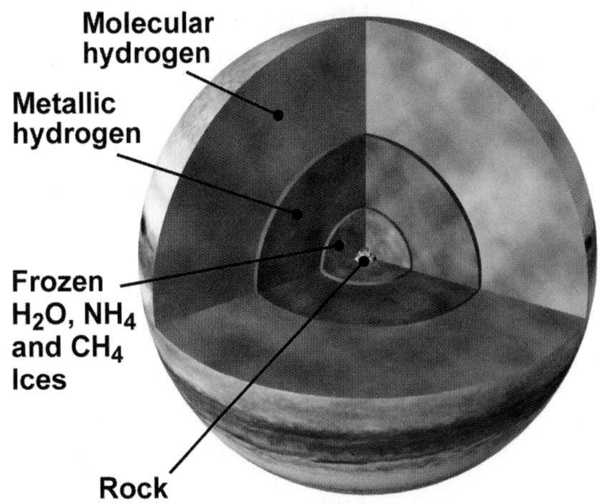

Atmospheres of the Outer Planets
(percentage by mole)

	Sun	Jupiter	Saturn	Uranus	Neptune
H_2	84	86.4	97	83	79
He	16	13.6	3	15	18
H_2O	-	0.1	-	-	-
CH_4	-	0.21	0.2	2	3
NH_3	-	0.07	0.03	-	-

Chemical similarity to the Sun

Chemically, the outer planets are more similar to the Sun than to Earth. All four are predominantly hydrogen and helium with a small percentage of water, ammonia, and methane.

The interiors of the gas planets

The diagram shows that the pressure increases with depth reaching more than a million atmospheres. At this high pressure, molecular hydrogen is literally squeezed apart into separate protons and electrons. Because the temperature is low, this "cold plasma" acts more like a liquid metal than a hot gas, and is referred to as *metallic hydrogen*.

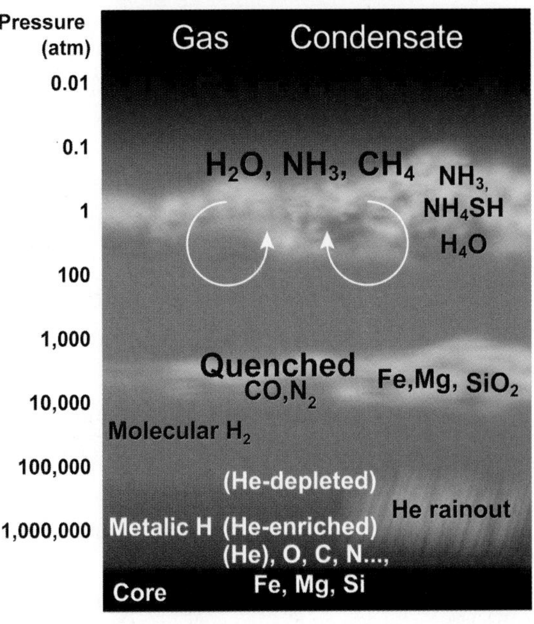

Why Jupiter and Saturn are different from Uranus and Neptune

Jupiter and Saturn have less helium in their atmospheres than the Sun because helium is soluble in metallic hydrogen. Scientists believe a cold, liquid helium "rain" occurs in the transition region between the molecular hydrogen and metallic hydrogen. Over time, this depletes helium form the atmospheres. The effect is strongest on Jupiter since Jupiter has a larger mass and therefore more metallic hydrogen compared to molecular hydrogen. Saturn has a smaller amount of metallic hydrogen since it has only a third the mass of Jupiter and therefore lower core pressure. Uranus and Neptune are smaller than Saturn and pressures are not high enough for metallic hydrogen to form.

Modeling the interior of a planet

Ideal gas laws don't apply at high pressure

How do we know what the inside of Jupiter or Saturn is like? We cannot easily go ourselves to find out so scientists build models. A model for the interior of a gas planet needs to include (at a minimum) variables of composition, temperature and pressure. One such model we already know is the ideal gas law. The ideal gas law relates pressure, temperature, and composition for gases that are *weakly interacting*. That means the individual molecules are far apart. At the enormous pressures inside a planet, molecules are squeezed tightly together so the ideal gas law does not apply.

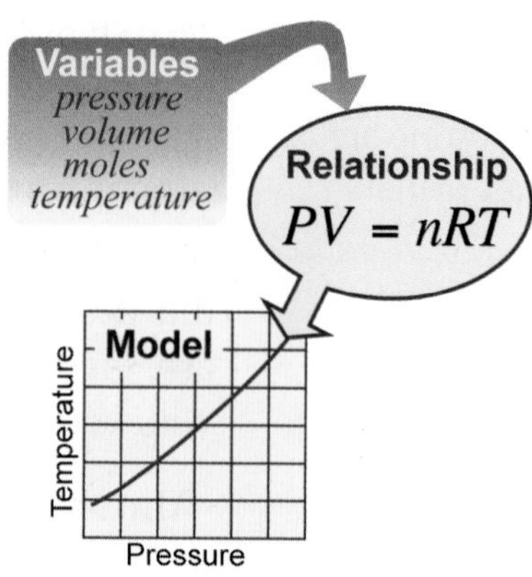

A model for the phases of matter in Jupiter

The diagram on the right shows the different phases of a hydrogen and helium mixture typical of the outer planets. This model is based on both calculations and on high pressure experiments on Earth. The ideal gas law describes only the lighter shaded region in the center left of the graph. At pressures above about a million atmospheres hydrogen is in its metallic phase. Jupiter's core falls in this region. However, the density and temperature at the core of Uranus fall in the liquid H_2 region so we expect the inner regions of Uranus to be liquid hydrogen.

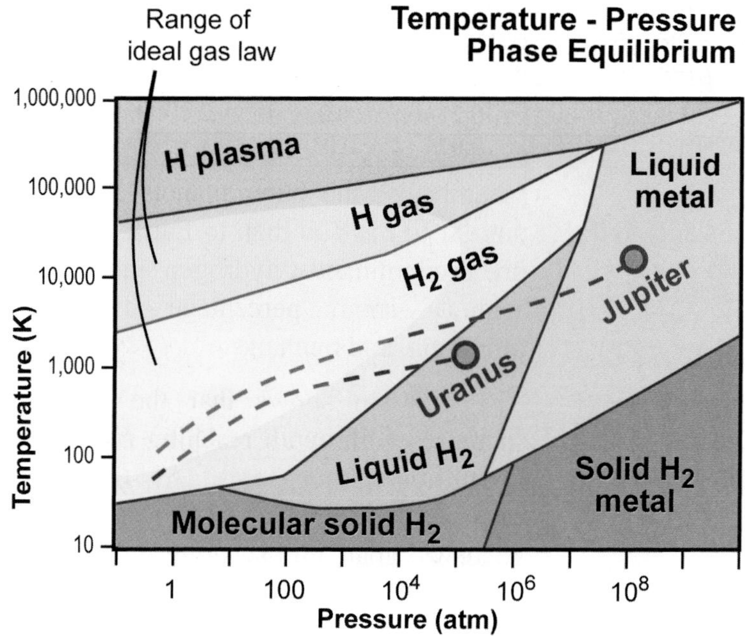

Matter can have multiple different liquid and solid phases

At high densities atoms and molecules are squeezed tightly together and their interactions become complex. Hydrogen, like other substances, exhibits distinct phases other than just solid, liquid, and gas. The diagram shows that there are two different liquid phases and two solid phases depending on the temperature and pressure. Molecular solid H_2 is an electrical insulator. At temperatures below 20K and pressures above a million atmospheres solid molecular hydrogen changes into an electrical conductor, its solid metal phase. This phase transition occurs when the intense pressure forces electrons to become dissociated from individual molecules.

21.3 The Possibility of Life Elsewhere

What about life on other planets?

Earth is a relatively warm, wet, and lush oasis compared to the rest of the solar system. At least, we humans think so, along with the rest of the plants and animals evolved to live here. We know that life arose on Earth at least 3 billion years ago when the planet was quite different than it is today. For example, there was very little (or no) oxygen on the primordial Earth. The universe is vast beyond imagining, and people have always wondered if there is life on other worlds. As a prerequisite to searching for extraterrestrial life it seems prudent to ask some basic questions.

- What are the physical and chemical conditions under which life can exist?
- Do those conditions exist outside of Earth? If so, where?
- How would we recognize alien life if we found it?

What are the possible range of conditions where life can exist?

What are the extreme boundaries where life exists here on Earth? Kilometers below the surface, on the sunless floor of the deep ocean, volcanic activity occasionally creates features called *hydrothermal vents*. These vents are jets of acidic, superheated water that exist in absolute darkness, at incredible pressure, and contain concentrations of poison gas and heavy metals. The hot water coming from a hydrothermal vent is rich in dissolved minerals, most notably sulfides. When it comes in contact with cold ocean water, some minerals precipitate, forming a black chimney-like structure around each vent called a *black smoker*.

Courtesy of Spiess, Macdonald, et al, 1980.

Chemo-synthetic life needs no sunlight

Despite the extreme conditions, a variety of *chemosynthetic* life exists at hydrothermal vents. Chemosynthetic bacteria use energy directly from heat and convert toxic minerals such as hydrogen sulfide into carbon compounds. One such reaction creates formaldehyde (CH_2O) from dissolved carbon dioxide, oxygen, and hydrogen sulfide.

Chemosynthesis
(example: hydrogen sulfide reaction)

$$CO_2 + O_2 + 4H_2S \rightarrow CH_2O + 4S + 3H_2O$$

Chemosynthetic bacteria support a food chain of other animals. Some creatures, such as gastropod snails, eat the bacteria directly. Tubeworms, host chemosynthetic bacteria in their tissues in exchange for organic compounds that the bacteria create from the vent chemicals and seawater. This is life at temperatures well over 300°C, at extreme pressures and totally without sunlight or organic matter.

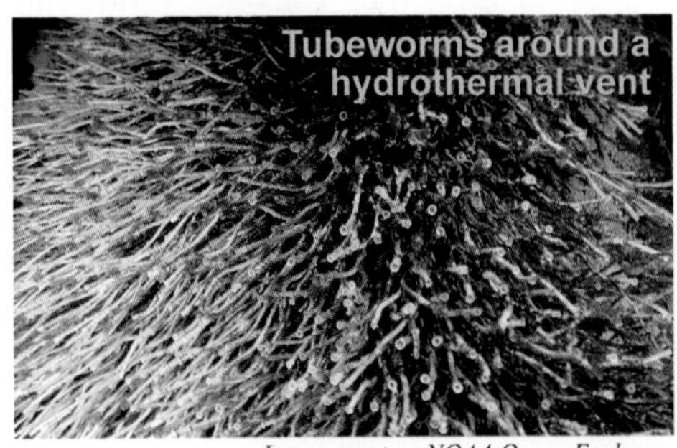

Image courtesy NOAA Ocean Explorer

Life deep underground

We once believed life was confined to the surface

Until recently, most scientists believed that life was restricted to the surface of planet Earth. The oceans, the air, the ground, and the soil are alive animals, plants, and microorganisms. Below the surface, in deep bedrock there had been little expectation for life. Deep underground the temperature is hot, toxic chemicals are abundant, background radiation is significant, and there is no sunlight.

Recently discovered species of bacteria live deep underground

Geologists and microbiologists have proved that life is plenty tough enough to survive deep underground. In the late 1980s, researchers found microbes living in rock 500 meters below the surface in South Carolina. Between 1998 and 2006, Princeton professor, Dr. Tullis Onstott and other researchers have discovered active, living colonies of bacteria 2.8 kilometers underground. In some cases, these bacteria have apparently remained prisoners of the deep for millions of years, making such colonies actual living fossils. "This is a revolution that's expanding the limits under which we know life to exist," says Onstott. "I think we're recognizing that life is a bit more tenacious than we had given it credit for, and there's a much broader realm of possibilities where life can survive." More recently researchers have discovered microorganisms in deep granite formations. These rocks completely lack organic materials. The new discoveries suggest deep subsurface life may be quite common.

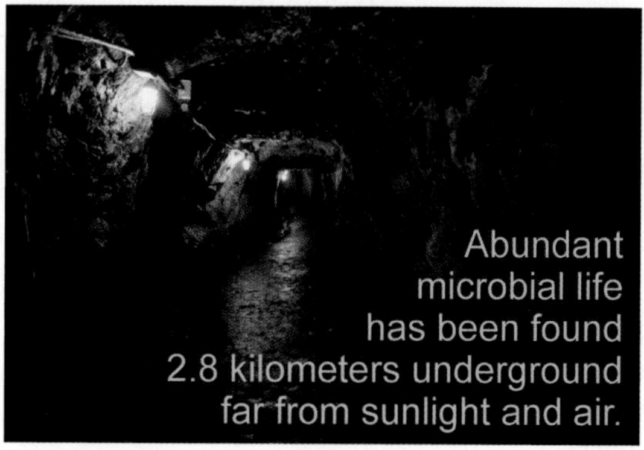
Abundant
microbial life
has been found
2.8 kilometers underground
far from sunlight and air.

Bacillus infernus is an example of thermophylic bacteria

Deep bacteria are classed as *thermophiles*, or heat lovers, because their environment has a temperature of 75°C (167°F). One particular species was recently named Bacillus infernus, for its high-temperature survival. Similar thermophilic microbes colonize volcanically heated springs, such as those in Yellowstone National Park, and scalding geysers on the ocean floor.

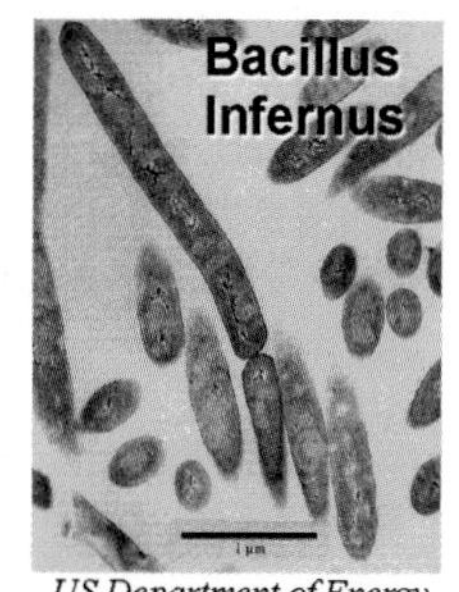
Bacillus
Infernus

US Department of Energy

Implications for ancient and extraterrestrial life

Biologists say the discovery of deep thermophilic bacteria has broad implications, stretching backward in time and outward into of space. If bacteria and archaea can survive far below ground without any contact with the surface, then it's possible that life could have even started there. Earth's surface was quite catastrophic when the planet was young. Deep rocks would have provided protection from sterilizing temperatures, meteor impacts and radiation. Today, many scientists believe life may be hiding beneath the dusty red surface of Mars and other extraterrestrial habitats. The surface of Mars is dry but it is very likely that liquid water exists in the warmer interior of the planet.

Evidence for life on Mars

The Viking mission (1976)

If you had to travel 75 million kilometers and could only look at one tiny spot for evidence of life, how would you do it? In 1976, before there were ATM's or CD-ROMs, NASA launched the Viking mission. During one year, a pair Viking spacecraft landed on Mars and carried out some basic chemical tests for life.

Viking on Mars 1976

Photo courtesy NASA

Viking carried Four chemical experiments to detect the signature of life

A *mass spectrometer* looked for organic compounds, such as alcohols or amino acids in Martian soil. The results were negative and Viking found martian soils to contain less carbon than lifeless lunar soils returned by the Apollo program.

The *gas exchange* experiment looked for O_2, CO_2, H_2, CH_4, and N_2 given off by a soil sample that was treated with a liquid complex of organic and inorganic nutrients. The *labeled release* experiment inoculated a sample of Martian soil with a drop of dilute aqueous nutrient solution tagged with radioactive ^{14}C. The *pyrolytic release* experiment exposed Martian soil to light and CO_2 that was tagged with ^{14}C. Any photosynthetic activity would incorporate the tagged carbon and be detected by heating the sample after a few weeks incubation.

Viking was inconclusive

The results from Viking were inconclusive. No life was detected, but life could not be ruled out because the samples were small and the instrumentation of the time was not as sophisticated as today.

The Mars Phoenix mission (2008)

In May of 2008, the NASA Phoenix mission landed on Mars to continue the search for life. Among Phoenix accomplishments was the a definitive observation of water in the Martian soil. One morning there was frost on the sides of a trench the lander had scraped previously. A second chemical finding were several salts and perchlorate (ClO_4) in Martian soil. Perchlorate is an oxidant not a powerful one. Perchlorates are found naturally on Earth at such places as Chile's hyper-arid Atacama Desert. While ambiguous as a signature of life, some microorganisms on Earth are fueled by processes that involve perchlorates, and some plants concentrate the substance.

Photos courtesy NASA

Europa

Jupiter's sixth moon is unusual

In 1997 the Galilleo spacecraft flew within 200 kilometers of Europa, the sixth moon of Jupiter. The smallest of the Jupiter's four major moons, Europa was an enigma. The most reflective object in the Solar System, Europa's surface was criss-crossed with strange lines. There were almost no craters yet Ganymede (Jupiter's largest moon) had plenty of craters. Something about Europa provided an active process for erasing craters from its surface.

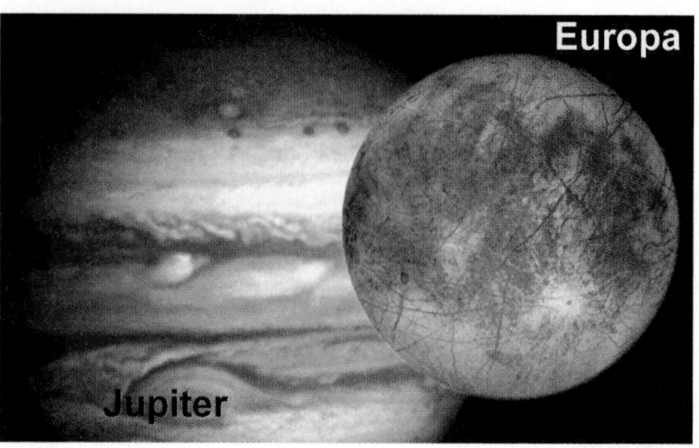

Europa has a thick crust of water ice

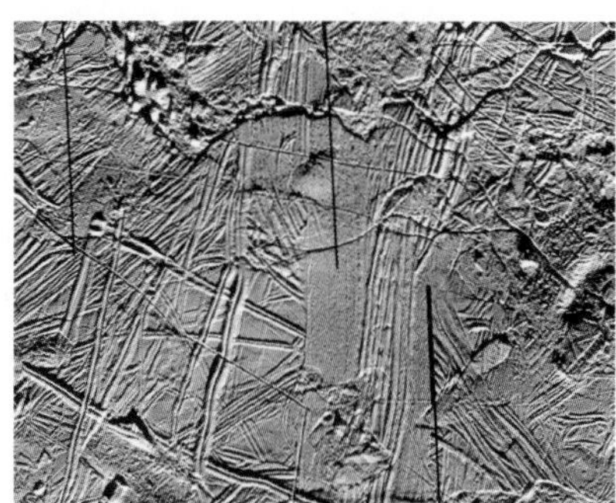

Europa's surface from 200 km altitude

Orbital photographs show a chaotic terrain unlike any other in the Solar System. Scientists believe that the cracks are in a layer of water ice that may be 100 kilometers thick. The observed scarcity of craters is due to periodic cracks in the ice caused by Jupiter's gravity. Upwelling liquid water freezes in the -160°C temperature to make new surface. Beneath Europa's ice lie the deepest water oceans in the Solar System, estimated at 200 km. For comparison, the deepest oceans on Earth are less than 10 kilometers deep.

Europa may have the deepest oceans in the Solar System

How does water stay liquid so far from the Sun? The answer is that Europa is fairly close to Jupiter for a moon. Jupiter's gravity creates strong tides that pull Europa's inner oceans up and down with each orbit. This repetitive massaging creates frictional heating that keeps the oceans liquid.

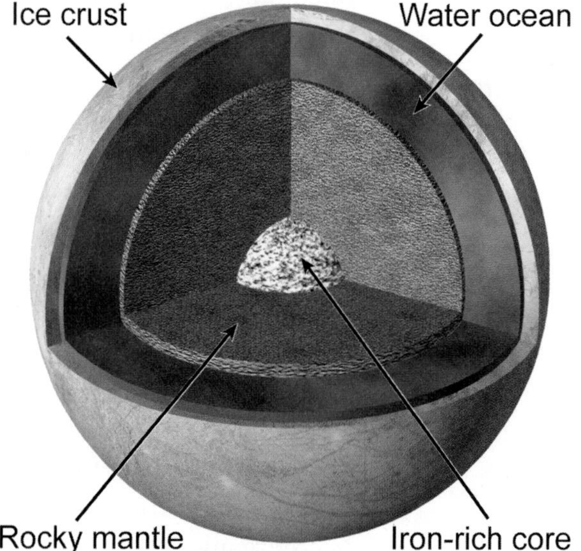

The potential for life on Europa

It is very likely that conditions near the base of Europa's oceans are similar to those on the sea floor of Earth. Chemosynthetic or thermophylic bacteria could conceivably thrive there. After Mars, Europa is currently considered by many scientists to be the next most likely candidate for extraterrestrial life.

Titan

Titan is larger than Mercury

Titan is Saturn's sixth moon from Saturn and is larger than the planet, Mercury. The only moon with a real atmosphere, the pressure at Titanian sea level is one and a half times that on Earth. Titan's atmosphere is 98.4% nitrogen with the remaining 1.6% being of methane and trace amounts of many organic compounds including ethane, diacetylene, methylacetylene, acetylene, and propane. The moon itself is composed of water ice and rocky material. Much as with Venus until the Space Age, the dense, cloudy atmosphere over Titan prevented any glimpse of its surface until the arrival of the Cassini–Huygens mission in 2004.

Titan is a rich world more like a planet than a moon

In January 2005, after traveling through space for 7 years, the Cassini-Huygens spacecraft reached Saturn and a probe descended through the clouds to see the surface for the first time. We now know that Titan is more like a planet, with oceans, rivers, mountains, and even weather! Titan's rivers and oceans are the only large, stable bodies of surface liquid known to exist anywhere other than Earth. Titan even has volcanoes that may erupt ammonia ice slush. he shores of a cold ethane ocean on Titan (artists rendering from Huygens probe images)

The potential for life on Titan

With the exception of the extreme cold (-180°C), Titan's abundant organic liquids and dense nitrogen atmosphere are seen by scientists as similar to the early Earth. Even with the cold, some consider Titan a potential site for finding extraterrestrial life. At a minimum we know Titan contains a prebiotic environment rich in complex organic chemistry. If there are warmer underground features on Titan the possibility of life there becomes greatly enhanced. Humans will not know until some future space mission when a person might be able to set foot on this distant but intriguing world.

Chapter 21 Review

Conceptual Questions

Section 21.1

1. The element hydrogen is
   a. More abundant than helium in the universe
   b. Less abundant than helium in the universe
   c. Always found combined with carbon as methane in molecular clouds
   d. Always found combined with nitrogen as ammonia in molecular clouds

2. The Big Bang theory is supported by which of the following observations?
   a. Stars like the Sun are about average in size.
   b. Distant galaxies are all receding from us.
   c. The Sun carries out nuclear fusion reactions in its core.
   d. Stars tend to be grouped into large collections called galaxies.

3. Are chemical reactions likely to occur in the Sun? Why or why not?

4. Research two examples of plasmas you might find on Earth. What gasses are involved? What is the average temperature?

5. A person tells you that they believe combustion of flammable gasses make heat in the Sun just as a fireplace makes heat on Earth. Write one paragraph agreeing or disagreeing and state at least 2 reasons to support your position.

6. Explain why a single nuclear fusion reaction like the one below does not occur, or occurs so rarely that it is completely unimportant in the Sun.

$$^{1}H + {}^{1}H + {}^{1}H + {}^{1}H = {}^{4}He + 2\nu_e + 2e^{+}$$

7. Describe the origin of the element, oxygen as it appears on Earth. Where do oxygen atoms come from? Are they made on Earth?

8. Name at least three major substances in the interstellar medium. Identify them as elements or compounds.

9. Which of the following phases of matter is most abundant among the visible matter in our universe? Explain where it is usually found.
   a) solid b) liquid c) gas d) plasma

Section 21.2

10. Suppose there were planets around a star that was larger and hotter than the Sun. Would you expect to find an Earth-like planet closer or farther away form the star? Give the reasoning behind your answer.

11. Explain how the "frost line" affected the formation of the inner and outer planets.

12. List three of the most important differences between the inner and outer planets.

13. Which planet has the highest average surface temperature? Is this planet the closest planet to the sun?

14. What is the meaning of the term *accretion* in the context of the formation of the solar system?

15. What happens to all the energy absorbed by a planet from the Sun?

16. What would happen if a planet absorbed *all* the energy it receives from the sun?

17. Explain the difference between thermal radiation and sunlight. Is there a difference?

18. Why does most sunlight pass though Venus atmosphere while thermal radiation is partly absorbed?

19. What gives Mars its reddish color?

20. Mar's atmosphere is similar in composition to Venus (mostly CO_2). Why is Mars so cold while Venus is so hot?

Section 21.3

21. Does all life on Earth require sunlight? Give two examples to support your answer.

22. Where do scientists believe the deepest liquid water oceans in the solar system can be found

23. If a moon has no craters, what does this tell us about its geology?

24. Research three hydrocarbon compounds that are liquid at -180°C. Are any of these present on Titan?

Section 21.1 Quantitative Problems

25. Write down a possible nuclear reaction that could make magnesium in a star from lighter elements. The rules are:
    a. The total of protons + neutrons must balance on both sides of the reaction.
    b. The reaction must involve only 2.
    c. A neutron can become a proton according to the reaction $n \rightarrow p + \nu_e + e^+$ where ν_e is an electron neutrino and e^+ is a positron (antimatter).

26. The Sun has a mass of 2.0×10^{30} kg. We know from calculated models that the Sun's core, where nuclear reactions occur, is about 50% of its mass, or 1.0×10^{30} kg. From the energy output we know that the Sun fuses 3.8×10^{28} protons per second into helium. Calculate how long it will take the Sun to use up 25% of the hydrogen in its core. Express your answer in years. (Hint: The mass of a proton is 1.67×10^{-27} kg)

Section 21.2 Quantitative problems

27. Make a graph of the surface temperatures of the inner planets versus the distance form the Sun. Fit a smooth curve to Mercury, Earth, and Mars. You may ignore Venus as an anomaly for the purposes of the graph.
    a. Estimate the radius of an orbit that would make the average surface temperature 100°C.
    b. Estimate the radius of an orbit that would make the average surface temperature 0°C.
    c. This region is known as the "life zone" for potential orbits of habitable planets. Where is Earth in the life zone of the Sun? Is our planet in the middle or near one or the other edge?

28. The inverse square law describes how the intensity *(I)* of solar radiation drops off with distance from the Sun or any other star. The *luminosity (L)* is the total energy output of the star in watts and *R* is the distance from from the star to the planet in meters.

$$I = \frac{L}{4\pi R^2}$$

    a. Calculate the intensity (w/m^2) of sunlight at the top of Earth's atmosphere. For the Sun, L = 3.8×10^{26} watts and the distance between the sun and Earth is 150×10^9 meters.
    b. Suppose Earth were orbiting Alpha Centaurii A, the nearest star to Earth. This star has a luminosity of 5.7×10^{26} watts. Calculate the intensity of light at Earth's orbit around Alpha Centaurii A and discuss whether water would exist on Earth's surface or not. Compare your intensity to that for Venus in arriving at your answer.

29. The power of thermal radiation increases with temperature to the fourth power according to the Stefan-Boltzman law .$I = \sigma T^4$ where I is the total power emitted (w), *R* is the area of the emitting surface (m^2), *T* is the temperature in K, and *s* is the Stefan-Boltzman constant, $s = 5.67 \times 10^{-8}$ J/ s·m^2·K^4.
    a. The surface area of the Sun is 6.18×10^{18} square meters. The temperature of teh Sun's surface is 5,780K. Calculate the power radiated by the Sun.
    b. Suppose the temperature of the surface of the Sun were to increase by 5%. What change does that make in the power radiated by the Sun? Express your answer as a percent.

Section 21.3

30. Europa has a mass of 4.2×10^{22} kg and a radius of 1,550 kilometers.
    a. Calculate the volume of Europa
    b. Calculate Europa's average density
    c. Assume Europa is rock (5,500 kg/m^3) and water (1,000 kg/m^3). What percentage of the moon is rock and what percentage is water?

GLOSSARY

absolute zero - the lowest possible temperature, at which the energy of molecular motion is essentially zero, or as close to zero as allowed by quantum theory

accuracy - describes how close a measurement is to the true value.

acid - a chemical that dissolves in water to create more H^+ ions than there are in neutral water

activated complex - high energy state where bonds are being broken and reformed. Also referred to as the activated complex

activation energy - the energy necessary to break bonds in the reactants so a chemical reaction can occur. Units are joules per mole (J/mol) or kilojoules per mole (kJ/mol) - The minimum amount of energy required for molecules to react

active site - a small region of an enzyme molecule that is responsible for "binding" or holding on to the substrate molecules

activity - the rate of decay of a radioactive sample

actual yield - the amount obtained in lab by actual experiment

addition polymerization - type of polymer formed by adding monomers together using their double bonds

addition reaction - a chemical reaction that adds atoms to the carbon atoms forming multiple bond(s) in a molecule

adhesion - the property of a liquid that causes it to stick to surfaces

alcohol - a hydrocarbon with a hydroxyl group, -OH, attached

aldehyde - a functional group in which the central carbon makes a double bond with oxygen and then is bonded to hydrogen on one side and to another hydrocarbon

alkane - a hydrocarbon containing only single bonds. A saturated hydrocarbon

alkene - a hydrocarbon containig one or more double bonds

alkyne - a hydrocarbon containing one or more triple bonds

alloy - solid material made up of two or more elements (usually metals) that is evenly mixed on the microscopic level

alloying element - One of the elements that make up a certain alloy

alpha decay - happens when a nucleus decays by releasing a helium nucleus

alpha helix - a common form of the secondary structure of a protein that forms a spiral or twist similar to a slinky or spiral staircase. A single protein chain is held together by hydrogen bonding attractions

alpha radiation - the radiation associated with alpha decay

aluminum - a metallic element, atomic number 13

amide linkage - type of bond formed between an amine group ($-NH_2$) and a carbonyl (C=O) group

amine - an organic molecule containing an amino group, NH_2

amino acid - chiral molecules that have an amino group on one end and a carboxyl group on the other. There are 20 different naturally occurring amino acids

ammonia - NH_3

ammonium - the ion NH_4^+

ammonium nitrate - the compound NH_4NO_3

amorphous - a microstructure that does NOT have any ordered pattern

ampere - SI unit of electrical current

amphoteric - a substance is amphoteric if it can act as either an acid or a base under different circumstances. Water is amphoteric

anode - the electrode at which oxidation occurs

antioxidant - a molecule that reacts easily with free radicals. Some sources of antioxidants include brightly colored fruits and vegetables, vitamin E, and chocolate

aqueous - dissolved in water- indicated by (aq).

aqueous equilibrium - when the amount of dissolved solutes remains constant over time

Aqueous solution - An aqueous solution is any solution where the solvent is water. Molecules, atoms, or ions in aqueous solutions are identified by the symbol *(aq)* after a chemical formula.

aromatic hydrocarbon - a hydrocarbon containing one or more cyclic ring structures

Atmosphere (atm) - a large unit of pressure equal to 101,325 Pa, - the average pressure of air at sea level, also equal to 14.7 pounds per square inch (14.7 psi).

atmospheric pressure - the pressure we feel from the air around us - 101,325 Pa or 14.7 psi (average)

atom economy - the fraction of reactant material that ends up in the final product

atomic mass unit (amu) - a mass unit equal to 1.66×10^{-24} grams or 1.661×10^{-27} kg

atomic number (Z) - the number of protons in a nucleus, unique to each element

atomic radius - the distance from the center of an atom to its "outer edge."

atomic structure - the organization of neutrons, protons, and electrons in an atom

ATP- adenosine triphosphate. A molecule that carries chemical energy from spontaneous chemical reactions to non spontaneous reactions

average - you calculate the average of a set of measurements by adding up all their values and dividing by the total number of measurements.

Avogadro's number - the number of atoms contained in one mole. It is 6.02×10^{23}

baking soda - a household variety of the compound sodium bicarbonate $NaHCO_3$

balanced chemical equation - a chemical equation that satisfies the law of conservation of mass, - when the total number of atoms of each element are the same on both the reactant side and the product side of a chemical equation

barometer - an instrument that measures atmospheric pressure

base - a chemical that dissolves in water to create less H^+ ions than there are in pure water (or equivalently, more OH^- ions)

bent - a shape formed when a central atom has two unshared pairs of electrons and is bonded to two other atoms

benzene - an aromatic hydrocarbon C_6H_6, with alternating single and double bonds between the 6 carbon atoms in the ring

beryllium - alkali earth element with atomic number 4

beta decay - when an unstable nucleus releases an electron

beta radiation - the radiation resulting from beta decay

bimolecular- an elementary process involving the collision of two molecules

binary alloy - An alloy made up of only two different elements

binary phase diagram - a graph showing the phase of an alloy at different temperatures and compositions

binding energy - is the process of chemical change

boiling point - the temperature at which a substance changes phase from liquid to gas. For example, the boiling point of water is 100°C

Boltzmann's constant (k) - relates the kinetic energy in molecular motion to temperature. Boltzmann's constant has the value, $k = 1.381 \times 10^{-23}$ J/K

boron - non-mettalic element with atomic number 5

bravais lattices - fourteen groups of crystal structures used to help group the many different types of crystals

British thermal unit (btu) - large unit of heat used in the U.S., 1 btu = 1055 joules

brittle - a property of materials that are hard and require little energy to fracture, materials that break or shatter relatively easily

bromine - element with atomic number 35 - one of the halogens

Bronsted-Lowry definition - a different way to look at what defines acids and bases. Acids are compounds that donate protons (H+). Bases are compounds that accept protons

Brownian motion - the erratic, jerky movement of tiny particles suspended in water, due to the random impacts of individual molecules in thermal motion

buffer - a solution that resists small changes in pH by chemical action

buffer capacity - describes the amount of excess acid or base a buffer can neutralize without changing the pH of the solution

calcium (Ca) - element with atomic number 20

calcium carbonate - the compound $CaCO_3$

calorie - older unit of heat, 1 calorie = 4.184 joules

Calvin cycle - a sequence of chemical reactions that makes energy rich sugars from CO_2 by using the high energy molecules ATP and NADPH formed in the light dependent reactions

capillary action - an effect where liquid is pulled up a thin tube by adhesion

carbohydrate - short and long chain polymers of sugar molecules. All organisms use carbohydrates as a main source of energy

carbon - element with atomic # 6

carbon cycle - the process showing how carbon moves between the atmosphere, the land, and the ocean. Carbon changes it's chemical form as it moves through the cycle. Some of the common chemical forms of carbon are CO_2, $C_6H_{12}O_6$, and H_2CO_3

carbon dioxide - the compound CO_2

carbon-14 - a radioactive isotope of carbon

carbonyl group - formed when a carbon atom is double bonded to an oxygen atom (C=O). Common organic reference

carboxylic acid - an organic molecule containing a carboxyl group, COOH

catalyst - a substance that speeds up the rate of a chemical reaction, by providing a pathway with a lower activation energy

cathode - the electrode at which reduction occurs

cathode ray - a stream of particles produced by high voltage placed across a vacuum tube. Later cathode rays were discoved to be electrons, the first sub-atomic particle to be discovered.

cell-emf - abbreviation for electromotive force of a cell

cellular respiration - a chemical reaction that breaks down glucose molecules in the presence of oxygen and releases energy that is used to carry out life processes

Celsius scale - a temperature scale with 100 degrees between the freezing point and the boiling point of water; water freezes at 0°C and boils at 100°C.

chain reaction - is the process of chemical change

chemical bond - a relatively strong connection between two atoms

chemical change - a change that affects the structure or composition of the molecules that make up a substance, typically turning one substance into another substance with different physical properties - a result of chemical reaction.

chemical engineering - The application of chemistry to solving human problems, such as creating materials, fuels, medicines, or processes

chemical equation - A chemical equation is an expression that describes the changes that happen in a chemical reaction - typically of the form: *reactants* → *products*.

chemical formula - a combination of element symbols and subscripts that tells you the ratio of elements in a compound. For example, H_2O is the chemical formula for water and tells you there are 2 hydrogen (H) atoms for every one oxygen (O) atom

chemical property - property that can only be observed when one substance changes into a different substance - such as iron's tendency to rust.

chemical reaction - a process that rearranges the atoms in any substance(s) to produce one or more different substances, the process of chemical change

chlorine - element with atomic number 17, a halogen

chlorophyll - pigment molecule that absorbs sunlight and uses it to excite electrons in it's molecular structure, thereby storing energy to aid in photosynthesis

closed system - a system that is not allowed to exchange matter with irs surroundings, often the products and reactants in a chemical reaction, a chemical or a mixture of chemicals in a closed container, a system in which only energy is allowed to be exchanged with the surroundings

coefficient - the number that tells you how many of each molecule participate in the balanced reaction

cohesion - the property of a liquid that causes it to hold together

colligative property - physical property of a solution that depends only on the number of dissolved solute particles not on the type (or nature) of the particle itself

common ion - an ion which is produced by two or more different chemicals in the same solution

compound - a substance containing more than one element in which atoms of different elements are chemically bonded together

concentrated - a solution containing a lot of solute compared to solvent.

concentration - describes the amount of each solute compared to the total solution

conclusion - a stated decision whether the results of experiments or observations confirm an idea or hypothesis, or not.

condensation - a phase change from gas to liquid, a substance in its gas phase may condense at a temperature below its boiling point.

condensation polymerization - type of polymerization reaction that links monomers through the loss of a small molecule such as water

condensed matter - matter in which atoms or molecules are closely packed together and strongly interacting. Both liquids and solids are condensed matter

conduction - the flow of heat energy through the direct contact of matter

conservation of mass - in any chemical reaction, the total mass remains the same: the total mass of reactants equals the total mass of products

control variables - variables that are kept constant.

conversion factor - a ratio of two different units that has a value of 1. For example, 3.785 liters/1 gallon is a conversion factor. The numbers are different but the actual physical quantity is the same because 1 gallon is the same volume as 3.785 liters.

copolymer - a molecule made from repeatedly bonding more than one type of monomer together

core - the center of the earth, sun or other planet, typically where temperature and pressure are very high

coulomb (C) - SI unit of electrical charge

covalent bond - a chemical bond that consists of one shared electron

cracking - a process used to break long chain hydrocarbons down into smaller hydrocarbon fragments. This process requires high temperatures and a catalyst

crystal - a piece of crystalline matter in which the microstructure is uniform and continuous over the entire piece

crystal structure - system which describes the distances and angles between atoms in a crystal

crystalline - a microstructure with a repeating, ordered pattern

cytochrome protein- any iron containing protein with heme groups, these proteins are import as catalysts in oxidation-reduction reactions in the electron transport chain

Dalton's law of partial pressures - the total pressure in a mixture of gasses is the sum of the partial pressures of each individual gas in the mixture

daughter nuclide - the nucleus resulting from a nuclear reaction

decay - the process during which a nucleus undergoes spontaneous change

dehydrogenation reaction - remove hydrogen, H2 from the alkane causing formation of double bonds. This process forms unsaturated alkenes

Deionized water - water without ions, purified using filtration methods

density - a property of a substance that describes how much matter the substance contains per unit volume - typical units are grams per cubic centimeter (g/cm3)

dew point - the temperature at which air is saturated with H_2O vapor (Rh = 100%)

diffusion - the spread of molecules through their surroundings through constant collisions with neighboring molecules

dilute - a solution containing relatively little solute compared to solvent

dimensional analysis - using conversion factors to convert between units

dipole-dipole attraction - the attractions between the positive part of one polar molecule and the negative part of another polar molecule

dislocation - a missing half-plane of atoms in a crystal. This is a 1-D defect

dissolved - when molecules of solute are completely separated from each other and dispersed into a solution

distillation - a process of separating, concentrating or purifying liquids by boiling and condensing them again.

Distilled water - water purified by heating water to steam and condensing the steam into clean or sterilized container

DNA - long polymer chain of nucleotides that stores and transmits genetic information. Nucleotides - a molecular unit composed of a 5 carbon sugar, a phosphate group, and a nitrogenous base

ductile - Describes materials that require much energy to fracture

electric charge - a fundamental property of matter than comes in positive and negative.

electric current - is the flow of electric charge

electrochemical cell - device in which redox reactions take place

electrode - the part of a cell where oxidation and reduction reactions occur

electrolysis - the result of a chemical reaction in an electrolytic cell

electrolyte - solute capable of conducting electricity when dissolved in an aqueous solution, also - a conductive solution, or the solution in which the electrodes are immersed

electrolytic cell - an electrochemical cell in which chemical reactions result from the application of electrical current

electromagnetic spectrum - the complete range of electromagnetic waves, including visible light

electromotive force (emf) - the difference in the electrical potential between the anode and the cathode of an electrochemical cell

electron - a tiny particle that fills the outer volume of an atom. Electrons have negative charge and are responsible for chemical bonds

electron configuration - a description of which orbitals contain electrons for a particular atom.

electron transport chain - a sequence of biologically important chemical reactions that transfers high-energy electrons from the Krebs cycle to make ATP molecules

electron volt - a unit of energy equal to 1.602×10^{-19} joules

electronegativity - a value between 0 and 4 that describes the relative "pull" of an element for electrons from other atoms. High numbers mean stronger attraction for electrons., the ability for an atom to attract another atom's electrons when bonded to that other atom.

electroneutrality -The property of charges to balance resulting in a charge neutral system

element - a unique type of atom. All atoms of the same element are similar to each other and different from atoms of other elements

element symbol - a one or two letter abbreviation for each element

elementary steps - a series of simple reactions that represent the overall progress of the chemical reaction at the molecular level

empirical formula - the simplest ratio of atoms in a substance

endothermic reaction - a chemical reaction that requires an input of energy to go from reactants to products. Endothermic reactions absorb energy, - a reaction that absorbs energy. $\Delta H>0$

energy - energy is a measure of a system's ability to change or create change in other systems, measured in joules

energy barrier - The energy needed to initiate a chemical reaction

energy level - for an electron in an atom, an energy level is the set of quantum states that have approximately the same energy

enthalpy - ΔH, the heat energy of a chemical change measured in joules per mole (J/mole) or kilojoules per mole (kJ/mole) at standard temperature and pressure

enthalpy of formation - the change in energy when one mole of a compound is assembled from pure elements

entropy, S - is a measure of the disorder or randomness of a system

enzymes- biological catalysts, responsible for speeding up the chemical reactions inside our bodies

equilibrium - a "balance" in a chemical system. At equilibrium the rate of the forward reaction is equal to the rate of the reverse reaction, and the concentration of reactants and products remain constant over time

equilibrium constant - numeric value of the equilibrium expression

equilibrium expression - is a special ratio of product concentrations to the reactant concentrations. Each concentration is raised to the power of the corresponding coefficient in the balanced equation

equilibrium position - the favored direction of a reversible reaction. Determined by each set of concentrations for the reactant(s) and product(s) at equilibrium

equivalence point - in a titration, the point at which the moles of H^+ from the acid and OH^- from the base are exactly equal

error - the unavoidable difference between a real measurement and the unknown true value of the quantity being measured.

ester- an organic molecule containing a -COOR group, is an ester. Esters are responsible for many smells and tastes in fruits and flowers

ether - a molecule where oxygen is bonded to two carbon groups. The two carbon groups are often hydrocarbon chains

eutectic point - the temperature and composition where an alloy's melting point is the lowest

evaporation - a phase change from liquid to gas at a temperature below the boiling point.

excess reactant - the reactant left over after the reaction is complete. The amount of reactant the did not react to form product

exothermic reaction - a chemical reaction that releases energy in going from reactants to products. Endothermic reactions have negative enthalpy - a reaction that releases energy. $\Delta H<0$

experiment - a situation specially set up to observe how something happens or to test a hypothesis.

experimental variable - the single variable that is changed to test its effect.

exponent - a number written in scientific notation consists of a value multiplied by a power of ten, such as $500 = 5 \times 10^2$. The exponent is the power of ten. In the example the exponent is 2. Exponents are negative for numbers that are less than 1. For example, 0.05 is the same as 5×10^{-2}, - the power of ten in scientific notation,

Fahrenheit scale - a temperature scale with 180 degrees between the freezing point and the boiling point of water; water freezes at 32 °F and boils at 212 °F.

fat - or lipids, are nonpolar molecules made from fatty acids and a glycerol molecule. Fats molecules are an efficient way to store energy for many types of animals. Fats are not soluble in water

first law of thermodynamics - energy can neither be created nor destroyed. The total energy in an isolated system remains constant; all the energy lost by one system must be gained by the surroundings or another system

fission products - is the process of chemical change

fission reaction - is the process of chemical change

flow - The ability of a liquid to move and change shape under a force, like gravity (weight)

fluorine - a reactive element with atomic number 9 - the lightest of the halogens

force - an action such as a push or a pull that has the ability to change the motion of an object, such as to start it moving, stop it, or turn it.

formula mass - the mass of one mole of a compound with a given chemical formula

free radical - a molecule or atom that is highly reactive due to its having one or more unpaired valence electrons

frequency - the rate at which an oscillation repeats - one hertz (Hz) is a frequency of 1 oscillation per second

fusion reaction - is the process of chemical change

gallium - a soft metallic element that has a low melting point, atomic number 31

galvanic cell - another name for voltaic cell

gamma decay - when a nucleus decays releasing electromagnetic energy

gene - segment of DNA that contains the code for a specific protein

geologists - scientists that study the matter making up the earth

geometric isomers - are molecules with the same sequence of atoms, but different placement of groups around a double bond

germanium - a mettaloid element with atomic number 32

glass - a solid structure with no long-range order for bond lengths and angles

global warming - the warming of the surface of the earth by gas molecules in the atmosphere which trap heat and reflect it back to the surface of the earth

glycolysis - (gly-KOHL-ih-sis) first step in cellular respiration. During glycolysis one molecule of glucose is broken in half yielding two 3 carbon molecules (of pyruvic acid)

graduated cylinder - a measuring instrument used to measure volume

grain boundary - the interface between two grains in a polycrystalline solid crystal. This is a 2-D defect

gram - SI unit of mass, = 1/1000 kilogram

green chemistry - The practice of chemical science in a manner that is safe, sustainable and produces minimal waste

group - a column of the periodic table - all the elements in a group have similar chemical properties

half-life - the time it takes for half of the atoms in a sample to decay

half-reactions - the oxidation and the reduction parts of a redox reaction

halogens - group 17 elements including chlorine, fluorine, bromine, etc.

hardness - The ease of deforming a material by a small amount

heat - thermal energy, energy due to temperature, the total energy in random molecular motion contained in matter

heat of fusion - the energy required to change the phase of one gram of a material from liquid to solid or solid to liquid at constant temperature, constant pressure, and at the melting point

heat of solution - the energy absorbed or released when a solute dissolves in a particular solvent

heat of vaporization - the energy required to change the phase of one gram of a material from liquid to gas or gas to liquid at constant temperature, constant pressure, and at the boiling point

helium - the second lightest element, atomic number 2, the lightest of the noble gasses

heterogeneous mixture - is a mixture that is not uniform, different samples may have different compositions

homogenous mixture - is a mixture taht is uniform throughout, any sample has the same composition as any other sample

homopolymer - a molecule made from repeatedly bonding the same monomer together

Hydration - the process of molecules with any charge separation to collect water molecules around them. While not "chemically bonded" a hydrated molecule does hold fairly tightly to its "private collection" of water molecules

hydrocarbon - a molecule made almost entirely from carbon and hydrogen atoms

hydrogen - the lightest element, atomic number 1

hydrogen bond - an intermolecular bond that forms between a hydrogen atom in one molecule and the negatively charged portion of another molecule (or another part of the same molecule) - in water, the attractions between the partially positive hydrogen from one molecule to the partially negative oxygen on an adjacent molecule

hydrogenation reaction - a type of addition reaction where hydrogen gas, H_2, is added to a double or triple bond in a hydrocarbon, causing the structure to become saturated

hydronium ion - the H_3O^+ ion forms when H^+ bonds to a complete water molecule. Hydronium ions are what give acids their unique properties. Any reference to "H^+" in aqueous solution usually means H_3O^+

hypothesis - a tentative explanation for something, or a tentative answer to a question.

ideal gas - a gas made from molecules that have no volume or interactions with each other. This gas doesn't exist in reality, but is the kind of gas assumed by the various gas laws including the ideal gas law. Gases with temperatures well above their boilng point (such as air) or at low pressures are good approximations of an ideal gas

igneous rock - formed from crystallized magma, covers more than 90%of the earths crust

indicator - a chemical that turns different colors at different values of pH. Indicators may be used to determine pH directly, or in reactions with solutions of known pH

inquiry - the process of learning through asking questions.

insoluble - not dissolvable in a particular solvent

intensity of radiation - the amount of energy that flows per unit area per second

interatomic forces - bond atoms together into molecules or ions.

intermediate - chemical species that is formed during the elementary steps, but is not present in the overall balanced equation

intermolecular attractions - the attractions between molecules

intermolecular forces - act between molecules, typically much weaker than the forces acting within molecules (interatomic forces)

inverse square law - the intensity of radiaiton emitted from a point source decreases inversely over the square of the distance

iodine - an element, one of the halogens, atomic number 53

ion - an atom or small molecule with an overall positive or negative charge due to an imbalance of protons and electrons

ion product constant - In dilute aqueous solutions, $[H^+] \times [OH^-] = 1.0 \times 10^{-14}$ and the value $K_w = 1.0 \times 10^{-14}$ at 25°C is known as the ion product constant for water

ionic bond - an attraction between oppositely charged ions. This attraction occurs with all nearby ions of opposite charge

ionic compound - a compound in which positive and negative ions attract each other to keep matter together. Salt (NaCl) is a good example.

ionization - the process of removing one or more electrons from an atom, creating a positive ion and a free electron

ionization energy - the energy required to completely remove an electron

ionizing radiation - radiation with high enough energy to remove electrons from their orbits

irreversible change - is a chemical change that rearranges atoms into different substances

isolated system - neither matter nor energy can be exchanged with surroundings

isomer - a specific structure of a molecule. This term is only used when a chemical formula could represent more than one molecule, the same group of atoms bonded into different structures are called isomers

isotopes - atoms or elements that have the same number of protons in the nucleus, but different number of neutrons

joule - the fundamental SI unit of energy (and heat)

Kelvin scale - a temperature scale that starts at absolute zero and has the same size degrees as Celsius degrees. $T_{Kelvin} = T_{Celsius} + 273$

ketone - a functional group in which the central carbon makes a double bond with oxygen and then is bonded to two hydrocarbons one on each side

kilogram - SI unit of mass

kinetic energy - energy of motion

kinetic molecular theory - the theory that explains the observed thermal and physical properties of matter in terms of the average behavior of a collection of atoms and molecules

Krebs cycle - a series of energy extracting chemical reactions that break pyruvic acid down into $CO_2(g)$, during cellular respiration

latent heat - thermal energy that is absorbed or released by a phase change

lava - molten rock or magma released from the vent(s) of a volcano

law of conservation of energy - energy can never be created or destroyed, just converted from one form into another

law of conservation of mass - the total mass of reactants (starting materials) and the total mass of products (materials produced by the reaction) is the same

law of mass action - general description of any equilibrium reaction

Le Chateliers principle - states that when a "change" is placed on a system at equilibrium, the system will shift in a direction that partially offsets the "change". The change can be defined as a change in temperature, concentration, volume or pressure

Lewis dot diagrams - a diagram showing one dot for each valence electron an atom has. These dots surround the element symbol for the atom

limiting reactant - the reactant present in the least amount. The reactant that "runs out" first

lipid - a molecule that typically falls into the category of fat or steroid

liter - an SI unit of volume equal to a cube 10 centimeters on a side, or 1,000 cm^3

lithium - the lightest metal in the alkali metals group, atomic number 3

logarithm - the logarithm of a number (A) is another number (B) such that $10^B = A$. For example, the logarithm of 1,000 = 3 since $10^3 = 1,000$. Logarithms are positive for numbers greater than one and negative for numbers less than one. For example, the logarithm of 0.01 is -2 since $10^{-2} = 0.01$. Logarithms are often abbreviated "log" so the expression 2 = log(100) means "2 equals the logarithm of 100"

London dispersion attraction - the attraction that occurs between non-polar molecules due to temporary slight polarizations that occur when the normally equal distribution of electrons is shifted

lone pairs - electrons that are paired up in a Lewis dot diagram, but are not shared between atoms in a covalent bond

macronutrients - elements needed in large quantity by your body. Note: Often only Na, Mg, P, S, Cl, K, and Ca are listed here because they come from mineral sources.The other elements (C,H,N, and O) are found in organic matter

macroscopic - on the scale that can be directly seen and measured, from a bacteria up to the size of a planet.

magma - molten rock found below the surface of the upper mantle

malleable - means a substance can be hammered into thin sheets without cracking.

mantissa - a decimal number that multiplies the power of ten in scientific notation

mass - measures how much matter there is, units of grams or kilograms (SI)

mass number (A) - the number of protons plus neutrons in the nucleus

matter - material, has mass and takes up space.

measurement - information that describes a physical quantity with both a number and a unit

melting point - the temperature at which a substance changes phase from solid to liquid. For example, the melting point of water is 0°C

meniscus - the curved top of a column of liquid

metallic bond - an attraction between metal atoms that loosely involves many electrons

metallic glass - a special form of glass made by cooling metal extremely quickly

metamorphic rock - is created from igneous and sedimentary rock under conditions of heat and pressure

micelle - a collection of phospholipid molecules that form a small sphere which allows the nonpolar tails to be away from the aqueous solvent in the center of the sphere

microscopic - to a chemist, on the scale of atoms, 10^{-9} meters and smaller.

microstructure - the spatial arrangement of atoms and molecules in matter

milliliter - an SI unit of volume equal to a cube 1 cm on a side, or 1 cm^3

minerals - a natural non-organic solid with an organized internal structure, where the atoms have a definite pattern

mixture - matter that contains more than one substance.

Mohs hardness scale - A hardness scale describing how easy it is to scratch one material with another

molality, m - unit of concentration used when temperature varies, moles of solute per kilogram of solvent

molar volume - 22.7 liters per mole at 0°C and 1 atmosphere of pressure, - the amount of space occupied by a mole of gas at standard temperature and pressure

molarity - the number of moles of solute per liter of solution

mole - the number 6.02×10^{23}, the average atomic mass in grams is the mass of one mole of atoms

mole ratio - A ratio comparison between substances in a balanced equation. The ratio is obtained from the coefficients in the balanced equation. The ratio allows for the conversion of one substance to another substance by using molar equivalent amounts

molecular formula - the exact number and types of atoms in a molecule

molecule - a neutral group of atoms that are covalently bonded together

monomer - a small molecule which is a building block of larger molecules called polymers

monosaccharide - a single sugar unit. Mono - meaning one and saccharide meaning sugar molecule

NADH - nicotinamide adenine dinucleotide; is a molecule capable of carrying high-energy electrons and transferring them to another pathway

NADPH- molecule that carries two high energy electrons and stores sunlight as chemical energy

nanometer (nm) - is 10^{-9} meters

natural laws - the unwritten rules that govern everything in the universe.

neon - a noble gas with atomic number 10

Nernst equation - the mathematical equation that relates the electrical potential of a cell to the standard reduction potential and the state of the reaction as given by the concentration of reactants and products

network covalent - a type of large molecule, usually made from hundreds to billions of atoms, in which each atom is covalently bonded to multiple neighboring atoms, forming a web of connections

neutral - an atom or molecule is neutral when it has zero total electric charge - in the context of acids and bases, neutral means the pH = 7.0 which also means the concentrations of H^+ and OH^- ions are equal

neutralization - a reaction in which the pH of an acid is raised by combining with a base, or the pH of a base is lowered by combining with an acid. Complete neutralization results in a pH of 7, the same as neutral water

neutron number (N) - the number of neutrons in a nucleus

nitrogen - an element that makes up 78% of earth's atmosphere, atomic number 7

nitrogen cycle - the process that shows how nitrogen moves from the atmosphere to the ocean, land and living systems

nitrogen fixation - the ability to use nitrogen gas from the atmosphere and convert it into a nitrogen compound such as ammonia or nitrate

nitrogenous base - nitrogen containing bases, one component of nucleotides

non spontaneous - a reaction that does not occur in the indicated direction

non-ionizing radiation - radiation that cannot cause ionization of atoms

non-polar covalent bond - a bond formed between two atoms in which electrons are shared equally or almost equally between the two atoms

non-viscous - having a low viscosity

nuclear energy - the energy released during nuclear reactions

nuclear power plant - a facility that harnesses nuclear reactions to produce commercial electricity

nuclear reactions - involve the nuclei and may change one element into another

nucleotides - a molecular unit composed of a 5 carbon sugar, a phosphate group, and a nitrogenous base

nucleus - the tiny core of an atom that contains all the protons and neutrons. The nucleus is extraordinarily small - about 1/10,000 the diameter of the atom

objective - describing only what actually occured without opinion or bias

octet rule - elements transfer or share electrons in chemical bonds to reach a stable configuration of 8 valence electrons. The light elements H, Li, Be, and B have helium as the closest noble gas so the preferred state is 2 valence electrons instead of 8

ohm (Ω) - the unit of electrical resistance

Ohm's law - the relationship between voltage, current and electrical resistance

oligosaccharides - two to ten monosaccharides linked together in a chain

open system - matter and energy can be exchanged with the surroundings

optical isomers - are formed when the carbon atom has four different groups attached to it. These isomers form "mirror images" of one another

orbital - group of quantum states that have similar spatial shapes. The orbitals are labeled s, p, d and f

organic molecule - a molecule primarily made from carbon and hydrogen, but often with some oxygen, nitrogen, one of the halides, or some other non-metal atoms

oxidation - a chemical reaction that increases the charge of an atom or ion by giving up electrons - loss of electrons - element charge increases

oxidation number - gives the number of electrons that an element has lost or gained in forming a chemical bond with another element

oxidizing agent - the element that oxidizes another element

oxygen - an element with atomic number 8

parent compound - the longest continuous chain of carbons in an organic compound. The parent compound tells us the base alkane name

parent nuclide - the original nucleus involved in a nuclear reaction

partial hydrogenation - a chemical reaction where only some of the unsaturated fat is hydrogenated. In this case some double bonds remain in the structure, causing the formation of transfats in place of the naturally occurring cis configuration

Pascal (Pa) - the SI unit of pressure, a very small unit of pressure equal to one newton of force per square meter of area (1 N/m^2).

Pauli exclusion principle - two electrons in the same atom may never be in the same quantum state

peptide bond - a bond formed between the amino group and the carboxyl group of two amino acids. The peptide bond is formed through condensation polymerization or the loss of water molecules

percent yield - the ratio of the amount of product actually obtained by experiment (actual yield) as compared to the amount of product calculated theoretically (theoretical yield) multiplied by 100

period - a row of the periodic table

periodic - repeating at regular intervals. The periodic table is named for this because the rows are organized by repeated patterns found in both the atomic structure and the properties of the elements

periodic table - a graphical chart of information on the elements that groups the elements in rows and columns according to their chemical properties

petroleum - a blend of hydrocarbons, formed from prehistoric organic matter, containing mainly straight and branched chain alkanes

pH scale - a measurement of the H^+ ion concentration that tells whether a solution is acid or base. Pure water has a pH of 7. Solutions with pH <7 are acidic. Solutions with pH >7 are basic

phase change - occurs when a substance changes how its molecules are organized without changing the individual molecules themselves. Examples are changing from solid to liquid or liquid to gas

phase equilibrium diagram - shows the relationship between the phases of matter, temperature and pressure

phases - the physical forms of matter: solid, liquid, and gas (and plasma)

phospholipid - a lipid that has a phosphate group attached, giving it a polar head and a nonpolar tail. The phosphate group is the polar region of the molecule and the hydrocarbon chain is the nonpolar region

phosphorous cycle - the process that shows how phosphorous moves between the land, the ocean, and living systems

photodissociation - the breaking of a chemical bond in a molecule caused by the absorption of a photon

photoionization - when a molecule absorbs a photon and ejects an electron causing it to become an ion

photon - the smallest possible quantity (or quanta) of light. Light exists in discrete bundles of energy called photons

photosynthesis - the chemical reaction that occurs almost exclusively in plants and algae and combines CO_2 and H_2O to form glucose $C_6H_{12}O_6$ and oxygen. The reaction is endothermic and it driven by sunlight

physical change - a change in physical properties, such as shape, phase or temperature; for example, grinding, melting, boiling, dissolving, heating or cooling.

physical property - property such as mass, density or color that you can measure or see through direct observation.

Planck's constant - defines the scale of energy at which quantum effects must be considered. It is equal to $h = 6.626 \times 10^{-34}$ joule-seconds (J.s)

plasma - a hot, energetic phase of matter in which the atoms are broken apart into positive ions and negative electrons that move independent of each othe

pleated sheet - a secondary structure of protein that forms a folded or pleated pattern similar to an old fashioned fan. Formed by two or more side by side protein chains held together by hydrogen bonding

point defects - single atom defects in a crystal. These are 0-D defects

polar - a molecule is polar when there is a charge separation that makes one side (or end) of the molecule more positive or negative than the other side (or end)

polar covalent bond - a bond formed between two atoms in which electrons are unequally shared

polarization - uneven distribution of positive and negative charge

polyatomic ion - a small molecule with an overall positive or negative charge

polymer - from "poly" many and "meros" parts. A long chain molecule formed by connecting small repeating units with covalent bonds - a molecule built up from many repeating units of a smaller molecular fragment

polymerization - a reaction that assembles a polymer through repeated additions of smaller molecular fragments, - chemical reaction that joins monomers to produce a polymer

polysaccharide - a long chain, greater than ten, monosaccharides connected together. Poly- means many and saccharide means sugar units. Examples are starch, cellulose and glycogen molecules

positive, negative - the charge on a proton is defined to be positive and the charge on an electron is defined to be negative

positron - a particle that has the same mass as the electron and positive charge

positron emission - when a nucleus decays by releasing a positron

precipitate - an insoluble compound that forms in a chemical reaction in aqueous solution

precision - describes how close measured values are to each other.

pressure - force per unit area, units of Pa (N/m^2) or psi (lbs/in^2) or atm, a expansive force per unit area that acts equally in all directions within a liquid or a gas.

primary structure - sequence of amino acids in a protein chain

principal quantum number - a number that specifies the quantum state and is related to the energy level of the electron

procedure - detailed instructions on how to do an experiment or make an observation

product - a substance that is created or released in a chemical reaction

protein - biologically functional polymers of amino acids, typically very large molecules with 100 or more amino acids

proton - a tiny particle in the nucleus that has a positive charge

quantum state - a specific combination of values of variables such as energy and position that is allowed by quantum theory

quantum theory - a theory of physics and chemistry that accurately describes the universe on very small scales, such as the inside of an atom

rad - radiation with high enough energy to remove electrons from their orbits

radiation - the transmission of energy through space

radioactivity - a process by which the nucleus of an atom spontaneously changes itself by emitting particles or energy, - property of some element to break up and release energy

random - scattered equally among all possible choices with no organized pattern

rate constant - a number that descrubes how fast a radioactive nuclide decays

rate determining step - The “slow” elementary step in the reaction mechanism determines the overall rate (or speed) of the chemical reaction

rate of decay - the number of decays per unit time

reactant - a substance that is used or changed in a chemical reaction, - the starting materials or substances in a chemical reaction

reaction mechanism- a proposed sequence of elementary steps that leads to product formation. Must be determined using experimental evidence

reaction profile : a graph showing the progress of a reaction with respect to the energy changes that occur during a collision

reaction rate - the speed at which a chemical reaction occurs, the change in concentration, of a reactant or product, per unit time

reactivity - the tendency of elements to form chemical bonds. A reactive element forms bonds easily therefore tends to have many reactions

redox - abbreviation for oxidation reduction

reducing agent - the element that reduces another element

reduction - a chemical reaction that decreases the charge of an atom or ion by accepting electrons, - gain of electrons. Element charge decreases

region of electron density - an area represented by shared or unshared electrons around an atom

relative humidity - the actual partial pressure of water vapor in air divided by the saturation vapor pressure at the same temperature

rem - unit of radiation dose that is weighted for its effect on body tissue, stande for Radiation Equivalent Man

repeatable - others who do the experiment or make the observation the same way obtain the same results or observations.

R-group - in an organic formula the R- group can stand for an H, hydrogen atom, or a hydrocarbon side chain, of any length. It can also refer to any group of atoms

RNA - a polymer composed of nucleotides that contain a 5 carbon sugar called ribose, a phosphate group, and a nitrogenous base. This molecule transmits the information stored in DNA to cell "machinery."

salinity - is a measure of how salty water is. The concentration of salt in ocean water is a measure of it's salinity

salt - an ionic compound in which the positive ion comes from an acid and the negative ion comes from a base - an ionic compound that dissolves in water to produce ions

salt bridge - an electrical connection between the oxidation and the reduction half-cells of an electrochemical cell

saturated - when a solution has dissolved all the solute it can possibly hold. Saturation occurs when the amount of dissolved solute gets high enough that the rate of "undissolving" matches the rate of dissolving

saturated hydrocarbon - a hydrocarbon containing only single carbon to carbon bonds

scale - a typical size that shows a certain level of detail.

scientific method - a process of learning that constantly poses questions and tentative answers that can be evaluated by comparison with objective evidence

scientific notation - a method of writing numbers as a base times a power of ten

second law of thermodynamics - energy (heat) spontaneously flows from higher temperature to lower temperature (basic interpretation of 2nd law)

sedimentary rock -is formed from the compression of small pieces of igneous rock

significant - a difference between two results is only significant when it is substantially greater than the error in either result.

significant figures - a way of writing data that tells the reader how precise a measurement is.

solar nebula - disk shaped region of dust and gases around a forming star. The sun's solar nebula is where the earth formed

solubility - the amount of a solute that will dissolve in a particular solvent at a particular temperature and pressure

solute - any substance in a solution other than the solvent

solution - a mixture that is homogeneous on the molecular level

solvent - the substance that makes up the biggest percentage of the mixture and is usually a liquid

specific heat - the quantity of energy, measured in J/g°C, it takes per gram to raise the temperature one degree Celsius

spectrometer - a device that measures the spectrum of light

spectroscopy - the science of analyzing matter using electromagnetic emission or absorption spectra

spectrum - a representation that analyzes a sample of light into its component energies or colors, can be a picture, graph or table of data

speed of light, c - a constant speed at which all electromagnetic radiation travels through vacuum, including visible light. The speed of light in vacuum is 3×10^8 m/s

spontaneous - a reaction that occurs in the indicated direction without any energy input

spontaneous reaction - A chemical reaction that happens without the need for external energy input

standard reduction potential - the potential of a cell measured under standard conditions of temperature, pressure and concentration

standard temperature and pressure - (STP) - conditions of one atmosphere of pressure and 0°C

steroid - a molecule with four carbon rings that is biologically active as either a hormone, vitamin, drug or poison

stoichiometric equivalent - the mole amounts of each substance in a balanced chemical equation are proportionate

stoichiometry - STOY-KEE-AHM-EH-TREE - is the study of the amounts of substances involved in a chemical reaction. The amounts can be studied in moles or in mass relationships

stratosphere - the second layer of gases after the troposphere. This region of gases extends from 10 to 50 km above the earth

strong acid/base - dissociates completely (or almost completely) usually yielding 1 mole of H+ or OH- ions for every mole of acid or base dissolved

structural isomer - molecule with the same number and type of atoms as another molecule but a different bonding pattern

substance - a kind of matter that can't be separated into other substances by physical means such as heating, cooling, filtering, drying, sorting or dissolving.

substitution reaction - a reaction in which one or more hydrogen atoms on an alkane are removed and replaced by different atoms. The name substitution reaction comes from the fact that a different atom is "substituted" for one or more of the hydrogen atoms

substrate - a reactant molecule that is able to be bound by a particular enzyme

supersaturation - when a solution contains more dissolved solute than it can hold. Supersaturated solutions are always unstable and the excess solute becomes "undissolved", often rapidly

surface tension - a force created by intermolecular attraction in liquids, such as hydrogen bonding in water. Surface tension acts to pull a liquid surface into the smallest possible area, for example, pulling a droplet of water into a sphere - a property of liquids to resist having their surface broken. The energy needed to remove a certain amount of area on a liquid surface, measured in J/m^2

surfactant - a substance that reduces the surface tension of a liquid

surroundings - everything outside the "system"

system - an interrelated group of matter and energy that we choose to investigate

Tap water - drinking water from the sink or "tap."

temperature - a measure of the average kinetic energy of atoms or molecules, units of degrees Fahrenheit (°F) degrees Celsius (°C) or Kelvins (K)

tensile strength - A measure of the energy needed to break a material in two

tertiary structure - protein structure that forms from the bending or folding of the secondary alpha helix. Tertiary structures form specific globular proteins. The structures are held together by intermolecular attractions, hydrophobicity of nonpolar regions, and ion-dipole interactions

tetrahedral - a shape formed when a central atom is bonded to four other atoms that all point into the corners of a tetrahedron

theoretical yield - the calculated amount produced if everything reacts perfectly

theory - an explanation that is supported by evidence.

thermal conductor - a material that conducts heat easily

thermal equilibrium - a condition where the temperatures are the same and heat no longer flows

thermal insulator - a material that resists the flow of heat

thermistor - electronic sensor for measuring temperature by changes in resistance

thermochemical equation - the equation that gives the chemical reaction and the energy information of the reaction

thermocouple - electronic sensor for measuring temperature that produces a temperature dependent voltage

thermometer - an instrument that measures temperature

titration - a laboratory process to determine the precise volume of acid or base of known concentration which exactly neutralizes a solution of unknown pH

trace amount - refers to a very small quantity

trace elements - elements which are needed in very small quantities to maintain optimum health. Too much can be toxic. Not enough can be fatal or cause disease

transition state: high energy state where bonds are being broken and reformed, also referred to as the activated complex

transpiration - is the release of water from the leaves of plants as they carry out photosynthesis

triglyceride - is a fat or lipid molecule used for energy storage. It is made of a glycerol molecule and three long chain fatty acids

trigonal planar - a flat, triangular geometry typical of molecules with three regions of electron density surrounding a central atom

trigonal pyramidal - a 3-D shape formed when a central atom has one unshared pair of electrons and is bonded to three other atoms, the central atom is not in the same plane as the surrounding three

triple point - the temperature and pressure at which the solid, liquid, and gas phases of a substance can all exist in equilibrium together

troposphere - the layer of gases closest to the earth, extending up to approximately 10 km above the earths surface

unbalanced chemical equation - a chemical equation that does not satisfy the law of conservation of mass. The number of each type of atom on the reactant side of the equation does not equal the same number for each atom on the product side

unimolecular - an elementary step where only one reactant is involved

unsaturated hydrocarbon - a hydrocarbon containing multiple bonds between the carbon atoms

valence electrons - electrons in highest unfilled energy level. These are the electrons that make chemical bonds

van der Waals attractions - another term used to describe the attractions between molecules. Most sources consider these identical to the broad term "intermolecular attractions". However, some people only associate the term van der Waals attractions with the London dispersion type of intermolecular attraction

variable - a quantity that is measured or changed in an experiment or observation.

viscosity -the resistance of a liquid to flow under an applied force

viscous - having a high viscosity

volcano - opening in the earth were magma, gases and other materials can escape from inside the earth

volt - the unit of voltage

voltage -the electrical potential difference

voltaic cell - electrochemical cell in which chemical reactions generate electricity

volume - an amount of space having length, width and height

VSEPR - an acronym that stands for Valence Shell Electron Pair Repulsion, a theory that states the shapes of molecules are dictated, in part, by the repulsion of the shared electrons and the unshared pairs of electrons

water - the compound H_2O

water cycle - the process showing how water molecules move between the atmosphere, the land, and the ocean

wavelength - the distance (separation) between any two successive peaks (or valleys) of a wave

weak acid/base - only partially dissociates in solution, typically only a few percent (or less) of molecules dissociate to yield H^+ or OH^- ions

weak base - a chemical that only partially dissociates in solution, typically only a few percent (or less) to produce OH- ions

weight - a force (push or a pull) that results from gravity acting on mass, measured in newtons in the SI system of units and pounds or ounces in the English system of units

INDEX

A

D

E

F

I

J

K

L

M

N

O

P

S

U

V

W

X

Y

Z

Table 1: SI Base Units

Quantity	Unit	Symbol
length	meter	m
mass	kilogram	kg
time	second	s
electric current	ampere	A
thermodynamic temperature	kelvin	K
amount of substance	mole	mole
luminous intensity	candela	cd

Table 2: SI Derived Units

Derived quantity	Name	Symbol
area	square meter	m^2
volume	cubic meter	m^3
speed, velocity	meter per second	m/s
acceleration	meter per second squared	m/s^2
mass density	kilogram per cubic meter	kg/m^3
specific volume	cubic meter per kilogram	m^3/kg
current density	ampere per square meter	A/m^2
magnetic field strength	ampere per meter	A/m
amount-of-substance concentration	mole per cubic meter	mol/m^3

Table 3: Fundamental Constants of Nature

Quantity	Symbol	Value	Units
Absolute zero	0 K	-273.15	°C
Strength of gravity at Earth's surface	g	9.8	N/kg or m/sec^2
Avogadro's number	N_A	6.02×10^{23}	atoms/mole
Boltzman's constant	k	1.381×10^{-23}	J/K
Earth: mass	m_{earth}	5.98×10^{24}	kg
Earth: radius	r_{earth}	6.38×10^{6}	m
Moon: mass	m_{moon}	7.36×10^{22}	kg
Moon: radius	r_{moon}	1.74×10^{6}	m
Sun: mass	m_{sun}	1.99×10^{30}	kg
Sun: radius	r_{sun}	6.96×10^{8}	m
Electron mass	m_e	9.109×10^{-31}	kg
Proton mass	m_p	1.673×10^{-27}	kg
Neutron mass	m_n	1.675×10^{-27}	kg
Planck's constant	h	6.626×10^{-34}	J · sec
Stefan-Boltzmann constant	s	5.670×10^{-8}	$W/m^2/kg^2$
Speed of light in vacuum	c	2.998×10^{8}	m/sec
Universal gravitational constant	G	6.670×10^{-11}	$N \cdot m^2/kg^2$
Elementary charge	e	1.602×10^{-19}	coulombs (C)
Speed of sound at 20°C	v_s	343	m/sec
Gas constant for air	R	287.1	J/kg · K
Coulomb's law constant	K	9×10^{9}	$N \cdot m^2$/coulombs $(C)^2$

Table 4: Conversions

Time
1 hour = 60 minutes = 3,600 seconds
1 day = 1,440 minutes = 8.64 10^4 seconds
1 year = 365.25 days = 3.156 10^7 seconds

Length
1 meter (m) = 100 centimeters (cm) = 39.37 inches
1 kilometer (km) = 1,000 meters = 0.6215 miles (mi)
1 foot (ft) = 12 inches (in.) = 2.54 centimeters (cm)
1 miles (mi) = 5,280 feet (ft) = 1,609 kilometers (km)

Speed
1 m/sec = 2.237 mi/hr = 3.600 km/hr
1 mi/hr = 0.4470 m/sec = 1.609 km/hr

Area
1 m^2 = 10,000 cm^2 = 1.76 ft^2 = 1,5550 in^2
1 in^2 = 6.452 cm^2 = 0.000645 m^2
1 acre = 43,560 ft^2 = 4,048 m^2

Volume
1m^3 = 1,000 liters = 10^6 cm^3 = 35.32 ft^3
= 264.2 gallons
1 gallon = 3.785 liters = 0.003785 m^3

Angles and angular speed
1 radian = 57.3
1 degree = 0.01745 radians
1 rad/sec = 9.549 rev/min (rpm)

Mass
1 kilogram (kg) = 1,000 grams (g)
1 atomic mass unit (amu) = 1.660×10^{-27} kg

Force
1 newton (N) = 0.2248 pounds (lbs)
1 pound (lb) = 4.448 newtons (N)

Pressure
1 pascal (Pa) = 1N/m^2
= 1.450×10^{-4} lbs/in^2 (psi)
1 lb/in^2 (psi) = 6,895 N/m^2 (Pa)
1 atmosphere (atm) = 14.70 lb/in^2 (psi)
= 101,300 N/m^2 (Pa)

Energy
1 kilowatt · hour (kwh) = 3.6×106 joules (J)
1 British thermal unit (Btu) = 1,054 joules (J)
1 electron volt (eV) = 1.602×10^{-19} joule (J)
1 J = 1 N · m = 9.484×10^{-4} Btu
= 2.778×10^{-7} kwh

Power
1 watt (w) = 1 J/sec
= 0.001341 horsepower (hp)
1 horsepower (hp) = 745.7 watts (W)

Table of the Elements

Element Name	Symbol	Atomic Number	Atomic Mass	Element Name	Symbol	Atomic Number	Atomic Mass
Actinium	**Ac**	89	[227]	Neodymium	**Nd**	60	144.24
Aluminum	**Al**	13	26.98	Neon	**Ne**	10	20.18
Americium	**Am**	95	[243]	Neptunium	**Np**	93	[237]
Antimony	**Sb**	51	121.76	Nickel	**Ni**	28	58.69
Argon	**Ar**	18	39.95	Niobium	**Nb**	41	92.91
Arsenic	**As**	33	74.92	Nitrogen	**N**	7	14.01
Astatine	**At**	85	[210]	Nobelium	**No**	102	[259]
Barium	**Ba**	56	137.33	Osmium	**Os**	76	190.23
Berkelium	**Bk**	97	[247]	Oxygen	**O**	8	16.00
Beryllium	**Be**	4	9.01	Palladium	**Pd**	46	106.42
Bismuth	**Bi**	83	208.98	Phosphorus	**P**	15	30.97
Bohrium	**Bh**	107	[264]	Platinum	**Pt**	78	195.08
Boron	**B**	5	10.81	Plutonium	**Pu**	94	[244]
Bromine	**Br**	35	79.90	Polonium	**Po**	84	[210]
Cadmium	**Cd**	48	112.41	Potassium	**K**	19	39.10
Cesium	**Cs**	55	132.91	Praseodymium	**Pr**	59	140.91
Calcium	**Ca**	20	40.08	Promethium	**Pm**	61	[145]
Californium	**Cf**	98	[251]	Protactinium	**Pa**	91	231.04
Carbon	**C**	6	12.01	Radium	**Ra**	88	[226]
Cerium	**Ce**	58	140.12	Radon	**Rn**	86	[220]
Chlorine	**Cl**	17	35.45	Rhenium	**Re**	75	186.21
Chromium	**Cr**	24	52.00	Rhodium	**Rh**	45	102.91
Cobalt	**Co**	27	58.93	Roentgenium	**Rg**	111	[272]
Copper	**Cu**	29	63.55	Rubidium	**Rb**	37	85.47
Curium	**Cm**	96	[247]	Ruthenium	**Ru**	44	101.07
Darmstadtium	**Ds**	110	[271]	Rutherfordium	**Rf**	104	[261]
Dubnium	**Db**	105	[262]	Samarium	**Sm**	62	150.36
Dysprosium	**Dy**	66	162.50	Scandium	**Sc**	21	44.96
Einsteinium	**Es**	99	[252]	Seaborgium	**Sg**	106	[266]
Erbium	**Er**	68	167.26	Selenium	**Se**	34	78.96
Europium	**Eu**	63	151.96	Silicon	**Si**	14	28.09
Fermium	**Fm**	100	[257]	Silver	**Ag**	47	107.87
Fluorine	**F**	9	19.00	Sodium	**Na**	11	22.99
Francium	**Fr**	87	[223]	Strontium	**Sr**	38	87.62
Gadolinium	**Gd**	64	157.25	Sulfur	**S**	16	32.07
Gallium	**Ga**	31	69.72	Tantalum	**Ta**	73	180.95
Germanium	**Ge**	32	72.61	Technetium	**Tc**	43	[98]
Gold	**Au**	79	196.97	Tellurium	**Te**	52	127.60
Hafnium	**Hf**	72	178.49	Terbium	**Tb**	65	158.93
Hassium	**Hs**	108	[277]	Thallium	**Tl**	81	204.38
Helium	**He**	2	4.00	Thorium	**Th**	90	232.04
Holmium	**Ho**	67	164.93	Thulium	**Tm**	69	168.93
Hydrogen	**H**	1	1.01	Tin	**Sn**	50	118.71
Indium	**In**	49	114.82	Titanium	**Ti**	22	47.87
Iodine	**I**	53	126.90	Tungsten	**W**	74	183.84
Iridium	**Ir**	77	192.22	Ununbium	**Uub**	112	[285]
Iron	**Fe**	26	55.85	Ununhexium	**Uuh**	116	[292]
Krypton	**Kr**	36	83.80	Ununoctium	**Uuo**	118	[294]
Lanthanum	**La**	57	138.91	Ununpentium	**Uup**	115	[288]
Lawrencium	**Lr**	103	[262]	Ununquadium	**Uuq**	114	[289]
Lead	**Pb**	82	207.20	Ununseptium	**Uus**	117	[293]
Lithium	**Li**	3	6.94	Ununtrium	**Uut**	113	[284]
Lutetium	**Lu**	71	174.97	Uranium	**U**	92	238.03
Magnesium	**Mg**	12	24.31	Vanadium	**V**	23	50.94
Manganese	**Mn**	25	54.94	Xenon	**Xe**	54	131.29
Meitnerium	**Mt**	109	[268]	Ytterbium	**Yb**	70	173.06
Mendelevium	**Md**	101	[258]	Yttrium	**Y**	39	88.91
Mercury	**Hg**	80	200.59	Zinc	**Zn**	30	65.41
Molybdenum	**Mo**	42	95.94	Zirconium	**Zr**	40	91.22

A value in square brackets is the atomic mass of the most stable isotope.